COPLANAR MICROWAVE INTEGRATED CIRCUITS

COPLANAR MICROWAVE INTEGRATED CIRCUITS

INGO WOLFF
IMST GmbH Kamp-Lintfort, Germany

WILEY-INTERSCIENCE
A JOHN WILEY & SONS, INC., PUBLICATION

Published by John Wiley & Sons, Inc., Hoboken, New Jersey
Published simultaneously in Canada

Library of Congress Cataloging-in-Publication Data:

Wolff, Ingo.
Coplanar microwave integrated circuits / Ingo Wolff.
p. cm.
Includes bibliographical references and index.
ISBN-13: 978-0-471-12101-5
ISBN-10: 0-471-12101-0
1. Microwave integrated circuits. I. Title.

TK7876.W64 2006
621.381'32–dc22

2005056821

10 9 8 7 6 5 4 3 2 1

CONTENTS

Preface **xi**

1 Introduction **1**

References, 9

2 Transmission Properties of Coplanar Waveguides **11**

2.1 Rigorous, Full-Wave Analysis of Transmission Properties, 11
2.1.1 The Coplanar Waveguide with a Single Center Strip and Finite Ground-Plane Width, 12
2.1.2 The Coplanar Waveguide with a Single Center Strip and Infinite Ground-Plane Width, 26
2.1.3 Coupled Coplanar Waveguides, 34
2.1.3.1 Scattering Matrix of Coupled Coplanar Waveguides, 36
2.1.3.2 Coupled Coplanar Waveguides and Microstrip Lines—A Comparison, 40
2.2 Quasi-Static Analysis of Coplanar Waveguides Using the Finite Difference Method, 46
2.2.1 Introduction, 46
2.2.2 The Finite Difference Method as Applied to the Analysis of Coplanar Waveguide Structures, 48
2.2.3 The Solution of Laplace's Equation for Planar and Coplanar Line Structures Using the Finite Difference Method, 48

2.2.4 Application of the Quasi-Static Techniques to the Analysis of Coplanar Waveguides, 55
2.2.5 Characteristic Parameters of Coplanar Waveguides, 63
2.2.6 The Influence of the Metalization Thickness on the Line Parameters, 72
2.2.7 The Influence of the Ground Strip Width on the Line Parameters, 74
2.2.8 The Influence of the Shielding on the Line Parameters, 75
2.2.9 Special Forms of Coplanar Waveguides, 76
2.2.10 Coplanar-like Waveguides, 80
2.2.11 Coupled Coplanar Waveguide Structures, 89
2.2.11.1 Analysis of the Characteristic Parameter Matrices, 90
2.2.11.2 Determination of the Scattering Matrix of Coupled Coplanar Waveguides, 92
2.3 Closed Formula Static Analysis of Coplanar Waveguide Properties, 95
2.3.1 Analysis of a Generalized Coplanar Waveguide with Supporting Substrate Layers, 95
2.3.1.1 Structure SCPW1, 98
2.3.1.2 Structure SCPW2, 100
2.3.1.3 Structure SCPW3, 100
2.3.1.4 Numerical Results, 100
2.3.2 Static Formulas for Calculating the Parameters of General Broadside-Coupled Coplanar Waveguides, 109
2.3.2.1 Analytical Formulas and Results for the General Broadside-Coupled Coplanar Waveguide, 110
2.3.2.2 Analysis of an Asymmetric Supported BSC-CPW, 115
2.3.2.3 Application of the GBSC-CPW as Single CPW, 117
2.3.2.4 Criteria for the Coplanar Behavior of the Structure, 118
Bibliography and References, 120

3 Coplanar Waveguide Discontinuities **145**

3.1 The Three-Dimensional Finite Difference Analysis, 145
3.2 Computation of the Electric Field Strength, 147
3.3 Computation of the Magnetic Field Strength, 150
3.3.1 Convergence and Error Discussion for the Analysis Technique, 152
3.4 Coplanar Waveguide Discontinuities, 154
3.4.1 Modeling the Discontinuities, 156
3.4.2 Extraction of the Model Parameters, 157
3.5 Description of Coplanar Waveguide Discontinuities, 161

3.5.1 The Coplanar Open End, 162
3.5.2 The Coplanar Waveguide Short-Circuited End, 167
3.5.3 The Gap in a Coplanar Waveguide, 169
3.5.4 The Coplanar Waveguide Step, 175
3.5.5 Air Bridges in Coplanar Waveguides, 183
3.5.6 The Coplanar Waveguide Bend, 192
3.5.7 The Coplanar Waveguide T-Junction, 202
3.5.7.1 Analysis of the Odd-Mode Excitation, 221
3.5.8 The Coplanar T-Junction as a Mode Converter, 225
3.5.9 The Coplanar Waveguide Crossing, 234
Bibliography and References, 241

4 Coplanar Lumped Elements 249

4.1 Introduction, 249
4.2 The Coplanar Interdigital Capacitor, 250
4.2.1 The Lumped Element Modeling Approach, 250
4.2.2 Enhancement of the Interdigital Capacitor Model for Application at Millimeter-Wave Frequencies, 269
4.3 The Coplanar Metal–Insulator–Metal (MIM) Capacitor, 272
4.4 The Coplanar Spiral Inductor, 276
4.4.1 Enhancement of the Inductor Model for Millimeter-Wave Frequencies, 290
4.4.2 Coupled Coplanar Rectangular Inductors, 291
4.5 The Coplanar Rectangular Spiral Transformer, 295
4.6 The Coplanar Thin-Film Resistor, 303
Bibliography and References, 304

5 Coplanar Element Library and Circuit Design Program 309

5.1 Introduction, 309
5.2 Modeling, Convergence, and Accuracy, 312
5.3 Overview on *Coplan for ADS*™, 315
5.3.1 Data Items, 317
5.3.2 Library Elements, 319
5.4 Cache Management, 321
5.5 Layout, 321
5.6 Coplanar Data Items, 322
5.6.1 Overview, 322
5.6.2 Description of the Data Items, 324
5.6.2.1 Coplanar Substrate Data Definition C_SUB, 325
5.6.2.2 Coplanar Line-Type Data Definition C_LINTYP, 327
5.6.2.3 Coplanar Coupled Lines Data Definition C_NL_TYP, 328
5.6.2.4 Coplanar Bridge-Type Data Definition C_AIRTYP, 331

5.6.2.5 Coplanar Grid Data Definition C_GRID, 333
5.6.2.6 Process (Foundry) Used for Fabrication C_PROCES, 335
5.6.2.7 Technological Data Definition (Default Foundry) C_TECH, 336
5.6.2.8 Layer Data Definition (Default Foundry) C_LAYER, 338
5.7 The Coplanar Components and Their Models, 339
5.7.1 Coplanar Waveguide RF-Port C_PORT, 341
5.7.2 Coplanar Transmission Line C_LIN, 344
5.7.3 Coplanar Inter-Metal via (No Step) Connection C_METIA, 345
5.7.4 Coplanar Resistively Loaded Transmission Line C_TFG, 347
5.7.5 Coplanar MIM-Capacitor to Ground C_CAPLIN, 349
5.7.6 Coplanar Open-Ended Transmission Line C_OPEN, 351
5.7.7 Coplanar Short-Circuited Transmission Line C_SHORT, 353
5.7.8 Gap in a Coplanar Transmission Line C_GAP, 354
5.7.9 Step in a Coplanar Transmission Line C_STEP, 355
5.7.10 Coplanar Waveguide Taper C_TAPER, 357
5.7.11 Coplanar Air Bridges C_AIR, 359
5.7.12 Bend in a Coplanar Transmission Line C_BEND, 360
5.7.13 T-Junction in Coplanar Transmission Lines C_TEE, 362
5.7.14 Crossing of Coplanar Transmission Lines C_CROSS, 364
5.7.15 Coplanar Interdigital Capacitor C_IDC, 366
5.7.16 Coplanar Rectangular Inductor C_RIND, 368
5.7.17 Coplanar Thin-Film Resistor C_TFR, 370
5.7.18 Coplanar Metal–Insulator–Metal Capacitor C_MIM, 371
Bibliography, 373

6 Coplanar Filters and Couplers **377**

6.1 Coplanar Lumped Element Filters, 377
6.1.1 The Coplanar Spiral Inductor as a Filter, 377
6.1.2 Design and Realization, 379
6.1.3 Results, 381
6.1.4 Phase-Shifting Filter Circuits, 386
6.2 Coplanar Passive Lumped-Element Band-Pass Filters, 388
6.2.1 Theoretical Background, 389
6.2.2 Properties of the Coplanar Hybrid Band-Pass Filters, 390
6.3 Special Coplanar Waveguide Filters, 392
6.3.1 The Coplanar Band-Reject Filter, 394
6.3.1.1 The Hybrid Band-Reject Filter, 394
6.3.1.2 The Monolithic Band-Reject Filter, 395
6.3.2 Coplanar Millimeter-Wave Filters, 398

6.4 Coplanar Edge-Coupled Line Structures, 404
6.4.1 Verification of Coupling Between Coupled Coplanar Waveguides, 405
6.4.2 End-Coupled Coplanar Line Structures, 409
6.4.3 Coplanar Waveguide End-Coupled to an Orthogonal Coplanar Waveguide, 411
6.5 Coupled Coplanar Waveguide Filters and Couplers, 414
6.5.1 Interdigital Filter Design, 414
6.5.2 Coplanar Waveguide Couplers, 420
6.6 Coplanar MMIC Wilkinson Couplers, 426
6.6.1 Conventional Wilkinson Couplers, 427
6.6.2 Wilkinson Couplers with Discrete Elements, 427
6.6.3 MMIC Applicable Wilkinson Couplers with Coplanar Lumped Elements, 429
6.6.4 Wilkinson Coupler in Coplanar Waveguide Technique for Millimeter-Wave Frequencies, 431
Bibliography and References, 434

7 Coplanar Microwave Integrated Circuits 439

7.1 Introduction, 439
7.1.1 The Effect of the Shielding on Modeling, 440
7.1.2 The Waveguide Properties, 441
7.2 Coplanar Transistors and Coplanar Switches, 444
7.2.1 Active Power Dividers and Combiners and Switches, 444
7.2.1.1 Power Dividers and Combiners, 444
7.2.1.2 Fundamental Coplanar Switch Circuits, 446
7.2.1.3 Results and Measurements, 447
7.2.1.4 Device Scaling, 450
7.2.1.5 Design and Realization of Coplanar RF Switches, 453
7.3 Coplanar Microwave Active Filters, 457
7.3.1 Introduction, 457
7.3.2 The Coplanar Active Inductor, 458
7.3.3 The First-Order Active Coplanar Band-Pass Filter, 460
7.3.4 The Fixed Center Frequency Second-Order Active Filter, 460
7.3.5 The Coplanar Active Tunable Filter, 463
7.4 Coplanar Microwave Amplifiers, 471
7.4.1 Coplanar Microwave Amplifiers in Waveguide Design, 471
7.4.1.1 Introduction, 471
7.4.1.2 Circuit Design and Technological Aspects, 472
7.4.1.3 Results and Comparison with Measurements, 475
7.4.2 Coplanar Lumped-Element MMIC Amplifiers, 477
7.4.2.1 Introduction, 477
7.4.2.2 MMIC Design and Results, 478

7.4.3 Influence of the Backside Metalization on the Design of a Coplanar Low-Noise Amplifier, 481
7.4.3.1 Modeling the Transistor and Its Noise Properties, 481
7.4.3.2 The Coplanar LNA Design, 484
7.4.3.3 Simulation Results, 484
7.4.3.4 Measurement Results, 485
7.4.4 Miniaturized Ka-band MMIC High-Gain Medium-Power Amplifier in Coplanar Waveguide Technique, 488
7.4.4.1 Introduction, 488
7.4.4.2 MMIC Design and Results, 488
7.5 Coplanar Electronic Circulators, 491
7.6 Coplanar Frequency Doublers, 495
7.6.1 Different Realization Concepts of FET Frequency Doublers, 495
7.6.1.1 The Single-Device FET Frequency Doubler, 495
7.6.1.2 The Balanced (Push–Push) FET Frequency Doubler, 495
7.6.1.3 The Wideband FET Frequency Doubler, 497
7.6.2 Realization of Coplanar Frequency Doublers, 497
7.6.2.1 The Coplanar Balanced Hybrid MIC Frequency Doubler, 498
7.6.2.2 The Coplanar Balanced Monolithic MIC Frequency Doubler, 500
7.6.3 A Coplanar Times Five Frequency Multiplier, 504
7.7 Microwave and Millimeter-Wave Oscillators in Coplanar Technology, 508
7.7.1 Coplanar Microwave Oscillators, 508
7.7.2 A 5-GHz Coplanar Voltage-Controlled Oscillator, 514
Bibliography and References, 518

Index **537**

PREFACE

This book combines the research results of a large research group under the leadership of the author and his colleagues at the University of Duisburg, Duisburg, Germany in the 1990s and later at the author's private research institute, the IMST GmbH, Kamp-Lintfort, Germany. Research subjects have been the materials, the technology, the design, and the realization of coplanar microwave integrated circuits. The author himself was responsible for the design and realization of this kind of circuit, the theoretical background, and the realization of simulating the various components, structures, and circuits. A large number of doctoral theses were elaborated in the research group under the author's guidance at that time. They are referenced in the bibliographies of the relevant chapters. The author has made intensive use of the results described in these dissertations when writing this book.

In the early years the research group was financed in the form of a collaborative research center (Sonderforschungsbereich) at the University of Duisburg by the German Research Foundation (Deutsche Forschungsgemeinschaft, DFG). The author thankfully acknowledges the great financial help given by the DFG in the form of this intensive research grant. In recent years the work has been continued at the private research institute of the author, the IMST GmbH, under various national and European research projects, funded by the State Government of the State Nordrhein-Westfalen, the German Federal Ministry of Education and Research (Bundesministerium für Bildung und Wissenschaft, BMBF), the European Community, and the European Space Agency (ESA). Also the results of research and development projects bilateral with industry companies and other research institutes shall be mentioned here. They also have been used in this book if they have been published in the open literature. The author is grateful for the huge support he and his research groups received from all of the mentioned partners.

Dr. Mohammed Abdo Tuko, an earlier scientist in the authors research group and now Professor at the Addis Ababa University, Ethiopia, corrected the English language of the first manuscript. The author thanks him for the intensive work he has contributed to this project.

Kamp-Lintfort
January 2006

INGO WOLFF

1

INTRODUCTION

In modern information and communication techniques, planar integrated microwave circuits play an important role. Such planar microwave circuits were used for the first time in the 1950s. They are produced with thin-film metallic strip lines on a plastic or ceramic substrate material, are cost-effective, and need reduced space as compared to, for example, waveguide circuits. Moreover, active elements like diodes and transistors can be easily integrated into the metallic planar waveguide structures. During the first 40 years of planar circuit development the so-called microstrip line that had been developed by ITT [1] was used primarily in planar microwave integrated circuit design. Active semiconductor elements as well as thin-film and thick-film capacitors and resistors have been integrated into the circuits using hybrid technologies.

With the development of modern microwave transistors like field effect transistors (MESFETs: metal-semiconductor field effect transistors) and heterostructure field effect transistors (HEMTs: high electron mobility transistors) on GaAs or InP materials, the application of hybrid and also of monolithic microwave integrated circuits has grown intensively over the last 25 years. Today, a broad class of analog and function block circuits is available to the microwave engineer in a frequency range from 0 to about 150 GHz. A wide range of literature has been published in international conference proceedings, in leading international journals, and in specialized books on the subject, such as references 2–6.

Coplanar Microwave Integrated Circuits, by Ingo Wolff.

Monolithic microwave integrated circuits (MMIC) offer the advantage of a cost-effective mass production, improved electrical parameters, smaller size and weight as well as improved reliability compared to the hybrid integrated circuits. The disadvantage of monolithic integrated circuits compared to the hybrid integrated ones is that a tuning, as it is possible for hybrid integrated circuits, is almost impossible after production. The design costs are normally very high, and the additional technology through-run that might be needed due to design errors is highly expensive. Therefore, accurate design tools are needed for an optimal "first shot" design result.

Looking closely to the technologies, which have been applied for the microwave integrated circuit design and production so far, a large part of all realized circuits (including possibly lumped elements) use a microstrip-based technology. Figures 1.1a to 1.1d show the most common forms of the microstrip line that have been used. Figure 1.1a shows the conventional microstrip line, which consists of a strip of width w and metalization thickness t on top of a substrate material of height h, which may be a dielectric material (plastic-based or ceramic) or a semi insulating semiconductor material (e.g., GaAs, InP). The backside of the substrate is completely covered by a metalization layer. The fundamental mode of the microstrip line is a quasi-TEM mode that has a dispersive behavior because at higher frequencies the electromagnetic field is more and more concentrated into the dielectric carrier material.

Figure 1.1b shows the so-called strip line where the strip of width w is inserted within a homogeneous dielectric material of relative permittivity ε_r shielded by two large conducting planes on top and bottom of the substrate material. The fundamental mode on this line is a true dispersion less TEM

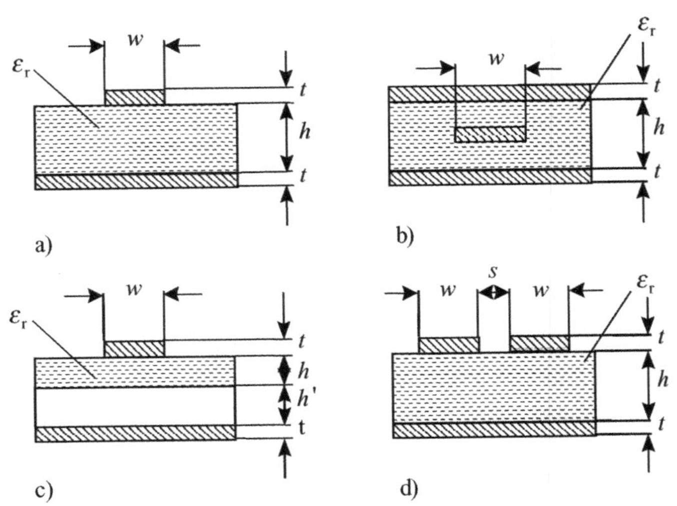

Fig. 1.1. Fundamental microstrip waveguides as they are used in microwave integrated circuits: (a) The conventional microstrip line, (b) the strip line, (c) the suspended microstrip line, and (d) the coupled microstrip lines.

mode, but this line is used only for special applications, such as in high-quality filter structures. This line is not commonly used for hybrid or monolithic integrated circuit applications because the implementation of active semiconductor elements cannot be easily realized.

The suspended microstrip line, which has a substrate material of reduced thickness separated from the ground metalization by an air region (Fig. 1.1c), is also normally only used for filter applications and only very seldom for circuit applications. The reduced substrate thickness leads to lower dielectric losses, which makes this line attractive for low-loss filters. Also, because of the small substrate height, the dispersion of this line is smaller than that in the case of the conventional microstrip line (Fig. 1.1a).

The coupled microstrip lines, shown in Fig. 1.1d, are often used in microwave integrated circuits, when couplers or filters are to be realized within the circuitry. The two lines can carry two fundamental quasi-TEM modes, the even and the odd mode, which have different effective dielectric constants (i.e., different phase velocities of their waves) and different dispersion properties because of the different field structures of the modes. This line structure often appears within a circuit if the circuit is not designed carefully enough and if two single microstrip lines come too close to each other. This leads to an unwanted parasitic coupling within microstrip circuits, which can be avoided only by leaving enough space between the two lines so that the coupling coefficient is reduced to an acceptable low value. This is one reason why microstrip-based circuits often need large space for their proper realization.

Figures 1.2a to 1.2d show an alternative line for the design of microwave integrated circuits—that is, coplanar waveguide structures. The coplanar strips

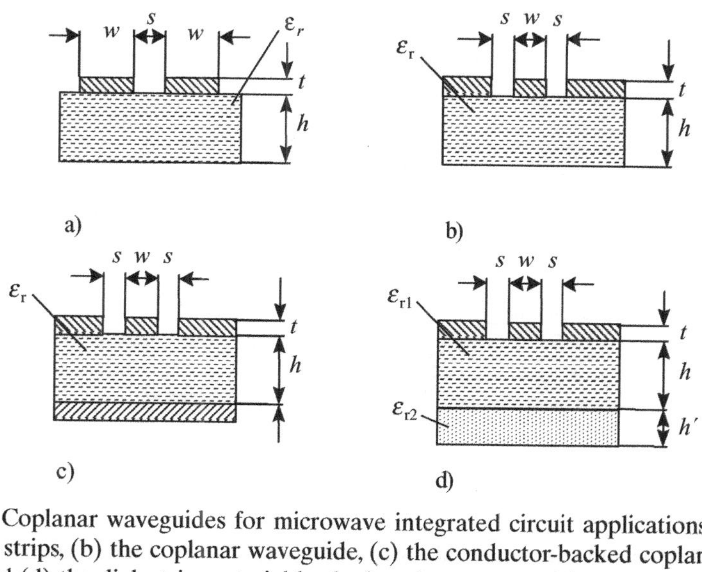

Fig. 1.2. Coplanar waveguides for microwave integrated circuit applications: (a) The coplanar strips, (b) the coplanar waveguide, (c) the conductor-backed coplanar waveguide, and (d) the dielectric-material-backed coplanar waveguide.

shown in Fig. 1.2a are normally used only in low radio-frequency (rf) circuits in conjunction with hybrid and/or lumped planar elements. For higher microwave frequencies, this line is not used in circuit design because it has a large stray field and does not define a solid common ground plane condition.

A true alternative to the microstrip line especially for applications in modern microwave integrated circuit design is the coplanar waveguide shown in Fig. 1.2b, which is the subject of this book. Alternative forms like the conductor-backed coplanar waveguide or the dielectric-material-supported coplanar waveguide are shown in Figs. 1.2c and 1.2d, respectively. Their properties are discussed in Chapter 2. The coplanar waveguide has the "hot" strip and the ground planes both on top of the dielectric carrier material and therefore forms a *real planar waveguide*. Because, in principle, it is a three-conductor line, it can carry two fundamental modes with zero cutoff frequency: (a) the so-called "even mode," which has equal potentials of the ground planes, and (b) the so-called "odd mode," which has ground potentials of different signs but equal magnitude.

Figure 1.3 shows the electric and the magnetic field distribution of (a) the even mode (coplanar waveguide mode) and (b) the odd mode (slotline mode). The even mode is a quasi-TEM mode with even symmetry with respect to the symmetry plane, its dispersion is very low (see also Chapter 2), and it is normally used for application in circuit design. The electric field lines begin (or end) at the center conductor and they end (or begin) on the two surrounding ground planes. The magnetic field lines enclose the center conductor. If current is transported on the center conductor (e.g., with direction into the paper plane as shown in Fig. 1.3a), the current densities in the ground planes have the

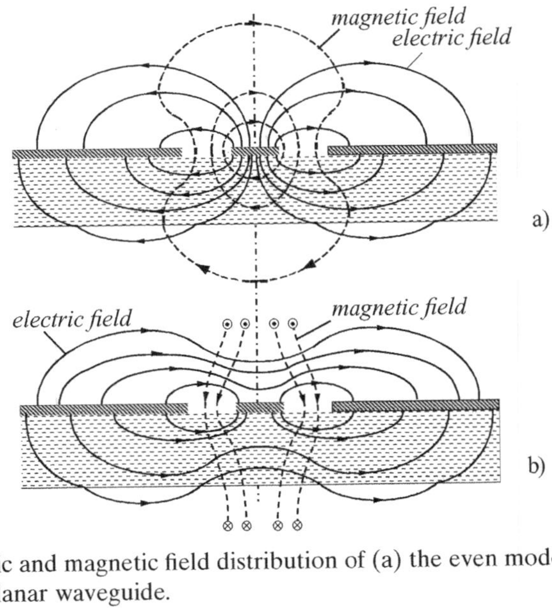

Fig. 1.3. Electric and magnetic field distribution of (a) the even mode and (b) the odd mode on a coplanar waveguide.

opposite direction. Because of the low dispersion of the fundamental "even mode," very broadband applications are possible, making this mode propagation applicable in microwave integrated circuits.

The electric field lines of the odd mode start on one ground plane and end on the other ground plane, which means that the potentials of the two ground planes have opposite signs. Not all of the electric field lines touch the center conductor. In the case of infinitely wide ground planes the odd mode, like a slot-line mode, is a hybrid mode and has magnetic field components in longitudinal direction and its dispersion can be considered large. If the ground plane width is finite, the magnetic field lines may be closed in the cross section enclosing the ground planes.

Despite its promising properties, the coplanar waveguide, up to now, has been used only seldom in commercial microwave integrated circuits. This is astonishing because in 1969 Wen [7] proposed the coplanar line as a possible microwave waveguide and in 1976 and 1977 Houdart [8, 9] demonstrated the big advantages of this waveguide in microwave circuit applications.

Tables 1.1 and 1.2 show two tables published in similar form by Houdart [8] in 1976. The tables show that he really recognized already at that time the broad application range of coplanar lines and components. He showed that the coplanar circuit approach is especially interesting for the realization of hybrid and monolithic microwave integrated circuits because it has several advantages compared to the microstrip line technique. An application of coplanar technologies to circuit design has been first described by Simon [10].

These advantages, as they are seen today (and as they already had been seen by Houdart 30 years ago), are as follows:

TABLE 1.1. Properties of Various Microwave Microcircuit Techniques as First Shown by Houdart [8]

	Microstrip Line	Suspended Strip Line	Slotline	Coplanar Waveguide
Characteristic impedance	25–95 Ω	40–130 Ω	40–130 Ω	30–140 Ω
Effective dielectric constant for $\varepsilon_r = 9.8$	≈6	≈2.4	≈5	≈5
Spurious modes	Low	High	Non-TEM propagation	Low
Integration level	High	Low	—	High
Technological difficulties	Ceramic holes edge plating	—	Double-side etching	—
Parallel components	Poor	Difficult	Easy	Easy
Series components	Easy (except distributed lines)	Easy (except distributed lines)	Difficult	Easy

TABLE 1.2a. Fundamental Lumped Elements and Filter Elements Realized in Coplanar Waveguide Technology

Circuit Element	Equivalent Circuit	Application
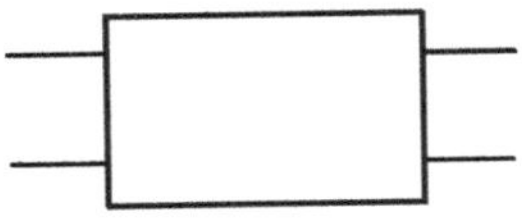	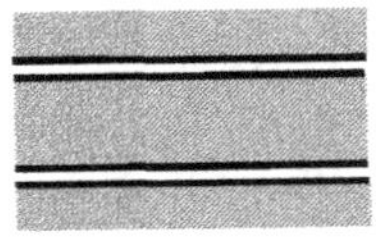	Transmission line
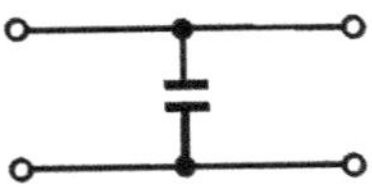		Stop-band filter
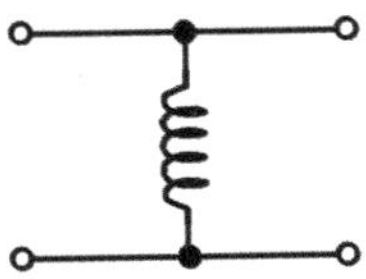	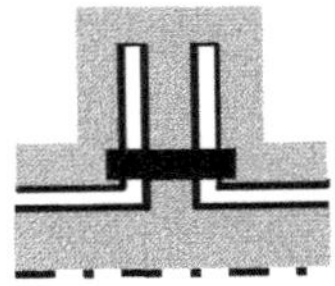	Pass-band filter
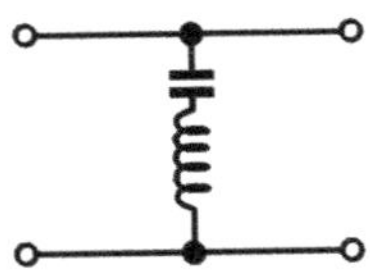		Stop-band elliptic filter

Source: After Houdart [8].

TABLE 1.2b. Fundamental Lumped Elements and Filter Elements Realized in Coplanar Waveguide Technology

Circuit Element	Equivalent Circuit	Application
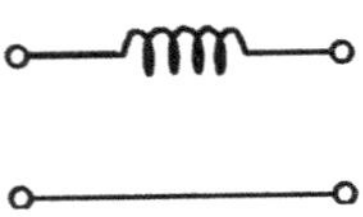	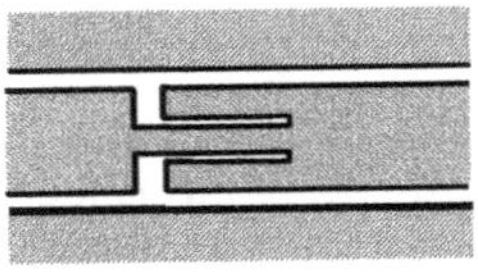	Stop-band filter
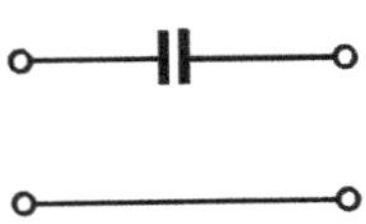	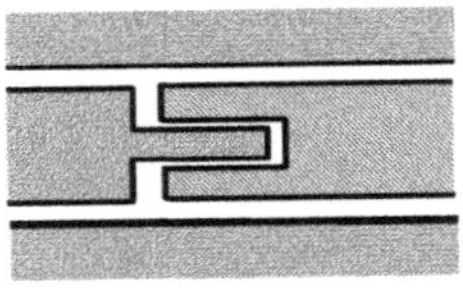	Pass-band filter
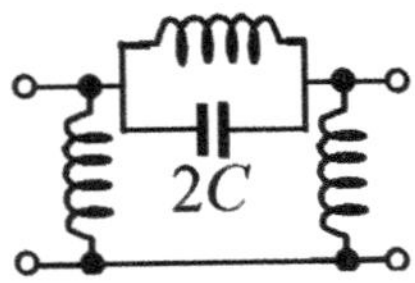	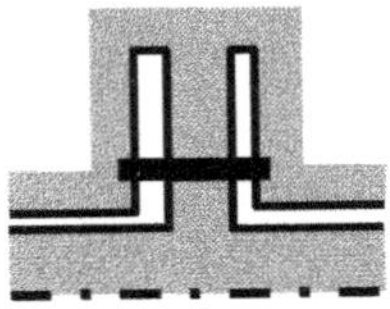	High-pass filter
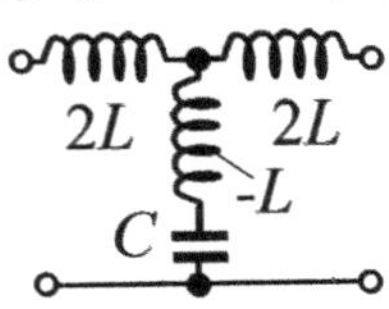	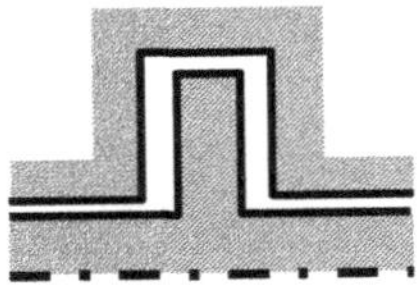	All-pass filter

Source: After Houdart [8].

- The available range of characteristic impedances is larger for the coplanar line (30–140 Ω) than for the microstrip line (25–95 Ω), for example.
- The coplanar-based microwave integrated circuit is a real planar circuit because the "hot" lines as well as the ground planes are located on the upper surface of the carrier material. This enables series and parallel implementation of active and passive lumped elements into the circuit without any via hole connections through the substrate material. Good ground contacts can be realized anywhere in the circuit, and the space saved from the elimination of via holes leads to a more condensed circuit design.
- No backside preparation and no substrate thinning are needed because the coplanar circuit in principle can work with arbitrarily thick substrate materials. Heat transfer problems can be solved using a flip chip technology when mounting the circuits into a housing. Together with the above-mentioned advantage of avoiding the via-holes, it means that three essential technology drawbacks, which might reduce the yield of the circuit production and which increase the costs, can be avoided.

  The coplanar technology provides the possibility to design highly condensed microwave integrated circuits, especially if additional use is made of a lumped element technique. Very small circuit layouts can be made up to highest frequencies. Because the fundamental coplanar waveguide does not use a conducting ground plane on the backside of the substrate material, the parasitic capacitances of the lumped circuit components like spiral inductors or interdigital capacitors are small compared to the microstrip case. This results in a much higher first resonant frequency of these components so that even at millimeter-wave frequencies (e.g., 40–60 GHz) a lumped element technique can be used in coplanar monolithic integrated circuits.
- The fundamental even mode of the coplanar waveguide is less dispersive than the fundamental mode of the microstrip line. This is especially true if the coplanar waveguides are carefully designed—that is, if small gap widths s are used. So, broadband circuits from low rf frequencies up into the millimeter-wave range can be realized. Because the coplanar waveguide has two geometrical design parameters for optimizing the waveguide with respect to the circuit requirements (line width w and gap width s), it has one more degree of freedom for the circuit designer than does the microstrip line.
- Finally, simple coplanar-based on-wafer measurement techniques are available for testing the coplanar circuits. On-wafer measurement results may be directly interpreted and transferred to the component or circuit properties, something that is not always true in the case of a microstrip-technology-based circuit or component.

For a long time, several disadvantages were claimed regarding the application of coplanar waveguides in integrated circuits. They shall be discussed here briefly:

- First it was claimed that the coplanar waveguide has higher losses compared to the microstrip line. As already mentioned above, there is one more geometrical parameter available for the design of a coplanar waveguide compared to the microstrip line so that, for instance, a 50-Ω line may be realized in many ways using different *w* and *s* values. Moreover, the losses of a 50-Ω line can be changed by, say, using a waveguide with a large center strip width. Therefore, by applying this technique, the losses of the coplanar waveguides can always be kept in the same order as those of the microstrip line.
- The second argument against coplanar circuits has been that a large part of the expensive semiconductor substrate (e.g., GaAs) is covered by the ground planes, and therefore coplanar circuits are not cost-effective. As will be shown in this book, coplanar circuits can be designed smaller in size than microstrip-based integrated circuits because additional ground planes on top of the substrate can reduce the coupling between adjacent lines. In fact, space reduction in the order of 30–50% is possible if coplanar circuits are used instead of microstrip-based circuits.

One of the disadvantages of the coplanar waveguide, which has already been mentioned above, is the fact that two fundamental modes can propagate on the line with zero cutoff frequency if the two ground planes are not held at the same potential. In this book it will be shown that different air-bridge techniques, which are able to sufficiently suppress the unwanted "odd mode" of the coplanar guide and which also do not incur an additional technology cost in the production of the circuits, have been developed for application in coplanar MMICs. In coplanar hybrid integrated circuits, this problem is a little bit more difficult because using (for example) bond wires as air bridges is not always easy, since a production of the bonded bridges with an accuracy and reproducibility required for high-quality circuits is difficult.

Finally, there is one main reason that, as the author of this book feels, kept the coplanar technique from being applied intensively: No accurate and flexible design basis was available for a long time. All available commercial circuit design software tools were specialized on the design of microstrip circuits, so the practicing engineer did not really dare to use the coplanar concept for his/her circuit design. Parallel to this book, the author and his research group have developed a software basis that can be implemented into the most common circuit design programs and that contains models for nearly all line structures, discontinuities, and lumped elements needed in a coplanar environment for circuit design. These design tools that have been intensively evaluated up to frequencies of 70 GHz should help the microwave engineer to realize that circuit design on the basis of coplanar waveguides can be much easier than in the microstrip case. At the end he will really enjoy the advantages and possibilities, which lie behind coplanar technology.

REFERENCES

1. D. D. Grieg and H. F. Engelmann, Microstrip—A new transmission technique for the kilomegacycle range, *Proc. IRE*, vol. 40, no. 12, 1952, pp. 1644–1650.
2. F. Ali, I. Bahl, and A. Gupta, *Microwave and Millimeter-Wave Heterostructure Transistors and Their Applications*, Norwood, MA: Artech House, 1989.
3. R. Goyal, *Monolithic Microwave Integrated Circuits: Technology & Design*, Norwood, MA: Artech House, 1989.
4. P. H. Ladbrooke, *MMIC Design GaAs FETs and HEMTs*, Norwood, MA: Artech House, 1989.
5. M. J. Howes and D. V. Morgan, *Gallium Arsenide, Materials, Devices, and Circuits*, Chichester: John Wiley & Sons, 1985.
6. L. E. Larson, *RF and Microwave Circuit Design for Wireless Communication*, Boston: Artech House, 1996.
7. C. P. Wen, Coplanar waveguides: A surface strip transmission line suitable for nonreciprocal gyromagnetic devices applications, *IEEE Trans. Microwave Theory Tech.*, vol. MTT-17, 1969, pp. 1087–1090.
8. M. Houdart, Coplanar lines: Application to broadband microwave integrated circuits, in: *Proceedings, 6th European Microwave Conference*, Rome, Italy, 1976, pp. 49–53.
9. M. Houdart, Coplanar lines: application to lumped and semilumped microwave integrated circuits, in: *Proceedings 7th European Microwave Conference*, 1977, pp. 450–454.
10. R. N. Simon, *Coplanar Waveguide Circuits Components and Systems*, New York: John Wiley & Sons, 2001.

2

TRANSMISSION PROPERTIES OF COPLANAR WAVEGUIDES

2.1 RIGOROUS, FULL-WAVE ANALYSIS OF TRANSMISSION PROPERTIES

In this chapter the full-wave propagation characteristics of coplanar waveguides shall be studied using rigorous analysis techniques like the spectral domain analysis that is known to be a fast and accurate computation technique, especially well-suited for the analysis of planar transmission line structures. Also the finite-difference time-domain (FDTD) analysis technique that is often applied to control the frequency-dependent transmission parameters of components and subsystems will be partly used. Using these techniques, it will be shown that dispersion of the coplanar waveguide mode—that is, the fundamental even mode on a coplanar waveguide (see Chapter 1), normally used in the circuit design—is small. As a result, approximate quasi-static methods can be applied in many cases and with high accuracy if CAD models for the analysis of coplanar circuits are developed.

First, a rigorous but simple spectral domain analysis approach will be used to compute the characteristics (effective dielectric constant as a measure for the phase velocity of wave propagation, characteristic impedance, and dielectric and ohmic losses) of coplanar waveguides, including their frequency dependence [250]. It includes the singularities of the currents on the strips and allows a computation of the characteristic impedances of individual strips. The formulation takes into account also the parasitic effects due to a finite ground

Coplanar Microwave Integrated Circuits, by Ingo Wolff.

plane width, which leads to changes of the waveguide impedances and propagation constants. Coplanar waveguides with a single center strip and with two or more coupled center strips will be discussed as examples.

In the second applied spectral domain technique, some additional effort has been put into the analysis techniques. That is, a method that is able to directly integrate the dielectric and conductor losses into the analysis is used [274]. Furthermore, this method considers also vertical current elements in the analysis and, therefore, can analyze real three-dimensional structures such as air bridges that are intensively used in coplanar integrated circuits.

The frequency-dependent computation of coplanar transmission line characteristics in spectral domain technique is well known and has been applied by a large number of authors [e.g., 7, 20, 35, 56, 65]. Since the task of this book is to prepare the basis for microwave integrated circuit design and not to describe field theoretical methods, these methods will not be discussed here; they are only applied to the coplanar waveguide structures, and the derived results are discussed.

Finally, in various sections also the finite-difference time-domain technique (FDTD) [360] is used to analyze the coplanar waveguide structures. The FDTD method is widely known in the mean time and is applied in many microwave design areas, so it must not be described here again.

2.1.1 The Coplanar Waveguide with a Single Center Strip and Finite Ground-Plane Width

As a first application of the described analysis techniques, coplanar waveguides with a single center strip (which is the conventional form of the coplanar waveguide) shall be considered. In this first examination, the ground planes of the coplanar waveguides are assumed to be of finite width, as shown in Fig. 2.1.1.

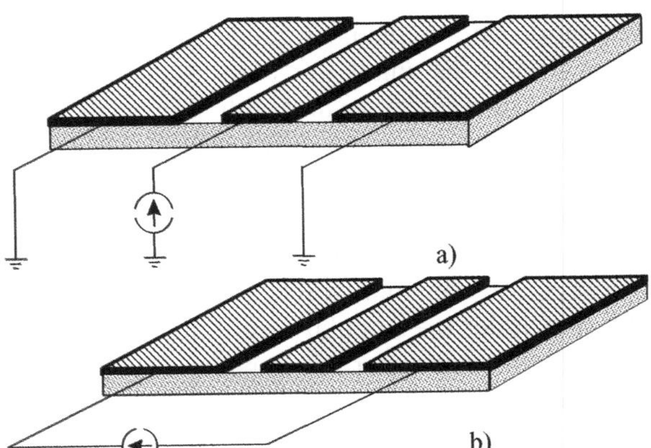

Fig. 2.1.1. Excitation (a) of the even mode (the coplanar waveguide mode) and (b) the odd mode (the slot-line mode) on a coplanar waveguide.

If the ground planes are of sufficient width, this assumption does not influence the properties of the fundamental even coplanar waveguide mode much (see discussion below), but it has a large effect on the odd mode and its properties, as will be shown in the next section. In the case of finite ground-plane width, it is not assured in the simulation that the ground planes always are at the same potential (i.e., $\varphi = 0$), as will be assumed (and guaranteed by air bridge technologies) in coplanar integrated circuits. The results that will be demonstrated in Section 2.1.1 are surely of high relevance for many applications in circuit design, but coplanar waveguides with finite ground plane widths are also used in various other applications. It will also be assumed that the coplanar waveguide in this first examination is enclosed in a metallic shielding that can be assumed to represent the package, which is always available in a real microwave integrated circuit.

The excitation of the two fundamental modes on a coplanar waveguide (called the even and the odd modes) is shown in Fig. 2.1.1. In the literature the even mode is often referred to as the *coplanar waveguide mode*, and the odd mode is often called the *slot-line mode.*

The electric and the magnetic field of the coplanar waveguide with finite ground plane width have been computed at a frequency of 1 GHz for both the even and the odd mode, and they are shown in Figs. 2.1.2 and Fig. 2.1.3. What is shown is a coplanar waveguide that is carried on a dielectric substrate material of dielectric constant $\varepsilon_0\varepsilon_r$ and height h. Above and below the substrate, a vacuum with the dielectric constant ε_0 is assumed. The metalization on top of the substrate consists of the center-strip conductor and the metalization of the two ground planes that are finite in width. One notices that the fields of the even mode (coplanar waveguide mode) are confined near the gaps between the conductors of the waveguide. The electric field lines are directed from the center conductor to the ground planes. The magnetic field lines surround the center conductor. On the other hand, the fields of the odd mode (slot-line mode) are more scattered in the space between the ground planes and they resemble the fields of an odd mode of two coupled strip lines or a slot line with a spacing of $w + 2s$. The electric field lines run from one of the ground planes to the other, nearly not touching the center conductor.

Both modes have a field distribution that is symmetrical with respect to the symmetry plane of the structure. The symmetry plane is a magnetic wall in the case of the even mode and an electric wall in the case of the odd mode. An introduction of an adequate wall into the symmetry plane would not disturb the field distributions that are shown in Fig. 2.1.2 and Fig. 2.1.3, respectively, for the even and the odd mode.

In monolithic microwave integrated circuits (MMICs), coplanar waveguides are frequently enclosed in a metallic shielding or they are conductor-backed, which leads to an additional parasitic (even) mode with a zero cutoff frequency. Its fields are shown in Fig. 2.1.4.

The field of the parasitic even mode (Fig. 2.1.4) is the most scattered of the three considered modes, and it propagates mostly in the air space above the

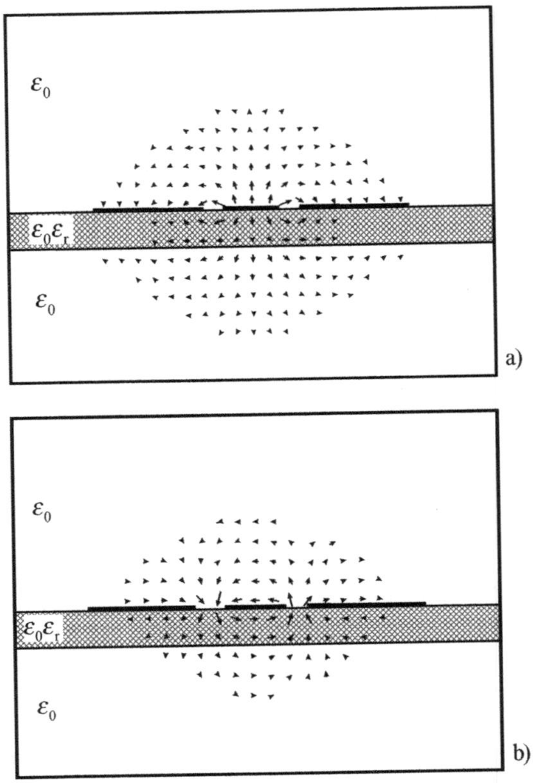

Fig. 2.1.2. The field distribution of the fundamental even mode (the coplanar waveguide mode) on a shielded coplanar waveguide with a single center strip. (a) The electric field and (b) the magnetic field.

conductors and below the substrate just like in a waveguide mode in a metallic waveguide. In the case where a coplanar circuit is conductor-backed or is enclosed inside a metallic package, this mode may form a cavity oscillation and may lead to a parasitic coupling between different parts of the circuit. To avoid such kind of parasitic coupling, a good knowledge of the propagation coefficients of these modes or the related cavity resonance frequencies is necessary. It may be derived from a full-wave analysis program like the one used here.

If the currents carried by the strip conductors of the two fundamental modes are calculated, it may be recognized that in the case of the even mode, the center conductor carries a current, which is the sum of the currents in the two outer ground planes in the opposite direction. In the case of the odd mode, the center conductor carries nearly no current. The current flows in the two outside ground planes in opposite directions.

The phase velocities of the fundamental modes on a coplanar waveguide are described by an effective dielectric constant using the same definition as in the case of a microstrip line, that is,

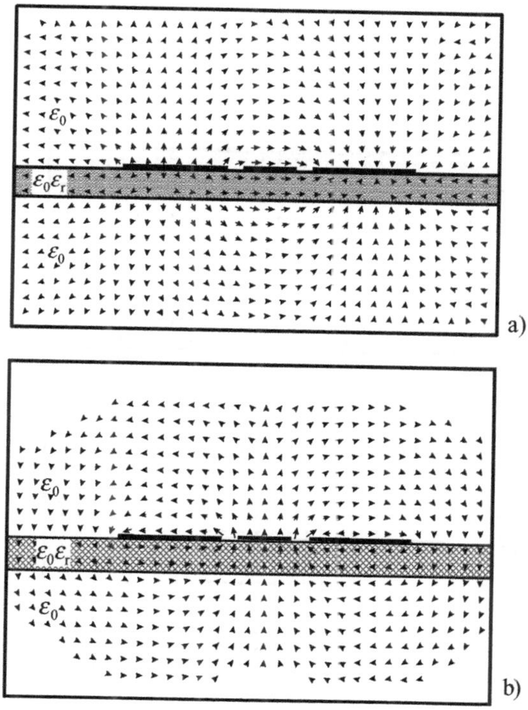

Fig. 2.1.3. The field distribution of the fundamental odd mode (the slot-line mode) on a shielded coplanar waveguide with a single center strip. (a) The electric field, (b) the magnetic field.

$$v_{ph} = \frac{c_0}{\sqrt{\varepsilon_{eff}}}. \tag{2.1.1}$$

The effective dielectric constants of the fundamental even and odd modes are given in Fig. 2.1.5 for different gap width (s) to substrate height (h) ratios as a function of frequency. These values are again calculated using the simple moment method as described briefly above, without considering losses within the line structure. The effective dielectric constant of the even mode, especially for small gap widths (i.e., s/h values), is less frequency-dependent than that of the odd mode.

If the coplanar waveguide is properly designed and a correct value of s/h is chosen, the dispersion of the effective dielectric constant of the even mode can be kept small (below 1%) for frequencies up to 40 GHz or even higher. On the other hand, the effective dielectric constant of the odd mode is strongly frequency-dependent. This is due to the fields of the odd mode (see Fig. 2.1.3) that are much more scattered in the space surrounding the conductors than those of the even mode. The odd mode is more sensitive to an increase of

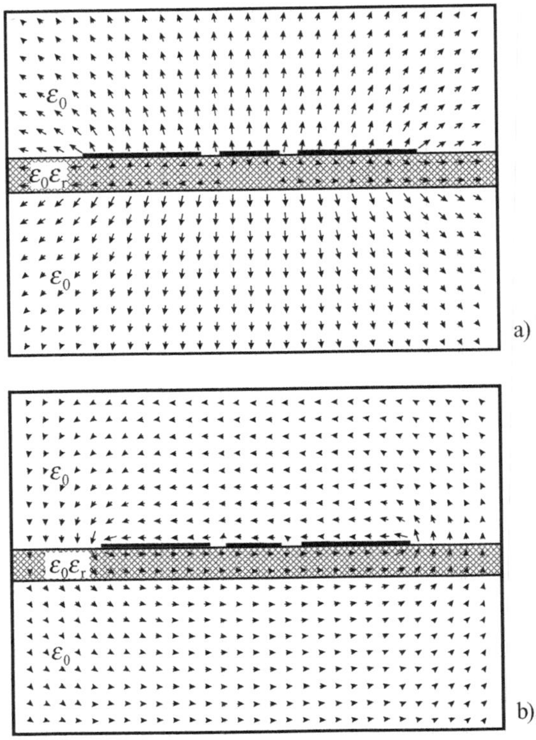

Fig. 2.1.4. The field distribution of the parasitic even mode on a coplanar waveguide with a single center strip. (a) The electric field, (b) the magnetic field.

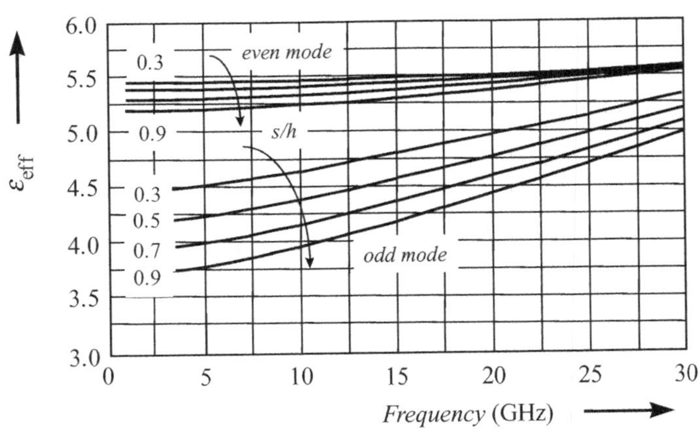

Fig. 2.1.5. Frequency dependence of the effective dielectric constant of the even and the odd mode on a coplanar waveguide with a single center strip, with the gap width s to substrate material h ratio as a parameter. s/h values = 0.3, 0.5, 0.7, and 0.9. $\varepsilon_r = 10$, $h = 635\,\mu m$.

frequency that leads to a concentration of the electromagnetic fields in the dielectric medium—that is, in the gaps between the strips. Larger gaps, which result in larger scattering of the electromagnetic field, also lead to a stronger dispersion of the effective dielectric constant, as can be clearly seen (from Fig. 2.1.5) as well for the even mode as for the odd mode. It should be pointed out again that the widths of the ground-plane strips are finite for the considered coplanar waveguide. In this case, the odd mode can propagate down to zero frequency because the two ground planes may have different potentials even at zero frequency. As a result, the effective dielectric constant of the odd mode is finite at zero frequency, as may be seen from Fig. 2.1.5 (compare also with Fig. 2.1.18, Section 2.1.2 for the case of an infinite ground plane width).

Figure 2.1.6 shows the computed power concentration ratio of the even and the odd mode on the considered coplanar waveguide. It is defined as the ratio of the power concentrated in the dielectric carrier material to the total power transported through the cross section of the waveguide. The frequency-dependent curves shown for the power concentration ratio confirm the well-known fact that the fields concentrate in the dielectric material and therefore near the slots of the coplanar waveguide for higher frequencies. The even mode propagates along three conductors (the center conductor of small width and the two ground planes of larger widths) while the odd mode, in principle, propagates only along the two ground conductors with spacing $w + 2s$. The center conductor is nearly not recognized by the odd mode. The power concentration ratio of the even mode for all frequencies is nearly equal to 0.5; that is, half of the transported power is concentrated in the air region above the substrate plane, and the other half is below the conductor plane in the sub-

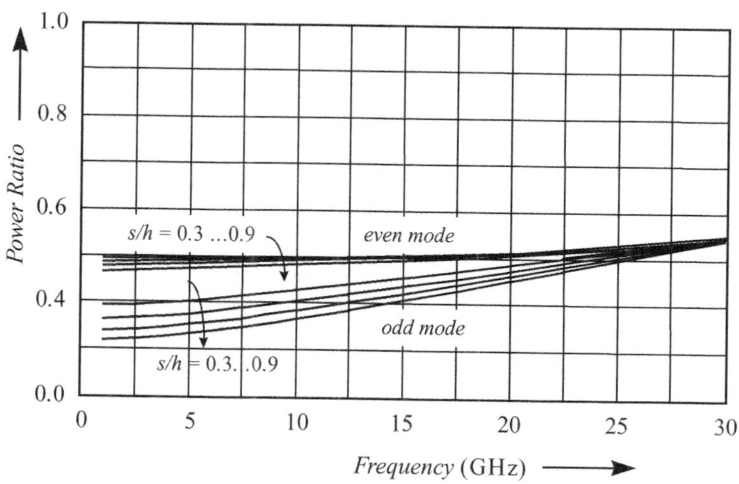

Fig. 2.1.6. The power concentration ratio of the power transported in the substrate and the power totally transported through the cross section of the coplanar waveguide in dependence on the frequency and with the slot width s to substrate height h ratio taken as a parameter. s/h values = 0.3, 0.5, 0.7, and 0.9. $\varepsilon_r = 10$, $h = 635\,\mu\text{m}$.

strate region. In a case of a very thin substrate and an air region below it, a small part of the field may also fill this air region.

In a well-designed case of a coplanar waveguide, the fields of the even mode are kept close to the gaps and this situation does not change much with frequency, especially not if the slot width s is small. If the height of the dielectric carrier material is large or if there is no air region under the substrate, then the effective dielectric constant of the even mode as a first approximation is given by

$$\varepsilon_{\text{eff}} \approx \frac{\varepsilon_r + 1}{2}. \tag{2.1.2}$$

From Fig. 2.1.5 it can also be observed that the values of the effective dielectric constants of the even mode and the odd mode (especially at higher frequencies) are very close to each other, so that a coupling between these two modes may occur and power may be converted from the even to the odd mode or vice versa in a microwave integrated circuit that is based on the coplanar waveguide as a transmission medium. The same is true with respect to the parasitic even mode. For circuit applications, the unwanted odd modes can be suppressed by adequate methods like air bridges as described in detail in Section 3.5.5. They provide equal potentials on both the ground planes so that the odd mode cannot be excited or will be suppressed if it is excited (e.g., at a line discontinuity). The parasitic mode, especially in conductor-backed coplanar circuits, cannot be controlled so easily in all cases.

Losses are claimed to be higher in coplanar waveguides, compared to the classical microstrip line. The computed attenuation coefficient α in dependence on the frequency is shown in Fig. 2.1.7. It is calculated using the simple

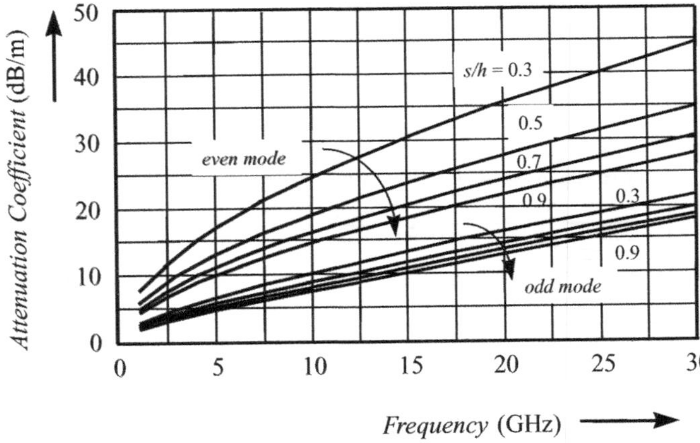

Fig. 2.1.7. The frequency dependence of the attenuation coefficient of the even and the odd mode on a coplanar waveguide for various slot width to substrate height ratios. s/h values = 0.3, 0.5, 0.7, and 0.9. $\varepsilon_r = 10$, $h = 635\,\mu\text{m}$.

moment method analysis described in the section above. In this method the computation of the losses is approximate, and it is done in a very simple way using the field distributions as calculated from the spectral domain analysis of the lossless structure. The dielectric losses (which, to a first approximation, can be neglected) and the conductor losses are then calculated using a perturbation technique. For the analysis of the conductor losses the surface resistance approach is applied.

From Fig. 2.1.7 it can be seen that the losses of the even mode of the coplanar waveguide are much higher compared to those of the odd mode. This is due to the fact that the electromagnetic field of the even mode is so closely concentrated in the gaps between the conductors that the current inside the center strip and the ground planes is heavily concentrated near the edges of the conductors, which leads to higher losses. Therefore, the losses increase with decreasing slot widths, as can be clearly seen from Fig. 2.1.7. For a low-loss design, therefore, large slot widths are needed. But this will possibly lead to higher dispersion, as shown in Fig. 2.1.5.

The problems that exist in the definition of the characteristic impedance for different propagation modes in the case of microstrip lines also exist for coplanar waveguides. This has been intensively discussed in the literature [119, 162, 192, 220]. Because the electromagnetic field of, say, an even mode is not really a TEM mode, a voltage between the electrodes and thereby a characteristic impedance of the line, in principle, cannot be defined. As can be seen from the above discussion of the effective dielectric constant (it means of the phase velocity of the propagating modes), the dispersion of the even mode is very low up to even high frequencies. This means that the even mode, to a good approximation, is a quasi-TEM mode, and therefore the problem of defining a characteristic line impedance is not so severe as in the case of the microstrip line. There are three possible definitions for the characteristic line impedance:

$$Z_{L1} = \frac{V}{I}, \qquad Z_{L2} = \frac{V^2}{2P}, \qquad Z_{L3} = \frac{2P}{I^2}, \tag{2.1.3}$$

where V is the voltage between the electrodes (center strip to ground plane), I is the current (e.g., in the center strip conductor), and P is the power transported along the line. All three definitions lead to different results of the characteristic impedances at higher frequencies.

In Fig. 2.1.8, the dependence of the characteristic line impedance of the even and the odd mode calculated using the simple spectral domain approach as described above and using the definition $Z_L = 2P/I$ are shown for different slot width s to substrate height h ratios. It can be observed that the dispersion of the characteristic impedance is much smaller for the even mode compared to the odd mode. Both values for the chosen geometrical parameters basically decrease with increasing frequency, but the dispersion of the even-mode characteristic impedance, to a first approximation, may be neglected up to frequencies of 40 GHz and for the line dimensions shown in Fig. 2.1.8. It is essential to mention that the odd-mode impedances are of the same order as

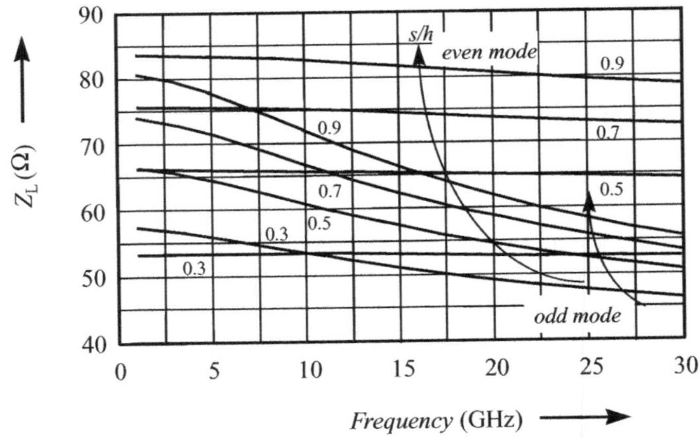

Fig. 2.1.8. Frequency dependence of the characteristic line impedance Z_L of a coplanar waveguide for varying gap width to substrate height ratios: $s/h = 0.3, 0.5, 0.7$, and 0.9. Substrate Al_2O_3, $\varepsilon_r = 10.0$, $h = 635\,\mu m$.

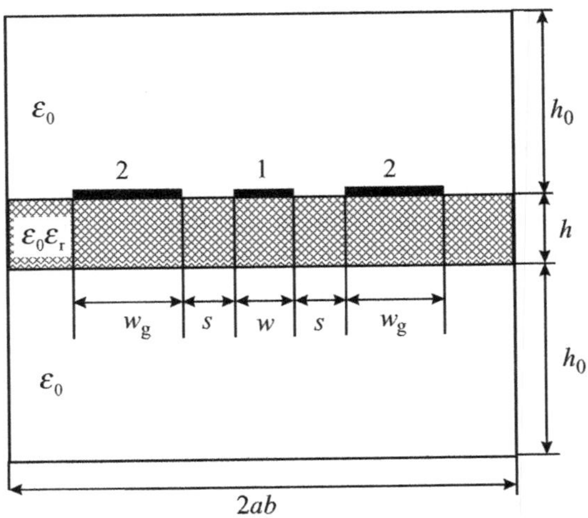

Fig. 2.1.9. A shielded coplanar waveguide with finite width of the ground planes. Substrate GaAs, $\varepsilon_r = 12.9$, $w = 75\,\mu m$, $s = 50\,\mu m$, $h = 410\,\mu m$.

those of the even mode impedances at low frequencies for the case of the waveguide considered here (with finite ground plane width).

The question arises as to how far the finite ground-plane width would have an influence on the line parameters of coplanar waveguides. In Fig. 2.1.9 the considered structure is shown again. Hoffmann [126], in his handbook, argues that when the ground-plane width w_g fulfills the condition $w_g \geq 0.5(2s + w)$, the effect of the ground width on the characteristic impedances of the even and the odd mode can be neglected.

The effect of the ground-plane width on the characteristic parameters of the coplanar waveguide has been studied here using the simple moment method for the case of a shielded structure. As an example, a coplanar waveguide on GaAs substrate ($\varepsilon_r = 12.9$) with height $h = 410\,\mu\text{m}$, a center-strip width of $w = 75\,\mu\text{m}$, and a gap width of $s = 50\,\mu\text{m}$ (as shown in Fig. 2.1.9) is considered.

The results are computed at a frequency of 1 GHz. The propagation constant (effective dielectric constant) of the coplanar waveguide with the mentioned dimensions has been computed using the current distributions of the three separate conductors, and the results are given in Fig. 2.1.10. One observes that the parameters of the odd mode are strongly dependent on the width of the ground planes. This can be explained by the field distribution of the odd mode. This has already been shown in Fig. 2.1.3 and has been discussed above. The electromagnetic field lines of the odd mode begin on one of the ground planes and end on the other one. They nearly do not touch the center strip. The field spreads over a wide area of the ground planes so that a variation of the ground-plane width also leads to a large variation of the propagation characteristics of this mode.

As can be seen from Fig. 2.1.10, the effective dielectric constant of the odd mode strongly decreases with increased values of w_g because in the case of a large ground-plane width, the electric field concentration in air is much higher than in the case of small ground-plane width. The effective dielectric constant of the even mode, which is of more interest to the circuit designer, is less affected by the width of the ground planes because the electromagnetic field is concentrated in the area around the gaps.

In any case there is an influence of the ground-plane width on the attenuation coefficient of the coplanar waveguide as is shown in Fig. 2.1.11. For both

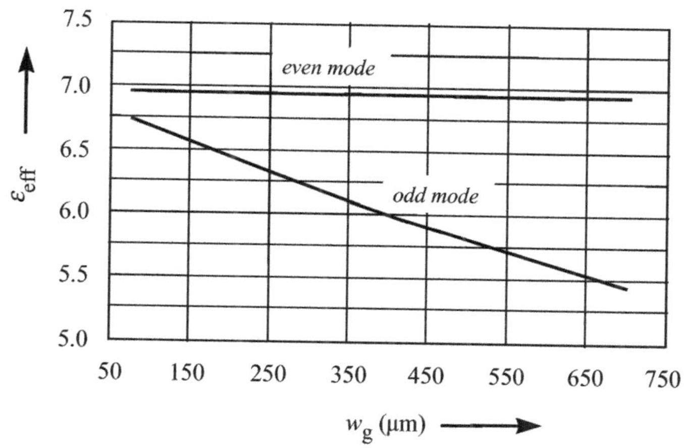

Fig. 2.1.10. Dependence of the effective dielectric constant of a coplanar waveguide on the width of the ground planes for the even mode and the odd mode. Line parameters: $w = 75\,\mu\text{m}$, $s = 50\,\mu\text{m}$, $h = 410\,\mu\text{m}$. Substrate GaAs: $\varepsilon_r = 12.9$, $f = 1\,\text{GHz}$.

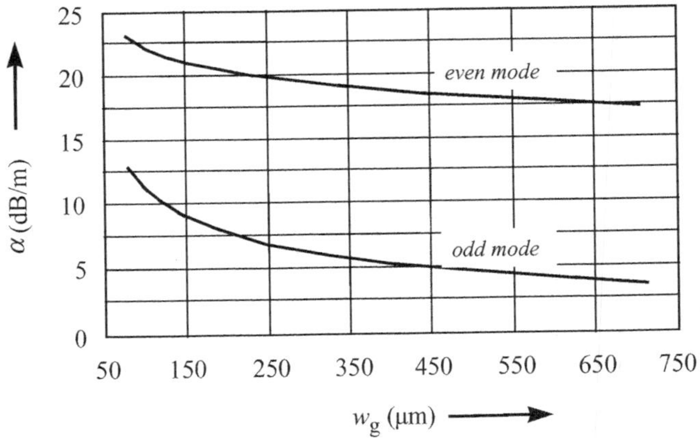

Fig. 2.1.11. Dependence of the attenuation coefficient of the even mode and the odd mode on a coplanar waveguide on the finite ground plane width w_g. Line parameters: $w = 75\,\mu m$, $s = 50\,\mu m$. Substrate GaAs, $h = 410\,\mu m$, $\varepsilon_r = 12.9$, $\rho_{Au} = 2.38 \times 10^5\,\Omega \cdot mm$, $\rho_{GaAs} = 1 \times 10^7\,\Omega \cdot mm$, $\tan\delta_{GaAs} = 0.0002$, $f = 1\,GHz$.

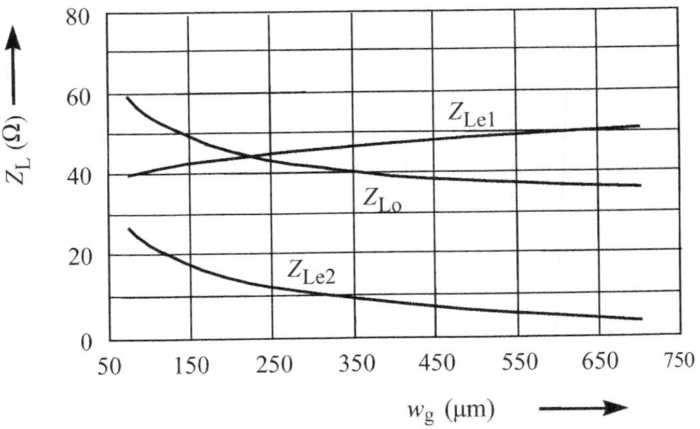

Fig. 2.1.12. Dependence of the characteristic impedance of the even mode and the odd mode on a coplanar waveguide on the finite ground plane width w_g. Line parameters: $w = 75\,\mu m$, $s = 50\,\mu m$, $h = 410\,\mu m$. GaAs: $\varepsilon_r = 12.9$, $f = 1\,GHz$.

the even and the odd mode, the attenuation coefficient α of the coplanar waveguide decreases with increasing width of the ground planes because the resistance per unit length of the waveguide is reduced by a larger ground-plane width. In the case of the even mode a width $w_g > 500\,\mu m$ must be ensured for the coplanar waveguide under consideration in order to get an attenuation coefficient that is nearly independent of the ground-plane widths.

The dimensions of the coplanar waveguide shown in Fig. 2.1.9 have been chosen so that the characteristic impedance of the even mode in the case of infinite ground plane width (Z_{Le1}) should be $50\,\Omega$. As Fig. 2.1.12 shows, the characteristic impedance of the even mode approaches the 50-Ω value for a width w_g of the ground plane in the order of 500 μm. For a width of w_g = 250 μm, which still is larger than the value given by Hoffmann (see above), the characteristic impedance deviates by more than 10% from the 50-Ω value. Also shown in Fig. 2.1.12 is the characteristic impedance of the odd mode (Z_{Lo}) and the parasitic even mode (Z_{Le2}).

The electromagnetic field distribution of the different modes on a coplanar waveguide placed on a GaAs substrate material without shielding can be well-measured using modern measurement techniques and equipment like an electro-optical measurement system [307, 327, 328]. To excite the different modes, special coaxial to coplanar waveguide probes have been used. In the case of the surface wave mode (which is the equivalent of the parasitic waveguide mode in the case of the shielded coplanar waveguide; see also Section 2.1.2), a special coplanar waveguide with a short circuit across the line was used to guarantee the excitation of this mode (Fig. 2.1.14d) [327].

The measurement is performed using the electro-optical effect of the GaAs substrate and measuring the voltage across the substrate material from the backside of the substrate. In Fig. 2.1.13 the measured potential of the even mode, the odd mode, and the surface wave mode in the cross section of the coplanar waveguide are shown. Figure 2.1.13a shows the magnitude and phase of the even-mode potential.

The high value of the signal under the center strip can be clearly identified. The magnitude of the potential under the ground planes is 18–20 dB below the potential of the center strip. The phase difference between the ground-plane potential and the center strip potential is 180°. In the case of the odd mode (Fig. 2.1.13b) the ground planes are on a high potential level and the center-strip potential is nearly zero. A 180° phase shift is measured between the potentials of the two ground planes. For the surface mode it can be observed that the magnitudes of the ground-plane and the center-strip potentials are nearly identical and no phase differences exist between the potentials. Despite the metallization structure on top of the substrate, the surface wave behaves like a plane wave propagating along the air–dielectric interface.

Figures 2.1.14a to 2.1.14c show the field distribution of the even (a) and the odd mode (b) as well as of the surface wave mode (c) along the coplanar waveguide as measured with the electro-optical measurement technique [328]. As already mentioned above, the excitation of the different modes was realized using RF probes. In the case of the even mode, a ground–signal–ground probe was used. For the odd mode, only two probe heads were used and a signal ground distribution was applied to the two ground planes. The center strip was not excited. Finally, in the case of the surface wave mode, a coplanar waveguide was excited in the conventional even mode but a short circuit was placed across the coplanar waveguide after a certain distance behind the probe (Fig. 2.1.14d). Then, the field distribution shown in Fig. 2.1.14c was measured behind the short circuit.

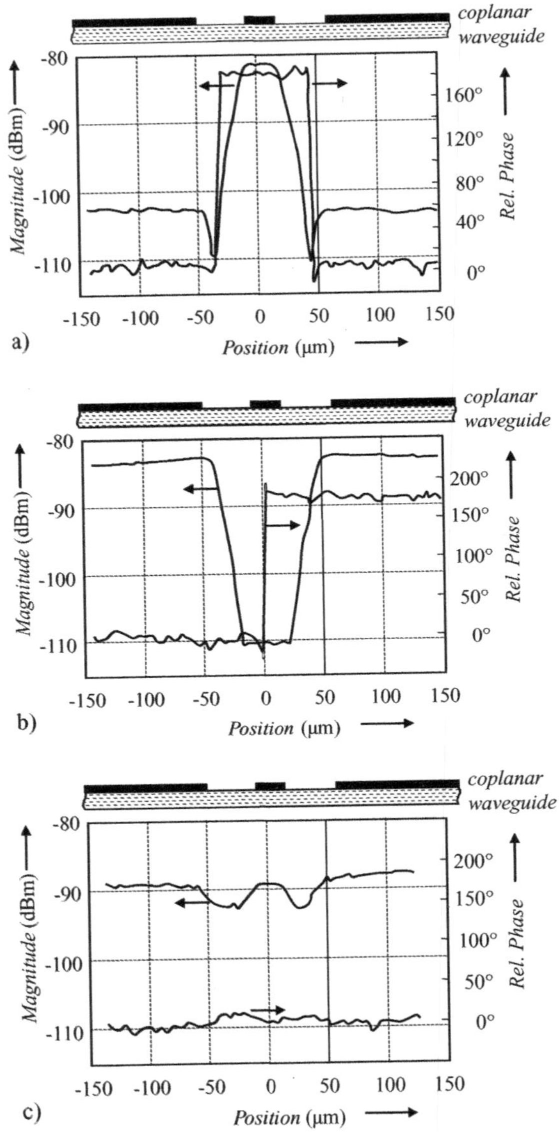

Fig. 2.1.13. Measured field distribution of the electric potential (magnitude and phase) for the even mode (a), the odd mode (b), and the surface wave mode (c) on the coplanar waveguide. Measurements have been performed using an electro-optical measurement technique [328].

As Fig. 2.1.15 shows, the field distribution of the even mode on a coplanar waveguide is almost frequency-independent. The figure shows the measured potential signal (magnitude) for a coplanar waveguide at a frequency of 6 GHz (a) and at a frequency of 1 GHz (b). As may be recognized from the figures, a)

b)

c)

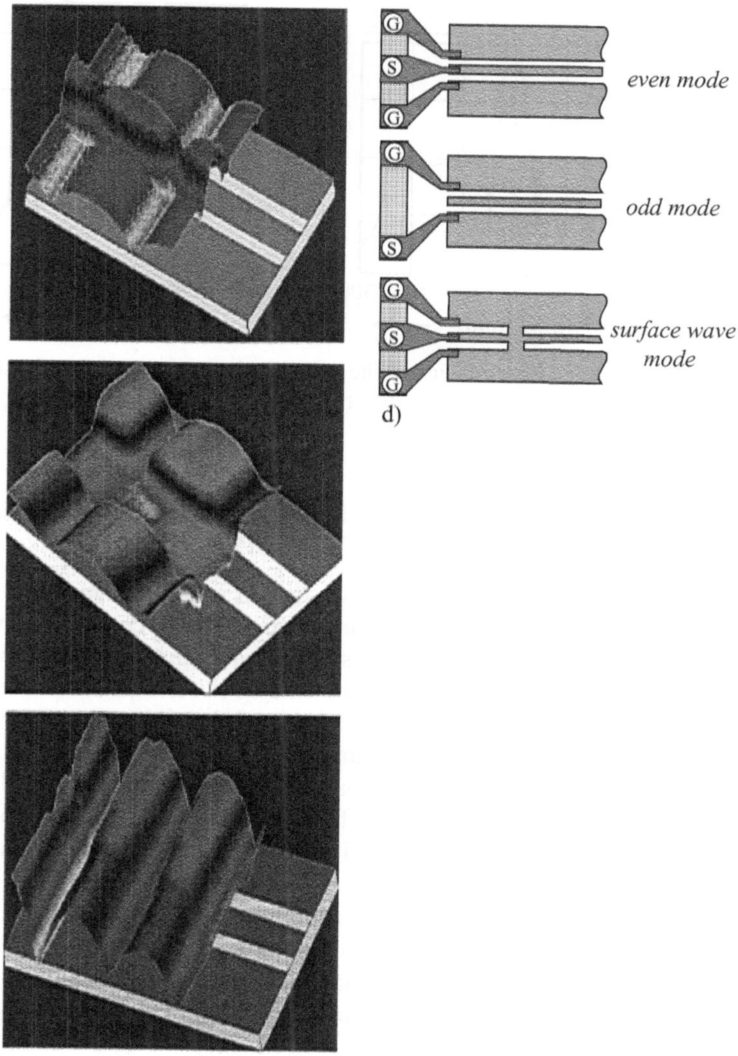

d)

Fig. 2.1.14. Measured field (potential) distribution along a coplanar waveguide for (a) the even mode, (b) the odd mode, and (c) the surface wave mode using an electro-optical measurement technique [328]. Frequency: 18 GHz. Shown area: 360 μm × 5500 μm for parts a and b and 370 μm × 5800 μm for part c. Part d shows how the different modes have been excited.

the field distribution is nearly identical. Only at the outside end of the ground planes, which are of finite width, some very small difference may be observed. The main fields near the center strip and in the gap region that determine the waveguide properties do not change much over the considered frequency range.

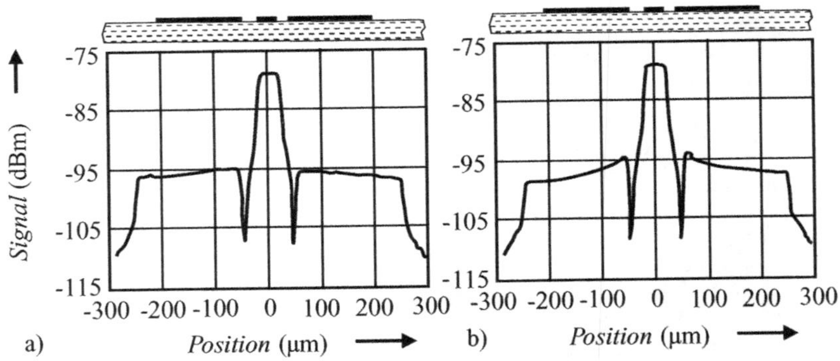

Fig. 2.1.15. Potential distribution under the metalization layer for a coplanar waveguide (a) at a frequency of 6 GHz and (b) at a frequency of 1 GHz [328].

2.1.2 The Coplanar Waveguide with a Single Center Strip and Infinite Ground-Plane Width

In this section, similar investigations as described in the previous section will be discussed, with the only difference that the considered coplanar waveguide has an infinitely wide ground plane. This also means that in the applied simulation technique (the second moment method as described at the beginning of the chapter), the ground planes on the left and the right side of the central strip at low frequencies are always assumed to be at the same electric potential. This especially influences the propagation of the odd mode at low frequencies that, under these conditions, has a nonzero cutoff frequency. This assumption also approximates a little bit better the conditions that are given in a microwave integrated circuit, whereby to avoid the propagation properties of the odd mode (slot-line mode), air bridges are used to keep the ground planes on one and the same electric potential (see also Section 3.5.5). The assumptions made here also include that the coplanar waveguide is considered to be an open structure; that is, there is no shielding assumed as in the case discussed in Section 2.1.1.

If the field distributions of the even and the odd mode are considered for an open environment surrounding the coplanar waveguides, results like the ones shown in Figs. 2.1.16 and 2.1.17 may be found. There is not much difference to be observed between the field distributions shown here and for the case of the shielded lines (Fig. 2.1.2 to Fig. 2.1.3). Only in the case of the open coplanar waveguide, the electric field lines do not end on the electric shielding as shown in Fig. 2.1.2a and especially in Fig. 2.1.3a. The parasitic waveguide mode, which has been discussed above (see Fig. 2.1.4) in the case of the open structure, is replaced by a surface wave propagating along the boundary between the dielectric substrate material and the air region. Like the parasitic waveguide mode, it has zero cutoff frequency and, therefore, may be excited inside a coplanar circuit together with the even and the odd modes.

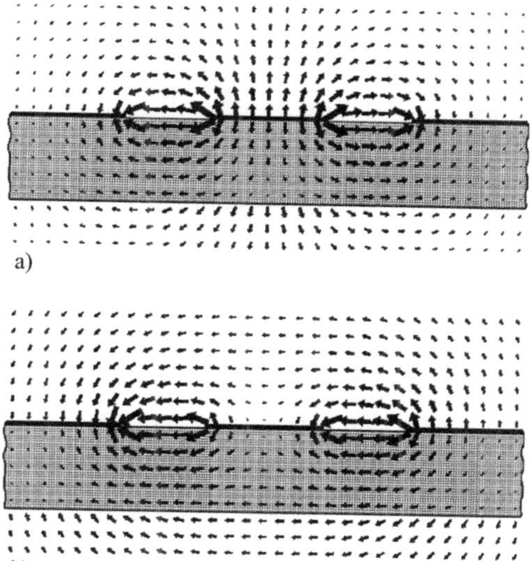

a)

b)

Fig. 2.1.16. Transversal electric field strength of (a) the even mode (the coplanar waveguide mode) and (b) the odd mode (the slot-line mode) on an open coplanar waveguide calculated for a frequency of 20 GHz. Line structure: $\varepsilon_r = 9.8$, $h = 250\,\mu m$, $s = 250\,\mu m$, $w = 350\,\mu m$.

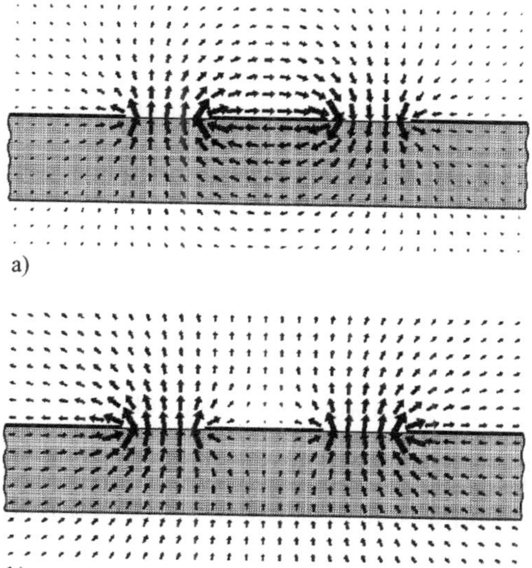

a)

b)

Fig. 2.1.17. Transversal magnetic field strength of (a) the even mode (coplanar waveguide mode) and (b) the odd mode (slot-line mode) on an open coplanar waveguide calculated for a frequency of 20 GHz. Line structure: $\varepsilon_r = 9.8$, $h = 250\,\mu m$, $s = 250\,\mu m$, $w = 350\,\mu m$.

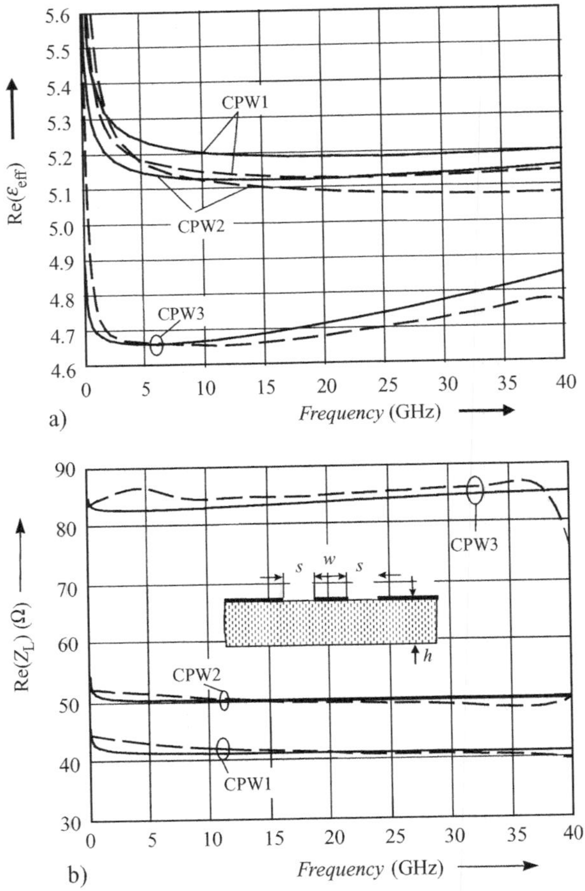

Fig. 2.1.18. The frequency-dependent real part of the effective dielectric constant (a) and the frequency-dependent characteristic impedance (b) of the even mode on coplanar waveguides without a metallic shielding and with parameters as shown in Table 2.1.1. (———) computed values, (– – –) measured values.

In the second moment method, which was briefly described in Section 2.1.1 and which is used for this analysis, the losses are directly included into the spectral domain analysis. Moreover, despite this method being also an approximating one, it makes it possible to calculate the influence of the dielectric and the conductor losses on the effective dielectric constant and characteristic impedance, whereas the first simple method (see Section 2.1.1) only delivers the effect of the losses on the attenuation coefficients (see discussion below). Figure 2.1.18 shows a comparison of the computed effective dielectric constants and the characteristic impedances of the even mode for three coplanar waveguides together with measurement results. The real part of the effective

TABLE 2.1.1. Geometrical Parameters and Material Parameters of the Analyzed Coplanar Waveguides

Waveguide	w (μm)	s (μm)	Z_0 (Ω)
CPW1	125	25	≈40
CPW2	125	50	≈50
CPW3	125	250	≈80

Substrate: Al_2O_3 ceramic material. Dielectric constant $\varepsilon_r = 9.8$, $\tan\delta = 0.0001$, substrate height $h = 250\,\mu m$, metalization thickness $t = 5\,\mu m$, material gold $\rho_{metal} = 2.4 \times 10^{-8}\,\Omega\cdot m$.

dielectric constant and the characteristic impedance is drawn because the analysis technique is a complex one when taking into account the losses of the structures, and therefore the effective dielectric constant and the characteristic impedance become complex.

Table 2.1.1 shows the geometrical parameters and material parameters of the coplanar waveguides that have been analyzed. The measured difference, for example, in the frequency dependence of the effective dielectric constant, as compared to Fig. 2.1.5, is that the effective dielectric constant increases strongly at low frequencies.

The reason for this behavior is the skin effect. For a metalization thickness of 5 μm, at frequencies of about 2–5 GHz, the skin depth is on the order of the metalization thickness, so that for lower frequencies the current density and thereby the electromagnetic field also penetrates the conducting material.

The magnetic field components form an inner inductance per unit line length that is added to the normal (outer) inductance per unit line length of the coplanar waveguide. Therefore the phase velocity and consequently the effective dielectric constant are changed. This effect becomes larger as the frequency decreases.

In a first approximation, this effect can be explained by the following equations: The propagation coefficient of the transmission line is defined approximately by

$$\gamma \approx \mathrm{j}\omega\sqrt{L'C'}\sqrt{1-\mathrm{j}\frac{R'}{\omega L'}} \Rightarrow \beta \approx \omega\sqrt{\frac{L'C'}{2}}\sqrt{1+\sqrt{1+\left(\frac{R'}{\omega L'}\right)^2}}. \tag{2.1.4}$$

If the effective dielectric constant is calculated from γ this leads to

$$\mathrm{Re}\{\varepsilon_{\mathrm{eff}}\} = -\mathrm{Re}\left\{\left(\frac{\gamma}{k_0}\right)^2\right\} \qquad \text{with } k_0 = \frac{\omega}{c_0}. \tag{2.1.5}$$

Since in the case $\omega \to 0$ the resistance per line length R' is nearly frequency-independent (dc resistances), for the effective dielectric constant the following result is derived:

$$\mathrm{Re}\{\varepsilon_{\mathrm{eff}}\} \approx \left(\frac{\omega \sqrt{\frac{L'C'}{2}} \sqrt{\frac{R'}{\omega L'}}}{k_0} \right)^2 \cong \frac{1}{\omega}, \tag{2.1.6}$$

and therefore the frequency dependence as shown in Fig. 2.1.18 results. This frequency dependence can also be measured as is shown in the same figure.

The complex characteristic line impedances in this case are calculated from the definition $Z_L = V^2/(2P^*)$ under the consideration of losses. The effect of the losses on the characteristic impedances of the even mode, as Fig. 2.1.18b shows, is not so severe. The characteristic impedance is mainly frequency-independent, and only a slight increase at very low frequencies may be observed.

Finally, Fig. 2.1.19 shows the frequency dependence of the characteristic impedances of the even mode (a) and the odd mode (b) for coplanar waveguides of different substrate material height. These results are again calculated using the moment method that considers the effect of the losses (see Section 2.1.1).

The first result, which may be drawn from these figures, is that using the $Z_L = V^2/(2P^*)$ definition the dispersion of the characteristic impedance of the even mode (coplanar waveguide mode) may be positive or negative, depending on the substrate height. For very small values of the substrate height, the effective dielectric constant increases with frequency, mainly because with increasing frequency a field concentration into the substrate occurs also at the backside of the substrate material with increasing frequency. This effect is small on the top of the substrate material, because the slot width of the assumed structure is small. With increased substrate height (e.g., $h = 500\,\mu\mathrm{m}$) there is no more electromagnetic field penetrating the substrate from the backside and the field concentration process may no longer occur at this side. On the other hand, for these structures, the influence of the losses that are considered in this investigation result in a small decrease of the characteristic impedance with increasing frequency.

The behavior of the characteristic impedance of the odd mode (slot-line mode) is different. A large dispersion may be observed, and a significant difference of the characteristic impedance compared to that calculated for the coplanar waveguide with finite ground-plane width (see Fig. 2.1.8) is found. Since here the assumption of an infinitely wide ground plane was made, the potential of both ground planes must be equal at low frequencies. Because the odd mode is a slot-line mode that needs different potentials on both ground planes (compare also with Fig. 2.1.1b), it cannot propagate on this line at low frequencies. Its cutoff frequency now is finite. Therefore, the characteristic impedance reduces to zero at very low frequencies, as shown in Fig. 2.1.19b. Because in coplanar integrated circuits all ground planes are kept on the same potential using an air bridge technology (see also Section 3.5.5 and the dis-

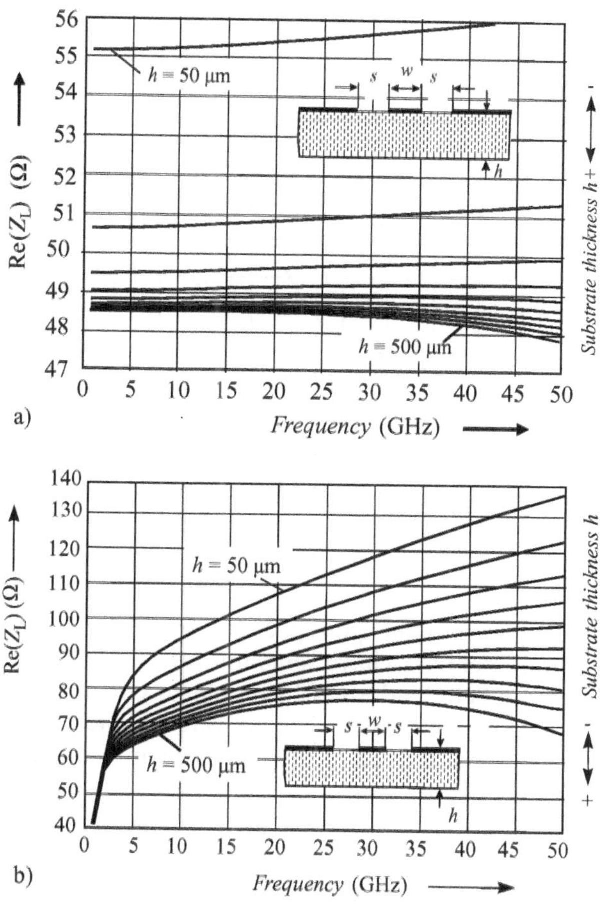

Fig. 2.1.19. Dispersion of the characteristic impedance for (a) the even mode (coplanar waveguide mode) and (b) the odd mode (slot-line mode), plotted against the frequency and with the substrate material height as a parameter ($h = 50$–$500\,\mu m$ in steps of $50\,\mu m$). Line parameters: $\varepsilon_r = 12.9$ (GaAs), $\tan\delta = 0.002$, $s = 50\,\mu m$, $w = 75\,\mu m$.

cussion there), the assumption of an infinite ground plane is the realistic one for the circuit designer. It means that the high dispersion of the odd-mode characteristics and especially the cutoff at low frequencies must be taken into account if components under use of the odd mode shall be designed (compare with the discussion on mode converters in Section 3.5.8).

As already mentioned above, in the case of an open coplanar waveguide, besides the fundamental even and odd mode an additional wave propagation, in a form similar to a surface wave, is possible along the dielectric–air interface even if a metalization is available in this surface. This additional mode may couple to the fundamental even mode of the coplanar waveguide, and it

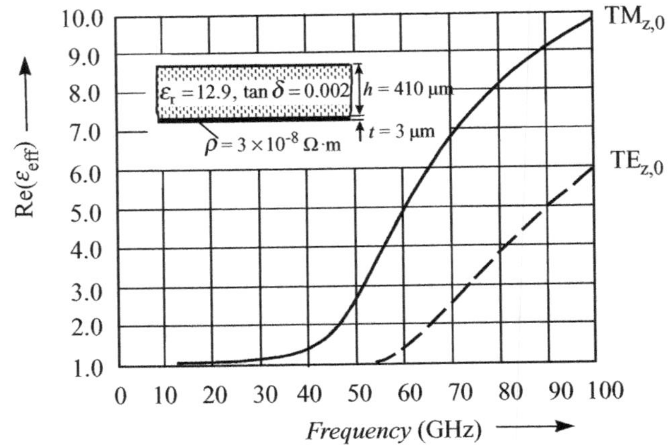

Fig. 2.1.20. Dispersion characteristic of the real part of ε_{eff} for the fundamental TM$_{z,0}$ and the TE$_{z,0}$ surface wave mode. Substrate GaAs, $\varepsilon_r = 12.9$, $\tan\delta = 0.002$. Substrate thickness $h = 410\,\mu$m, specific conductivity $\rho = 3 \times 10^8\,\Omega\cdot$m.

may even radiate power into the open space (compare also references 235 and 274). It is, therefore, essential to have some information available on this parasitic mode if a circuit is to be designed in coplanar circuit technology.

Using the advanced spectral domain analysis technique (see Section 2.1.1), which considers also the influence of the dielectric and conductor losses, the properties of a surface wave along a dielectric material air interface with or without a metalization on the backside of the dielectric slab (Fig. 2.1.20) may be investigated. There is not much difference in the value of the effective dielectric constant for these two cases if the surface wave is considered, because a possible backside metalization does not influence much the phase velocity of the surface wave.

Because of the considered losses, the propagation constant is complex and it may be described by an effective dielectric constant and an attenuation coefficient. In Fig. 2.1.20 the real part of the complex effective dielectric constant (that defines the phase velocity of the wave) is shown for the case of a dielectric slab material (GaAs) that is metalized on the backside. The dispersion characteristics of the fundamental TM$_{z,0}$ and the TE$_{z,0}$ mode are shown. The fundamental surface wave mode TE$_{z,0}$ has a cutoff frequency of zero, whereas the next-higher-order mode TE$_{z,0}$ has a cutoff frequency near 53 GHz for the structure that is considered here. From Fig. 2.1.20 it can be seen that the effective dielectric constant increases strongly with frequency and for frequencies higher than 60 GHz, it comes into the order of the effective dielectric constant of the fundamental coplanar waveguide mode (even mode).

If the losses are analyzed, three different cases can be considered, because a possible lossy backside metalization may have an influence on the losses. In

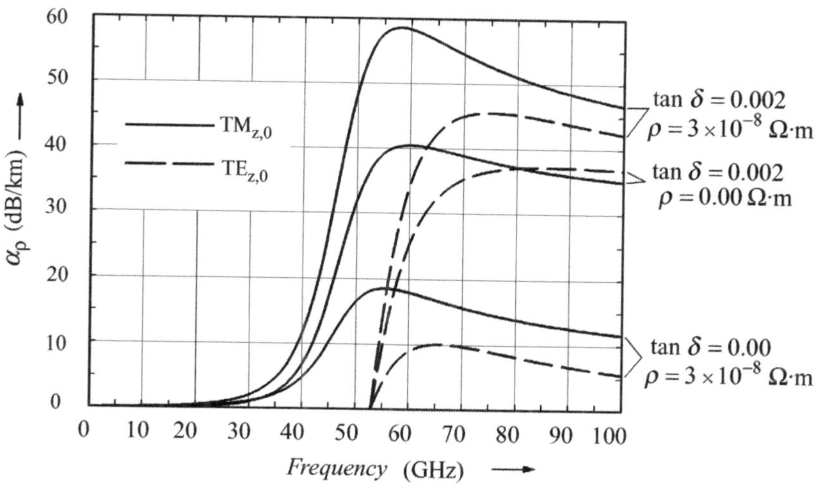

Fig. 2.1.21. Frequency dependence of the attenuation coefficients of the two fundamental surface waves along a dielectric slab substrate. Structure parameters: See Fig. 2.1.20.

Fig. 2.1.21 the attenuation coefficient of the two modes is shown for three different cases: (1) the case, where a lossy dielectric material and a metalization on the backside of the substrate is considered, (2) the case of a lossy dielectric material with an ideal backside metalization ($\rho = 0.0\Omega \cdot$m), and (3) the case where the substrate material is loss less ($\tan\delta = 0$), but a lossy backside metalization is available. Because the surface wave is propagating mainly in the dielectric–air interface, the losses of the surface wave are low and the effect of the backside metalization on the attenuation coefficient is of secondary importance. For the considered case, the main influence on the attenuation coefficient is taken by the dielectric losses of the substrate material because these losses are high ($\tan\delta = 0.002$). If substrate materials with low dielectric losses are considered, the influence of the conductor losses from the backside metalization may be dominant [235, 274]. The strong differences between the three cases, especially the strong influence of the dielectric losses, can be observed easily.

To investigate the influence of the substrate thickness on the effective dielectric constant of the surface wave, an Al_2O_3 substrate is considered in Fig. 2.1.22. The figure shows the real part of the effective dielectric constant for the $TM_{z,0}$ mode with respect to frequency and for different substrate material heights. With increasing substrate height the dispersion of the effective dielectric constant is increased so that already at low frequencies the effect of the surface wave may be recognizable in a circuit on MIC or MMIC basis.

Finally, Fig. 2.1.23 shows the measured effective dielectric constant and attenuation coefficient for the three different modes: the even mode, the odd

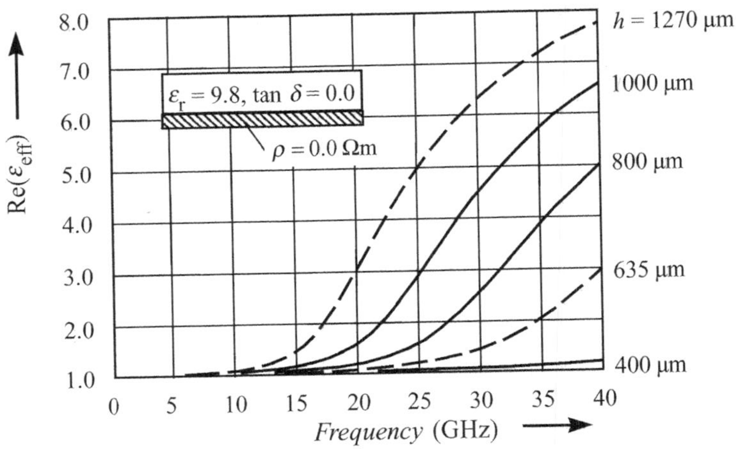

Fig. 2.1.22. Dispersion characteristic of the effective dielectric constant of the fundamental $TM_{z,0}$ surface wave mode for different substrate thickness. Material: Al_2O_3.

mode, and the microstrip like surface wave mode on a coplanar waveguide. The measurement technique used here was again an electro-optical measurement technique [328] that determines the field distribution of the different modes using a laser optical signal and the electro-optical effect of the investigated GaAs substrate material.

As can be seen from the figures, the principal dependence of the effective dielectric constant and the attenuation coefficient on the frequency (as has already been described above in the theoretical analysis) is also recognizable in the measurement results. If the measurement results are compared with simulation results of the moment method in detail, a good agreement can be found.

2.1.3 Coupled Coplanar Waveguides

Coupled transmission lines have multiple applications in components like filters, directional couplers, interdigital capacitors, and planar spiral inductors (see Sections 4 and 6). Therefore, the proper knowledge of the frequency-dependent transmission properties of coupled coplanar waveguides is essential for the circuit design. Besides this aspect of the coupled coplanar lines, another aspect is also essential in many cases of circuit design: that is, the unwanted coupling between neighboring line structures and the definition of a minimum distance which must be kept between the lines so that unwanted coupling is small enough for the proper performance of the circuit (see Section 2.1.3.2). Two different forms of coupled coplanar waveguides shall be discussed here. Both of them are of high relevance for the circuit designer and are shown in Figs. 2.1.24a and 2.1.24b. The structure shown in Fig. 2.1.24a

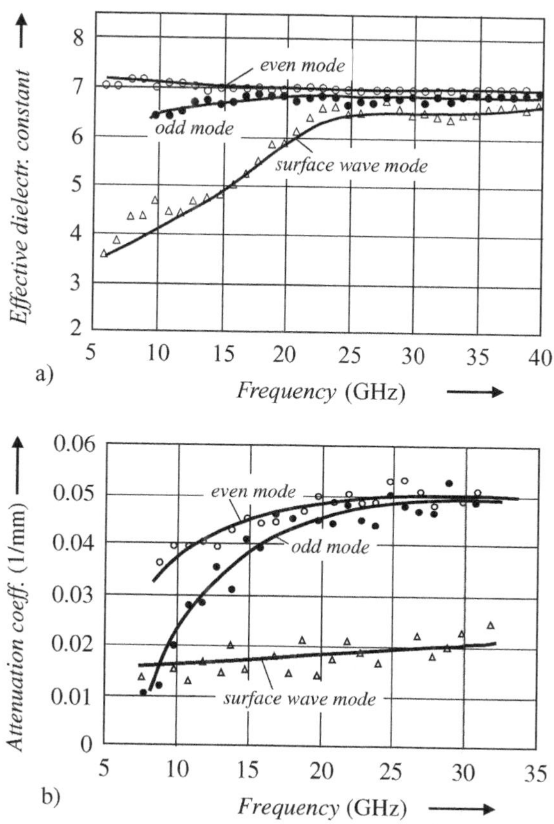

Fig. 2.1.23. Measured frequency dependence of the effective dielectric constant (a) and of the attenuation coefficient (b) for the even mode, the odd mode, and the surface wave mode on a coplanar waveguide on GaAs substrate material. Measurement technique: Electro-optical effect of the substrate material [328]. Symbols: Measured values. Lines: medium value.

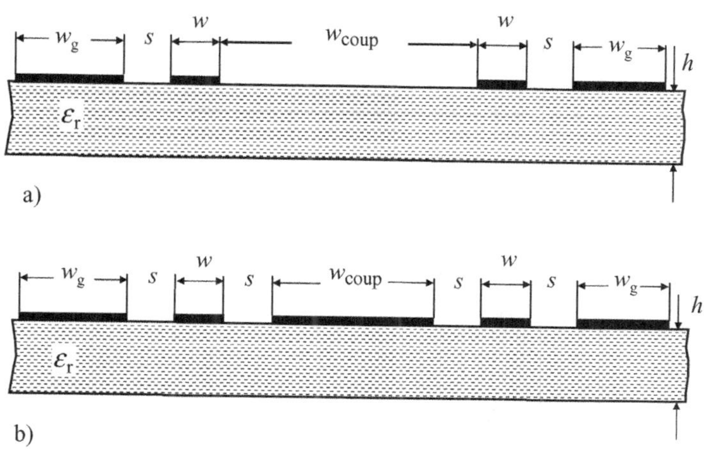

Fig. 2.1.24. The cross section of a coplanar waveguide with two coupled center strips (a) and of two coupled coplanar waveguides (b).

consists of two strips of widths w_g (which are the ground strips) and two (or possibly more) center strips of widths w that are used for signal transmission. There is no additional ground plane between these coupled center strips in the considered case. Transmission lines of this kind are used in interdigital capacitors and spiral inductors, as will be shown later in Chapter 4. An alternative form of coupled coplanar waveguides is shown in Fig. 2.1.24b where an additional ground plane of width w_{coup} is brought between the two "hot" center strips. Transmission lines of this kind are, for example, used in couplers. On the other hand, these structures represent two closely spaced single coplanar waveguides that inevitably bring an unwanted coupling in a circuit layout.

2.1.3.1 Scattering Matrix of Coupled Coplanar Waveguides. Figure 2.1.25 shows a system of multiply coupled transmission lines that are the coplanar waveguides described above.

For the circuit designer parameters such as signal transmission, coupling and isolation are essential for analysis and design. These parameters must be known for the relevant fundamental even mode as well as for the mode conversion into the unwanted odd mode or vice versa. If a correct description of all possible couplings on a line system like the one shown in Fig. 2.1.25 is to be given, all scattering parameters of such a system, in consideration of different propagating modes, must be known [248]. A method to derive the scattering parameters of multiply coupled microstrip lines from the calculated transmission line parameters has been described in reference 48. This method shall be expanded here for the application to coplanar waveguide structures. In the previous section it has been mentioned that the voltage power defini-

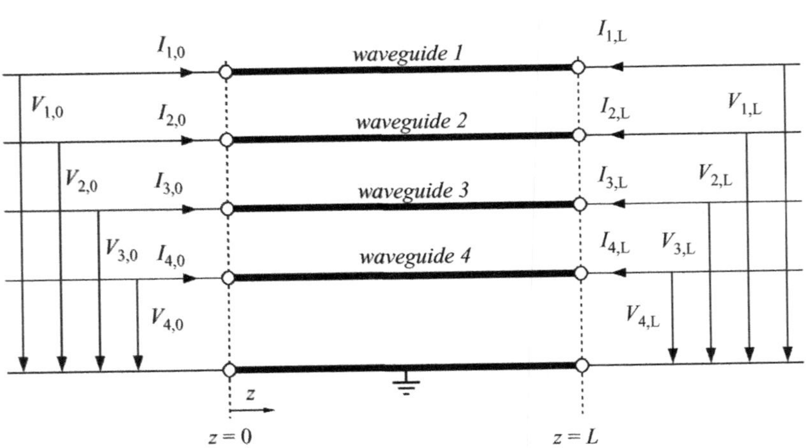

Fig. 2.1.25. Schematic representation of transmission line currents and transmission line voltages on a multiply coupled line structure of length L.

tion of the characteristic impedance is most advantageous for the single coplanar waveguide even if there are not big differences of the values determined with the other two methods. For multiply coupled line systems, the definition of the characteristic impedance is much more complicated because a wave impedance matrix $\boldsymbol{Z}_L$ must be determined in this case. The elements of this impedance matrix are the characteristic impedances of the single strips carrying a special propagating mode. The possibilities to calculate wave impedance matrices of coupled line systems have been discussed in the literature for a long time [192, 213]. Here a method based on references 28 and 162, which has also been published in reference 243, shall be used.

If one special eigensolution for the open, lossy, and coupled line structure is considered, it can be shown, using the reciprocity theorem, that the adjungated eigenvectors of voltage and current have the property

$$\boldsymbol{V}^m \cdot (\boldsymbol{I}^n)^{*T} = 0 \qquad \text{for } \gamma_m \neq \gamma_n. \tag{2.1.7}$$

The elements of $\boldsymbol{V}^m$ and $\boldsymbol{I}^n$ are the voltages of the different strips with respect to a defined reference point and the longitudinal (z-direction) currents within the strips for mode m and mode n, respectively. If the transported power of mode m is calculated from the transversal electric and magnetic fields using the Poynting vector

$$P^\nu = \frac{1}{2} \iint_{A_{tr}} \left[\boldsymbol{E}_{tr}^\nu \times (\boldsymbol{H}_{tr}^\nu)^* \right] \cdot \boldsymbol{u}_z \, dA, \tag{2.1.8}$$

with $\boldsymbol{u}_z$ the unit vector in z-direction the result is

$$P^m = \frac{1}{2} \boldsymbol{V}^m \cdot (\boldsymbol{I}^m)^{*T}, \qquad \text{with } m = 1, \ldots, N-1, \tag{2.1.9}$$

where N is the number of strips forming the coupled line system on the substrate material. If Eqs. (2.1.8) and (2.1.9) are combined, using diagonal matrices of the size $[(N-1) \times (N-1)]$: $\boldsymbol{P}$, $\boldsymbol{V}$, and $\boldsymbol{I}$, we obtain

$$\boldsymbol{P} = \text{diag}\{P^1, \ldots, P^i, \ldots, P^{N-1}\} = \frac{1}{2} \boldsymbol{V} \cdot \boldsymbol{I}^{*T} \tag{2.1.10}$$

Using these definitions, the wave impedance matrix of the coupled line system can be calculated after the propagation coefficients have been determined following the steps listed below:

Step 1: Calculation of the power transported by each mode and definition of the equivalent diagonal element of the matrix $\boldsymbol{P}$.

Step 2: Calculation of the slot voltages of the coplanar structure and definition of the diagonal elements of the matrix $\boldsymbol{V}$.
Step 3: Determination of $\boldsymbol{I} = 2(\boldsymbol{V}^{-1} \cdot \boldsymbol{P})^{*T}$.

Using the so-defined matrices, each element of the $\boldsymbol{Z}_L$ matrix can be calculated from

$$Z_L^{[m,n]} = \frac{V^{[m,n]}}{I^{[m,n]}}. \tag{2.1.11}$$

At the beginning of the evaluation process of the scattering parameters for a line section as shown in Fig. 2.1.24, the calculation of the eigenvalues γ_v (of the mode currents $\boldsymbol{I}^{\mathrm{mode},v}$) and the characteristic wave impedance matrix [162] must be performed. To do this, the four transmission line voltages V_1 to V_4 shown in Fig. 2.1.25 (as an example) will be determined. Using these voltages, the power transported by the different wave modes is calculated. In Fig. 2.1.25, a line system with four transmission lines (as an example), on which four fundamental modes can propagate, is shown. These four modes form a complete system of TEM modes so that each TEM field distribution on the line can be represented by a superposition of these four modes. The relation between the transmission line currents and the mode currents at the beginning (index "0") and at the end (index "L") of the transmission line can then be written as

$$\begin{pmatrix} \boldsymbol{I}_0^{\mathrm{strip}} \\ \boldsymbol{I}_L^{\mathrm{strip}} \end{pmatrix} = \begin{pmatrix} \boldsymbol{M}_\mathrm{I} & 0 \\ 0 & \boldsymbol{M}_\mathrm{I} \end{pmatrix} \cdot \begin{pmatrix} \boldsymbol{I}_0^{\mathrm{mode}} \\ \boldsymbol{I}_L^{\mathrm{mode}} \end{pmatrix}, \tag{2.1.12}$$

with

$$\boldsymbol{M}_\mathrm{I} = \left(\boldsymbol{I}^{\mathrm{mode},1}, \ldots, \boldsymbol{I}^{\mathrm{mode},4}\right) \cdot \mathrm{diag}\left(\frac{1}{I_1^{\mathrm{mode},1}}, \ldots, \frac{1}{I_1^{\mathrm{mode},4}}\right). \tag{2.1.13}$$

Under the assumption of a TEM approximation, the two transmission line equations

$$V_{i,L}^{\mathrm{mode},v}(L) = V_{i,0}^{\mathrm{mode},v} \cosh(\gamma_v L) - Z_L^{[v,i]} I_{i,0}^{\mathrm{mode},v} \sinh(\gamma_v L), \tag{2.1.14}$$

$$I_{i,L}^{\mathrm{mode},v}(L) = I_{i,0}^{\mathrm{mode},v} \cosh(\gamma_v L) - \frac{V_{i,0}^{\mathrm{mode},v}}{Z_L^{[v,i]}} \sinh(\gamma_v L) \tag{2.1.15}$$

are valid for the mode voltages $V_i^{\mathrm{mode},v}$ and the mode currents $I_i^{\mathrm{mode},v}$. Both equations define a relation between the mode voltages and mode currents at the beginning (index "0") and at the end (index "L") of the line. These equations will be used in the next step to set up a relation between the transmission line voltages and the mode currents. To evaluate this relationship, a series of excitation and open-end experiments are established for all modes at both ends of the line. An application of the superposition technique that defines the transmission line voltage as the sum of the mode voltages leads to

$$\left.\begin{array}{l} \boldsymbol{V}^{\text{mode}} = \boldsymbol{Z}_{\text{mode}} \cdot \boldsymbol{I}^{\text{mode}} \\ V_i^{\text{strip}} = \sum_{\nu=1}^{4} V_i^{\text{mode},\nu} \end{array}\right\} \Rightarrow \boldsymbol{V}^{\text{strip}} = \boldsymbol{M}_{\text{V}} \cdot \boldsymbol{I}^{\text{mode}}, \tag{2.1.16}$$

with

$$\boldsymbol{V}^{\text{strip}} = \begin{pmatrix} \boldsymbol{V}_0^{\text{strip}} \\ \boldsymbol{V}_L^{\text{strip}} \end{pmatrix}, \qquad \boldsymbol{V}^{\text{mode}} = \begin{pmatrix} \boldsymbol{V}_0^{\text{mode}} \\ \boldsymbol{V}_L^{\text{mode}} \end{pmatrix}, \tag{2.1.17}$$

which relates the transmission line voltages and the mode currents. Introducing Eqs. (2.1.12) and (2.1.16) finally delivers the relation between the transmission line voltages and currents on the coupled coplanar waveguides in the form

$$\boldsymbol{V}^{\text{strip}} = \boldsymbol{M}_{\text{V}} \cdot \begin{pmatrix} \boldsymbol{M}_{\text{I}} & 0 \\ 0 & \boldsymbol{M}_{\text{I}} \end{pmatrix}^{-1} \cdot \boldsymbol{I}^{\text{strip}} = \boldsymbol{Z} \cdot \boldsymbol{I}^{\text{strip}}. \tag{2.1.18}$$

The direct conversion of the so-calculated impedance matrix $\boldsymbol{Z}$ does not directly lead to the wanted scattering matrix $\boldsymbol{S}$, which describes the structure with respect to their two fundamental even and odd modes. For the determination of the scattering matrices, all strip voltages and strip currents of the structure must be reduced to the even-mode and odd-mode components (as shown in Fig. 2.1.26) of the two coplanar waveguides I and II, respectively. The reduction of the currents is shown in the lower part of the figure:

$$\begin{pmatrix} \boldsymbol{X}_0^{\text{cop}} \\ \boldsymbol{X}_L^{\text{cop}} \end{pmatrix} = \begin{pmatrix} \boldsymbol{M}_{X,\text{cop}} & 0 \\ 0 & \boldsymbol{M}_{X,\text{cop}} \end{pmatrix} \cdot \begin{pmatrix} \boldsymbol{X}_0^{\text{strip}} \\ \boldsymbol{X}_L^{\text{strip}} \end{pmatrix}, \tag{2.1.19}$$

where X stands for V (voltage) or I (current), respectively, and the two transformation matrices for the voltages and the currents are given by

$$\boldsymbol{M}_{\text{V,cop}} = 0.5 \begin{pmatrix} -1 & 2 & 0 & 0 \\ 2 & 0 & 0 & 0 \\ 0 & 0 & 2 & -1 \\ 0 & 0 & 0 & 2 \end{pmatrix} \quad \text{and} \quad \boldsymbol{M}_{\text{I,cop}} = 0.5 \begin{pmatrix} 0 & 2 & 0 & 0 \\ 2 & 1 & 0 & 0 \\ 0 & 0 & 2 & 0 \\ 0 & 0 & 1 & 2 \end{pmatrix}. \tag{2.1.20}$$

If Eq. (2.1.18) is inserted into Eq. (2.1.19), this leads to the impedance matrix Z_{cop}, which, after normalization (element by element) $Z_{\text{cop,norm}}^{[m,n]} = Z_{\text{cop}}^{[m,n]} / (Z_{L,m} Z_{L,n})^{1/2}$ by the line impedances $Z_{L,m}$ and $Z_{L,n}$ of the connecting lines, defines the scattering matrix

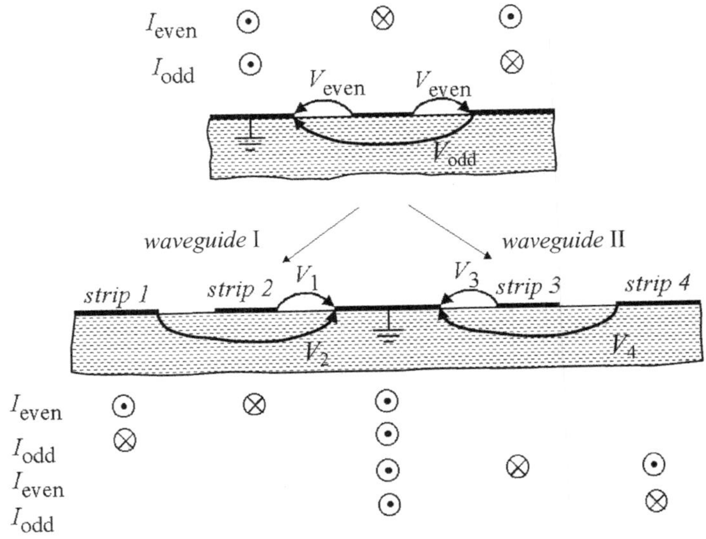

Fig. 2.1.26. Schematic representation of the reduction of the strip currents and voltages to their components with respect to the coplanar waveguides I and II.

$$\boldsymbol{S} = (\boldsymbol{Z}_{\text{cop,norm}} + \boldsymbol{U})^{-1} \cdot (\boldsymbol{Z}_{\text{cop,norm}} - \boldsymbol{U}), \tag{2.1.21}$$

where $\boldsymbol{U}$ is the unit matrix. The line impedances $Z_{L,m}$ and $Z_{L,n}$ of the connecting lines are those of the coplanar waveguide mode (even mode) and the slot-line mode (odd mode) calculated, for example, from a voltage power relationship as defined in Eq. (2.1.3). The used voltages are V_{even} and V_{odd}, as shown in the upper part of Fig. 2.1.26.

2.1.3.2 Coupled Coplanar Waveguides and Microstrip Lines—A Comparison. In this section we will discuss how large the coupling between two coupled coplanar waveguides (as shown in Fig. 2.1.27) will be in comparison to that of two coupled microstrip lines [274]. These investigations can lead to criteria as to what distance two coplanar waveguides must be placed from each other in a circuit design, so that the coupling between them is negligibly small. As a design basis in practical circuit design, the rule $w_{\text{coupl}} \geq 2s + w$ is frequently used. This design rule will be compared to accurate, frequency-dependent scattering parameter calculations.

Furthermore, the coupling coefficient between two coupled coplanar waveguides and two microstrip lines shall be compared to show that a more condensed circuit layout is possible in the case of coplanar technology-based integrated circuits. Figure 2.1.27 shows the structure that is to be analyzed. At the four ends of two coupled waveguides, four ports are defined. Port 1 and port 3 are connected to the coplanar waveguide I, whereas ports 2 are 4 are

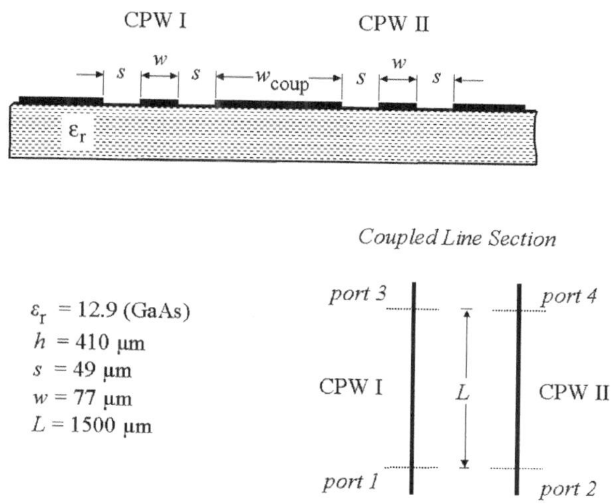

Fig. 2.1.27. Geometry and port definition of two coupled coplanar waveguides.

connected to waveguide II. If the scattering parameters of the two coupled coplanar waveguides are to be analyzed with the existence of the even and the odd mode on each line, the structure shown in Fig. 2.1.27 must be described by eight ports: four ports describing the even mode propagation and four ports for the odd mode propagation.

Figure 2.1.28a shows the frequency-dependent magnitude of the reflection coefficient $|S_{11}^{ee}|$ at port 1, the isolation $|S_{21}^{ee}|$, and the coupling coefficient $|S_{41}^{ee}|$ for the technically relevant even mode between port 1 and ports 2 and 4, respectively. Figure 2.1.28b shows the mode conversion parameters $|S_{mn}^{oe}|$ for conversion from the even mode to the odd mode between port 1 and ports 2, 3, and 4, respectively. The coupled coplanar waveguides satisfy the above-mentioned condition: $w_{coup} \geq 2s + w$ (see dimensions of the structure given in Fig. 2.1.27).

The figure also shows that the mentioned design rule fulfills all requirements for the circuit design; that is, the input reflection coefficient $|S_{11}^{ee}|$ (Fig. 2.1.28a) for all considered frequencies is lower than −48 dB, the isolation $|S_{21}^{ee}|$ is always better than −30 dB and the coupling coefficient $|S_{41}^{ee}|$ has a maximum value of only −47.6 dB for frequencies higher than 20 GHz. This is a value that is below a well-measurable value in microwave integrated circuits. A similar good behavior may be found for the conversion of the even mode into the unwanted odd mode Fig. 2.1.29b). The coupling parameter $|S_{21}^{oe}|$ is always below −20 dB. The design rule $w_{coup} \geq 2s + w$ therefore may be claimed as being too pessimistic, and smaller coupling width w_{coup} therefore may be allowed.

To discuss the integration density that can be used in coplanar circuit design, the width w_{coup} of the ground plane between the two coplanar waveguides has been varied between 77 μm and 450 μm, keeping all other line parameter to the values shown in Fig. 2.1.27. Figure 2.1.29 shows (a) the mag-

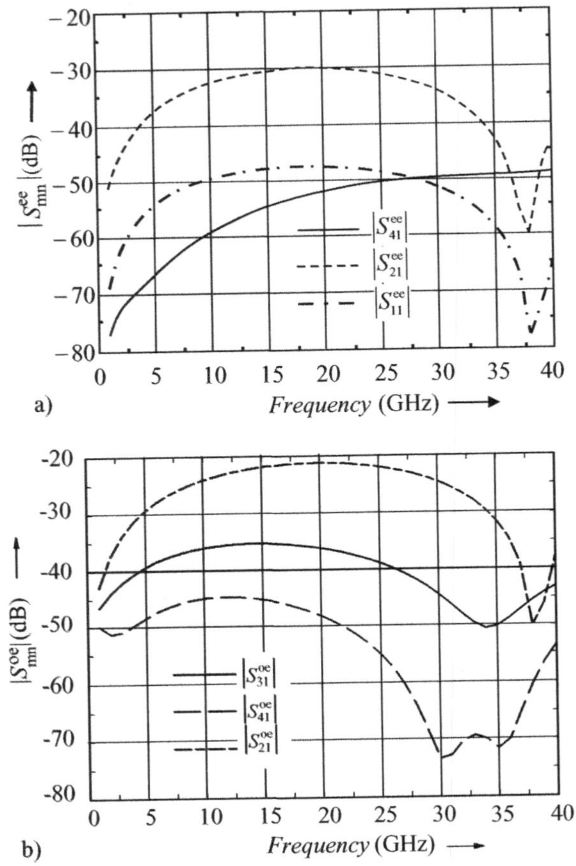

Fig. 2.1.28. Frequency dependence of the coupling coefficient $|S_{41}^{ee}|$, the isolation $|S_{21}^{ee}|$, and the input reflection coefficient $|S_{11}^{ee}|$ for (a) the fundamental even mode of two coupled coplanar waveguides and (b) the even-mode to odd-mode conversion scattering parameters. Coupling width $w_{coup} = 175\,\mu m$.

nitude of the coupling coefficient for the even mode and (b) the magnitude of the isolation between the even mode and the odd mode at port 1 and port 2 with respect to dependence on the frequency and the parameter w_{coup}. Both figures show the strong dependence of the scattering parameters on the frequency and on the coupling width between the lines. It can be observed that the coupling coefficient $|S_{41}^{ee}|$ even for the smallest assumed coupling width $w_{coup} = 77\,\mu m$ is still below -40 dB for all considered frequencies. On the other hand, the coupling between the even mode at port 1 and the odd mode at port 2 increases to a maximum value of -17 dB for this small value of the coupling width. Nevertheless, it can be seen that even for such a small coupling width which leads to a value $w_{coup}/(w + 2s) = 0.44$, a decoupling between the two coplanar waveguides acceptable for circuit design may be realized.

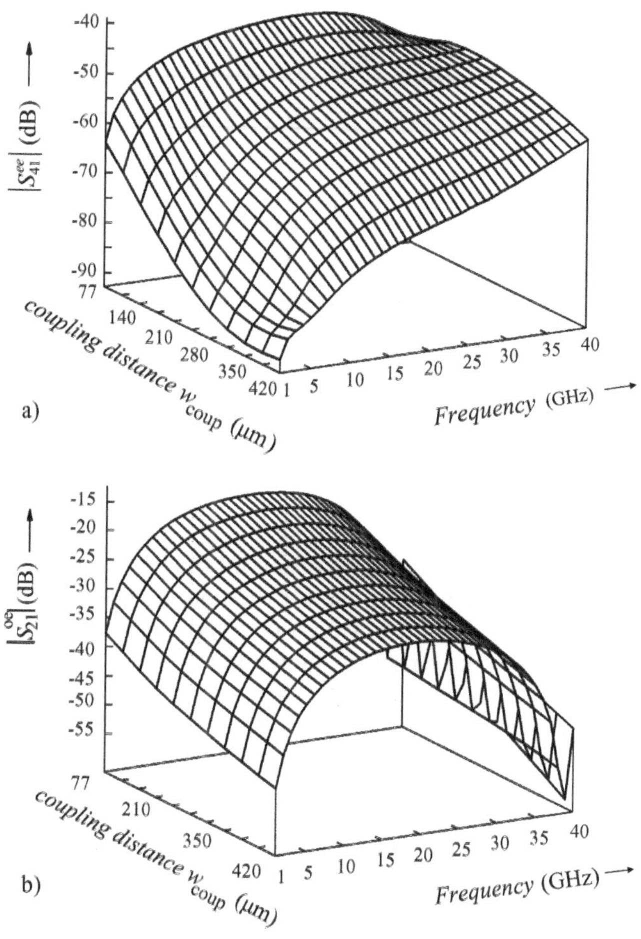

Fig. 2.1.29. The frequency dependence of (a) the coupling coefficient $|S_{41}^{ee}|$ and (b) the isolation $|S_{21}^{oe}|$ for different values of the coupling width w_{coup}. Structure and geometrical parameters as shown in Fig. 2.1.27.

Thus, the considered coupling between the two coplanar waveguides is just about small enough for application in circuit design. Experimental results have shown that similar results are obtained for two coupled coplanar waveguides with different strip width w.

It has been frequently claimed that coplanar microwave integrated circuits enable a higher integration density compared to microstrip-line-based technologies. One of the reasons for this argument, as has already been mentioned, is the presence of an additional ground plane between the two coupled coplanar waveguides that reduces the coupling as compared to the case of two parallel microstrip lines. Looking at the above-demonstrated results for the two

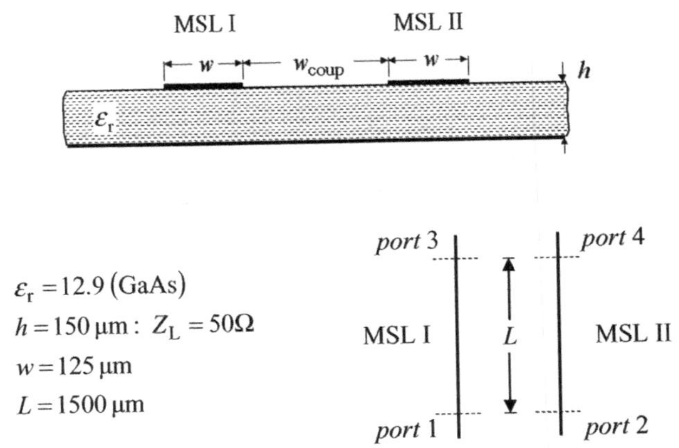

Fig. 2.1.30. Parameters of the coupled microstrip lines and an equivalent four-port structure.

coupled coplanar waveguides, it can be observed that these two lines are characterized by a very small coupling, even if the coupling width between them is small. To make a comparison between coupled coplanar waveguides and coupled microstrip lines, the structure shown in Fig. 2.1.30 is analyzed.

A disadvantage of this coupled line structure compared to the coplanar structure already becomes evident during the design of the structure; that is, if the substrate thickness h is chosen to be, say, $h = 150\,\mu m$, the strip width w for a 50-Ω impedance is fixed to $w = 125\,\mu m$. However, in the case of the coplanar line, multiple combinations of slot width values s and strip width values w can lead to a 50-Ω line. In other words, in the design process the coplanar line has one more degree of freedom.

Figure 2.1.31 shows the frequency-dependent eigen reflection coefficient $|S_{11}|$, the isolation $|S_{21}|$, and the coupling coefficient $|S_{41}|$ for the coupled microstrip lines. If a value $w_{coup} = 275\,\mu m$ is chosen, a direct comparison can be made to the coplanar structure shown in Fig. 2.1.27 because both structures then have the same lateral dimensions, that is, $4s + 2w + w_{coup}$ (Fig. 2.1.27) = $525\,\mu m = 2w + w_{coup}$ (Fig. 2.1.31). The values of the eigenreflection coefficient and the isolation are acceptable for the circuit designer, but the coupling coefficient of −12 dB at 40 GHz would be too high if such a coupling is an unwanted coupling in a circuit. At lower frequencies (up to 10 GHz), the coupling value is acceptable for a circuit layout.

If again a criterion for the integration density of a microwave circuit shall be derived for the case of the microstrip line, Fig. 2.1.32 can be used. The figure shows the magnitude of the coupling coefficient $|S_{41}|$ in dependence on the frequency and for variable values of the coupling width w_{coup} between the lines. It can be seen that the dependence of the coupling coefficient on the coupling width is much lower than in the case of the coupled coplanar waveguides (compare to Fig. 2.1.29a).

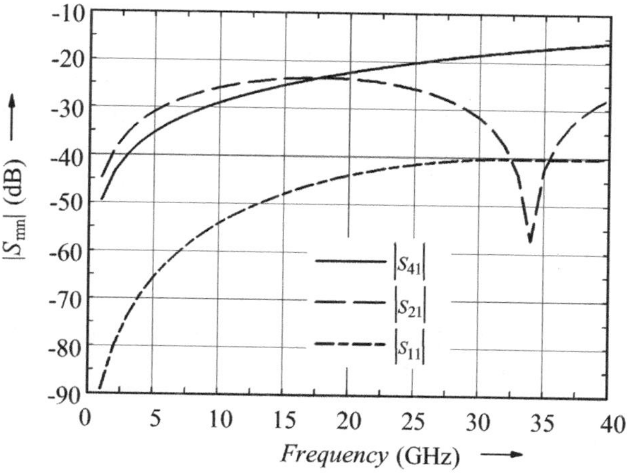

Fig. 2.1.31. The frequency dependence of the input reflection coefficient $|S_{11}|$, the coupling coefficient $|S_{41}|$, and the isolation $|S_{21}|$ of two coupled microstrip lines. Coupling width between the two lines: $w_{coup} = 275\,\mu m$.

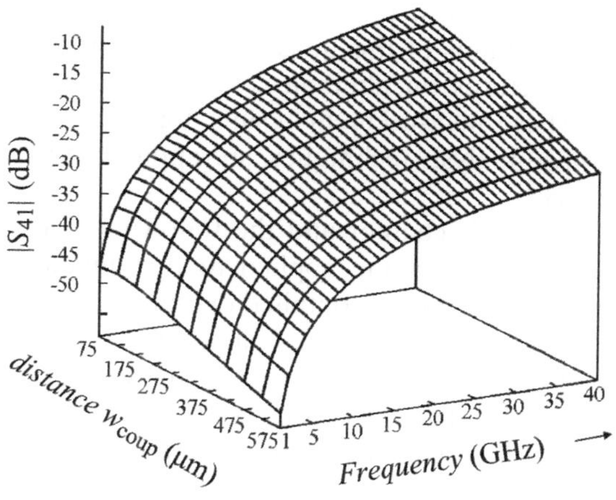

Fig. 2.1.32. Frequency dependence of the coupling coefficient $|S_{41}|$ between two coupled microstrip lines with variable coupling width w_{coup}. Geometrical parameters and structure as shown in Fig. 2.1.30.

Only for a coupling width of $w_{coup} > 425\,\mu m$, $|S_{41}|$ is below -20 dB over the whole frequency range. This is a value that must be required in circuit design to avoid an unwanted coupling inside a microwave circuit. Observe that even for a value $w_{coup} = 575\,\mu m$ the coupling coefficient $|S_{41}|$ still has a maximum value (at 40 GHz) of -24.5 dB that is much higher than the equivalent coefficient for all considered coupling widths in the case of the coplanar waveguide.

Consequently, this means that coplanar circuits can be designed much more compact than microstrip-technology-based circuits. This is a result of the close confinement of the electromagnetic field to the gap region of the coplanar waveguide. On the other hand, the stray field of the microstrip line covers a much broader area, and the design engineer must keep larger distances between the line structures to avoid unwanted coupling effects in microstrip-based circuits.

2.2 QUASI-STATIC ANALYSIS OF COPLANAR WAVEGUIDES USING THE FINITE DIFFERENCE METHOD

2.2.1 Introduction

The aim of this chapter is to show that because of the special properties of coplanar waveguides as they have been described in Section 2.1, a quasi-static analysis in many cases is sufficient to describe the electromagnetic properties of coplanar lines and components for application in microwave integrated circuits. This is, dependent on the geometrical parameters of the structures, valid up to very high frequencies in the millimeter-wave range (50–60 GHz). To control the application limits of this quasi-static analysis technique, accurate measurements of structures on alumina ceramics and on gallium-arsenide material are used. These measurement results, together with computational results from full-wave analysis techniques, are applied to verify the applicability of this approach.

The method used in this chapter and the following chapters is the static finite difference analysis technique applied to two-dimensional and three-dimensional problems [248]. First, a detailed overview of the applied theoretical technique and of the resulting numerical accuracy of the applied method is given. Then the application of the method to analyze coplanar waveguides on multilayer carrier materials is discussed, and the derived results as well as the influence of the geometrical and electrical parameters on the propagation properties of the waveguides are investigated for the fundamental coplanar waveguide mode. Furthermore, some special forms of coplanar waveguides are considered.

In many microwave circuits, closely coupled waveguides are needed to realize couplers, filters, and the like. For the analysis of such circuits, the calculation of the line parameters on the basis of simple closed formula approximations normally are not sufficient to predict the circuit properties with the needed accuracy. This is because in many analysis techniques of this kind, only the coupling of the next nearby line is considered and coupling effects to other lines are not taken into account. Therefore in the third part of this chapter the complete parameter matrices of multiply coupled coplanar waveguides will be analyzed using the quasi-static finite difference approach.

The single and the coupled coplanar lines can be considered as two-dimensional, longitudinally homogeneous structures. The line discontinuities (such as line bends, T-junctions, crossings, etc.), however, belong to the group of three-dimensional components that have to be analyzed using a three-dimensional analysis technique. The modeling of these three-dimensional components will be discussed in Chapter 3. Because all these components are of small geometrical size, they can be described using equivalent circuits derived from quasi-static electric and magnetic field computations.

The needed field computations will again be performed using a three-dimensional quasi-static finite difference technique. In this connection it is very essential also to consider the air-bridge technique that is needed in all discontinuity structures to suppress the influence of the unwanted fundamental odd mode on the coplanar structures. They can be described using equivalent circuits derived from quasi-static electric and magnetic field computations. Different buildup techniques of these air bridges are also considered and analyzed using the three-dimensional finite difference technique (see Section 3.5.5).

Technological progress in the area of monolithic microwave integrated circuits (MMICs) permits us to realize circuits with more and more reduced geometrical size. Moreover, because of the need for new frequency bands for communication applications (for example), the working frequencies of the circuits are shifting to higher frequencies. Hence, two requirements are always very important in microwave circuit design: (1) the increase of the component density inside the circuits and (2) the increase of the resonant frequencies of the used lumped components such as spiral inductors, interdigital and MIM (metal–insulator–metal) capacitors, and lumped resistors. Because of the missing backside metallization and thereby the reduced parasitic ground capacitances in coplanar technology (if applied in this way), the resonant frequencies of lumped elements can be shifted to very high frequencies (>50 GHz).

Components such as line discontinuities can be described by equivalent circuits with discrete elements like capacitors, inductors, and resistors. These equivalent circuits, again, are determined using the three-dimensional finite difference analysis technique. Included into the analysis are all effects that are essential for monolithic integration of the components such as the metallization thickness and the influence of possibly available air bridges in different realization forms. Measurements are used to verify the numerical results.

A *coplanar component design library* has been developed, as will be described in Chapter 5. It bases on the finite difference analysis technique. An extensive measurement evaluation of more than 100 coplanar elements and comparison to simulation results up to frequencies of 60 GHz will be described in the following chapters.

2.2.2 The Finite Difference Method as Applied to the Analysis of Coplanar Waveguide Structures

As already mentioned in the introduction, coplanar waveguides belong to the class of quasi-TEM lines if they are used in their fundamental even mode (coplanar waveguide mode). As has been shown in Section 2.1, the dispersion of this mode is low. Therefore, an analysis based on a quasi-static computation technique will deliver results accurate enough for the design of coplanar microwave integrated circuits even for application at higher frequencies. Only in the millimeter-wave range, care must be taken that the influence of dispersion is considered. In these cases, possibly full-wave analyses as they have been described in Section 2.1 must be used for the design basis.

The quasi-static analysis technique assumes that the longitudinal electric and magnetic field components on the line structure are always small compared to the transversal ones and that the transversal current-density components in the conducting strips can be neglected against the longitudinal ones. Under these assumptions, it is uniquely possible to describe the waveguides by frequency-independent parameters that are defined by the capacitance per line length C', the inductance per line length L', the longitudinal resistance per line length R', and the transversal conductance per line length G'.

The capacitance per unit line length C' can be derived from the charge distribution on the line conductors and the electric field distribution between the line electrodes using the solution of Laplace's equation $\Delta\varphi = 0$. An analytical solution of this equation (because of the inhomogeneous material distribution) is found only with difficulties (compare also Section 2.3). Therefore a numerical solution can be used with advantage. A method that seems to be especially suited under the above-defined conditions is the quasi-static finite difference analysis technique [4, 31, 248] that will be applied under consideration of the boundary conditions on a metallic shielding, enclosing the coplanar waveguide structure. The following sections describe the applied method in detail, considering additionally techniques for the efficient use of the method in the case of coplanar waveguides and components.

2.2.3 The Solution of Laplace's Equation for Planar and Coplanar Line Structures Using the Finite Difference Method

An essential step in the numerical analysis of a given boundary value problem is the discretization of the region that is to be analyzed. It is also a criterion for the accuracy and the efficiency of the finite difference method [66, 112, 126, 147, 128, 163, 164]. Therefore, it is part of the task to find an optimum discretization scheme so that besides a low numerical expense a relative high accuracy of the calculated data can be achieved. Additionally, a discretization technique that can be applied to a variety of problems occurring in circuit design and analysis should be used. This is not possible without introducing certain restrictions with respect to the problems to be analyzed. One of these restrictions that seems to be reasonable with respect to the planar nature of the structures used in planar circuits is the assumption that material boundaries do only occur in one direction (horizontal direction). Additionally, in the analysis technique used here, the number of parallel material layers is restricted to four, which is large enough for technically interesting planar waveguide structures. Using these criteria, the following assumptions must be used to discretize planar waveguide structures like those in Figs. 1.1 and 1.2:

- The ratio between the different geometrical sizes of the structures may become large. Therefore, it is necessary to use a nonequidistant discretization to reduce the numerical expense to an acceptable amount.
- At the strip-line edges where the electromagnetic field is highly concentrated, or at positions where field singularities may occur, a refined discretization must be used to reach a better solution for the field distribution.
- For the sake of a good convergence, the ratio of the applied mesh sizes should not exceed the value five in the case of a nonequidistant discretization. A good solution is doubling (or dividing by 2) the mesh sizes at the mesh boundaries. This allows calculating the potential values in the mesh boundary easily.
- The choice of a quadratic mesh cell reduces the computational algorithm and thereby reduces computation time.
- The discretization should be chosen in such a way that the material boundaries and the conductor edges lie on the mesh nodes.

Considering the above-mentioned criteria, the discretization scheme (as qualitatively shown in Fig. 2.2.1) is used for the analysis of coplanar microwave structures. The structures are enclosed in a conducting shielding so that Dirichlet walls (electric walls) with constant electric potential limit the analysis space to the interior of this shielding. In the case of magnetic field analyses, the electric walls may be replaced by magnetic walls (Neumann walls). The region inside the shielding is discretized below and above the conducting strips using mesh cells with sizes that increase with the distance from the strips. Because the metallization thickness normally is much smaller than the other geometrical line parameters, the mesh in the area near the metallic strips is once more refined in the y-direction and possibly also in the x-direction (Fig. 2.2.1). If the line structure is symmetric, as shown in Fig. 2.2.1, a magnetic wall in the symmetry plane may be introduced, thus reducing the numerical expense for analyzing the field by a factor of two.

After finishing the discretization scheme, Laplace's equation must be written in finite difference form. Using the solution of the two-dimensional Laplace equation,

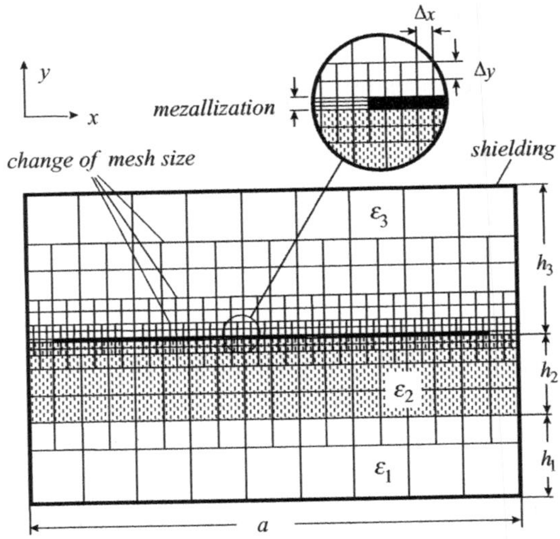

Fig. 2.2.1. The discretization scheme for planar line structures considering the metalization thickness ($\Delta x = \Delta y$).

$$\Delta\varphi = \frac{\partial^2\varphi}{\partial x^2} + \frac{\partial^2\varphi}{\partial y^2} = 0, \tag{2.2.1}$$

the electric field strength $\boldsymbol{E}$ can be calculated as follows:

$$\boldsymbol{E} = -\text{grad}\ \varphi = -\frac{\partial\varphi}{\partial x}\boldsymbol{u}_x - \frac{\partial\varphi}{\partial y}\boldsymbol{u}_y, \tag{2.2.2}$$

where $\boldsymbol{u}_x$ and $\boldsymbol{u}_y$ are the unit vectors in x- and y-coordinate directions, respectively. The boundary conditions, assuming ideal conducting strips and shielding, can be written as follows:

Electric wall (Dirichlet wall):

$$\boldsymbol{E}_t = 0 \rightarrow \varphi = \text{const.}, \tag{2.2.3a}$$

Magnetic wall (Neumann wall):

$$E_n = 0 \rightarrow \frac{\partial\varphi}{\partial n} = \text{const.}, \tag{2.2.3b}$$

Dielectric interface:

$$\varepsilon_1 E_{n1} = \varepsilon_2 E_{n2} \rightarrow \varepsilon_1\left(\frac{\partial\varphi}{\partial n}\right)_{\text{I}} = \varepsilon_2\left(\frac{\partial\varphi}{\partial n}\right)_{\text{II}}, \tag{2.2.3c}$$

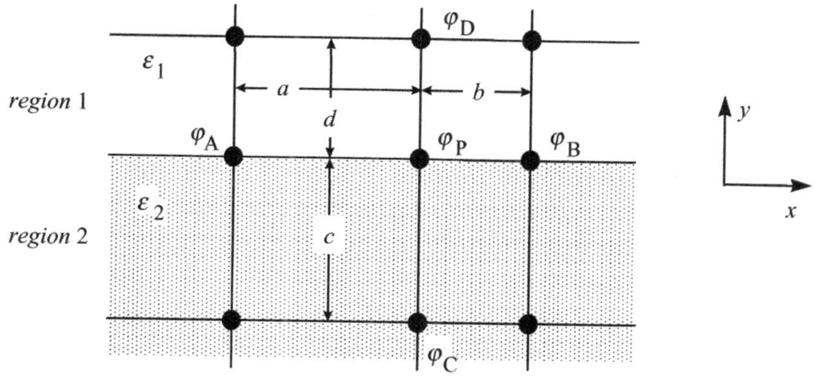

Fig. 2.2.2. General discretization mesh for developing the difference equation.

where $\boldsymbol{E}_t$ and $\boldsymbol{E}_n$ are the tangential and the normal electric field strengths, respectively.

If the differential equation (2.2.1) is replaced by an adequate difference equation, the potential φ_P in a central point P (Fig. 2.2.2) must be substituted by the potentials of the surrounding mesh nodes. This is achieved by developing these potentials into series expansions with respect to the potential in P as follows:

$$\varphi_A = \varphi_P - a\left(\frac{\partial\varphi}{\partial x}\right)_{\mathrm{I}} + \frac{a^2}{2}\left(\frac{\partial^2\varphi}{\partial x^2}\right)_{\mathrm{I}} - \frac{a^3}{6}\left(\frac{\partial^3\varphi}{\partial x^3}\right)_{\mathrm{I}} + \frac{a^4}{24}\left(\frac{\partial^4\varphi}{\partial x^4}\right)_{\mathrm{I}} \mp \ldots, \quad (2.2.4)$$

$$\varphi_D = \varphi_P + d\left(\frac{\partial\varphi}{\partial y}\right)_{\mathrm{I}} + \frac{d^2}{2}\left(\frac{\partial^2\varphi}{\partial y^2}\right)_{\mathrm{I}} + \frac{d^3}{6}\left(\frac{\partial^3\varphi}{\partial y^3}\right)_{\mathrm{I}} + \frac{d^4}{24}\left(\frac{\partial^4\varphi}{\partial y^4}\right)_{\mathrm{I}} \mp \ldots, \quad (2.2.5)$$

where a, b, c, and d are the mesh sizes as shown in Fig. 2.2.2 for a mesh point P on an interface between two different dielectric media. If it is assumed that the mesh sizes a, b, c and d are small, the series expansions shown in Eq. (2.2.4) and Eq. (2.2.5) may be truncated after the second differential term. The weighted sum of Eq. (2.2.4) and Eq. (2.2.5) together with Laplace's equation (2.2.1) then gives a relation between φ_P, φ_A, and φ_D:

$$\frac{d}{a}\varphi_A + \frac{a}{d}\varphi_D - \left(\frac{d}{a} + \frac{a}{d}\right)\varphi_P + d\left(\frac{\partial\varphi}{\partial x}\right)_{\mathrm{I}} - a\left(\frac{\partial\varphi}{\partial y}\right)_{\mathrm{I}} = 0. \quad (2.2.6)$$

Similar equations can be derived for the other mesh nodes in region I and region II (Fig. 2.2.2):

$$\frac{d}{b}\varphi_B + \frac{b}{d}\varphi_D - \left(\frac{d}{b} + \frac{b}{d}\right)\varphi_P - d\left(\frac{\partial\varphi}{\partial x}\right)_{\mathrm{I}} - b\left(\frac{\partial\varphi}{\partial y}\right)_{\mathrm{I}} = 0, \quad (2.2.7)$$

$$\frac{c}{b}\varphi_B+\frac{b}{c}\varphi_C-\left(\frac{c}{b}+\frac{b}{c}\right)\varphi_P-c\left(\frac{\partial\varphi}{\partial x}\right)_{\text{II}}+b\left(\frac{\partial\varphi}{\partial y}\right)_{\text{II}}=0, \tag{2.2.8}$$

$$\frac{c}{a}\varphi_A+\frac{a}{c}\varphi_C-\left(\frac{c}{a}+\frac{a}{c}\right)\varphi_P+c\left(\frac{\partial\varphi}{\partial x}\right)_{\text{II}}+a\left(\frac{\partial\varphi}{\partial y}\right)_{\text{II}}=0. \tag{2.2.9}$$

If Eq. (2.2.6) is added to Eq. (2.2.7) and Eq. (2.2.8) is added to Eq. (2.2.9), the following equations result for the mesh shown in Fig. 2.2.2:

$$\frac{d}{a}\varphi_A+\frac{d}{b}\varphi_B+\frac{(a+b)}{d}\varphi_D-\left(\frac{a}{d}+\frac{d}{a}+\frac{b}{d}+\frac{d}{b}\right)\varphi_P=(a+b)\left(\frac{\partial\varphi}{\partial y}\right)_{\text{I}}, \tag{2.2.10}$$

$$\frac{c}{a}\varphi_A+\frac{c}{b}\varphi_B+\frac{(a+b)}{c}\varphi_C-\left(\frac{a}{c}+\frac{c}{a}+\frac{b}{c}+\frac{c}{b}\right)\varphi_P=-(a+b)\left(\frac{\partial\varphi}{\partial y}\right)_{\text{II}}. \tag{2.2.11}$$

Applying the boundary condition given in Eq. (2.2.3c) to Eqs. (2.2.10) and (2.2.11), a relation between the electric potential φ_P in the central point P and the potentials in the surrounding mesh nodes can be derived as a linear combination of the form

$$\left(\frac{\varepsilon_1}{d}+\frac{\varepsilon_2}{c}+\frac{d\varepsilon_1+c\varepsilon_2}{ab}\right)\varphi_P-\left(\frac{d\varepsilon_1+c\varepsilon_2}{a(a+b)}\right)\varphi_A-\left(\frac{d\varepsilon_1+c\varepsilon_2}{b(a+b)}\right)\varphi_B$$
$$-\frac{\varepsilon_2}{c}\varphi_C-\frac{\varepsilon_1}{d}\varphi_D=0. \tag{2.2.12}$$

The general discretization scheme shown in Fig. 2.2.2 and its describing equation (2.2.12) can be reduced to simpler forms in the case of special mesh and node configurations, as shown in Table 2.2.1.

If Eq. (2.2.12) is applied to all mesh nodes, a linear system of equations for the electric potentials in all mesh nodes results, which can be written as a matrix equation of the form

$$\boldsymbol{M}\varphi+\boldsymbol{B}=0, \tag{2.2.13}$$

where φ is the vector of the node potentials, $\boldsymbol{M}$ is the symmetrical coefficient matrix defined by Eq. (2.2.12), and $\boldsymbol{B}$ is the vector of the electric potentials in the boundary nodes. The system matrix $\boldsymbol{M}$ is of band structure; that is, it has nonzero elements only near the principal diagonal.

For the solution of Eq. (2.2.13) with respect to the electric node potentials, a Gauss–Seidel iteration technique or another adequate method is the possible candidate (see, e.g., references 89, 112, and 197). The disadvantage of the Gauss–Seidel technique is its bad convergence behavior in the case of high node numbers, as shown for a microstrip line in Fig. 2.2.3.

TABLE 2.2.1. Simplified Algorithms for Calculating the Electrical Potentials in Special Node Arrangements

Case	Arrangement	Equation
Homogeneous medium standard mesh $a = b = c = d$	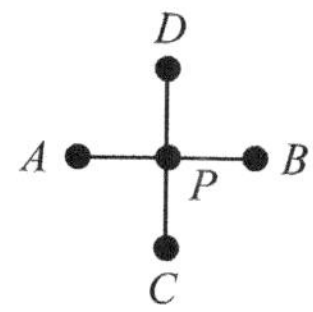	$4\varphi_P - \varphi_A - \varphi_B - \varphi_C - \varphi_D = 0$
Homogeneous medium change of mesh size with five nodes $a = b = d$ $c = 2a$	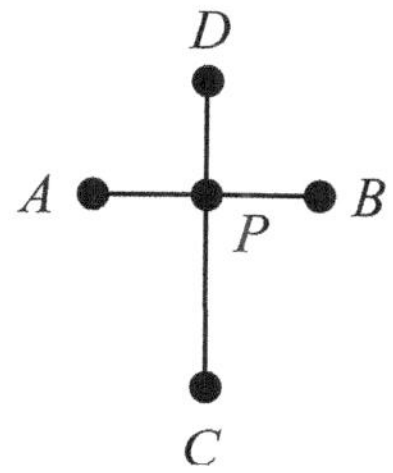	$9\varphi_P - 3\varphi_A - 3\varphi_B - \varphi_C - 2\varphi_D = 0$
Homogeneous medium change of mesh size with six nodes $a = b = d$ $c = 2a$	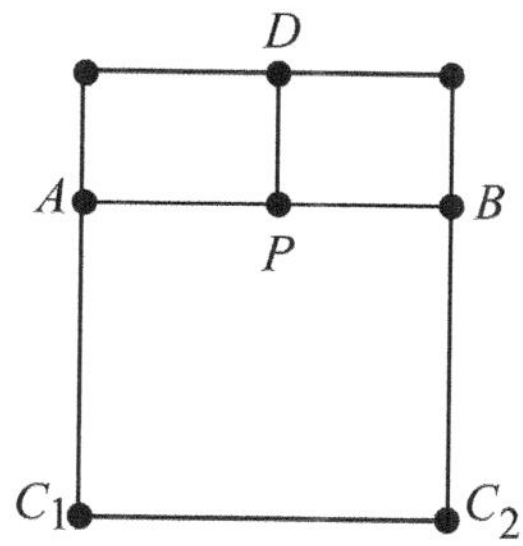	$18\varphi_P - 6\varphi_A - 6\varphi_B - \varphi_{C1} - \varphi_{C2} - 4\varphi_D = 0$
Inhomogeneous medium $a = b = c$ $k = d/c$	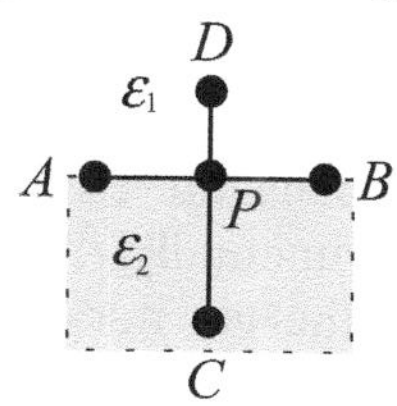	$\left\{\left(\frac{2}{k}+2k\right)\varepsilon_1+4\varepsilon_2\right\}\varphi_P-2\varepsilon_2\varphi_C - \frac{2}{k}\varepsilon_1\varphi_D-(k\varepsilon_1+\varepsilon_2)(\varphi_A+\varphi_B)=0$
Electric Wall $a = b = c$	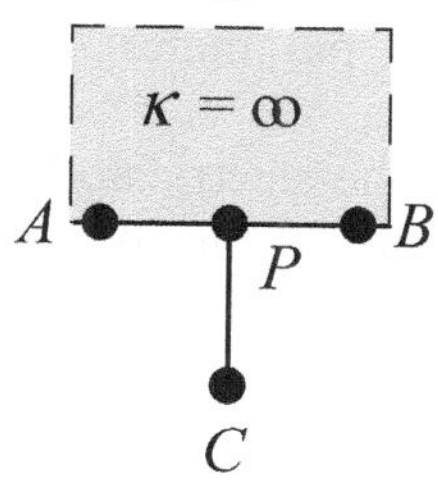	$\varphi_A = \varphi_B = \varphi_P =$ constant
Magnetic wall $a = b = c$	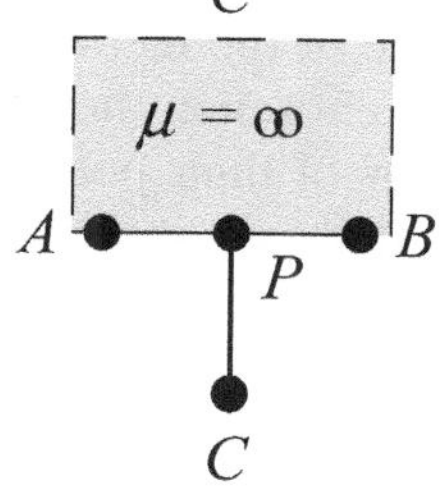	$4\varphi_P - \varphi_A - \varphi_B - 2\varphi_C = 0$

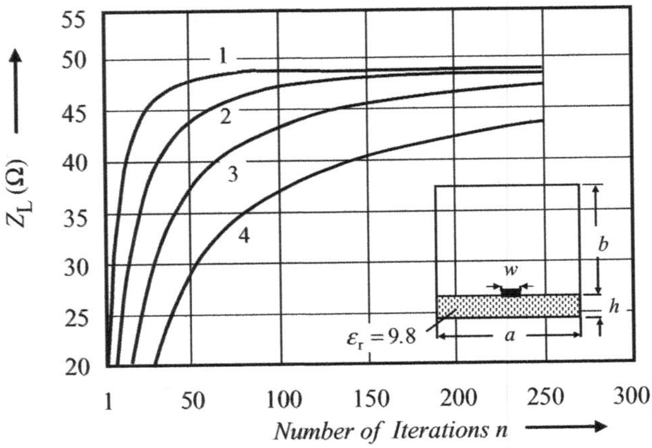

Fig. 2.2.3. Convergence of the Gauss–Seidel iteration technique for solving the linear equation system for the case of a shielded microstrip line (characteristic impedance Z_L), which is dependent on the number of mesh nodes (parameter) and the number of iteration steps. Dimensions of the shielding: $a = 20w$, $b = 10w$, $h = w$. Curve 1, $w = 5\Delta x \Rightarrow 2062$ mesh nodes; 2, $w = 10\Delta x \Rightarrow 4750$ mesh nodes; 3, $w = 20\Delta x \Rightarrow 12{,}000$ mesh nodes; 4, $w = 40\Delta x \Rightarrow 34{,}000$ mesh nodes.

A good alternative to the Gauss–Seidel technique is the successive overrelaxation method (see reference 3), which is also an iterative method and solves the equation system using an iteration scheme:

$$\varphi_n^{(\nu+1)} = \varphi_n^{(\nu)} - k r_n^{(\nu)}, \tag{2.2.14}$$

where k is a relaxation factor for optimizing the convergence with values $0 \le k \le 2$, which must be determined experimentally. $r_n^{(\nu)}$ is the nth element of the residuum vector $\boldsymbol{r}^{(\nu)} = \boldsymbol{M} \cdot \boldsymbol{\varphi}^{(\nu)} + \boldsymbol{B}$ in the νth iteration step. Figure 2.2.4 shows the improved convergence of the successive overrelaxation technique, which is dependent on the convergence parameter k and the number of iterations n.

Also, the choice of the start values for the electric node potentials has a big influence on the convergence of the computations. For example, if the characteristic impedance of a microstrip line is to be calculated, the electric potential of the shielding may be set to zero and the potential of the strip to 1 V at the beginning of the iteration process.

Figure 2.2.5 shows how the convergence of the method behaves for different values of all the other node potentials chosen at the beginning of the iteration. These nodes start with a value of φ_0 of their potential between 0.05 V and 0.9 V. It can be seen from the results that fastest convergence is found for a starting value of node potentials between 0.05 V and 0.1 V. This is because the electric field is mainly concentrated near the strip of the waveguide structure.

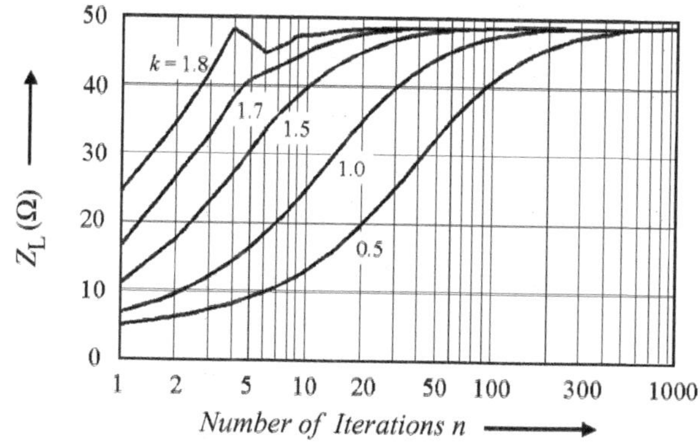

Fig. 2.2.4. Convergence of the successive overrelaxation method, which is dependent on the convergence parameter k and the number of iterations n for the example shown in Fig. 2.2.3, curve 2.

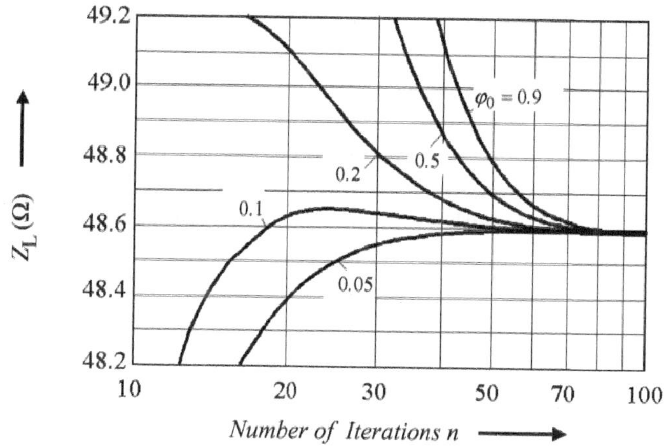

Fig. 2.2.5. Convergence of the iteration technique in dependence on the starting values of the node potentials for the example shown in Fig. 2.2.3, curve 2.

2.2.4 Application of the Quasi-Static Techniques to the Analysis of Coplanar Waveguides

The fundamental coplanar waveguide has already been defined in Chapter 1: It is a planar waveguide with all electrodes on one side of the substrate material. On this waveguide, a quasi-TEM mode can propagate if at least one of the electrodes has finite cross-sectional dimensions. Figure 2.2.6 shows the coplanar waveguide as it is used in most applications. In principle, it is a planar three-strip line [23, 24, 50, 78, 150].

Several advantages and disadvantages of coplanar waveguides compared with microstrip lines have already been discussed in the previous chapters.

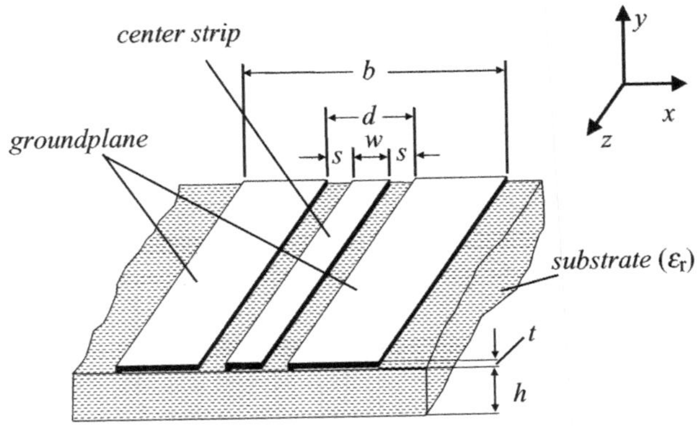

Fig. 2.2.6. The symmetrical coplanar three-strip line (coplanar waveguide CPW).

Besides the aspects discussed there, further specialties have to be mentioned here as the basis for the analysis technique used in this chapter: The first aspect is the dispersion of the line that has already been discussed in Section 2.1. For a true TEM-mode propagation on a waveguide, a homogeneous medium is needed in the cross section of the line. In the case of the coplanar waveguide, the field carrying space of the line has a piecewise homogeneous medium only. This, as is well known, leads to different phase velocities of possible waves in the different media. Because of the boundary conditions in the plane between the media, only one mode with a common phase velocity in all media can propagate along the line, so that a hybrid mode is built up which has not only transversal field components, but also longitudinal electric and magnetic field components [18]. In the case of the fundamental coplanar mode, these field components are small compared to the transversal components, especially at low frequencies. But with increasing frequency the influence of these longitudinal components becomes larger. This means that the wave propagation along the line can no longer be described using frequency-independent characteristic parameters (impedance, phase velocity). This frequency dependence of the line parameters, which is called the dispersion of the line parameters, is strongly dependent on the field distribution and the geometrical parameters of the considered line [13, 17–19, 49, 64].

Figure 2.2.7 shows the distribution of the electrical field strength on a microstrip line and a coplanar waveguide, as they have been computed using the above-described static finite difference technique. If the microstrip line is compared to the coplanar waveguide with respect to the above-discussed aspects, the following results can be deduced:

- The field of the coplanar waveguide is mainly concentrated inside the slots between the ground planes and the center strip and, therefore, is only slightly changing with increasing frequency. The field of the microstrip line, on the other hand, has a large stray field in the air region above the

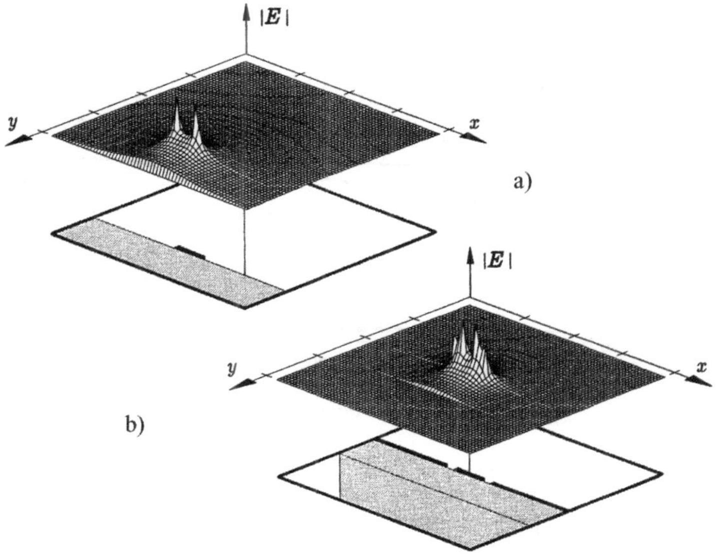

Fig. 2.2.7. Magnitude of the electrical field strength of the fundamental quasi-TEM mode in the cross section for (a) a microstrip line and (b) a coplanar waveguide.

substrate and is concentrated more and more inside the substrate material with increasing frequency.

- The coplanar waveguide with (assumed) zero thickness metallization performs a magnetic wall on the upper side of the substrate (refer also to Section 2.3). Therefore, neglecting the low value of the field in the air region below the substrate, nearly the same part of the electromagnetic field is concentrated in the air region and inside the substrate material. As a consequence, with increasing frequency and to a first approximation, the electromagnetic field of the coplanar waveguide is concentrated to an equal amount in the air and the substrate region within the slot area. In the case of the microstrip line, however, an increase of the frequency leads to a reduction of the field part in the air region and to a concentration of the field inside the dielectric substrate material.
- The capacitance per unit line length, and hence the characteristic impedance of the microstrip line, is dependent on the ratio w/h of strip width w and substrate height h (and, of course, on the dielectric constant of the substrate material). Therefore, for the realization of low characteristic impedance values, lines with large strip widths w are needed. In the case of the coplanar line, the characteristic impedance is mainly independent of the substrate height and can be adjusted by variation of the strip width w to slot width s ratio w/s. Therefore lines with small cross-sectional dimensions can be realized, and as a consequence, the assumption of a quasi-TEM wave propagation is valid in a larger frequency range.

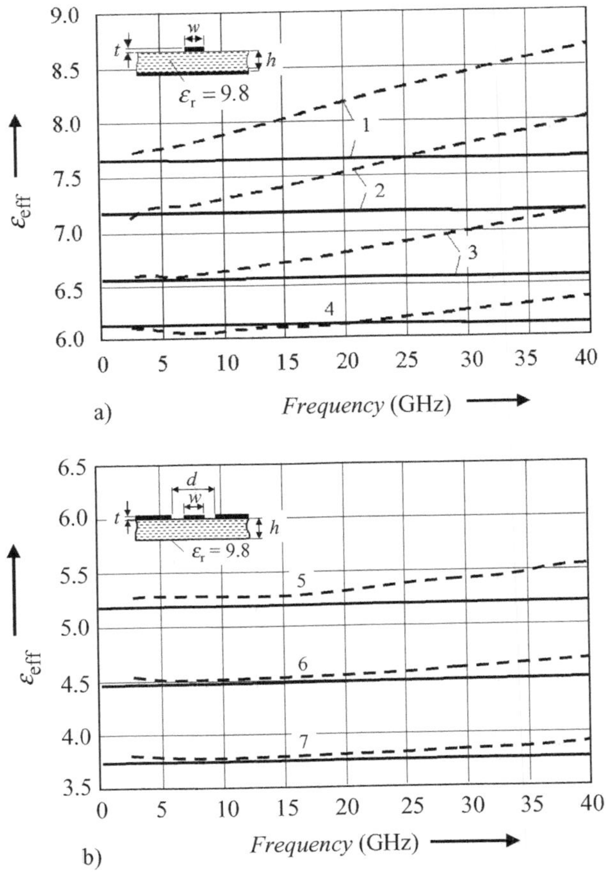

Fig. 2.2.8. Measured (– – –) and calculated (———) dispersion of the effective dielectric constant of (a) a microstrip line and (b) a coplanar waveguide. Metalization thickness: $t = 5\,\mu\text{m}$.
Parameters:

Curve:	1	2	3	4	5	6	7
w (μm):	1031	547	225	62.5	438	375	188
h (μm):	250	250	250	250	625	250	125
Z_L (Ω):	20	35	50	80	40	50	80

Using the above-mentioned arguments, it may be concluded that the dispersion of the coplanar line is much lower than that of the microstrip line of comparable geometrical size. To prove this assumption, the phase velocity, described by the effective dielectric constant $\varepsilon_{eff} = (c_0/v_{ph})^2$, with v_{ph} the phase velocity of the wave propagating on the line, has been measured for various microstrip lines and coplanar waveguides. The results are shown in Fig. 2.2.8. The solid lines are the results calculated with the static finite difference technique as described above (for the applied method see below). The theoretical results consider the metalization thickness of $t = 5\,\mu m$, which is essential especially in the case of coplanar waveguides. For frequencies lower than 2.5 GHz the accuracy of the applied measurement techniques is low due to calibration. Therefore these results are omitted in Fig. 2.2.8.

Figure 2.2.8 shows that the dispersion of the microstrip line is much larger than that of the coplanar waveguide. Furthermore, it can be observed that the dispersion of the microstrip line is strongly dependent on the ratio w/h whereas in the case of the coplanar waveguide, only a small dependence on the d/h ratio can be observed. It is also interesting to note that the dispersion of the coplanar waveguide, even for large values of the s/h ratio (see Fig. 2.2.8b, curve 7), is not larger than 4% in the considered frequency range of up to 40 GHz.

From the results of various measurements, it has been found that the dispersion of coplanar waveguides on substrates with different material heights ($h = 125$, 250, and $625\,\mu m$) is smaller than 4% in the frequency range of up to 40 GHz if the following conditions are fulfilled: $s/h \leq 2$ and $d/h \leq 5$.

If the characteristic parameters like the propagation coefficient γ and the characteristic line impedance Z_L of a quasi-TEM mode on a planar waveguide are to be determined, only the capacitance, the inductance, the resistance, and the conductance per unit line length must be known (see reference 126):

$$Z_L = \sqrt{\frac{R' + j\omega L'}{G' + j\omega C'}}, \tag{2.2.15}$$

$$\gamma = \alpha + j\beta = \sqrt{(R' + j\omega L')(G' + j\omega C')}. \tag{2.2.16}$$

The losses of the planar waveguides normally are so small that they generally do not change the fundamental field distribution compared to the lossless case. If it is considered that the inductance per unit line length L' of a lossless TEM-mode waveguide can be computed from the capacitance per unit line length C_0' of the same waveguide, replacing the dielectric substrate material by air [126], Eqs. (2.2.15) and (2.2.16) in the case of a lossless waveguide can be reduced to

$$Z_L = \sqrt{\frac{L'}{C'}} = \sqrt{\frac{\varepsilon_0 \mu_0}{C_0' C'}} = \frac{1}{c_0 \sqrt{C_0' C'}}, \tag{2.2.17}$$

$$\beta = \frac{2\pi f}{c_0} \sqrt{\varepsilon_{\text{eff}}}, \tag{2.2.18}$$

with

$$\varepsilon_{\text{eff}} = \frac{C'}{C_0'}, \tag{2.2.19}$$

where ε_{eff}, the effective dielectric constant, is the relative permittivity of a material which homogeneously fills the planar waveguide and which then has the same phase velocity as the considered waveguide with inhomogeneous material distribution in its cross section (e.g., the coplanar waveguide). This means that it is good enough to calculate the capacitance per unit line length of the waveguide, if the characteristic impedance Z_L and the effective dielectric constant ε_{eff} are to be determined.

To compute the capacitance per unit line length of the coplanar waveguide using the finite difference technique described above, first the potential distribution in the cross section of the line has to be determined. If the electric potential of the inner conductor is chosen to be φ_i and the potential of the ground electrodes is φ_g, the capacitance per unit line length of the waveguide can be calculated from

$$C' = \frac{Q'}{(\varphi_i - \varphi_g)}, \tag{2.2.20}$$

where Q' is the charge per unit line length on the center strip. As shown in Fig. 2.2.9, the charge per unit line length can be determined by computation of the normal electric field component of the electrical field strength from the computed electric potential along the contour of the center strip of the coplanar waveguide by [5]

$$Q' = \oint D_n ds = \varepsilon_0 \oint \varepsilon_r E_n ds = \varepsilon_0 \sum_i \varepsilon_r \frac{\Delta\varphi_i}{\Delta x_i} \Delta x_j, \tag{2.2.21}$$

$$j = \begin{cases} 2 & \text{inside the metallization layer,} \\ 1 & \text{in all other cases.} \end{cases}$$

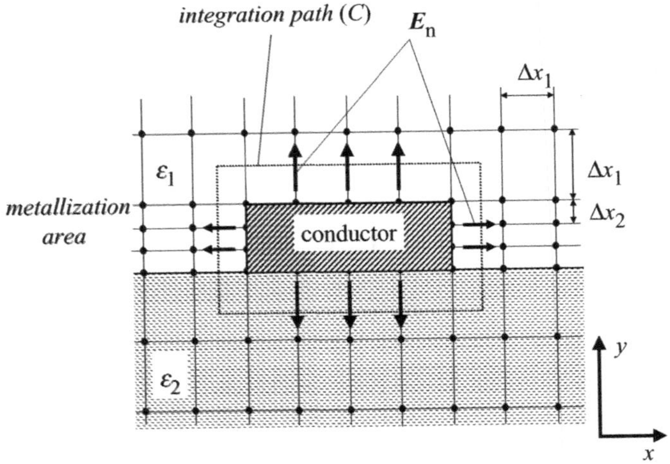

Fig. 2.2.9. Calculating the charge per unit line length on the inner conductor.

Principally, three different loss mechanisms contribute to the attenuation coefficient α of planar waveguides: the conductor loss described by the attenuation coefficient α_c, the dielectric loss described by α_d, and the cross-conductance loss described by α_g. The reason for the conductor losses is the finite conductivity of the metalization layers, along with the dielectric losses that are polarization losses of the dielectric substrate material. The reason for the cross-conductance losses is a possible finite conductivity of the dielectric carrier material. For applications in the microwave area, mainly substrate materials with very low conductivity and small loss factor tan δ_ε are used (possible exception is silicon, Si). Therefore the values α_d and α_g can be assumed to be negligibly small compared to α_c if substrate materials like plastic materials, ceramics, or GaAs semiconductors are used.

The calculation of the conductor losses is performed using a perturbation technique under the assumption that the losses are small. This means that the static field distribution in the cross section of the waveguide is determined without considering the losses, and then this field is used to compute the surface current density along the conductor contour [26]. The attenuation coefficient is then calculated from this surface current density using the surface resistance derived from skin effect theory.

Using the assumption of TEM-mode propagation on the waveguide, the surface current density can be derived from the normal component of the electrical field strength $\boldsymbol{E}_n$, which is a measure for the tangential magnetic field strength $\boldsymbol{H}_t$ in each point under the assumption made. The current in each point of the metallic contour, therefore, can be written as

$$I_i = \oint_{C_i} \boldsymbol{H}_t \cdot ds = \sqrt{\frac{\varepsilon_0 \varepsilon_r}{\mu_0}} \oint_{C_i} E_n \, ds, \tag{2.2.22}$$

$$i = \begin{cases} 1 & \text{inner strip,} \\ 2 & \text{ground conductor,} \end{cases}$$

where the integration must be performed along a closed curve enclosing the considered point. The so-calculated current along the metallic electrode contour then delivers the needed surface current density function $g(s)$ for computing the conductor losses. Figure 2.2.10 shows the magnitude of such a current distribution on a coplanar line. The total current through the conductor strip is then found by integrating along the conductor contour.

Under the assumption of an exponential decay of the transported power along the line length, the attenuation coefficient α_c can be written as [125]

$$\frac{\alpha_c}{\text{dB}} = 4.343 \frac{R_f'}{Z_L}, \tag{2.2.23}$$

$$\frac{R_f'}{\Omega} = 8.24 \times 10^{-3} \sqrt{\mu_r (f/\text{GHz})(\rho/\rho_{\text{Cu}})} \left\{ \oint_{C_1} \frac{g_1^2(s_1)}{I_z^2} ds_1 + \oint_{C_2} \frac{g_2^2(s_2)}{I_z^2} ds_2 \right\}, \tag{2.2.24}$$

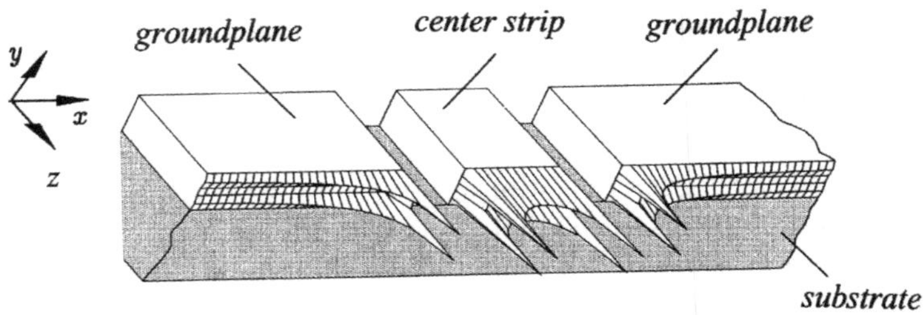

Fig. 2.2.10. The surface current density distribution in the cross section of a symmetrical coplanar waveguide with finite metalization thickness.

where R'_f is the resistance per unit line length of the waveguide, f is the frequency, ρ is the specific resistance of the used conductor material, and $\rho_{Cu} = 1.72 \times 10^{-6}\,\Omega\cdot\text{cm}$ is the specific resistance of copper. g_1 and g_2 are the surface current density functions as they have been defined above for the inner conductor and the ground metalization, respectively. They can be determined from the numerically computed electric potential distribution using Eq. (2.2.22). The integrals are to be calculated along the closed contours of the conductors. Z_L is the characteristic impedance of the waveguide and I_z is the total longitudinal current through the cross section of the center strip.

The above-described method assumes that the metalization thickness t is at least three times the value of the skin depth δ. This means that the used equations are only valid for frequencies higher than a limit frequency f_{lim}, which is defined by the specific resistance ρ of the used metallic material on the one side and by the metalization thickness t on the other side:

$$f_{\text{lim}}/\text{GHz} = 39.313\frac{(\rho/\rho_{\text{Cu}})}{(\text{t}/\mu\text{m})^2}. \tag{2.2.25}$$

For a gold conductor on a ceramic substrate material ($t = 5\,\mu\text{m}$) the limiting frequency is about 2.5 GHz. Below this frequency the dc current resistance per unit line length R'_{dc} can be used instead of R'_f with good accuracy:

$$R'_{\text{dc}} = \rho\frac{b-d+w}{wt(b-d)}. \tag{2.2.26}$$

If a linear function is assumed for the frequency region $f \le f_{\text{lim}}$ (what only is an approximation of the real situation), the following approximate formula for calculating the resistance per unit line length can be found for the total frequency range:

$$R'_f = \begin{cases} R'_{dc} + \left\{R'_f(f = 1\,\text{GHz})\sqrt{f_{\text{lim}}/\text{GHz}} - R'_{dc}\right\}\dfrac{f}{f_{\text{lim}}} & \text{for } f \le f_{\text{lim}}, \\ R'_f(f = 1\,\text{GHz})\sqrt{f/\text{GHz}} & \text{for } f \ge f_{\text{lim}}, \end{cases} \tag{2.2.27}$$

TABLE 2.2.2. Comparison Between the Results of Different Quasi-static Computation Methods for the Characteristic Impedance and the Effective Dielectric Constant of Various Coplanar Waveguides ($\varepsilon_r = 9.8$, $t = 0$, $d = 625\,\mu m$)

		Z_L (Ω)			ε_{eff}		
$h(\mu m)$	w/d	FDM	FEM	CM	FDM	FEM	CM
125	0.08	114.81	116.0	117.68	4.100	4.048	3.993
125	0.2	89.96	90.0	92.56	3.854	3.785	3.767
125	0.32	76.35	76.8	78.59	3.737	3.698	3.665
125	0.4	69.35	69.9	71.35	3.701	3.670	3.637
125	0.6	54.78	55.1	56.24	3.721	3.684	3.676
125	0.8	31.30	31.3	42.17	3.900	3.875	3.877
125	0.92	31.74	31.2	32.03	4.135	4.114	4.134
250	0.08	105.28	106	106.82	4.874	4.809	4.847
250	0.2	81.07	81.2	82.7	4.743	4.685	4.719
250	0.32	68.32	68.8	69.81	4.666	4.624	4.645
250	0.4	61.97	62.6	63.35	4.633	4.585	4.614
250	0.6	49.26	49.6	50.34	4.602	4.566	4.589
250	0.8	37.82	38.0	38.5	4.652	4.605	4.650
250	0.92	29.61	29.3	29.85	4.751	4.705	4.760
625	0.08	101.31	102.0	102.15	5.275	5.213	5.300
625	0.2	77.24	76.9	78.25	5.240	5.189	5.272
625	0.32	64.70	65.0	65.64	5.218	5.165	5.254
625	0.4	58.55	59.0	59.42	5.206	5.165	5.245
625	0.6	46.47	46.7	47.15	5.188	5.141	5.231
625	0.8	35.85	35.9	36.29	5.190	5.141	5.235
625	0.92	28.31	27.9	28.41	5.212	5.165	5.254

with $R_f'(1\,\text{GHz})$ to be calculated from Eq (2.2.24) for a frequency of $f = 1\,\text{GHz}$.

To check the accuracy of the quasi-static analysis technique using the finite difference technique described above, the characteristic impedance and the effective dielectric constant of various coplanar waveguides have been calculated and the results have been compared to numerical results of other quasi-static analysis techniques using conformal mapping technique (CM) [126] and a finite element technique (FEM) [173]. Table 2.2.2 shows the results of this comparison. It can be observed that the agreement is mainly good, but in the case of very small substrate height some deviations can be found.

2.2.5 Characteristic Parameters of Coplanar Waveguides

Using the above-described finite difference technique for the analysis of the electric field, the parameters of the coplanar waveguide like the characteristic impedance, the effective dielectric constant, and the attenuation coefficient are calculated and compared to measurement results. In all computations it is assumed that the width of the ground planes is (quasi-) infinite. This means that the ground planes reach up to the shielding as it is defined, for example, in Fig. 2.2.1.

It also means that the ground planes are always on the same potential. Various coplanar waveguides with constant total slot width d = 625 μm (see Fig. 2.2.6) and different w/d ratios on substrates of different heights h are used to analyze the influence of the substrate height on the characteristic parameters. Alumina ceramic substrates with a dielectric constant of ε_r = 9.8 and heights of 125 μm, 250 μm, and 625 μm have been used for this investigation.

The measurement of the characteristic impedance Z_L has been performed measuring the reflection coefficient at the input of the line at low frequencies under the assumption of an ideal transition between two lines of different characteristic impedances using an on-wafer measurement equipment. Assuming that the influence of transition from the coplanar waveguide to the probe head may be neglected at low frequencies [193], a transition from the coplanar waveguide to a 50-Ω line may be assumed. The value of the effective dielectric constant ε_{eff} has been measured by comparison of the phase coefficient of two identical open-ended lines with different line lengths. During the measurements a 3-mm-thick air region was assured under the substrate material so that the same conditions as in the simulations could be guaranteed in the measurements as well. The calculated and the measured values of the characteristic impedance Z_L and the effective dielectric constant ε_{eff} in dependence on the ratio w/d are shown in Figs. 2.2.11 and 2.2.12. The deviations between the measured and calculated characteristic impedance are on the order of 5% and, therefore, within the order of the measurement accuracy. In the case of the effective dielectric constant the deviation is only about 2%. This is because a more accurate measurement technique has been used in this case. In Fig.

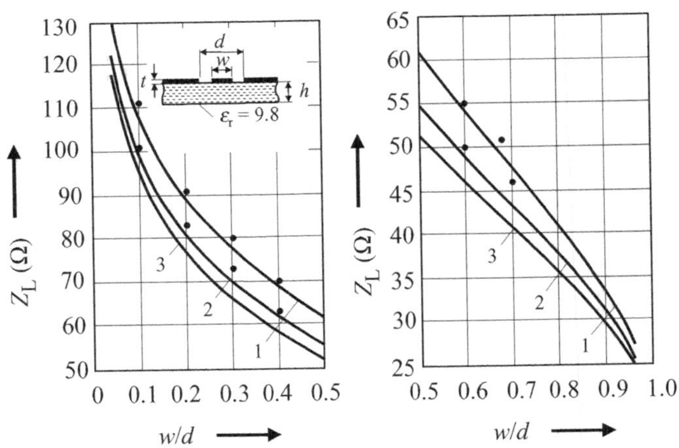

Fig. 2.2.11. Measured (•) and calculated (———) characteristic impedance of coplanar waveguides in dependence on the ratio w/d (d = 625 μm, t = 5 μm). Curve 1, h = 125 μm; curve 2, h = 250 μm; curve 3, h = 625 μm.

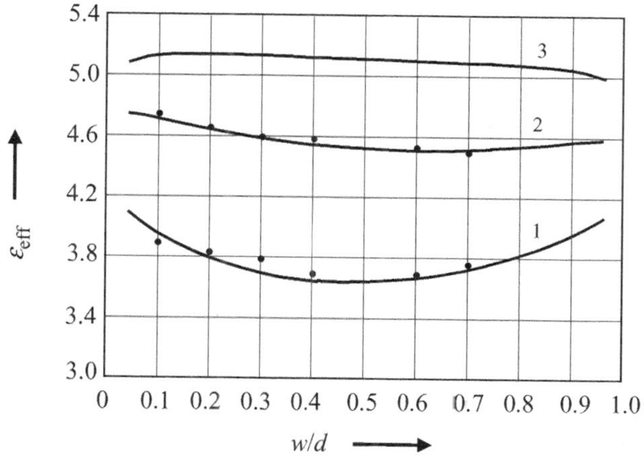

Fig. 2.2.12. Measured (•) and calculated (———) effective dielectric constant of coplanar waveguides in dependence on the ratio w/d ($d = 625\,\mu m$, $t = 5\,\mu m$). Curve 1, $h = 125\,\mu m$; curve 2, $h = 250\,\mu m$; curve 3, $h = 625\,\mu m$.

2.2.12, curve 3 shows a decrease of the effective dielectric constant for values $w/d \leq 0.1$ and $w/d \geq 0.9$, which (as will be shown later) is a result of the influence of the considered finite metalization thickness t.

As has been mentioned above, when calculating the attenuation coefficient, only the contribution of the conductor losses will be considered. To prove the accuracy of this approximation, the attenuation coefficients of different microstrip lines and coplanar waveguides have been computed and measured. The measurement results have been drawn from the measured reflection coefficients at the input of open-ended lines with different line lengths.

Figure 2.2.13 shows a comparison between the measured and calculated attenuation coefficients for (a) microstrip lines and (b) coplanar waveguides. While the agreement between measured and calculated values is quite good in the case of the microstrip line (Fig. 2.2.13a), deviations can be clearly observed especially at higher frequencies in the case of the coplanar waveguides (Fig. 2.2.13b). The reason for this, besides neglecting radiation losses and dielectric losses, can also be attributed to the assumed ideal metalization. One of the assumptions is that the gold strips on top of the substrate are of ideal rectangular cross section. In reality, this cross section, which is built of a thin chromium layer and a plated gold layer, is much more complex.

Surface roughness [1] and inhomogeneity in the line length direction are also not considered in the calculations. These effects are of much more influence in the case of the coplanar waveguide (compared to the microstrip line) because in this waveguide the strip current flows to a large extent along the strip edges (see Fig. 2.2.10).

In the case of the microstrip line, on the other hand, there is a much bigger part of the current that also flows in the center of the strip. Comparing Figs.

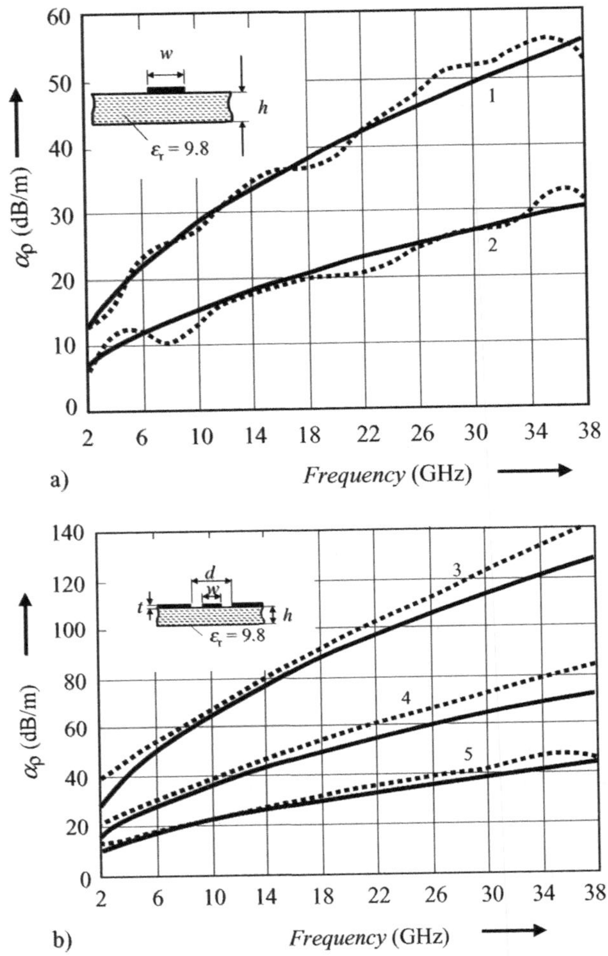

Fig. 2.2.13. Measured (– – –) and calculated (———) attenuation coefficients of (a) the microstrip line and (b) the coplanar line, which are dependent on the frequency ($h = 250\,\mu m$, $t = 5\,\mu m$).

Curve:	1	2	3	4	5
w (μm):	62.5	225	62.5	125	62.5
d (μm):	—	—	112.5	225	625
Z_L (Ω):	80	50	40	50	80

2.2.13a and 2.2.13b, it can also be observed that the losses of the microstrip line and the coplanar waveguide are of the same order if the coplanar waveguide is well-designed (see curve 5 of Fig. 2.2.13 as an example). It can be said that despite the assumed approximations, the agreement of measured and calculated results is good enough for circuit design applications. This indicates

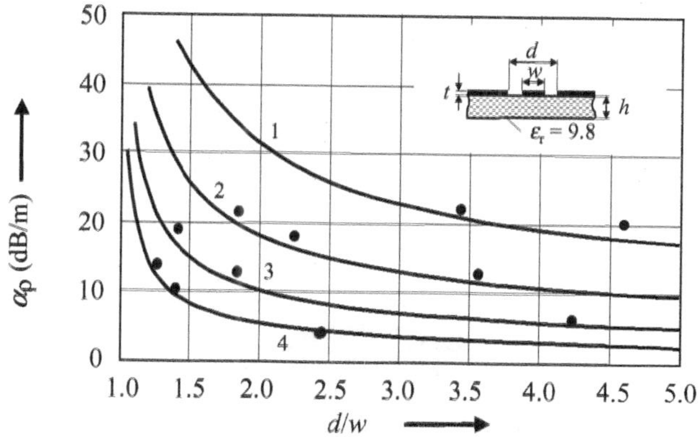

Fig. 2.2.14. Measured (•) and calculated (———) attenuation coefficient α of the coplanar waveguide in dependence on the normalized total slot width d/w ($h = 250\,\mu\text{m}$, $t = 5\,\mu\text{m}$). Curve 1, $w = 31.25\,\mu\text{m}$; curve 2, $w = 62.50\,\mu\text{m}$; curve 3, $w = 125.0\,\mu\text{m}$; curve 4, $w = 250.0\,\mu\text{m}$.

that the applied perturbation technique is an acceptable concept for the loss calculations in the case of the coplanar waveguides.

The attenuation coefficient of the coplanar waveguide is mainly determined by the width of the inner strip w and the metalization thickness t on the one side and by the slot width s on the other side. The dependence on the slot width can be explained by the fact that with reduced slot width the electromagnetic field is concentrated more and more inside the slot. Therefore, the current mainly flows at the edges of the conductors, and the area for current flow becomes smaller and smaller. As long as the electromagnetic field in the air region below the substrate is negligibly small, the substrate height h does not have a significant influence on the attenuation coefficient α. To demonstrate the dependence of the attenuation coefficient of the coplanar waveguide on the lateral dimensions, the attenuation coefficient for constant values of the strip width w and varying total slot width d (i.e., for different slot widths s) is depicted in Fig. 2.2.14.

The measured values of α are somewhat larger than the calculated ones because only the conductor losses, as described above, have been considered in the calculations. Nevertheless, there is an agreement between measurement and computation that is good enough for circuit design applications.

For the verification of the coplanar waveguide losses in MMIC circuit design, a 2-inch GaAs wafer, consisting of more than 100 test circuits, has been designed and fabricated (foundry: Daimler Benz Research Center, Ulm, Germany). A photograph of the wafer is depicted in Fig. 2.2.15. Besides the simple coplanar waveguide structures, waveguide discontinuities (see Chapter 3) and lumped elements (see Chapter 4) in coplanar environment have also been included on the wafer.

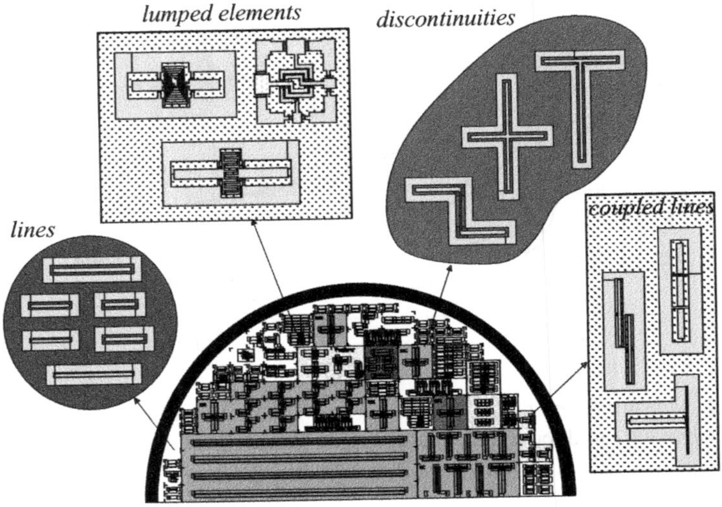

Fig. 2.2.15. GaAs wafer with coplanar test circuits.

The left and right half are identical to ensure that each circuit can be measured. The following substrate parameters are used for all structures and comparative simulations: $h = 450\,\mu m$, $\varepsilon_r = 12.9$, $t = 3\,\mu m$ (galvanically enlarged metalization layer); metalization conductivity parameters are $\sigma = 41\,S/\mu m$, $t_2 = 0.48\,\mu m$ (gate metalization layer).

The verification measurements have been performed with an on-wafer probe in a frequency range from 400 MHz to up to 67 GHz. Here, only the verification of the coplanar waveguide structures with respect to their dispersion characteristics and losses will be discussed. The verification of the other components that are also on the wafer will be discussed subsequently in Chapters 3 and 4.

The coplanar waveguides investigated here have ground strips of finite width and variable spaces between ground and electrical sidewalls of the shielding (see Fig. 2.2.1). The coplanar line configuration has also been used to test any undesired influences of the top cover, the sidewalls, and the bottom metalization. It was found that the influence of the top and bottom walls can be neglected if the distance to the upper electrical wall of the shielding is larger than twice the substrate height and if the distance to the bottom metalization is equal to the substrate height (compare it with the case of a microstrip line in which the distance to the upper shielding must be larger than 10 times the substrate height).

In the verification, very long coplanar waveguides have been used to prove the calculation of the line losses. On a GaAs material the substrate losses can be neglected to a first approximation. The main losses result from the (gold) metalization. Four long lines of different geometrical sizes are depicted on each half of the wafer in Fig. 2.2.15. Also, 40-, 50-, 60-, and 70-Ω configurations

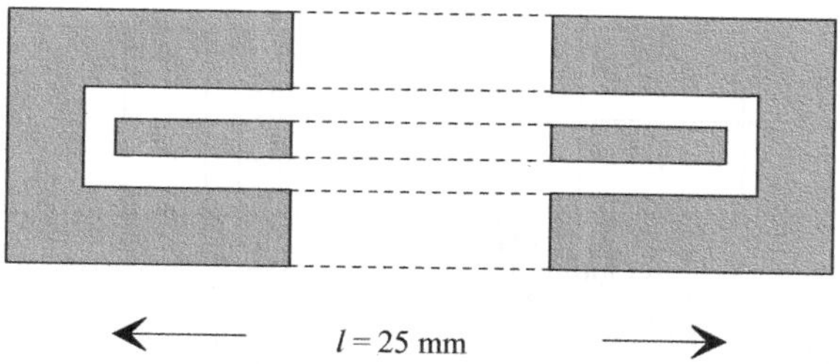

Fig. 2.2.16. Used test structure on GaAs substrate for comparison of the measured and simulated losses of coplanar waveguides. 40-Ω line: $w = 100\,\mu m$, $s = 35\,\mu m$. 50-Ω line: $w = 100\,\mu m$, $s = 75\,\mu m$. 60-Ω line: $w = 40\,\mu m$, $s = 57\,\mu m$. 70-Ω line: $w = 40\,\mu m$, $s = 103\,\mu m$.

(see Table 2.2.2) with a length of 25 mm have been realized. Figure 2.2.16 shows the used test structures. When comparing measurement and simulation results, it must be made sure that the influence of the probe heads and their positions on the coplanar waveguide is taken into account correctly. Also, because the ground plane surrounds the coplanar waveguide at its ends, as well, the influence of the end capacitance (coplanar waveguide open end; see Section 3.5.1) must be considered in the simulation. The test structure given in Fig. 2.2.16, therefore, has been simulated as a series connection of (1) an open-ended coplanar waveguide, (2) a short piece of coplanar waveguide ($l = 30\,\mu m$), (3) an input port, (4) a coplanar waveguide of length 24,940 μm, (5) an output port, (6) a 30-μm-long coplanar waveguide, and (7) an open-ended coplanar waveguide.

Figures 2.2.17 and 2.2.18 depict, as examples, the measured and simulated scattering parameters for the 50-Ω and the 70-Ω coplanar waveguides over a frequency range of nearly 60 GHz. Note that all scattering parameters are normalized to 50-Ω impedance. In both cases the simulated and measured magnitudes agree well up to the highest frequencies as shown in the figures. The measured and simulated results are nearly congruent in the depiction. This means especially that the losses of the waveguides are well-simulated. The phase angles of S_{21} and S_{11} show a similar good agreement over the same frequency ranges. Because of the large lengths of the waveguides, dispersion effects cannot be well-detected from the phase distribution, but they are present as will be shown shortly.

As has already been mentioned in the previous sections, coplanar waveguides of constant characteristic line impedance may be built with different geometrical sizes. Table 2.2.3 shows five coplanar waveguides of different geometrical parameters all having a characteristic impedance of nearly 50 Ω.

The following compromise conditions must be considered in the application of coplanar waveguides: On the one hand a coplanar waveguide with small geometrical center strip width w and small slot width s (example, CPW5 in Table 2.2.3) has high losses, but its dispersion is low. On the other hand, coplanar waveguides of large center-strip widths (e.g., CPW1 in Table

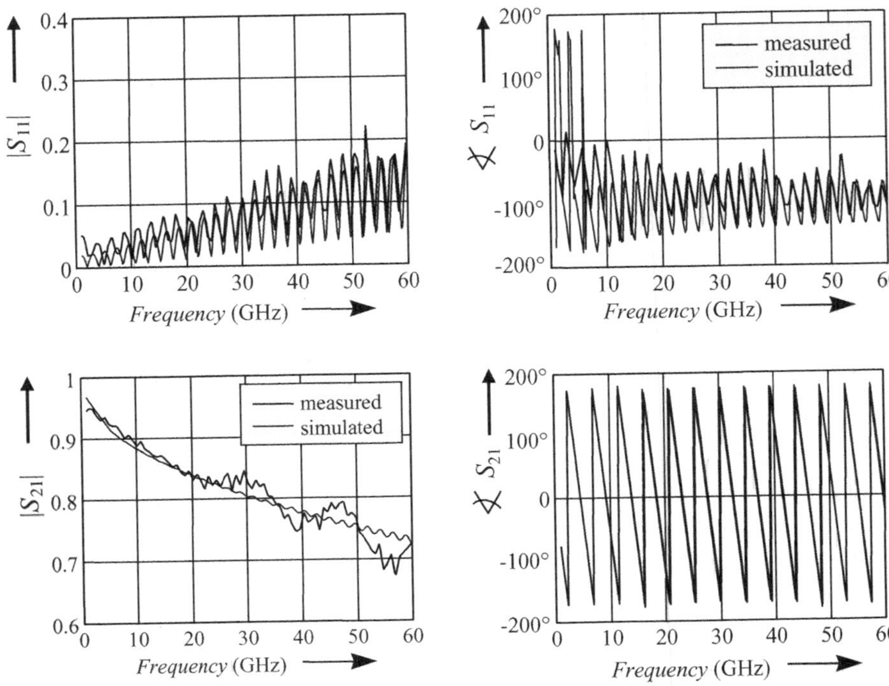

Fig. 2.2.17. Comparison between measurement (thick lines) and simulation (thin lines) of the scattering parameters of a long (25 mm) 50-Ω coplanar waveguide, plotted against the frequency.

2.2.3) have reduced losses but much higher dispersion of its characteristic parameters.

To measure the dispersion effects of the lines, "short" coplanar waveguides with a line length of 1 mm have been used as test structures as shown in Fig. 2.2.19. The scattering parameters are measured and simulated in a similar way as described above for the "long" lines. The results for two examples—one waveguide with a broad center strip (CPW1, Table 2.2.3) and one waveguide with a medium center strip (CPW3, Table 2.2.3)—are shown in Fig. 2.2.20 and Fig. 2.2.21, respectively. Observe the expanded scale for the magnitude measurements.

From both figures it can be seen that over the frequency range of interest there is a certain deviation between the measured and the simulated results. It should be remembered that the simulation technique is a quasi-static one, and it does not take into account the dispersion of the coplanar waveguide. The influence of this dispersion can be clearly seen in the results shown in Figs. 2.2.20 and 2.2.21. It can also be observed, that the agreement between measurement and simulation is better for the case of the CPW3, which has a smaller center strip width and also a smaller slot width. This means that the

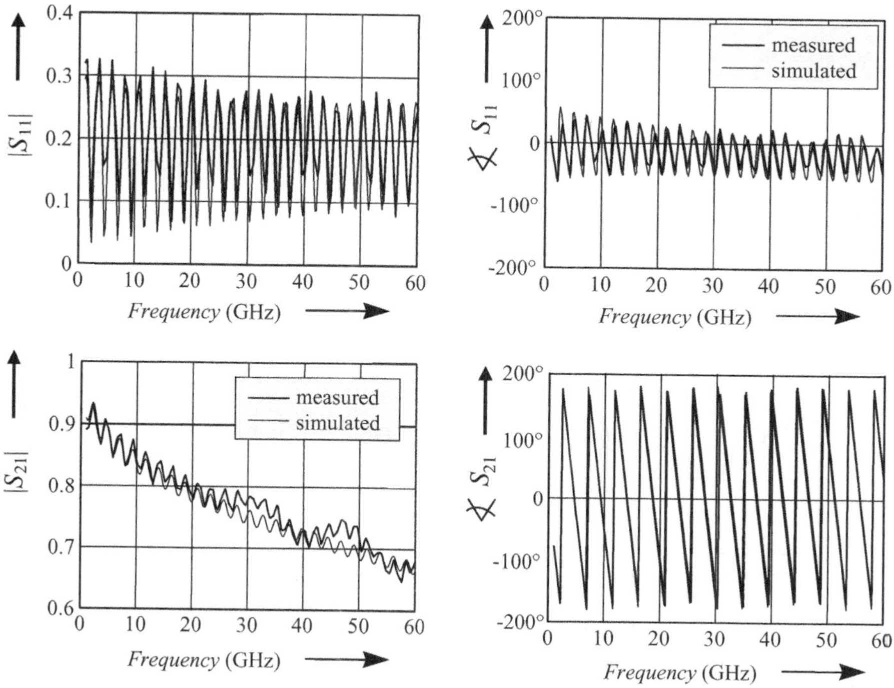

Fig. 2.2.18. Comparison between measurement (thick lines) and simulation (thin lines) of the scattering parameters of a long (25 mm) 70 Ω coplanar waveguide, plotted against the frequency.

TABLE 2.2.3. 50-Ω Coplanar Waveguides of Different Geometrical Sizes Together with Their Characteristic Line Impedance and Effective Dielectric Constant as Calculated from the Quasi-Static Finite Difference Model

Identification	w (μm)	s (μm)	$Z_{L,\text{stat}}$ (Ω)	ε_{eff}
CPW1	100	75	49.8	6.66
CPW2	75	56	49.3	6.64
CPW3	50	37	49.3	6.56
CPW4	25	19	49.5	6.35
CPW5	10	10	49.3	5.81

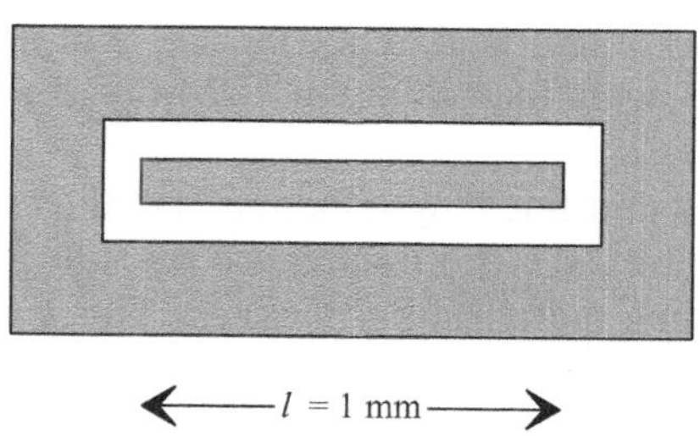

Fig. 2.2.19. Coplanar waveguide test structure to measure the dispersion characteristics.

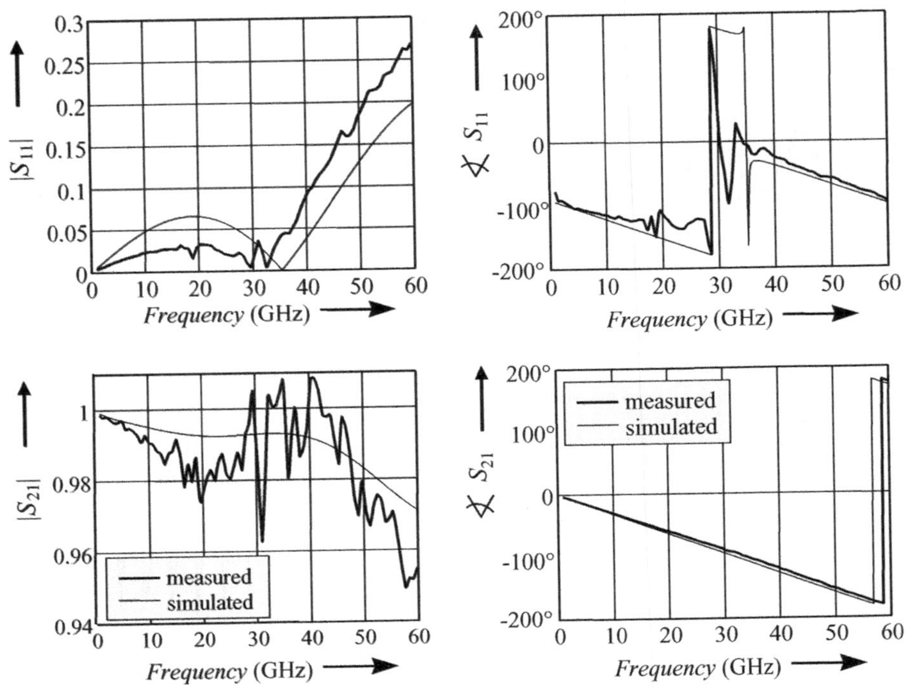

Fig. 2.2.20. Comparison between measurement (thick lines) and simulation (thin lines) of the scattering parameters of a short (1-mm) 50-Ω coplanar waveguide (CPW1, Table 2.2.2), plotted against the frequency.

dispersion of this line is smaller and that the agreement between simulation and measurement should be better.

2.2.6 The Influence of the Metalization Thickness on the Line Parameters

The influence of the metalization thickness t on the line parameters [12, 20, 21, 74, 111, 146] is significant only if t is on the order of the other geometrical line dimensions. This can be the case in monolithic microwave integrated circuits, where strip and slot widths of 5–20 μm and a metalization thickness of 2–5 μm are used. The characteristic impedance Z_L and the effective dielectric constant ε_{eff} decrease with increasing metalization thickness. This can be explained using Fig. 2.2.22.

The capacitance per unit line length increases in the case $t > 0$ compared to the case $t = 0$ by a value $\Delta C'$. The inductance per unit line length, however, decreases with increasing metalization thickness. This leads [according to Eq. (2.2.17)] to a decrease of the characteristic impedance Z_L with increasing metalization thickness. The effective dielectric constant ε_{eff} is defined [Eq (2.2.19)] by the quotient of C' and C'_0, and the percentage increase of the capacitance

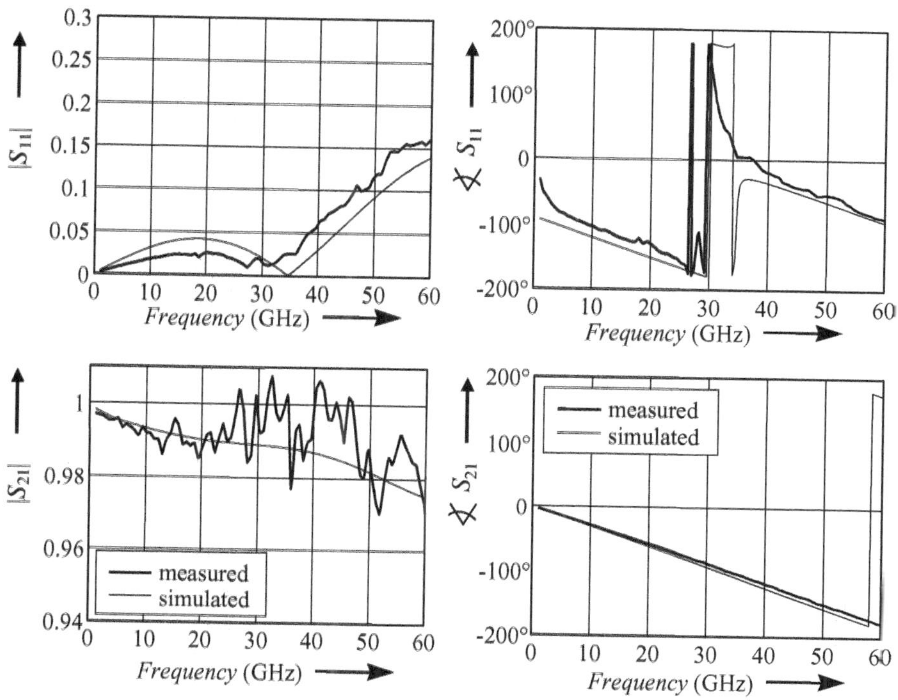

Fig. 2.2.21. Comparison between measurement (thick lines) and simulation (thin lines) of the scattering parameters of a short (1-mm) 50-Ω coplanar waveguide (CPW3, Table 2.2.2), plotted against the frequency.

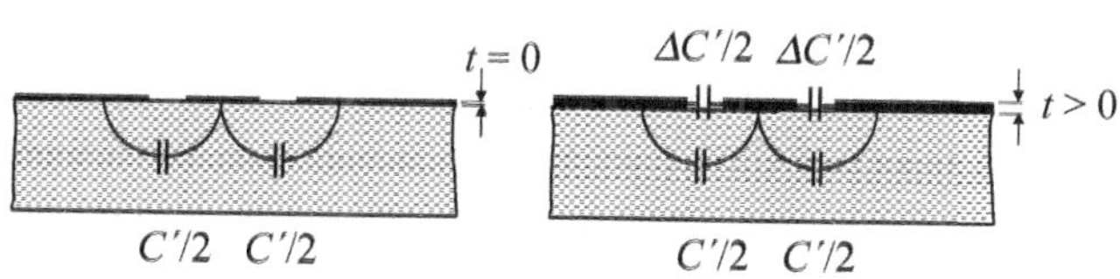

Fig. 2.2.22. The influence of the metalization thickness on the line parameters of the coplanar waveguide.

per line length C_0' with the metalization thickness t is higher compared to that of C'. Therefore, ε_{eff} decreases with increasing metalization thickness t.

In many approximation techniques used to calculate the characteristic parameters of planar waveguides, the influence of the metalization thickness t on the characteristic impedance and the effective dielectric constant is considered only by assuming an effective broadening of the line width [74]. Such an approximation is valid only for very small t/w and t/s values. It means that this technique can only be applied in cases where the influence of the metalization thickness is small.

TABLE 2.2.4. Comparison Between Different Methods for Calculating the Line Parameters of Coplanar Waveguides, in Dependence on the Metalization Thickness t ($w = 15\,\mu m$, $d = 35\,\mu m$, $h = 100\,\mu m$, $\varepsilon_r = 12.9$)

	Z_L (Ω)			ε_{eff}		
t (μm)	FDM	FEM	[74]	FDM	FEM	[74]
0.5	48.9	48.5	47.7	6.708	6.574	6.588
1.0	48.0	47.9	45.9	6.525	6.345	6.340
1.5	47.1	46.9	—	6.358	6.157	—
2.0	46.3	46.0	—	6.204	6.007	—
2.5	45.6	45.1	—	6.060	5.834	—
3.0	44.8	44.1	—	5.925	5.723	—
3.5	44.2	43.3	—	5.798	5.584	—
4.0	43.5	42.7	—	5.678	5.485	—
4.5	42.9	42.0	—	5.563	5.358	—
5.0	42.3	41.3	—	5.456	5.236	—

Table 2.2.4 compares the characteristic impedance and the effective dielectric constant of a coplanar waveguide as calculated from the finite difference technique, a finite element technique [173], and values obtained from the above-mentioned approximation technique [74]. Table 2.2.4 shows that this approximation technique is applicable only to a metalization thickness of up to 1 μm.

The dependence of the line parameters on the metalization thickness t is also shown in Fig. 2.2.23, where the characteristic impedance and the effective dielectric constant of various 50-Ω coplanar waveguides are depicted with respect to their dependence on the metalization thickness t. For very low strip widths w the influence of the technology on the strip cross section must also be considered because the cross section possibly is no longer a definite rectangle.

2.2.7 The Influence of the Ground Strip Width on the Line Parameters

In all calculations of this chapter, up to now, the width of the ground plane (see Fig. 2.2.6) has been assumed to be infinite. In real circuits, however, components may be placed closely together, to keep the needed substrate size small (compare also the discussions in Section 2.1). As a result, coplanar waveguides often have only a very small space available for their ground planes. The ground plane width has only a small influence on the line parameters of the fundamental even mode (coplanar waveguide mode) as long as it is essentially larger than the total slot width d (refer also to the discussion in Section 2.1.3.2). However, when its width decreases to a certain limit, the influence on the line parameters can no longer be neglected [146]. The minimum ground plane width needed, to a first approximation, is dependent on the strip line width w. Figure 2.2.24 shows the dependence of the characteristic impedance

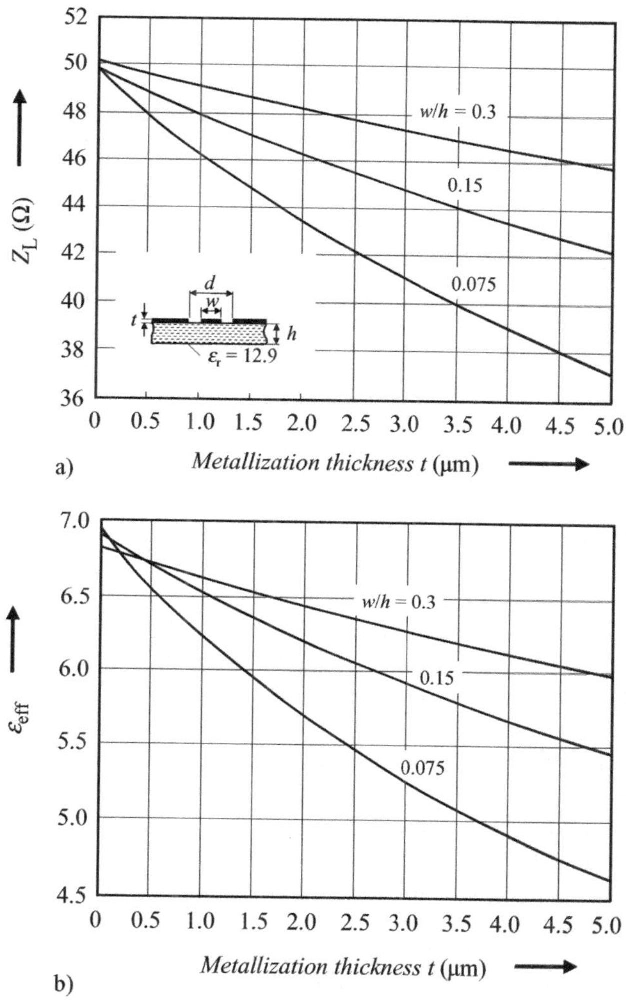

Fig. 2.2.23. Characteristic impedance (a) and effective dielectric constant (b) of three 50-Ω coplanar waveguides in dependence on the metalization thickness t ($w/d = 3/7$, $h = 100\,\mu\text{m}$).

of two coplanar waveguides on the ground plane width together with some measurement results. Coplanar lines with ground planes of unequal widths and unequal slot widths additionally have the tendency to excite the unwanted odd mode on the waveguide and, therefore, are used very seldom in circuit design practice.

2.2.8 The Influence of the Shielding on the Line Parameters

According to Section 2.2.3, the theoretical analysis of the coplanar waveguides using the finite difference technique is applied to a structure that is enclosed

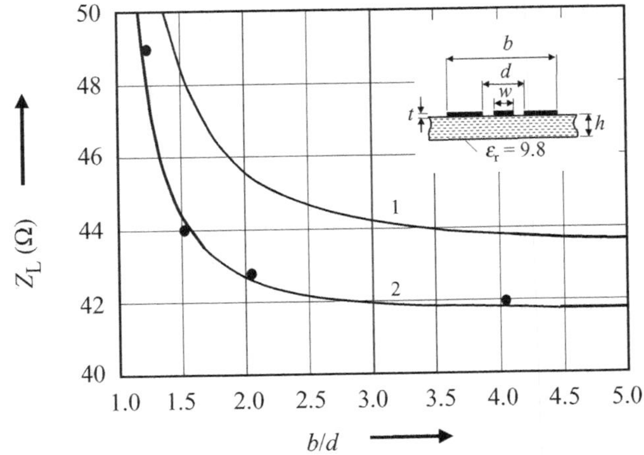

Fig. 2.2.24. Influence of the ground plane width on the characteristic impedance of a coplanar waveguide ($w/d = 5/7$, $h = 250\,\mu m$, $t = 5\,\mu m$). Curve 1, $w/h = 0.25$; curve 2, $w/h = 1.0$; • measurements for curve 2.

in a metallic shielding. This shielding, depending on its geometrical size, may have an essential influence on the calculated line parameters [146]. The discussion of these influences leads to an estimation of the influence of a real shielding on coplanar lines and components if the circuit is brought into a package. Furthermore, it leads to acquiring experiences as to how far the shielding has to be put away from the coplanar waveguide when working with the numerical analysis technique.

The influence of the shielding on the calculation results heavily depends on the geometrical parameters of the coplanar line, and two different examples will be considered here to demonstrate this fact. The results of the investigations are shown in Fig. 2.2.25. The characteristic impedance decreases with decreasing shielding size because the capacitance per unit line length is increased. If the width and the height of the shielding are at least four times as large as the total slot width d (Fig. 2.2.6), the influence of the shielding on the characteristic impedance is negligibly small.

The effective dielectric constant of the coplanar line at first increases with a decreasing shielding size but then has a minimum value in dependence on the parameters a and b (see inset of Fig. 2.2.25). The reason for this behavior is that part of the electromagnetic field below the substrate material that with decreasing values a and b at first increases compared to the part in the substrate material, but then vanishes for very small values of the spacing.

2.2.9 Special Forms of Coplanar Waveguides

Besides the symmetrical coplanar waveguide that has been intensively discussed in the previous chapters some special forms of coplanar waveguides do

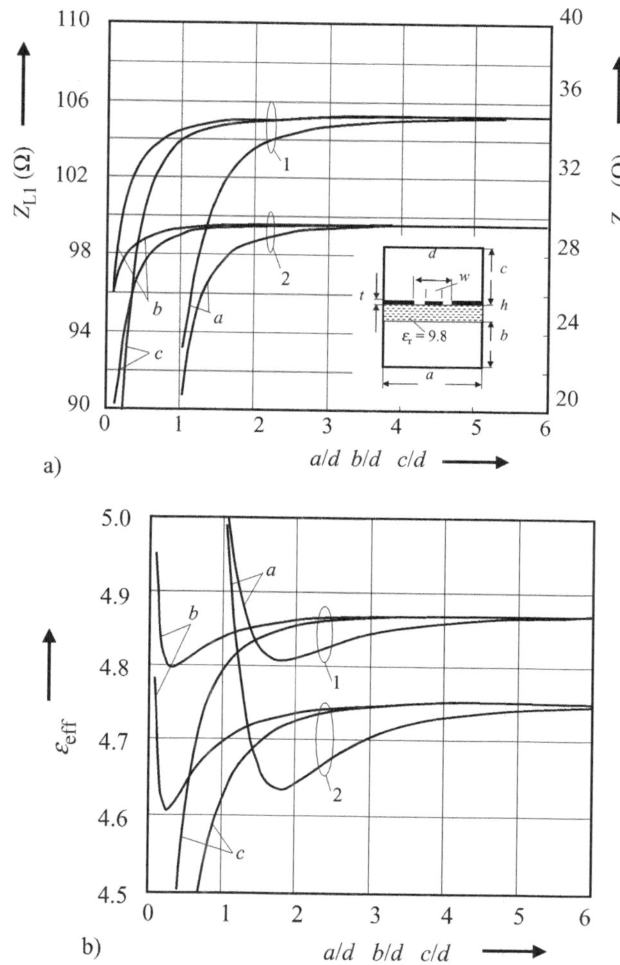

Fig. 2.2.25. Influence of the shielding on the characteristic impedance (a) and the effective dielectric constant (b) of the coplanar waveguide ($h = 250\,\mu m$, $d = 625\,\mu m$, $t = 0$). Curve 1, $w = 50\,\mu m$; curve 2, $w = 575\,\mu m$.

exist, but are not used so frequently as the symmetrical ones [49, 93, 97, 145, 190]. If, for example, the backside of the coplanar waveguide substrate is metalized, a totally new waveguide structure is produced.

The conductor-backed coplanar waveguide (CBCPW) [145], depending on the substrate height and the geometrical line parameters, is able to propagate waves different from those of the normal coplanar waveguide (see also the discussion in Section 2.3). For large slot widths and relatively large w/h ratios, the field distribution of this line is similar to that of a microstrip line (Fig. 2.2.26). In other cases it is a typical coplanar waveguide field. Figure 2.2.26 shows the typical current and field distributions of these two cases.

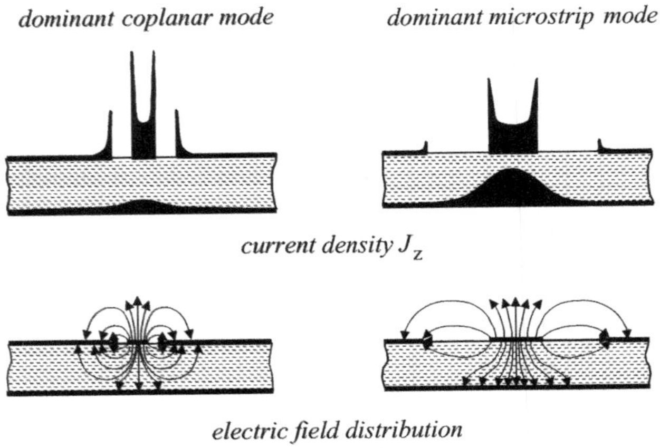

Fig. 2.2.26. Current and electric field distribution of a coplanar waveguide with a backside metalization of the substrate material. Shown are the two dominant cases: the coplanar mode (left) and the microstrip mode (right).

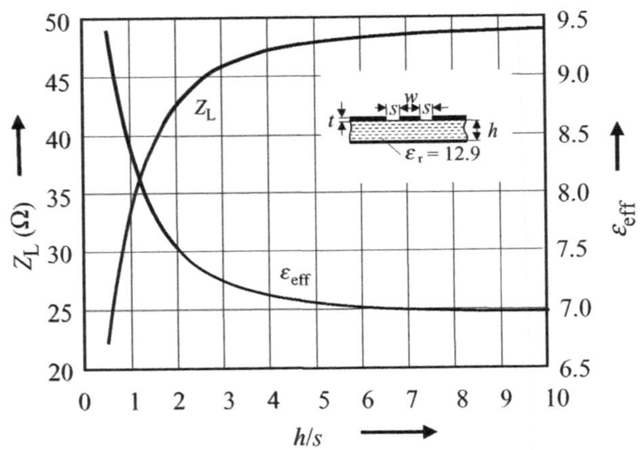

Fig. 2.2.27. Dependence of the line parameters of the backside metalized coplanar waveguide on the substrate height ($w/s = 1.5$, $t = 0$).

An interesting property of this waveguide mode is that, with increasing frequency, its electromagnetic field is more and more concentrated into the slots and thereby a mode change from the microstrip to the coplanar mode may occur. In this process the microstrip mode loses more and more its dominant role and the field changes to that of the coplanar mode. This frequency-dependent field distribution is also the reason for a much higher dispersion of such a waveguide.

Figure 2.2.27 shows the dependence of the line parameters of such a conductor backed coplanar waveguide on the substrate height h. With decreasing substrate height, the part of the electromagnetic field inside the substrate

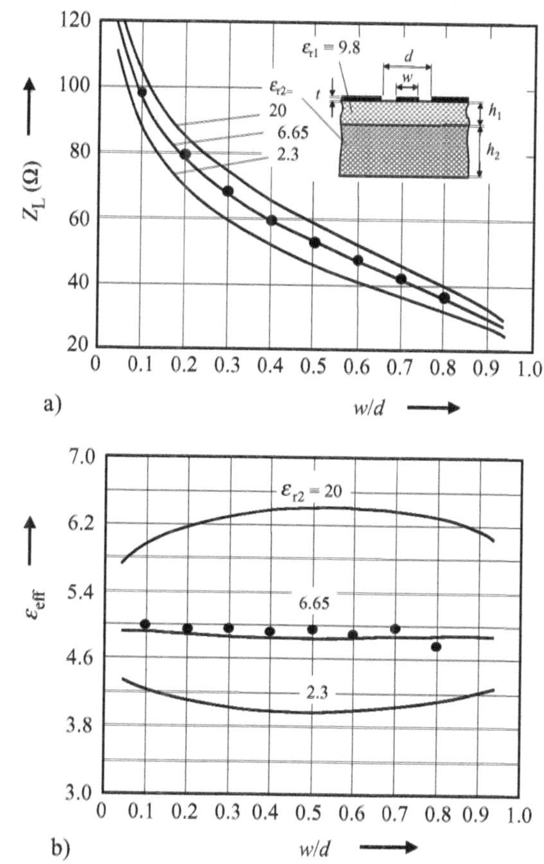

Fig. 2.2.28. Measured (•) and calculated (———) characteristic impedance (a) and effective dielectric constant (b) of different coplanar waveguides on a two-layer substrate material in dependence on the normalized strip line width w/d ($h_1 = 125\,\mu m$, $h_2 = 1600\,\mu m$, $d = 625\,\mu m$, $t = 5\,\mu m$).

material increases so that the effective dielectric constant becomes larger. Also, the attenuation coefficient of the line increases in this case.

Another special form of the coplanar waveguide can be built using more than one layer for the substrate material. In this case, the space below the metalization plane is no longer homogeneous. The properties of these lines and their dispersion strongly depend on the height of the substrate materials and their permittivities (see also the discussion in Section 2.3).

Figure 2.2.28 shows, as an example, the line parameters of three different coplanar waveguides with a two-layer substrate material together with some measured results. The measured dispersion of these lines in the frequency range from 2.5 GHz to 38 GHz was smaller than 4% of the static values.

2.2.10 Coplanar-like Waveguides

In monolithic microwave integrated circuits (MMICs) for application at millimeter-wave and submillimeter-wave frequencies, new transmission lines with low loss and low dispersion are needed. In this section, two new coplanar-like transmission lines that fulfill the above-mentioned requirements are analyzed using the finite difference time domain (FDTD) analysis and the quasi-static finite difference method (FDM) [314, 330]. The additional requirement for these new waveguide structures is that they can be produced in a compatible manner as the semiconductor elements that are monolithically integrated into the circuit layout.

Two coplanar-like transmission lines with an elevated center conductor, produced by air-bridge technologies (see Section 3.5.5), are analyzed and investigated with respect to their applicability for the above-mentioned purpose. Using field theoretical analysis techniques, it will be shown that these lines show the wanted properties up to the very high frequencies. The two waveguide structures that are analyzed are shown in Fig. 2.2.29. Figures 2.2.29a and 2.2.29b show a coplanar-like waveguide with an elevated center conductor as it was first proposed for application in sampling circuits and subpicosecond transmission lines by Bhattacharya et al. [314]. Figure 2.2.29d shows a similar structure where the ground planes, in an area near the center conductors, are elevated into the same height as the center conductor. Koßlowski [243] used this form of a coplanar waveguide for the first time in a similar form as a capacitively loaded waveguide (see Fig. 4.2.22).

Both types of coplanar-like waveguides can be easily produced using conventional air-bridge technology (Section 3.5.5) that is normally available in MMIC production techniques. The elevated center strip is carried by buttresses that are placed in a certain distance (10–200 μm or even larger) under the center strip along the transmission line.

The four line structure elements shown in Fig. 2.2.29 have been analyzed using the finite difference time domain (FDTD) technique (see Section 2.1) and the quasi-static finite difference method (FDM as described in this chapter) to calculate the electromagnetic field distribution, the current density distribution inside the conductors, the effective dielectric constant, the characteristic impedance, and the losses of these new transmission media.

As a first result of the analysis, Fig. 2.2.30 shows the simulated electric and magnetic field distribution of the waveguide structure shown in Fig. 2.2.29a. It can be observed clearly that the main electric field occurs under the elevated center conductor and near the edges of the center conductor. Both field areas are in an air region. This fact leads to a reduction of the effective dielectric constant, as will be demonstrated below.

The current density distribution inside the center conductor and the ground plane of the structure shown in Fig. 2.2.29a is depicted in Fig. 2.2.31. Because of the center conductor elevation, the current density distribution inside the conducting areas are asymmetrical with respect to the z-coordinate as can be seen from the figure.

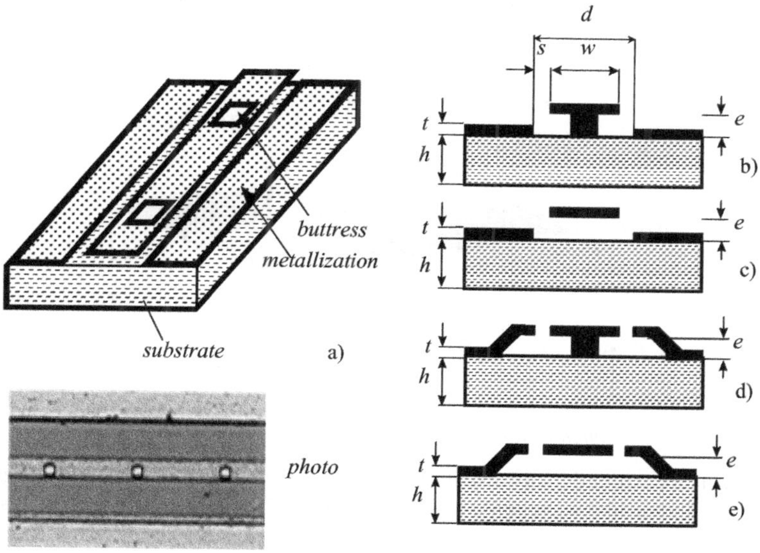

Fig. 2.2.29. (a) A quasi-coplanar waveguide in a three-dimensional figure. (b) Cross section for the line with an elevated center conductor and buttress. (c) Same as part b but without buttress. (d) Cross section with an elevated center conductor and elevated ground planes, with buttress. (e) Same as part d but without buttress.

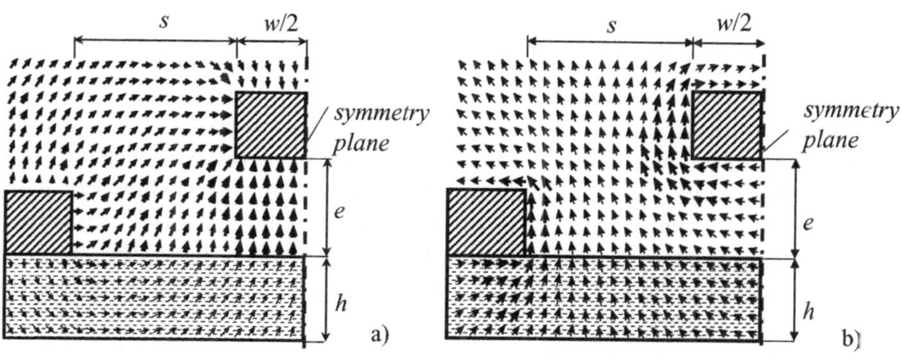

Fig. 2.2.30. Electromagnetic field distribution in a coplanar-like waveguide with elevated center conductor. (a) The electric field and (b) the magnetic field as simulated using a finite difference time domain (FDTD) technique. (The figures show only the left side of the symmetrical waveguide.)

Also, it can be observed that, compared to a conventional coplanar waveguide, the maximum values of the current density at the edges of the conducting areas here are smaller because the distance between ground plane and elevated center conductor is enlarged. This should lead to a reduced loss and attenuation of the considered waveguide structure.

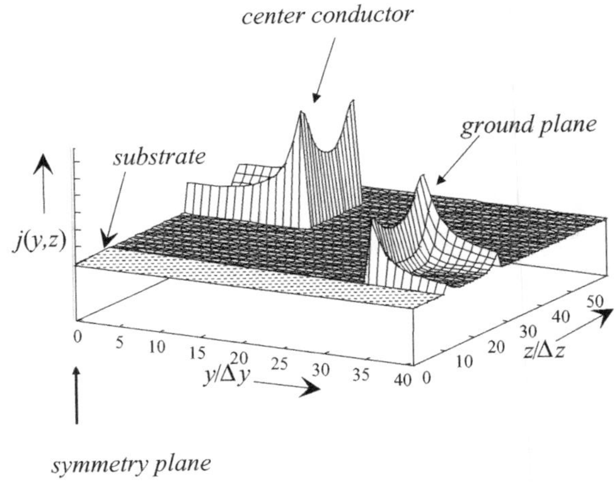

Fig. 2.2.31. The current density distribution inside the center conductor and the ground plane of a coplanar-like waveguide as shown in Fig. 2.2.29a. (Shown is only one side of the symmetrical waveguide structure.)

The results of the analysis with respect to the effective dielectric constant and the attenuation parameter α are shown in Fig. 2.2.32a and 2.2.32b, respectively. Figure 2.2.32a shows that the effective dielectric constant of the special waveguide under consideration (see figure legend) is reduced considerably if the height of the center conductor above the substrate is 3 μm, a value that can be easily realized. The figure also shows that the effective dielectric constant is nearly frequency-independent up to highest frequencies if, as has been done in the analysis, the structure is considered to be lossless.

The losses of the waveguide have been analyzed using the field distribution of the lossless waveguide and analyzing the current density as shown in Fig. 2.2.31. They are reduced by a factor of two to three compared to the conventional coplanar waveguide as depicted in Fig. 2.2.32b. This is due to the smaller current densities in the edges of the center conductor as they can be deduced from the FDTD analysis.

These results are taken together in Table 2.2.5, where a conventional coplanar waveguide (elevation height $e = 0$) and the new structure with an elevation of the center conductor of 3 μm are compared up to a frequency of 300 GHz. The capacitance per line length is four times smaller for the waveguide with elevated center conductor, whereas the inductance per line length is nearly independent of the elevation height. This means that, in the final analysis, the phase velocity of the electromagnetic waves on the new structure is half the value of the conventional waveguide. The attenuation coefficient is at least divided by the factor of two.

If the structure shown in Fig. 2.2.29a (with the geometrical parameters as given in Table 2.2.5) is analyzed using a full-wave FDTD technique, which also

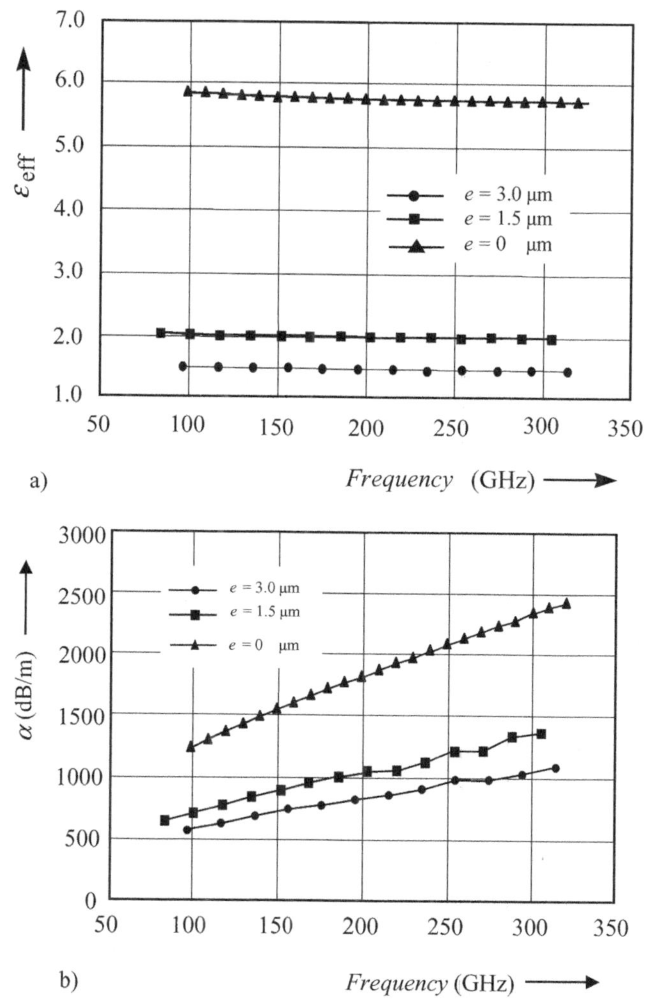

Fig. 2.2.32. The effective dielectric constant (a) and the attenuation coefficient (b) of the line structure shown in Fig. 2.2.29a, plotted against the frequency. Parameters of the waveguide: $w = 8\,\mu m$, $s = 5\,\mu m$, $t = 2\,\mu m$, elevation height $e = 3\,\mu m$, substrate GaAs, $\varepsilon_r = 12.8$. Conductivity of the metalization: $\sigma = 43.5 \times 10^6\,S/m$.

TABLE 2.2.5. Comparison of the Characteristic Parameters of the New Coplanar-like Waveguide as Shown in Fig. 2.2.29a ($e = 3\,\mu m$) and the Conventional Coplanar Waveguide ($e = 0\,\mu m$)[a]

	$e = 3\,\mu m$	$e = 0\,\mu m$	Advantages of the Waveguide with Elevated Center Conductor
C' (pF/m)	48	192	Four times smaller
L' (nH/m)	341	330	Nearly equal
v_{ph} (10^8 m/s)	2.47	1.25	Nearly doubled
α (dB/m)	1000	2300	At least divided by two

[a] All values for a frequency of 300 GHz. Waveguide parameters: $w = 8\,\mu m$, $s = 5\,\mu m$, $t = 2\,\mu m$, $\varepsilon_r = 12.8$, $\sigma = 43.5 \times 10^6\,S/m$.

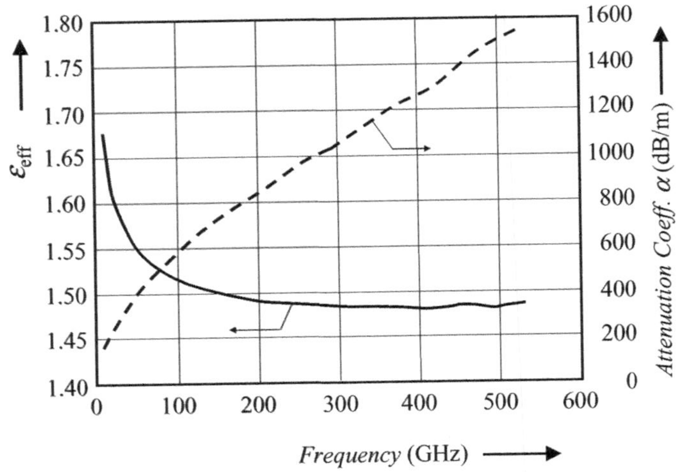

Fig. 2.2.33. Dispersion behavior of the effective dielectric constant due to the skin effect at low frequencies.

considers the current density fields inside the conductor, a large dispersion of the effective dielectric constant at low frequencies can be observed. This is shown in Fig. 2.2.33. The reason for this dispersive behavior of the effective dielectric constant is the field that penetrates the conductor regions at low frequencies (skin effect).

This effect is especially large in the case of small geometrical line parameters, as they have been assumed in the analyzed waveguide example. For structures with large line widths and slot widths (e.g., 20–50 μm) and larger metalization thickness (e.g., 3–5 μm), this effect is not so large and occurs only at very low frequencies (e.g., 1–2 GHz).

Due to the above-mentioned results of the FDTD analysis, it appears that the electromagnetic field of the new waveguide structures is nearly of a TEM mode and that the line properties also can be analyzed using the simple quasi-static finite difference method (FDM) as described at the beginning of this chapter. A comparison between the results of the FDTD analysis and a FDM calculation for the same line used in Fig. 2.2.33 shows (Table 2.2.6) that both results are nearly identical, and therefore the much simpler quasi-static analysis technique is well-suited for analyzing the more complex structures shown in Fig. 2.2.29b.

Figure 2.2.34 shows the distribution of equipotential lines for the electric field in a waveguide structure defined in Fig. 2.2.29b. The figure shows that with increasing center conductor width the electric field is concentrated more and more in the air-gap between the ground planes and the center conductor.

TABLE 2.2.6. Comparison of an FDTD and an FD analysis of a Waveguide as Shown in Fig. 2.2.29a)[a]

Elevation	Parameter	FD Technique	FDTD Technique	Difference (%)
$e = 3\,\mu m$	ε_{eff}	1.451	1.452	0.044
	Z_L (Ω)	84.73	84.92	0.22
$e = 0\,\mu m$	ε_{eff}	5.57	5.58	0.16
	Z_L (Ω)	41.37	41.43	0.14

[a] Parameters of the waveguide: $w = 8\,\mu m$, $s = 5\,\mu m$, $t = 2\,\mu m$, elevation height $e = 3\,\mu m$, substrate GaAs. Analysis frequency: 5 GHz.

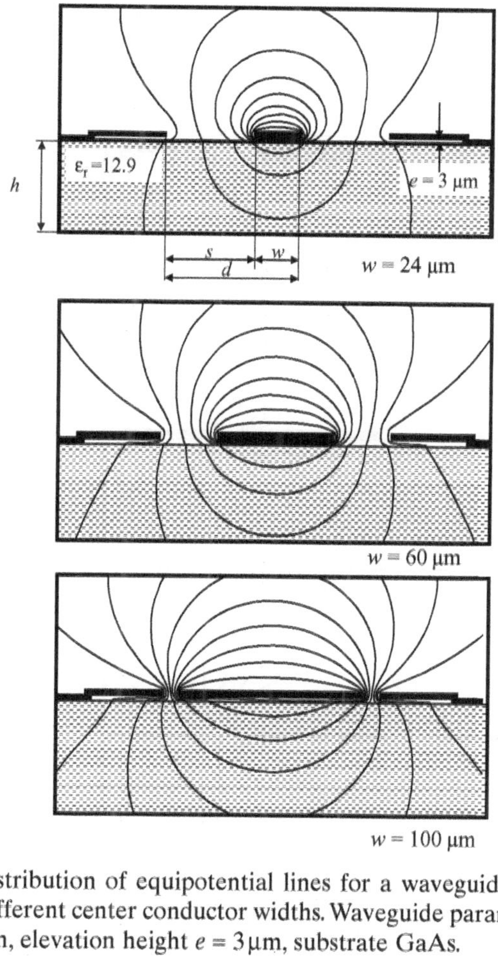

Fig. 2.2.34. The distribution of equipotential lines for a waveguide as shown in Fig. 2.2.29b for three different center conductor widths. Waveguide parameters: $d = 120\,\mu m$, $h = 350\,\mu m$, $t = 3\,\mu m$, elevation height $e = 3\,\mu m$, substrate GaAs.

Moreover, in the case of a waveguide with a large center conductor width a considerable field is concentrated in the substrate region under the center conductor, thus increasing the effective dielectric constant with increasing width w.

A similar effect can be observed with respect to its dependence on the elevation height of the center conductor, as is shown in Fig. 2.2.35. For a relatively large value of the elevation height (e.g., $3\,\mu m$), various equipotential lines are closed under the center conductor in the air-gap region. This means that a high electric field is concentrated in this area. For smaller values of e (e.g., $2\,\mu m$ or $1\,\mu m$) the equipotential lines and thereby the electric field are more and more shifted into the substrate material area so that the effective

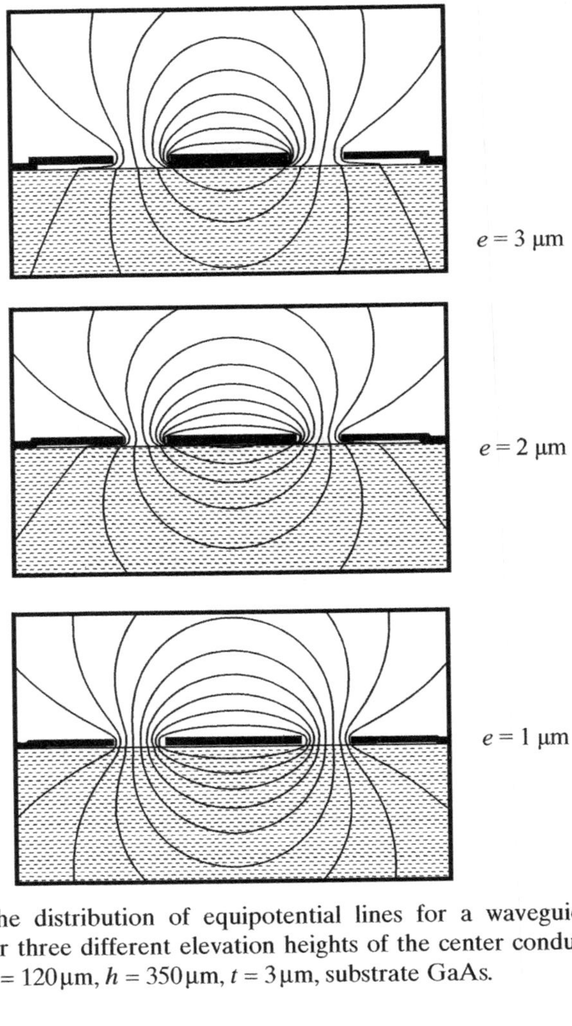

Fig. 2.2.35. The distribution of equipotential lines for a waveguide as shown in Fig. 2.2.29b for three different elevation heights of the center conductor. Waveguide parameters: $d = 120\,\mu m$, $h = 350\,\mu m$, $t = 3\,\mu m$, substrate GaAs.

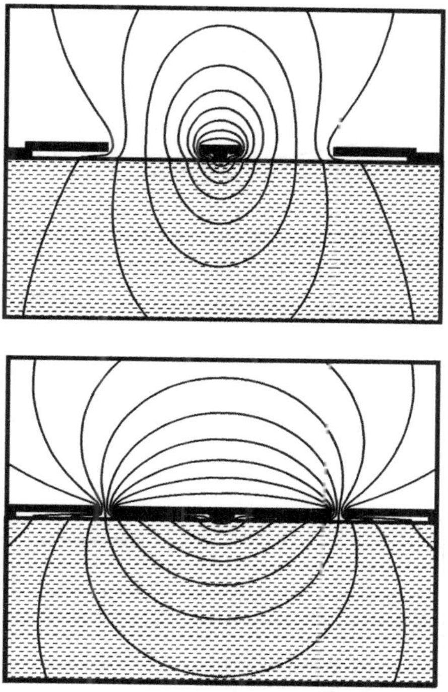

Fig. 2.2.36. The influence of a buttress on the electric field distribution near the center conductor demonstrated for two different waveguides with elevated center conductor and two different widths of the center conductors.

dielectric constant is increased with decreasing values of the elevation height, as already documented in Fig. 2.2.32a.

As shown in Fig. 2.2.36, the influence of the buttress under the center conductor leads to an increase of the electric field inside the dielectric carrier material and therefore to an increase of the effective dielectric constant. This means it has a similar influence as that of a capacitance placed at the position of the buttress. From this description, it is possible to develop a distributed waveguide model for the coplanar-like waveguide with elevated center conductor and stepwise-introduced buttresses. The influence of the buttresses is described by capacitances at the position of the buttresses along the line direction.

Using the same techniques (FD analysis), the waveguide structures shown in Figs. 2.2.29d and 2.2.29e can also be analyzed. The dependence of the effective dielectric constant and the characteristic impedance of the line structures shown in Fig. 2.2.29d and Fig. 2.2.29e on the w/d-ratio is shown in Fig. 2.2.37a and Fig. 2.2.37b, respectively. From Fig. 2.2.37a it can be seen that the buttresses have a large influence especially on the value and the w/d-dependence of the effective dielectric constant. For a constant thickness of the buttresses

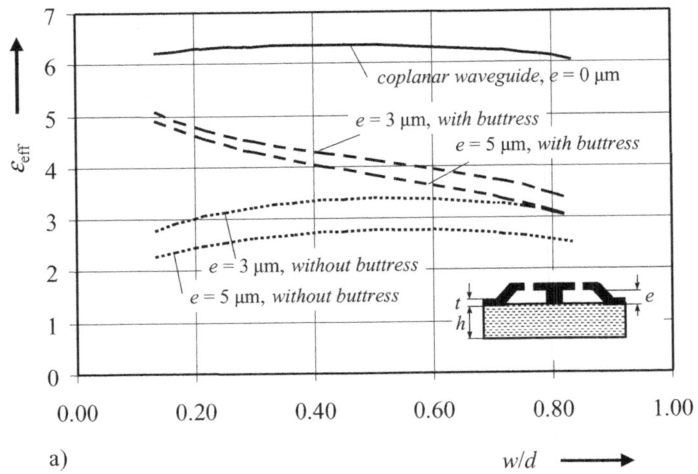

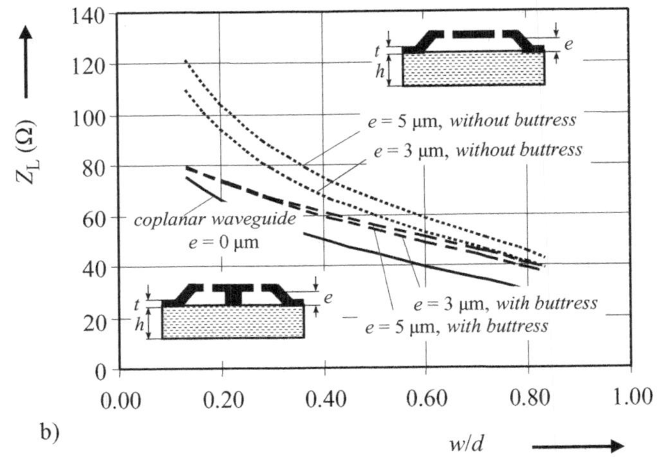

Fig. 2.2.37. The effective dielectric constant (a) and the characteristic impedance (b) of a waveguide structures as shown in Figs. 2.2.29b and 2.2.29c. Waveguide parameters: $d = 120\,\mu\text{m}$, $t = 3\,\mu\text{m}$, $b_{\text{buttress}} = 10\,\mu\text{m} = \text{const.}$, $h = 350\,\mu\text{m}$, $\varepsilon_r = 12.8$.

(here: 10 µm), the effective dielectric constant of waveguides with wide center conductors are nearly of the same value as those of the waveguides without buttresses. With decreasing values of the *w*/*d* ratio the values of the effective dielectric constant increases and reaches the values of the regular coplanar waveguide for very small values of *w*/*d*.

Figure 2.2.37b shows another advantage of the coplanar waveguide structure with elevated center conductor. That is, the spectrum of the available characteristic impedances is much larger than that of a conventional coplanar waveguide. For the waveguide without buttresses, characteristic impedances between 140 Ω and 40 Ω may be realized, whereas a conventional coplanar waveguide with equivalent parameters allows only realizing impedances between 80 Ω and 35 Ω.

The investigations briefly described above show that the new coplanar-like transmission lines are good candidates for designing millimeter-wave and submillimeter-wave monolithic integrated circuits. They have a low effective dielectric constant that enlarges the dimensions of the circuit elements at these high frequencies and reduces the losses. In addition, the dispersion of the line properties is low so that circuits can be designed in a quasi-static manner even at wave–wave frequencies. The technology to produce these lines is compatible to the standard MMIC production technique.

2.2.11 Coupled Coplanar Waveguide Structures

In this chapter, coupled coplanar line structures shall be investigated using the quasi-static analysis method that can be used in microwave integrated circuits to develop components like filters and couplers (compare also Chapter 4). As already has been mentioned in Section 2.1.3, two different kinds of coupled coplanar waveguides may be considered (see Fig. 2.1.24). The structure shown in Fig. 2.2.38 is often used for filter and coupler design and shall be discussed here. The coplanar structure shown in Fig. 2.2.38 is different from a coupled microstrip line structure. If coupled microstrip lines are considered, each of the strip lines has the same distance to the ground plane. In the coplanar case, the different strips not only may have their own strip width and distance to the neighboring strip, but they also have their own distance from the right and the left ground plane. Additionally, there may be one or more other strips between the considered line and the ground planes. Vice versa, in the case of coupled microstrip lines there is always a direct way, for example, for the electric field lines from the strip to the ground.

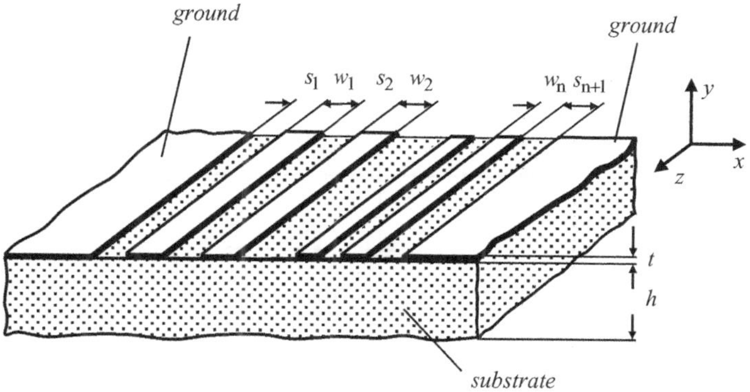

Fig. 2.2.38. Cross section of a multiple-strip-line system in coplanar waveguide technology.

The coupled line structure shown in Fig. 2.2.38 will be considered as homogeneous in z-direction. Furthermore, quasi-TEM mode propagation is assumed on the coupled line system. Losses shall be small. Under these conditions the propagation properties of this line system can be described by using voltages and currents between the strips and on the strips, respectively [41, 42, 76]. They are the solutions of the telegraphist's equations. The per unit line length parameter matrices which are the elements of these equations will be calculated here, using the above-described finite difference method (compare also Section 2.1.3 and the applied full-wave analysis).

2.2.11.1 ***Analysis of the Characteristic Parameter Matrices.*** A multiply coupled line section of infinitesimal short length dz is considered in Fig. 2.2.39. It consists of n coupled strip lines and one ground plane. Quasi-TEM wave propagation is assumed on this line system. The capacitances and inductances per unit line length of each strip line are given by C'_{ii} and L'_{ii}, respectively. The parameters C'_{ij} and L'_{ij} with $i \neq j$ vice versa characterize the coupling capacitances and inductances between the different strip lines, respectively. The inductance matrices of the coupled coplanar line system can be deduced from the capacitance matrices as in the case of the single coplanar waveguide. Therefore, only the capacitance matrix is derived here. In many cases it has been reported in the literature (e.g., references 36 and 60) that for an analysis of coupled lines, only the capacitances C'_{ii} of the single strips and the elements C'_{ij} of those lines are needed, which are direct neighbors. This is not true in the case of coupled coplanar lines as shown in Fig. 2.2.38, because the different single lines do not have their own ground planes. For this reason the coupling of a strip, which may be in the next position to the considered strip

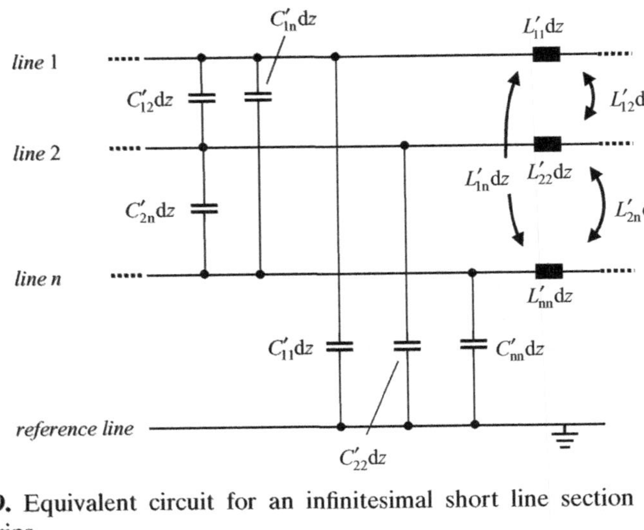

Fig. 2.2.39. Equivalent circuit for an infinitesimal short line section of n coupled lossless strips.

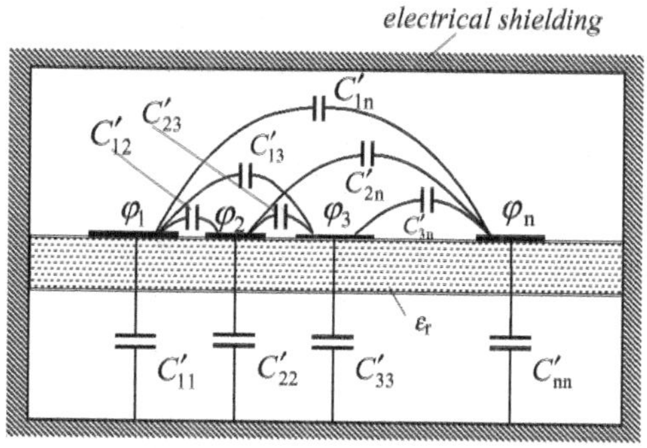

Fig. 2.2.40. The partial capacitances per unit line length of multiply coupled coplanar strips.

line (or which may even be in a more distant position), still has a recognizable influence on the line properties. Therefore the total capacitance matrix is needed for an accurate description of the coupled coplanar line system.

The static finite difference method, which is discussed in this chapter, only can analyze shielded structures. Ground planes are not necessarily connected to the shielding, and they therefore are considered in the same way as the center-strip conductors in this investigation.

Figure 2.2.40 shows a system of n coupled coplanar strips on a homogeneous, isotropic, and lossless substrate material. The relative dielectric constant of the substrate is ε_r. The distances between the outer strips and the electrical shielding shall be much larger than the geometrical size of the coupled line structure. The potentials of the strips are defined as $\varphi_1, \varphi_2, \dots, \varphi_n$ and the shielding has the potential $\varphi_0 = 0$ V. If the potentials of the strips are calculated using the finite difference method, the charge distribution can be analyzed using Eq. (2.2.21). The following equation system describes the connection between the charge per unit line length on the strip conductors and their potentials:

$$
\begin{aligned}
Q'_1 &= C'_{11}\varphi_1 + C'_{12}(\varphi_1 - \varphi_2) + C'_{13}(\varphi_1 - \varphi_3) + \dots + C'_{1n}(\varphi_1 - \varphi_n),\\
Q'_2 &= C'_{21}(\varphi_2 - \varphi_1) + C'_{22}\varphi_2 + C'_{23}(\varphi_2 - \varphi_3) + \dots + C'_{2n}(\varphi_2 - \varphi_n),\\
Q'_3 &= C'_{31}(\varphi_3 - \varphi_1) + C'_{32}(\varphi_3 - \varphi_2) + C'_{33}\varphi_3 + \dots + C'_{3n}(\varphi_3 - \varphi_n), \qquad (2.2.28)\\
&\vdots\\
Q'_n &= C'_{n1}(\varphi_n - \varphi_1) + C'_{n2}(\varphi_n - \varphi_2) + C'_{3n}(\varphi_n - \varphi_3) + \dots + C'_{nn}\varphi_n
\end{aligned}
$$

can be found. The elements $C'_{\nu\mu}$ are the partial capacitances per unit line length as shown in Fig. 2.2.36. For these capacitances the relationship $C'_{\nu\mu} = C'_{\mu\nu}$, $\nu \neq \mu$ is valid.

From the equation system (2.2.36), a method to determine the partial capacitances can be derived. If the potential of one strip is defined to be 1 V and all other potentials including the potential of the shielding is set to 0 V, then the partial capacitances can directly been calculated from Eq. (2.2.28). For the special case that $\varphi_1 = 1\,\text{V}$ and all other potentials are zero, Eq. (2.2.28) delivers

$$\begin{aligned} Q_1' &= (C_{11}' + C_{12}' + C_{13}' + \ldots + C_{1n}')\varphi_1, \\ Q_2' &= -C_{21}'\varphi_1, \\ Q_3' &= -C_{31}'\varphi_1, \\ &\vdots \\ Q_n' &= -C_{n1}'\varphi_1. \end{aligned} \tag{2.2.29}$$

If all partial capacitance coefficients shall be determined, n different calculations are needed. Considering geometrical symmetries reduces the calculation expense drastically, because not only fewer potential configurations are needed for the analysis, but also the calculation of the different coefficients is much easier.

The described method for the calculation of the partial capacitances is also valid if the lines are not lossless, but the losses are small. The method can be applied, as long as the fundamental field distribution and the characteristic propagation parameters are not changed essentially by the losses. Under these conditions the conductor losses can be calculated from the determined surface current densities. The elements of the resistance per unit line length matrix then again can be calculated using the method described above [see Eq. (2.2.22) and p. 61 ff]. In these calculations the subdiagonal elements of the matrix may no longer be neglected against the main-diagonal elements, because the currents in the different strips heavily depend on all currents of all other strips.

2.2.11.2 Determination of the Scattering Matrix of Coupled Coplanar Waveguides. In the analysis of a coupled n-line system of length l besides the determination of the propagating wave modes, the calculation of the scattering matrix with respect to defined $2n$ reference planes (ports) is essential. In this section it shall be shown [248] how this scattering matrix can be derived from the above-described calculations (compare also Section 2.1.3.1 for the case of the full-wave analysis).

In the previous section it was shown how the capacitance per unit line length matrix $\boldsymbol{C}'$, the inductance per unit line length matrix $\boldsymbol{L}'$, and the resistance per unit line length matrix $\boldsymbol{R}'$ can be determined. Using these matrices, the impedance per unit line length matrix $\boldsymbol{Z}' = \boldsymbol{R}' + \mathrm{j}\omega\boldsymbol{L}'$ and the admittance per unit line length matrix $\boldsymbol{Y}' = \mathrm{j}\omega\boldsymbol{C}'$ can be computed. Using these matrices, the telegraphist's equations of the coupled multiline system shown in Fig. 2.2.41 is given by [140]:

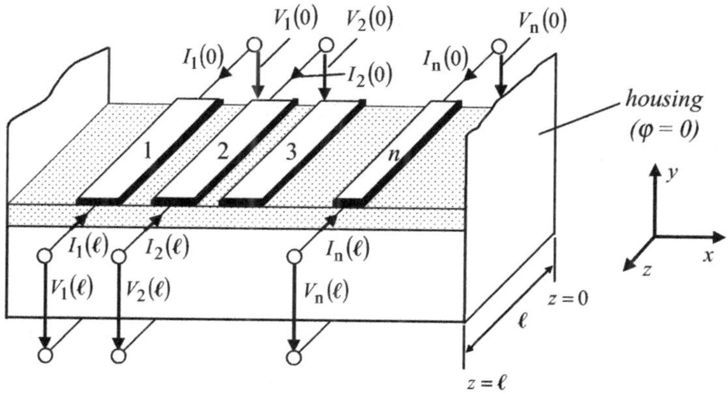

Fig. 2.2.41. Definition of the line currents and the line voltages of a coupled line system.

$$\frac{d}{dz}\begin{bmatrix} \boldsymbol{V}(z) \\ \boldsymbol{I}(z) \end{bmatrix} = \begin{bmatrix} 0 & \boldsymbol{Z}' \\ \boldsymbol{Y}' & 0 \end{bmatrix}\begin{bmatrix} \boldsymbol{V}(z) \\ \boldsymbol{I}(z) \end{bmatrix}. \tag{2.2.30}$$

The solutions of the equation system (2.2.30) are the equations that for a coupled line section of length ℓ have the following form [42]:

$$\begin{bmatrix} \boldsymbol{V}(0) \\ \boldsymbol{I}(0) \end{bmatrix} = \begin{bmatrix} \boldsymbol{M}_v \mathbf{cosh}(\Gamma\ell)\boldsymbol{M}_v^{-1} & \boldsymbol{M}_v \mathbf{sinh}(\Gamma\ell)\boldsymbol{M}_i^{-1} \\ \boldsymbol{M}_i \mathbf{sinh}(\Gamma\ell)\boldsymbol{M}_v^{-1} & \boldsymbol{M}_i \mathbf{cosh}(\Gamma\ell)\boldsymbol{M}_i^{-1} \end{bmatrix}\begin{bmatrix} \boldsymbol{V}(\ell) \\ \boldsymbol{I}(\ell) \end{bmatrix} \tag{2.2.31}$$

where the diagonal matrices $\mathbf{cosh}(\Gamma\ell)$ and $\mathbf{sinh}(\Gamma\ell)$ are given by

$$\mathbf{cosh}(\Gamma\ell) = \begin{bmatrix} \cosh(\gamma_1\ell) & & 0 \\ & \ddots & \\ 0 & & \cosh(\gamma_n\ell) \end{bmatrix},$$
$$\mathbf{sinh}(\Gamma\ell) = \begin{bmatrix} \sinh(\gamma_1\ell) & & 0 \\ & \ddots & \\ 0 & & \sinh(\gamma_n\ell) \end{bmatrix} \tag{2.2.32}$$

and

$$\Gamma = \begin{bmatrix} \gamma_1 & & 0 \\ & \ddots & \\ 0 & & \gamma_n \end{bmatrix}. \tag{2.2.33}$$

$\boldsymbol{\Gamma}$ is the diagonal matrix of the n propagation coefficients describing the n possible TEM modes on the coupled coplanar structure. The matrices $\boldsymbol{M}_i$ and $\boldsymbol{M}_v$ are quadratic $n \times n$ current and voltage transformation matrices that connect the strip currents and strip voltages (Fig. 2.2.37) to the wave variables $\boldsymbol{a}$ and $\boldsymbol{b}$ as follows:

$$\boldsymbol{V} = \boldsymbol{M}_v(\boldsymbol{a}+\boldsymbol{b}), \tag{2.2.34}$$

$$\boldsymbol{I} = \boldsymbol{M}_i(\boldsymbol{a}-\boldsymbol{b}). \tag{2.2.35}$$

It may be shown [42] that the voltage transformation matrix $\boldsymbol{M}_v$ and the diagonal matrix $\boldsymbol{\Gamma}$ can be calculated as the solution of the complex eigenvalue problem:

$$\boldsymbol{Z}'\boldsymbol{Y}'\boldsymbol{M}_v = \boldsymbol{M}_v\boldsymbol{\Gamma}. \tag{2.2.36}$$

The current transformation matrix then is given by

$$\boldsymbol{M}_i = \boldsymbol{Y}'\boldsymbol{M}_v\boldsymbol{\Gamma}. \tag{2.2.37}$$

Using the two transformation matrices, the characteristic wave impedance and wave admittance matrices of the coupled line system can be determined:

$$\boldsymbol{Z}_L = \boldsymbol{M}_v\boldsymbol{M}_i^{-1}, \tag{2.2.38}$$

$$\boldsymbol{Y}_L = \boldsymbol{M}_i\boldsymbol{M}_v^{-1}. \tag{2.2.39}$$

If the scattering matrix of the coupled line system shall be calculated, first the impedance matrix must be known. Rearranging Eq. (2.2.31), the impedance matrix directly may be derived as

$$\begin{bmatrix} V(0) \\ V(\ell) \end{bmatrix} = \underbrace{\begin{bmatrix} \boldsymbol{M}_v\mathbf{coth}(\Gamma\ell)\boldsymbol{M}_i^{-1} & \boldsymbol{M}_v(\mathbf{sinh}(\Gamma\ell))^{-1}\boldsymbol{M}_i^{-1} \\ \boldsymbol{M}_v(\mathbf{sinh}(\Gamma\ell))^{-1}\boldsymbol{M}_i^{-1} & \boldsymbol{M}_v\mathbf{coth}(\Gamma\ell)\boldsymbol{M}_i^{-1} \end{bmatrix}}_{\boldsymbol{Z}} \begin{bmatrix} I(0) \\ I(\ell) \end{bmatrix}. \tag{2.2.40}$$

Using well-known transformation relations, the scattering matrix can be found as

$$\boldsymbol{S} = (\boldsymbol{Z}+Z_L\boldsymbol{U})^{-1}\cdot(\boldsymbol{Z}-Z_L\boldsymbol{U}), \tag{2.2.41}$$

where $\boldsymbol{U}$ is the unit matrix and Z_L is the normalizing characteristic impedance. A verification of the theoretical results discussed here by measurements will be given in Section 4.4.1.

2.3 CLOSED FORMULA STATIC ANALYSIS OF COPLANAR WAVEGUIDE PROPERTIES

As already discussed in the introduction, the coplanar waveguide is able to transport two fundamental waves with cutoff frequency zero: the so-called even mode and the odd mode, the electromagnetic fields of which are shown in Fig. 1.3. Only the even mode is used for application in coplanar circuit design; the odd mode is commonly suppressed by an adequate air-bridge technology, as described in Section 3.5.5. The fundamental even mode of the coplanar waveguide is a quasi-TEM mode with a low dispersion, as already shown in Section 2.1. This means that static analysis techniques are suited to calculate at least approximately, for example, the capacitance per line length and thereby the phase velocity and the characteristic impedance of the waveguide for application in circuit design. In this chapter, accurate and simple approximate analytic formulas are presented for calculating the quasi-static TEM parameters of some supported coplanar waveguide structures [175, 246]. These include the open, covered, and dielectric overlay coplanar waveguides. The formulas have been designed for use in CAD programs and are only valid whenever the supporting material is of lower dielectric constant. Comprehensive comparisons have been made by using a rigorous spectral domain hybrid mode analysis. Accuracy of the developed formulas has been found to be better than 1% for most of the operating ranges of physical dimensions, available dielectric materials ($\varepsilon_r < 20$), and low frequencies ($f < 20\,\text{GHz}$).

2.3.1 Analysis of a Generalized Coplanar Waveguide with Supporting Substrate Layers

A generalized structure of the coplanar waveguide assuming an additional supporting dielectric layer under the main substrate and/or a dielectric cover on top of the waveguide—and, alternatively, a metallic shielding on top of the line (which might model the metallic housing of the circuit)—will be considered here. All these structures are of advantage in (M)MIC design and application. Using these alternatives, three different structures will be considered in the following (Fig. 2.3.1):

- The open and supported coplanar waveguide (Fig. 2.3.1a) with a dielectric supporting material (ε_{r1}, h_1) below the main dielectric substrate (ε_{r2}, h_2), where the space above the coplanar strips can be filled with either another dielectric (ε_{r3}) or with air.
- The covered and supported CPW, which is similar to the open coplanar waveguide except that a metallic top covers the structure (Fig. 2.3.1b).
- The third configuration that consists of four layers. The main substrate (ε_{r2}, h_2), the supporting dielectric material (ε_{r1}, h_1), the overlay dielectric material (ε_{r4}, h_4), and another layer of dielectric material (ε_{r3}) or air (Fig. 2.3.1c).

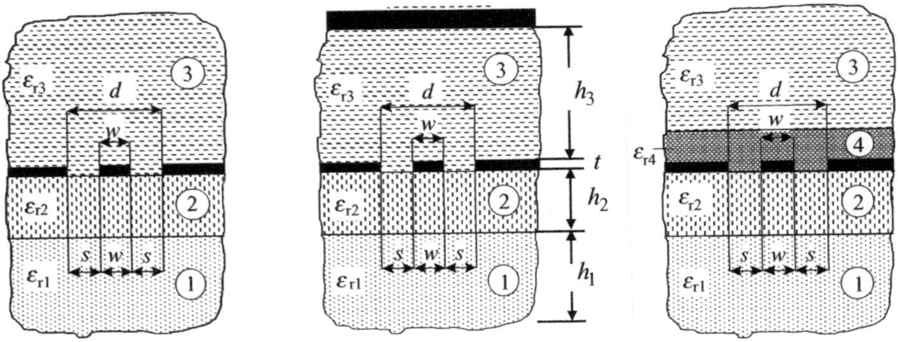

Fig. 2.3.1. Supported coplanar waveguide structures to be analyzed: (a) Open and supported coplanar waveguide, (b) covered and supported coplanar waveguide, and (c) overlay supported coplanar waveguide.

For the sake of brevity, these structures will be referred to in the following as SCPWs (supported coplanar waveguides) and will be numbered as SPCW1 (Fig. 2.3.1a), SPCW2 (Fig. 2.3.1b), and SPCW3 (Fig. 2.3.1c). If the relative dielectric constants ε_{r1}, ε_{r3}, and ε_{r4} are chosen to be equal to one, the general coplanar waveguide (CPW) in open or covered form results, depending on whether a metallic shielding on top is chosen or not. With the three models shown in Fig. 2.3.1, nearly all cases that are of interest in microwave circuit design can be modeled.

The approach for analysis used here is similar to that used in reference 68. It starts with finding an exact expression for the characteristic impedance Z_L^a, replacing all dielectric materials by air and using the conformal mapping technique. As the next step, an effective dielectric constant is defined which describes the correct phase velocity of the even mode on the coplanar waveguide (considering the dielectric materials) by using some approximate techniques as will be described below. This effective dielectric constant is also used to describe the correct value of the characteristic impedance for the coplanar waveguide filled with the dielectric materials. The last step is then to evaluate the error of the assumed expressions by comparison with rigorous numerical results.

The calculation is started with the expression for the characteristic impedance Z_L^a of the air-filled coplanar waveguide, which may be written as

$$Z_L^a = \frac{1}{c_0 C_t'^a}, \tag{2.3.1}$$

where $c_0 = 2.9979 \times 10^8$ m/s is the velocity of light in vacuum, and $C_t'^a$ is the total capacitance per unit length of the structure when replacing all dielectric materials by air.

To calculate the effective dielectric constant of the considered coplanar lines, filling factors q_ν ($\nu = 1, 2, 3, 4$) that correctly describe the influence of the considered dielectric material and their material parameters on the line parameters are introduced:

$$\varepsilon_{\text{eff}} = q_1\varepsilon_{r1} + q_2\varepsilon_{r2} + q_3\varepsilon_{r3} + q_4\varepsilon_{r4}. \tag{2.3.2}$$

Equation (2.3.2) is limited to a coplanar waveguide of, at most, four dielectric layers with q_1, q_2, q_3, q_4 representing the filling factors of the four dielectric regions 1, 2, 3, and 4, respectively. The value of the characteristic impedance and the phase velocity in the presence of the dielectric materials is then calculated by using the following relations:

$$Z_L = \frac{Z_L^a}{\sqrt{\varepsilon_{\text{eff}}}}, \qquad v_{\text{ph}} = \frac{c_0}{\sqrt{\varepsilon_{\text{eff}}}}. \tag{2.3.3}$$

The expression for the total capacitance per unit length $C_t'^a$ as well as the filling factors q_1, q_2, q_3, and q_4 can be obtained in terms of the corresponding air filled basic capacitances per unit length, as shown in Figs. 2.3.2a to 2.3.2c, respectively. They correspond to the regions of the original supported coplanar waveguides shown in Figs. 2.3.1a to 2.3.1c, respectively, and are obtained by replacing all dielectric interfaces in the original structure by magnetic walls. Although the assumption of the boundaries being magnetic walls is hardly verified for all cases, especially for large slots and very small cover heights, it has been proven to yield excellent results for practical line dimensions as has been shown in references 68, 94, and 109.

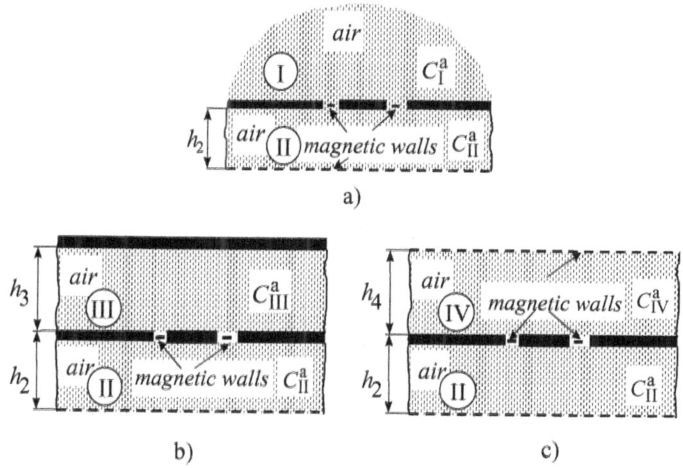

Fig. 2.3.2. Air-filled coplanar waveguides corresponding to the waveguides shown in Fig. 2.3.1.

With reference to Fig. 2.3.2, four basic capacitances per unit line length are defined as $C_{\mathrm{I}}'^a$, $C_{\mathrm{II}}'^a$, $C_{\mathrm{III}}'^a$, and $C_{\mathrm{IV}}'^a$ representing the electric fields in the regions I, II, III, and IV, respectively.

The expressions for all these basic capacitances are available in the literature [68, 70, 94, 146]; they will not be derived separately here. They have been obtained by conformal mapping techniques and can be rewritten as follows:

$$C_i'^a = 2\varepsilon_0 \frac{K(k_i)}{K(k_i')} \qquad (i = \mathrm{I}, \mathrm{II}, \mathrm{III}, \text{and IV}), \tag{2.3.4}$$

where

$$k_{\mathrm{I}} = \frac{w}{w+2s}, \qquad k_{\mathrm{II}} = \frac{\sinh\left(\dfrac{\pi w}{4h_2}\right)}{\sinh\left(\dfrac{\pi(w+2s)}{4h_2}\right)},$$

$$k_{\mathrm{III}} = \frac{\tanh\left(\dfrac{\pi w}{4h_3}\right)}{\tanh\left(\dfrac{\pi(w+2s)}{4h_3}\right)}, \qquad k_{\mathrm{IV}} = \frac{\sinh\left(\dfrac{\pi w}{4h_4}\right)}{\sinh\left(\dfrac{\pi(w+2s)}{4h_4}\right)}, \tag{2.3.5}$$

with $K(k)$ and $K(k')$ the complete elliptic integral of the first kind and its complement, and $k_i' = \sqrt{1-k_i^2}$.

Accurate expressions for the ratio $K(k)/K(k')$ are available in reference 14 and are given below:

$$\frac{K(k)}{K(k')} = \begin{cases} \dfrac{1}{\pi}\ln\left[2\left(1+\sqrt{k}\right)/\left(1-\sqrt{k}\right)\right], & 0.5 \le k^2 \le 1 \\ \dfrac{\pi}{\ln\left[2\left(1+\sqrt{k'}\right)/\left(1-\sqrt{k'}\right)\right]}, & 0.0 \le k^2 \le 0.5 \end{cases}. \tag{2.3.6}$$

The value of the total capacitance per unit length $C_t'^a$ as well as the filling factor q_1 to q_4 can be written in terms of the above values of the basic capacitances. This will be done for each structure shown in Fig. 2.3.1 separately (in accordance with their practical importance) as follows.

2.3.1.1 Structure SCPW1. As shown in Fig. 2.3.1a, the structure SCPW1 consists of three layers only. Therefore $q_4 = 0$. With reference to Figs. 2.3.1a and 2.3.2a, the following exact values can be obtained:

$$C_t'^a = 2C_{\mathrm{I}}'^a,$$

$$q_3 = \frac{C_{\mathrm{I}}'^a}{C_t'^a} = \frac{1}{2}. \tag{2.3.7}$$

In order to estimate the values of the filling factors q_1 and q_2, a formula for the effective dielectric constant of a conventional coplanar waveguide that was first given by Veyres and Fouad-Hanna [68] and verified later by Ghione and Naldi [109] will be used. This formula can be written as a function of the air-filled capacitances as follows:

$$\varepsilon_{\text{eff}} = 1 + 0.5(\varepsilon_r - 1)\frac{C_{\text{II}}^{\prime a}}{C_{\text{I}}^{\prime a}}. \tag{2.3.8}$$

This effective dielectric constant can also be written as a function of the filling factor of the main substrate as follows:

$$\varepsilon_{\text{eff}} = (1 - q_m) + \varepsilon_r q_m. \tag{2.3.9}$$

A simple comparison between Eqs. (2.3.8) and (2.3.9) shows that the filling factor in this case is not a function of the relative dielectric constant of the main substrate and is only a function of the structure physical dimensions as given below:

$$q_m = 0.5\frac{C_{\text{II}}^{\prime a}}{C_{\text{I}}^{\prime a}} = \frac{C_{\text{II}}^{\prime a}}{C_t^{\prime a}}. \tag{2.3.10}$$

Because the assumption in Eq. (2.3.10) is verified for any air–dielectric interface, then it can be suggested that it may also be valid for any other two dielectric interfaces (when replacing the air under the main dielectric substrate by another dielectric material). Therefore,

$$q_2 = q_m. \tag{2.3.11}$$

It should be pointed out that the conclusion derived from Eq. (2.3.10) is not generally correct because the filling factor should also be a function of the type of dielectric interface. Accordingly, the obtained value of the filling factor q_2 given by Eqs. (2.3.10) and (2.3.11) should be considered as an approximate value, and comparison with rigorous numerical results will be required (see below) to find out whether further improvement of these formulas is necessary or not.

The value of the filling factor q_1 can then be determined from the values of q_3 and q_2 given by Eqs. (2.3.7), (2.3.10), and (2.3.11) in addition to the following well-known relation for the three filling factors:

$$q_1 + q_2 + q_3 = 1. \tag{2.3.12}$$

The following value for the filling factor q_1 is finally obtained:

$$q_1 = (C_{\text{I}}^{\prime a} - C_{\text{II}}^{\prime a})/C_t^{\prime a}. \tag{2.3.13}$$

The values of Z_L and ε_{eff} of the SCPW1 structure can then be calculated by using Eqs. (2.3.1) to (2.3.6) in association with Eqs. (2.3.7), (2.3.10), (2.3.11), and (2.3.13).

2.3.1.2 *Structure SCPW2.* Reference is now made to Figs. 2.3.1b and 2.3.2b. This structure also consists of three layers, and therefore $q_4 = 0$. The following exact equations can be obtained:

$$C_t'^a = C_{\text{I}}'^a + C_{\text{III}}'^a,$$
$$q_3 = C_{\text{III}}'^a / C_t'^a. \tag{2.3.14}$$

The assumed values for q_1 and q_2 of Eqs. (2.3.10), (2.3.11), and (2.3.13) are used also in this case. The values of Z_L and ε_{eff} are calculated by using Eqs. (2.3.1) to (2.3.6), (2.3.10), (2.3.11), and (2.3.13). It should be pointed out here that the value of $C_t'^a$ in this case is different from that in the case of the structure SCPW1, which was given by Eq. (2.3.7).

2.3.1.3 *Structure SCPW3.* In this case, the reference is to Fig. 2.3.1c and 2.3.2c. The following exact relation can be obtained:

$$C_t'^a = 2C_{\text{I}}'^a. \tag{2.3.15}$$

The following additional approximate expressions can also be assumed in the same way that resulted in Eqs. (2.3.8) to (2.3.13), that is,

$$q_4 = C_{\text{IV}}'^a / C_t'^a,$$
$$q_3 = \left(C_{\text{I}}'^a - C_{\text{IV}}'^a\right) / C_t'^a. \tag{2.3.16}$$

The same approximate values of Eqs. (2.3.10), (2.3.11), and (2.3.13) for q_1 and q_2 are used. Z_L and ε_{eff} are then calculated by Eqs. (2.3.1)–(2.3.6), (2.3.10), (2.3.11), (2.3.13), (2.3.15), and (2.3.16).

The conventional coplanar waveguide (shown in Fig. 1.2b) can be considered as a limiting case for the structure SCPW1 (when both $\varepsilon_{r1} = \varepsilon_{r3} = 1$). In this case the values of Z_L and ε_{eff} converge to those of Veyres and Fouad-Hanna [68].

2.3.1.4 *Numerical Results.* Two assumptions have been made during the derivation. These are: the modeling of the two slots as magnetic walls as well as the assumed approximate values for q_1 and q_2 in the case of the SCPW1 and the SCPW2 structure, in addition to those of q_3 and q_4 for the case of the SCPW3 structure.

The assumption of modeling the two slots as magnetic walls is always verified in order to ensure proper behavior of the structure as coplanar transmission line. The assumed values of the filling factors are logical. Moreover,

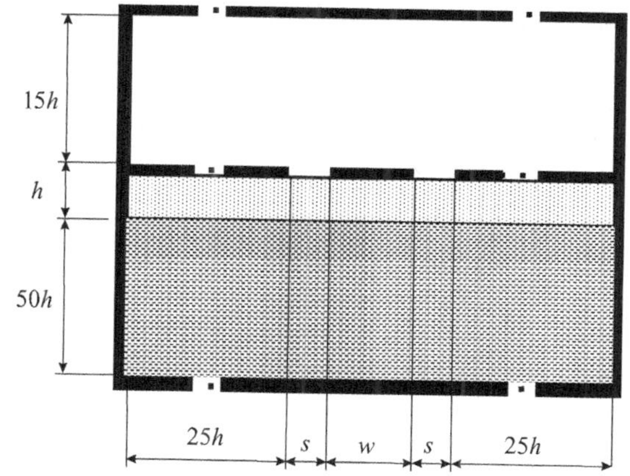

Fig. 2.3.3. Dimensions of the shielding box for the spectral domain analysis, used for accuracy comparison.

they lead to the correct limits in the case of SCPW1 when $h_2 \to 0$, $h_2 \to \infty$, or $\varepsilon_{r1} = \varepsilon_{r2} = \varepsilon_{r3}$. Comprehensive comparisons with results from a rigorous spectral domain analysis (see Section 2.1) for a wide range of physical dimensions and dielectric constants have been carried out to prove the validity of the assumed approximations.

The calculations have been made at a frequency of 1 GHz, and the number of spectral terms are truncated to 4000 and the dimensions of the shielding box are selected as shown in Fig. 2.3.3 in order to avoid an effect of both the top and the bottom covers as well as the lateral sidewalls. The substrate thickness has been chosen to be 200 μm.

The first group of numerical results is presented to assess for the validity of the presented formulas. Comprehensive comparisons with the results that are obtained by a rigorous spectral domain hybrid mode approach at low frequencies (1 GHz), as described in Section 2.1, have shown that the error of the derived formulas is less than 1% for most of the applicable range of physical dimensions and available dielectric materials.

The accuracy decreases as the spacing between the coplanar ground planes ($w + 2s$, Fig. 2.3.3) increases. Moreover, it is more sensitive to the increase of the slot width s than to a corresponding increase of the strip width w. Some of the comparisons are displayed in Tables 2.3.1 and 2.3.2, respectively.

Table 2.3.1 shows a comparison with respect to the characteristic impedance of the structure SCPW1 (see Fig. 2.3.1) with the parameters $h_2 = 200\,\mu\text{m}$, $\varepsilon_{r1} = \varepsilon_{r3} = 1$, and $\varepsilon_{r2} = 2.25$, 12.9, and 20.0, respectively. It should be pointed out here again that in this case the presented expressions converge to those of Veyres and Fouad-Hanna [68]. It should also be pointed out that Ghione and Naldi [109] have verified Veyres and Fouad-Hanna's expressions for a single dielectric material ($\varepsilon_{r2} = 10$) by comparison with the upper and lower values that are obtained by spectral domain variational analysis. The main purpose

TABLE 2.3.1. Comparison of the Derived Formula Results for the Characteristic Impedance (in Ohms) of the Conventional Coplanar Waveguide with Results from a Rigorous Spectral Domain Analysis at a Frequency of 1 GHz

		Derived Formula Results for Z_L (Ω)			Spectral Domain Analysis for Z_L (Ω)		
a/b	b (μm)	$\varepsilon_{r2} = 20$	$\varepsilon_{r2} = 12.9$	$\varepsilon_{r2} = 2.25$	$\varepsilon_{r2} = 20$	$\varepsilon_{r2} = 12.9$	$\varepsilon_{r2} = 2.25$
0.2	50	54.49	67.95	140.75	55.76	68.28	141.40
	170	57.52	70.29	142.86	57.52	70.27	142.97
	230	59.00	72.13	144.43	59.02	71.95	143.95
	350	62.89	76.44	147.85	62.60	75.93	146.30
0.4	50	42.04	51.47	106.57	42.22	51.69	106.99
	170	43.88	53.60	108.47	43.86	53.56	108.32
	230	45.28	55.21	109.82	45.20	55.05	109.34
	350	48.46	58.82	112.62	48.24	58.44	111.41
0.6	50	33.32	40.80	84.45	33.48	40.99	84.83
	170	34.87	42.59	86.04	34.86	42.56	85.84
	230	35.99	43.87	87.12	35.93	43.76	86.68
	350	38.41	46.63	89.24	38.26	46.36	88.38
0.8	50	25.68	31.45	65.09	25.86	31.66	65.51
	170	26.81	32.45	66.62	26.80	32.71	66.03
	230	27.56	33.61	66.98	27.53	33.54	66.59

Parameters: $h_2 = 200\,\mu\text{m}$, $\varepsilon_{r1} = \varepsilon_{r3} = 1$, $\varepsilon_{r2} = 2.25$, 12.9, and 20.0.

of Table 2.3.1 is then to extend the validity of Veyres and Fouad-Hanna's assumption for a wider range of dielectric materials and give the user of these formulas the safety of accuracy over the wide spectrum of dielectric constants tested in Table 2.3.1 ($\varepsilon_{r2} = 1$ up to $\varepsilon_{r2} = 20$).

Table 2.3.2 shows a similar comparison for the SCPW1 structure but with the presence of a supporting dielectric material and $\varepsilon_{r3} = 1$. Three cases are displayed. These are: GaAs ($\varepsilon_{r2} = 12.9$) supported by quartz ($\varepsilon_{r1} = 3.78$), GaAs ($\varepsilon_{r2} = 12.9$) supported by alumina ($\varepsilon_{r1} = 10.0$), and a hypothetical substrate ($\varepsilon_{r2} = 20.0$) supported by alumina ($\varepsilon_{r1} = 10.0$). From the result it may be observed that the agreement between the rigorous spectral domain analysis results and the results from the approximate formulas is quite good (error below 1%) for the small slot widths and reasonable line widths.

Even for the large slot width of 200 μm and a strip width of 20 μm, the agreement is still within accuracy with an error below 1%. Only for larger strip widths (e.g., 800 μm), the error increases to about 2–3%. In the following, some typical results for the parameters of various coplanar waveguides shall be discussed using the results from the above given formulas for the effective dielectric constant and the characteristic impedance of the waveguides.

As a first result, the effective dielectric constant of a single coplanar waveguide on a substrate material of height $h = 200\,\mu\text{m}$ without backside metal-

TABLE 2.3.2. Comparison of the Derived Formula Results for the Characteristic Impedance (in Ohms) of the Open Supported Coplanar Waveguide with Results from a Rigorous Spectral Domain Analysis at a Frequency of 1 GHz

		Static Approximations Z_L (Ω)			Spectral Domain Analysis Z_L (Ω)		
s (μm)	w (μm)	$\varepsilon_{r2} = 20$ $\varepsilon_{r1} = 10$	$\varepsilon_{r2} = 12.9$ $\varepsilon_{r1} = 3.78$	$\varepsilon_{r2} = 12.9$ $\varepsilon_{r1} = 10$	$\varepsilon_{r2} = 20$ $\varepsilon_{r1} = 10$	$\varepsilon_{r2} = 12.9$ $\varepsilon_{r1} = 3.78$	$\varepsilon_{r2} = 12.9$ $\varepsilon_{r1} = 10$
20	20	45.51	55.96	55.91	45.85	56.37	56.33
	60	33.24	40.90	40.79	33.38	41.08	40.98
	120	27.68	34.11	33.90	27.12	34.16	33.96
	200	24.52	30.30	29.98	24.50	30.28	29.92
	800	19.01	23.73	22.60	18.83	23.72	22.47
60	20	61.82	76.09	76.84	62.01	76.32	76.09
	60	45.81	56.46	56.07	45.83	56.50	56.14
	120	37.73	46.62	46.01	37.67	45.56	46.00
	200	32.95	40.88	39.96	32.81	40.72	39.89
	800	24.41	30.99	28.69	24.05	30.56	28.43
100	20	70.56	86.97	86.36	70.60	87.03	86.48
	60	53.26	65.80	64.98	53.18	65.71	64.97
	120	44.09	53.65	53.92	43.90	54.44	53.42
	200	38.47	47.91	46.39	38.19	47.60	46.22
	800	27.98	35.76	32.61	27.48	35.16	32.25
200	20	83.77	103.81	101.75	83.24	103.18	101.48
	60	65.28	81.22	78.85	64.80	80.66	78.94
	120	54.87	68.62	65.81	54.31	67.95	65.44
	200	48.95	60.58	57.26	47.50	59.80	56.22
	800	34.42	44.60	39.47	33.66	43.65	38.95

Parameters: $h_2 = 200\,\mu\text{m}$, $\varepsilon_{r1} = 3.78$ and 10, $\varepsilon_{r3} = 1$, $\varepsilon_{r2} = 12.9$ and 20.0.

ization and with air above the waveguide structure is shown in Fig. 2.3.4. The dielectric constant of the substrate material is $\varepsilon_r = 12.9$ (GaAs). As has already been discussed in Sections 2.1 and 2.2, the results of the static formula also show that the effective dielectric constant of a coplanar waveguide with a small slot width ($s = 10\,\mu\text{m}$) has a relatively high value (here: $\varepsilon_{\text{eff}} \approx 7$) and it is nearly independent of the center strip width. For the case of a larger gap between the center strip and the ground planes (e.g., 400 μm), the effective dielectric constant is reduced (here: $\varepsilon_{\text{eff}} \approx 6$) and it decreases to values close to 5 for increased center-strip width. The reason for this is that the electric stray field in the air region is largely increased if the gap width is increased and in addition if the center strip is wider.

As a consequence, the characteristic impedance of a coplanar waveguide with a small gap between the center strip and the ground planes has a low

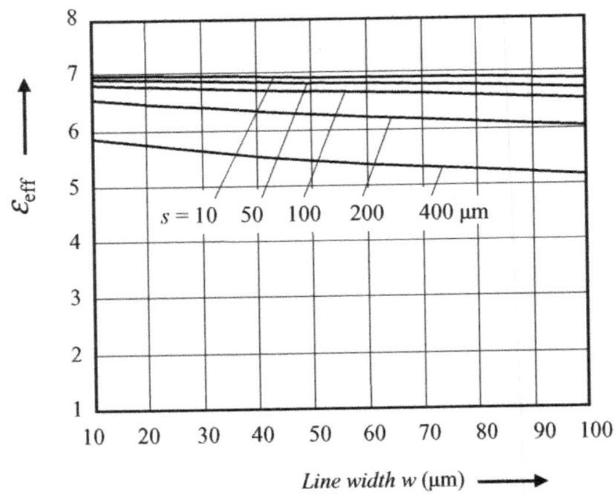

Fig. 2.3.4. Effective dielectric constant of a single coplanar waveguide on a GaAs substrate ($\varepsilon_r = 12.9$) of height $h = 200\,\mu m$ without backside metalization in dependence on the center strip width w and with the gap width s as a parameter.

value. As may be seen from Fig. 2.3.5, characteristic impedances between $30\,\Omega$ and $100\,\Omega$ are realizable on GaAs substrate if gap widths and strip widths between $10\,\mu m$ and $100\,\mu m$ are used. It may also be observed that there is nearly no difference between the characteristic impedances for waveguides on the 200-µm-thick substrate and the 400-µm-thick substrate. The reason for this is that the electric field is concentrated in the slot area and does not touch the backside of the substrate material even for the substrate of 200-µm thickness and for waveguides with 100-µm gaps.

Figure 2.3.6 shows the dependence of the characteristic impedance on the center-strip width and with the dielectric constant of the substrate material as a parameter. Two effects may be observed: (1) The absolute value of the characteristic impedance decreases with increasing dielectric constant, and (2) the slope of the curves in dependence on the center strip width becomes steeper with increasing dielectric constant.

If a coplanar waveguide with a special characteristic impedance is to be designed, the dependencies given in Fig. 2.3.7 are of good use for a first estimation of the center-strip width and the adjoined gap width that are needed to realize a special characteristic impedance. The figures are given for a dielectric constant of 12.9 (GaAs) and a substrate height (which, however, does not have a big influence, according to the discussion above) of $200\,\mu m$. The two figures give a different scaling of the strip width and the gap width.

It can be observed that a 50-Ω line may be realized in a large variety of structures. So, for example, a center-strip width of about $100\,\mu m$ requires a gap width of about $65\,\mu m$ (Fig. 2.3.7a). A 50-Ω line may, however, also be realized using a center-strip width of $60\,\mu m$ and a slot width of $40\,\mu m$ (Fig. 2.3.7b) or the combination of a strip width of $15\,\mu m$ and a slot width of $10\,\mu m$, and so

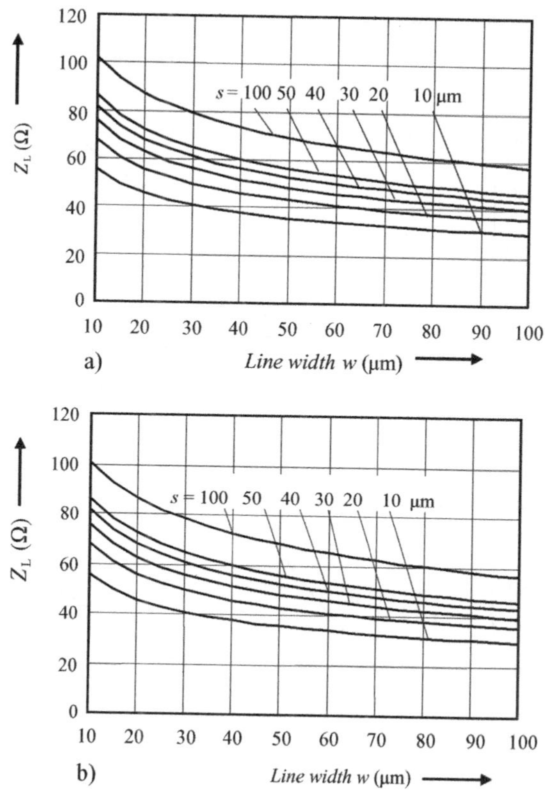

Fig. 2.3.5. Characteristic impedance of a single coplanar waveguide on a GaAs substrate ($\varepsilon_r = 12.9$) of height $h = 200\,\mu m$ (a) and $h = 400\,\mu m$ (b) without backside metalization in dependence on the center strip width w and with the gap width s as a parameter.

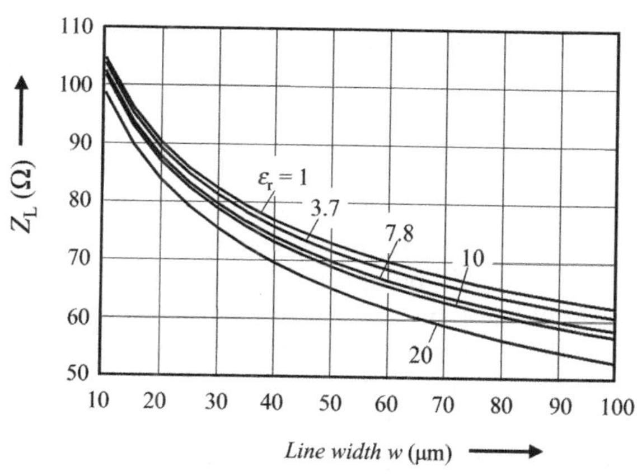

Fig. 2.3.6. Characteristic impedance of a single coplanar waveguide ($s = 100\,\mu m$) on a substrate of height $h = 200\,\mu m$ without backside metalization, plotted against the center strip width w and with the dielectric constant of the substrate material as a parameter.

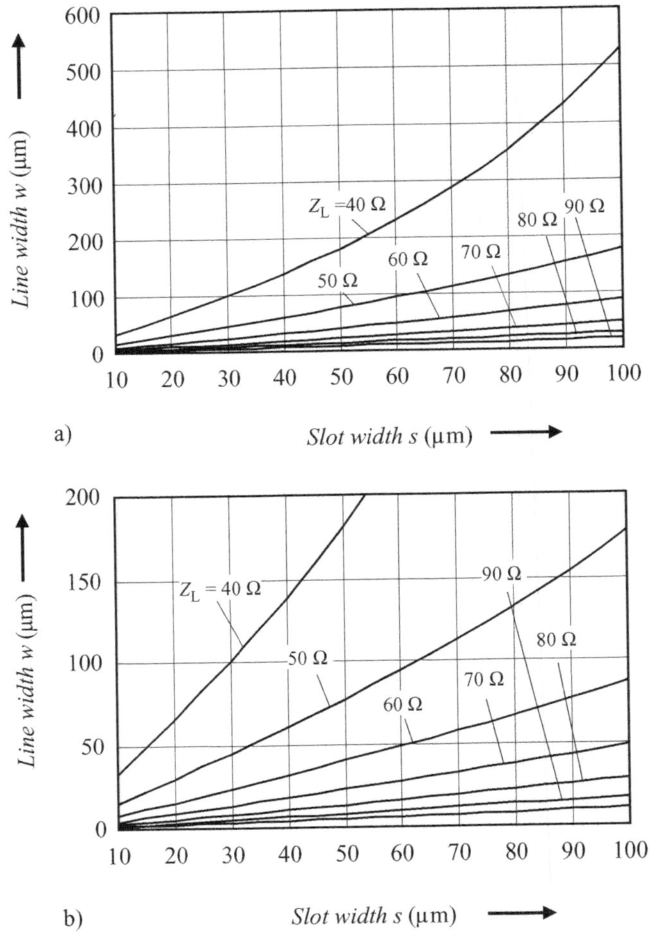

Fig. 2.3.7. The dependence of the center strip width on the slot width for a constant characteristic impedance of a coplanar waveguide. Substrate material GaAs; $\varepsilon_r = 12.9$, $h = 200\,\mu\text{m}$.

on. It all depends on the requirements of the circuits and the used frequency for a particular circuit design. So, if for example, lines with low losses are needed, a large strip width will be optimal. On the other hand in this case, the signal frequency should not be too high because with a larger center-strip width a larger slot width is needed, and this leads to higher dispersion of the coplanar waveguide (see Section 2.1). In any case, a compromise can be found depending on the requirements of the special circuit design task.

In a microwave integrated circuit environment, coplanar waveguides are finally used in a metalized package. The top shielding of this package may influence the properties of the coplanar waveguide. Therefore, an investigation of this influence is made using the metalized shielding in a height h_3 above

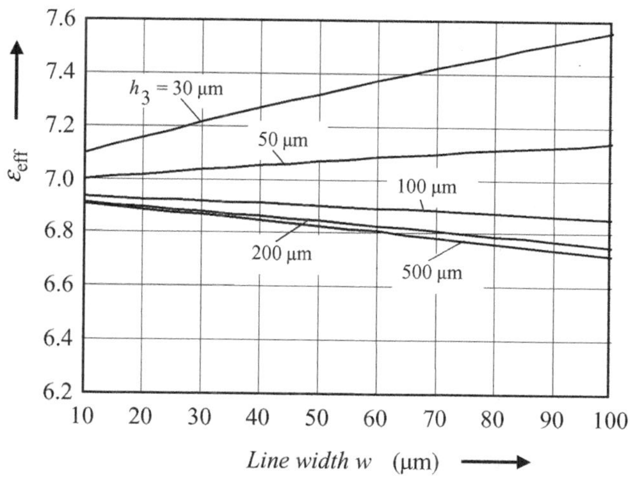

Fig. 2.3.8. The effective dielectric constant of a coplanar waveguide, plotted against the center strip width and with the height of a shielding plane above the substrate as a parameter.

the upper substrate plane of the waveguide. Figure 2.3.8 shows the influence of this shielding plane on the effective dielectric constant of the coplanar waveguide. The effective dielectric constant is drawn against the center-strip width with the height of the shielding as a parameter. It may be observed that two different effects occur in dependence on the height h_3: If h_3 is very small, the effective dielectric constant increases with increasing strip width. The electric field lines are pressed into the dielectric material and increase the effective dielectric constant if the strip width is increased. On the contrary, if h_3 is large, the electric field lines have enough space between the upper substrate plane and the shielding plane to fill the air space and thereby reduce the effective dielectric constant if the strip width increases.

Also of large interest, especially in the case of the coplanar waveguide with small gap width, is the influence of the metalization thickness on the waveguide properties. This has already been intensively discussed in Section 2.2.6 using the finite difference analysis technique. Here the influence of the metalization thickness is considered by introducing an additional capacitance per unit line length:

$$\Delta C' = 1.75 \frac{\varepsilon_0 t}{s} \tag{2.3.17}$$

parallel to the capacitance C'^{a}_{II} (see Fig. 2.3.2). The influence of the metalization thickness on the effective dielectric constant is shown in Fig. 2.3.9. Chosen are three different coplanar waveguides on GaAs substrate ($\varepsilon_r = 12.9$, $h = 200\,\mu m$) of characteristic impedance $50\,\Omega$ but with different center strip widths. If the results shown in Fig. 2.3.9 are compared to those that are shown in

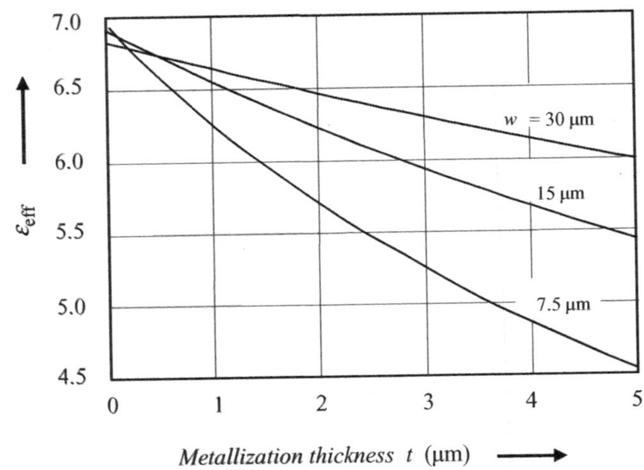

Fig. 2.3.9. The influence of the metalization thickness on the effective dielectric constant for three single 50-Ω coplanar waveguides on a GaAs substrate ($\varepsilon_r = 12.9$) of height $h = 200\,\mu\text{m}$. Parameter: center strip width w.

Fig 2.2.23b, which have been calculated using the most accurate finite difference technique, it may be concluded that the described method is of a good accuracy.

Coplanar waveguides are normally used with thick substrate material because it is a big advantage of the coplanar technology that it needs no backside preparation. In some special cases, thin substrate may be of interest, for example, for better heat transfer. For such a case, a backside metalization may be of interest. Under this condition, it must be known how big the influence of a finite substrate thickness on the characteristic impedance and the effective dielectric constant is. The formulas derived in this section are also able to analyze this case. An evaluation of the characteristic impedance in dependence on the substrate thickness is shown in Fig. 2.3.10. All lines that are analyzed have a 50-Ω impedance in the case of an infinite substrate thickness, but they are of different center-strip and slot width. The chosen substrate material is GaAs with a relative dielectric constant of 12.9. Figure 2.3.10 shows that all lines, independent of their strip width, indeed have the 50-Ω impedance for a substrate height larger than 250 μm, independent of whether the backside is metalized or not. In the case of the nonmetalized backside (single coplanar waveguide, SCPW), the influence of the substrate height for values larger than 200 μm is negligibly small. If the substrate height is reduced, the characteristic impedances of coplanar waveguides that are not metalized on the backside increase, because electric field lines enter the air space under the substrate material and the effective dielectric constant is decreased. On the contrary, the characteristic impedance of the conductor-backed coplanar waveguide (CBCPW) decreases with decreasing substrate height because electric field

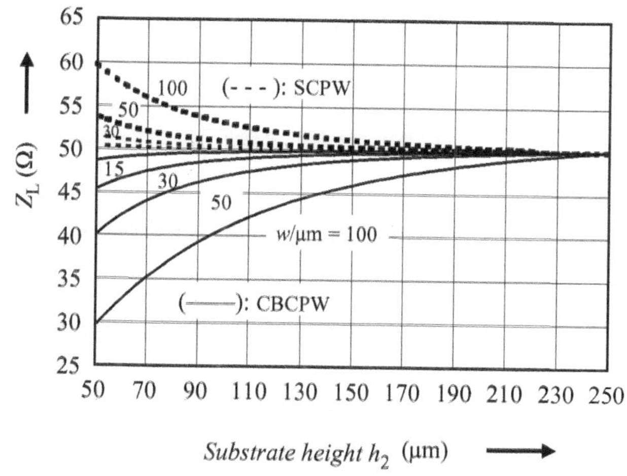

Fig. 2.3.10. Influence of the substrate height on the characteristic impedance of a single coplanar waveguide without backside metalization (SCPW) and a conductor-backed coplanar waveguide (CBCPW). All coplanar waveguides are 50-Ω lines in the case of infinite substrate thickness. Substrate material GaAs ($\varepsilon_r = 12.9$).

lines from the center conductor may end on the metalization on the substrate backside.

This means that more electric field lines are concentrated in the dielectric substrate material and the effective dielectric constant increases. Also, because these field lines no longer end on the ground planes placed on top of the substrate material, this effect is similar to an increase of the slot width which also leads to a decrease of the characteristic impedance. A more detailed discussion of the effects connected with conductor-backed coplanar waveguides, especially with possible mode transfer, will be given in the next section.

2.3.2 Static Formulas for Calculating the Parameters of General Broadside-Coupled Coplanar Waveguides

The configuration that is dealt with in this section is shown in Fig 2.3.11. It is similar to the one that has been introduced and analyzed numerically by Hatsuda [35] (by using the finite difference method) under the name of symmetrical two-strip-conductor coplanar-type strip line. In this section, fast and exact analytic formulas are presented for the quasi-static TEM parameters of general broadside-coupled coplanar waveguides (with general dielectric interface, GBSC-CPW). They may be of interest in coplanar circuit design whenever close coupling between two coplanar waveguides (e.g., in couplers or in lumped element transformers) is needed (compare Section 4.5, Fig. 4.5.8). The approach used here is based on isolating the odd and the even modes that may exist on the structure by assuming an electric wall in the case of the odd mode

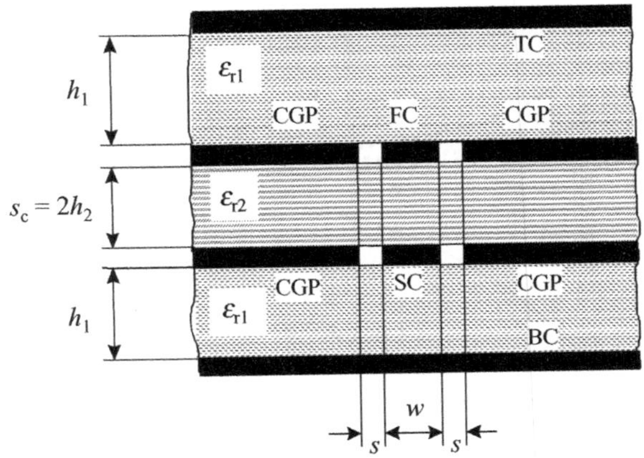

Fig. 2.3.11. Cross-sectional view of the general broadside-coupled coplanar waveguide (GBSC-CPW).

and a magnetic wall for the even mode. The cross section of each mode is then divided into two regions, with the field in each region represented by a capacitance whose expressions have been taken from the literature. Numerical results are presented in order to investigate various properties of the structure. Analytic formulas and numerical results are also presented for an asymmetrical broadside-coupled coplanar waveguide as well as for the single CPW that results from connecting the two coupled strips of the GBSC-CPW at the input and output ports. High speed of computation and exactness justify the use of these formulas in (M)MIC-CAD programs. Criteria are also obtained to ensure the coplanar behavior of the structure. Investigations on the dispersion characteristics of the broadside-coupled coplanar waveguide in reference 257 show that up to high frequencies dispersion may be neglected and that the quasi-static analysis is adequate for designing microwave and millimeter-wave circuits using this kind of waveguide.

2.3.2.1 Analytical Formulas and Results for the General Broadside-Coupled Coplanar Waveguide. The configuration that shall be considered in this section is shown in Fig. 2.3.11. It consists of two coupled strips placed face to face on a dielectric layer of thickness s_c and relative dielectric constant ε_{r2} (center dielectric material); FC and SC denote these two strips, respectively. They are placed near four coplanar ground planes (CGP) at a distance (slot width) equal to *s*.

Top and bottom metallic plates TC and BC may cover the structure, respectively, positioned at a distance h_1 from the surfaces of the dielectric material. The spacing between the center dielectric and the metallic top and bottom covers may be filled by another dielectric material whose thickness is h_1 and whose relative dielectric constant is ε_{r1}. Let us call them the upper and lower dielectric materials, respectively. This structure supports two fundamental

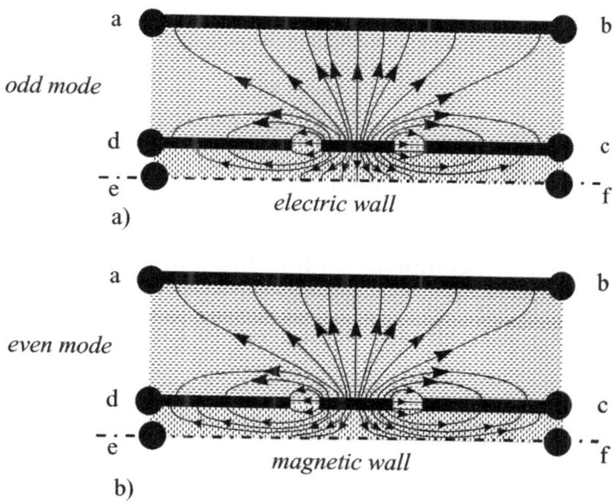

Fig. 2.3.12. The even (a) and the odd (b) mode of the general broadside-coupled coplanar waveguide (GBSC-CPW).

modes, namely, the odd and the even mode with respect to the symmetry plane shown in Fig. 2.3.12. They can be isolated by assuming an electric wall for the odd mode and a magnetic wall for the even mode, as shown in Figs. 2.3.2.12a and 2.3.2.12b, respectively. The analytic formulas of the odd and even mode parameters of the structure can be obtained as follows.

A. The Odd-Mode. The analytic expression for the odd-mode capacitance, Fig. 2.3.12a, can be obtained by modeling the two slots as magnetic walls. This assumption is always verified as long as the structure behaves as a coplanar waveguide. Criteria to ensure this behavior will be discussed in Section 2.3.2.5. An electrical wall is placed at the lower bound of the center dielectric material to ensure the field distribution of the odd mode.

The total odd mode capacitance per unit length can then be considered as the sum of two components, C'_{o1} and C'_{o2}, representing the electric field in the upper and the middle dielectric materials, respectively. The expressions for those two components (C'_{o1} and C'_{o2}) have been derived [94] using conformal mapping technique by mapping each of the regions into a half-plane first and then into a parallel-plate configuration. The resulting odd mode capacitance per unit length can be obtained by rewriting the results of reference 94 in accordance with our physical dimensions as follows:

$$C'_o = C'_{o1} + C'_{o2} \tag{2.3.18}$$

with

$$C'_{oi} = 2\varepsilon_0\varepsilon_{ri}\frac{K(k_{oi})}{K(k'_{oi})}, \qquad i = 1, 2, \tag{2.3.19}$$

where

$$k_{oi} = \frac{\tanh\left(\frac{\pi w}{4h_i}\right)}{\tanh\left(\frac{\pi(w+2s)}{4h_i}\right)}. \tag{2.3.20}$$

$K(k)$ and $K(k')$ are the complete elliptic integral of the first kind and its complement, respectively. Furthermore, $k' = \sqrt{1-k^2}$. Accurate expressions for the ratio $K(k)/K(k')$ have already been given in Eq. (2.3.6) and must not be repeated here.

B. The Even Mode. The total even mode (Fig. 2.3.12b) capacitance per unit line length can be derived in the same way; the only difference is that a magnetic wall is assumed at the lower bound of the center dielectric. The result in this case is as follows:

$$C'_e = C'_{e1} + C'_{e2}, \tag{2.3.21}$$

where

$$C'_{e1} = C'_{o1} \tag{2.3.22}$$

and

$$C'_{e2} = 2\varepsilon_0 \varepsilon_{r2} \frac{K(k_{e2})}{K(k'_{e2})} \tag{2.3.23}$$

with

$$k_{e2} = \frac{\sinh\left(\frac{\pi w}{4h_2}\right)}{\sinh\left(\frac{\pi(w+2s)}{4h_2}\right)}. \tag{2.3.24}$$

Odd and even mode characteristic impedances, effective dielectric constants, and phase velocities of wave propagation can be calculated by using the above given equations and the following well-known dependencies:

$$Z_{L(o,e)} = \frac{1}{c_0 \sqrt{C'_{(o,e)} C'^{a}_{(o,e)}}}, \tag{2.3.25}$$

$$\varepsilon_{\text{eff}(o,e)} = \frac{C'_{(o,e)}}{C'^{a}_{(o,e)}}, \tag{2.3.26}$$

$$v_{(o,e)} = \frac{c_0}{\sqrt{\varepsilon_{\text{eff}(o,e)}}}. \tag{2.3.27}$$

In these equations, $c_0 = 2.9979 \times 10^8\, m/s$ is the velocity of light in vacuum and $C'^{a}_{(o,e)}$ is the odd- and even-mode capacitance when replacing the dielectric materials by air.

Calculated odd- and even-mode characteristic impedances and effective dielectric constants by using the derived formulas are plotted in Figs. 2.3.13a and 2.3.13b, respectively, for $\varepsilon_{r1} = 1$ (air), $\varepsilon_{r2} = 12.9$ (GaAs), $h_1 = 1000\,\mu$m, $s = 20\,\mu$m, $w = 50$–$400\,\mu$m, and $s_c = 50$, 100, and 200$\,\mu$m. The results generally demonstrate relatively low values for the odd- and even-mode characteristic impedances and relatively high values of the odd and even mode effective dielectric constants. In order to get a deeper insight into the properties of this coupled structure, the coupling coefficient C_c (at the center frequency) and the mode velocities ratio v_e/v_o are plotted in Figs. 2.3.14a and 2.3.14b, respectively. The materials are: $\varepsilon_{r1} = 1$ (air), 3.78 (quartz), and = 10 (Al_2O_3); $h_1 = 5000\,\mu$m; $\varepsilon_{r2} = 12.9$ (GaAs); $s_c = 50\,\mu$m; $s = 50$–$400\,\mu$m; and $w = 50$–$400\,\mu$m. The following properties (some of them are typical of all broadside-coupled MIC structures, while the others are special for the coplanar structure discussed here) are observed:

1. The increase of the coupling coefficient C_c is always associated with a corresponding increase in the mode velocities ratio v_e/v_o.
2. The decrease of the center substrate thickness, $s_c = 2h_2$, results in the increase of both C_c and v_e/v_o.
3. The increase of slot width s results in the increase of both C_c, and v_e/v_o.
4. The increase of the strip width w of the coupled strips increase both C_c and v_e/v_o.
5. The increase of the relative dielectric constant ε_{r2} increases both C_c and v_e/v_o of the center substrate.
6. Increasing the relative dielectric constants ε_{r1} of the upper and lower dielectric material results in decreasing both C_c and v_e/v_o. However, a larger decrease is observed in v_e/v_o than in C_c. For example, in the case of $\varepsilon_{r1} = 10$ (alumina) and $\varepsilon_{r2} = 12.9$ (GaAs), the mode velocities ratio v_e/v_o is less than 1.1 while a good coupling coefficient of 0.791 can be still achieved. This type of interface of two dielectric materials of nearly equal dielectric constants is sometimes desirable in particular circuit applications. Moreover, the presence of the lower dielectric material will serve as support for the middle GaAs substrate, which can be thin and fragile in the case of (M)MIC applications. This will result in improving the mechanical strength as well as the average power-handling capability of the whole structure. However, the presence of the upper dielectric material will not permit the insertion of series and parallel lumped passive and active elements, which is considered as one of the most important

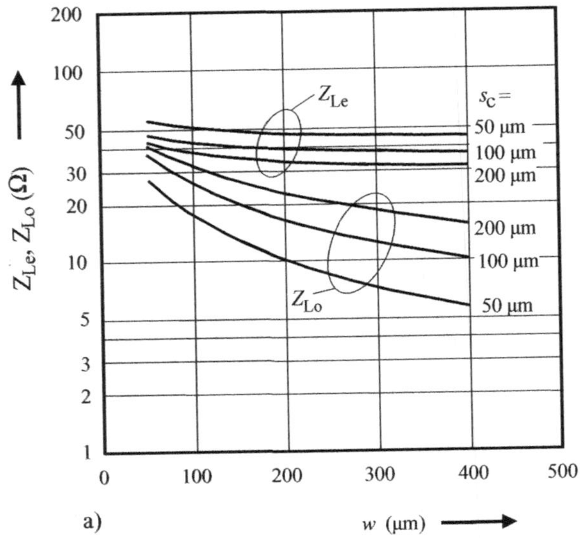

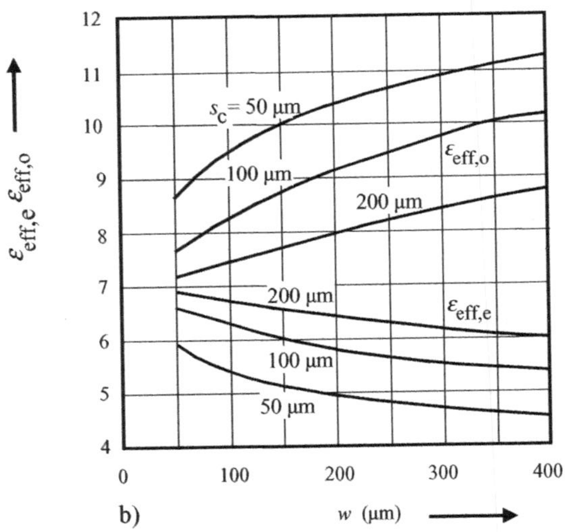

Fig. 2.3.13. The even and the odd mode characteristic impedance (a) and the even and the odd mode effective dielectric constant (b) of the broadside-coupled coplanar waveguide, plotted against typical physical dimensions of the waveguide. Parameters: $s_g = 20\,\mu\text{m}$, $h = 1000\,\mu\text{m}$, $\varepsilon_{r1} = 1$, $\varepsilon_{r2} = 12.9$.

advantages of CPW structures over the microstrip configuration. One solution is to remove, partially, the upper dielectric material from some part of the circuit to permit for such an insertion. The other is to derive new analytic formulas that can deal with such new CPW structure where only the upper dielectric material is replaced by air.

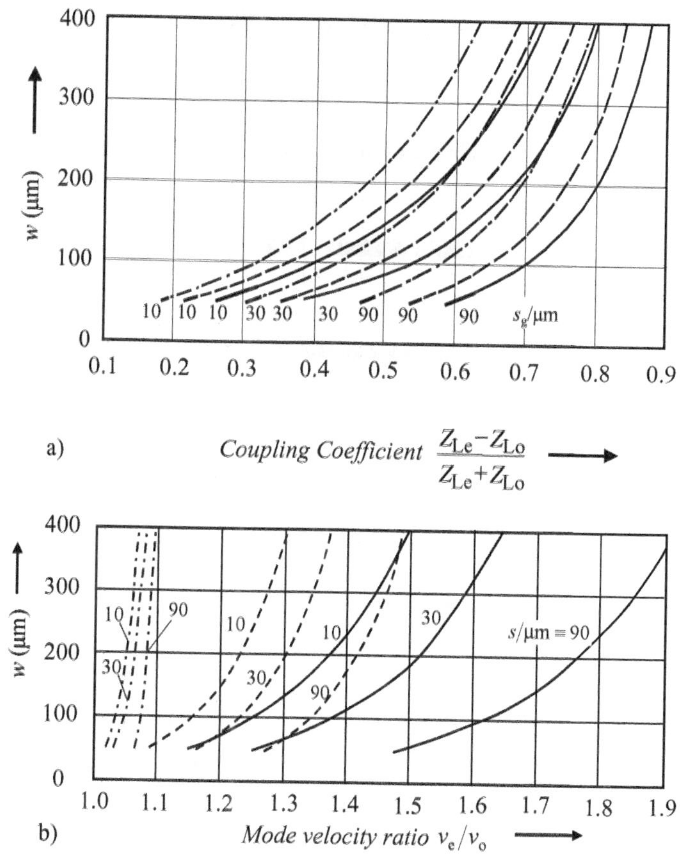

Fig. 2.3.14. (a) Variation of the coupling coefficient C_c of the GBSC-CPW and (b) variation of the mode velocities ratio v_e/v_o, plotted against typical values of the physical dimensions. Parameters: $s_c = 50\,\mu\text{m}$, $h_1 = 5000\,\mu\text{m}$, ——— $\varepsilon_{r1} = 1$, $\varepsilon_{r2} = 12.9$, – – $\varepsilon_{r1} = 3.78$, $\varepsilon_{r2} = 12.9$, –·– $\varepsilon_{r1} = 10.0$, $\varepsilon_{r2} = 12.9$.

2.3.2.2 ***Analysis of an Asymmetric Supported BSC-CPW.*** The new structure that is shown in Fig. 2.3.15a may be called as an asymmetrical supported broadside-coupled coplanar waveguide (ASBSC-CPW) and can be analyzed by considering it as asymmetrical coupled lines. In this case, two modes will be propagating on the lines but with unequal mode characteristic impedances seen by each of the lines.

The mode characteristic impedances and effective dielectric constants are calculated using the self- and mutual capacitances per unit length of the ASBSC-CPW C'_{11}, C'_{12}, C'_{22}, L'_{11}, L'_{12}, and L'_{22} as well as Eq. (5) of reference 105 after replacing the suffixes L and R (standing for left and right lines, respectively) by the suffixes 1 and 2 (standing for the first and second strip, respectively). It should be pointed out, here, that Eq. (5a) of reference 105 should correctly be read as follows:

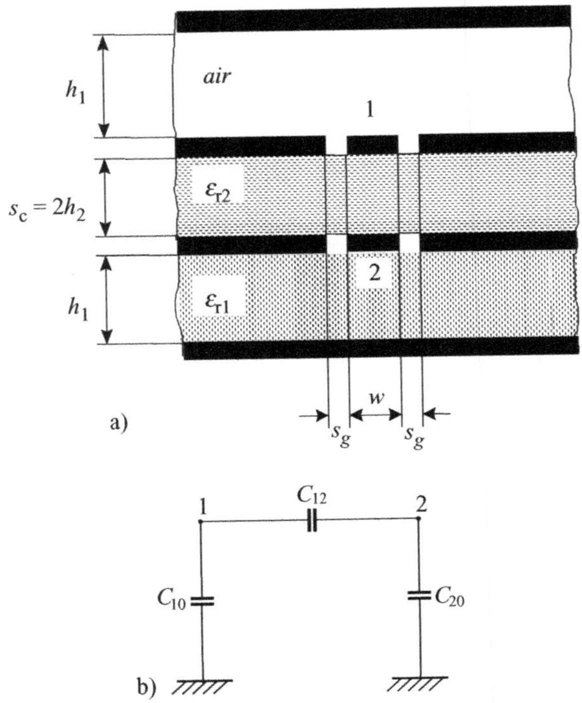

Fig. 2.3.15. Cross-sectional view of (a) the ACBSC-CPW and (b) the equivalent circuit showing the self- and mutual capacitances of the structure.

$$\varepsilon_{\mathrm{eff}(C,\Pi)} = c_0^2\left(L'_{11}C'_{11} + L'_{22}C'_{22} - 2L'_{12}C'_{12} \pm \lambda\right)/2, \tag{2.3.28}$$

$$Z_{C1} = \frac{c_0}{\sqrt{\varepsilon_{\mathrm{eff,C}}}}\left(L'_{11} - L'_{12}/R_{\Pi}\right), \tag{2.3.29}$$

$$Z_{\Pi 1} = \frac{c_0}{\sqrt{\varepsilon_{\mathrm{eff},\Pi}}}\left(L'_{11} - L'_{12}/R_{C}\right), \tag{2.3.30}$$

$$Z_{C2} = -R_C R_{\Pi} Z_{C1},$$

$$Z_{\Pi 2} = -R_C R_{\Pi} Z_{\Pi 1}, \tag{2.3.31}$$

where

$$\lambda = \sqrt{4\left(L'_{12}C'_{22} - L'_{11}C'_{12}\right)\left(L'_{12}C'_{11} - L'_{22}C'_{12}\right) + \left(L'_{22}C'_{22} - L'_{11}C'_{11}\right)^2}$$

and

$$R_{(C,\Pi)} = \frac{\left(L'_{22}C'_{22} - L'_{11}C'_{11}\right) \pm \lambda}{2\left(L'_{12}C'_{22} - L'_{11}C'_{12}\right)}. \tag{2.3.32}$$

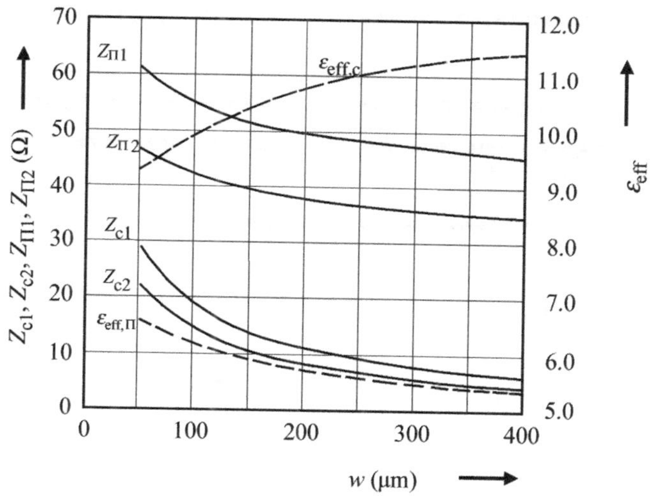

Fig. 2.3.16. Variation of the mode (C and Π) characteristic impedances and effective dielectric constants of the ASBSC-CPW with typical values of physical dimensions (see text).

In this case, the expressions for the self- and mutual capacitance per unit length C'_{11}, C'_{12}, and C'_{22} can be written with reference to Fig. 2.3.15 and by using Eqs. (2.3.18)–(2.3.24) as follows:

$$C'_{12} = 0.5(C'_{o2} - C'_{e2}), \tag{2.3.33}$$

$$C'_{11} = C'^{a}_{o1} + C'_{e2} + C'_{12}, \tag{2.3.34}$$

$$C'_{22} = C'_{o1} + C'_{e2} + C'_{12}. \tag{2.3.35}$$

Calculated values for Z_{C1}, Z_{C2}, $Z_{\Pi 1}$, $Z_{\Pi 2}$, $\varepsilon_{\text{eff}(C)}$, and $\varepsilon_{\text{eff}(\Pi)}$ are plotted in Fig. 2.3.16 for $\varepsilon_{r2} = 12.9$, $h_1 = 1000\,\mu\text{m}$, $s = 20\,\mu\text{m}$, $w = 50$–$400\,\mu\text{m}$, and $s_c = 50$, 100, and 200 μm, where the upper dielectric is air while the lower dielectric is quartz ($\varepsilon_{r1} = 3.78$).

2.3.2.3 Application of the GBSC-CPW as Single CPW. Two connections for the GBSC-CPW to external input source are suggested. Fig. 2.3.17a shows the normal way of connection as two coupled lines to be used as building block for (M)MICs, especially as directional couplers that permit tight coupling or for the transmission of electromagnetic power between two coplanar waveguides deployed on different surfaces. Fig. 2.3.17b also shows the use of the GBSC-CPW as a single CPW [35]. This can be achieved by connecting the two coupled strips to the same potential. In this case, only the even mode will be excited along the strips.

The design parameters of the resultant single CPW can be then written as follows:

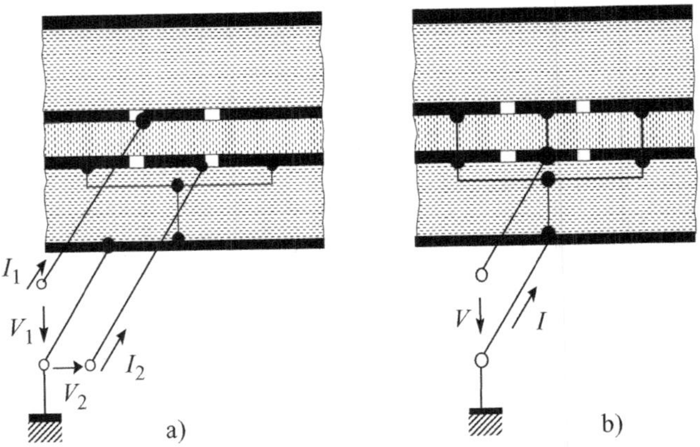

Fig. 2.3.17. Two possible connections for different use of the GBSC-CPW.

$$C'_s = 2C'_e, C'^a_s = 2C'^a_e, \tag{2.3.36}$$

$$Z_{L,s} = 0.5\left(c_0\sqrt{C'_s C'^a_s}\right)^{-1} \tag{2.3.37}$$

and

$$\varepsilon_{\text{eff},s} = C'_s / C'^a_s \tag{2.3.38}$$

where C'_e and C'^a_e have already been given by Eqs. (2.3.23) and (2.3.24). Typical numerical results for the characteristic impedance as well as for the effective dielectric constant as compared to those of the conventional CPW with finite substrate thickness are displayed in Table 2.3.3. It is observed that this configuration gives nearly half the characteristic impedance of the conventional CPW, while its effective dielectric constant is nearly the same.

This means that, by using the connections of Fig. 2.3.17 with the proposed GBSC-CPW, it is possible to obtain a single CPW with the same value of the characteristic impedance but with better slot width manufacturing tolerance allowances. It should be pointed out that it is possible to obtain a similar effect on the impedance in the presence of a near by backed ground plane. However, the decrease in the characteristic impedance will be always associated with an increase in the effective dielectric constant in this case and the possibility to excite the unwanted microstrip-like mode on the conductor-backed coplanar waveguide (compare Section 2.2.9).

2.3.2.4 Criteria for the Coplanar Behavior of the Structure. If the slot width s or the top cover height h_1 increase, their effects on the characteristics of the structure can be ignored. Moreover, increasing the slot width s up to a certain limit will cause some electric field lines to cross the dielectric interface (at the slot) to reach the electric wall (in case of the odd mode) or the ground

TABLE 2.3.3. Comparison Between the Design Parameters of the Conventional CPW (Line 1) and the GBSC-CPW When Used as Single CPW (Line 2)

			Z_L/Ω		ε_{eff}	
h (μm)	s (μm)	w (μm)	Line 1	Line 2	Line 1	Line 2
50	10	50	41.29	20.83	6.37	6.25
	10	400	33.54	17.50	5.30	4.97
	30	50	63.87	32.55	5.86	5.64
	30	400	49.09	25.83	4.69	4.23
	50	50	81.88	42.12	5.46	5.16
	50	400	60.96	32.56	4.31	3.78
200	10	50	35.87	17.95	6.90	6.89
	10	400	26.08	13.18	6.30	6.17
	30	50	49.70	24.90	6.83	6.80
	30	400	34.84	17.68	6.08	5.90
	50	50	58.84	29.51	6.75	6.71
	50	400	41.14	20.95	5.90	5.69

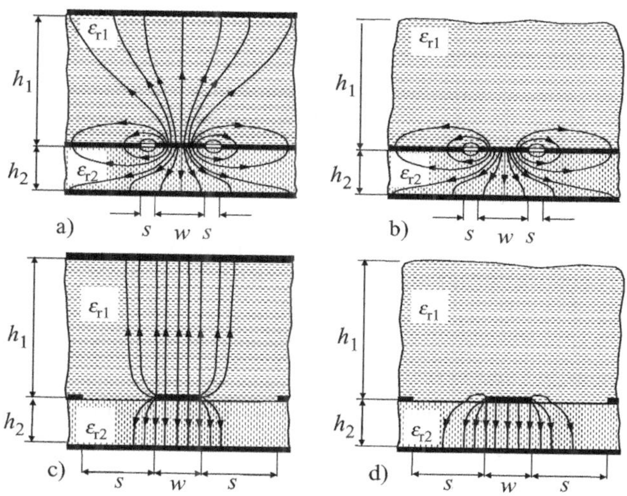

Fig. 2.3.18. The four limiting cases of the GBSC-CPW under an odd-mode excitation.

planes (in the case of the even mode). In this case, the assumption of modeling the slot width as a magnetic wall will not be verified. This effect (due to the increase of the value of s) is more critical in the case of the odd mode than in the case of the even mode. Thus the discussion will be concentrated on the odd mode that is identical to the case of a single coplanar waveguide with top cover and metallic-backed ground plane [145]. Four limiting cases are shown in Figs. 2.3.18a–2.3.18d, respectively.

Case 1: h_1 is comparable to h_2, and s is smaller than a critical slot width s_c. In this case the odd mode configuration behaves as a single coplanar structure with top cover and metallic ground plane [145].

Case 2: h_1 is much larger than h_2, but s is still smaller than s_c. In this case the odd mode configuration behaves as a single coplanar structure with metallic-backed ground plane [94].

Case 3: h_1 is comparable to h_2, and s is much larger than s_c. In this case the odd mode configuration behaves as a covered microstrip line.

Case 4: h_1 is much larger than h_2, and s is much larger than s_c. In this case the odd mode configuration behaves as an open microstrip line.

BIBLIOGRAPHY AND REFERENCES

1. S. P. Morgan, Effect of surface roughness on eddy current losses at microwave frequencies, *J. Appl. Phys.*, vol. 20, 1949, pp. 352–362.
2. G. J. Habetier and E. L. Wachspress, Symmetric successive overrelaxation in solving difference equations, *Math. Comput.*, vol. 15, no. 73, Jan. 1961, pp. 356–362.
3. C. G. Broyden, On convergence criteria for the method of successive overrelaxation, *Math. Comput.*, vol. 18, no. 85, Jan. 1964, pp. 136–141.
4. H. E. Green, The numerical solution of some important transmission-line problems, *IEEE Trans. Microwave Theory Tech.*, vol. MTT-13, Sept. 1965, pp. 676–692.
5. M. V. Schneider, Computation of impedance and attenuation of TEM-line by finite difference methods, *IEEE Trans. Microwave Theory Tech.*, vol. MTT-13, 1965, pp. 793–800.
6. K. S. Yee, Numerical solution of initial boundary value problems involving Maxwell's equations in anisotropic media, *IEEE Trans. Antennas Propag.*, vol. AP 14, 1966, pp. 302–307.
7. R. F. Harrington, *Field Computation by Moment Methods,* New York: Macmillan, 1968, Sections 1–3 and 1–7.
8. R. A. Pucel, D. J. Massé, and C. P. Hartwig, Losses in microstrip, *IEEE Trans. Microwave Theory Tech.*, vol. MTT-16, 1968, pp. 342–350.
9. H. R. Schwarz, Tridiagonalization of a symmetrical band matrix, *Numer. Math.*, vol. 12, 1968, pp. 231–241.
10. C. P. Wen, Coplanar waveguide: A surface strip transmission line suitable for nonreciprocal gyromagnetic device applications, *IEEE Trans Microwave Theory Tech.*, vol. MTT-17, 1969, pp. 1087–1090.
11. M. V. Schneider, Dielectric loss in integrated microwave circuits, *Bell Syst. Tech. J.*, vol. 48, 1969, pp. 2325–2332.
12. M. A. R. Hunston and J. R. Weale, Variation of microstrip impedance with strip thickness, *Electronics Lett.*, vol. 5, 1969, pp. 697–698.
13. J. S. Hornsby and A. Gopinath, Numerical analysis of a dielectric-loaded waveguide with a microstrip line—Finite difference method, *IEEE Trans. Microwave Theory Tech.*, vol. MTT-17, 1969, pp. 684–690.

14. W. Hilber, From approximation to exact relations for characteristic impedances, *IEEE Trans. Microwave Theory Tech.*, vol. MTT-17, 1969, pp. 259–265.

15. C. P. Wen, Attenuation characteristics of coplanar waveguides, *Proc. IEEE*, vol. 58, no. 1, 1970, pp. 141–142.

16. E. J. Denlinger, A frequency dependent solution for microstrip transmission lines, *IEEE Trans. Microwave Theory Tech.*, vol. MTT-19, 1971, pp. 30–39.

17. G. K. Grünberger and H. Meinke, Experimenteller und theoretischer Nachweis der Längsfeldstärken in der Grundwelle der Mikrowellen-Streifenleitung, *Nachrichtentechn. Z.*, vol. 24, 1971, pp. 364–368.

18. G. Kowalski and R. Pregla, Dispersion characteristics of shielded microstrips with finite thickness, *Arch. Elektron. Übertragungstechn.*, vol. 25, 1971, pp. 193–196.

19. R. Mittra and T. Itoh, A new technique for the analysis of the dispersion characteristics of microstrip lines, *IEEE Trans. Microwave Theory Tech.*, vol. MTT-19, Jan. 1971, pp. 47–56.

20. R. Horton, B. Easter, and A. Gopinath, Variation of microstrip losses with thickness of strip, *Electronics Lett.*, vol. 7, 1971, pp. 490–491.

21. E. Yamashita and K. Atsuki, Analysis of thick-strip transmission lines, *IEEE Trans Microwave Theory Tech.*, vol. MTT-19, 1971, pp. 120–123.

22. T. Hatsuda, Computation of coplanar stripline characteristics by relaxation method, *Electronics Commun. Japan*, vol. 54, no. 11, 1971, pp. 76–82.

23. T. Hatsuda, Computation of the characteristics of coplanar-type strip lines by the relaxation method, *IEEE Trans. Microwave Theory Tech.*, vol. MTT-20, 1972, pp. 413–416.

24. T. Hatsuda, Relaxation method computation of coplanar type strip lines characteristics, *Rev. Electrical Commun. Lab.*, vol. 20, no. 9/10, 1972, pp. 825–836.

25. M. K. Krage and G. I. Haddad, Frequency dependent characteristics of microstrip transmission lines, *IEEE Trans. Microwave Theory Tech.*, vol. MTT-20, Oct. 1972, pp. 678–688.

26. W. Wiesbeck, Berechnung der Dämpfung ungeschirmter Streifenleitungen, *Wiss. Ber. AEG-TELEFUNKEN*, vol. 45, no. 4, 1972, pp. 162–166.

27. A. E. Luna, Parallel coplanar strips on a dielectric substrate, in: *Report: Government, Republic of Germany*, 1973, Report no. AD 772–882.

28. K. D. Marx, Propagation modes, equivalent circuits and characteristic terminations for multiconductor transmission lines with inhomogeneous dielectrics, *IEEE Trans. Microwave Theory Tech.*, vol. MTT-21, July 1973, pp. 450–457.

29. A. A. Yashin, Capacitance of a coplanar microstrip line when allowance is made for the effect of a screening plane, *Telecommun. Radio Eng.*, vol. 27/8, no. 11, 1973, pp. 114–115.

30. H. J. van Linde, High-order finite-difference methods for Poisson's equation, *Math. Comput.*, vol. 28, no. 126, April 1974, pp. 369–391.

31. J. H. C. van Heuver, Conduction and radiation losses in microstrip, *IEEE Trans. Microwave Theory Tech.*, vol. MTT-22, 1974, pp. 841–844.

32. L. J. Linner, A method for the computation of the characteristic immittance matrix of multiconductor striplines with arbitrary widths, *IEEE Trans. Microwave Theory Tech.*, vol. MTT-22, no. 11, 1974, pp. 930–937.

33. G. Kowalski, Coplanar printed meander lines, *Arch. Elektron. Übertragungstech.*, vol. 28, no. 6, 1974, pp. 257–262.
34. J. B. Knorr and K. Kuchler, Analysis of coupled slots and coplanar strips on dielectric substrate, *IEEE Trans. Microwave Theory Tech.*, vol. MTT-23, no. 7, 1975, pp. 541–548.
35. T. Hatsuda, Computation of coplanar type strip line characteristics by relaxation method and its application to microwave circuits, *IEEE Trans. Microwave Theory Tech.*, vol. MTT-23, no. 10, 1975, pp. 795–802.
36. W. P. Ou, Design equations for an interdigital coupler, *IEEE Trans. Microwave Theory Tech.*, vol. MTT-23, 1975, pp. 253–255.
37. Y. Fujiki, M. Suzuki, and T. Kitazawa, Higher order modes in coplanar type transmission lines, *Electronics and Commun. Japan*, vol. 58, no. 2, 1975, pp. 74–81.
38. V. Rizzoli, A unified variational solution to microstrip array problems, *IEEE Trans. Microwave Theory Tech.*, vol. MTT-23, no. 2, 1975, pp. 223–234.
39. A. A. Yashin, Capacitance of a screened coplanar microstrip line, *Telecommun. Radio Eng.*, vol. MTT-30, no. 6, 1975, pp. 121–123.
40. J. B. Knorr and K. D. Kuchler, Analysis of coupled slots and coplanar strips on dielectric substrate, *IEEE Trans. Microwave Theory Tech.*, vol. MTT-23, July 1975, pp. 541–547.
41. R. Briechle, Übertragungseigenschaften gekoppelter, verlustbehafteter Mehrleitersysteme mit geschichtetem Dielektrikum, *Frequenz*, vol. 29, March 1975, pp. 69–79.
42. R. Briechle, Analyse von Kammleitungsfiltern und ähnlichen Baugruppen aus gekoppelten Mehrleitersystemen, *Frequenz*, vol. 29, April 1975, pp. 94–100.
43. A. Taflove and M. E. Brodwin, Numerical solution of steady state electromagnetic scattering problems using the time dependent Maxwell's equations, *IEEE Trans. Microwave Theory Tech.*, vol. MTT-23, 1975, pp. 623–630.
44. T. Kitazawa, Y. Hayashi, and M. Suzuki, A coplanar waveguide with thick metal coating, *IEEE Trans. Microwave Theory Tech.*, vol. MTT-24, no. 9, 1976, pp. 604–608.
45. A. Farrar and A. T. Adams, Computations of propagation constants for the fundamental and higher order modes in microstrip lines, *IEEE Trans. Microwave Theory Tech.*, vol. MTT-24, July 1976, pp. 456–460.
46. H. Fritzsche, Capacitances of coplanar microstrip lines in integrated circuits, *Siemens Forsch. Entwicklungsberichte*, vol. 5, no. 2, 1976, pp. 72–75.
47. T. J. Simpson and B. Tseng, Dielectric loss in microstrip lines, *IEEE Trans. Microwave Theory Tech.*, vol. MTT-24, 1976, pp. 106–108.
48. D. Pavlidis and H. L. Hartnagel, The design and performance of three line microstrip couplers, *IEEE Trans. Microwave Theory Tech.*, vol. MTT-24, March 1976, pp. 631–640.
49. E. Müller, *Wellenwiderstand und mittlere Dielektrizitätskonstante von koplanaren Zwei- und Dreidrahtleitungen auf einem dielektrischen Träger und deren Beeinflussung durch Metallwände*, Doctoral Thesis, University of Stuttgart, Germany, 1977.
50. L. Rothe, Eine Methode zur Ermittlung der Feldverteilung von Streifenleitungen, *Nachrichtentech. Elektron.*, vol. 27, 1977, pp. 387–390.

51. I. Sakagami, K. Ono, S. Yajima, N. Nagai, and K. Hatori, An analysis of coplanar type strip transmission lines by conformal mapping, *Trans. Inst. Electron. Commun. Eng. Japan E*, vol. E60, no. 3, 1977, p. 5.

52. Y. Noguchi and N. Okamoto, Analysis of characteristics of the shielded coplanar waveguide by conformal mapping method, *Trans. Inst. Electron. Commun. Eng. Japan E*, vol. E60, no. 8, 1977, pp. 423–424.

53. M. Kumar and B. N. Das, Field and potential distribution of coplanar strips, *J. Inst. Electronics Telecommun. Eng. India*, vol. 23, no. 3, 1977, pp. 137–140.

54. P. K. Saha, Dispersion in shielded planar transmission lines on two layer composite substrate, *IEEE Trans. Microwave Theory Tech.*, vol. MTT-25, no. 11, 1977, pp. 907–911.

55. E. Mueller, Measurement of the effective relative permittivity of unshielded coplanar waveguides, *Electronics Lett.*, vol. 13, no. 24, 1977, pp. 729–730.

56. J. B. Davies and D. Mirshekar Syahkal, Spectral domain solution of arbitrary coplanar transmission line with multilayer substrate, *IEEE Trans. Microwave Theory Tech.*, vol. MTT-25, no. 2, 1977, pp. 143–146.

57. M. Ikeuchi, K. Inoue, H. Sawami, and H. Niki, Finite element analysis of the coplanar type striplines, *Trans. Inst. Electronics Commun. Eng. Japan*, part E, vol. E61, no. 5, 1978, p. 10.

58. R. H. Jansen, High speed computation of single and coupled microstrip parameters including dispersion, high order modes, loss and finite strip thickness, *IEEE Trans. Microwave Theory Tech.*, vol. MTT-26, Feb. 1978, pp. 75–82.

59. R. H. Jansen, Unified user oriented computation of shielded, covered and open planar microwave and millimeter wave transmission line characteristics, *Proc. IEE Microwave Optics Acoustics*, vol. 3, Jan. 1978, pp. 14–22.

60. J. S. Wong, Microstrip tapped-line filter design, *IEEE Trans. Microwave Theory Tech.*, vol. MTT-27, Jan. 1979, pp. 44–50.

61. A. Gopinath, A comparison of coplanar waveguide and microstrip for GaAs monolithic integrated circuits, in: *1979 IEEE MTT S International Microwave Symposium Digest*, April 30–May 2, 1979, Orlando, FL, 1979, pp. 109–111.

62. J. P. Becker and D. Jäger, Electrical properties of coplanar transmission lines on lossless and lossy substrates, *Electronics Lett.*, vol. 15, no. 3, 1979, pp. 88–90.

63. D. Mirshekar-Syahkal and J. B. Davies, Accurate solution of microstrip and coplanar structures for dispersion and for dielectric and conductor losses, *IEEE Trans. Microwave Theory Tech.*, vol. MTT-27, July 1979, pp. 694–699.

64. W. Schumacher and R. Lehmann, Berechnung der Felder von Streifen- und Schlitzleitungen, der Stromverteilung auf den Leitern der Schlitzleitung und deren Auswirkung auf die Leitungsdämpfung, *Arch. Elektron. Übertragungstech.*, vol. 33, 1979, pp. 417–420.

65. W. Schumacher, *Zur Berechnung der Leitungswellenlänge zwei und drei schichtiger planarer Leitungen mit einer Momentenmethod*, Doctoral Thesis, University Karlsruhe, Germany, 1979.

66. L. Frnell, Solution of Poisson's equations on a nonuniform grid, *J. Comput. Phys.*, vol. 53, 1980, pp. 408–425.

67. J. Majer, Graphical method for conduction loss calculation of TEM transmission lines, *Electronics Lett.*, vol. 16, 1980, pp. 638–639.

68. C. Veyres and V. Fouad-Hanna, Extension of the application of conformal mapping techniques to coplanar lines with finite dimensions, *Int. J. Electronics*, vol. 48, no. 1, 1980, pp. 47–56.
69. K. Koshiji, E. Shu, and S. Miki, An analysis of coplanar waveguides with finite conductor thickness computation and measurement of characteristic impedance, *Trans. Inst. Electronics Commun. Eng. Japan*, Part E, vol. E64, no. 8, 1981, p. 61.
70. V. Fouad-Hanna, Finite boundary corrections to coplanar stripline analysis, *Electronics Lett.*, vol. 16, no. 15, 1980, pp. 604–606.
71. V. V. Nikolskiy and O. A. Golovanov, Application of autonomous multimodal blocks to analysis of slotted, high Q and coplanar lines, *Radio Eng. Electronic Phys.*, vol. 25, no. 6, 1980, pp. 35–40.
72. V. Fouad-Hanna and D. Thebault, Analysis of asymmetrical coplanar waveguides, *Int. J. Electronics*, vol. 50, no. 3, 1981, pp. 221–224.
73. Y. Hayashi, T. Kitazawa, and S. Sasaki, Analysis of coplanar strip lines on an anisotropic substrate using Galerkin's method, *Trans. Inst. Electronics Commun. Eng. Japan*, Part E, vol. E64, no. 7, 1981, p. 14.
74. K. C. Gupta, R. Garg, and R. Chadha, *Computer Aided Design of Microwave Circuits*, Boston: Artech House, 1981.
75. D. Mirshekar-Syahkal and J. B. Davies, Accurate analysis of tapered planar transmission lines for microwave integrated circuits, *IEEE Trans. Microwave Theory Tech.*, vol. MTT-29, no. 2, 1981, pp. 123–128.
76. J. Siegl, V. Tulaja, and R. Hoffmann, General analysis of interdigitated microstrip couplers, *Siemens Forsch. Entwicklungsber.*, vol. 10, Oct. 1981, pp. 228–236.
77. G. Mur, Absorbing boundary conditions for the finite difference approximation of the time domain electromagnetic field equations, *IEEE Trans. Electromagn. Compat.*, vol. EMC 23, 1981, pp. 377–382.
78. R. A. Pucel, Design consideration for monolithic microwave circuits, *IEEE Trans. Microwave Theory Tech.*, vol. MTT-29, 1981, pp. 513–534.
79. Y. C. Shih and T. Itoh, Analysis of conductor backed coplanar waveguide, *Electronics Lett.*, vol. 18, no. 12, 1982, pp. 538–540.
80. T. Kitazawa and Y. Hayashi, Quasi static characteristics of coplanar waveguide on a sapphire substrate with its optical axis inclined, *IEEE Trans. Microwave Theory Tech.*, vol. MTT-30, no. 6, 1982, pp. 920–922.
81. R. Sorrentino and G. Leuzzi, Full wave analysis of integrated transmission lines on layered lossy media, *Electronics Lett.*, vol. 18, 1982, no. 14, pp. 607–609.
82. A. Gopinath, Losses in coplanar waveguides, *IEEE Trans. Microwave Theory Tech.*, vol. MTT-30, no. 7, 1982, pp. 1101–1104.
83. Y. Fukuoka and T. Itoh, Analysis of slow wave phenomena in coplanar waveguide on a semiconductor substrate, *Electronics Lett.*, vol. 18, no. 14, 1982, pp. 589–590.
84. B. Janiczak, Analysis of multimodes coplanar lines for microwave integrated circuits applications, in: *Conference, Proceedings 13th European Microwave Conference*, Nuernberg, D, 5–8 Sept. 1983, pp. 425–430.
85. S. M. Riad, A. A. Riad, F. W. Stephenson, M. Ahmad, A. M. Shaarawi, and K. Razzaghi, The effect of processing parameters on the characteristic impedance value of thick film coplanar waveguides, in: *Special issue on the 1983 Int. Micro-*

electronics Symposium, Philadelphia, USA, Oct. 31–Nov. 2, 1983, International Journal for Hybrid Microelectronics, vol. 6, no. 1, 1983, pp. 317–321.

86. B. Janiczak, Behaviour of guided modes in systems of parallelly located transmission lines on dielectric substrates, *Electronics Lett.*, vol. 19, no. 19, 1983, pp. 778–779.
87. R. Sorrentino, A. Silbermann, and G. Leuzzi, Characteristics of coplanar waveguides for monolithic microwave integrated circuit applications, in: *Proceedings of MELECON '83, Mediterranean Electrotechnical Conference*, Athens, GR, May 24–26, 1983, vol. 1, 1983, pp. B2.09.1–B2.09.2.
88. M. Kitlinski and B. Janiczak, Dispersion characteristics of asymmetric coupled slot lines on dielectric substrates, *Electronics Lett.*, vol. 19, no. 3, 1983, pp. 91–92.
89. I. N. Bronstein and K. A. Semendjajew, *Taschenbuch der Mathematik*, Leipzig: Harri Deutsch, 1983.
90. B. D. Milovanovic and V. V. Jankovic, Fast accurate parameters computation of coupled microstrip and coplanar strip lines, in: *Proceedings of MELECON '83. Mediterranean Electrotechnical Conference, Athens*, GR, May 24–26, 1983, vol. 1, 1983, pp. B5.06.1–B5.06.2.
91. A. M. Pavio, Hybrid mode technique yields waveguide dispersion analysis, *Microwave Syst. News, MSN*, vol. 13, no. 4, 1983, pp. 106–111.
92. G. Leuzzi, A. Silbermann, and R. Sorrentino, Mode propagation in laterally bounded conductor backed coplanar waveguides, in: *1983 IEEE MTT S International Microwave Symposium Digest*, Boston, May 31–June 3, 1983, pp. 393–395.
93. D. A. Rowe and B. Y. Lao, Numerical analysis of shielded coplanar waveguides, *IEEE Trans. Microwave Theory Tech.*, vol. MTT-31, no. 11, 1983, pp. 911–915.
94. G. Ghione and C. Naldi, Parameters of coplanar waveguides with lower ground plane, *Electronics Lett.*, vol. 19, no. 18, 1983, pp. 734–735.
95. M. N. Malyshev and I. G. Mironenko, Planar lines based on layered dielectric, *Radioelectronics Commun. Syst.*, vol. 26, no. 1, 1983, pp. 24–29.
96. S. K. Koul and B. Bhat, Simplified analysis of stripline, microstripline and coplanar strips, with anisotropic substrates for MIC and SAW applications, in: *1983 IEEE MTT S Int. Microwave Symposium Digest*, Boston, May 31–June 3, 1983, pp. 236–238.
97. Y. Fukuoka, Y. Shih, and T. Itoh, Analysis of slow-wave coplanar waveguide for monolithic integrated circuits, *IEEE Trans. Microwave Theory Tech.*, vol. MTT-31, 1983, pp. 567–573.
98. K. Koshiji, E. Shu, and S. Miki, Simplified computation of coplanar waveguide with finite conductor thickness, *IEE Proceedings, Part H (Microwaves, Optics and Antennas)*, vol. 130, no. 5, 1983, pp. 315–321.
99. D. P. Kasilingam and D. B. Rutledge, Surface wave losses of coplanar transmission lines, in: *1983 IEEE MTT S Int. Microwave Symposium Digest*, Boston, May 31–June 3, 1983, pp. 113–116.
100. S. N. Arkhanov and E. P. Timofeev, The field structure of the natural modes in screened slotted and coplanar lines, *Radioelectronics Commun. Syst.*, vol. 26, no. 8, 1983, pp. 60–62.
101. S. K. Koul and B. Bhat, Generalized analysis of microstrip like transmission lines and coplanar strips with anisotropic substrates for MAC, electrooptic modulator,

and SAW application, *IEEE Trans. Microwave Theory Tech.*, vol. MTT-31, no. 12, 1983, pp. 1051–1059.

102. Y. Fukuoka, Y. C. Shih, and T. Itoh, Analysis of slow wave coplanar waveguide for monolithic integrated circuits, *IEEE Trans. Microwave Theory Tech.*, vol. MTT-31, no. 7, 1983, pp. 567–573.
103. B. J. Janiczak, Analysis of coplanar waveguide with finite ground planes, *Arch. Elektron. Uebertragungstech.*, vol. 38, no. 5, 1984, pp. 341–342.
104. M. Ikeuchi, Boundary element analysis of shielded microstrip lines with dielectric layers, *Trans. Inst. Electronics Commun. Eng. Japan*, Part E, vol. E67, no. 11, 1984, pp. 585–590.
105. S. S. Bedair, Characteristics of some asymmetrical coupled transmission lines, *IEEE Trans. Microwave Theory Tech.*, vol. MTT-32, no. 1, 1984, pp. 108–110.
106. H. Lee and V. K. Tripathi, Generalized spectral domain analysis of planar structures having semi infinite ground planes, in: *1984 IEEE MTT S Int. Microwave Symposium Digest*, San Francisco, May 29–June 1, 1984, pp. 327–329.
107. A. M. Lerer, Losses in the conductors of coplanar waveguides, *Radio Eng. Electronic Phys.*, vol. 29, no. 7, 1984, pp. 69–74.
108. B. J. Janiczak, On the analysis of multimodes coplanar lines for microwawe integrated circuits applications, *Mikrowellen Mag.*, vol. 10, no. 2, 1984, pp. 157–165.
109. G. Ghione and C. Naldi, Analytical formulas for coplanar lines in hybrid and monolithic MICs, *Electronics Lett.*, vol. 20, no. 4, 1984, pp. 179–181.
110. K. Koshiji and E. Shu, Effect of inner conductor offset in a coplanar waveguide, *IEEE Trans. Microwave Theory Tech.*, vol. MTT-32, no. 10, 1984, pp. 1387–1390.
111. V. F. Hanna and D. Thebault, Theoretical and experimental investigation of asymmetric coplanar waveguides, *IEEE Trans. Microwave Theory Tech.*, vol. MTT-32, 1984, pp. 1649–1651.
112. D. Braess, The convergence rate of a multigrid method with Gauss–Seidel relaxation for the Poisson equation, *Math. Comput.*, vol. 42, no. 166, April 1984, pp. 505–519.
113. D. F. Williams and S. E. Schwarz, Reduction of propagation losses in coplanar waveguide, in: *1984 IEEE MTT S International Microwave Symposium Digest*, San Francisco, May 29–June 1, 1984, pp. 453–454.
114. V. Fouad-Hanna and D. Thebault, Theoretical and experimental investigation of asymmetric coplanar waveguides, *IEEE Trans. Microwave Theory Tech.*, vol. MTT-32, no. 12, 1984, pp. 1649–1651.
115. A. B. Omar and K. Schünemann, Space domain decoupling of LSE and LSM fields in generalized planar guiding structures, *IEEE Trans. Microwave Theory Tech.*, vol. MTT-32, Dec. 1984, pp. 1626–1632.
116. A. Nakatani and N. G. Alexopoulos, A generalized algorithm for the modeling of the dispersive characteristics of microstrip, inverted microstrip, stripline, slotline, finline, and coplanar waveguide circuits on anisotropic substrates, in: *1985 IEEE MTT S Int. Microwave Sympos. Digest*, St. Louis, June 4–6, 1985, pp. 457–459.
117. T. Kitazawa and Y. Hayashi, Analysis of asymmetrical coplanar waveguide and coplanar strip lines with anisotropic substrate, *Electronics Lett.*, vol. 21, no. 21, 1985, pp. 986–987.

118. T. Kitazawa and K. Mittra, Quasi-static characteristics of asymmetrical and coupled coplanar type transmission lines, *IEEE Trans. Microwave Theory Tech.*, vol. MTT-33, no. 9, 1985, pp. 771–778.

119. D. Bhattacharya, Characteristic impedance of coplanar waveguide, *Electronics Lett.*, vol. 21, no. 13, 1985, p. 57.

120. T. Kitazawa and Y. Hayashi, Coupled coplanar waveguide with anisotropic substrate, *Electronics Lett.*, vol. 21, no. 25–26, 1985, pp. 1197–1198.

121. C. Seguinot, Kadiri. M. El, P. Kennis, P. Pribetich, and J. P. Villotte, Time domain response of MIS coplanar waveguides for MMICs, *Electronics Lett.*, vol. 21, no. 25–26, 1985, pp. 1185–1186.

122. A. Nakatani and N. G. Alexopoulos, Toward a generalized algorithm for the modeling of the dispersive properties of integrated circuit structures on anisotropic substrates, *IEEE Trans. Microwave Theory Tech.*, vol. MTT-33, no. 12, 1985, pp. 1436–1441.

123. N. Alexopoulos, Integrated circuit structures on anisotropic substrates, *IEEE Trans. Microwave Theory Tech.*, vol. MTT-33, no. 10, 1985, pp. 847–881.

124. B. J. Janiczak, Multiconductor planar transmission line structures for high directivity coupler applications, in: *1985 IEEE MTT S. International Microwave Symposium Digest*, St. Louis, 4–6 June 1985, pp. 215–218.

125. W. K. Gwarek, Analysis of an arbitrarily shaped planar circuit a time domain approach, *IEEE Trans. Microwave Theory Tech.*, vol. MTT-33, 1985, pp. 1067–1072.

126. R. Hoffmann, *Microwave Integrated Circuit Handbook*, Boston: Artech House, 1985.

127. M. J. Berger, Stability of interfaces with mesh refinement, *Math. Comput.*, vol. 45, no. 172, Oct. 1985, pp. 301–318.

128. W. J. Goedherr and J. H. H. M. Potters, A compact finite difference scheme on a non-equidistant mesh, *J. Comput. Phys.*, vol. 61, 1985, pp. 269–279.

129. A. Sawicki and K. Sachse, Analysis and applications of modified coupled coplanar lines, *Electronics Lett.*, vol. 22, no. 14, 1986, pp. 746–747.

130. T. Kitazawa, Y. Hayashi, and R. Mittra, Asymmetrical coupled coplanar type transmission lines with anisotropic substrates, *IEE Proceedings, Part H (Microwaves, Optics and Antennas)*, vol. 133, no. 4, 1986, pp. 265–270.

131. T. Kitazawa and Y. Hayashi, Quasistatic characteristics of a coplanar waveguide with thick metal coating, *IEE Proceedings, Part H (Microwaves, Optics and Antennas)*, vol. 133, no. 1, 1986, pp. 18–20.

132. R. W. Jackson, Coplanar waveguide vs, microstrip for millimeter wave integrated circuits, in: *1986 IEEE MTT S Int. Microwave Symp. Digest*, Baltimore, June 2–4, 1986, pp. 699–702.

133. D. Mirshekar-Syahkal, Dispersion in shielded coupled coplanar waveguides, *Electronics Lett.*, vol. 22, no. 7, 1986, pp. 358–360.

134. M. S. Leong, P. S. Kooi, S. Prakash, and A. L. Satya, Effect of a conducting enclosure on the characteristic impedance of coplanar waveguides, *Microwave J.*, vol. 29, no. 8, 1986, pp. 105–106, 108.

135. A. J. H. Hallett, The convergence of accelerated overrelaxation iterations, *Math. Comput.*, vol. 47, no. 175, July 1986, pp. 219–223.

136. V. M. Hietala, Y. R. Kwon, and K. S. Champlin, Low loss slow wave propagation along a microstructure transmission line on a silicon surface, *Electronics Lett.*, vol. 22, no. 14, 1986, pp. 755–756.

137. D. H. Choi and W. J. R. Hoefer, The finite difference time domain method and its application to eigenvalue problems, *IEEE Trans. Microwave Theory Tech.*, vol. MTT-34, 1986, pp. 1464–1470.

138. H. Baudrand, M. Kaddour, and M. Ahmadpanah, Bias variable characteristics of coupled coplanar waveguide on GaAs substrate, *Electronics Lett.*, vol. 23, no. 4, 1987, pp. 171–172.

139. J. Fang, X. Zhang, and K. K. Mei, Dispersion characteristics of microstrip lines in the vicinity of a coplanar ground, *Electronics Lett.*, vol. 23, no. 21, 1987, pp. 1142–1143.

140. F. E. Gardiol, *Lossy Transmission Lines*, Boston: Artech House, 1987.

141. R. Majidi Ahy, B. A. Auld, K. J. Weingarten, D. M. Bloom, and M. Riaziat, Electro optic sampling measurement of coplanar waveguide (coupled slot line) modes, *Electronics Lett.*, vol. 23, no. 24, 1987, pp. 1262–1263.

142. T. S. Bird, Mutual coupling in finite coplanar rectangular waveguide arrays, *Electronics Lett.*, vol. 23, no. 22, 1987, pp. 1199–1201.

143. T. Kitazawa and Y. Hayashi, Quasistatic and hybrid mode analysis of shielded coplanar waveguide with thick metal coating, *IEE Proc., Part H (Microwaves, Optics and Antennas)*, vol. 134, no. 3, 1987, pp. 321–323.

144. M. Riaziat, I. J. Feng, B. A. Auld, and R. Majidi Ahy, Single mode operation of coplanar waveguides, *Electronics Lett.*, vol. 23, no. 24, 1987, pp. 1281–1283.

145. G. Ghione and C. U. Naldi, Coplanar waveguides for MMIC application: Effect of upper shielding, conductor backing, finite-extent ground planes, and line-to-line coupling, *IEEE Trans. Microwave Theory Tech.*, vol. MTT-35, 1987, pp. 260–267.

146. T. Kitazawa and Y. Hayashi, Variational method for coplanar waveguide with anisotropic substrates, *IEE Proc., Part H (Microwaves, Optics and Antennas)*, vol. 134, no. 1, 1987, pp. 7–10 (4 pages, 5 figures, 1 table, 12 references).

147. P. Luchini, An adaptive-mesh finite-difference solution method for the Navier–Stokes equations, *J. Comput. Phys.*, vol. 68, 1987, pp. 183–306.

148. F. Brito, H. Baudrand, and M. Ahmadpanah, Capacitance calculation in coplanar lines on semiconductor substrates, in: *MIOP '87, Mikrowellentechnologie und Optoelektronik, Kongressmesse fuer Hoechstfrequenztechnologie*, Digest, Wiesbaden, D, 19–21 Mai, 1987, vol. 1, 1987, pp. 4B.3.1–4B.3.9.

149. A. N. Sychev, Calculation of the parameters of unbalanced planar and coplanar striplines, *Telecommuni. Radio Eng.*, vol. 43, no. 3, 1988, pp. 123–127.

150. P. Pribetich, C. Seguinot, and P. Kennis, Propagation characteristics of coplanar transmission lines laid on semiconductor substrates, *Alta Frequenza*, vol. LVII, no. 7, 1988, pp. 417–430.

151. H. Shigesawa, M. Tsjui, and A. A. Oliner, Conductor backed slot line and coplanar waveguide: dangers and full wave analyses, in: *1988 IEEE MTT International Microwave Symposium Digest*, 25–27 May 1988, New York, vol. 1, 1988, pp. 199–202.

152. R. Majidi Ahy, K. J. Weingarten, M. Riaziat, D. M. Bloom, and B. A. Auld, Electrooptic sampling measurement of dispersion characteristics of slot line and copla-

nar waveguide (coupled slot line) even and odd modes, in: *1988 IEEE MTT International Microwave Symposium Digest*, 25–27 May 1988, New York, vol. 1, 1988, pp. 301–304.

153. R. Delrue, C. Seguinot, P. Pribetich, and P. Kennis, The effects of a dielectric capacitor layer and metallization on the propagation parameters of coplanar waveguide for MMIC, *IEEE Trans. Microwave Theory Tech.*, vol. MTT-36, no. 8, 1988, pp. 1285–1288.

154. D. Bourreau and P. Guillon, Characteristics of coplanar waveguides with metal coating on multilayer substrate: Application to broadband LiNbO(ind 3):Ti traveling wave light modulators/switch, in: *1988 IEEE MTT International Microwave Symposium Digest*, 25–27 May 1988, New York, vol. 2, 1988, pp. 1079–1082.

155. J. Fang and K. K. Mei, A super absorbing boundary algorithm for solving electromagnetic problems by time domain finite difference method, in: *1988 IEEE AP S International Symposium Digest*, Syracuse, NY., June 1988, pp. 472–475.

156. X. Zhang, J. Fang, K. K. Mei, and Y. Liu, Calculation of the dispersive characteristics of microstrips by the time domain finite difference method, *IEEE Trans. Microwave Theory Tech.*, vol. MTT-36, 1988, pp. 263–267.

157. W. K. Gwarek, Analysis of arbitrarily shaped two dimensional microwave circuits by finite difference time domain method, *IEEE Trans. Microwave Theory Tech.*, vol. MTT-36, 1988, pp. 738–744.

158. X. Zhang and K. K. Mei, Time domain finite difference approach to the calculation of the frequency dependent characteristics of microstrip discontinuities, *IEEE Trans. Microwave Theory Tech.*, vol. MTT-36, 1988, pp. 1775–1787.

159. R. K. Hoffmann, *Integrierte Mikrowellenschaltungen*, Berlin: Springer-Verlag, 1988 (in German).

160. S. Koike, N. Yoshida, and I. Fukai, Transient analysis of microstrip line on anisotropic substrate in three dimensional space, *IEEE Trans. Microwave Theory Tech.*, vol. MTT-36, 1988, pp. 34–43.

161. T. Shibata, T. Hayashi, and T. Kimura, Analysis of microstrip circuits using three dimensional full wave analysis in the time domain, *IEEE Trans. Microwave Theory Tech.*, vol. MTT-36, 1988, pp. 1064–1070.

162. L. Wiemer and R. H. Jansen, Reciprocity related definition of strip characteristic impedance for multiconductor hybrid mode transmission lines, *Microwave Opt. Tech. Lett.*, no. 1, 1988, pp. 22–25.

163. S. Koshizuka, Y. Oka, S. Kondo, and Y. Togo, Interpolating matrix method: a finite difference method for arbitrary arrangement of mesh points, *J. Comput. Phys.*, vol. 75, 1988, pp. 444–468.

164. N. M. Wigley, An efficient method for subtracting off singularities at corners for Laplace's equation, *J. Comput. Phys.*, vol. 78, 1988, pp. 369–377.

165. A. Benghalia, M. Ahmadpanah, and H. Baudrand, Accurate two dimensional approach for capacitance calculation in microcoplanar MES transmission lines, *Electronics Lett.*, vol. 24, no. 16, 1988, pp. 996–998.

166. S. S. Bedair and I. Wolff, Dividing the total capacitance: An approach for deriving closed form expressions for the computer-aided design of various microwave integrated circuits, in: *Proceedings MIOP'88 Conference*, paper 7A.1, Wiesbaden 2–4 March, 1988.

167. C. Paiva, A. M. Barbosa, and A. L. Topa, A new analytical method for the study of layered anisotropic waveguides, in: *MELECON '89: Mediterranean Electrotechnical Conference Proceedings*. Integrating Research, Industry and Education in Energy and Communication Engineering, 11–13 April 1989, Lisbon, Portugal, pp. 691–694.

168. G. C. Liang, Y. W. Liu, and K. K. Mei, Analysis of coplanar waveguide by the time domain finite difference method, in: *1989 IEEE MTT S International Microwave Symposium Digest*, 13–15 June, 1989, Long Beach, CA, vol. 3, 1989, pp. 1005–1008.

169. M. Drissi, V. Fouad-Hanna, and J. Citerne, Analysis of radiating end effects of symmetric and asymmetric coplanar waveguide using integral equations technique, in: *1989 IEEE MTT S International Microwave Symposium Digest*, 13–15 June 1989, Long Beach, CA, vol. 2, 1989, pp. 791–794.

170. N. Fache and D. De Zutter, Circuit parameters for single and coupled microstrip lines by a rigorous full wave space domain analysis, *IEEE Trans. Microwave Theory Tech.*, vol. MTT-37, no. 2, 1989, pp. 421–425.

171. D. Marcuse, Electrostatic field of coplanar lines computed with the point matching method, *IEEE J. Quantum Electronics*, vol. 25, no. 5, pt. 1, 1989, pp. 939–947.

172. C. T. Chang and G. A. Garcia, Crosstalk between two coplanar waveguides, *Arch. Elektron. Uebertragungstech.*, vol. 43, no. 1, 1989, pp. 55–58.

173. P. Waldow, Feldtheoretische Berechnung vielfach gekoppelter Leitungsanordnungen, *Mikrowellen & HF Magazin*, vol. 15, 1989, pp. 449–453.

174. M. F. Iskander and T. S. Lind, Electromagnetic coupling of coplanar waveguides and microstrip lines to highly lossy dielectric media, *1989 IEEE MTT S International Microwave Symposium*, 13–15 June 1989, Long Beach, CA, *IEEE Trans. Microwave Theory Tech.*, vol. MTT-37, no. 12, 1989, pp. 1910–1917.

175. S. S. Bedair and I. Wolff, Fast and accurate analytic formulas for calculating the parameters of a general broadside coupled coplanar waveguide for (M)MIC applications, *IEEE Trans. Microwave Theory Tech.*, vol. MTT-37, no. 5, 1989, pp. 843–850.

176. R. R. Boix and M. Horno, Modal quasistatic parameters for coplanar multiconductor structures in multilayered substrates with arbitrary transverse dielectric anisotropy, *IEE Proc., Part H (Microwaves, Optics and Antennas)*, vol. 136, no. 1, 1989, pp. 76–79.

177. W. Schroeder and I. Wolff, A new hybrid mode boundary integral method for analysis of MMIC waveguides with complicated cross section, in: *IEEE 1989 MTT S International Microwave Symposium Digest*, 13–15 June 1989, Long Beach, CA, vol. 2, 1989, pp. 711–714.

178. M. F. Iskander and T. S. Lind, Electromagnetic coupling of microstrip lines and coplanar waveguides to multilayer lossy media, in: *1989 IEEE MTT S International Microwave Symposium Digest*, 13–15 June 1989, Long Beach, CA, vol. 1, 1989, pp. 175–178.

179. W. Mertin, K. D. Herrmann, and E. Kubalek, Electron beam testability of monolithic microwave integrated circuits (MMIC), *Electron and Optical Beam Testing of Integrated Circuits. Proceedings of the Second European Conference*, Duisburg, D, October 1–4, 1989, *Microelectronic Eng.*, vol. 12, no. 1/4, 1990, pp. 287–293.

180. G. C. Liang, Y. W. Liu, and K. K. Mei, Full wave analysis of coplanar waveguide and slotline using the time domain finite difference method, *1989 IEEE MTT S*

International Microwave Symposium, 13–15 June 1989, Long Beach, CA, *IEEE Trans. Microwave Theory Tech.*, vol. 37, no. 12, 1989, pp. 1949–1957.

181. P. Pribetich, C. Seguinot, and P. Kennis, Modeling of propagation characteristics of coplanar transmission lines laid on semiconductor substrates, in: *MIOP '89, Microwaves and Optronics, 4th Exhibition and Conference for Ultra High Frequency Technology, Conference Proceedings*, Sindelfingen, D, 28.2. 2.3.1989, 1989, pp. 5A.1.1–5A.1.6.

182. T. Itoh, Overview of quasi planar transmission lines, *IEEE Trans. Microwave Theory Tech.*, vol. MTT-37, no. 2, 1989, pp. 275–280.

183. M. Tsuji, H. Shigesawa, and A. A. Oliner, Printed circuit waveguides with anisotropic substrates: A new leakage effect, in: *1989 IEEE MTT S International Microwave Symposium Digest*, 13–15 June 1989, Long Beach, CA, vol. 2, 1989, pp. 783–786.

184. D. Kinowski, F. Huret, P. Pribetich, and P. Kennis, Spectral domain analysis of coplanar superconducting line laid on multilayered GaAs substrate, *Electronics Lett.*, vol. 25, no. 12, 1989, pp. 788–789.

185. T. Kitazawa, Variational method for planar transmission lines with anisotropic magnetic media, *IEEE Trans. Microwave Theory Tech.*, vol. MTT-37, no. 11, 1989, pp. 1749–1754.

186. I. Wolff, D. Kiefer, and S. S. Bedair, Considering some undesired effects due to dense packaging in supported coplanar waveguide MMICs by using combined methods, in: 1989 *IEEE MTT-S International Microwave Symposium Digest*, 1989, pp. 657–660.

187. F. Alessandri, U. Goebel, F. Melai, and R. Sorrentino, Theoretical and experimental characterization of integrated millimeter wave structures, in: *MELECON '89: Mediterranean Electrotechnical Conference Proceedings*. Integrating Research, Industry and Education in Energy and Communication Engineering, 11–13 April 1989, Lisbon, Portugal, 1989, pp. 710–713.

188. F. Alessandri, U. Goebel, F. Melai, and R. Sorrentino, Theoretical and experimental characterization of nonsymmetrically shielded coplanar waveguides for millimeter wave circuits, *1989 IEEE MTT S International Microwave Symposium*, 13–15 June 1989, Long Beach, CA, *IEEE Trans. Microwave Theory Tech.*, vol. MTT-37, no. 12, 1989, pp. 2020–2027.

189. C. J. Railton and J. P. McGeehan, Analysis of microstrip discontinuities using the finite difference time domain technique, in: *1989 IEEE MTT S International Microwave Symposium Digest*, Long Beach, CA, June 1989, pp. 1009–1012.

190. R. N. Simons, G. E. Ponchak, and K. S. Martzaklis, Channelized coplanar waveguide: discontinuities, junctions, and propagation characteristics, in: *1989 IEEE MTT-S International Microwave Symposium Digest*, 1989, pp. 915–918.

191. S. Nam, H. Ling, and T. Itoh, Time domain method of lines applied to the uniform microstrip line and its step discontinuity, in: *1989 IEEE MTT S International Microwave Symposium Digest*, Long Beach, CA, June 1989, pp. 997–1000.

192. V. K. Tripathi and H. Lee, Spectral domain computation of characteristic impedances and multiport parameters of multiple coupled microstrip lines, *IEEE Trans. Microwave Theory Tech.*, vol. MTT-37, Jan. 1989, pp. 215–221.

193. M. Naghed and I. Wolff, Ein Koplanar-Streifenleitungsübergang für die Messungen mit Hilfe des Spitzenmeßplatzes, *Mikrowellen & HF Mag.*, vol. 15, 1989, pp. 261–263.

194. H. Shigesawa, M. Tsuji, and A. A. Oliner, A new mode coupling effect on coplanar waveguides of finite width, in: *1990 IEEE MTT S International Microwave Symposium Digest*, 8–10 May 1990, Dallas, TX, vol. 3, 1990, pp. 1063–1066.

195. M. F. Iskander and T. S. Lind, Analysis of coupling characteristics of coplanar waveguides and microstrip lines to multi-layer dielectric media, in: *1990 International Symposium Digest. Antennas and Propagation. Institute of Electrical and Electronics Engineers*. Merging Technologies for the 90's, 7–11 May 1990, Dallas, TX, vol. 1, 1990, pp. 302–305.

196. K. S. Kong, C. W. Kuo, T. Kitazawa, and T. Itoh, Analysis of the superconducting coplanar waveguide by combining spectral domain method and phenomenological equivalence method, *Electronics Lett.*, vol. 26, no. 19, 1990, pp. 1558–1560.

197. G. Wittum, Mehrgitterverfahren, *Spektrum der Wissenschaften*, April 1990, pp. 78–90.

198. S. S. Gevorgian and I. G. Mironenko, Asymmetric coplanar strip transmission lines for MMIC and integrated optic applications, *Electronics Lett.*, vol. 26, no. 22, 1990, pp. 1916–1918 (3 pages, 8 references).

199. T. Kitazawa and T. Itoh, Asymmetrical coplanar waveguide with finite metallization thickness containing anisotropic media, in: *1990 IEEE MTT S International Microwave Symposium Digest*, 8–10 May 1990, Dallas, TX, vol. 2, 1990, pp. 673–676.

200. Y. C. Shih and M. Maher, Characterization of conductor backed coplanar waveguide using accurate on wafer measurement techniques, in: *1990 IEEE MTT S International Microwave Symposium Digest*, 8–10 May 1990, Dallas, TX, vol, 1990, pp. 1129–1132.

201. D. S. Phatak, N. K. Das, and A. P. Defonzo, Dispersion characteristics of optically excited coplanar striplines: comprehensive full wave analysis, *IEEE Trans. Microwave Theory Tech.*, vol. MTT-38, no. 11, 1990, pp. 1719–1730.

202. W. Heinrich, Full wave analysis of conductor losses on MMIC transmission lines, *IEEE Trans. Microwave Theory Tech.*, vol. MTT-38, no. 10, 1990, pp. 1468–1472.

203. Y. Qian and E. Yamashita, Additional approximate formulas and experimental data on micro coplanar striplines (MMICs), *IEEE Trans. Microwave Theory Tech.*, vol. MTT-38, no. 4, 1990, pp. 443–445.

204. N. C. Chia, C. W. Yu, and H. C. Chun, Full wave analysis of coplanar waveguides by variational conformal mapping technique, *IEEE Trans. Microwave Theory Tech.*, vol. MTT-38, no. 9, 1990, pp. 1339–1344.

205. C. W. Yen and A. O. Jeremiah, Impedance calculations for modified coplanar waveguides, *Int. J. Electronics*, vol. 68, no. 5, 1990, pp. 861–875.

206. I. D. Robertson and A. H. Aghvami, Multi level transmission line circuits for MMICs, in: *IEE Colloquium on 'Components for Novel Transmission Lines'*, 26 March 1990, London, vol. No. 048, 1990, pp. 3/1–3/4.

207. M. Riaziat, R. Majidi Ahy, and I. J. Feng, Propagation modes and dispersion characteristics of coplanar waveguides, *IEEE Trans. Microwave Theory Tech.*, vol. MTT-38, no. 3, 1990, pp. 245–251.

208. M. Horno, F. L. Mesa, F. Medina, and R. Marques, Quasi TEM analysis of multilayered, multiconductor coplanar structures with dielectric and magnetic anisotropy including substrate losses, *IEEE Trans. Microwave Theory Tech.*, vol. MTT-38, no. 8, 1990, pp. 1059–1068.

209. M. Tsuji, H. Shigesawa, and A. A. Oliner, Theory and experiments of mode coupling and power leakage on coplanar waveguides of finite width, in: *1990 Asia Pacific Microwave Conference APMC '90*, 18–21 Sept. 1990, Tokyo, Japan, *IEICE Trans. (Japan)*, vol. E74, no. 5, 1991, pp. 1264–1269.

210. B. A. Neganov, The singular integral method of representing fields in eigenwave problems for shielded stripline slotline microwave structures, *Sov. J. Commun. Technol. Electronics*, vol. 35, no. 1, 1990, pp. 60–70.

211. H. Y. Lee and T. Itoh, Experimental and theoretical characterizations of very thin coplanar waveguide and coplanar slow wave structures, in: *1990 IEEE MTT S International Microwave Symposium Digest*, 8–10 May 1990, Dallas, TX, vol. 1, 1990, pp. 175–178.

212. D. M. Sheen, S. M. Ali, M. D. Abouzahra, and J. A. Kong, Application of the three dimensional finite difference time domain method to the analysis of planar microstrip circuits, *IEEE Microwave Theory Tech.*, vol. MTT-38, 1990, pp. 849–857.

213. N. Fache and D. De Zutter, New high frequency circuit model for coupled lossless and lossy waveguide structures, *IEEE Trans. Microwave Theory Tech.*, vol. MTT-38, March 1990, pp. 252–259.

214. G. Kibuuka, *Computation of lumped and semi-lumped elements in microstrip and coplanar techniques based on spectral domain analysis of planar lines*, Doctoral Thesis, Duisburg University, Duisburg, 1990.

215. M. E. Goldfarb and A. Platzker, Losses in GaAs microstrip, *IEEE Trans. Microwave Theory Tech.*, vol. MTT-38, 1990, pp. 1957–1963.

216. T. K. Lee, H. Ling, and T. Itoh, Boundary element characterization of coplanar waveguides, *IEEE Microwave Guided Wave Lett.*, vol. 1, no. 12, 1991, pp. 385–387 (3 pages, 8 references).

217. Y. C. Shih, Broadvol characterization of conductor backed coplanar waveguide using accurate on wafer measurement techniques, *Microwave J.*, vol. 34, no. 4, 1991, pp. 95–96,98,100,102–103,105.

218. A. K. Ganguly and C. M. Krowne, Characteristics of microstrip transmission lines with high dielectric constant substrates, *IEEE Trans. Microwave Theory Tech.*, vol. MTT-39, no. 8, 1991, pp. 1329–1337.

219. S. N. Radcliffe, A. J. Smith, and T. P. Young, Computer modelling of coplanar waveguide, in: *IEE Colloquium on 'Field Analysis of Microwave Devices and Circuits'*, 21 Jan. 1991, London, vol. 015, 1991, pp. 8/1–8/5.

220. A. Janhsen and V. Hansen, Determination of the characteristic impedance of single and coupled lines in layered dielectric media, in: *1991 IEEE MTT S International Microwave Symposium Digest*, 10–14 June 1991, Boston, vol. 2, 1991, pp. 765–768.

221. K. Wu and R. Vahldieck, Field distribution and dispersion characteristics of fundamental and higher order modes in miniature hybrid MIC (MHMIC) considering finite conductor thickness and conductivity, in: *1991 IEEE MTT S International Microwave Symposium Digest*, 10–14 June 1991, Boston, vol. 3, 1991, pp. 995–998.

222. C. N. Chang, W. C. Chang, and C. H. Chen, Full wave analysis of multilayer coplanar lines, *IEEE Trans. Microwave Theory Tech.*, vol. MTT-39, no. 4, 1991, pp. 747–750.

223. W. H. Haydl, T. Kitazawa, J. Braunstein, R. Bosch, and M. Schlechtweg, Millimeterwave coplanar transmission lines on gallium arsenide, indium phosphide and quartz with finite metalization thickness, in: *1991 IEEE MTT S International Microwave Symposium Digest*, 10–14 June 1991, Boston, vol. 2, 1991, pp. 691–694.

224. M. Tsuji, H. Shigesawa, and A. A. Oliner, New interesting leakage behavior on coplanar waveguides of finite and infinite widths, in: *1991 IEEE MTT S International Microwave Symposium Digest*, 10–14 June 1991, Boston, MA, vol. 2, 1991, pp. 563–566.

225. L. Zhu and E. Yamashita, New method for the analysis of dispersion characteristics of various planar transmission lines with finite metallization thickness, *IEEE Microwave Guided Wave Lett.*, vol. 1, no. 7, 1991, pp. 164–166.

226. T. Kitazawa, C. W. Kuo, K. S. Kong, and T. Itoh, Planar transmission lines with finitely thick conductors and lossy substrates, in: *1991 IEEE MTT S International Microwave Symposium Digest*, 10–14 June 1991, Boston, MA, vol. 2, 1991, pp. 769–772.

227. P. Pribetich, C. Seguinot, and P. Kennis, Systematic determination of the propagation characteristics of coplanar lines on semiconductor substrate, *IEEE Trans. Microwave Theory Tech.*, vol. MTT-39, no. 7, 1991, pp. 1083–1089.

228. R. Bromme and R. H. Jansen, Systematic investigation of coplanar waveguide MIC/MMIC structures using a unified strip/slot 3D electromagnetic simulator, in: *1991 IEEE MTT S International Microwave Symposium Digest*, 10–14 June 1991, Boston, vol. 3, 1991, pp. 1081–1084.

229. M. Drissi, V. Fouad-Hanna, and J. Citerne, Analysis of coplanar waveguide radiating end effects using the integral equation technique, *IEEE Trans. Microwave Theory Tech.*, vol. MTT-39, no. 1, 1991, pp. 112–116.

230. W. von. Wendorff, M. Stopka, and D. Jäger, Electrooptic sampling of microwave signals on MSM coplanar lines, in: *21st European Microwave Conference, Workshop Proceedings*, Stuttgart, D, 13 September 1991, pp. 52–57.

231. S. Gupta, J. F. Whitaker, and G. A. Mourou, Subpicosecond pulse propagation on coplanar waveguides: Experiment and simulation, *IEEE Microwave and Guided Wave Lett.*, vol. 1, no. 7, 1991, pp. 161–163.

232. K. Wu and R. Vahldieck, Hybrid mode analysis of homogeneously and inhomogeneously doped low loss slow wave coplanar transmission lines, *IEEE Trans. Microwave Theory Tech.*, vol. MTT-39, no. 8, 1991, pp. 1348–1360.

233. I. Wolff and M. Rittweger, Finite difference time domain analysis of planar microwave circuits, *Arch. Elektrotech.*, vol. 74, 1991, pp. 189–201.

234. C. N. Chang, Y. C. Wong, and C. H. Chen, Hybrid quasistatic analysis for multilayer coplanar lines, *IEE Proc., Part H (Microwaves, Antennas and Propagation)*, vol. 138, no. 4, 1991, pp. 307–312.

235. T. Becks and I. Wolff, Improvements of spectral domain analysis techniques for arbitrary planar circuits, in: *Directions in Electromagnetic Wave Modeling*, edited by H. L. Bertoni and L. B. Felsen, New York: Springer Verlag, 1991, pp. 339–346.

236. Yi Ke Jeng, I. Sheng Tsai, and H. C. Chun, Dispersion and leakage characteristics of coplanar waveguides, *IEEE Trans. Microwave Theory Tech.*, vol. MTT-40, no. 10, 1992, pp. 1970–1973.

237. U. D. Keil, D. R. Dykaar, A. F. J. Levi, R. F. Kopf, L. N. Pfeiffer, S. B. Darack, and K. W. West, High speed coplanar transmission lines, *IEEE J. Quantum Electronics*, vol. 28, no. 10, 1992, pp. 2333–2342.

238. J. R. Kessler and P. D. Coleman, Impedance and complex propagation constant measurements at W band on planar circuits, *Int. J. Infrared Millimeter Waves*, vol. 13, no. 4, 1992, pp. 397–424.

239. Y. Qian, E. Yamashita, and K. Atsuki, Modal dispersion control and distortion suppression of picosecond pulses in suspended coplanar waveguides, *IEEE Trans. Microwave Theory Tech.*, vol. MTT-40, no. 10, 1992, pp. 1903–1909.

240. M. A. Magerko, L. Fan, and K. Chang, Multiple dielectric structures to eliminate moding problems in conductor backed coplanar waveguide MIC, *IEEE Microwave Guided Wave Lett.*, vol. 2, no. 6, 1992, pp. 257–259.

241. T. S. Lind and M. F. Iskander, On the coupling characteristics of coplanar waveguides and microstrip lines to multilayer dielectric media, *IEEE Trans. Electromagn. Compat.*, vol. 34, no. 2, 1992, pp. 117–123.

242. F. L. Mesa, G. Cano, F. Medina, R. Marques, and M. Horno, On the quasi TEM and full wave approaches applied to coplanar multistrip on lossy dielectric layered media, *IEEE Trans. Microwave Theory Tech.*, vol. MTT-40, no. 3, 1992, pp. 524–531.

243. S. Koßlowski, *A contribution to the design of monolithic integrated microwave circuits on GaAs substrate in coplanar technology*, Doctoral Thesis, Duisburg University, 1992.

244. T. Rozzi, G. Gerini, and P. Sewell, Discontinuities and radiation in planar circuits, in: *9th MM 92 Conference*, Brighton, GB, 14–15 Oct 1992, pp. 17–22.

245. W. H. Haydl, Experimentally observed frequency variation of the attenuation of millimeter wave coplanar transmission lines with thin metallization, *IEEE Microwave Guided Wave Lett.*, vol. 2, no. 8, 1992, pp. 222–324.

246. S. S. Bedair and I. Wolff, Fast, accurate and simple approximate analytic formulas for calculating the parameters of supported coplanar waveguides for (M)MIC's, *IEEE Trans. Microwave Theory Tech.*, vol. MTT-40, no. 1, 1992, pp. 41–48.

247. M. Rittweger, *Simulation transienter elektrodynamischer Ausbreitungsphänomene zur Analyse der Übertragungseigenschaften von Systemen der Mikro und Millimeterwellentechnik*, Doctoral Thesis, Duisburg University, Duisburg, Germany, 1992.

248. M. Naghed, *Analyse koplanarer Mikrowellenstrukturen mit der Methode der quasistatischen Finiten Differenzen*, Doctoral Thesis, Duisburg University, Duisburg, Germany, 1992.

249. K. K. Mei and J. Fang, Superabsorption a method to improve absorbing boundary conditions, *IEEE Trans. Antennas Propag.*, vol. 40, 1992, pp. 1001–1010.

250. I. Wolff, Finite difference time domain simulation of electromagnetic fields and microwave circuits, *Int. J. Numer. Modelling*, vol. 5, Aug. 1992, pp. 163–182.

251. K. K. M. Cheng and J. K. A. Everard, A new technique for the quasi-TEM analysis of conductor backed coplanar waveguide structures, *IEEE Trans. Microwave Theory Tech.*, vol. MTT-41, no. 9, 1993, pp. 1589–1592.

252. M. Gillick, I. D. Robertson, and J. S. Joshi, An analytical method for direct calculation of E & H field patterns of conductor backed coplanar waveguides, *IEEE Trans. Microwave Theory Tech.*, vol. MTT-41, no. 9, 1993, pp. 1606–1610.

253. Y. Chen and B. Becker, Analysis of bilateral coplanar waveguides printed on anisotropic substrates for use in monolithic MICs, *IEEE Trans. Microwave Theory Tech.*, vol. MTT-41, no. 9, 1993, pp. 1489–1493.

254. R. W. Jackson, Circuit model for substrate resonance coupling in grounded coplanar waveguide circuits, *IEEE Trans. Microwave Theory Tech.*, vol. MTT-41, no. 9, 1993, pp. 1641–1645.

255. W. Heinrich, Design of coplanar waveguides for mm wave MMIC's, in: *MIOP 92, Mikrowellen und Optronik, 7. Kongreßmesse f. Höchstfrequenztechnik*, Sindelfingen, D, 25–27. Mai, 1993, pp. 76–80.

256. M. Gillick, I. D. Robertson, and J. S. Joshi, Direct analytical solution for the electric field distribution at the conductor surfaces of coplanar waveguides, *IEEE Trans. Microwave Theory Tech.*, vol. MTT-41, no. 1, 1993, pp. 129–135.

257. C. Nguyen, Dispersion characterisitics of the broadside coupled coplanar waveguide, *IEEE Trans. Microwave Theory Tech.*, vol. MTT-41, no. 9, 1993, pp. 1630–1633.

258. C. C. Tien, C. K. C. Tzuang, and S. T. Peng, Effect of finite width backside plane on overmoded conductor backed coplanar waveguide, *IEEE Microwave Guided Wave Lett.*, vol. 3, no. 8, 1993, pp. 259–261.

259. C. C. Tien, C. K. C. Tzuang, and J. Monroe, Effect of lateral walls on the propagation characteristics of finite width conductor backed coplanar waveguides, *Electronics Lett.*, vol. 29, no. 15, 1993, pp. 1357–1358.

260. M. R. Lyons, J. P. K. Gilb, and C. A. Balanis, Enhanced dominant mode operation of a shielded multilayer coplanar waveguide via substrate compensation, *IEEE Trans. Microwave Theory Tech.*, vol. MTT-41, no. 9, 1993, pp. 1564–1567.

261. M. Helal, J. F. Legier, E. Paleczny, P. Pribetich, and P. Kennis, Experimental characterization of under skin depth thickness strips of planar lines for MMIC applications, in: *MIOP '93, Mikrowellen und Optronik, 7. Kongreßmesse f. Höchstfrequenztechnik*, Sindelfingen, D, 25–27. May, 1993, pp. 373–377.

262. E. Drake, F. Medina, and M. Horno, Improved quasi TEM spectral domain analysis of boxed coplanar multiconductor microstrip lines, *IEEE Trans. Microwave Theory Tech.*, vol. MTT-41, no. 2, 1993, pp. 260–267.

263. Liu Yaozhong and T. Itoh, Leakage phenomena in multilayered conductor backed coplanar waveguides, IEEE *Microwave Guided Wave Lett.*, vol. 3, no. 11, 1993, pp. 427–427.

264. W. Heinrich, Quasi TEM description of MMIC coplanar lines including conductor loss effects, *IEEE Trans. Microwave Theory Tech.*, vol. MTT-41, no. 1, 1993, pp. 45–52.

265. W. T. Lo, C. K. C. Tzuang, S. T. Peng, C. C. Tien, C. C. Chang, and J. W. Huang, Resonant phenomena in conductor backed coplanar waveguides (CBCPW's), *IEEE Trans. Microwave Theory Tech.*, vol. MTT-41, no. 12, 1993, pp. 2099–2108.

266. C. C. Tien, C. K. C. Tzuang, S. T. Peng, and C. C. Chang, Transmission characteristics of finite width conductor backed coplanar waveguide, *IEEE Trans. Microwave Theory Tech.*, vol. MTT-41, no. 9, 1993, pp. 1616–1624.

267. M. Gillick and I. D. Robertson, Ultra low impedance CPW transmission lines for multilayer MMIC's, in: *1993 IEEE Microwave and Millimeter Wave Monolithic Circuits Symp.*, Atlanta, GA, June 14–15, 1993, pp. 127–130.

268. M. Gillick and I. D. Robertson, Ultra low impedance CPW transmission lines for multilayer MMIC's, in: *1993 IEEE MTT S Int. Microwave Symp. Digest*, Vol. 1, Atlanta, GA, June 14–18, vol. 1, 1993, pp. 145–148.

269. G. Merceur Maze, S. Tadjini, and J. L. Bonnefoy, Analysis of a CPW on electric and magnetic biaxial substrate, *IEEE Trans. Microwave Theory Tech.*, vol. MTT-41, no. 3, 1993, pp. 457–461.

270. A. A. Omar and Y. L. Chow, Coplanar waveguide with top and bottom shields in place of air bridges, *IEEE Trans. Microwave Theory Tech.*, vol. MTT-41, no. 9, 1993, pp. 1559–1563.

271. G. Ghione, A CAD oriented analytical model for the losses of general asymmetric coplanar lines in hybrid and monolithic MICs, *IEEE Trans. Microwave Theory Tech.*, vol. MTT-41, no. 9, 1993, pp. 1499–1510.

272. R. Kulke and T. Sporkmann, Coplanar waveguide elements for a European CAD environment, in: *Proc. 23rd European Microwave Conference*, Madrid, Spain, Sept. 6–9, 1993, vol. 23, pp. 209–211.

273. M. S. Islam, E. Tuncer, and D. P. Neikirk, Calculation of conductor loss in coplanar waveguide using conformal mapping., *Electronics Lett.*, vol. 29, no. 13, 1993, pp. 1189–1191.

274. T. Becks, *Elektrodynamische Simulation von passiven dreidimensionalen Komponenten in (M)MIC Schaltungen mit dem Spektralbereichsverfahren*, Doctoral Thesis, Duisburg University, Duisburg, Germany, 1993.

275. X. P. Lin and K. Naishadham, A new computationally efficient method for the analysis of planar transmission lines and complex MMIC elements, in: *NAECON 1994, 1994 IEEE National Aerospace and Electronics Conf.*, Vol. 2, Dayton, OH, May 23–27, 1994, pp. 983–990.

276. S. S. Gevorgian, Basic characteristics of two layered substrate coplanar waveguides, *Electronics Lett.*, vol. 30, no. 15, 1994, pp. 1236–1237.

277. S. Hofschen and I. Wolff, Berechnung der Wellenausbreitungseigenschaften supraleitender koplanarer Wellenleiter mit Hilfe der Methode der zweidimensionalen Finiten Differenzen im Zeitbereich, in: *Supraleitung und Tieftemperaturtechnik, zum Statusseminar, Berichte zu F&E Projekten aus dem Förderbereich Physikalische Technol. des Bundesministeriums für Forschung und Technol.*, Weimar, D, 13–14 June 1994, pp. 137–140.

278. W. A. Artuzi Jr., and T. Yoneyama, Characterization and measurements of laterally shielded coplanar waveguide at millimeter wavelengths, *IEEE Trans. Microwave Theory Tech.*, vol. 42, no. 1, 1994, pp. 150–153.

279. R. Kulke, T. Sporkmann, D. Köther, I. Wolff, and P. Pogatzki, Coplanar elements support circuit designs to 67 GHz, Part 1, *Microwaves & RF*, vol. 33, no. 13, 1994, pp. 103–104, 106, 108–109, 112–114, 116.

280. P. Pogatzki and O. Kramer, A coplanar element library for the accurate CAD of (M)MICs, *Microwave Eng. Eur.*, no. Dec./Jan., 1994, pp. 41, 42, 45–46.

281. M. Y. Frankel, R. H. Voelker, and J. N. Hilfiker, Coplanar transmission lines on thin substrates for high speed low loss propagation, *IEEE Trans. Microwave Theory Tech.*, vol. 42, no. 3, 1994, pp. 396–402.

282. D. Mirshekar-Syahkal and J. Danneel, Criteria for single mode operation of packaged coplanar waveguide circuits, in: Modelling, Design and Application of MMIC's, *IEE Colloquium*, London, GB, June 17, 1994, IEE Colloquium, vol. 92, 1994, pp. 8/1–8/4.

283. S. Alexandrou, C. C. Wang, M. Curie, R. Sobolewski, and T. Y. Hsiang, Loss and dispersion at subterahertz frequencies in coplanar waveguides with varying ground plane widths, in: *Technol. for Optical Fiber Communications*, Los Angeles, CA, Jan 25, 1994, *Proc.*, *SPIE*, vol. 2149, 1994, pp. 108–118.

284. R. R. Kumar, S. Aditya, and D. Chadha, Modes of a shielded conductor backed coplanar waveguide, *Electronics Lett.*, vol. 30, no. 2, 1994, pp. 146–148.

285. K. K. M. Cheng and I. D. Robertson, Numerically efficient spectral domain approach to the quasi TEM analyss of supported coplanar waveguide structures, *IEEE Trans. Microwave Theory Tech.*, vol. 42, no. 10, 1994, pp. 1958–1965.

286. S. Hofschen and I. Wolff, Improvements of the 2-D-FDTD method for the simulation of small CPW's on GaAs using time series analysis, in: *1994 IEEE MTT-S International Microwave Sympos. Digest*, San Diego, CA, May 1994, pp. 39–42.

287. X. Shujun and R. Vahldieck, Signal propagation in conductor backed superconductor coplanar waveguides, in: *Proceedings, 24th European Microwave Conference*, Cannes, F, Sept. 5–8, 1994, vol. 1, pp. 384–389.

288. A. Görür, C. Karpuz, and M. J. Lancaster, Modified coplanar meander transmission line for MMICs, *Electronics Lett.*, vol. 30, no. 16, 1994, pp. 1317–1318.

289. K. M. Rahman and C. Nguyen, Spectral domain analysis of eigenmodes in a shielded multilayer coplanar waveguide for microwave and millimeter wave integrated circuits, *Int. J. Infrared Millimeter Waves*, vol. 15, no. 7, 1994, pp. 1297–1314.

290. K. Wu, Y. Xu, and R. G. Bosisio, Theoretical and experimental analysis of channelized coplanar waveguides (CCPW) for wideband applications of integrated microwave and millimeter wave circuits, *IEEE Trans. Microwave Theory Tech.*, vol. 42, no. 9, part 1, 1994, pp. 1651–1659.

291. H. Klingbeil and W. Heinrich, Calculation of CPW A.C. resistance and inductance using a quasi static mode matching approach, *IEEE Trans. Microwave Theory Tech.*, vol. 42, no. 6, 1994, pp. 1004–1007.

292. A. Görür, A novel coplanar slow wave structure, *IEEE Microwave and Guided Wave Lett.*, vol. 4, no. 3, 1994, pp. 76–88.

293. J. Y. Ke and C. H. Chen, Dispersion and attenuation characteristics of coplanar waveguides with finite metallization thickness and conductivity, *IEEE Trans. Microwave Theory Tech.*, vol. 43, no. 5, 1995, pp. 1128–1135.

294. W. Tongqing and K. Wu, Dynamic analysis of uniplanar guided wave structures with trapezoidal conductor profile and microshielding enclosure, *IEICE Trans. Electronics*, vol. E78 C, no. 8, 1995, pp. 1100–1105.

295. A. K. Rastogi, N. J. McEwan, I. U. Khairuddin, and A. Z. Jakal, Effect of substrate thickness and metallization thickness on dispersion characteristics of CPW, *Int. J. Infrared Millimeter Waves*, vol. 16, no. 8, 1995, pp. 1393–1406.

296. L. Yaozhong and T. Itoh, Four layered coplanar waveguide with double side conductor backing, in: *Electrical Performance of Electronic Packaging, IEEE 4th Topical Meeting*, Portland, OR, Oct. 2–4, 1995, pp. 188–190.

297. A. K. Rastogi, A. Z. Jakal, N. J. McEwan, and I. U. Khairuddin, Loss minimisation and sensitivity analysis for coplanar waveguide, *Int. J. Infrared Millimeter Waves*, vol. 16, no. 10, 1995, pp. 1789–1810.

298. P. C. Hsu and C. Nguyen, New multilayer planar transmission lines for microwave and millimeter wave integrated circuits, *IEEE Trans. Microwave Theory Tech.*, vol. 43, no. 8, 1995, pp. 1809–1813.

299. Y. Liu, K. Cha, and T. Itoh, Non leaky coplanar (NLC) waveguides with conductor backing, *IEEE Trans. Microwave Theory Tech.*, vol. 43, no. 5, 1995, pp. 1067–1072.

300. K. M. Rahman and C. Nguyen, On the computation of complex modes in lossless shielded asymmetric coplanar waveguides, *IEEE Trans. Microwave Theory Tech.*, vol. 43, no. 12, pt. 1, 1995, pp. 2713–2716.

301. S. C. Wu and H. Grebel, Phase shifts in coplanar waveguides with patterned conductive top covers, *J. Phys. D: Appl. Phys.*, vol. 28, no. 2, 1995, pp. 437–439.

302. K. K. M. Cheng and I. D. Robertson, Quasi TEM analysis of V shaped conductor backed coplanar waveguide, *IEEE Trans. Microwave Theory Tech.*, vol. 4, no. 8, 1995, pp. 1992–1994.

303. T. Roy, T. K. Sarkar, and S. Madhavan, Surface integral formulation for calculating conductor and dielectric losses of various transmission structures, *IEEE Trans. Microwave Theory Tech.*, vol. 43, no. 1, 1995, pp. 176–185.

304. D. Budimir, Q. H. Wang, A. A. Rezazadeh, and I. D. Robertson, V shaped CPW transmission lines for multilayer MMICs, *Electronics Lett.*, vol. 31, no. 22, 1995, pp. 1928–1930.

305. R. Kulke, T. Sporkmann, D. Köther, I. Wolff, and P. Pogatzki, Modeling and analysis aid coplanar designs, part 2, *Microwaves & RF*, vol. 34, no. 1, 1995, pp. 89–90, 92, 95–96.

306. R. Kulke, T. Sporkmann, D. Köther, I. Wolff, and P. Pogatzki, Examine the applications of coplanar circuits, part 3, *Microwaves & RF*, vol. 34, no. 2, 1995, pp. 112, 114, 116–117.

307. G. David, W. Schroeder, D. Jäger, and I. Wolff, 2D electro optic probing combined with field theory based multimode wave amplitude extraction: A new approach to on wafer measurement, in: *1995 IEEE MTT S Internat. Microwave Symposium, Digest*, vol. 3, Orlando, FL, May 16–20, 1995, vol. 3, pp. 1049–1052.

308. L. Duvillaret, J. M. Lourtioz, and L. Chusseau, Absolute voltage measurements on III V integrated circuits by internal electro optic sampling, *Electronics Lett.*, vol. 31, no. 1, 1995, pp. 23–24.

309. W. Tongqing and W. Ke, Dynamic analysis of uniplanar guided wave structures with trapezoidal conductor profile and microshielding enclosure, *IEICE Trans. Electronics*, vol. E78 C, no. 8, 1995, pp. 1100–1105.

310. G. David, W. Schroeder, D. Jäger, and I. Wolff, 2D electro optic probing in combination with field theoretical multimode extraction technique: A new approach to on wafer measurements, in: *1995 IEEE MTT S International Microwave Symposium Digest*, Orlando, May 1995, pp. 1049–1052.

311. C. L. Holloway and E. F. Kuester, A quasi closed form expression for the conductor loss of CPW lines, with an investigation of edge shape effects, *IEEE Trans. Microwave Theory Tech.*, vol. 43, no. 12, pt. 1, 1995, pp. 2695–2701.

312. T. Q. Deng, M. S. Leong, and P. S. Kooi, Accurate and simple closed form formulas for coplanar waveguide synthesis, *Electronics Lett.*, vol. 31, no. 23, 1995, pp. 2017–2019.

313. S. Gevorgian, L. J. P. Linner, and E. L. Kollberg, CAD models for shielded multilayered CPW, *IEEE Trans. Microwave Theory Tech.*, vol. 43, no. 4, part I, 1995, pp. 772–779.

314. U. Bhattacharya, S. T. Allen, and M. J. Rodwell, DC-715 GHz sampling circuits and subpico-second nonlinear transmission lines using elevated coplanar waveguide, *IEEE Microwave Guided Wave Lett.*, vol. 5, Feb. 1995, pp. 50–52.

315. K. K. M. Cheng, Analysis and synthesis of coplanar coupled lines on substrates of finite thicknesses, *IEEE Trans. Microwave Theory Tech.*, vol. 44, no. 4, 1996, pp. 636–639.

316. N. H. Zhu, W. Qiu, E. Y. B. Pun, and P. S. Chung, Analysis of two layer planar transmission lines with the point matching method, *Int. J. Electronics*, vol. 80, no. 1, 1996, pp. 99–105.

317. C. Rieckmann, A. Jostingmeier, and A. S. Omar, Application of the eigenmode transformation technique for the analysis of planar transmission lines, *1996 IEEE MTT S International Microwave Symposium/Exhibition: "Bridging the Spectrum"*, 18–20 June 1996, San Francisco, CA, *IEEE Trans. Microwave Theory Tech.*, vol. 44, no. 12, pt. 2, 1996, pp. 2479–2486.

318. R. R. Kumar, S. Aditya, and D. Chadha, Dielectric loaded shielded edge coupled conductor backed coplanar waveguide structures, *Arch. Elektron. Übertragungstech.*, vol. 50, no. 6, 1996, pp. 384–388.

319. A. C. Polycarpou, M. R. Lyons, and C. A. Balanis, Finite element analysis of MMIC waveguide structures with anisotropic substrates, *IEEE Trans. Microwave Theory Tech.*, vol. 44, no. 10, pt. 1, 1996, pp. 1650–1663.

320. D. Kremer and R. Pregla, MOLCAR a very efficient and accurate design tool for the hybrid analysis of multilayered waveguide structures, in: *Integrated Optics Devices: Potential for Commercialization*, San Jose, CA, Feb. 12–14, 1996, *Proc. SPIE*, vol. 2997, 1996, pp. 255–262.

321. S. G. Garcia, T. M. Bao Hung, R. G. Martin, and B. G. Olmedo, On the application of finite methods in time domain to anisotropic dielectric waveguides, *IEEE Trans. Microwave Theory Tech.*, vol. 44, no. 12, pt. 1, 1996, pp. 2195–2206.

322. F. K. Jean, Quasi TEM analysis of coplanar waveguides with an inhomogeneous semiconductor substrate, *IEEE Trans. Microwave Theory Tech.*, vol. 44, no. 9, 1996, pp. 1586–1589.

323. N. Cam, Unsymmetrical broadside coupled coplanar waveguides for loose and tight coupling MIC and MMIC applications, *Int. J. Infrared Millimeter Waves*, vol. 17, no. 8, 1996, pp. 1321–1328.

324. R. Schmidt, *Vollwellenanalyse von verlustbehafteten koplanaren Leitungen und Leitungsdiskontinuitäten*, Doctoral Thesis, 1996, pp. 1–134.

325. A. H. Hamade, A. B. Kouki, and F. M. Ghannouchi, A CAD suitable approach for the analysis of nonuniform MMIC and MHMIC transmission lines, *IEEE Trans. Microwave Theory Tech.*, vol. 44, no. 9, 1996, pp. 1614–1617.

326. G. E. Ponchak, S. Robertson, F. Brauchler, J. East, and L. P. B. Katehi, Finite width coplanar waveguide for microwave and millimeter wave integrated circuits, in:

ISHM 96, 1996 International Symposium on Microelctronics, Minneapolis, MN, Oct. 8–10, 1996, *Proc. SPIE*, vol. 2920, 1996, pp. 517–521.

327. G. David, *Höchstfrequenz Charakterisierung von monolithisch integrierten Mikrowellenbauelementen und schaltungen durch zweidimensionale elektrooptische Feldverteilungsmessungen*, Doctoral Thesis, Duisburg University, Duisburg, Germany, 1996.

328. G. David, R. Tempel, I. Wolff, and D. Jäger, In circuit electro-optic field mapping for function test and characterization of MMICs, in: *1996 IEEE MTT S International Microwave Symposium Digest*, June 1996, pp. 1533–1536.

329. T. Q. Deng, M. S. Leong, P. S. Kooi, and T. S. Yeo, Synthesis formulas for coplanar lines in hybrid and monolithic MICs., *Electronics Lett.*, vol. 32, no. 24, 1996, pp. 2253–2254.

330. S. Hofschen and I. Wolff, Simulation of an elevated coplanar waveguide using 2 D FDTD, *IEEE Microwave Guided Wave Lett.*, vol. 6, no. 1, 1996, pp. 28–30.

331. Z. R. Bu, V. F. Fusco, J. A. C. Stewart, Y. Wu, H. S. Gamble, B. M. Armstrong, and N. B. Buchanan, Characteristics of trenched coplanar waveguide for SiMMIC applications, in: *1997 IEEE MTT S International Microwave Symposium Digest*, 8–13 June 1997, Denver, CO, vol. 2, 1997, pp. 735–738.

332. T. Q. Deng, M. S. Leong, P. S. Kooi, and T. S. Yeo, Synthesis formulas simplify coplanar waveguide design., *Microwaves & RF*, vol. 36, no. 3, 1997, pp. 84, 86, 91–92, 94, 96, 98.

333. K. K. M. Cheng, Characteristics parameters of symmetrical triple coupled CPW lines, *Electronics Lett.*, vol. 33, no. 8, 1997, pp. 685–687.

334. P. H. Ic, K. L. Yong, and K. P. Han, Dispersion characteristics of a broadside coupled coplanar waveguide, *Electronics Lett.*, vol. 33, no. 11, 1997, pp. 965–966.

335. K. K. M. Cheng, Effect of conductor backing on the line to line coupling between parallel coplanar lines, *IEEE Trans. Microwave Theory Tech.*, vol. 45, no. 7, 1997, pp. 1132–1134.

336. E. Carlsson and S. Gevorgian, Effect of enhanced current crowding in a CPW with a thin ferroelectric film, *Electronics Lett.*, vol. 33, no. 2, 1997, pp. 145–146.

337. V. Fouad-Hanna, O. Picon, and S. Visan, Finite difference time domain characterization of coplanar waveguide structures, *Arch. Elektron. Übertragungstech.*, vol. 51, no. 1, 1997, pp. 40–47.

338. S. Gevorgian, T. Martinsson, A. Deleniv, E. Kollberg, and I. Vendik, Simple and accurate dispersion expression for the effective dielectric constant of coplanar waveguides, *IEE Proc. Microwaves Antennas Propag.*, vol. 144, no. 2, 1997, pp. 145–148.

339. J. S. Ko, B. K. Kim, and K. Lee, Simple modeling of coplanar waveguide on thick dielectric over lossy substrate, *IEEE Trans. Electron Devices*, vol. 44, no. 5, 1997, pp. 856–861.

340. Y. Gao, *A near field measurement system for measuring electric and magnetic fields in planar high frequency circuits*, Doctoral Thesis. Fortschrittberichte VDI, Reihe 21: Elektrotechnik, vol. 244, 1998, pp. 1–158.

341. N. Baghlani and S. N. Safavl, Analysis of a coplanar waveguide with finite metallization thickness on an anisotropic multilayer substrate using method of lines, in: *IEEE Antennas and Propagation Society International Symposium. 1998 Digest*,

in conjunction with *USNC/URSI National Radio Science Meeting*, 21–26 June 1998, Atlanta, GA, vol. 3.

342. W. S. Jyh and D. L. Yu, Propagation characteristics of the slotline first higher order mode, *IEEE Trans. Microwave Theory Tech.*, vol. 46, no. 11, pt. 1, 1998, pp. 1774–1781.
343. M. Birk, H. Kibbel, C. Warns, A. Trasser, and H. Schumacher, All silicon nonlinear transmission line, in: *43rd International Wissenschaftliches Kolloquium*, vol. 3, Ilmenau, D, 21–24 Sept. 1998, pp. 334–339.
344. W. L. Jae, P. H. Ic, H. Y. Tae, and K. P. Han, Quasi static analysis of conductor backed coupled coplanar waveguide., *Electronics Lett.*, vol. 34, no. 19, 1998, pp. 1861–1862.
345. V. Milanovic, M. Ozgur, D. C. DeGroot, J. A. Jargon, M. Gaitan, and M. E. Zaghloul, Characterization of broad band transmission for coplanar waveguides on CMOS silicon substrates, *IEEE Trans. Microwave Theory Tech.*, vol. 46, no. 5, pt. 2, 1998, pp. 632–640.
346. M. N. Choong and S. K. Young, Coplanar waveguides on silicon substrate with thick oxidized porous silicon (OPS) layer, *IEEE Microwave Guided Wave Lett.*, vol. 8, no. 11, 1998, pp. 369–371.
347. A. Jrad, P. Ferrari, J. W. Tao, C. Fuchs, A. Dominjon, G. Angenieux, and J. L. Coutaz, Choice of CPW characteristic impedance for lossy nonlinear transmission lines synthesis, *Electronics Lett.*, vol. 35, no. 12, 1999, pp. 985–986.
348. M. Ribo, J. de la Cruz, and L. Pradell, Circuit model for mode conversion in coplanar waveguide asymmetric series impedances, *Electronics Lett.*, vol. 35, no. 21, 1999, pp. 1851–1853.
349. M. Ribo and L. Pradell, Circuit model for mode conversion in coplanar waveguide asymmetric shunt impedances, *Electronics Lett.*, vol. 35, no. 9, 1999, pp. 713–715.
350. A. K. Rastogi and S. Mishra, Coplanar waveguide characterization with thick metal coating, *Int. J. Infrared Millimeter Waves*, vol. 20, no. 3, 1999, pp. 505–520.
351. G. Ghione, M. Goano, G. Madonna, C. Omegna, M. Pirola, S. Bosso, D. Frassati, and A. Perasso, Microwave modeling and characterization of thick coplanar waveguides on oxide coated lithium niobate substrates for electro optical applications, in: *1999 IEEE MTT S International Microwave Symposium Digest*, 13–19 June, Anaheim, CA, vol. 3, 1999, pp. 1007–1010.
352. J. F. Shao and S. W. Bai, Analysis of asymmetric coplanar waveguide with conductor backing, *IEEE Trans. Microwave Theory Tech.*, vol. 47, no. 2, 1999, pp. 238–240.
353. M. Kunze and W. Heinrich, Modified FD formulation for conductor loss calculation in MMIC coplanar waveguides, in: *MIOP 99, 29th European Microwave Conf., Conf. Proc.*, vol. 2, Munich, D, Oct. 5–7, 1999, pp. 423–426.
354. G. E. Ponchak, RF transmission lines on silicon substrates, in: *GAAS 99 Conf. Proc., 1999 European Gallium Arsenide and Related III V Compounds, Applications Symp.*, Munich, D, 4–5 Oct. 1999, pp. 414–417.
355. C. Karpuz and A. Gorur, Slow wave phenomena in modified coplanar waveguides, *Electronics Lett.*, vol. 35, no. 4, 1999, pp. 309–311.
356. M. Tanabe, M. Nishitsuji, Y. Anda, and Y. Ota, A low impedance coplanar waveguide using an SrTiO(ind 3) thin film for GaAs power MMIC's, *IEEE Trans. Microwave Theory Tech.*, vol. 48, no. 5, 2000, pp. 872–874.

357. F. Schnieder and W. Heinrich, Low dispersive coplanar waveguides and thin film microstrip lines for sub mm wave monolithic integration, in: *THz Conference 2000, 8th International Conference on Terahertz Electronics*, Darmstadt, D, 28–29 Sept. 2000, pp. 165–168.

358. L. Vietzorreck, Modeling of transmission lines and passive elements for multilayer circuits, in: *2000 Topical Meetings on Silicon Monolithic Integrated Circuits in RF Systems. Digest of Papers*, 26–28 April 2000, Garmisch, Germany, pp. 23–24.

359. N. H. Huynh, *Verbesserung der Effizienz der FDTD Methode für die Analyse von koplanaren MMICs*, Doctoral Thesis, 2000, Germany, pp. 1–104.

360. A. Taflove and S. C. Hagness, *Computational Electromagnetics, The Finte Difference Time Domain Method*, Boston, London, 2000.

361. H. S. Gamble, O. L. Kam, S. H. Raza, B. M. Armstrong, S. J. N. Mitchell, Y. Suidong, V. F. Fusco, and J. A. C. Stewart, Coplanar waveguides on SOI and OPS substrates, in: *MEMS Design, Fabrication, Characterization, and Packaging*, 30 May–1 June 2001, Edinburgh, UK, *Proceedings of the SPIE*, The International Society for Optical Engineering, vol. 4407, 2001, pp. 363–371.

362. J. H. Chung, Coplanar waveguide dispersion characteristics including anisotropic substrates, *IEEE Trans. Microwave Theory Tech.*, vol. 49, no. 2, 2001, pp. 362–368.

363. G. E. Ponchak, J. Papapolymerou, and E. M. Tentzeris, Coupling between finite ground coplanar waveguides embedded in polyimide layers for 3D MMICs on Si, in: *GAAS 2001 Conference Proceedings, 9th European Gallium Arsenide and Other Semiconductors Applications Symposium*, London, 24–25 Sept. 2001, pp. 291–294.

364. R. N. Simon, *Coplanar Waveguide Circuits Components and Systems*, New York: John Wiley & Sons, 2001.

3

COPLANAR WAVEGUIDE DISCONTINUITIES

3.1 THE THREE-DIMENSIONAL FINITE DIFFERENCE ANALYSIS

In Chapter 2, basically two-dimensional line structures have been investigated using different analysis techniques. Especially in Section 2.2, it was shown how a quasi-static finite difference technique can be used with remarkable advantage to analyze coplanar waveguides that are homogeneous along the wave propagation direction. For these structures, a two-dimensional analysis technique was adequate. If a real coplanar circuit design is to be considered, the problems become more complex because, besides the lines, three-dimensional components must be investigated. A first class of components that are of essential influence in the circuit design are the waveguide discontinuities, such as open and shorted ends, impedance steps, line gaps, waveguide bends, T-junctions, and crossings. In the case of the coplanar waveguide, there is a very special discontinuity element that is needed in a large number inside circuits: the air bridges that connect the two ground planes of waveguides to assure equal potentials of the ground planes. These elements are investigated in this chapter. The results presented in this chapter are mainly taken from investigations of Naghed [61] and additionally from various German and European research projects under the leadership of the author.

For the analysis and simulation of coplanar waveguide discontinuities, a three-dimensional field analysis is needed in principle to accurately describe the frequency-dependent properties of these components. Quasi-static and

Coplanar Microwave Integrated Circuits, by Ingo Wolff.

full-wave analyses may be used. Because of the small size (compared to the wavelength) of the discontinuities and the reduced dispersion of the coplanar waveguide (compared to microstrip lines), the accuracy of a quasi-static technique is good enough even up to frequencies of about 50 GHz. The full-wave analysis techniques, like the finite difference time domain (FDTD) technique, are of high numerical expense so that they are, normally, not directly applicable in computer-aided circuit design. They are, however, often used to control the results generated by the static analyses in critical cases such as, for example, at higher frequencies.

The quasi-static analysis technique assumes a pure TEM-propagation inside the discontinuity. This is also the wanted situation inside a microwave circuit. Using appropriate methods like the air-bridge technology (see Section 3.5.5), it may be assumed that inside the real circuit only a quasi-TEM mode can propagate and that an excitation of the fundamental odd mode or possibly even higher-order modes (e.g. in conductor backed circuits) is avoided.

Full-wave analysis techniques are normally used to compute the scattering matrices of the considered components. Here the three-dimensional quasi-static analysis technique is used to create an equivalent circuit with resistors, capacitors, and inductors describing the properties of the discontinuities. The elements of the equivalent circuit are deduced from the quasi-static field analysis. This circuit is finally used as a model for describing the components in a circuit design software.

The additional advantage of this technique is that the elements of the equivalent circuit describe the fundamental physical properties like power dissipation and storage of electric and magnetic field energy inside the component. The dependence of these elements on the geometrical and electrical parameters of the circuit can clearly be identified and can be used to optimize the circuit layout.

As has already been explained in detail in Section 2.2, the applied quasi-static analysis technique makes use of a numerical solution of Laplace's equation for the electric potential inside the considered component. Analogous to the two-dimensional analysis that has been described in Chapter 2.2, a three-dimensional finite difference analysis technique may also be used. To realize this, the investigated discontinuity is connected to some feeding coplanar waveguides, and this structure is then analyzed inside an electric or magnetic shielding. Those walls of the shielding where the feeding lines of the structure end are assumed to be magnetic walls. All other walls are defined to be electric walls (Fig. 3.1.1). All walls must be far away from the discontinuity structure so that the influence of the walls on the discontinuity properties can be considered negligibly small. This also guarantees that the electromagnetic field distribution on the coplanar waveguide at the magnetic walls—that is, at the input and output ports—is a pure TEM mode. The metalization on top of the substrate material that forms the discontinuity is considered to be of infinite conductivity. Its thickness is assumed to be different from zero. Using the cri-

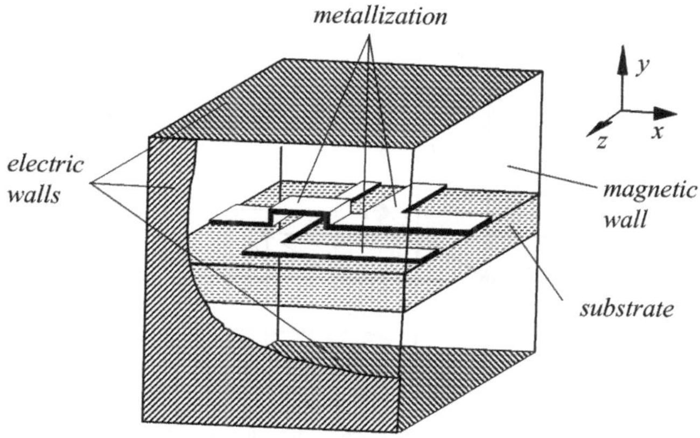

Fig. 3.1.1. Three-dimensional coplanar discontinuity inside a shield of electric and magnetic walls.

teria described in detail in Section 2.2 performs the discretization of the three-dimensional space inside the shielding. Starting with a very fine discretization near the metalization plane, the mesh is chosen to become more and more coarse in the y-direction. As in the two-dimensional case, only quadratic meshes are used. In planes where the mesh size is changed (by a factor of 2; see Fig. 3.2.2), nodes must be considered that do not have six (as in the general case) but only five neighboring nodes. In the computation process of the node potentials, the potential of the missing node will be generated by an interpolation technique. Finally, the electric and magnetic field distribution is computed in the discretized space from the solution of Laplace's equation as demonstrated below.

3.2 COMPUTATION OF THE ELECTRIC FIELD STRENGTH

As is well known, the electric field strength of a TEM-wave can be computed from the static electric potential φ. The potential itself is a solution of Laplace's equation. If this solution is to be determined numerically, Laplace's equation must be developed into finite difference form. For this purpose, a fundamental discretization as shown in Fig. 3.2.1 is considered. Points A to F and P are nodes of a discretized three-dimensional space. Moreover, the points A, B, E, F, and P are in the boundary plane between two dielectric materials with different dielectric constants ε_{r1} and ε_{r2}.

In an analogous way as described in Section 2.2.3, the potential of the central point P can be described as a linear combination of the other six node potentials [41], that is,

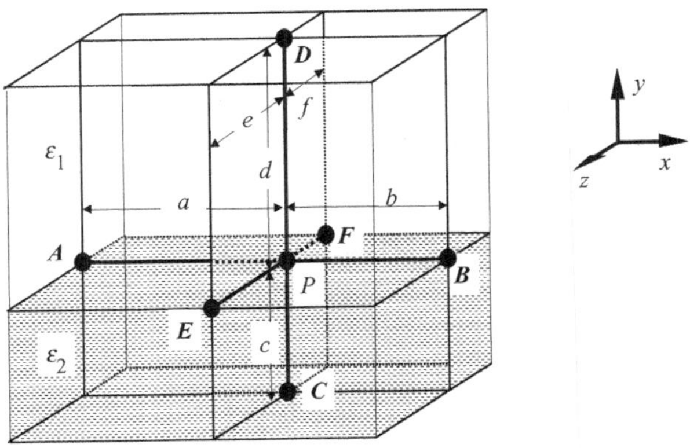

Fig. 3.2.1. Three-dimensional mesh for the development of Laplace's difference equation.

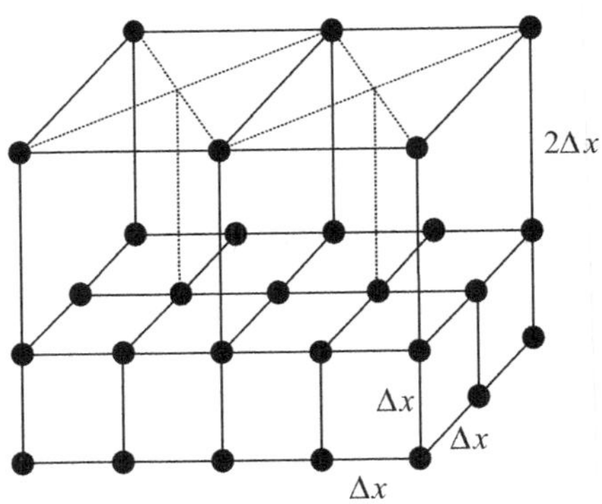

Fig. 3.2.2. Position of the mesh nodes in a plane where the mesh size changes.

$$\left(\frac{d\varepsilon_1 + c\varepsilon_2}{ab} + \frac{\varepsilon_1}{d} + \frac{\varepsilon_2}{c} + \frac{d\varepsilon_1 + c\varepsilon_2}{ef}\right)\varphi_{\mathrm{P}} = \frac{d\varepsilon_1 + c\varepsilon_2}{a(a+b)}\varphi_{\mathrm{A}}$$
$$+ \frac{d\varepsilon_1 + c\varepsilon_2}{b(a+b)}\varphi_{\mathrm{B}} + \frac{\varepsilon_2}{c}\varphi_{\mathrm{C}} + \frac{\varepsilon_1}{d}\varphi_{\mathrm{D}} + \frac{d\varepsilon_1 + c\varepsilon_2}{e(e+f)}\varphi_{\mathrm{E}} + \frac{d\varepsilon_1 + c\varepsilon_2}{f(e+f)}\varphi_{\mathrm{F}}. \quad (3.2.1)$$

If the potential distribution is known on the metalized structure, the potential of all other nodes considering the boundary conditions on the electric and the magnetic walls may be computed in the total space inside the shielding using Eq. (3.2.1). By application of, for example, the Gauss–Seidel iteration technique or the overrelaxation technique (see Section 2.2.3) and by application of a heavily nonequidistant discretization mesh, the numerical expense with respect to the computation time and the required memory capacity can be kept small enough for practical applications. The applied discretization scheme is similar to that shown in Fig. 2.2.2. Only the change of the mesh size must be extended to the third dimension. In Fig. 3.2.2 such a change of the mesh size is shown.

For the computation of the electric field strength, well-defined boundary conditions at the shielding are needed. Furthermore, the potentials of the metalized structure must be known. Because the applied method is static, only those metalized planes may have different potentials that are galvanically separated. This condition, of course, reduces the application of the method to a special class of problems, but it is, at the same time, the precondition for the propagation of the even coplanar waveguide mode in coplanar microwave circuits. Therefore, the condition is fulfilled for most practical problems that are to be solved in coplanar circuit design.

The three space components of the electric field strength are computed from $\boldsymbol{E} = -\text{grad}(\varphi)$ using the partial derivatives of the potential function. These derivative values are simply replaced by the equivalent difference values of the two potentials in the neighbored mesh nodes. In Fig. 3.2.3 the static electric field components of a coplanar waveguide step are shown as an example. The field components are computed in the metalization plane of the structure. As will be shown below, this electric field distribution can be used to extract the capacitive model parameters for the equivalent circuit describing the discontinuity.

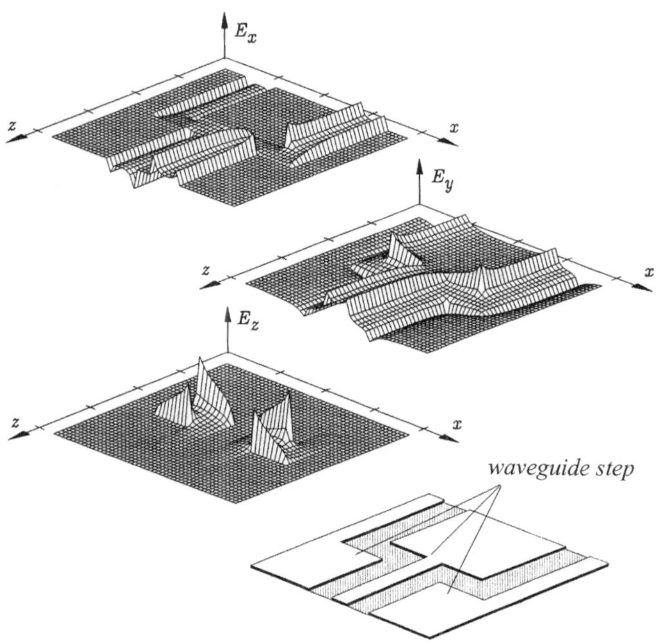

Fig. 3.2.3. The space components of the electric field strength on a coplanar waveguide step, computed in the metalization plane.

3.3 COMPUTATION OF THE MAGNETIC FIELD STRENGTH

The general assumption for the analysis of the discontinuities made above was that they are small compared to the wavelength. Under this assumption the geometrical current distribution inside the discontinuity will only change to a negligible amount due to its time dependence. This again means that the magnetic field inside the structure can be described using a stationary current density in the metalized structure. In several publications (e.g., references 8 and 14) the inductive effects of planar line discontinuities are analyzed using this current distribution. The analyzed structure is divided into elementary cells [8], and the current distribution is calculated using arbitrary two-dimensional test functions fulfilling the continuity condition. This method has an acceptable efficiency in the case of microstrip discontinuities because normally the metalized areas of such structures are small. In the case of coplanar discontinuities, however, the expense for analyzing discontinuities is too large because of the widespread ground planes.

An alternative method that is used here for the coplanar discontinuities makes use of the analogy of the static electric and the static magnetic potential distribution inside the structures [45]. The used analogy is explained considering the coplanar structure in Fig. 3.3.1. The structure consists of a two-dimensional metalized region I and an air-filled region II (slot region) that is positioned between the two parts of region I (center conductor and ground conductor) in the xz-plane. It is assumed that the metalized plane has zero thickness. The total space surrounding this planar structure is assumed to be filled with a material of permeability $\mu_r = 1$. The structure shown in Fig. 3.3.1 is a model for a coplanar structure if the magnetic field is to be analyzed because the dielectric substrate materials do not have influence on this analy-

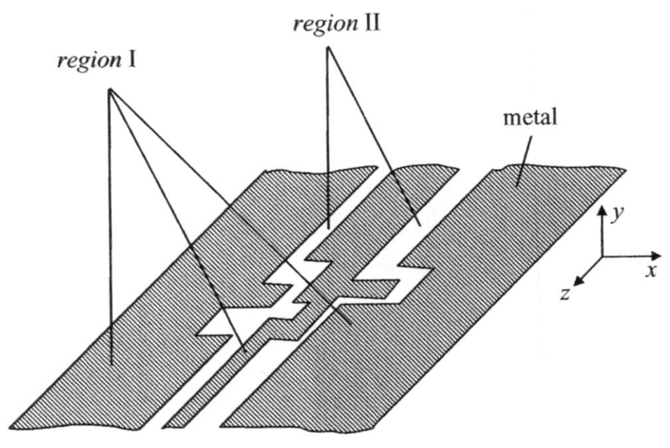

Fig. 3.3.1. Structure for discussing the analogy between electrostatic and magnetostatic field computation.

sis if a TEM field is assumed. Because of the symmetry of the structure, it is sufficient to consider only one half-space (e.g., the upper half-space) of the structure.

If the electrostatic field is computed for the structure shown in Fig. 3.3.1, the following conditions must be fulfilled for the potential φ and the electric field strength $\boldsymbol{E}$:

$$\partial\varphi/\partial x = 0 \rightarrow E_x = 0 \quad \text{in region I } (y=0), \tag{3.3.1a}$$

$$\partial\varphi/\partial z = 0 \rightarrow E_z = 0 \quad \text{in region I } (y=0), \tag{3.3.1b}$$

$$\partial\varphi/\partial y = 0 \rightarrow E_y = 0 \quad \text{in region II } (y=0), \tag{3.3.1c}$$

$$\left.\begin{array}{l}\text{curl } \boldsymbol{E} = 0\\ \text{div } \boldsymbol{E} = 0\end{array}\right\} \rightarrow \Delta\varphi = 0 \quad \text{for } y \geq 0. \tag{3.3.2}$$

For the static magnetic potential Ψ and the magnetic field strength $\boldsymbol{H}$ the analogous equations are

$$\partial\Psi/\partial x = 0 \rightarrow H_x = 0 \quad \text{in region II } (y=0), \tag{3.3.3a}$$

$$\partial\Psi/\partial z = 0 \rightarrow H_z = 0 \quad \text{in region II } (y=0), \tag{3.3.3b}$$

$$\partial\Psi/\partial y = 0 \rightarrow H_y = 0 \quad \text{in region I } (y=0), \tag{3.3.3c}$$

$$\left.\begin{array}{l}\text{curl } \boldsymbol{H} = 0\\ \text{div } \boldsymbol{H} = 0\end{array}\right\} \rightarrow \Delta\Psi = 0 \quad \text{for } y \geq 0. \tag{3.3.4}$$

If Eqs. (3.3.1) and (3.3.2) are compared with Eqs. (3.3.3) and (3.3.4), respectively, it may be observed that the boundary conditions for the electric field in region I are the same as for the magnetic field in region II and vice versa. This is also true for the magnetic field of a stationary surface current density in the metalized region I, as long as the symmetry of the structure is maintained.

This analogy means that the magnetic field of the coplanar structures can be analyzed in the same way as the electric field. In the analysis, only region I (metalized area) must be replaced against region II (slot area) and at the same time the electric potential φ is replaced by the magnetic potential Ψ. If it is assumed that the magnetic potential is constant in region II (magnetic walls in the slot areas, compare with the discussion in Section 2.3), the magnetic field can be derived from Laplace's equation for the magnetic potential Ψ. Equation (3.2.1) can be used in an analogous way if all relative dielectric constants are replaced by 1 and the electric potential φ is replaced by the magnetic potential Ψ. Also, the shielding that has been discretized for the electric field analysis can be used for the magnetic field computation if magnetic walls are replaced by electric walls and vice versa. The magnetic field strength can then be derived from

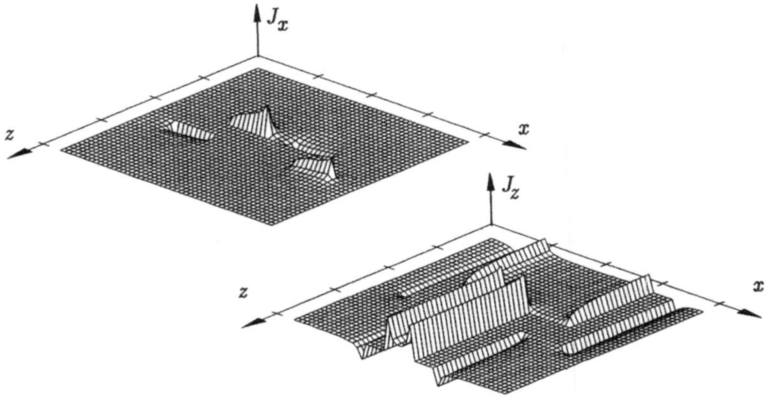

Fig. 3.3.2. The x- and z-components of the surface current density on a coplanar waveguide step.

$$H = -\text{grad}\,\Psi, \tag{3.3.5}$$

that is, from the partial derivatives of the potential function Ψ.

From the knowledge of the tangential magnetic field, the surface current density inside the metalization can also be computed using well-known methods. In Fig. 3.3.2 the x- and the z-components of the surface current density on a coplanar waveguide step are shown.

The typical current distribution J_z over the cross section of the coplanar structure can be well-observed. On the feeding lines of the discontinuity, only a z-component of the current density may be found. An x-component is found only very near to the discontinuity. Finally, the magnetic field distribution in the slot area will be used later to determine the inductive components of the equivalent circuit (see Section 3.4.2).

3.3.1 Convergence and Error Discussion for the Analysis Technique

In Section 2.2.3 the convergence behavior of the Gauss–Seidel technique and the overrelaxation technique has been discussed for the two-dimensional analysis of the electrostatic potential φ. Equivalent investigations for the three-dimensional problems have shown that an optimum relaxation factor k again on the order of $k = 1.8$ can be found for the overrelaxation method. Smaller values of k lead to longer computation times, whereas the choice of larger values leads to instabilities of the method. The choice of the starting values for the potential influences the convergence of the method, but due to the big variety of the analyzed three-dimensional structures, they cannot be optimized very easily.

The choice of the mesh and its size, to a first approximation, mainly has an influence on the accuracy of the computation. Nevertheless, there is also an influence on the convergence. If the mesh is largely nonequidistant, the convergence rate is reduced and under certain conditions this can lead to instabilities of the technique. Therefore, for the analysis of the coplanar structures in this book, cubic mesh cells are used. If the cell size is changed, the cell length is changed by a factor of 2 (see Fig. 3.2.2); only in the area near the metalization, the mesh length in the y-direction (perpendicular to the substrate plane) is different from those lengths in the x- and z-direction. But, care is taken that even in this area the ratio of the cell lengths is not larger than five.

To investigate the influence of the mesh size on the convergence and the accuracy of the analysis technique, the coplanar waveguide step shown in Fig. 3.2.3 was analyzed using different mesh sizes. Figure 3.3.3 shows the relative error of the analysis referenced to the final ($n = 150$) result of the finest mesh size ($s/\Delta x = 16$, where s is the slot width of the coplanar waveguide), plotted against the number of iterations n and the normalized discretization size $s/\Delta x$. It may be observed that with reduced mesh size the final error decreases, but the convergence of the method is reduced heavily. The reason for the slow convergence is the large number of mesh nodes that are established by a too small discretization size or by the choice of a too large dimension of the shielding compared to the structure's dimensions.

Two main groups of errors may be identified for the finite difference method. In the first group are those that are related to geometrical or material-specific properties of the structure and that prevent a unique numerical computation of the electric or magnetic field in certain regions. Metallic edges and the field singularities in their surrounding must be mentioned in this connection. These errors may be compensated by an adequate approximation technique [30].

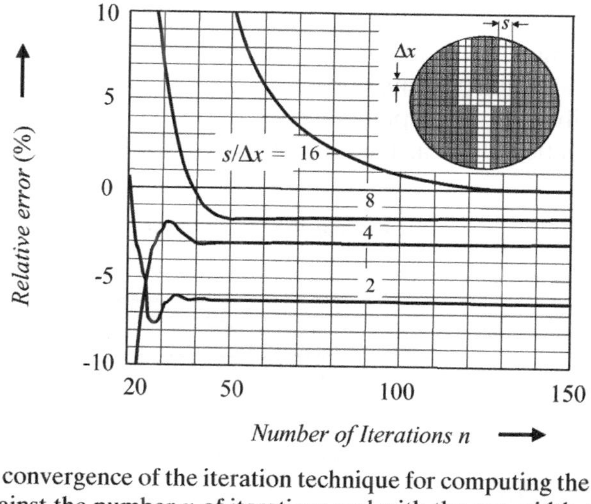

Fig. 3.3.3. The convergence of the iteration technique for computing the electric potential, plotted against the number n of iterations and with the gap width s normalized to the mesh size Δx as a parameter.

The second group of errors depends on the chosen analysis technique and its properties. One of these errors is, for example, the cutoff error [31] that occurs if Laplace's differential equation is approximated by a difference equation. On the other hand, an error occurs if the space that is to be analyzed has a complex structure and cannot be correctly resolved by the applied mesh structure. Consequently, these errors may be essential. In this connection, the inaccuracy resulting from a nonequidistant mesh plays an important role. These errors can only be avoided by using a finer mesh size that leads to a higher numerical expense or by the selection of more complex computation algorithms. Also, to the second group of errors belong those errors that result from the finite size of the shielding. Their influence will be analyzed in the next sections.

3.4 COPLANAR WAVEGUIDE DISCONTINUITIES

A coplanar waveguide discontinuity is an abrupt change of the geometrical parameters and/or the material parameters in a homogeneous coplanar waveguide. Here only those discontinuities will be considered that result from a change in geometrical parameters of the strip and ground-plane metalization. In Fig. 3.4.1 some of the typical coplanar waveguide discontinuities such as the open end, the shorted end, the waveguide step, the bend, and the crossing are shown. At low frequencies these discontinuities may be considered as ideal interconnections between the different coplanar waveguides. At higher frequencies, however, their properties are changing. If a fundamental even mode (quasi-TEM mode) is incident on one of the coplanar waveguides, then at the discontinuity its electric and magnetic fields will be scattered and a mixture of the fundamental even and odd mode together with hybrid modes will be excited in the space surrounding the discontinuity. Hybrid modes are characterized by showing all six field components of the electromagnetic field.

If the reflection and transmission properties of a discontinuity are to be correctly described, in principle, a three-dimensional full-wave analysis of the electromagnetic field near the discontinuity must be executed [12, 23, 26,

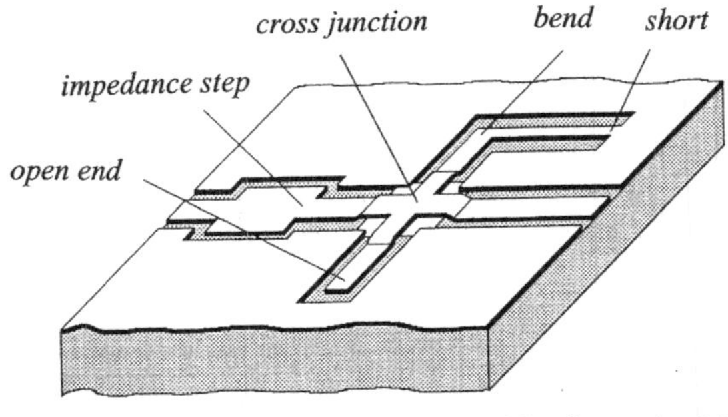

Fig. 3.4.1. Typical coplanar waveguide discontinuities.

35–37]. These analysis techniques however are of large numerical expense. If it is considered that the geometrical size of the discontinuities normally is very small compared to the wavelength of the electromagnetic fields, simple models of high accuracy can be developed if an accurate quasi-static electric *and* magnetic field analysis is used. The development of such models and the description of the model parameters [1, 41, 61] will be the subject of the following sections. Later, the most essential coplanar waveguide discontinuities will be discussed in detail in dependence on their geometrical parameters. Table 3.1 shows a summary of the most essential coplanar waveguide discontinuities as they are used in coplanar circuit design together with their equivalent circuit models.

TABLE 3.1. The Most Essential Coplanar Waveguide Discontinuities for Circuit Design and Their Equivalent Circuits

Coplanar Waveguide Discontinuities		Ideal Waveguide Circuit	Equivalent Circuit
Waveguide step	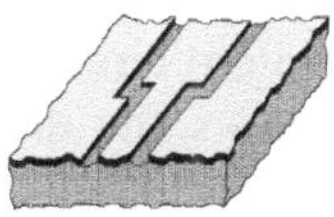	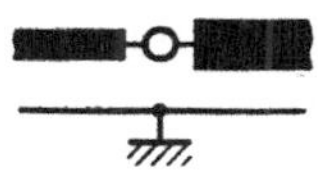	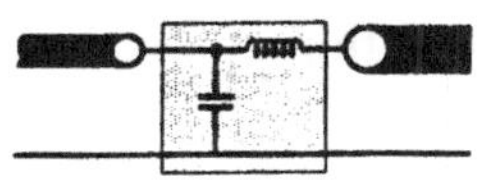
Shorted end	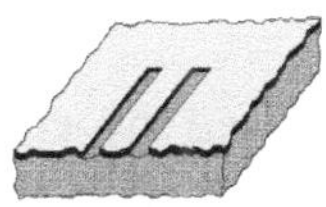		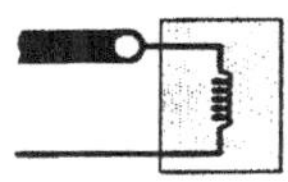
Open end			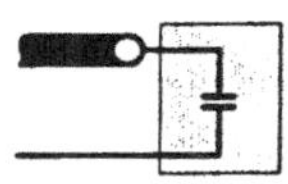
Waveguide bend	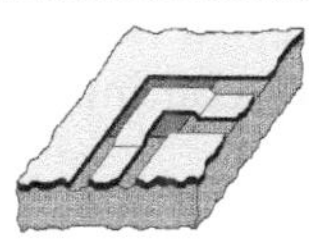	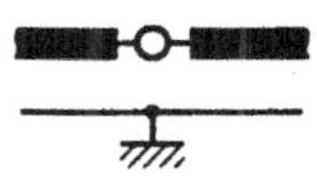	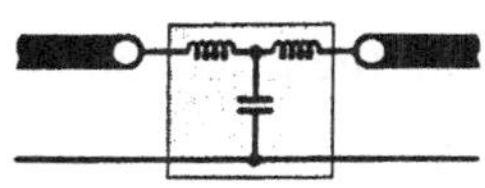
Waveguide T-junction		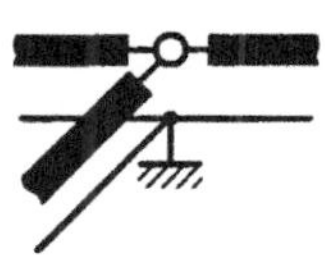	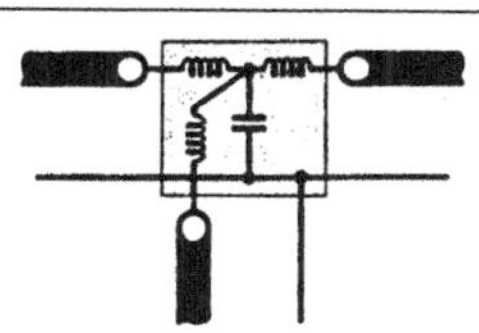
Waveguide crossing		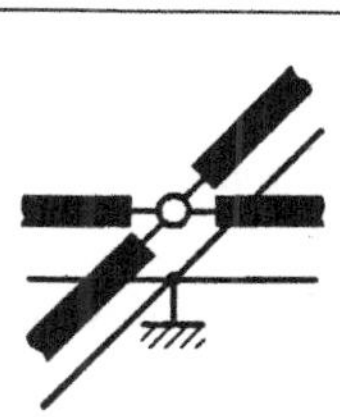	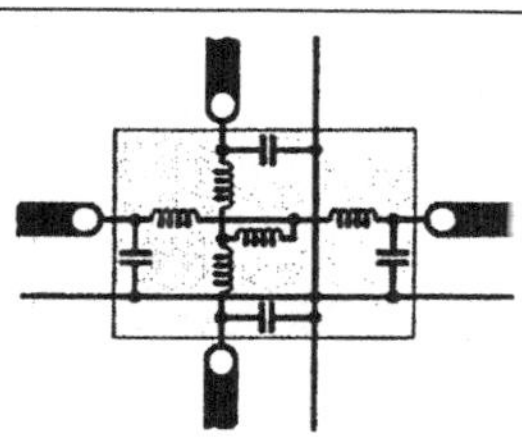

3.4.1 Modeling the Discontinuities

It is assumed that the linear geometrical size of the discontinuities is much smaller than the wavelength of the electromagnetic field. Under this condition, the frequency-dependent properties of the discontinuities can be described using an equivalent circuit description (model) with ideal discrete components like resistors, inductances, and capacitors. To find this model, as a first step, the reference planes (ports) that describe the geometrical size of the discontinuity must be defined. At these reference planes the discontinuities are connected to homogeneous coplanar waveguides. It is assumed that in the reference planes, the field disturbances that have been excited by the discontinuity have decayed to a nonmeasurable value. This then defines the geometrical size of the discontinuity to the area between the reference planes. At the same time, the geometrical size of the shielding that is used in the numerical analysis must be large enough so that the field disturbances may not be recognized in the shielding planes (e.g. in the magnetic walls).

In Fig. 3.4.2a the reference planes, the discontinuity area, and the feeding lines are shown for a general discontinuity. Figure 3.4.2b shows the principal description of this discontinuity by an equivalent circuit model. In the models that will be derived for different discontinuities, the disturbance of the electric field will be described using equivalent capacitances. The value of these capacitances will be determined using the electric field and surface charge distribution in the discontinuity area.

The magnetic field disturbances are caused by changes in the current density distribution—for example, in the form of a cut in magnitude or changes of the current density directions. These disturbances will be described by inductances in the equivalent circuits. It will be shown in the next section how these inductances can be determined from the distribution of the magnetic flux in the metalization plane (upper substrate plane).

Conductor losses may be described by resistors, but because of the small size of the discontinuities, the losses of the discontinuities, neglecting possible

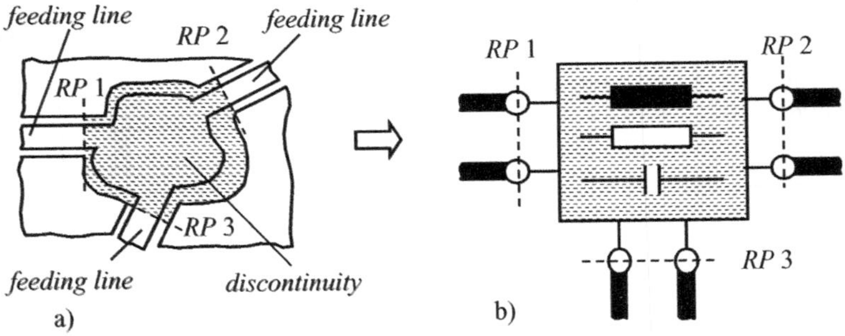

Fig. 3.4.2. Modeling of the coplanar waveguide discontinuity: (a) general discontinuity, (b) modeling of the discontinuity by a general equivalent circuit model. *RP* stands for reference plane.

radiation losses (see below), are negligibly small. Therefore in this chapter only reactive equivalent circuits are used as models for the discontinuities.

Besides the above-mentioned reactive effects inside the waveguide discontinuities, other processes must be investigated and correctly described by the equivalent circuits. There are two main effects that have to be considered additionally: The first effect is that a part of the power transported towards a discontinuity will be radiated from the discontinuity into the free space if the circuit is open. If the circuit is enclosed into a conducting package, the power is radiated into the air space above the substrate (and also partly into the substrate) that leads to surface currents in the package. The radiated power in any case is lost from the circuit. An adequate resistor inside the equivalent circuit model can describe it. The second effect is that at the discontinuity surface waves (see Section 2.1) may be excited. This effect then leads to unwanted coupling to other waveguides or components inside the circuit. Both mentioned effects are strongly frequency-dependent and cannot be simulated by the applied quasi-static simulation technique as described in this chapter.

3.4.2 Extraction of the Model Parameters

The model parameters of the coplanar waveguide discontinuities are estimated numerically. For this purpose the electric and the magnetic field distribution inside the discontinuity is computed using the quasi-static finite difference technique. The fields can be calculated from the static electric and magnetic potentials.

In a first step the capacitive elements of the equivalent circuits shall be determined. Figure 3.4.3 shows a coplanar waveguide step as an example. The

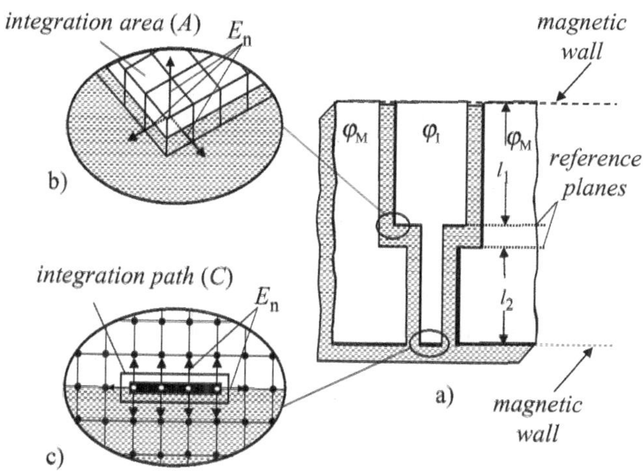

Fig. 3.4.3. Estimation of the charges for calculating the capacitance of a coplanar waveguide step. (Assumed values of the potentials: center strip φ_I is 1 V ground plane φ_M is 0 V).

electric field distribution outside the metalized area is analyzed using the three-dimensional method described in Section 3.2. It is assumed that the line lengths l_1 and l_2 (see Fig. 3.4.3) of the feeding coplanar waveguides are known. By the disturbance of the electric field in the discontinuity area the surface charge of the metalized structure is changed, compared to that of the homogeneous waveguide area. This change of the surface charge distribution is a cause for the additional capacitive effect of the discontinuity. To estimate this additional charge, in a first step, the total charge Q_{total} on the center strips of the discontinuity shown in Fig. 3.4.3 is computed by integration of the normal electric field component over the total area A of the center strip (Fig. 3.4.3b), that is,

$$Q_{\text{total}} = \oint\!\!\oint_A \boldsymbol{D}\cdot\boldsymbol{n}\, dA = \varepsilon_0\varepsilon_r \oint\!\!\oint_A E_n\, dA. \tag{3.4.1}$$

In the next step the charges per unit line length Q_1' and Q_2' on the two coplanar waveguides of length l_1 and l_2 (Fig. 3.4.3a) are calculated using the normal electric field strength and integrating along a contour in the magnetic walls of the shielding (i.e., in the ports of the discontinuity where the field disturbances have decayed down, see discussion above), as shown in Fig. 3.4.3c, that is,

$$Q_i' = \oint_C D_n\, ds = \varepsilon_0\varepsilon_r \oint_C E_n\, ds, \qquad i = 1, 2 \tag{3.4.2}$$

The additional charge Q_{add} that is stored in the discontinuity region, therefore, is given by the difference of the total charge Q_{total} and the charge on the two uniform coplanar waveguides. If the potential difference (i.e., the voltage) between the center strip and the ground planes, $V = \varphi_{\text{I}} - \varphi_{\text{M}}$, is taken, the additional capacitance due to the discontinuity is

$$C_{\text{add}} = \frac{Q_{\text{add}}}{V} = \frac{Q_{\text{total}} - l_1 Q_1' - l_2 Q_2'}{V}. \tag{3.4.3}$$

The accuracy of the so-calculated additional capacitance is heavily dependent on the chosen lengths of the two coplanar waveguides—that is, on the definition of the reference planes. Also, the distance of the electric walls (side walls of the shielding: see inset of Fig. 3.4.4) has an influence on the computation results.

To investigate these influences, Fig. 3.4.4 shows the relative error of the computation in dependence on the shielding dimensions. The reference value is the additional capacitance for very large values of the shielding dimensions; that is, all distances of the shielding from the discontinuity structure have been chosen larger than 12 times the total line width d of the coplanar waveguide with the larger dimensions (Fig. 3.4.4, inset).

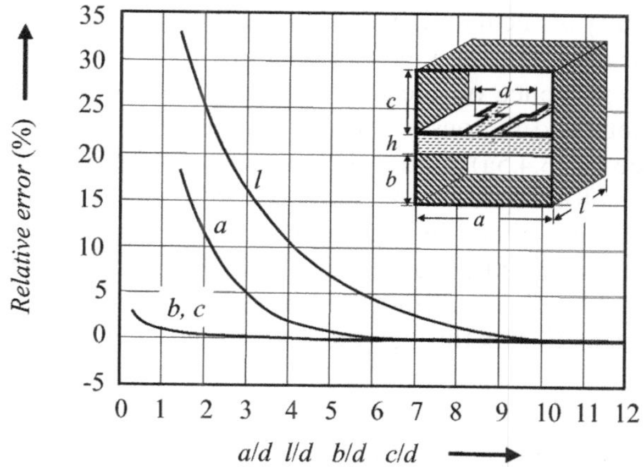

Fig. 3.4.4. Influence of the shielding on the equivalent capacitance computation for a coplanar waveguide step.

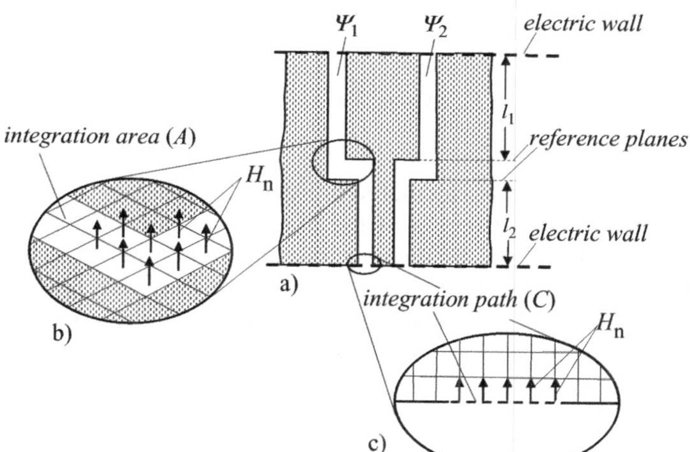

Fig. 3.4.5. Calculation of the total magnetic flux through the discontinuity discussed at the example of a coplanar waveguide step (used potentials: $\Psi_1 = -1A$, $\Psi_2 = 1A$).

The results that are shown in Fig. 3.4.4 show a large dependence of the calculated additional capacitance especially on the position of the magnetic walls (reference planes). There is also an influence of the electric walls (side walls) of the shielding. The top and bottom shielding nearly has no influence on the calculated capacitance of the structure.

If the inductive elements of the equivalent circuit are to be determined, the magnetic field distribution in the metalization plane must be known. Figure 3.4.5 shows, again using the waveguide step as an example, how the inductance of the discontinuity is calculated. In the analysis of the inductance it is assumed that the metalization thickness is very small ($t \to 0$) and that the substrate material has no influence on the magnetic field (i.e., $\mu_r = 1$). The slot areas between the metalized regions are set to the constant potentials Ψ_1 and Ψ_2, respectively (see Fig. 3.4.5a). Under these conditions the magnetic potential is determined in the total space above the metalization layer using the finite difference method as described in Section 3.3. From the potential the magnetic field strength can be computed using Eq. (3.3.5). If the component of the magnetic field normal to the metalization plane is integrated over the slot region A_{slot}, the total magnetic flux through the discontinuity region can be calculated as follows:

$$\Phi_{\text{total}} = \iint_{A_{\text{slot}}} \boldsymbol{B} \cdot \boldsymbol{n}\, dA = \mu_0 \iint_{A_{\text{slot}}} H_n\, dA. \tag{3.4.4}$$

The magnetic flux per unit line length connected to the two uniform coplanar waveguides of lengths l_1 and l_2 (see Fig. 3.4.5a) can be calculated, again using the normal magnetic field component and integrating along the integration path C across the slot width as shown in Fig. 3.4.5c:

$$\Phi_i' = \int_{C_i} B_n\, ds = \mu_0 \int_{C_i} H_n\, ds, \qquad i = 1, 2. \tag{3.4.5}$$

The total magnetic flux as calculated in Eq. (3.4.4) consists of two parts. The first part is the flux in the region of the homogeneous coplanar waveguides that can be calculated from the flux per unit line length times the line lengths l_1 and l_2. The second part is the additional magnetic flux Φ_{add} created by the field disturbances of the magnetic field in the discontinuity region. This part, therefore, can be calculated from

$$\Phi_{\text{add}} = \Phi_{\text{total}} - \Phi_1' l_1 - \Phi_2' l_2. \tag{3.4.6}$$

As long as the magnetic field is symmetrical with respect to the metalization plane, the above-calculated magnetic field is identical to a magnetic field of a current I inside the center strip. The value of this current depends on the magnetic potential as

$$I = 2 \int_{\text{I}}^{\text{II}} \boldsymbol{H} \cdot d\boldsymbol{s} = 2(\Psi_{\text{II}} - \Psi_{\text{I}}). \tag{3.4.7}$$

The integral has to be evaluated along the electric wall (Fig. 3.4.5) and the upper edge of the center conductor (from region I to region II). Factor 2 considers the fact that the integral is taken only along one-half of the conductor contour and therefore only delivers half of the current.

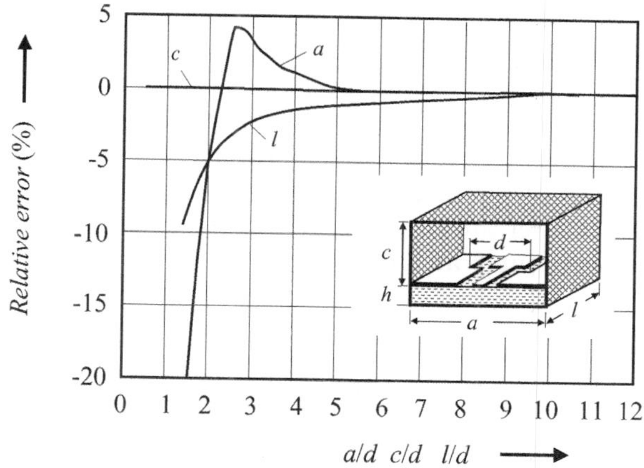

Fig. 3.4.6. Influence of the shielding on the computation of the equivalent inductance for a coplanar waveguide step.

The equivalent, additional inductance describing the influence of the discontinuity now can be calculated from the additional magnetic flux Φ_{add} and the current I as

$$L_{\text{add}} = \frac{\Phi_{\text{add}}}{I}. \tag{3.4.8}$$

Just as in the case of the capacitance computation, in the inductance calculation there is an influence of the shielding on the computation result. Figure 3.4.6 shows this influence for the used example of a coplanar waveguide step. In this case the influence of the top and the bottom walls is extremely small. Again the influence of the distances that have been defined between the reference planes and the ports (electric walls, Fig. 3.4.5a) is the largest one.

In the case of an equivalent circuit for a complex discontinuity that contains more than one inductance, multiple configurations of the magnetic fluxes in the slot regions must be used to estimate these inductances. Examples for such more complex structures will be given in the next sections, where, among other discontinuities, the coplanar waveguide T-junction and crossing are described.

3.5 DESCRIPTION OF COPLANAR WAVEGUIDE DISCONTINUITIES

In this section the most essential coplanar waveguide discontinuities as they are used in microwave integrated circuit design will be described using the above-discussed analysis technique and their equivalent circuits. The simulation results are verified using measurements between 45 MHz and 40 GHz. Also, simulation results from full-wave analyses such as the moment method and the finite difference time domain technique (see Section 2.1) are used for verification. The dependence of the model parameters on the different geometrical parameters will be studied systematically and intensively to demonstrate the properties of the discontinuities in a circuit design. All examples discussed in this chapter are of a geometrical size that is normally used in monolithic microwave integrated circuits.

3.5.1 The Coplanar Open End

If a homogeneous coplanar waveguide abruptly ends at some length l, a coplanar waveguide open end is formed [4, 18, 19, 33]. Different forms of the coplanar open ends exist, as shown in Fig. 3.5.1. In Fig 3.5.1a the open end of the coplanar waveguide is indeed an abrupt ending of the total waveguide. The center conductor, as well as each ground plane, ends in one geometrical position. Figure 3.5.1b shows a different form of the open end where only the center conductor ends. The ground planes of the homogeneous coplanar waveguide form a short, so that a finite gap exists between the centre conductor

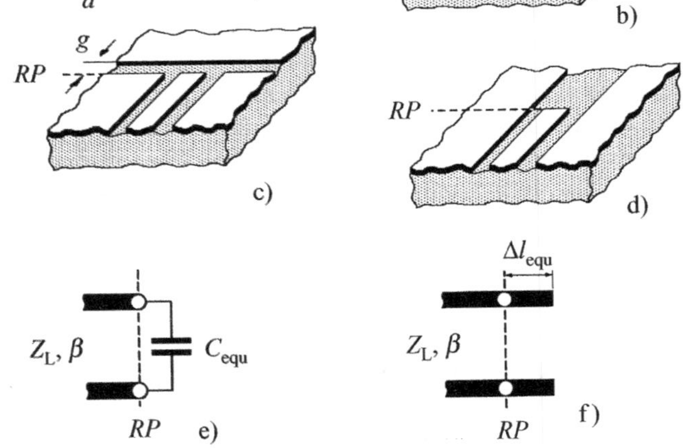

Fig. 3.5.1. Four different forms of coplanar waveguide open ends (a)–(d) and two possible equivalent circuit models describing the dominant capacitive effect of the discontinuity (e) and (f). *RP* defines the reference planes at the ports.

and the ground plane at the end of the waveguide. This form of the coplanar waveguide open end is the most frequently used in coplanar microwave integrated circuit design. Two other forms of coplanar open ends are shown in Fig. 3.5.1c and Fig. 3.5.1d. They are sometimes used in filter structures.

An ideal waveguide open end has a reflection coefficient of one; that is, the incident wave on the waveguide is totally reflected. In an actual open end, a part of the wave is transmitted into the open space behind the discontinuity. This leads to two effects: The first effect is an additional stray field at the end of the waveguide, and the second effect may be radiation into free space along the surface of the substrate (surface wave modes) or into the substrate. The first effect is dominant in microwave integrated circuit design; that is, at the end of the open-ended coplanar waveguide an additional electric stray field is built up that stores reactive electric energy.

Figures 3.5.1e and 3.5.1f show two possible models for the coplanar open end, as they have already been used for the microstrip open end. In the first model, an equivalent circuit containing just one capacitor connected to the end of the coplanar line is used. The second model uses an additional piece of homogeneous waveguide of length Δl_{equ} added to the line length l of the original coplanar waveguide. The equivalent line length can be calculated from the equivalent capacitance C_{equ} of the open end and the capacitance per unit line length C' of the uniform coplanar waveguide:

$$\Delta l_{equ} = C_{equ}/C'. \tag{3.5.1}$$

The computation of the equivalent capacitance C_{equ} and the capacitance per unit line length C' is performed using the technique described in Section 3.4.2. The potential of the center conductor φ_1 is set to 1 V and the metalized ground plane potential φ_M is set to 0 V, and the capacitances are calculated using the technique described in the above-mentioned section. The accuracy of the method was first tested with an example of a microstrip open end, and the results were compared to the closed formula solution of Kirschning, Jansen, and Koster [18] that was deduced from a hybrid mode moment method analysis and that is known to be of good accuracy.

Figure 3.5.2 shows the comparison of the two solutions for different line width to substrate height ratios w/h. The agreement between both solutions is found always to be good. The deviation between the two solutions for high ε_r values is on the order of maximum 5%. Considering that the accuracy of the Kirschning formula is claimed to be 2.5%, this result shows that the quasi-static technique used here is of a similar accuracy.

Next, coplanar waveguide open-end structures as shown in Figs. 3.5.1a and 3.5.1b will be analyzed. Figure 3.5.3 shows the computed electric field components E_x, E_y, and E_z calculated in the metalization plane at the open end of the coplanar waveguide positioned in the middle of the field analysis area. The E_x- and E_y-components show clearly a maximum value at the line end, and a small stray field of these components can be observed. The E_z-compo-

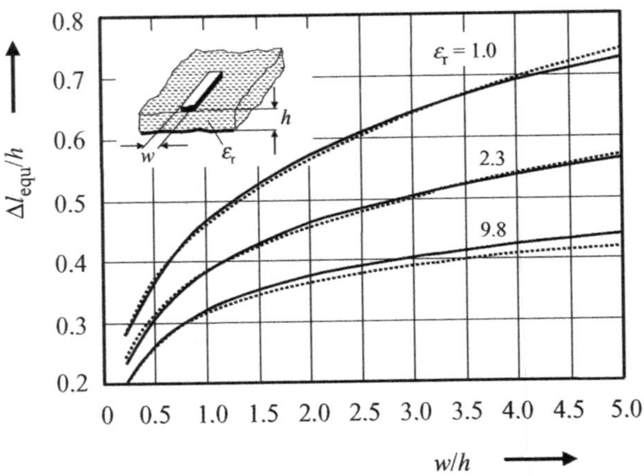

Fig. 3.5.2. Normalized equivalent line length of a microstrip open end, plotted against the line width to substrate height ratio w/h. (———) Quasi-static finite difference solution as described in Section 3.4.2; (. . .) using the closed formula description of Kirschning et al. [18].

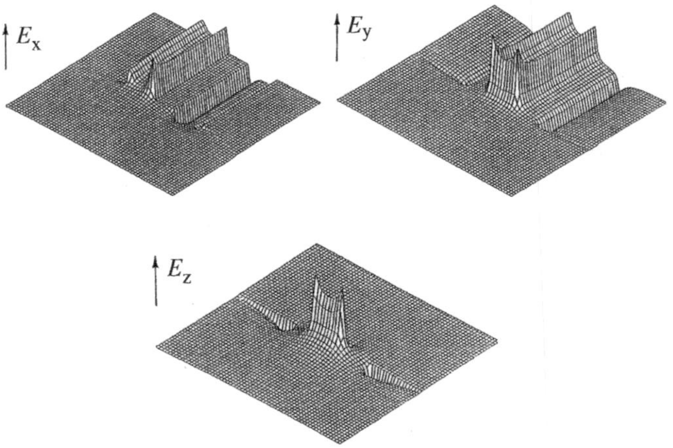

Fig. 3.5.3. The three electric field components at the end of an open-ended coplanar waveguide (used coordinate system: see Fig. 3.3.1).

nent has a recognizable value only at the end of the line. It forms a strong stray field from the waveguide end into the substrate area.

The resulting equivalent capacitances for the structure 3.5.1a are shown in Fig. 3.5.4a, plotted against the center line width w to total slot width d (see inset in Fig. 3.5.4a) ratio w/d. The parameter of the different curves shown in the figure is the normalized substrate height h/d. Figure 3.5.4b shows the equivalent capacitance for the open-end structure of Fig. 3.5.1b, plotted against the

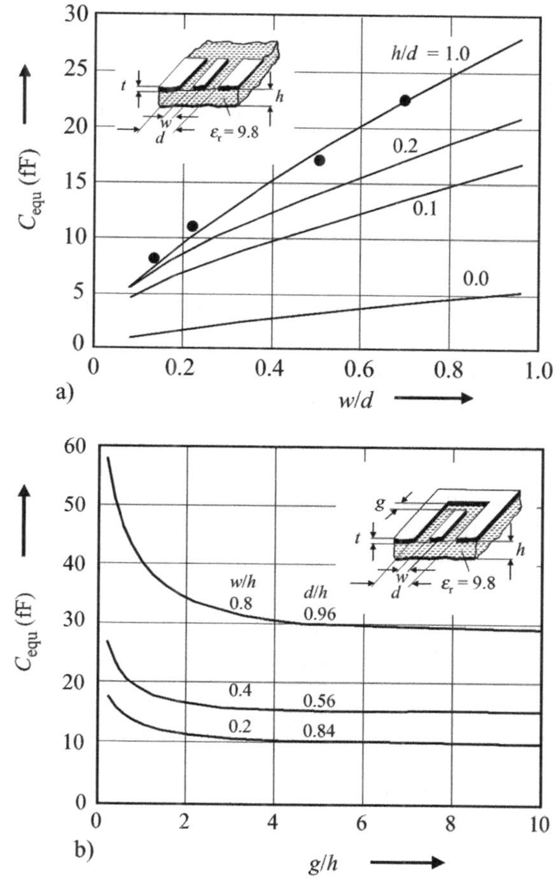

Fig. 3.5.4. (a) Measured (•) and calculated (———) equivalent capacitance of a coplanar waveguide open end as shown in the inset, plotted against the normalized center line width w/d ($w = 625\,\mu m$, $t = 5\,\mu m$). (b) Calculated equivalent capacitance of an open coplanar waveguide as shown in the inset, plotted against the normalized gap width g/h between the center conductor and the ground plane ($h = 625\,\mu m$, $t = 0\,\mu m$).

normalized gap width g/h at the end of the open waveguide (see inset, Fig. 3.5.4b). Parameters of the different curves in this case are the normalized center line width w/h and the normalized total slot width d/h. The measurement results shown in Fig. 3.5.4a have been performed using structures on a ceramic substrate that have been measured at frequencies between 45 MHz and 26.5 GHz. No real variation of the measurement results compared to the simulations could be found in the above-mentioned frequency range. That means that the open end of the coplanar waveguide indeed can be simulated using one constant capacitance C_{equ}. As may be expected, the equivalent capacitance C_{equ} increases with increasing center line width and decreasing total slot width. Both dependencies may be easily explained by the physics of the end capacitance and its electrical field. The equivalent capacitance also is dependent on the h/d ratio of the coplanar line. Starting from a value $h/d = 0$, there is a large increase of the equivalent capacitance with increasing h/d value because (e.g., for a constant value of d) the substrate height h is increased, and more and more electric field is stored in the dielectric substrate material of high permittivity.

For values $h/d > 1$ there is nearly no more increase of the equivalent capacitance with increasing h/d value because nearly all electric field lines below the metalization plane now are inside the dielectric material. The discussion of the dependence on the total slot width d for constant values of the substrate height h is similar. With decreasing values of the total slot width (assuming the center conductor width w is kept constant), the electric field is more and more concentrated inside the dielectric material and therefore the equivalent capacitance is increased.

Figure 3.5.4b shows simulated results for the open end that, as already mentioned above, is most frequently used in coplanar microwave integrated circuits because it partly avoids radiation of energy into free space at the end of the coplanar waveguide.

Compared to the structure shown in Fig. 3.5.1a or Fig. 3.5.4a, there will be an additional capacitance between the center conductor and the ground plane at the end of the waveguide. To find the dependence of the equivalent capacitance on this gap width, the results in Fig. 3.5.4b are shown. It may be observed that the equivalent capacitance heavily increases for very small gap widths g. To avoid an additional capacitive effect at the end of the line, the gap width g should at least be equal to (or larger than) the total slot width d of the coplanar waveguide.

The description of the open-end effect in coplanar waveguide by the equivalent capacitance C_{equ} may be interpreted as the description of a parasitic effect that normally is not intended in a circuit design. Nevertheless, as will be shown later in examples of intelligent circuit design, this parasitic capacitive effect may be used to realize, for example, lumped element filters in a very compact form. The big advantage of the coplanar technology as compared to the microstrip technology is that the designer has access to the value of this parasitic capacitance by designing the gap width g, for example. This is not possible in the case of a microstrip open end where the distance between the strip conductor and the ground plane is always the constant substrate height h. Insofar in this case the end effect indeed must be taken to be parasitic because in most cases, it cannot be changed and used to the advantage of the circuit designer.

This section will be concluded by comparison of broadband measurements on different coplanar open-end structures and the adequate simulation results. For this purpose the open ends have been produced on GaAs substrate material, and their reflection coefficients have been measured using an on-wafer measurement setup in a frequency range from 470 MHz to 67 GHz. In

Fig. 3.5.5a an example of an open-end test structure as defined in Fig. 3.5.1a is shown. Figure 3.5.5b shows the measured input reflection coefficients and the simulated results. Discrepancies can be observed for the magnitude of the reflection coefficient (but consider the applied scale of the figure!) for frequencies higher than 40 GHz, whereas the agreement of the phase up to highest frequencies is very good. The reason for the deviation between the measured and simulated magnitudes is that the structure shown in Fig. 3.5.5a radiates power at the open end that is not simulated by the used quasi-static simulation technique as has been described at the beginning of this chapter.

3.5.2 The Coplanar Waveguide Short-Circuited End

The coplanar waveguide short-circuited end, as a direct galvanic connection between the center conductor and the ground plane, is shown in two different forms in Figs. 3.5.6a and Fig. 3.5.6b. This discontinuity is used in many subsystems such as filters or matching networks. In contrast to the case of

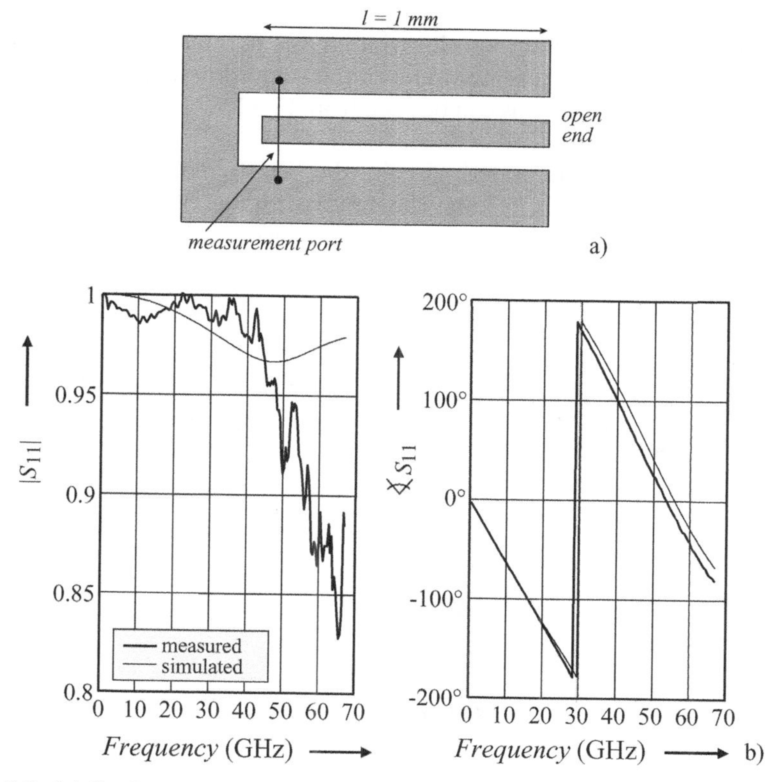

Fig. 3.5.5. (a) Coplanar open-end test structure and (b) comparison between measured and simulated input reflection coefficient, plotted against frequency. Characteristic impedance of the coplanar waveguide: 50 Ω.

the microstrip short end that needs a via-hole connection through the substrate material (and that can be realized only with high technological expense), the coplanar short end can be realized very easily. Since in a well-designed coplanar microwave integrated circuit the slot width between the center conductor and the ground plane normally is very small, the coplanar short end in most of the cases used in practice is nearly an ideal short. The small field disturbances that may occur at the end of the shorted coplanar waveguide can be described using an equivalent inductance L_{equ} as shown in Fig. 3.5.6c.

The equivalent inductance is calculated using the magnetic field distribution in the slots of the coplanar waveguide as explained in detail in Section 3.4. The substrate material that is assumed to have the relative permeability $\mu_r = 1$ does not influence the magnetic field distribution so that the short end and its equivalent inductance are independent of the substrate height. Therefore, the equivalent inductance is only dependent on the width w of the centre conductor and the slot width s. The influence of the metalization thickness t, because of the applied analysis technique (see Section 3.4), cannot be considered in the calculations of L_{equ}.

The computed values of the equivalent inductances for different coplanar waveguides are shown in Fig. 3.5.7, plotted against the line width to total slot width ratio w/d and with the total slot width d as a parameter. The inductance sharply decreases with increasing values of the center line width w or decreasing values of the slot width s as can be seen from Fig. 3.5.7 (assuming w = const. for the last argument).

Inside a coplanar integrated circuit, the short-circuited end is often designed as shown in Fig. 3.5.6b. The finite width g of the ground-plane metalization does not have a measurable influence on the properties of the

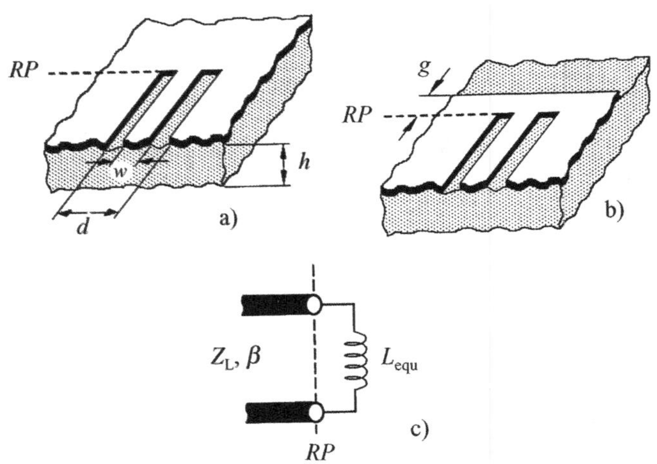

Fig. 3.5.6. (a, b) Two different forms of a coplanar waveguide short end and (c) the model of the discontinuity using an equivalent inductance L_{equ}. RP stands for reference plane.

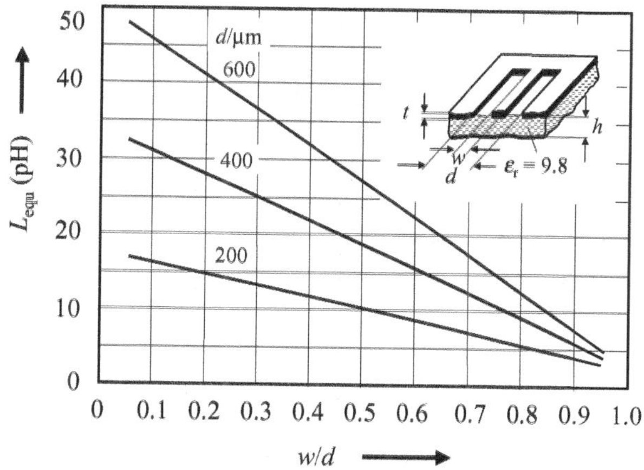

Fig. 3.5.7. The equivalent inductance L_{equ} of a coplanar waveguide short end, plotted against the line width to total slot width ratio *w/d*.

short end, if its value is at least equal to or larger than the slot width *s*. This condition can be easily fulfilled in practical circuit design. Therefore, no special investigations are needed for this special form of the coplanar short end.

3.5.3 The Gap in a Coplanar Waveguide

In filters and resonator structures a small gap in the center conductor of the coplanar waveguide is used as a coupling element. The resonant frequencies of resonator structures and the transmission properties of filters, on a large scale, are influenced by these coupling elements, so a design of these subsystems is not possible without an accurate knowledge of the gap properties. Therefore, in this section a gap in the center conductor of a coplanar waveguide as shown in Fig. 3.5.8a will be analyzed and described by a capacitive equivalent circuit, as shown in Fig. 3.5.8b.

The equivalent circuit is identical to the one used in microstrip technology [2, 33]. It consists of a π-circuit with capacitances C_g, C_{p1}, and C_{p2}. The capacitance C_g represents the coupling between the two center conductors of the coplanar waveguides. The two capacitances C_{p1} and C_{p2} describe the electric stray field from the center conductors to the ground plane. Again, all capacitances of the equivalent circuit are determined using the method described in Section 3.3, but two different steps in the analysis process have to be considered if all three capacitances are to be determined.

Two different potential distributions will be used to determine the three capacitances. The two distributions are shown in Fig. 3.5.9. Figure 3.5.9a shows the so-called even case, where the potentials φ_1 and φ_2 of the two ports of the discontinuity are fixed to the same value $\varphi_1 = \varphi_2 = 1\,\text{V}$ and the ground planes are set to a potential $\varphi_0 = 0\,\text{V}$. The equivalent circuit for this case contains only the two parallel shunt capacitances C_{p1} and C_{p2}.

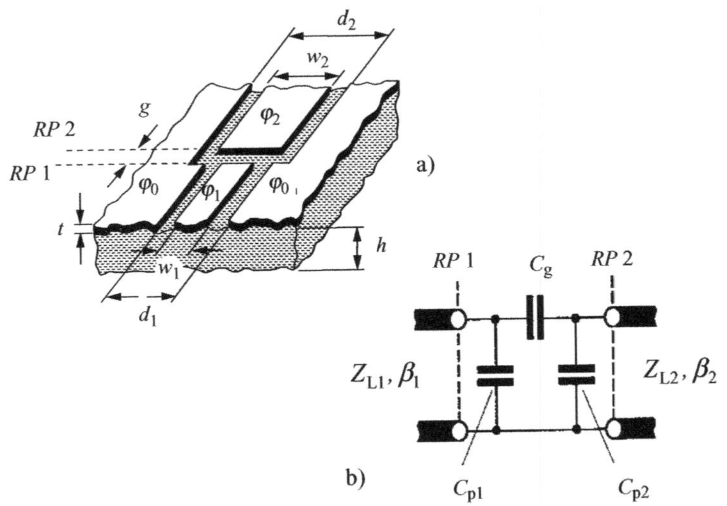

Fig. 3.5.8. (a) Gap in the center conductor of a coplanar waveguide. (b) Equivalent circuit for the discontinuity.

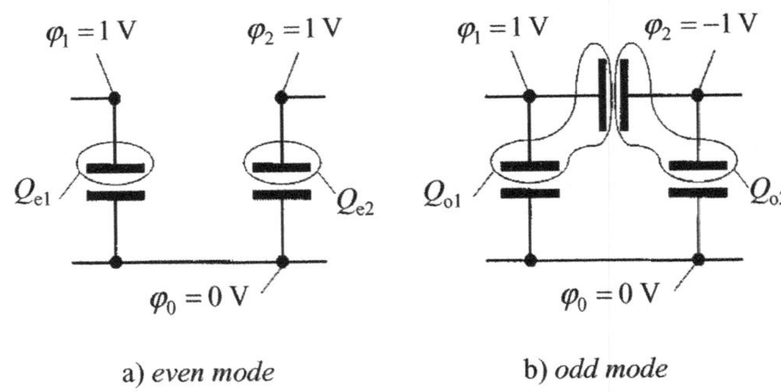

Fig. 3.5.9. The equivalent circuit of a coplanar waveguide gap under (a) the even potential and (b) the odd potential condition.

In Fig. 3.5.9b the odd case of the potential distribution is shown; that is, $\varphi_1 = 1$ V, $\varphi_2 = -1$ V, and the potential of the ground plane is again $\varphi_0 = 0$ V. The adjungated equivalent circuit contains all three capacitances. For both cases the resulting potential distribution on the discontinuity is computed using the three-dimensional finite difference analysis technique (Section 3.1), and the charges Q_{e1}, Q_{e2} on the electrodes as shown in Fig. 3.5.9a as well as the charges Q_{o1}, Q_{o2} on the electrodes as shown in Fig. 3.5.9b are computed. Using these charges, the elements of the equivalent circuit can be determined as follows:

$$C_{p1} = \frac{Q_{e1}}{(\varphi_1 - \varphi_0)} - l_1 C_1', \tag{3.5.2}$$

$$C_{p2} = \frac{Q_{e2}}{(\varphi_2 - \varphi_0)} - l_2 C_2', \tag{3.5.3}$$

and

$$C_g = \frac{Q_{o1} - Q_{e1}}{\varphi_1 - \varphi_2} - l_1 C_1' = \frac{Q_{o2} - Q_{e2}}{\varphi_1 - \varphi_2} - l_2 C_2', \tag{3.5.4}$$

where l_1 and l_2 are the lengths of the two coplanar waveguides between the reference planes and the gap discontinuity. C_1' and C_2' are the capacitances per unit line length of the two uniform coplanar waveguides, respectively.

The relation between the three capacitances has been determined using Eqs. (3.5.2) to (3.5.4) and the electric field disturbances in the discontinuity region that can be observed from Fig. 3.5.10. In this figure, the electric field normal to the substrate and the metalization plane is shown for the even- and the odd-mode potential distribution on the center conductors. It can be clearly seen that the main field disturbances occur very near to the gap discontinuity and that immediately behind the gap the typical electric field distribution of the coplanar waveguides is regenerated at some distance from the discontinuity. For the gap discontinuity, if the gap width g is small, it may be essential to consider the influence of the metalization thickness t on the capacitance C_g. This can be easily done using the method described in Section 2.2.6.

To determine up to what frequencies the equivalent circuit model is valid, several gap discontinuities have been fabricated on ceramic substrate and their scattering parameters have been measured up to frequencies of 40 GHz. The

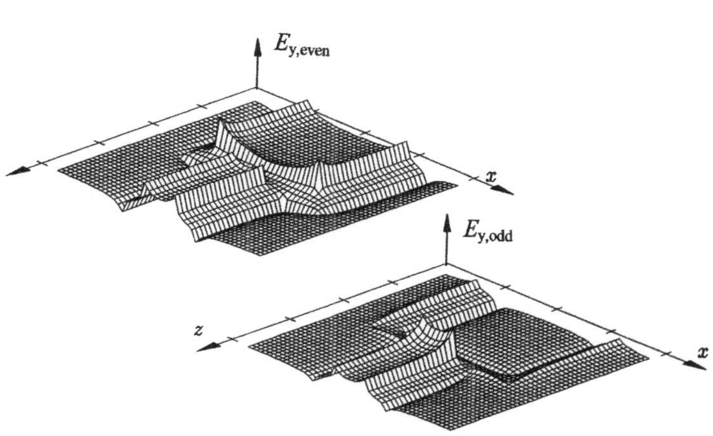

Fig. 3.5.10. The electric field component normal to the substrate and the metalization plane for (a) the case of the even potential distribution and (b) the case of the odd potential distribution on the center conductors of the two coplanar waveguides in a gap discontinuity.

results of theses measurements in comparison to the simulation results are shown in Fig. 3.5.11. The simulated scattering parameters have been derived from the equivalent circuit representation. The figure shows a quite good agreement between measurement and simulation results for the magnitudes up to frequencies of 30 GHz. The very good agreement of the phases even for higher frequencies is a hint as to the high quality of the derived equivalent circuit representation.

The dependence of the three capacitances C_g, C_{p1}, and C_{p2} on the normalized gap width g/h is shown in Figs. 3.5.12 and Fig. 3.5.13. Also shown in the figures are measured results for various coplanar gap discontinuities on ceramic substrate. The measurements have been performed as scattering parameter measurements using a vector network analyzer. The measured

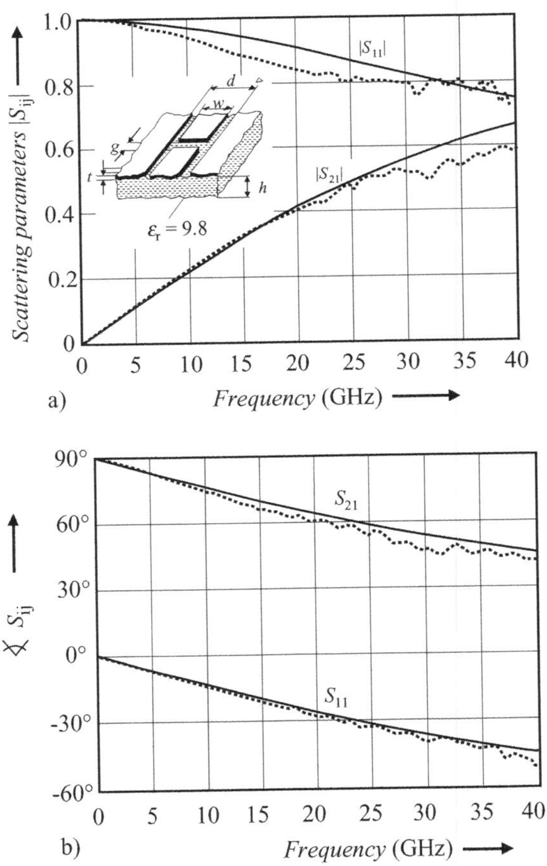

Fig. 3.5.11. Measured (. . .) and simulated (———) magnitude (a) and phase (b) of the scattering parameters of a gap in the center conductor of a coplanar waveguide, plotted against frequency. Parameters: $w = 350\,\mu m$, $d = 675\,\mu m$, $g = 25\,\mu m$, $h = 625\,\mu m$, $t = 5\,\mu m$, calculated equivalent circuit parameters: $C_g = 36.5\,fF$, $C_{p1} = C_{p2} = 1.7\,fF$.

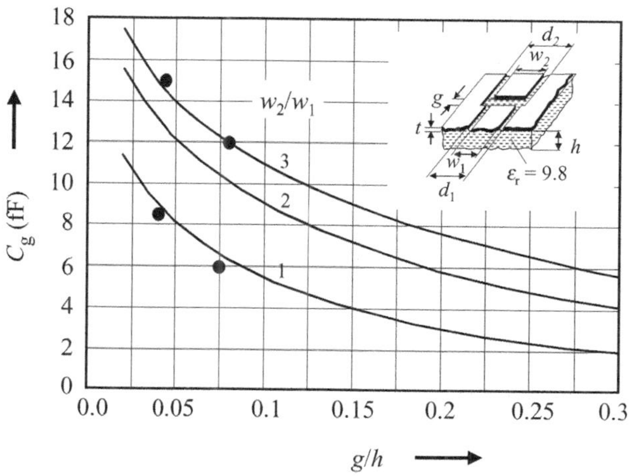

Fig. 3.5.12. Measured (•) and simulated (———) coupling capacitances C_g of gaps in different coplanar 50-Ω waveguides in dependence on the gap width g. ($w_1/d_1 = w_2/d_2 = 6/11$, $d_1/h = 0.352$, $h = 625\,\mu m$, $t = 5\,\mu m$).

scattering parameters are then converted to the capacitance values using the equivalent network and an optimization routine for parameter fitting. The line width parameters w_1/d_1 and w_2/d_2 have been chosen in such a way that the two coplanar waveguides, despite their different center strip widths, have a characteristic impedance of 50 Ω.

Figure 3.5.12 shows that the coupling capacitance C_g decreases with increasing gap width, as would be expected. It also decreases with a decreasing w_2/w_1 ratio. Also, this effect can be directly explained from the geometry of the gap structure.

More interesting is the dependence of the parallel capacitances C_{p1} and C_{p2} on the gap width g, as shown in Fig. 3.5.13. Especially for the capacitance C_{p1}, it can be seen that the value may become negative for a gap between two coplanar waveguides of unequal line widths w_1 and w_2. The reason for this is the fact that the electric field lines of the center conductor with the larger line width (w_2) end also on the ground-plane region of the waveguide with the smaller center conductor width (w_1). As a consequence, the field lines of this conductor are pushed back along the line, and near the discontinuity the line capacitance per unit line length is smaller than that of the undisturbed line. This means that the additional capacitance of the discontinuity becomes negative.

To test the broadband capabilities of the derived model for the coplanar waveguide gap, several gap-coupled resonators have been built on GaAs substrate material of height $h = 450\,\mu m$. Two of them, one of a 40-Ω coplanar waveguide and the other of a 70-Ω coplanar waveguide, are shown in Fig. 3.5.14 together with their geometrical parameters.

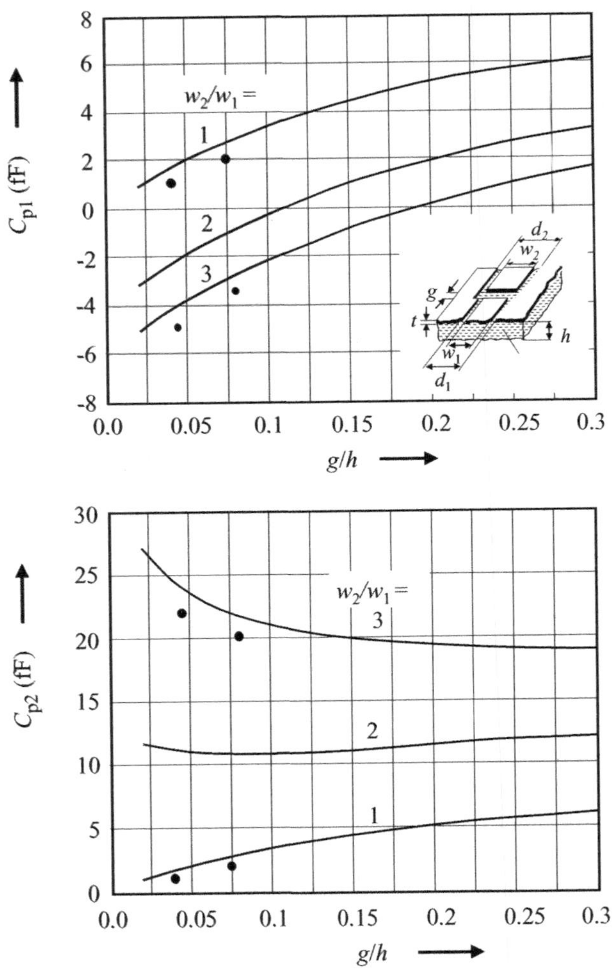

Fig. 3.5.13. Measured (•) and simulated (———) capacitances C_{p1} and C_{p2} of gaps in different coplanar 50-Ω waveguides, plotted against gap width g ($w_1/d_1 = w_2/d_2 = 6/11$, $d_1/h = 0.352$, $h = 625\,\mu m$, $\varepsilon_r = 9.8$, $t = 5\,\mu m$).

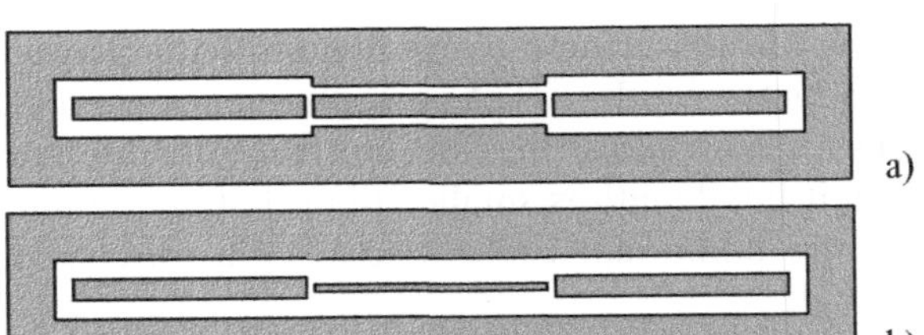

Fig. 3.5.14. Gap-coupled coplanar resonators as test structures. Feed lines are 50-Ω coplanar waveguides. (a) 40-Ω coplanar resonator, length $l = 1000\,\mu m$, total slot width $d = 240\,\mu m$, gap width $g = 25\,\mu m$. (b) 70-Ω coplanar waveguide resonator, length $l = 1000\,\mu m$, total slot width $d = 202\,\mu m$, gap width $g = 25\,\mu m$. Substrate GaAs, $\varepsilon_r = 12.9$, $h = 450\,\mu m$.

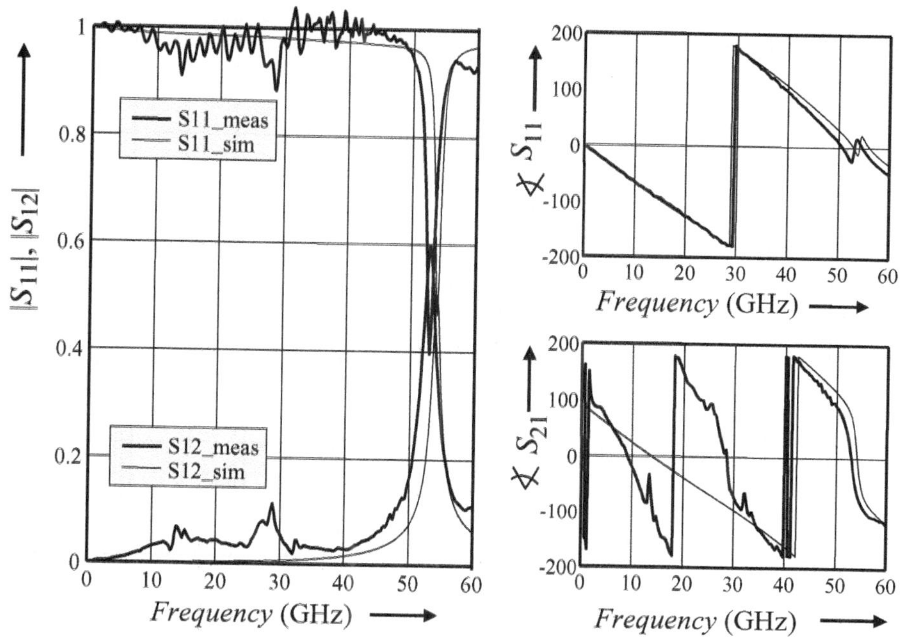

Fig. 3.5.15. Comparison between measured and simulated scattering parameters of the gap-coupled 40-Ω coplanar test resonator.

Figures 3.5.15 and 3.5.16, respectively, show the comparison between measurement and simulation (including the two coupling gaps) for the 40-Ω resonator and for the 70-Ω resonator. The test circuit behavior is like a $\lambda/2$-resonator of 1000-μm length. The measured resonant frequencies are $f_{r,\text{meas}} = 53\,\text{GHz}$ and $f_{r,\text{meas}} = 54.2\,\text{GHz}$. The deviation between measurement and simulation results from the calculated ε_{eff}, which is a static value for the applied FD-simulator ($\varepsilon_{\text{eff,static}} = 6.66$). The actual ε_{eff} is frequency-dependent and can be simulated using a spectral domain analysis technique and is found to be $\varepsilon_{\text{eff}}(f = 60\,\text{GHz}) = 7.106$. The theoretical resonant frequencies for these values are $f_r(\varepsilon_{\text{eff}} = 6.66) = 58.39\,\text{GHz}$, $f_r(\varepsilon_{\text{eff}} = 7.106) = 56.65\,\text{GHz}$ without considering the influence of the coupling gaps. Figure 3.5.16 shows the measurement results for the 70-Ω test-resonator. The difference between the measured and the simulated resonant frequency is still a little bit larger in this case, because the slot width of the 70-Ω coplanar waveguide is larger than that of the 40-Ω resonator (see Fig. 3.5.14) and this leads to an increased dispersion of the effective dielectric constant.

3.5.4 The Coplanar Waveguide Step

Figure 3.5.17 shows a discontinuity where, in a defined area, the center conductor width of a coplanar waveguide changes from a value w_1 to a value w_2.

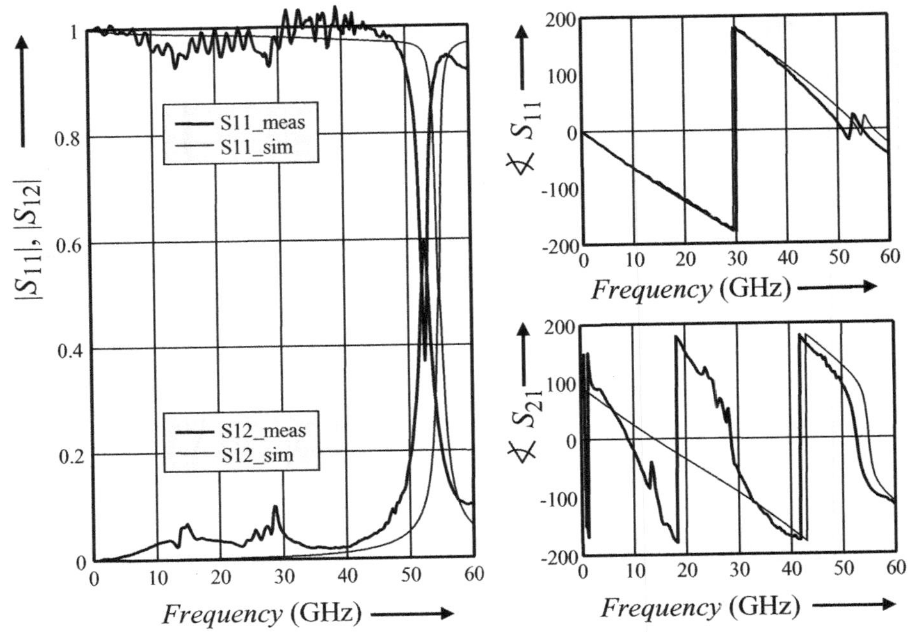

Fig. 3.5.16. Comparison between measured and simulated scattering parameters of the gap-coupled 70-Ω coplanar test resonator.

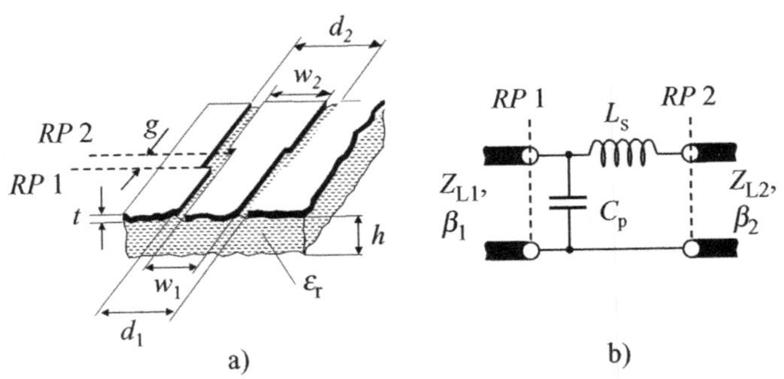

Fig. 3.5.17. The coplanar waveguide step (a) and its equivalent circuit (b).

In the same area, the total slot width is also changed from d_1 to d_2. This discontinuity often is called an impedance step. In coplanar technology, however, this discontinuity is not necessarily an impedance step because two coplanar waveguides of the same characteristic impedance but with different center conductor widths may be designed as has already been discussed in various sections. In most cases this discontinuity, however, is used as an impedance step. Nevertheless, because of the reason given above, this continuity shall be called here a coplanar waveguide step.

The scattering of the fundamental coplanar quasi-TEM mode at the step discontinuity will be described using the equivalent circuit shown in Fig. 3.5.17b. Similar circuits have been used for modeling the microstrip step in the literature [e.g., references 3, 10, 11, and 20]. The inductance L_s is a measure for the interruption in the current density at the step from the wide to the narrow center conductor width. The capacitance C_p describes the disturbances of the electric field near the step. The way as to how to determine the capacitive and the inductive elements of the equivalent circuit has been described in detail in Sections 3.3 and 3.4 and does not need to be repeated here.

The dependence of the parallel capacitance C_p and the inductance L_s on the geometrical line parameters is shown in Figs. 3.5.18 and 3.5.19, respectively. For all the shown results the dimensions of the waveguides have been chosen in such a way that the characteristic impedances of the two waveguides are equal. The distance g between the two waveguides (see Fig. 3.5.17) in all simulated cases is kept constant to 50 μm.

Several test structures of different sizes on GaAs substrate material have been used to measure the broadband response of coplanar steps. A first example is shown in Fig. 3.5.20. A coplanar waveguide step from a 50-Ω waveguide to a 40-Ω waveguide and back to a 50-Ω waveguide is used to measure the scattering parameters of the structure over a frequency range of 60 GHz. The geometrical parameters of the lines are given in the figure legend.

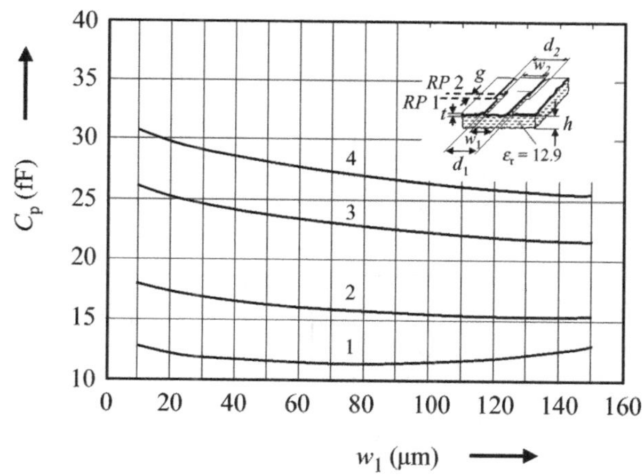

Fig. 3.5.18. Equivalent capacitance C_p of a coplanar waveguide step, plotted against the center conductor width w_1 (h = 400 μm, d_1 = 170 μm, g = 50 μm, t = 5 μm).

Curve Number:	1	2	3	4
w_2 (μm):	150	250	350	400
d_2 (μm):	350	350	390	420
Z_{L2} (Ω):	50	35	26	22

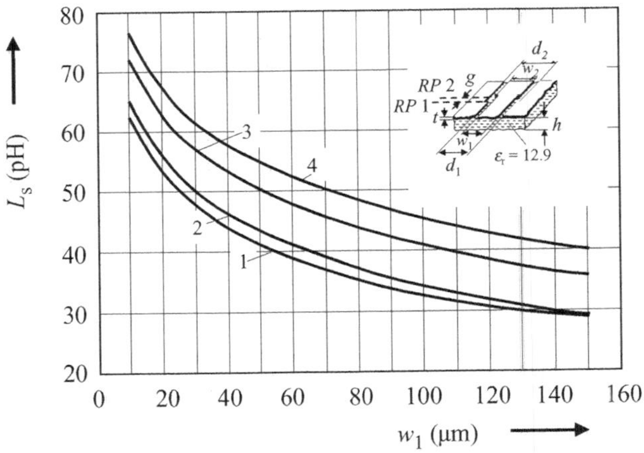

Fig. 3.5.19. Dependence of the equivalent inductance for a coplanar waveguide step on the center conductor width w_1 (all other parameters as in Fig. 3.5.18).

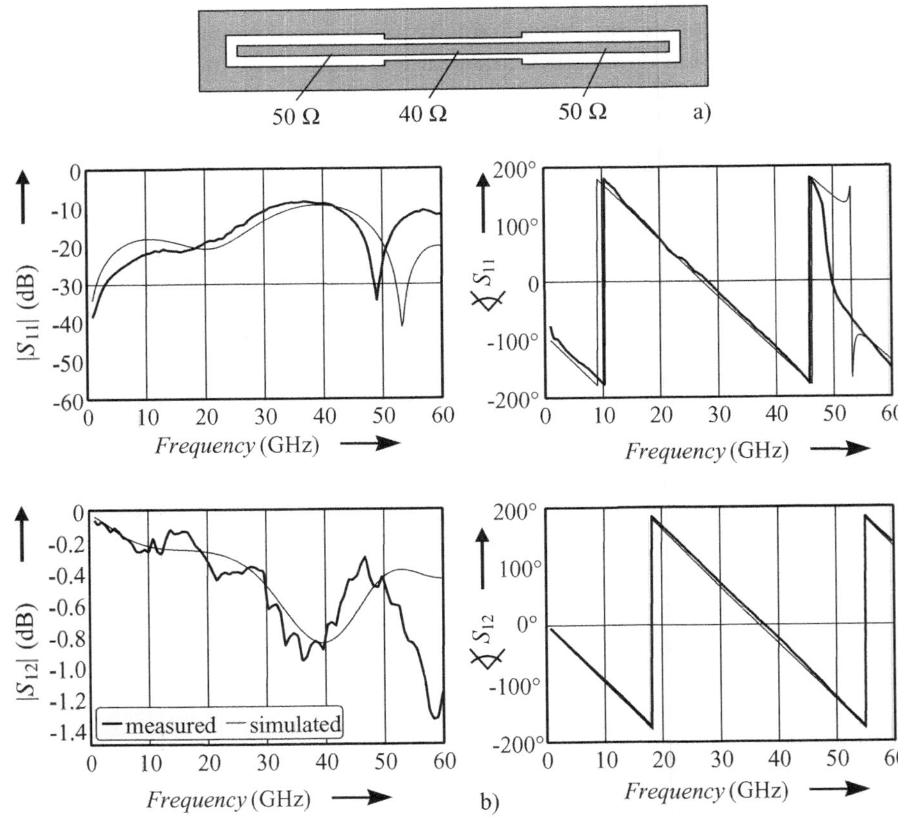

Fig. 3.5.20. (b) Comparison of measured and simulated scattering parameters of a coplanar test structure with two waveguide steps as shown in part (a). Geometrical parameters: For 50-Ω waveguide, $w = 100\,\mu m$ and $s = 75\,\mu m$; for 40-Ω waveguide, $w = 100\,\mu m$ and $s = 35\,\mu m$. Step parameter $g = 75\,\mu m$. Substrate GaAs, $\varepsilon_r = 12.9$, $h = 450\,\mu m$.

The simulated characteristic impedances of the lines are 48.4 Ω at low frequencies and 49.13 Ω (from a spectral domain analysis) at 60 GHz for the 50-Ω line ($\varepsilon_{\text{eff}} = 6.66$ at $f = 0$ and 7.202 at 60 GHz). For the 40-Ω line the equivalent parameters are: $Z_L = 39.4\,\Omega$, $\varepsilon_{\text{eff}} = 6.66$ at $f = 0$, and 39.88 Ω and $\varepsilon_{\text{eff}} = 7.106$ at $f = 60$ GHZ. As Fig. 3.5.20a shows, the step in the waveguide is not placed in the center strip, but in the ground plane structure. Figure 3.5.20b shows the measured and simulated scattering parameters, separately for magnitude and phase.

The results show a good agreement up to frequencies of about 45 GHz. For higher frequencies the dispersion of the coplanar waveguides affects the results. The simulated results have been calculated on the basis of the described static analysis technique, so that the waveguide dispersion cannot be taken into account. Figure 3.5.21 shows similar results for a waveguide step

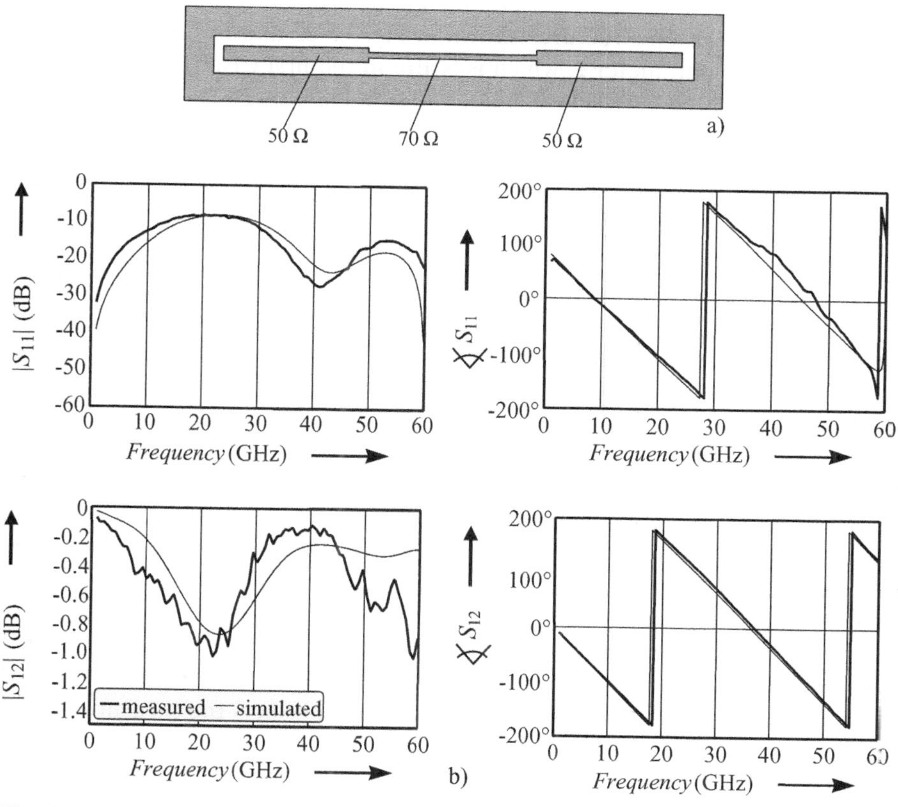

Fig. 3.5.21. (b) Comparison of measured and simulated scattering parameters of a coplanar test structure with two waveguide steps as shown in part (a). Geometrical parameters: For 50-Ω waveguide, w = 100 μm and s = 75 μm; for 70-Ω waveguide, w = 40 μm and s = 103 μm; total width of the structure, b is 650 μm. Length of the 70-Ω section; l, is 1000 μm, step parameter g is 75 μm. Substrate: GaAs, $\varepsilon_r = 12.9$, h = 450 μm.

from 50Ω to 70Ω and back to 50Ω, where, in this case, the width of the center conductor has been changed at the discontinuities.

The geometrical parameters of the structure are given again in the figure inscription. The measured and simulated results are in a quite good agreement up to 60GHz; only the insertion loss shows some deviations at higher frequencies. This better agreement results from the lower dispersion of the structure, which is especially dependent on the gap width s of the coplanar waveguides.

A good structure for testing the accuracy of the gap model is a series connection of three coplanar waveguides of the same characteristic impedance $Z_L \approx 50\Omega$ but with different geometrical line parameters on GaAs substrate material. It is also a good structure to test the influence of the step parameter g (Figure 3.5.17) on the properties of a coplanar waveguide step. Such structures are shown in Figs. 3.5.22 and 3.5.23.

In both figures three series-connected 50-Ω coplanar waveguides are shown that are of different geometrical parameters (see figure legends). The difference between the step structures shown in Fig. 3.5.22 and in Fig. 3.5.23 is the value of the gap parameter g, which is 50μm in the first case (Fig. 3.5.22) and 250μm in the second case (Fig. 3.5.23). The simulated quasi-static characteristic impedances of the used two waveguides are $Z_L = 49.8\Omega$ and $Z_L = 49.5\Omega$, respectively—that is, nearly the same value. The simulated effective dielectric constants are a little bit more different due to the different geometrical parameters (see Figs. 3.5.22 and 3.5.23): They are $\varepsilon_{eff} = 6.66$ and $\varepsilon_{eff} = 6.35$, respectively.

Figures 3.5.22b and 3.5.23b show the comparison of the measured and the simulated magnitude and phase angles of the scattering parameters, plotted against frequency. The results show a good agreement of the simulated and measured results over a frequency range of 50GHz for the magnitude as well as for the phase angle. The results also show that the reflection coefficient is always well below −10dB over the total frequency range for the step with the smaller gap parameter g (50μm, Fig. 3.5.22), and the insertion loss is not higher than 0.4dB. In the case of the step with the higher g value (g = 250μm, Fig. 3.5.23) the results are not so satisfactory. The reflection coefficient increases to values higher than −10dB at a frequency of about 45GHz, and the insertion loss becomes higher than 0.5dB at frequencies above 40GHz. These results show that the step parameter g can be used to optimize the transmission behavior of coplanar waveguide steps and (and this is very essential for the circuit design) that the derived model clearly predicts the measured scattering parameters over a large frequency range.

As has been demonstrated above, the coplanar waveguide step can be designed with a reflection coefficient smaller than −10dB over a large frequency range, which means that it is not a big discontinuity in the waveguide structure. The small reflection coefficient may still be improved using a linear taper between the waveguide sections [13]. The resulting reflection coefficient is dependent on the line widths and the taper length l. In the case of linear

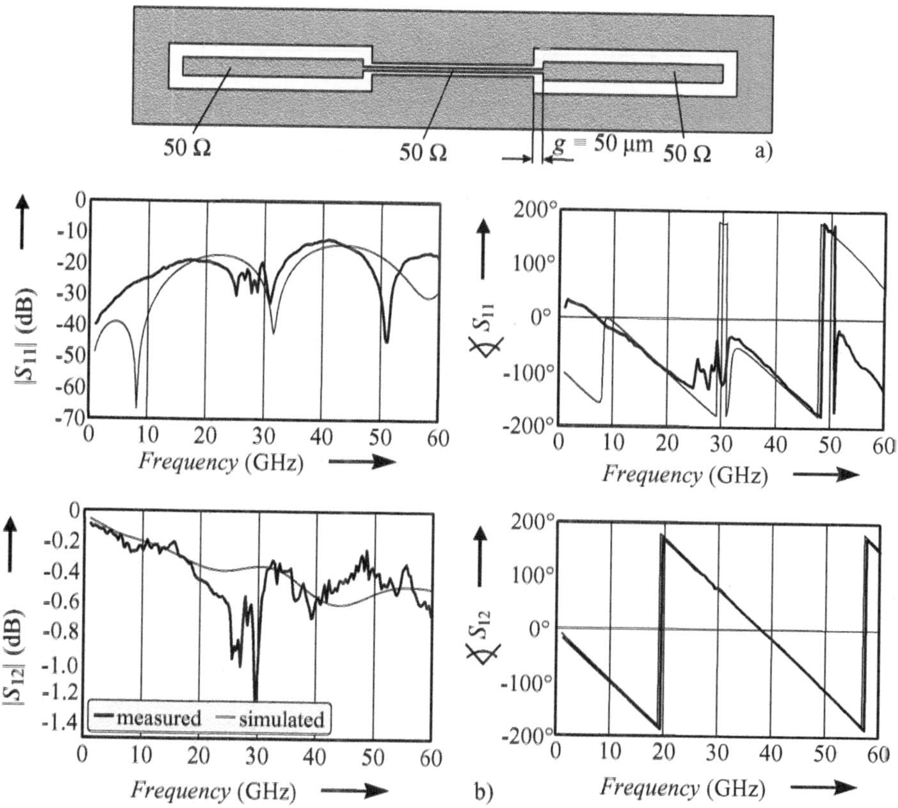

Fig. 3.5.22. (b) Comparison of measured and simulated scattering parameters of a coplanar test structure with two waveguide steps as shown in part (a). Geometrical parameters: For first 50-Ω waveguide, $w = 100\,\mu m$ and $s = 75\,\mu m$; for second 50-Ω waveguide (center line section), $w = 25\,\mu m$ and $s = 19\,\mu m$; total width of the structure, b, is $650\,\mu m$. Length of the center 50-Ω section, l, is $900\,\mu m$; step parameter, g, is $50\,\mu m$. Substrate GaAs; $\varepsilon_r = 12.9$, $h = 450\,\mu m$.

coplanar tapers with an angle $\alpha < 45°$, as shown in Fig. 3.5.24, the reflection may become negligibly small. This is demonstrated in Fig. 3.5.25, where measurement results of a 50-Ω–50-Ω–50-Ω waveguide step and a linear coplanar 50-Ω–50-Ω–50-Ω waveguide taper are compared. All geometrical waveguide parameters are identical for both structures (Fig. 3.5.24a). From a comparison of the measured results, it may be observed that especially for low frequencies the reflection coefficient of the taper is lower whereas the transmission coefficients of both structures are nearly identical over the total measured frequency range.

If the results of the taper are compared to the measurement results that are shown in Fig. 3.5.22 for the step construction with a step parameter $g = 50\,\mu m$, no big improvements may be found in using a taper construction instead of a

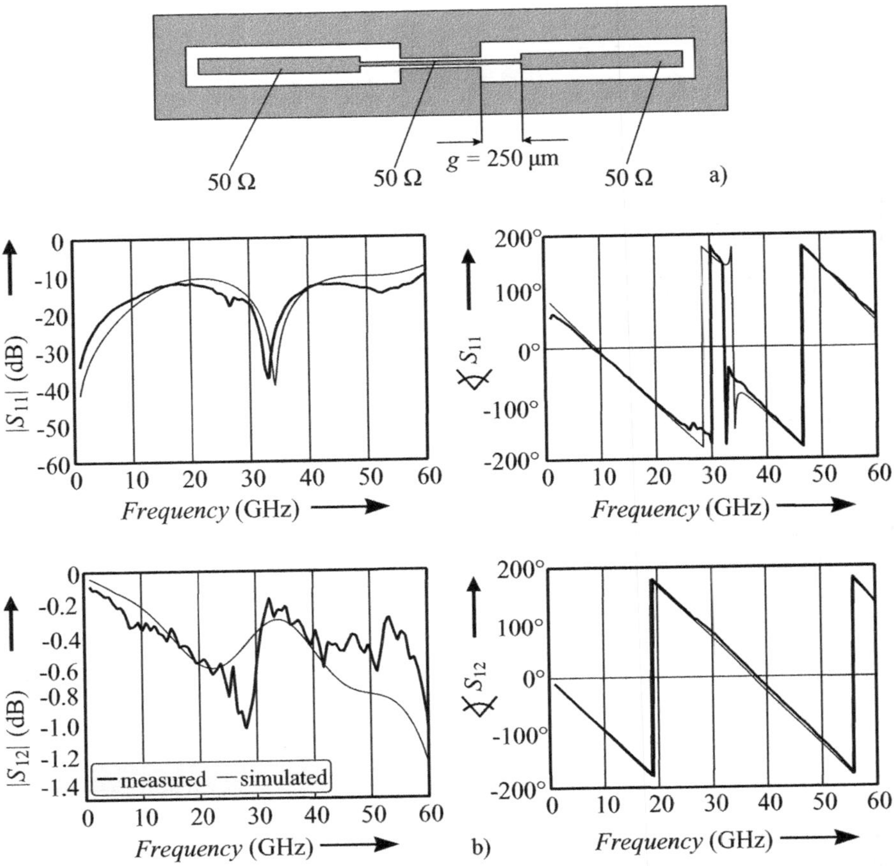

Fig. 3.5.23. (b) Comparison of measured and simulated scattering parameters of a coplanar test structure with two waveguide steps as shown in part (a). Geometrical parameters: For first 50 Ω waveguide, $w = 100\,\mu m$ and $s = 75\,\mu m$; for second 50-Ω waveguide (center line section), $w = 25\,\mu m$ and $s = 19\,\mu m$; total width of the structure, b, is $650\,\mu m$. Length of the center 50-Ω section, l, is $500\,\mu m$; step parameter g is $250\,\mu m$. Substrate GaAs; $\varepsilon_r = 12.9$, $h = 450\,\mu m$.

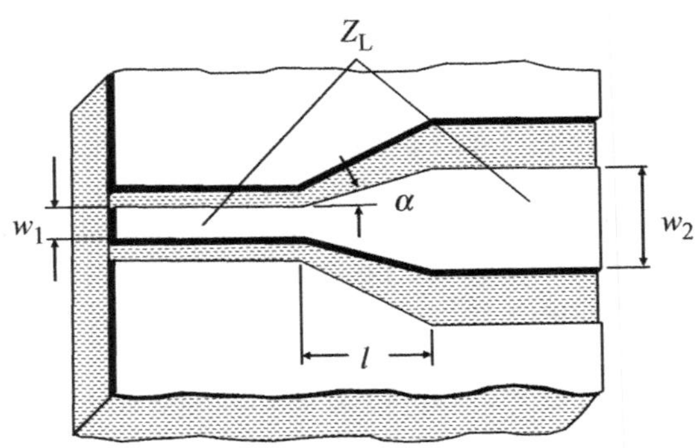

Fig. 3.5.24. Coplanar waveguide taper between two coplanar waveguides of the same characteristic impedance.

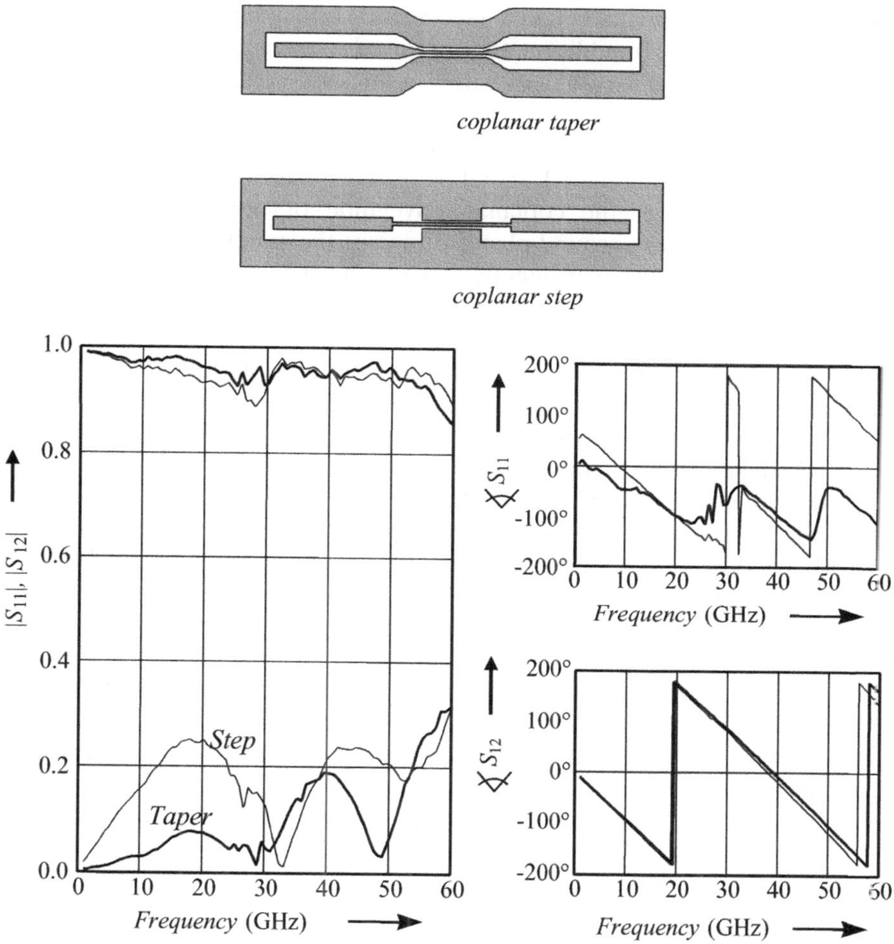

Fig. 3.5.25. Comparison between measured scattering parameters of a coplanar waveguide step (thin lines) and a linear coplanar taper (thick lines). All line sections have a 50-Ω characteristic impedance. Geometrical parameters of the line sections are the same as in Figs. 3.5.22 and 3.5.23. Transition length $g = 250\,\mu m$. Substrate GaAs, $\varepsilon_r = 12.9$, $h = 450\,\mu m$.

step. In the case of the step shown in Fig. 3.5.22, the total length of the transition is even much smaller compared to the taper construction.

3.5.5 Air Bridges in Coplanar Waveguides

Air bridges are indispensable to monolithic microwave integrated circuit design in coplanar waveguide technique. They have to ensure the biasing of active areas on the chip and they are necessary to maintain the ground-plane potential across the wafer. They are also indispensable as connecting elements in the design of interdigital couplers (see Section 6.5) and planar spiral inductors (see Section 4.3). In addition, they are used to connect coupled slot lines or to connect coplanar waveguides with slot lines (see, e.g., Sections 3.5.7 and 3.5.8).

Air bridges are fundamental components mainly used to suppress multimode propagation along the RF signal paths by equalizing the ground potentials on both sides of the coplanar waveguide or component. Potential differences between the ground planes often are the result of different propagation times of the electromagnetic wave along the ground plane/slot line structures. Different propagation times arise at discontinuities and if the ground plane structure is not symmetrically placed inside a circuit layout. Therefore the air bridge, which itself is a discontinuity, often is used in direct connection with other discontinuities (see discussion in Section 3.5.6 and following sections), and it is of tremendous help to guarantee that only the fundamental coplanar TEM mode propagates inside a circuit.

On the other hand, air bridges represent frequency-dependent discontinuities to the RF transmission lines causing losses and phase shifting. These parasitic effects depend on the physical size and the location of the air bridges inside the circuit. Neglecting these effects may lead to large deviations of the design goals from the measured data. Moreover, these deviations will increase with the number of air bridges in a circuit.

Typically, the size of an air bridge in an MMIC is very small. Its height h_{br} above the substrate is about $3\,\mu m$, and the length bw (Fig. 3.5.26) ranges from $10\,\mu m$ up to $50\,\mu m$. There are at least two principal ways to build an air bridge in an MMIC fabrication process: Either the inner conductor is galvanic-plastically built as a bridge across the deposited short-circuiting strip or the

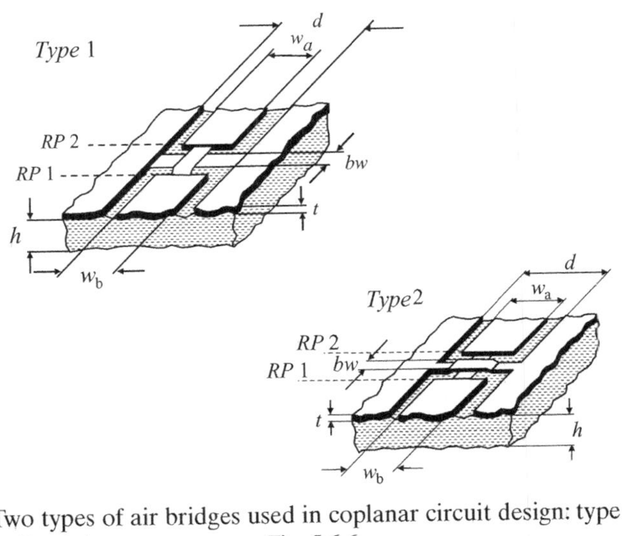

Fig. 3.5.26. Two types of air bridges used in coplanar circuit design: type 1 and type 2. For more details of the structures see Fig. 5.6.6.

inner conductor remains a thin and resistive evaporated layer (gate metallization layer) and the strip short-circuiting the outer conductors of the coplanar waveguide is built as a bridge crossing the center strip in an electroplating process.

To distinguish the two principally different types of air bridges, the first type is called type 1 and the second one type 2. Both types of air bridges are shown in Fig. 3.5.26.

For a first estimation of the air-bridge properties, the cross section of type 1 may be regarded as an air-filled microstrip line (Fig. 3.5.27a) while that of type 2 may be regarded as a very narrow covered coplanar waveguide (Fig. 3.5.27b). For both cases, theoretical results predict a considerable reduction of the characteristic impedance and the effective permittivity of the undisturbed coplanar waveguide in the air-bridge region. As mentioned earlier, the air-bridge height is typically very small (3 μm). Therefore, a significant additional capacitance per unit line length is located between the inner conductor of the coplanar waveguide and the short-circuiting strip for both kinds of air bridges.

Obviously, the characteristic impedance of the undisturbed coplanar waveguide (Z_L) is larger than that in the air-bridge region (Z_{ab}). A simple estimation results in

$$Z_L = \frac{1}{\sqrt{C'L'}} \geq Z_{ab}, \qquad (3.5.5)$$

where C' and L' denote the capacitance and inductance per unit length of the coplanar waveguide, respectively.

A coplanar waveguide built on GaAs substrate ($\varepsilon_r = 12.9$) with an inner conductor width of 75 μm and a slot width of 50 μm has a characteristic imped-

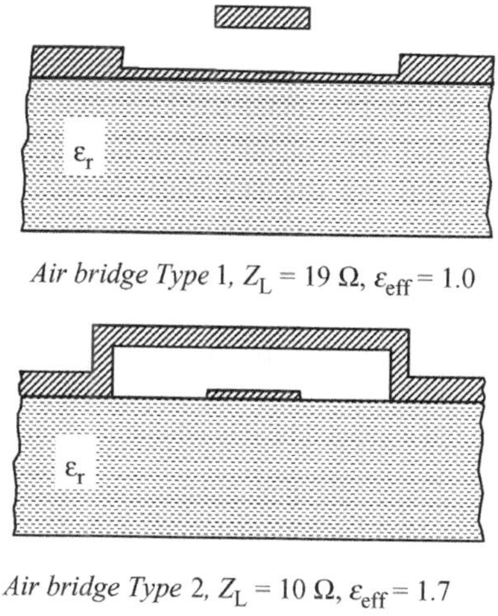

Fig. 3.5.27. The cross section of both types of air-bridge discontinuities.

ance that is near to 50 Ω and the effective permittivity is about 7. An air-bridge discontinuity built into such a line leads to a local change of the characteristic impedance and will reduce the effective permittivity of the transmission line. For the lateral dimensions given above and an air-bridge height of 3 μm the characteristic impedance at the location of the discontinuity caused by a type 1 air bridge and by a type 2 air bridge is reduced to ~19 Ω and ~10 Ω, respectively. The value of the effective permittivity is reduced to exactly 1.0 for type 1 and to about 1.7 for type 2 in the crossing region.

Due to these numerical data, both types can be expected to be a severe discontinuity in the coplanar waveguide structure. Fortunately, the total length of an air bridge is typically small (10–50 μm), and the influence of one single bridge on the transported RF power remains negligible. However, for practical MMIC designs a large number of air bridges is necessary, leading to an accumulation of attenuation and phase shifting effects. Additionally, it is important to keep in mind that due to the missing galvanic enlargement of the evaporated layer below the bridge, the short-circuiting conductor of type 1 and the inner conductor of type 2 are significantly more resistive than the normal coplanar waveguide structure.

This means that an additional resistance per unit length (R') of the coplanar waveguide has to be considered if the air bridge of type 2 is applied. This resistance per unit line length causes additional losses depending on the total length and on the inner conductor's width of the type 2 air bridge. The air bridge of type 1 is almost not concerned by this fact related to the RF power flow.

In order to make plain the effect of the reduced value of the effective permittivity in the air-bridge region, an experimental arrangement has been examined. A number of coplanar waveguides were built as resonant lines, open ended at both sides. The reference line was without any air bridge and had a resonant frequency near 15 GHz. Additional test lines included a number of 10 air bridges. Five air bridges were located near to both ends of the coplanar waveguides. Both types of air bridges were examined. The lengths of the air bridges were 10 μm and 50 μm, respectively. The measured results of the resonant frequencies of these resonators are given in Fig. 3.5.28.

There are two effects shifting the resonant frequency of such a line resonator. On the one hand, the effective permittivity is reduced in the air-bridge areas. This effect gives a frequency shift to higher frequencies. On the other hand, there are additional capacitances to ground, loading the line resonator. This effect results in a frequency shift to lower frequencies. For the special resonator test-circuit under consideration, the capacitive load effect is prevailing for air bridges of type 2. For type 1 air bridges the effect due to the reduced effective permittivity is the dominant one. However, for both types of air bridges the phase velocity of the transmission line is changed compared to a transmission line without air bridge. This has to be considered for the design of MMICs.

Both air-bridge types can be used in MMIC design, but a selection is necessary in each case. A type 1 air bridge yields best results with respect to low

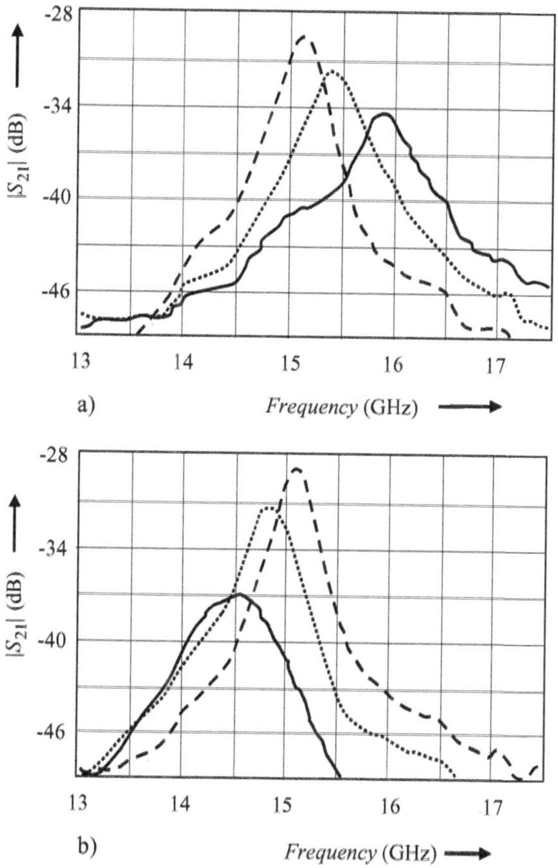

Fig. 3.5.28. Resonant frequency shift due to 10 air bridges of different lengths: bw = 10 μm (dotted line) and 50 μm (solid line) in comparison to a reference coplanar waveguide resonator without air bridges (dashed line). (a) Air bridge of type 1. (b) Air bridge of type 2.

RF losses on coplanar waveguides. The evaporated short-circuiting strip with higher resistance even with a length of only 10 μm is quite sufficient to suppress higher-order modes on the coplanar waveguide. Unfortunately, such a small strip may reduce the efficiency of an active circuit because the ohmic losses cannot be neglected if a biasing current has to be transported by this strip. For this case, a required number of type 2 air bridges should be introduced into the circuit to ensure a low resistive bias of the active areas.

Figure 3.5.29 shows the normal components of the quasi-static electric field distribution inside the two types of air bridges that have been computed using the finite difference technique as described in Section 3.2. The different properties of the two air-bridge structures may be observed from these field distributions.

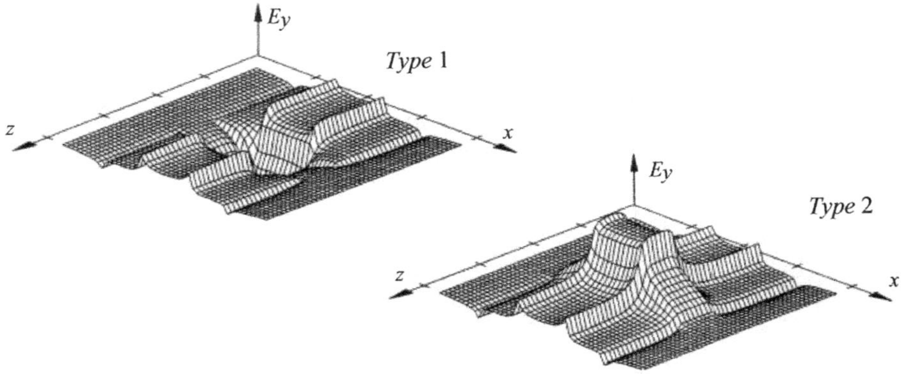

Fig. 3.5.29. Normal components of the electric field strength in air bridges of type 1 and type 2, respectively.

In reference 32 it has been suggested to reduce the distortion due to the low characteristic impedance of the air bridge discontinuity by reducing the center conductor's width. By this means, enlarging the value of the characteristic impedance by increasing the inductance per unit line length of the bridge is possible. On the other hand, reducing the width of the inner conductor will again lead to an increasing resistance per unit line length. Figure 3.5.30 shows the principle of these compensated air bridges as well as photos of their technological realizations.

To simulate this impedance matching effect, a 5-mm-long coplanar 50-Ω waveguide containing the two air-bridge structures as shown in Fig. 3.5.30 has been analyzed using the quasi-static finite difference technique. The result of the analysis is shown in Fig. 3.5.31. The figure shows the phase of the transmission coefficient of the waveguide without air bridge and for the cases where an air bridge is inserted into the waveguide in dependence on the bridge length bw. Parameter of the different curves is the width w_b of the center conductor in the compensation area. It can be observed from the figure that compensation is only possible for the type 1 air bridge.

Experimental investigations have shown that there is no significant advantage to compensate the low characteristic impedance of an air bridge by modifying the lateral dimensions of the conductor strip in the air bridge region in order to increase the inductance per unit line length. Although the characteristic impedance of the air-bridge discontinuity can be adjusted to 50 Ω, the disadvantage caused by additional losses due to the thinner strip prevails. So, matching the characteristic impedance normally is of no big advantage.

The renunciation of impedance matching contemporaneously means that the value of the effective permittivity will remain reduced for both types of air bridges. Furthermore, the above-described experimental investigations on the resonant frequency shift of coplanar waveguide resonators have shown that a considerable influence of these discontinuities must be taken into

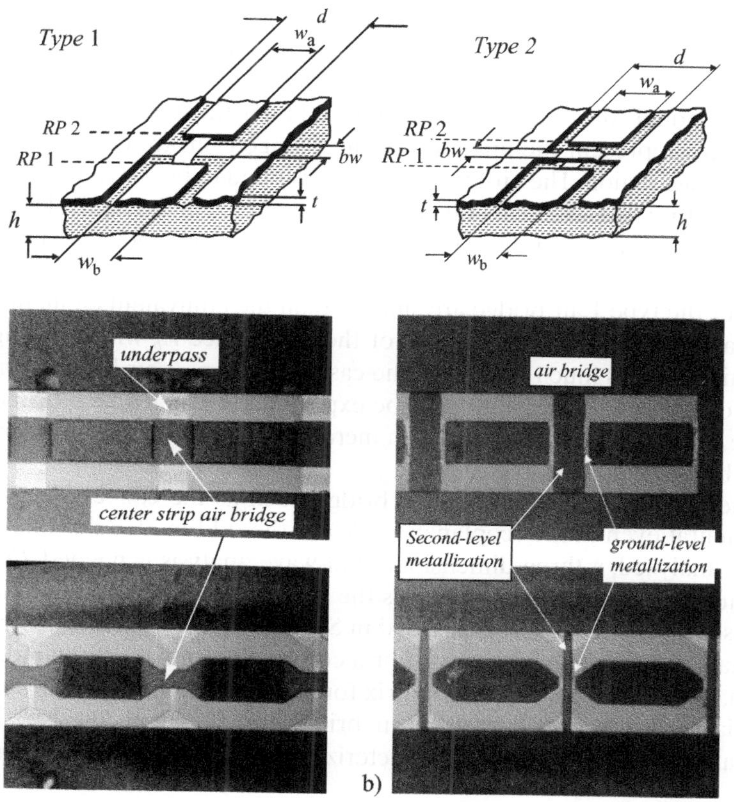

Fig. 3.5.30. Noncompensated and compensated air bridges of type 1 and type 2 and the technological realization of noncompensated and compensated air bridges. (a) Type 1, (b) type 2.

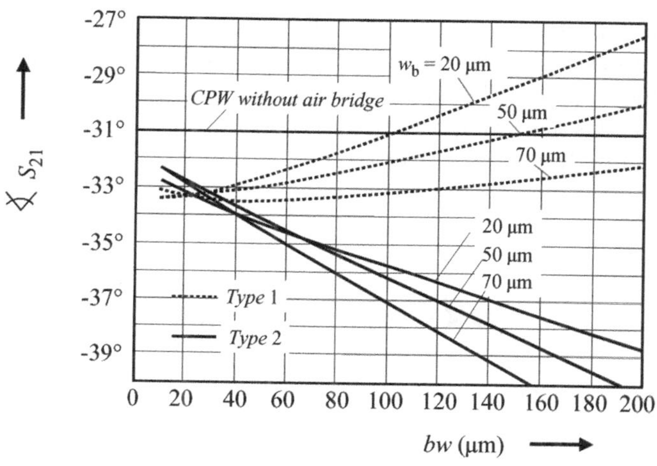

Fig. 3.5.31. Influence of the two types of air bridges on the phase of the transmission coefficient of a 50-Ω coplanar waveguide at 20 GHz (w_a = 70 μm, d = 170 μm, air-bridge height h_{br} = 2.5 μm, t = 2.5 μm. Substrate GaAs, ε_r = 12.9, h = 410 μm.

account and that therefore good models must be available that help to simulate the influence of the air bridges in circuit design.

Figure 3.5.32 shows an equivalent circuit model for the air bridges. A capacitance C_b models the additional capacitance under the air bridge, and the two inductances represent the influence of the changed surface current density in the air-bridge region. The air bridge can be assumed to be a symmetrical structure. Therefore two identical inductances are assumed in the equivalent circuit.

The dependence of the model parameters on the geometrical dimensions of the two discussed air-bridge types is shown in Fig. 3.5.33. Both elements C_b and L_b of the type 1 air bridge are smaller than the equivalent elements of the type 2 air bridge. Also, the increase of the inductance L_b with the length bw of the air bridge is much smaller in the case of the type 1 air bridge compared to that of the type 2 bridge. As may be expected, the capacitance C_b increases and the inductance L_b decreases with increasing width of the center strip w in the air bridge region.

If the mode conversion at an air bridge, such as the conversion from the fundamental even mode (coplanar mode) to the odd mode (slot line mode), is to be analyzed, a three-dimensional full-wave analysis is needed. One possible candidate for such an analysis is the three-dimensional moment method that bases on the technique described in Section 2.1.2 and that is described in full detail in references 59 and 80. For a correct description of the mode conversion, a generalized scattering matrix for the two considered modes must be defined; that is, the two-port (the air bridge together with the two feeding coplanar waveguides) must be characterized by an (4×4) element scattering matrix.

For the design engineer, the disturbance of the even mode (coplanar waveguide mode) that is caused by the air bridge as well as the suppression of the odd mode (slot-line mode) that may be reached is of interest. The disturbance of the even mode can be described by the reflection coefficient S_{11}^{ee} that describes the reflected even mode if the two-port is fed by the even mode. To study the suppression of the odd mode, the analysis of the transmission coefficient S_{21}^{oo} that describes the transmission of an odd mode into an odd mode is meaningful. This scattering parameter is a measure for the power of the odd mode that can be transmitted over the discontinuity.

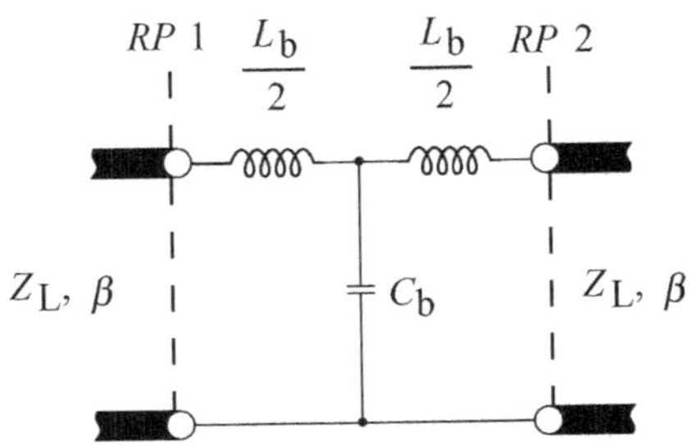

Fig. 3.5.32. Equivalent circuit for an air bridge in a coplanar waveguide.

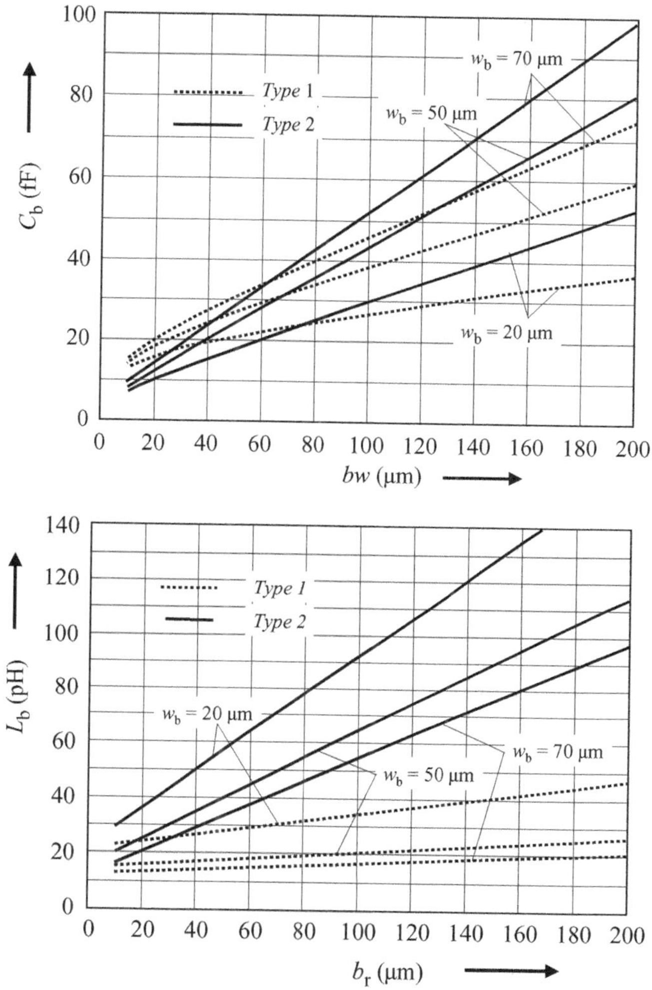

Fig. 3.5.33. The dependence of the equivalent circuit model elements for coplanar air bridges on the geometrical parameters ($w = 70\,\mu m$, $d = 170\,\mu m$, air-bridge height $h_{br} = 2.5\,\mu m$, $t = 2.5\,\mu m$).

In Fig. 3.5.34 the dependence of these two scattering parameters on the frequency is shown for the type 1 and the type 2 air bridges. Shown is a frequency range up to 60 GHz in that most of the coplanar microwave integrated circuits are used today. A comparison shows that both air bridges have a reflection coefficient S_{11}^{ee} below 1%. At higher frequencies, type 2 has a smaller reflection coefficient S_{11}^{ee} compared to that of the type 1 bridge. This is a result of the geometrical structures, especially of the air-bridge length. The type 2 air bridge investigated was only 50 µm long, whereas the type 1 air bridge had a length of 150 µm. The odd mode suppression of both air bridges is nearly identical

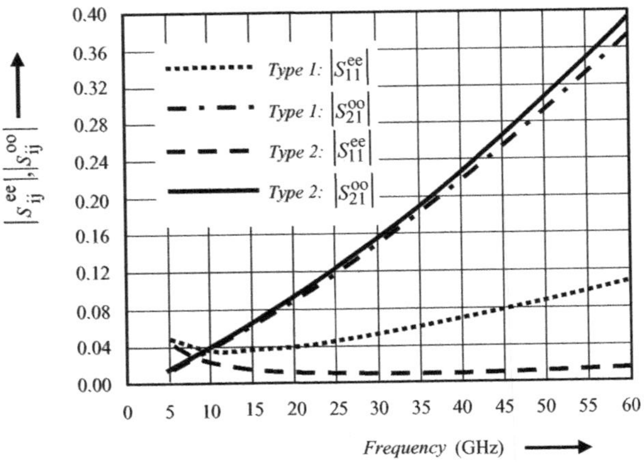

Fig. 3.5.34. The magnitude of the reflection coefficient S_{11}^{ee} and the magnitude of the transmission coefficient S_{21}^{oo} for a type 1 and a type 2 air bridge, plotted against the frequency.

but at higher frequencies slightly better for the type 1 bridge. As a result, it may be stated that both air bridges fulfill their tasks; that is, they have a small reflection coefficient for the even mode and excited odd modes are sufficiently suppressed.

3.5.6 The Coplanar Waveguide Bend

In microwave integrated circuit design, one of the essential requirements is the dense packaging of the layout. For this purpose, coplanar waveguides must be bended often. The waveguide bend, therefore, is an essential discontinuity often used in circuit design. The bend also is a discontinuity in that different propagation times of the wave along the ground planes and slot areas lead to a potential difference between the ground planes and therefore to the excitation of the odd mode on the coplanar structure.

As has already been discussed in detail in Section 3.5.5, these potential differences between the ground planes can only be avoided by using air bridges. Therefore, always a combination of the bend discontinuity and an air bridge structure will be considered here. Also in the circuit design, only this kind of coplanar bend is used. Figure 3.5.35 shows three different coplanar waveguide bends with integrated air bridges. The air-bridge structure used in Fig. 3.5.35a is the type 2 air bridge. In Fig. 3.5.35b the type 1 air bridge is used together with the bend. Finally, Fig. 3.5.35c shows a special form of the type 1 air bridge. In this structure the fundamental metallization layer of the structure (evaporated adhesive layer or gate metallization layer) connects the two ground planes in the total bend area, and the center conductor in this region is built as an air-bridge structure. This kind of coplanar bend will be called air-bridge bend in the future.

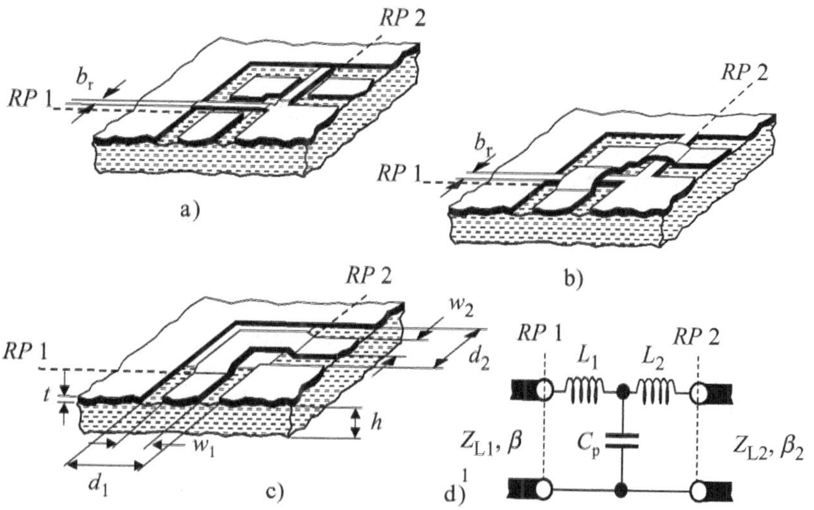

Fig. 3.5.35. (a) Coplanar waveguide right-angled bend with type 2 air bridges, (b) coplanar waveguide right-angled bend with type 1 air bridges, (c) coplanar waveguide right-angled air-bridge bend, and (d) equivalent circuit for the bends.

For modeling the bend discontinuity, the equivalent circuit shown in Fig. 3.5.35d will be used [5–7, 20]. The disturbances of the current density inside the discontinuity are described by the inductances L_1 and L_2. The capacitance C_p describes the changes of the electric field in the discontinuity region. Included into this capacitance is also the influence of the air bridges on the electric field distribution of the discontinuity. Analogue to the methods described in Section 3.2 and 3.3 a capacitance C_p and a total inductance L are determined using the quasi-static finite difference technique. The inductances L_1 and L_2 are then calculated from this total inductance L using the inductance per unit line length L_1' and L_2' of the coplanar waveguides as follows:

$$L_1 = L\frac{L_2'}{(L_1' + L_2')}, \tag{3.5.6}$$

$$L_2 = L\frac{L_1'}{(L_1' + L_2')}. \tag{3.5.7}$$

As already mentioned above, waveguide bends are used in integrated circuits to reduce the needed space of the circuit layout. The 90° bend requires smaller space than, say, a curved line that possibly has a smaller reflection coefficient and a higher transmission coefficient. If the coplanar air-bridge bend (Fig. 3.5.35c) is used, it is possible in principle to compensate the reflection behavior of the discontinuity by changing the size of the center conductor (see also the discussion in Section 3.5.5). Similar to the case of the microstrip bend, a better transmission property can be reached if the center conductor and the ground planes are truncated in the area of the discontinuity.

For the investigation of the equivalent circuit elements and their dependence on the geometrical parameters, a bend between two coplanar waveguides of different geometrical sizes has been used. A coplanar air bridge bend with constant total slot width $d_1 = d_2 = d$ (see Fig. 3.5.35c) but different center strip widths w_1 and w_2 was considered.

The results of the analysis are shown in Fig. 3.5.36 for the capacitance C_p and in Fig. 3.5.37 for the inductances L_1 and L_2. The values of the circuit elements are drawn in dependence on the center strip width w_1 and with constant values of w_2 as parameters. Under these conditions the capacitance C_p increases linearly with the center strip width w_1 of the first line.

The inductances L_1 and L_2, on the other hand, decrease with increasing values of w_1. From both figures it also follows that the capacitive effect of the discontinuity is dominant. The reason for this is that the disturbance of the electric field that is created by the relatively small air-bridge height (2.5 to 6 μm using conventional GaAs technologies) represents a large contribution to the capacitance C_p.

The value of C_p of course is also dependent on the applied type of air bridges. Therefore in Fig. 3.5.38 the capacitance C_p is shown for two different values of the center strip width w_2, plotted against the width w_1 for all three air-bridge types defined in Fig. 3.5.35. It may be observed that the capacitance is largest for the air-bridge bend discontinuity if the center strip width w_2 is large (upper curves in Fig. 3.5.38, $w_2 = 160\,\mu m$). For small values of w_2 there is not a big difference in the capacitances of the different air-bridge types (lower

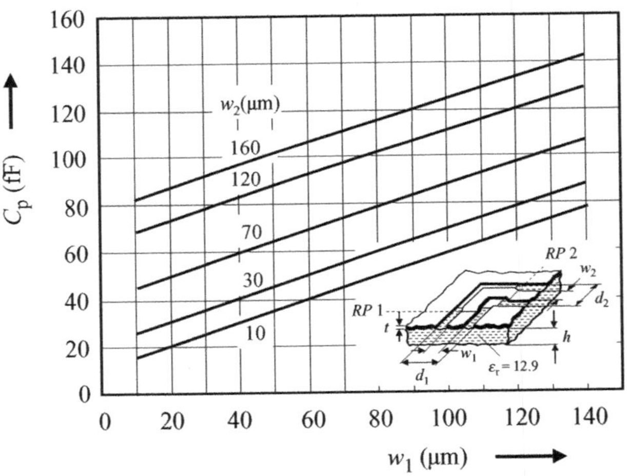

Fig. 3.5.36. The dependence of the equivalent capacitance C_p of the coplanar air-bridge bend on the geometrical parameters ($d_1 = d_2 = 170\,\mu m$. Substrate GaAs, $\varepsilon_r = 12.9$, $h = 400\,\mu m$, $t = 3\,\mu m$, air-bridge height $= 3\,\mu m$).

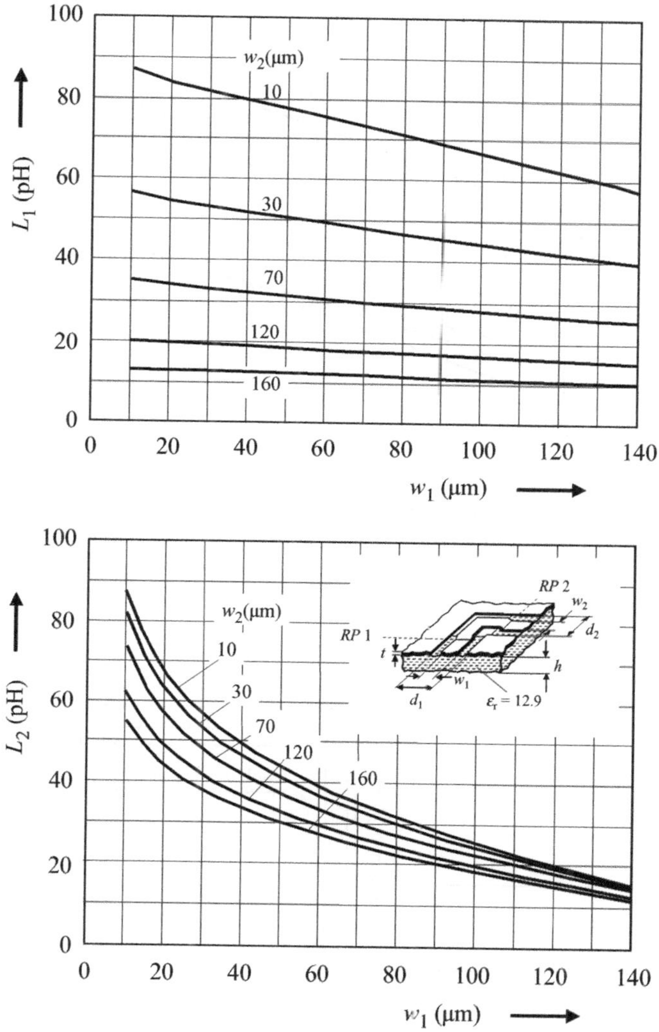

Fig. 3.5.37. The dependence of the equivalent inductances L_1 (a) and L_2 (b) of a coplanar air-bridge bend on the center strip width w. Geometrical parameters are the same as in Fig. 3.5.36.

curves in Fig. 3.5.38). Despite the fact that the air-bridge bend has the highest capacitance (that leads to a higher frequency dependence of the scattering matrix), it is often used in coplanar circuit design especially at lower frequencies because it suppresses more effectively the excitation of the fundamental odd mode on coplanar waveguides (see below).

To test the accuracy of the simulation for the coplanar bend with respect to its dependence on the frequency, several test structures on GaAs substrate have been used. The first one is shown in Fig. 3.5.39.

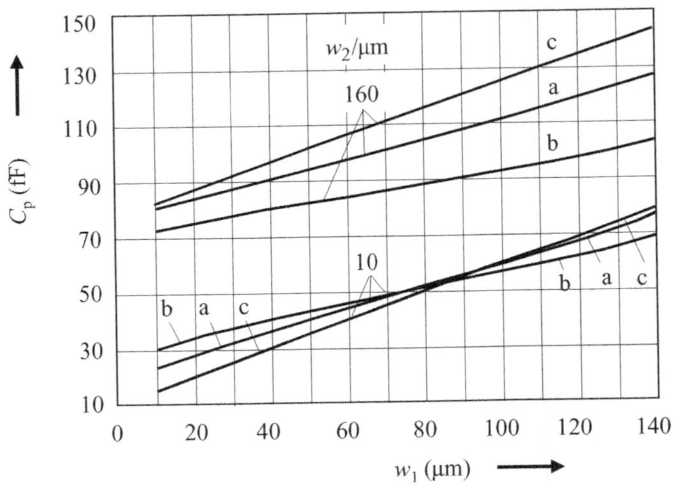

Fig. 3.5.38. The influence of the different air-bridge types on the equivalent capacitance of the coplanar right-angled bend. (a) air bridge type 1, (b) air bridge type 2, (c) air bridge bend. Geometrical parameters are the same as in Fig. 3.5.36.

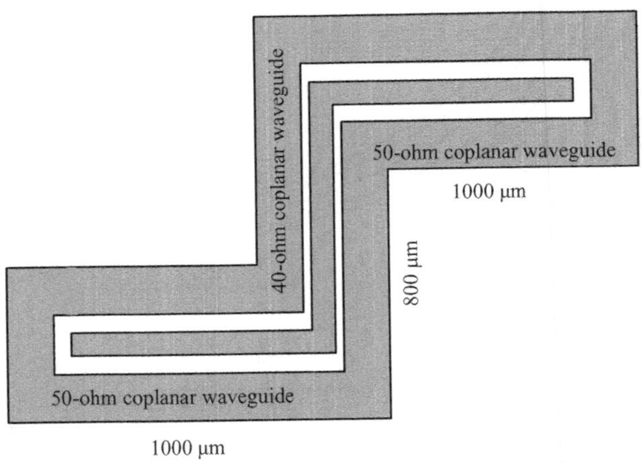

Fig. 3.5.39. Test structure with two coplanar bends without air bridges. Line parameters: For $Z_L = 49.8\,\Omega$, we have $w = 100\,\mu m$ and $s = 75\,\mu m$; for $Z_L = 40\,\Omega$, we have $w = 100\,\mu m$ and $s = 35\,\mu m$, and $t = 3\,\mu m$. Ground plane width: $200\,\mu m$. Substrate GaAs, $\varepsilon_r = 12.9$, $h = 450\,\mu m$.

It is a series connection of two asymmetrical bends between two feed lines of $50\,\Omega$ that are $1000\,\mu m$ long and a 40-Ω coplanar waveguide that connects the two bends. No air bridges are used in this first test structure. This means that large discrepancies between the simulation results (that base on the quasi-TEM analysis for the even coplanar waveguide mode) and the measurement results should be detected because the odd slot-line mode will be excited at the asymmetrical discontinuity.

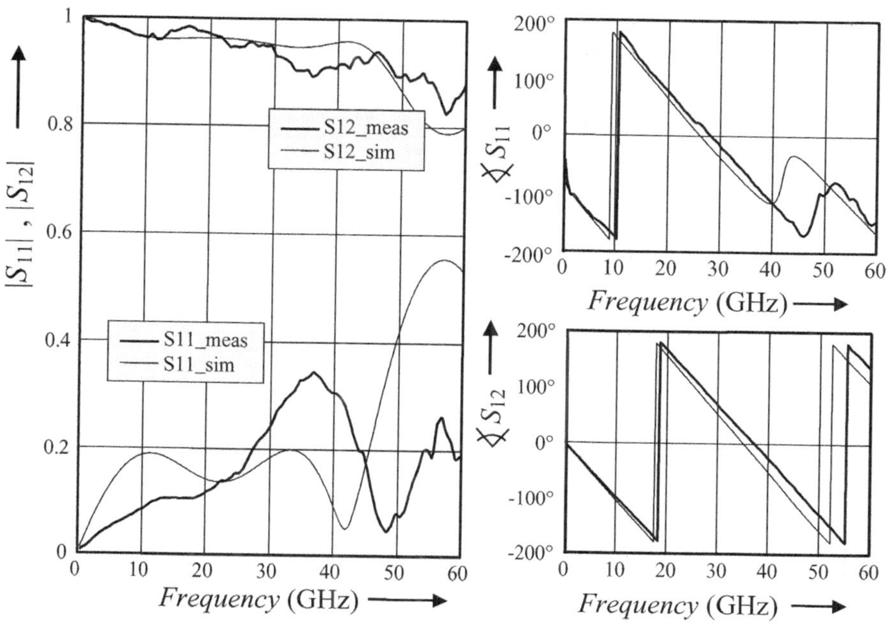

Fig. 3.5.40. Comparison of the simulated and the measured scattering parameters of the test structure shown in Fig. 3.5.39.

The simulated and the measured scattering parameters of the structure are shown in Fig. 3.5.40 for a frequency range from 45 MHz to 60 GHz. As expected, the agreement between the simulation results and the measured results is not very good. This is true for the magnitude of the scattering parameters as well as for the phase angles. The discrepancies increase with increasing frequency.

A similar test structure like the one in Fig. 3.5.39 is shown in Fig. 3.5.41. It depicts the same bend structure, but in this case with additional four air bridges that are placed directly at the bend ports. The used air bridge is of type 1 that connects the ground planes using the first-level metalization. The center strip forms the air bridge across this metal connection (see also Section 3.5.5).

Figure 3.5.42 shows the improvement of the comparison between simulation and measurement reached in this case. Again the simulated and measured scattering parameters up to 60 GHz are depicted. Now the agreement between simulation and measurement is much better. Only for frequencies higher than 40 GHz, some deviations in the magnitude of the scattering parameters can be observed. For the reflection coefficient, a small frequency shift may be seen at lower frequencies. The simulated phase is in good agreement with the measurements up to highest frequencies.

A very good test structure to prove the accuracy of the derived models for the coplanar bends is a coplanar meander line in that a large number of bends is used. Such a test structure, again built on GaAs substrate, is shown in Fig. 3.5.43. The meander line contains 12 coplanar waveguide bends with totally

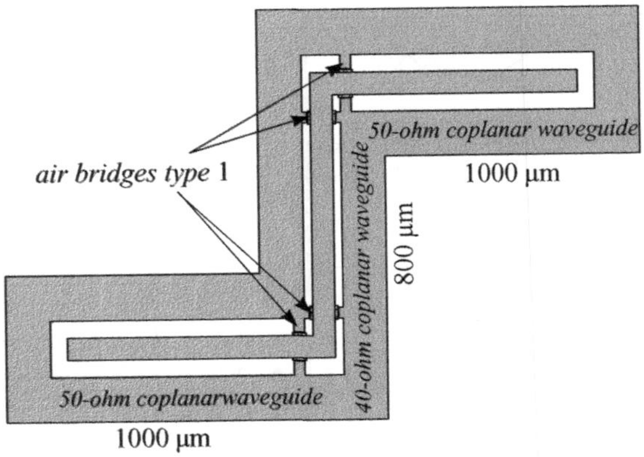

Fig. 3.5.41. Test structure with two coplanar waveguide bends as shown in Fig. 3.5.39 and additional four air bridges of type 1. Parameters: $wb = 50\,\mu m$, $bs = 14\,\mu m$, $bg = 6\ \mu m$ (see Fig. 5.6.6).

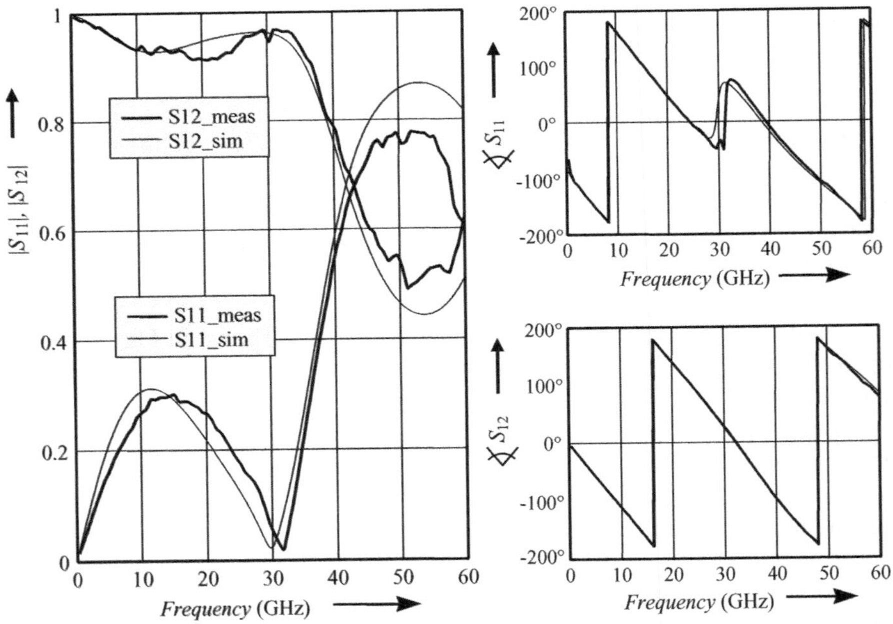

Fig. 3.5.42. Comparison of the simulated (thin lines) and measured (thick lines) scattering parameters for the test structure shown in Fig. 3.5.41, plotted against the frequency.

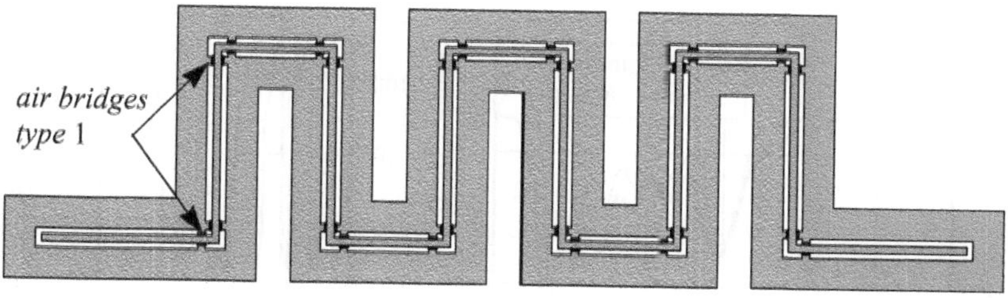

Fig. 3.5.43. Test structure with 12 coplanar waveguide bends and 24 air bridges of type 1. Geometrical parameters: $w = 50\,\mu m$, $s = 37\,\mu m$, ground plane width between the line sections = 200 µm. Total size of the structure: length $L = 6414\,\mu m$, height $H = 1764\,\mu m$, metalization thickness $t = 3\,\mu m$, metalization thickness evaporated layer $t_2 = 0.48\,\mu m$. Substrate GaAs, $\varepsilon_r = 12.9$, $h = 450\,\mu m$. Air bridges type 1: $wb = 50\,\mu m$, $bs = 14\,\mu m$, $bg = 6\,\mu m$; compare also Fig. 5.6.6 for the detailed construction of the air bridges.

24 air bridges of type 1. Further geometrical and electrical parameters are given in the figure inscription.

The structure, due to the series connection of the high number of discontinuities, shows a large number of resonance effects over the frequency range up to 60 GHz. It may be simulated directly as a series connection of the line sections and the bend discontinuities, because the coupling between the adjacent coplanar waveguides with the ground plane between them is very small. This situation is very different compared to a microstrip meander line, where the coupling between the line sections is essential and influences the transmission properties heavily.

The comparison between simulated and measured scattering parameters in Fig. 3.5.44 shows that the agreement is outstanding up to the highest frequency (60 GHz). All the ripples produced by the multiple resonances are reproduced well in magnitude (loss) and phase (effective dielectric constant). It should be pointed out that a similar test structure without air bridges did not show the strong resonance at 30 GHz. Thus, this effect may be interpreted to be due to the additional capacitances of the air bridges within the circuit.

A full-wave analysis on the basis of a three-dimensional spectral domain technique [62, 80] has been used to investigate the frequency-dependent transmission properties of the fundamental even mode (coplanar waveguide mode) and the excitation of the odd mode on the coplanar bend. Because the bend is a typical discontinuity with a large asymmetry in the ground-plane structure, an excitation of the odd mode (slot line mode) may be expected at the discontinuity.

In Fig. 3.5.45 the transmission coefficient S_{21}^{ee} of the even mode (analyzed using a full-wave moment method) is shown for the three discussed coplanar bends (types 1, 2, and air-bridge bend). Additionally, the transmission coefficient of the coplanar bend without any air-bridge structure is depicted. It can

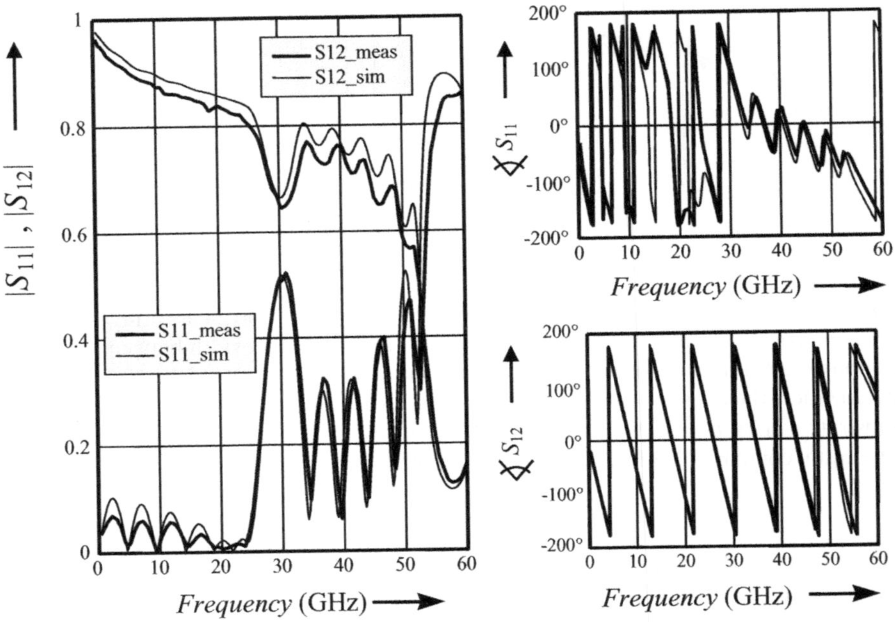

Fig. 3.5.44. Comparison of the simulated (thin lines) and measured (thick lines) scattering parameters of the coplanar meander line test structure shown in Fig. 3.5.43.

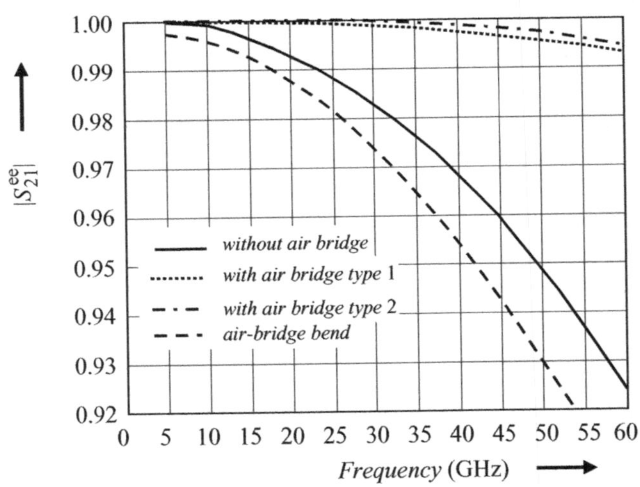

Fig. 3.5.45. Magnitude of the transmission coefficient $|S_{21}^{ee}|$ for a coplanar waveguide bend with different air-bridge structures. Geometrical parameters: $w = 75\,\mu m$, $s = 50\,\mu m$. Substrate GaAs, $\varepsilon_r = 12.9$, $h = 200\,\mu m$, air-bridge height $= 3\,\mu m$.

be seen that the bends with air bridges of types 1 and 2 are nearly ideal, and their transmission coefficient is higher than 0.99 for all frequencies up to 60 GHz. The air-bridges included into the discontinuity compensate the frequency dependence of the bend. The air-bridge bend has a much larger frequency dependence of its transmission coefficient; it decreases down to the value of 0.93 at 50 GHz.

The reason for this behavior is the length of the air bridge that, as already has been discussed above, behaves like a microstrip line with an effective dielectric constant of 1 and that, because of the small height above the ground metalization (3 μm), has a high capacitance. The transmission properties of the air-bridge bend are even worse than those of the coplanar bend without any air bridge.

To discuss the mode conversion from the even mode to the odd mode, Fig. 3.5.46 shows the transmission coefficient S_{21}^{oe} that characterizes the odd mode at port 2 excited by an incident even mode at port 1. The results are quite similar for the bends with type 1 and 2 air bridges. At 60 GHz a transmission coefficient of about 7% from the even mode to the odd mode may be observed for these bends. The air-bridge bend, however, has a much better behavior in the total frequency range. It suppresses the odd mode at port 2 excited by an incident even mode at port 1 down to a transmission coefficient of only 3.5% at 60 GHz. It also may be seen that the coplanar bend without any air bridge already excites the odd mode at very low frequencies to a considerable value. At higher frequencies the odd-mode excitation is so high that an integrated microwave circuit containing such a bend would no longer work properly.

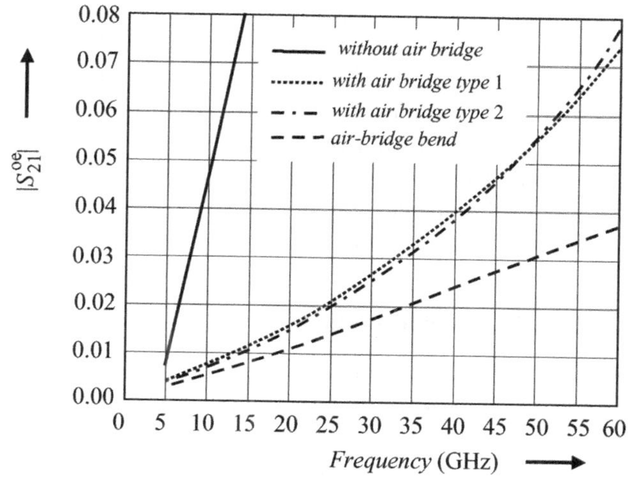

Fig. 3.5.46. Frequency dependence of the transmission coefficient $|S_{21}^{oe}|$ for coplanar bends with different air-bridge structures as well as for the bend without any air bridge. Geometrical parameters as in Fig. 3.5.34.

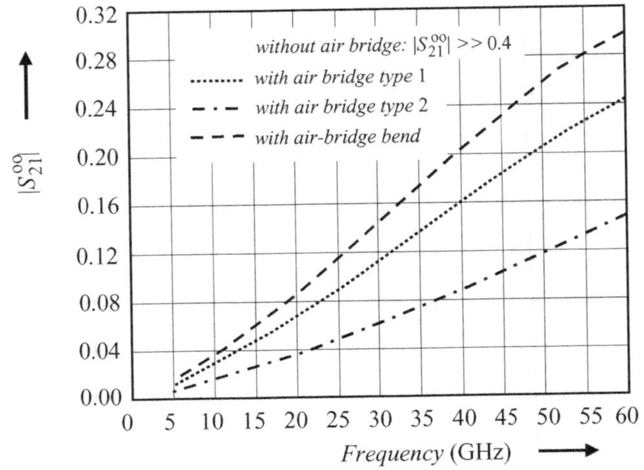

Fig. 3.5.47. Frequency dependence of the transmission coefficient $|S_{21}^{oo}|$ describing the transmission of an incident odd mode into an odd mode at the output of the bend. Geometrical parameters are the same as in Fig. 3.5.34.

Finally, in Fig. 3.5.47 the transmission properties of the bends are shown for the case that the structure is excited by an incident odd mode and the odd mode is measured at the output port. In this case the coplanar air-bridge bend has the worst properties. It transmits the unwanted mode with a transmission coefficient of about 30% at 60 GHz. Under this condition the type 2 air bridge shows the best properties together with the bend structure.

The discussion shows that a final decision for the choice of a coplanar air bridge bend is not easy. If a very small reflection coefficient for the even mode is needed in the circuit design, the bend with type 2 air bridges is optimal. If a high suppression of the odd mode is wanted, the air-bridge bend should be considered, taking into account that the frequency dependence of its scattering parameters is somewhat higher. Possibly at millimeter-wave frequencies the capacitive effect of this structure may lead to problems in a proper circuit design. For frequencies up to 30 GHz, it looks like the best solution.

3.5.7 The Coplanar Waveguide T-Junction

The most common connection between three coplanar waveguides is the T-junction. Figure 3.5.48a shows as an example the coplanar air bridge T-junction that is built similar to the air-bridge bend, as shown in Fig. 3.5.35c. Also air bridges of type 1 and type 2 may be used in connection with the T-junction. Fig. 3.5.48b shows the equivalent circuit that will be used to model the T-junction. It is similar to the equivalent circuit used for microstrip T-junctions [5, 16, 20].

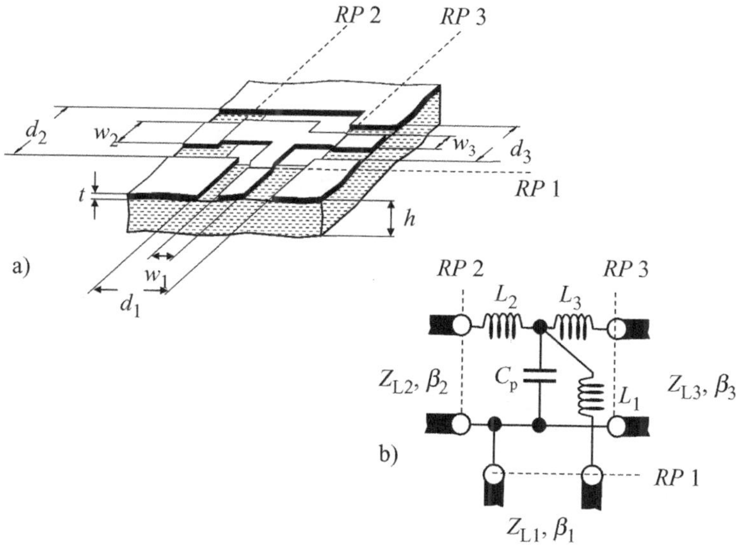

Fig. 3.5.48. (a) The coplanar waveguide T-junction of type air-bridge T-junction and (b) its equivalent circuit. *RP* stands for reference planes.

In contrast to the case of a microstrip T-junction where, depending on the used equivalent circuit, the reference planes are chosen at different positions on the feeding lines [20], for the case of the coplanar T-junction the reference planes are defined directly at the border between the discontinuity region and the feeding coplanar waveguides (Fig. 3.5.48a). The capacitance C_p of the equivalent circuit shown in Fig. 3.5.48b is calculated using the method as described in Section 3.4.2:

$$C_p = \frac{Q_{cc} - l_1 Q_1' - l_2 Q_2' - l_3 Q_3'}{\varphi_{cc} - \varphi_g} \tag{3.5.8}$$

where l_i (i = 1, 2, 3) are the lengths of the feeding coplanar waveguides, Q_i' (i = 1, 2, 3) are the charges per unit line length on the uniform coplanar waveguides, Q_{cc} is the total charge on the center conductor of the structure, and φ_{cc} and φ_g are the potentials of the center conductor and the ground plane, respectively.

The determination of the equivalent inductances L_i ($i = 1, 2, 3$) needs a little bit more expensive analysis because three different analysis steps are needed. The T-junction is excited in three different ways, so that a signal is transmitted always only from one port to a second port. The three different excitation conditions are simulated using the static magnetic potentials as described in Section 3.4.2. Figure 3.5.49 shows the considered structure where the static magnetic potentials Ψ_i ($i = 1, 2, 3$) in the slot areas A_i ($i = 1, 2, 3$) are defined.

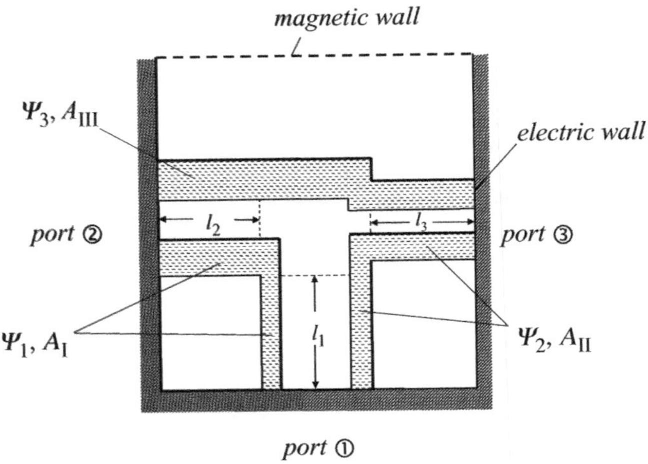

Fig. 3.5.49. Distribution of the static magnetic potential for the determination of the inductances in the equivalent circuit of a coplanar T-junction.

Using the values $\Psi_1 = +1$ A, $\Psi_2 = -1$ A, and $\Psi_3 = -1$ A, the sum of the inductances L_1 and L_2 can be determined as follows:

$$\begin{aligned} &\Psi_1 = +1\text{A}, \\ &\Psi_2 = -1\text{A} \rightarrow \Phi_{12} = \mu_0 \iint_{A_I} H_y dA \rightarrow L_1 + L_2 = \frac{\Phi_{12}}{4\text{A}} - l_1 L_1' - l_2 L_2', \qquad (3.5.49) \\ &\Psi_3 = +1\text{A}. \end{aligned}$$

The integral has to be taken over the area A_I, as shown in Fig. 3.5.49. L_1' and L_2' are the inductances per unit line length of the coplanar waveguides 1 and 2, respectively.

In an analogous way, the sum of the inductances L_1 and L_3 can be found from a current flow simulation between ports ① and ③ of the T-junction:

$$\begin{aligned} &\Psi_1 = +1\text{A}, \\ &\Psi_2 = -1\text{A} \rightarrow \Phi_{13} = \mu_0 \iint_{A_{II}} H_y dA \rightarrow L_1 + L_3 = \frac{\Phi_{13}}{4\text{A}} - l_1 L_1' - l_3 L_3', \qquad (3.5.10) \\ &\Psi_3 = +1\text{A}. \end{aligned}$$

The third equation that is needed to determine the three inductances is found from a current flow simulation from port ② to port ③ (Fig. 3.5.50) by

$$\Psi_1 = +1\text{A},$$

$$\Psi_2 = -1\text{A} \rightarrow \Phi_{23} = \mu_0 \iint_{A_{\text{III}}} H_y dA \rightarrow L_2 + L_3 = \frac{\Phi_{23}}{4\text{A}} - l_2 L_2' - l_3 L_3', \quad (3.5.11)$$

$$\Psi_3 = +1\text{A}.$$

The three inductances L_i (i = 1, 2, 3) can now be calculated combining the three equations (3.5.9)–(3.5.11). Using the combination of the above-simulated current flows, the surface current density distribution in the metalized structure can be simulated for the general case that a current is fed into one of the ports and is distributed to both of the other ports. The resulting surface current densities are shown in Fig. 3.5.50. A current signal of 2 A fed in one of the ports is distributed into two currents of value 1 A, each at the remaining ports. This distribution is forced upon the structure by the choice of the constant magnetic potentials in the slot areas.

To test the used model of the T-junction with respect to its applicability at different frequencies, the transmission properties of a symmetrical coplanar T-junction on GaAs substrate material have been measured and compared to simulation results. The scattering parameters of the T-junction have been measured up to 40 GHz using a vector network analyzer.

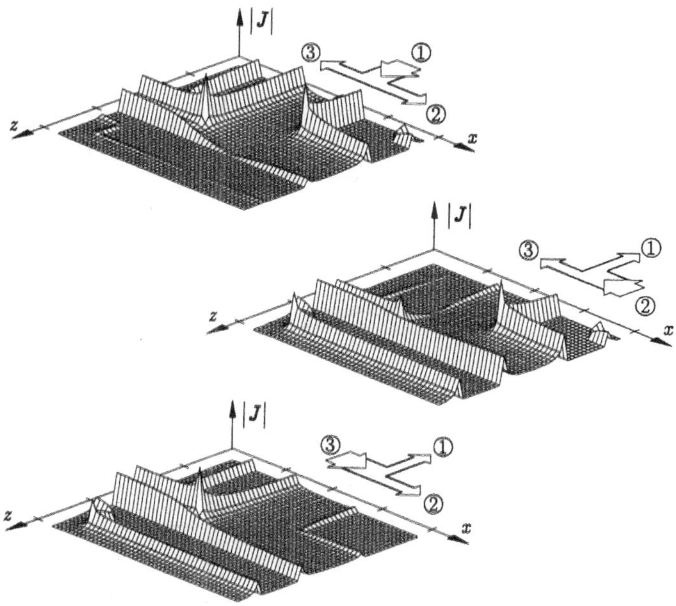

Fig. 3.5.50. Magnitude of the surface current density distribution on the metalized structure of an asymmetrical coplanar T-junction for three different excitations of the current at ports ① to ③.

Also the scattering parameters have been calculated from the equivalent circuit of the T-junction as shown in Fig. 3.5.48b. Figure 3.5.51 shows the results. A good agreement is found between the measured and simulated results over the whole frequency range. It may also be observed that the scattering parameters are strongly frequency-dependent. This frequency dependence may not be neglected in circuit design.

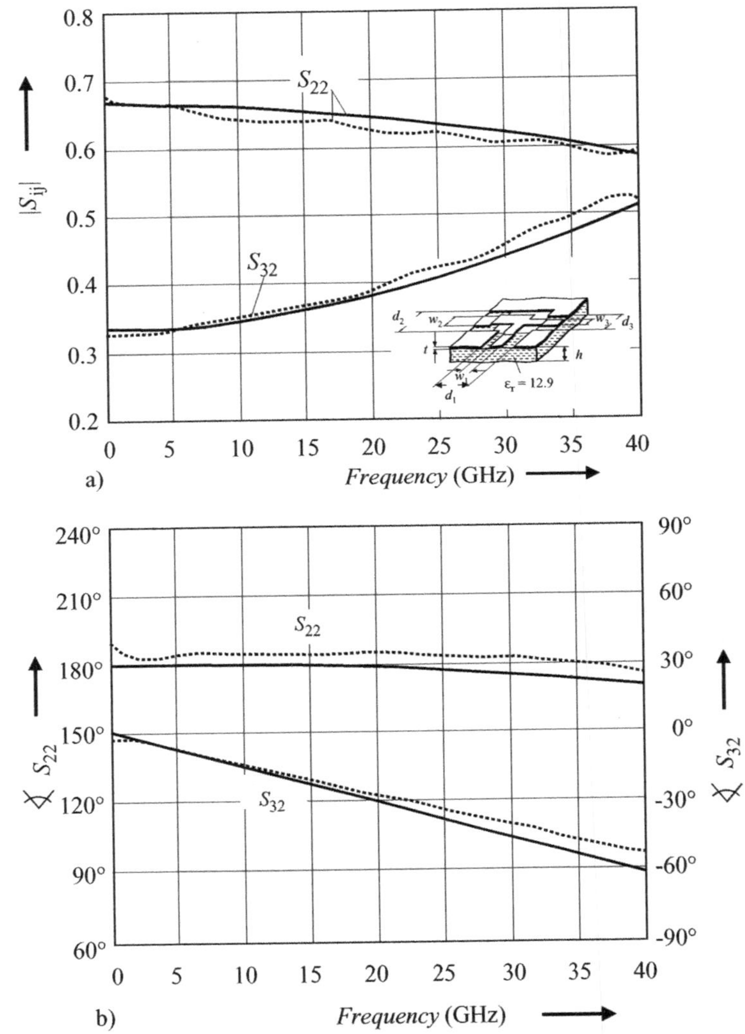

Fig. 3.5.51. Measured (. . .) and calculated (———) scattering parameters of a coplanar air-bridge T-junction in dependence on the frequency. Geometrical parameters: $w_1 = w_2 = w_3 = 75\,\mu m$, $d_1 = d_2 = d_3 = 175\,\mu m$, $Z_{L1} = Z_{L2} = Z_{L3} = 50\,\Omega$. Air bridge height = 3 μm. Substrate GaAs, $\varepsilon_r = 12.9$, $h = 400\,\mu m$. Model parameters: $L_1 = 13\,pH$, $L_2 = L_3 = 74\,pH$, $C_p = 114\,fF$.

The dependence of the equivalent circuit elements on the geometrical parameters is shown in Figs. 3.5.52 and 3.5.53 for coplanar T-junctions with identical and constant line widths of line 2 and line 3. The width of line 1 is varied. Line width $w_2 = w_3$ is taken as a parameter of the different curves. The assumed height of the air-bridge structure is 3 μm and the distance of the reference planes from the air bridge is 10 μm. It may be observed from the figures, that the capacitive effect of the air bridge is dominant. The inductance L_1 is decreasing with increasing values of w_1 and may become even negative (Fig. 3.5.53). It means that for these cases the inductance of the coplanar air bridge T-junction at port ① is smaller than that of a homogeneous coplanar waveguide of same line length. The inductances L_2 and L_3 are positive for all values of w_1.

In Fig. 3.5.54 a technological realization of a T-junction with type 1 air bridges (left figure) and an air bridge T-junction (right figure) as it has been used in a traveling wave amplifier design is shown. The T-junction with type 1 air bridges only has ground connections in the lower (gate) metalization layer at the three ports in form of thin strips. The T-junction itself is placed directly on the substrate material. At the three ports, the center strip of the feeding coplanar waveguides is formed in the form of an air bridge to cross the ground connections. The right figure shows the air bridge T-junction. It may be clearly recognized how the total center strip metalization in the region of the T-junction is lifted up into a height of about 3 μm above the substrate material and how the first level ground metalization connects the ground planes on the three sides of the T-junction construction.

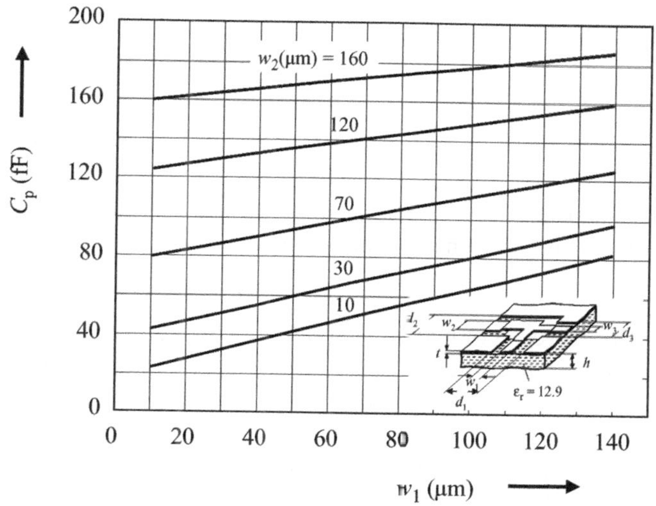

Fig. 3.5.52. The capacitance C_p of the equivalent circuit for the coplanar air-bridge T-junction, plotted against the line width w_1. Geometrical parameters: $d_1 = d_2 = 170$ μm, $t = 3$ μm, air-bridge height = 3 μm. Substrate GaAs, $\varepsilon_r = 12.9$, $h = 400$ μm.

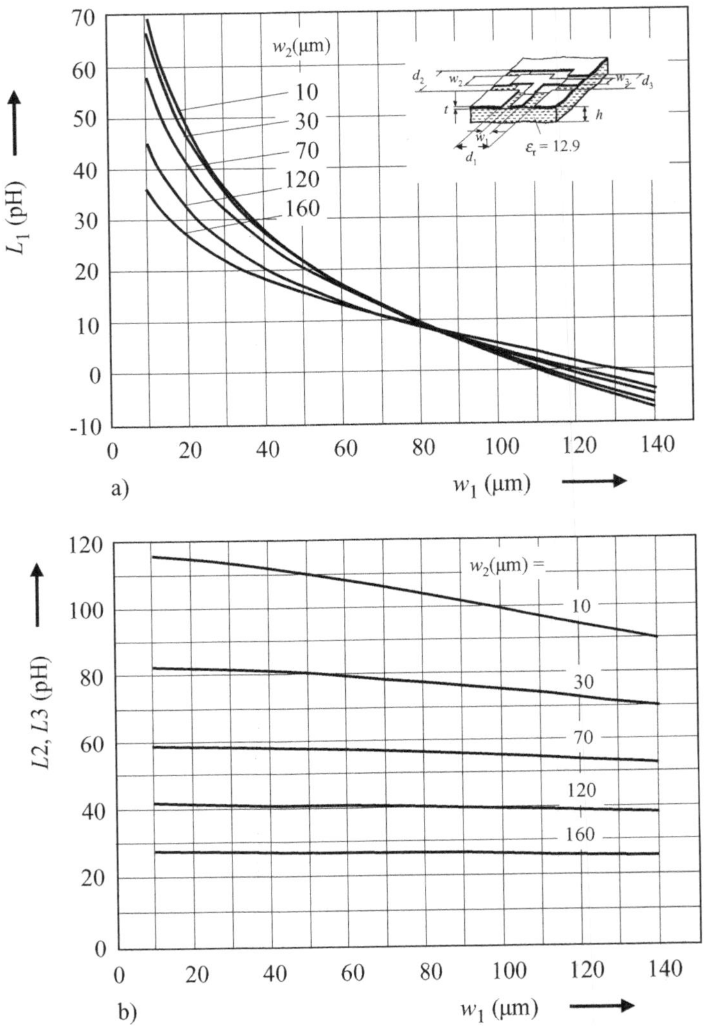

Fig. 3.5.53. The inductances L_1, L_2, and L_3 of a coplanar T-junction in dependence on the geometrical parameters. Other geometrical parameters as in Fig. 3.5.52.

Because the three connected coplanar waveguides are of different characteristic impedance, there is also an impedance step integrated into the T-junction construction. This kind of construction has, at least at lower frequencies where the additional capacitance does not influence the circuit design heavily, big advantages in the suppression of the unwanted odd mode (slotline mode). Mechanically, these T-junctions can be produced in very stable form.

To measure the field distribution of air bridges in coplanar microwave circuits, a T-junction circuit is fabricated as shown in Fig. 3.5.55. This coplanar

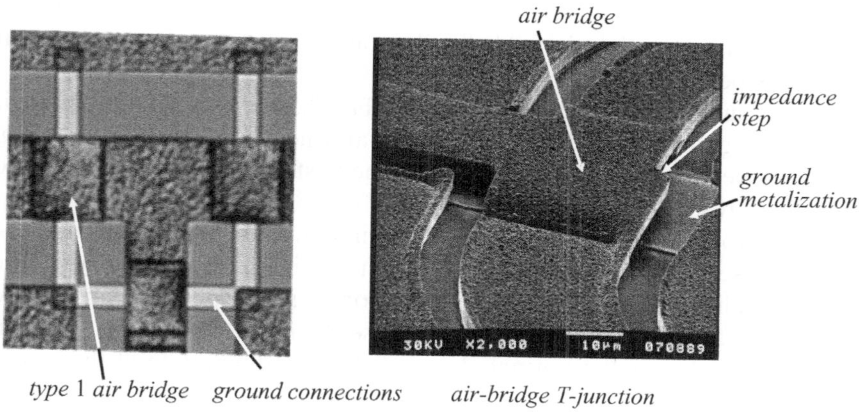

Fig. 3.5.54. Technological construction of two T-junctions. The T-junction with type 1 air bridge and ground connections in the gate metalization layer at each of the three ports (left) and the air bridge T-junction where the T-junction totally is formed as an air bridge and the area under the air bridge is metalized in the gate layer.

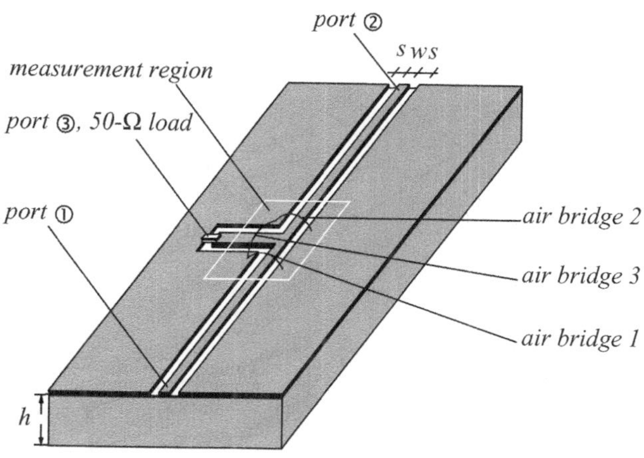

Fig. 3.5.55. Coplanar T-junction as a test structure for measuring the field distribution near the junction considering the influence of the air bridges.

T-junction has a conductor width of 500 μm and a gap width of 217 μm. It is fabricated on ceramic substrate with a dielectric constant of 9.8 and a thickness of 635 μm. Port 3 is terminated with a 50-Ω impedance, and port 2 is terminated with a 50-Ω coaxial load.

The three air bridges that have been used are placed to the T-junction one by another. The electric field normal to the substrate plane has been measured for all cases using an electric field probe [68]. The measured field distributions deliver a clear insight into the mode structure in the vicinity of the T-junction, as may be observed from the following measurement results. First, the three-dimensional normal electric field distribution of the T-junction without air bridges has been measured.

The measurement results are shown in Fig. 3.5.56. This measurement is taken in a region of 10,000 μm × 10,000 μm at a height of the field probe of 100 μm above the substrate. The measurement steps are 100 μm in *x*- and *y*-directions, respectively.

In order to analyze the measured results in detail, the field distribution in three cross sections are selected and shown in Fig. 3.5.57 (next page). The transmission line near the input port 1 transports a nearly pure coplanar mode (even mode) as shown in Fig. 3.5.57a, because the reflected signal is small. In this section the maximum normal electric field is in the middle of the center

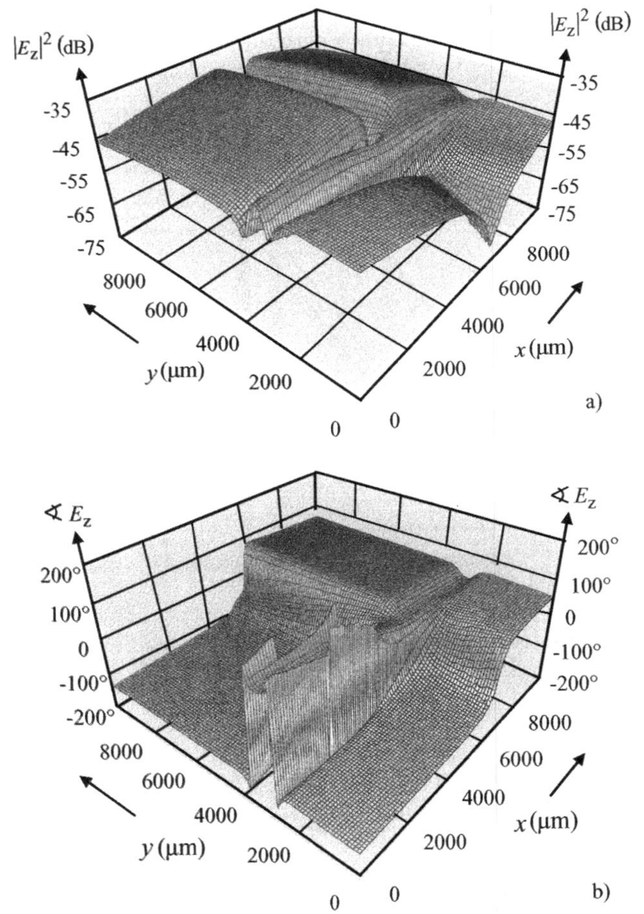

Fig. 3.5.56. Measured normal electric field distribution near a T-junction without air bridges. (a) Magnitude $|E_z|^2$ and (b) phase of E_z [96].

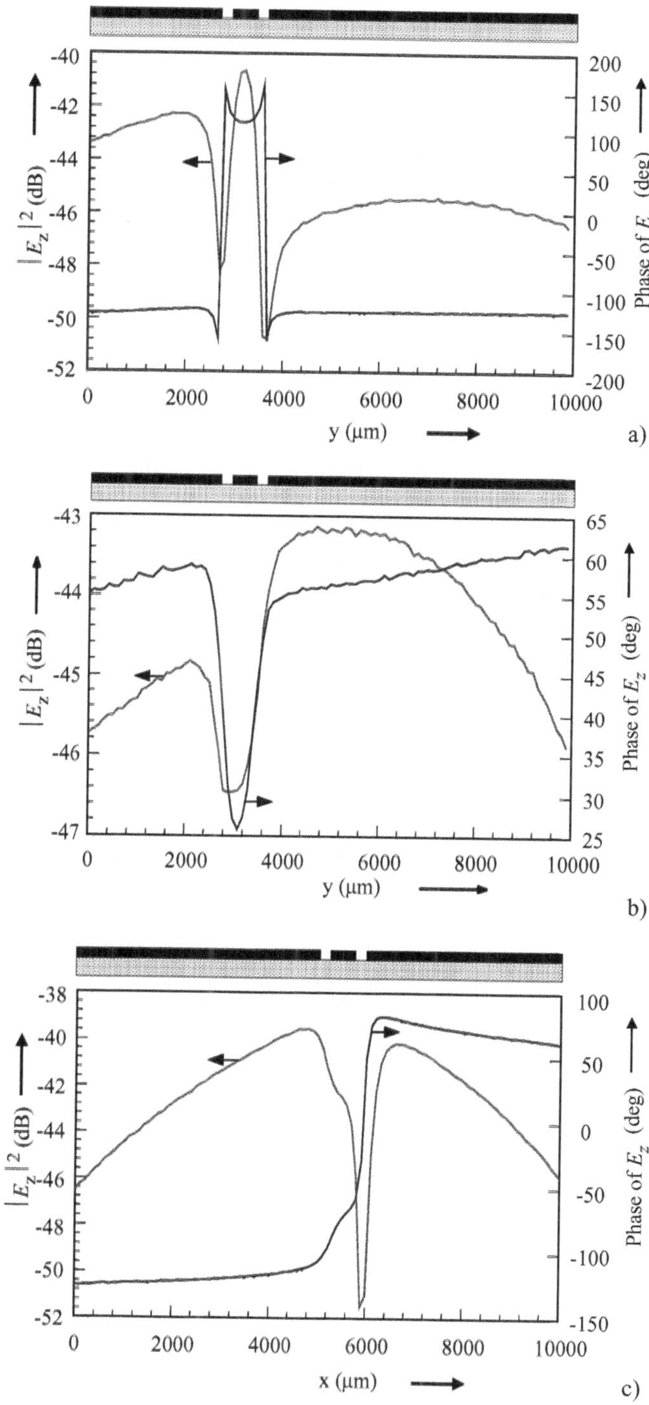

Fig. 3.5.57. Measured normal electric field distribution near a T-junction without air bridges at different cross sections: (a) $x = 0$ (input port ①), (b) $x = 10{,}000\,\mu m$ (port ②), and (c) $y = 10{,}000\,\mu m$ (port ③) [96].

conductor and the minimum of the electric field normal to the substrate is near to the slots.

The phases of the normal electric fields have about 120° difference between the center conductor and the ground planes. Near port 2 the amplitude and the phase of the electric field are almost constant as shown in Fig. 3.5.57b (consider the given scaling in this figure). This is the typical microstrip like mode or surface wave-like (SWL) mode field distribution. Because of the asymmetry of the T-junction discontinuity, such a microstrip-like mode can be excited as has already been discussed in Section 2.1.

Near port 3 the minimum normal electric field is measured in one slot area between the center conductor and the ground plane as shown in Fig. 3.5.57c. In the second slot, the field has a maximum. This field distribution is typical for a superposition of an even and an odd mode on the coplanar waveguide. The phase difference between the center conductor and the ground-plane field across the first slot amounts to 160°. There is only a small phase difference of about 40° between the center conductor and the ground-plane field across the second slot.

Figure 3.5.58 shows the three-dimensional normal electric field distribution of the T-junction with only air bridge 1 in place. Figure 3.5.59 shows the normal electric field at three different sections, which are the same as the sections shown in Fig. 3.5.57.

When air bridge 1 is placed at the T-junction port 1, there is principally no difference of the normal electric field distribution near port 1 and port 3 as shown in Figs. 3.5.59a and 3.5.59c, except that the magnitude of the normal electric field near port 1 is more symmetrical. However, near port 2 the maximum normal electric field now can be measured on the center conductor and the minimum can be measured in the slots, even though the difference between them is not big.

The phase of the normal electric field has now about a 60° difference between the center conductor and the ground plane as shown in Fig. 3.5.59b. That means that the even mode is more dominant now compared to the microstrip like mode, which is dominant in the field shown in Fig. 3.5.57b for the T-junction without any air bridge.

Figure 3.5.60 shows the normal electric field distribution of the T-junction with two air bridges, air bridge 1 and air bridge 2, in place. The normal electric field distributions at three different sections are shown in Fig. 3.5.61. The electromagnetic field near port 1 and port 2 are now nearly coplanar waveguide modes (even modes). The magnitude of the normal electric field has about 25 dB and 13 dB difference between the center conductor and the ground plane near port 1 and near port 2, respectively, as shown in Figs. 3.5.61a and 3.5.61b. The field distributions are also symmetrical, and at port 2 the phase difference between the center conductor and ground plane is about 110°. At port 3 the maximum magnitude of the normal electric field is measured in the middle of the center conductor as shown in Fig. 3.5.61c. The odd mode is suppressed by the two air bridges.

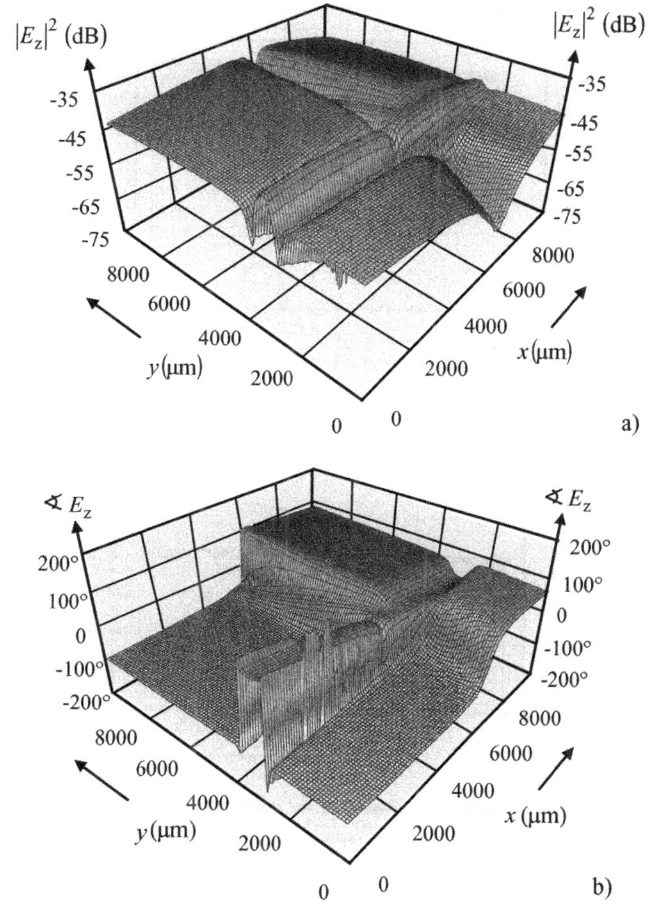

Fig. 3.5.58. Measured normal electric field distribution of the T-junction with air bridge number 1 in place. (a) Magnitude $|E_z|^2$ and (b) phase of E_z [96].

Figure 3.5.62 shows the normal electric field distribution of the T-junction with all three air bridges 1–3 in place. The normal electric field distributions at three different sections are shown in Fig. 3.5.63.

The coplanar modes at port 1 and port 2 are not much changed. However, at port 3 there is a phase difference of about 100° between the center conductor and the ground-plane field now. It may be observed that at all three ports the transmission modes are coplanar modes, even though the field is not an ideal even mode at port 3.

To demonstrate the accuracy and the limits of the quasi-static analysis technique as it has been described above and from which the equivalent circuits of the T-junctions are derived, coplanar waveguides with stub resonators, coupled by T-junctions without air bridges and with type 2 air bridges (see Fig.

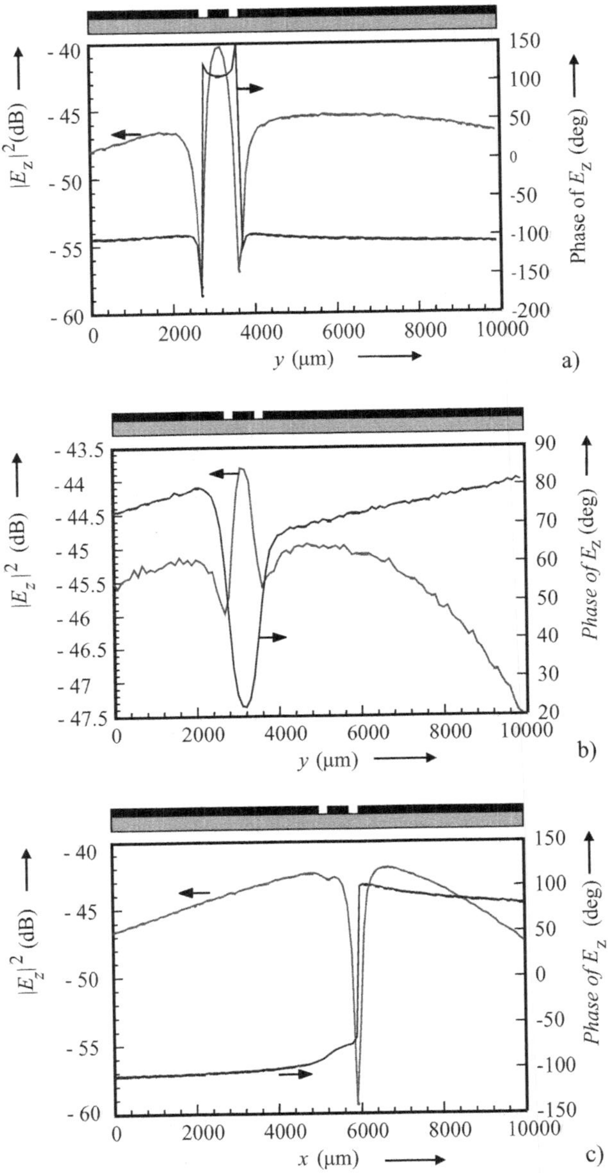

Fig. 3.5.59. Measured normal electric field distribution of the T-junction with air bridge number 1 in place at sections: (a) $x = 0$ (input port ①), (b) $x = 10{,}000\,\mu m$ (port ②) and (c) $y = 10{,}000\,\mu m$ (port ③) [96].

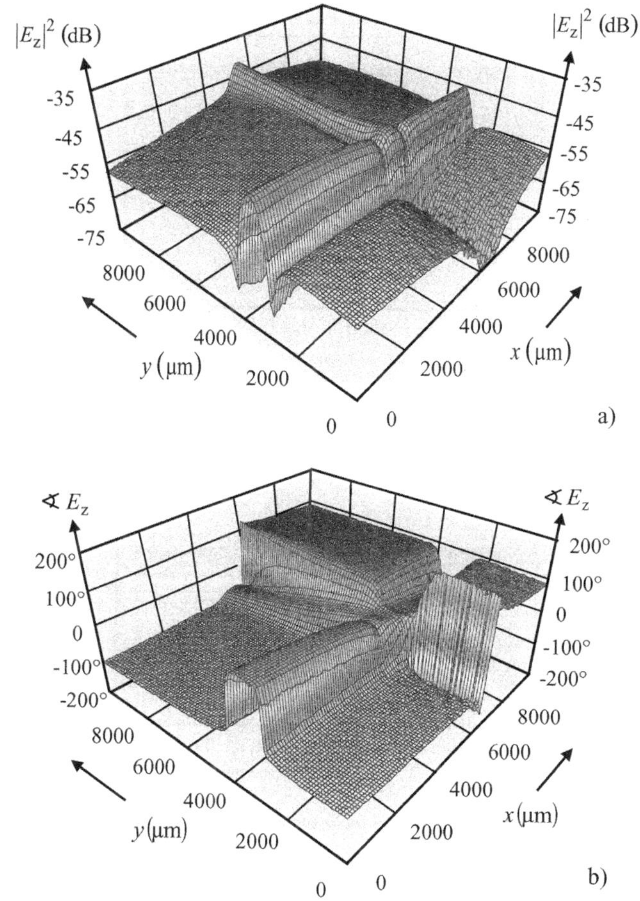

Fig. 3.5.60. Measured normal electric field distribution of T-junction with air bridge 1 and 2 in place: (a) Magnitude $|E_z|^2$ and (b) phase of E_z [96].

3.5.26), are used. Measurements with the other air-bridge types delivered similar results; therefore they are not presented here. Two different waveguides, both with a characteristic impedance of 50Ω but with different geometrical sizes, are used. In the first case a coplanar waveguide with a relatively wide slot and center strip width is used (for actual parameters see figure legend of Figs. 3.5.64 to 3.5.67); the second waveguide has a smaller slot width. Figures 3.5.65 to 3.5.67 show the simulated and the measured reflection coefficients of these structures fabricated on GaAs substrate material (substrate height 450 μm) over a large frequency range up to 60 GHz.

In Fig. 3.5.65 the simulated and measured scattering parameters of the measurement structure shown in Fig. 3.5.64a are compared. The used stub resonator is connected to the feed line using a T-junction without any air bridge.

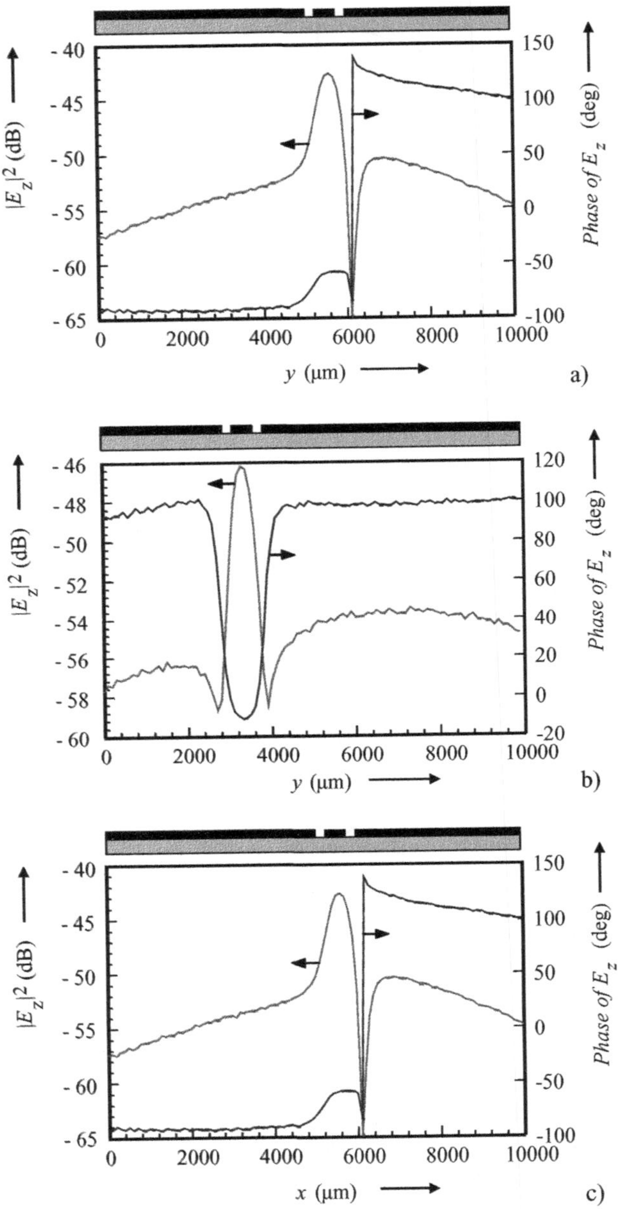

Fig. 3.5.61. Normal electric field distribution of T-junction with air bridge 1 and 2 in place at sections: (a) $x = 0$ (input put ①), (b) $x = 10{,}000\,\mu$m (port ②) and (c) $y = 10{,}000$ μm (port ③) [96].

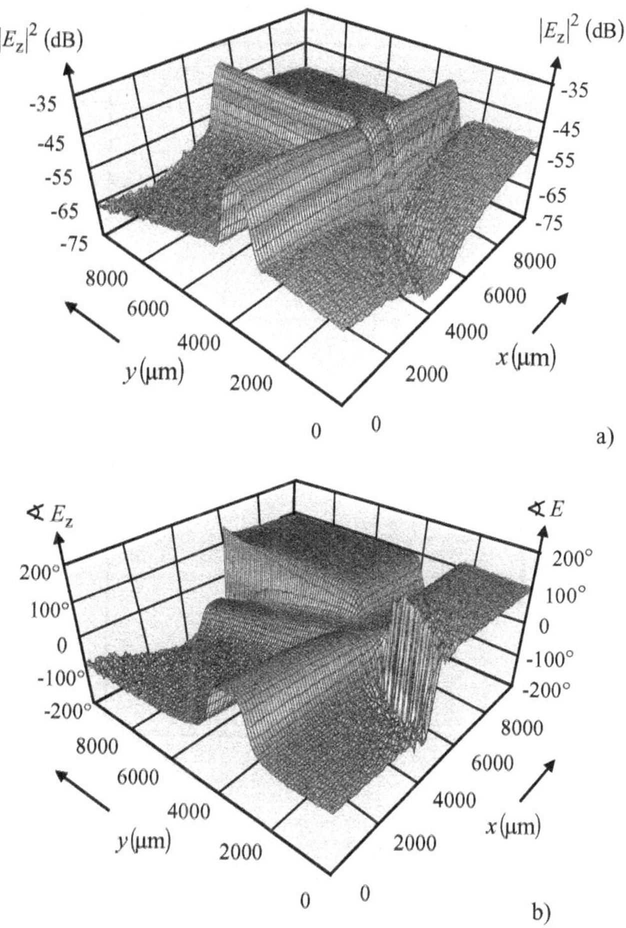

Fig. 3.5.62. Measured normal electric field distribution of a T-junction with all three air bridges 1–3 in place: (a) magnitude $|E_z|^2$ and (b) phase of E_z [96].

Under these conditions it must be expected that the odd coplanar waveguide mode (the slot-line mode) and eventually also a surface wave mode (see, e.g., Fig. 2.1.14) are excited at the asymmetrical T-junction. The influence of these modes on the measured scattering parameters can be clearly seen from Fig. 3.5.65. The agreement between the simulation results and the measured results is bad. This result is not astonishing considering the fact that the simulation has been performed on the assumption that the modes in the test structure are purely coplanar waveguide modes. The quasi-static analysis technique does not allow considering the influence of the odd mode and the surface wave mode as it has already been discussed in Chapter 2. But because in a microwave integrated circuit only the coplanar waveguide mode should exist

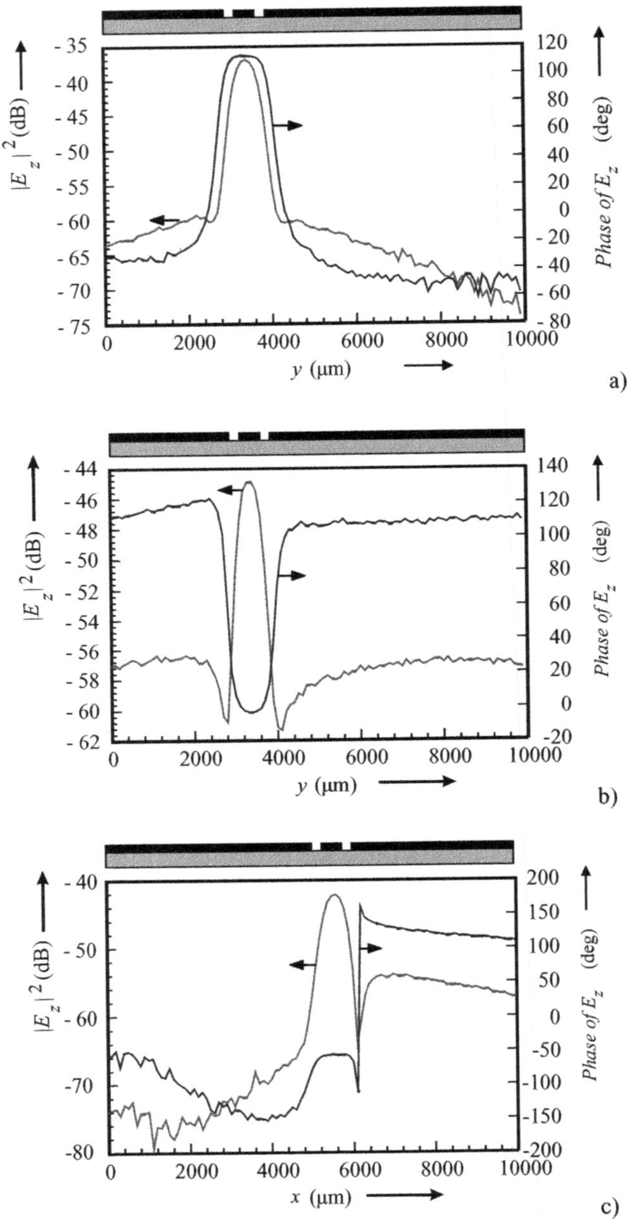

Fig. 3.5.63. Measured normal electric field distribution of a T-junction with three air bridges 1–3 in place at sections: (a) $x = 0$, (b) $x = 10{,}000\,\mu m$, and (c) $y = 10{,}000\,\mu m$ [96].

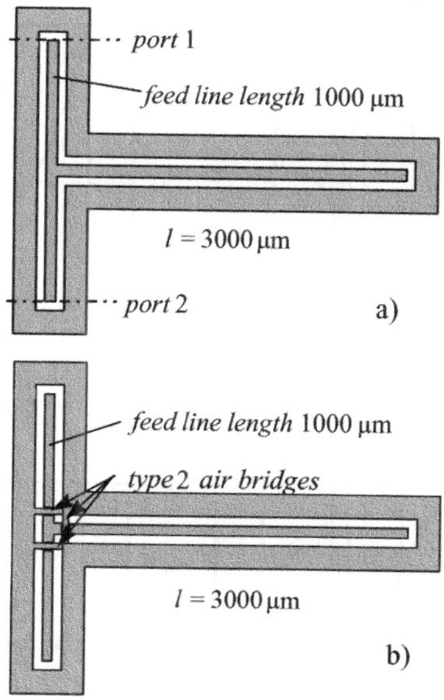

Fig. 3.5.64. Test structures for measuring the frequency response of the coplanar T-junction over a large frequency range, without (a) and with (b) type 2 air bridges. Line parameters: $Z_L = 49.8\,\Omega$, $w = 100\,\mu m$, $s = 75\,\mu m$, $t = 3\,\mu m$. Substrate GaAs, $\varepsilon_r = 12.9$, $h = 450\,\mu m$. Air-bridge parameters: $bg = 8\,\mu m$, $bw = 51\,\mu m$, $bs = 14\,\mu m$. Compare Fig. 5.6.6.

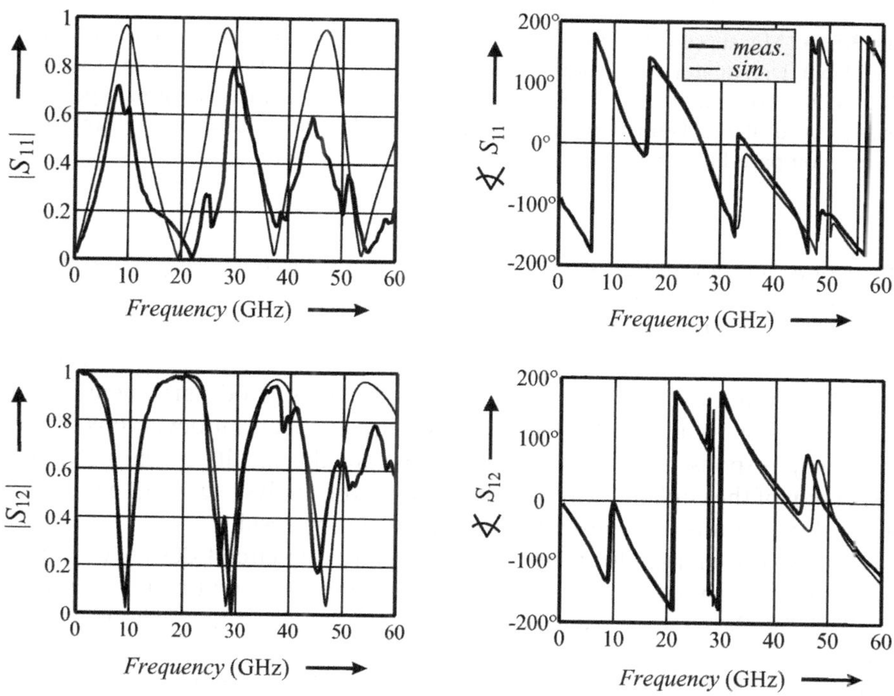

Fig. 3.5.65. Measured (thick lines) and simulated (thin lines) scattering parameters of a coplanar T-resonator with a stub length of $l = 3\,mm$, $w = 100\,\mu m$, $s = 75\,\mu m$, $t = 3\,\mu m$, $Z_L = 49.8\,\Omega$ (simulated). No air bridges are used at the T-junction. Substrate GaAs, $\varepsilon_r = 12.9$, $h = 450\,\mu m$.

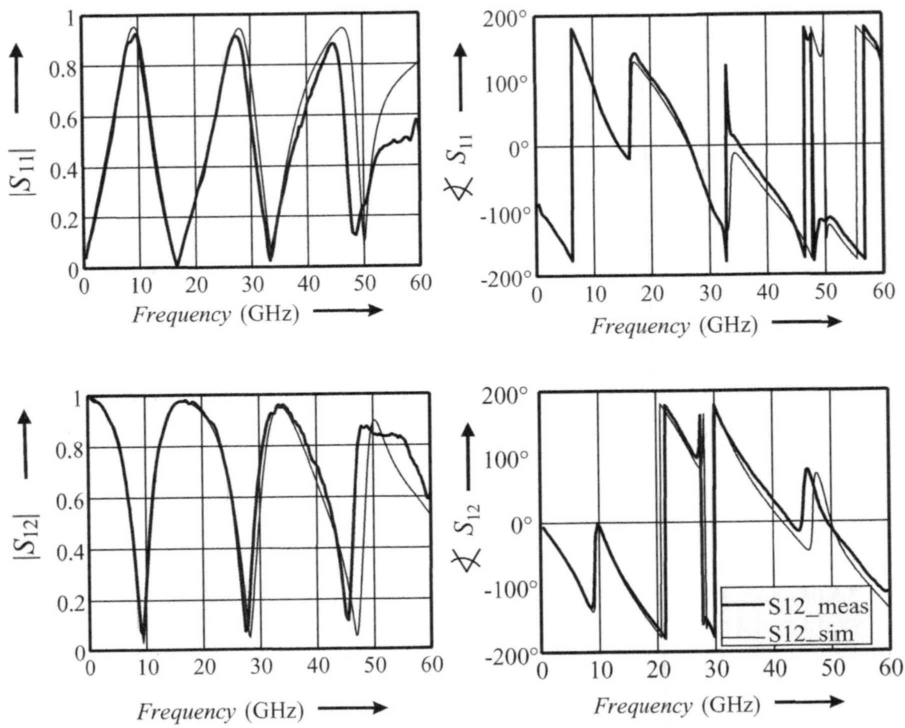

Fig. 3.5.66. Measured (thick lines) and simulated (thin lines) scattering parameters of a coplanar T-resonator with a stub length of $l = 3\,\text{mm}$, $w = 100\,\mu\text{m}$, $s = 75\,\mu\text{m}$, $t = 3\,\mu\text{m}$, $Z_L = 49.8\,\Omega$ (simulated). Substrate material: GaAs, $\varepsilon_r = 12.9$, $h = 450\,\mu\text{m}$. Type 2 air bridges used at the T-junction. Air bridge parameters: $bg = 8\,\mu\text{m}$, $bw = 50\,\mu\text{m}$, $bs = 14\,\mu\text{m}$. Compare Fig. 5.6.6.

(provided that adequate use is made of the air-bridge technology), the given test structure will never occur in a real circuit design.

Figure 3.5.66 shows how the influence of the air bridges improves the agreement between simulation and measurement results. The measured magnitudes and phase angles of the test structure are in a very good agreement up to frequencies of about 40 GHz. Above this frequency, certain deviations in the frequency response may be detected. These are the results of the available dispersion of the coplanar waveguides. Because the analysis technique is quasi-static, dispersion is not considered in the simulation results.

Because dispersion also is normally not wanted in circuit design, coplanar waveguide structures with smaller slot widths may be used at higher frequencies, keeping the characteristic impedance constant. Figure 3.5.67 shows a comparison between simulation and measurement results for the case of an equivalent test structure as shown in Fig. 3.5.64b, but a 50-Ω coplanar waveguide with smaller slot width is used in this case. It can be clearly seen that,

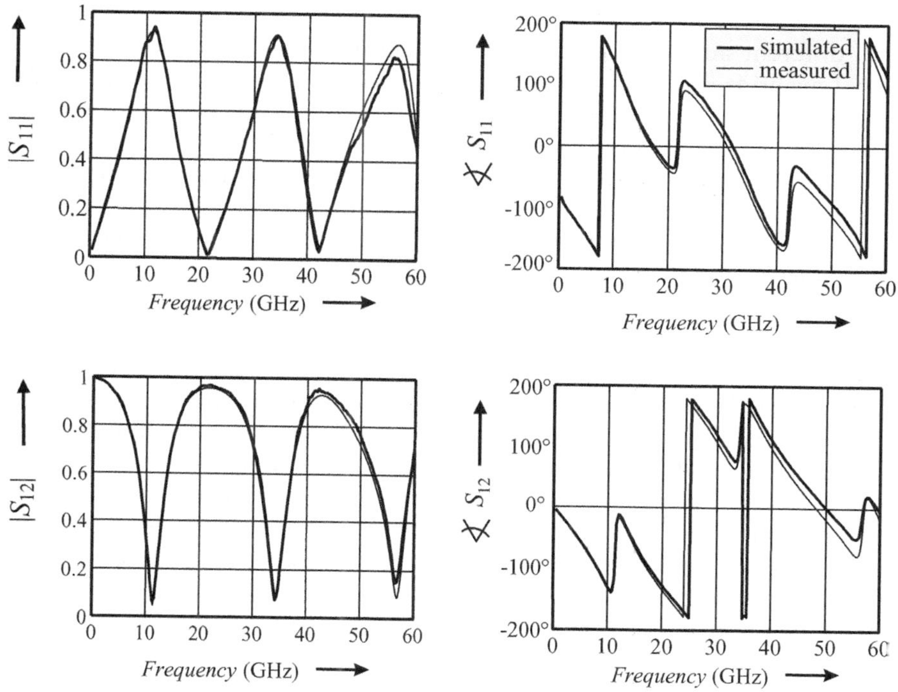

Fig. 3.5.67. Measured and simulated reflection coefficient of a coplanar T-resonator with a stub length of $l = 2.5\,\text{mm}$. $w = 50\,\mu\text{m}$, $s = 37\,\mu\text{m}$, $t = 3\,\mu\text{m}$, $Z_L = 49.3\,\Omega$ (simulated). Substrate GaAs, $\varepsilon_r = 12.9$, $h = 450\,\mu\text{m}$. Used air bridges (type 2): $bg = 8\,\mu\text{m}$ and $bw = 50\,\mu\text{m}$, $bs = 14\,\mu\text{m}$. Compare Fig. 5.6.6.

whereas for the structure with the larger geometrical dimensions the agreement between simulation and measurement degrades for frequencies beyond 40 GHz, the agreement for this structure with the smaller dimensions is quite good up to 60 GHz.

The above-given discussion clearly shows the limits of the quasi-static analysis technique if coplanar structures with large geometrical dimensions are used at very high frequencies. On the other hand, the minimization trends for MMIC circuits push the line geometry to values such as $d = w + 2s < 60\,\mu\text{m}$. In this case, static simulations, taking the effect of the metal thickness into account, may be valid up to frequencies of 100 GHz or even higher.

3.5.7.1 Analysis of the Odd-Mode Excitation. The quasi-static analysis allows only characterizing the properties of the coplanar T-junction with respect to the fundamental even mode (the coplanar mode). If information on the mode conversion properties of the junctions is to be derived, full-wave analysis techniques like the moment method [59, 62, 80] (see also Section 2.1) or the finite difference time domain (FDTD) technique [53, 54, 60] must be applied. To consider the odd-mode suppression capability of the T-junctions with different air-bridge structures, coplanar T-junctions with conventional type 2 air bridges and an air-bridge T-junction have been investigated. Only the results of the spectral domain analysis technique (which are in excellent agreement with those from the FDTD analysis) are reported here.

Figure 3.5.68 shows a comparison between the measured and simulated reflection coefficient S_{11}^{ee} at port ① and the transmission coefficient S_{21}^{ee} for even-mode signal transmission from port ① to port ② of a coplanar air-bridge T-junction on GaAs substrate material (ε_r = 12.9, h = 410 μm). The center conductor width w is 75 μm for all three lines and the slot width s is 50 μm. All feeding lines, therefore, have a characteristic impedance of 50 Ω. The air-bridge height is 3 μm; that is, it is the same T-junction that has been discussed in Fig. 3.5.51. As can be observed from Fig. 3.5.68 in comparison with Fig. 3.5.51, the simulated frequency dependencies of the scattering parameters derived from the quasi-static and the full-wave analysis agree very well.

In Fig. 3.5.69 the same results are shown for the coplanar T-junction with conventional type 2 air bridges. It may be clearly seen that the frequency dependence of this T-junction is much lower because of the reduced capacitance of the type 2 air bridges compared to the air-bridge T-junction.

The scattering parameters have been measured with an on-wafer measurement equipment applying a coplanar probe head. Therefore, only the even-mode (coplanar mode) parameters could be measured. The odd mode

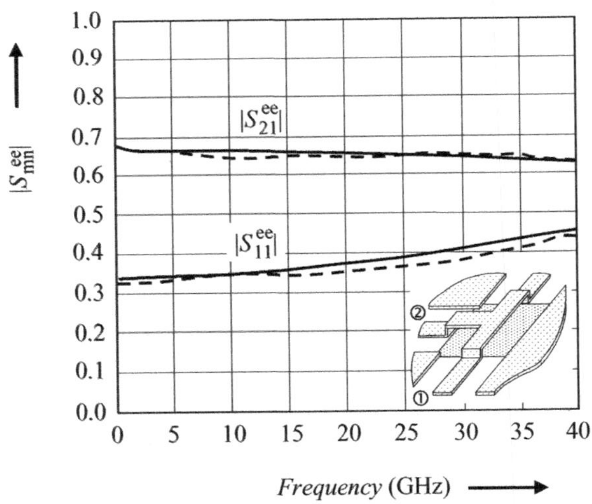

Fig. 3.5.68. Magnitude of the reflection coefficient S_{11}^{ee} and the transmission coefficient S_{21}^{ee} of the coplanar waveguide mode for a coplanar air bridge T-junction in dependence on the frequency (——— theory, – – – measured). Geometrical parameters: $w_1 = w_2 = w_3$ = 75 μm, $s_1 = s_2 = s_3$ = 50 μm, t = 3 μm, air-bridge height = 3 mm. Substrate GaAs, ε_r = 12.9, h = 400 μm.

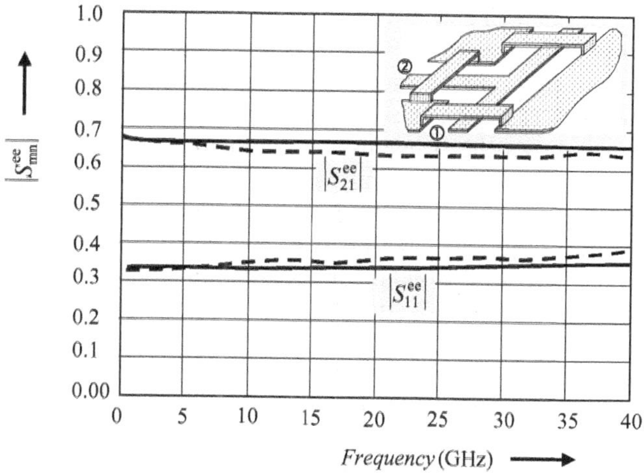

Fig. 3.5.69. Magnitude of the reflection coefficient $|S^{ee}_{11}|$ and the transmission coefficient $|S^{ee}_{21}|$ of the coplanar waveguide mode for a coplanar T-junction with conventional air bridges (type 2), plotted against the frequency (——— theory, – – – measured). Geometrical parameters: $w_1 = w_2 = w_3 = 75\,\mu m$, $s_1 = s_2 = s_3 = 50\,\mu m$, $t = 3\,\mu m$, air-bridge height $= 3\,mm$. Substrate GaAs, $\varepsilon_r = 12.9$, $h = 410\,\mu m$.

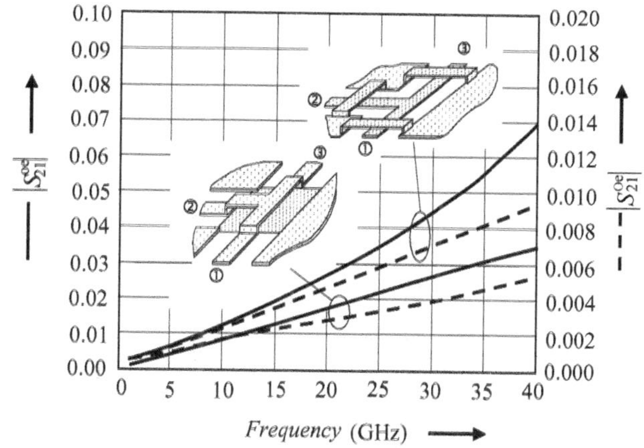

Fig. 3.5.70. Magnitude of the scattering parameter $|S^{oe}_{21}|$ describing the mode conversion from the even mode (coplanar waveguide mode) to the odd mode (slot-line mode) for two different air bridges and with two different sizes of the junctions: ——— $w = 75\,\mu m$, $s = 50\,\mu m$ (big size, left scale), – – – $w = 15\,\mu m$, $s = 10\,\mu m$ (small size, right scale).

(slot-line mode) is shorted by this kind of probe head, and no mode transfer properties can therefore be verified by measurements.

In Fig. 3.5.70 the mode conversion properties of the two different T-junctions that have been compared in Figs. 3.5.68 and 3.5.69 are depicted.

Two structures of each of the T-junctions with different sizes have been investigated. Both structures have 50-Ω feed lines and are built on GaAs substrate material, but in the first structure the center conductor width w is 75 μm and the slot width s is 50 μm, the second structure is smaller: w = 15 μm, s = 10 μm. The figure shows the transmission coefficient S_{21}^{oe}, which describes the transfer of power from the even mode (coplanar mode) to the odd mode (slot-line mode) between port ① and port ② for the investigated T-junctions, plotted against the frequency. The left scale is valid for the junctions of large size, whereas the right scale is valid for the junctions of small size.

It can be seen from Fig. 3.5.70 that the excitation of the odd mode is much higher in the case of the T-junction with conventional air bridges (type 2) compared to the air-bridge T-junctions. This is the same result as in the case of the coplanar bends (see Section 3.5.6). The T-junctions of the smaller size show a much lower mode conversion. The transmission coefficients of the smaller junctions are a factor five less than those of the bigger-size junctions.

Using the FDTD analysis, the influence of the air-bridge position on the excitation of the odd mode in a coplanar T-junction has been investigated in reference 49. Figure 3.5.71 shows the results of this investigation. V_1^{even} is the voltage of the coplanar incident mode at port ①. V_3^{odd} is the voltage of the odd (slot-line) mode at port ③. The feeding coplanar waveguides of the T-junction are assumed to be infinitely long. As can be observed from Fig. 3.5.71, the magnitude of the odd-mode voltage increases with increasing distance of the air bridges or bond wires from the T-junction region, especially at higher frequencies. This means that for an optimal suppressing of the unwanted odd mode at the output ports of the T-junction, the air bridge should be positioned as close as possible to the discontinuity region.

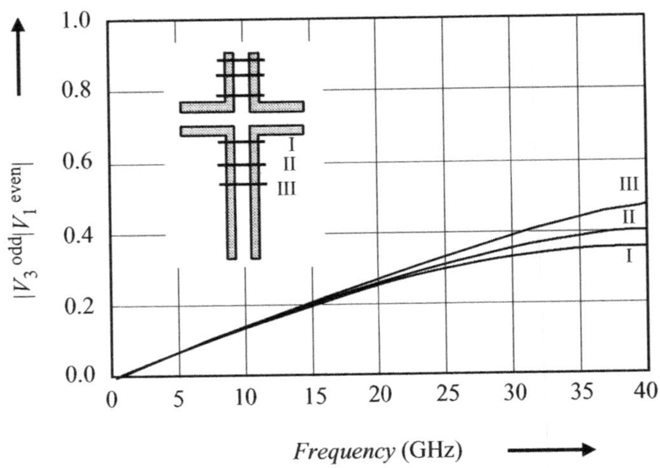

Fig. 3.5.71. Mode conversion from the even mode to the odd mode in a coplanar T-junction, plotted against the position of the air bridges or bond wires, respectively [60].

3.5.8 The Coplanar T-Junction as a Mode Converter

In Section 3.5.7 the coplanar T-junction has mainly been discussed under the aspect that an incident coplanar waveguide mode (even mode) is transferred to a coplanar waveguide mode at the output of the T-junction. It was the aim of the investigations to design a T-junction with an optimal suppression of the possibly excited odd mode on the connected coplanar waveguides. This is the normal way to use this component in a coplanar microwave integrated circuit.

In special circuit design problems, sometimes the even *and* the odd mode are used in a microwave circuit. Examples for such kind of circuit design may be found in references 24 and 25. The authors of these publications have demonstrated, for example, how a magic T-junction can be built in a condensed layout using the even and the odd mode of coplanar waveguides. In the same publications, two coplanar T-junctions have been presented which can be used for a mode conversion from the even coplanar mode to the odd coplanar mode.

Both T-junctions shall be investigated here theoretically using the advanced spectral domain analysis technique, as presented in Section 2.1.2 [80]. A measurement of the mode transfer properties is difficult because measurement techniques in monolithic circuit technology use on-wafer probe heads of coplanar structure that normally shorten the odd mode. Therefore, only the even-mode signals can be measured directly.

Two types of coplanar T-junctions, called type I and type II, are analyzed here. Figure 3.5.72 shows the first type (type I). In addition, the used dis-

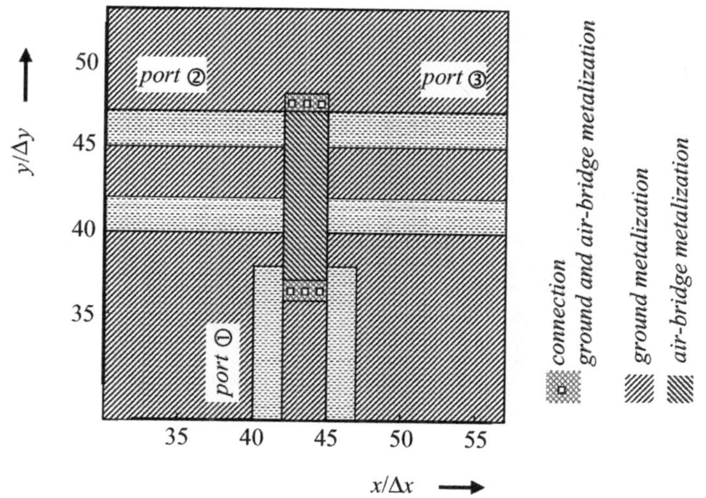

Fig. 3.5.72. Structure of the coplanar T-junction of type I for even-mode to odd-mode conversion. Shown is also the used discretization of the applied spectral domain analysis technique. Geometrical parameters: $w_1 = w_2 = w_3 = 75\,\mu m$, $s_1 = s_2 = s_3 = 50\,\mu m$, substrate GaAs, $h = 410\,\mu m$, $\varepsilon_r = 12.9$, air-bridge height $= 3\,\mu m$, metalization thickness $t = 0$. Mesh size for the analysis: $\Delta x = \Delta y = 25\,\mu m$.

cretization of the spectral domain technique is shown in the figure. The center conductor of the coplanar waveguide at port ① has been used as an air bridge of the same width, crossing the total coplanar waveguide between ports ② and ③. The ground planes of the coplanar waveguide at port ① are connected by a metalization layer under the air bridge. The theoretical analysis of this structure has been performed for GaAs substrate material using the geometrical parameters as shown in the inscription of Fig. 3.5.72. The characteristic impedances of the three connected coplanar waveguides are 50 Ω for the even mode. They are nearly frequency-independent in the considered frequency range up to 40 GHz. On the other hand, the characteristic impedance of the odd mode has values between 51 Ω and 81 Ω in the frequency range.

In Fig. 3.5.73 the investigated T-junction of type II is shown. In this structure, the center conductor of the coplanar waveguide at port ① is "through"-connected to the opposite ground plane of the structure. At the entrance of port ① the ground planes are connected by an air bridge of type 2 (see Section 3.5.5) and the center conductors of the coplanar waveguides at ports ② and ③ are connected by an air bridge of type 1.

It is assumed that an even coplanar waveguide mode is incident at port ① of the structures and that it is converted to an odd mode at ports ② and ③. Also of interest for the investigation is, how this structure transfers the odd mode from port ② to port ③. Therefore, in Fig. 3.5.74 the eigenreflection coefficients of the even mode at port ① and of the odd mode at port ② (which is equal to that at port ③) are shown.

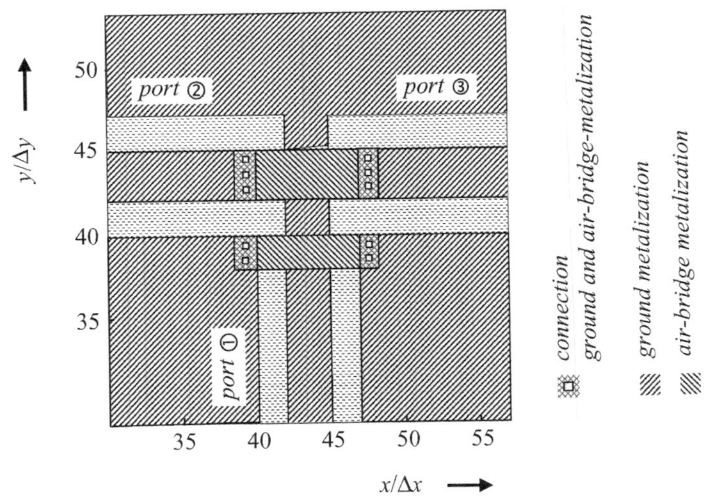

Fig. 3.5.73. Structure of the coplanar T-junction of type II for even-mode to odd-mode conversion. Shown is also the used discretization of the spectral domain analysis technique. Parameters: $w_1 = w_2 = w_3 = 75\,\mu m$, $s_1 = s_2 = s_3 = 50\,\mu m$, substrate material GaAs, $h = 410\,\mu m$, $\varepsilon_r = 12.9$, air-bridge height = 3 μm, metalization thickness $t = 0$. Mesh size for the analysis: $\Delta x = \Delta y = 25\,\mu m$.

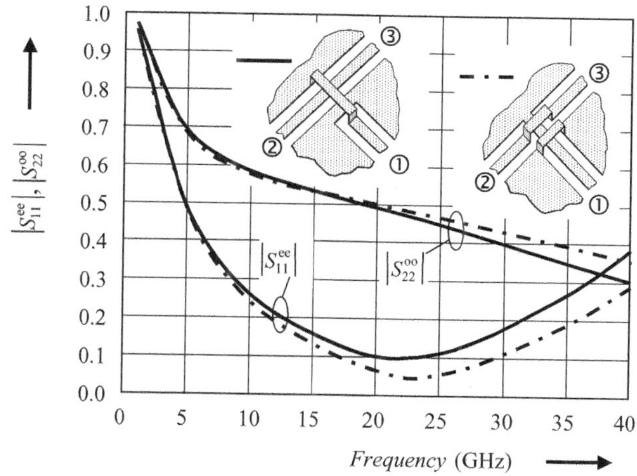

Fig. 3.5.74. Frequency dependence of the even-mode reflection coefficient at port ① and of the odd-mode reflection coefficient at port ② for the considered coplanar T-junctions as shown in Figs. 3.5.72 and 3.5.73.

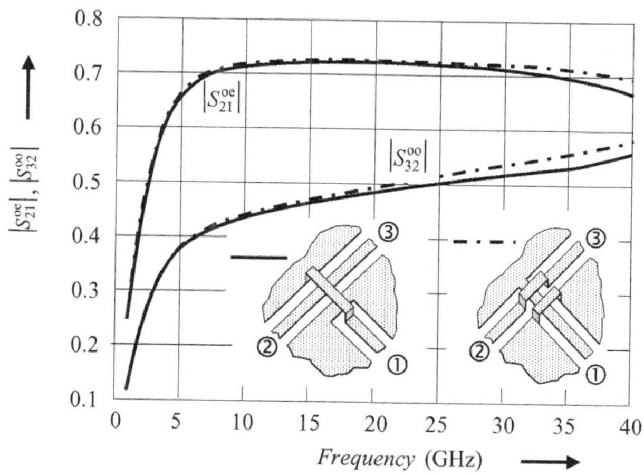

Fig. 3.5.75. Frequency dependence of the even-mode to odd-mode transmission coefficient between port ① and port ② and of the odd-mode to odd-mode transmission coefficient between port ② and port ③ for the considered coplanar T-junctions as shown in Figs. 3.5.72 and 3.5.73.

In Fig. 3.5.75 the transmission coefficients for transmitting an incident even mode at port ① to the odd mode at port ② and for transmitting an incident odd mode from port ② to an odd mode at port ③ are shown.

Both structures show similar reflection and transmission properties. Because of the high dispersion of the odd mode (see also Section 2.1), the frequency range for a good matching at port ① is limited. With respect to the matching properties, the type II T-junction is a little bit better than structure I. At very low frequencies the transmitted power nearly becomes zero. Also essential for the application of these mode transformers are the properties of the structures with respect to unwanted mode suppression. These properties are shown in Fig. 3.5.76.

Analyzing these properties, the type II junction shows a much better performance compared to that of structure I. The unwanted conversion from the even mode at port ① to the even mode at port ② or the odd mode at port ② to the even mode at port ③ is about five times larger in the case of the type I junction. But, looking for the absolute values of the scattering parameters, it must be stated that the unwanted mode suppression of both structures is excellent, and it is in any case good enough for circuit design applications.

A first investigation on the possibility to optimize the transmission properties of this kind of mode converters is demonstrated in Fig. 3.5.77. The figure shows the frequency dependence of the scattering parameters $|S_{11}^{ee}|$ and $|S_{32}^{eo}|$ on the distance a (see inset in Fig. 3.5.77) between the end of transmission line 1 and the ground plane as a parameter. The parameter a has been changed between 25 µm, 50 µm, and 75 µm. It can be observed that an improvement of the eigen reflection coefficient at port ① always leads to a decreased suppression of the unwanted mode, so a compromise must be found between the needed input reflection coefficient and the wanted mode suppression.

Another application of the T-junction structure for mode conversion is shown in Figs. 3.5.78 and 3.5.79, where connections between a coplanar waveguide and a slot line are shown. Slot lines are interesting waveguides

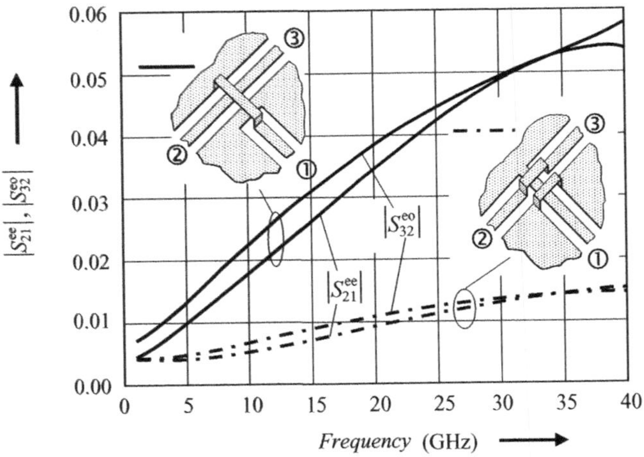

Fig. 3.5.76. Analyzed frequency dependence of the unwanted conversion from the coplanar even mode at port ① to the coplanar even mode at port ② as well as conversion of the odd mode incident at port ② into an even mode at port ③. Geometrical structures are the same as shown in Fig. 3.5.72 and 3.5.73.

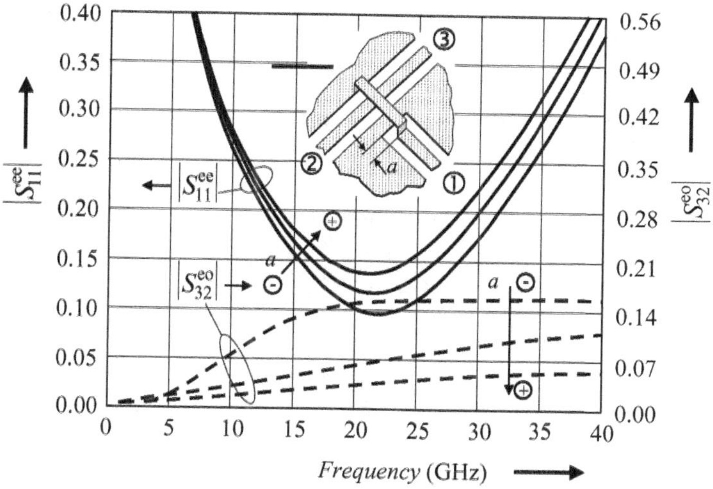

Fig. 3.5.77. Influence of the geometric parameter a (see inset) on the input reflection coefficient S_{11}^{ee} and the transmission coefficient S_{32}^{eo} which describes the unwanted mode suppression in the case of the type I junction. Geometrical parameters are the same as in Fig. 3.5.72.

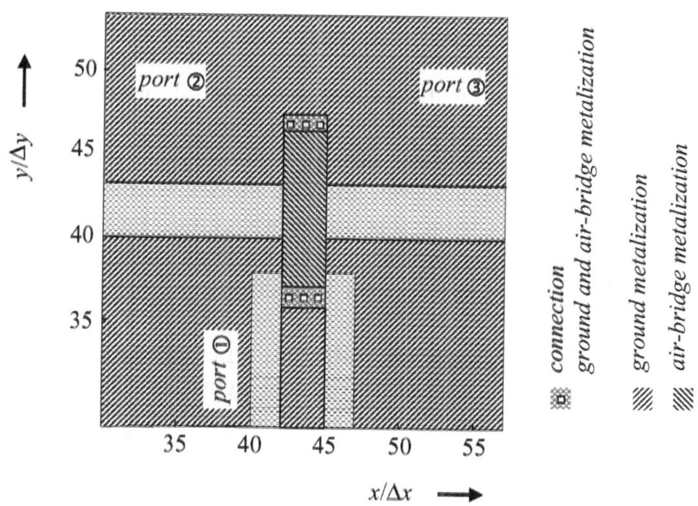

Fig. 3.5.78. Mode converter between a coplanar waveguide and a slot line, type I. Substrate: GaAs, $\varepsilon_r = 12.9$, $h = 410\,\mu m$, air-bridge height $= 3\,\mu m$. Discretization used for the analysis: $\Delta x = \Delta y = 25\,\mu m$.

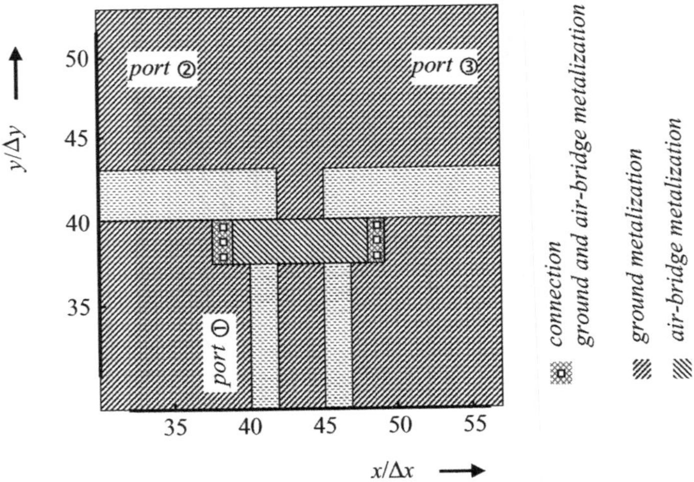

Fig. 3.5.79. Mode converter between a coplanar waveguide and a slotline, type II. Substrate: GaAs, $\varepsilon_r = 12.9$, $h = 410\,\mu m$, air-bridge height = $3\,\mu m$. Discretization used for the analysis: $\Delta x = \Delta y = 25\,\mu m$.

in uniplanar circuits [24, 25, 39] that use a combination of coplanar waveguide structures and slot lines. It is assumed that the fundamental even mode, the coplanar quasi-TEM mode, is incident at port ① of the structure and that a slot-line mode of TE mode propagates on the slot line. The effective conversion from the slot-line mode to the coplanar waveguide mode and vice versa is the aim of the mode converters shown in the figures.

The advantage of the uniplanar circuit technology is its use in parallel planar structures that propagate even and odd modes, so many new functions may be realized in a circuit using this mixed waveguide technology. The fundamental connecting components of this technology between the coplanar waveguide and the slot line are various T-junctions and other kinds of mode converters.

Figures 3.5.78 and 3.5.79 show two different types of mode converters that have been analyzed. The coplanar waveguide again has a center strip line width w of $75\,\mu m$ and a slot width s of $50\,\mu m$, so its characteristic impedance is about $50\,\Omega$. The slot width of the slot lines is $75\,\mu m$, so that the characteristic impedances of these lines because of the highly dispersive properties of the slot-line mode vary between $44\,\Omega$ and $64\,\Omega$ in the considered frequency range from 1 GHz to 40 GHz.

All used air bridges have a bridge height of $3\,\mu m$. In the type I mode converter (Fig. 3.5.78) the center strip at the end of the coplanar waveguide is formed as an air bridge crossing the slot line. A metalization layer at the end of the coplanar waveguide connects both ground planes of the coplanar waveguide. On the other hand, in type II of the mode converter (Fig. 3.5.79) the center conductor of the coplanar waveguide is "through"-connected to the opposite ground plane of the slot line, and the ground planes of the coplanar waveguide are connected by a type 2 air-bridge construction at the end of the coplanar waveguide.

In Fig. 3.5.80 the frequency dependence of the input reflection coefficients for the coplanar waveguide mode at port ① of the coplanar waveguide and for the slot-line mode at port ② of the slot line are shown for the type I converter as well as for the type II converter. Because of the high dispersion of the slot-line mode, matching at port ① (coplanar waveguide) can only be achieved with acceptable low reflection coefficients in a small frequency band. Because of the cutoff frequency of the slot-line mode that is always higher than zero ($f_c > 0$), the reflection coefficients for both modes (coplanar waveguide mode and slot line mode) increase to high values at low frequencies. For a frequency $f = 0$, the total power incident at port ① will be reflected back into the coplanar waveguide. A comparison of the two converter structures shows that structure II has some slight advantages concerning the reflection coefficient of the coplanar waveguide mode at higher frequencies. But there is no principal difference between these two structures.

An analysis of the transmission coefficients, depicted in Fig. 3.5.81, shows at low frequencies the already discussed properties. For very low frequencies, no power is transported from the coplanar waveguide to the slot line or vice versa. Both structures show very similar properties; only for frequencies higher than 20 GHz, some small deviations of the transmission properties may be observed.

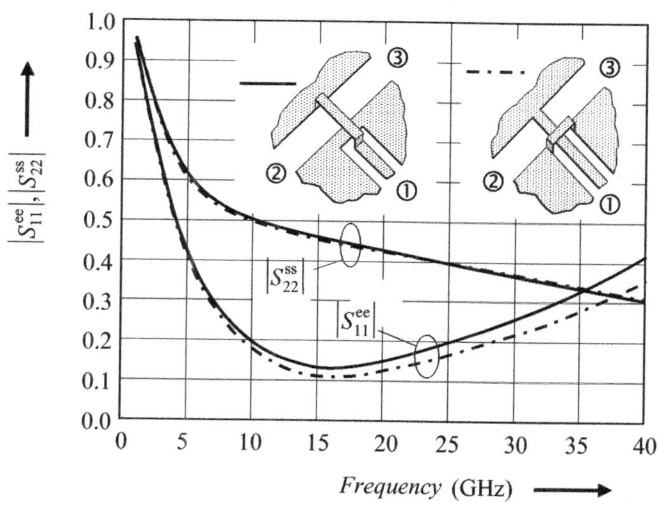

Fig. 3.5.80. Frequency dependence of the input reflection coefficients at port ① of the coplanar waveguide (even coplanar waveguide mode) and port ② of the slot line (slot-line mode). Geometrical parameters as in Figs. 3.5.78 and 3.5.79.

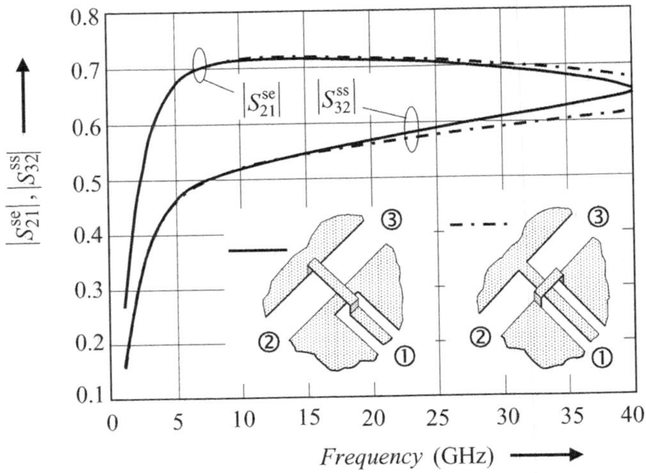

Fig. 3.5.81. Frequency dependence of the transmission coefficients from port ① of the coplanar waveguide (even coplanar waveguide mode) to port ② of the slot line (slot-line mode) and from port ② of the slot line (slot-line mode) to port ③ of the slot line (slot-line mode). Geometrical parameters are the same as in Figs. 3.5.78 and 3.5.79.

In Fig. 3.5.82 the properties of the structures concerning the suppression of the unwanted mode conversion—that is, the suppression of the slotline mode into an odd mode of the coplanar waveguide—is shown. The excitation of the odd mode on the coplanar waveguide, in the case where a slot-line mode is incident on the slotline, is unwanted in the described application.

Both types of converters do not have good properties with respect to this unwanted mode suppression. Especially at higher frequencies, high power is transported from the slot-line mode into the odd mode of the coplanar waveguide. Reasons for this property are the wide field distributions of the odd coplanar waveguide mode and of the slot-line mode. These wide field distributions that have already been discussed in Chapter 2 lead to a high coupling of both modes in the junction area.

This coupling can be reduced in the case of the type I coupler by increasing the air-bridge length a. By this way, the line end of the coplanar waveguide is shifted away from the slot line. However, as an investigation of the transmission properties and their dependence on this parameter (Fig. 3.5.83) shows, the reduction of the unwanted mode conversion, reached by this technique, unfortunately correlates directly with an increased input reflection coefficient of the even coplanar waveguide mode at port ①. If the distance a, shown in the inset of Fig. 3.5.83, is changed from 25 µm to 75 µm, an improvement of the unwanted mode suppression by about 13% can be reached at a frequency of 16 GHz. At the same time, however, the input reflection coefficient of the coplanar waveguide mode at port ① is increased by about 9%. Nevertheless, the results that have been analyzed using complex field theoretical simulation

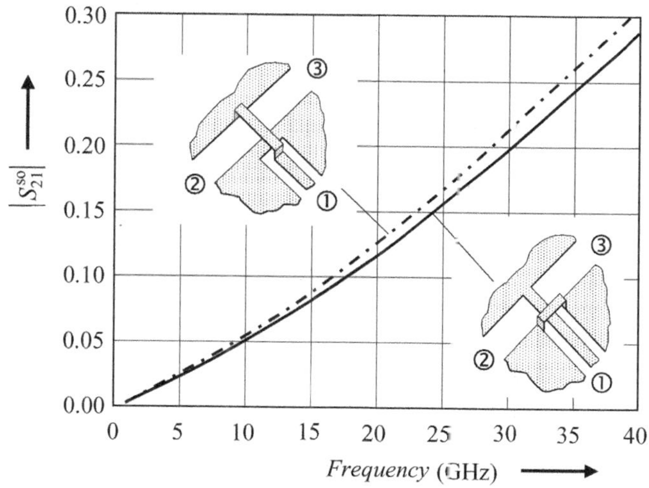

Fig. 3.5.82. Analyzed frequency dependence of the unwanted mode conversion from the odd coplanar waveguide mode at port ① to the slot-line mode at port ② or vice versa. Geometrical parameters are the same as in Figs. 3.5.78 and 3.5.79.

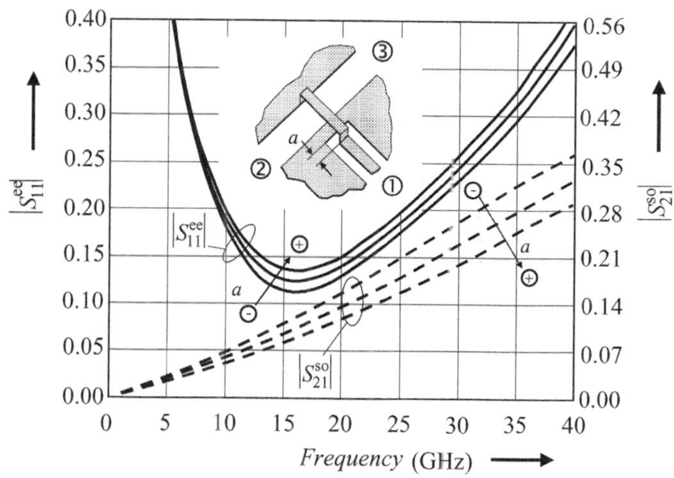

Fig. 3.5.83. Influence of the parameter a (see inset) on the frequency dependence of the input reflection coefficient S_{11}^{ee} and the unwanted mode conversion coefficient $|S_{21}^{so}|$ for the mode converter of type I. Geometrical parameters are the same as in Fig. 3.5.79.

techniques show the advantages of this kind of CAD methods for the circuit designer.

3.5.9 The Coplanar Waveguide Crossing

The coplanar waveguide crossing, as shown in Fig. 3.5.84a, leads to discussions similar to that described in the previous section for the coplanar T-junction. Also, for this component the three possible forms of air bridges may be used to assure a pure even-mode (coplanar-mode) propagation in the connected coplanar waveguides. An air-bridge crossing is shown in Fig. 3.5.84a which connects the four ground planes using the gate metalization layer under the crossing, which itself is constructed as an air bridge.

The equivalent circuit model of this component is shown in Fig. 3.5.84b. In the case of the air-bridge crossing, the capacitive effect of the crossing is simulated by four equivalent capacitances of equal values. In addition, five inductances describe the influence of the discontinuity on the current density distribution.

Figure 3.5.85 shows a technological realization of two crossings with different air bridges as used in GaAs monolithic integrated circuits. From the equivalent circuit parameters, it may be deduced that the air-bridge crossing (Fig. 3.5.85b) has a higher capacitance compared to that of the crossing with

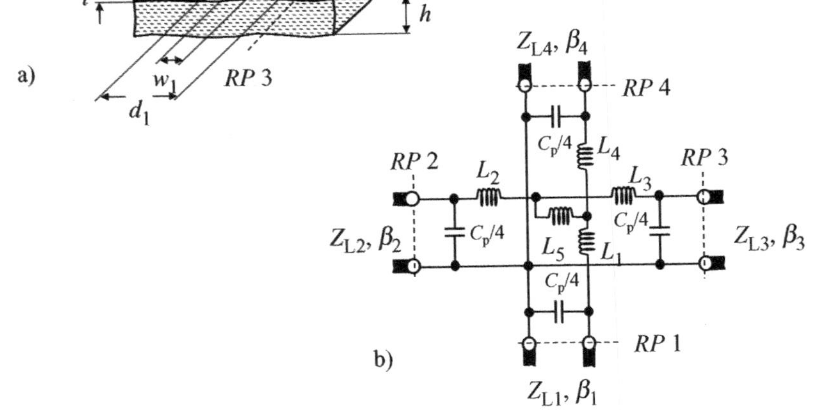

Fig. 3.5.84. (a) The coplanar waveguide crossing of type air-bridge crossing, and (b) the equivalent circuit of the coplanar crossing. *RP* stands for reference planes.

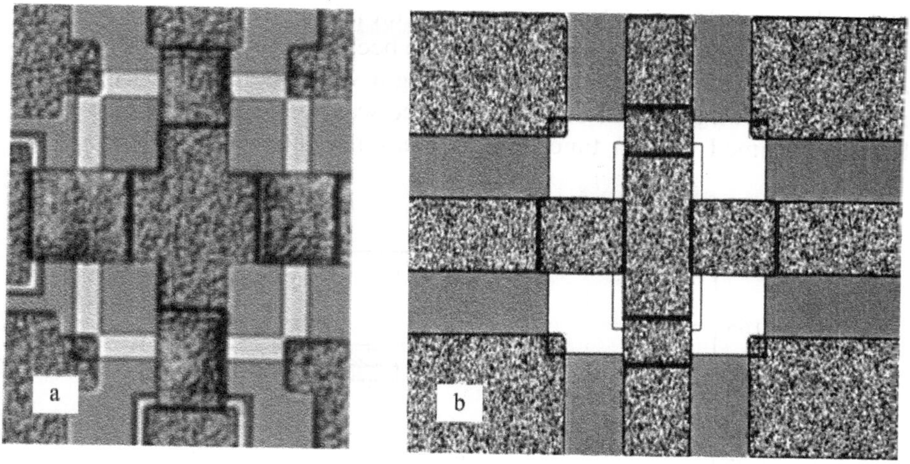

Fig. 3.5.85. (a) Technological realization of the coplanar crossing with type 1 air bridges. (b) technological realization of the air-bridge crossing.

type 1 air bridges (Fig. 3.5.85a). As has already been discussed in detail for the coplanar T-junction (Section 3.5.7), the air-bridge T-junction has the higher odd-mode suppression, and this holds also in the case of the coplanar crossing.

The capacitance C_p is determined in the same way as has been described in Section 3.4.2. If the five inductances have to be determined, five different analysis steps are needed. Similar to the case of the coplanar T-junction (Section 3.5.7), five different excitations of the current density distribution are simulated. From the resulting five equations the unknown inductances can be derived. The left side of Eq. (3.5.12) shows the ports between which the currents have been excited using the method described in Section 3.5.7. The right-hand side shows the resulting equation for determining the inductances.

Current between port ② and ③: $$\rightarrow L_2 + L_3 = \frac{\Phi_{23}}{4A} - l_2 L_2' - l_3 L_3', \qquad (3.5.12a)$$

Current between port ③ and ④: $$\rightarrow L_3 + L_4 + L_5 = \frac{\Phi_{34}}{4A} - l_3 L_3' - l_4 L_4', \qquad (3.5.12b)$$

Current between port ② and ④: $$\rightarrow L_2 + L_4 + L_5 = \frac{\Phi_{24}}{4A} - l_2 L_2' - l_4 L_4', \qquad (3.5.12c)$$

Current between port ① and ④: $$\rightarrow L_1 + L_4 = \frac{\Phi_{14}}{4A} - l_1 L_1' - l_4 L_4', \qquad (3.5.12d)$$

Current between port ① and ③: $$\rightarrow L_1 + L_3 + L_5 = \frac{\Phi_{13}}{4\mathrm{A}} - l_1 L_1' - l_3 L_3'. \qquad (3.5.12e)$$

For the verification of the equivalent circuit model, the scattering parameters of a symmetrical coplanar crossing have been determined from the equivalent circuit and have been measured using a vector network analyzer. The results are shown in Fig. 3.5.86. The measured scattering parameters that have been determined using a time domain measurement technique show good

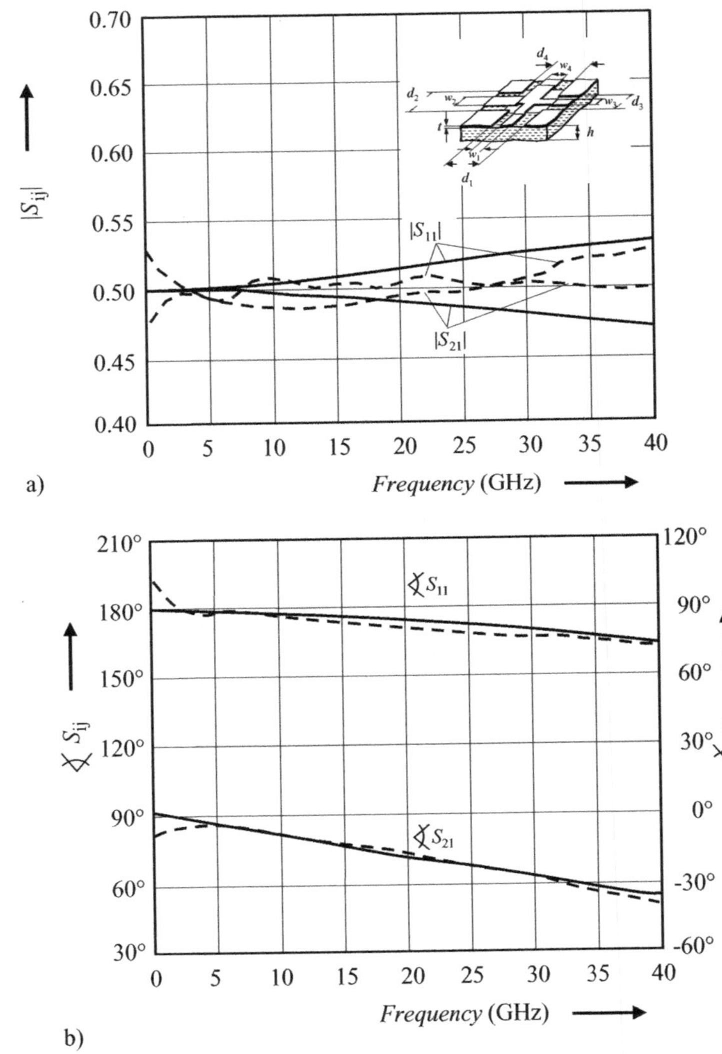

Fig. 3.5.86. Measured (- - -) and calculated (———) absolute value (a) and phase (b) of the scattering parameters of a symmetrical coplanar air-bridge crossing in dependence on the frequency. Geometrical parameters: $w_1 = w_2 = w_3 = w_4 = 75\,\mu\text{m}$, $d_1 = d_2 = d_3 = d_4 = 175\,\mu\text{m}$, substrate GaAs, $\varepsilon_r = 12.9$, $h = 410\,\mu\text{m}$, $t = 3\,\mu\text{m}$, air-bridge height $= 3\,\mu\text{m}$. Equivalent circuit parameters: $L_1 = L_2 = L_3 = L_4 = 53\,\text{pH}$, $L_5 = -22\,\text{pH}$, $C_p = 172\,\text{fF}$.

agreement with the simulated results up to high frequencies. This is true especially for the measured phases. The measured magnitudes show a small ripple in dependence on the frequency due to the used measurement technique.

Finally, the dependence of the equivalent circuit components on the center conductor width $w_1 = w_4$ is shown in Figs. 3.5.87–3.5.89 for constant values $w_2 = w_3$ as parameters.

The capacitance C_p increases with increasing center conductor widths w_1 and w_3, and the inductances L_1 to L_4 are decreasing with increasing center conductor widths. Astonishingly, the inductance L_5 has always a negative value for the considered geometrical data. It means that the total inductance is lower than the equivalent inductance of a homogeneous coplanar waveguide. This may lead to difficulties if, for example, the equivalent circuit is to be derived from measured scattering parameters by curve-fitting and the curve-fitting program is only able to work with positive values. In this case the inductance L_5 is taken to be zero, and the values of the other inductance might be different from those that are calculated from the magnetic potential analysis.

A coplanar test structure to measure the quality of the simulation technique for the coplanar crossing, with respect to dependence on the frequency, is shown in Fig. 3.5.90. Two open-ended stub resonators of 1500-μm length on GaAs substrate are used, and the scattering parameters between the two coplanar feed lines are measured and compared to the simulation results. The first test structure shown in Fig. 3.5.90 has no air bridges included.

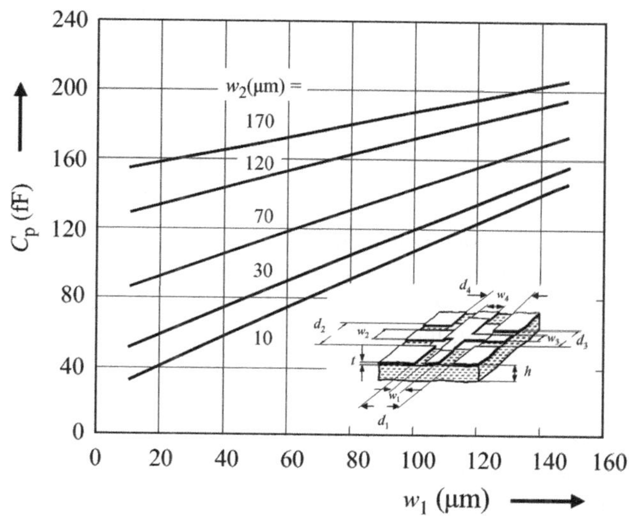

Fig. 3.5.87. Dependence of the equivalent capacitance C_p of a coplanar crossing on the geometrical parameters w_1 and w_2. Other geometrical parameters: $w_4 = w_1$, $w_3 = w_2$, $d_1 = d_2 = d_3 = d_4 = 170\,\mu m$, $t = 3\,\mu m$, air-bridge height = 3 μm, substrate GaAs, $\varepsilon_r = 12.9$, $h = 410\,\mu m$.

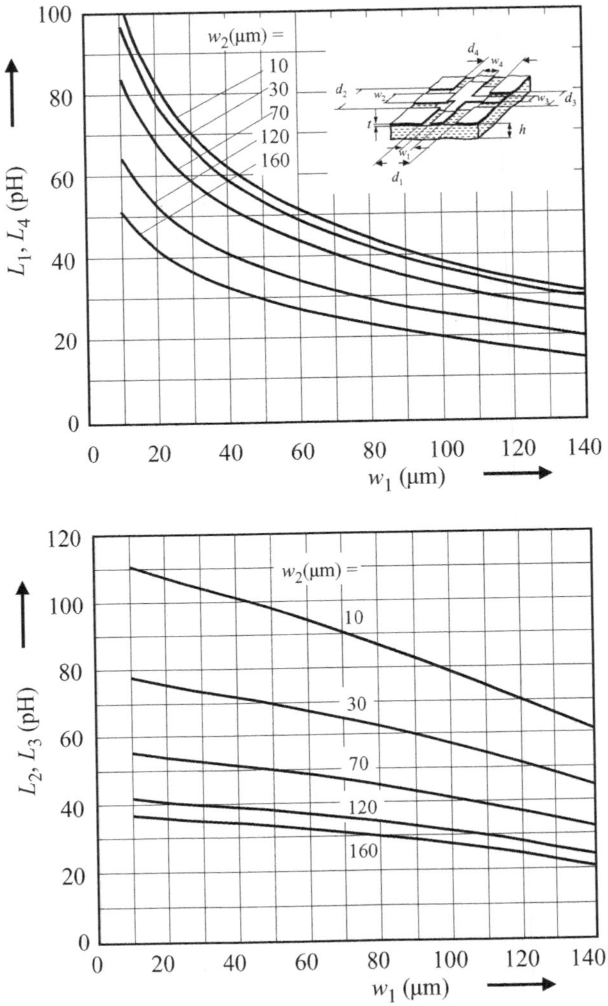

Fig. 3.5.88. Dependence of the equivalent inductances L_1, L_4 (a) and L_2, L_3 (b) of a coplanar crossing on the geometrical parameters w_1 and w_2. Other geometrical parameters: $w_4 = w_1$, $w_3 = w_2$, $d_1 = d_2 = d_3 = d_4 = 170\,\mu m$, $t = 3\,\mu m$, air-bridge height $= 3\,\mu m$, substrate GaAs, $\varepsilon_r = 12.9$, $h = 410\,\mu m$.

The measurement results for this test structure are shown in Fig. 3.5.91. Because no air bridges are included in the test structure, the odd slot-line mode is excited at the discontinuity and the agreement between measurement and simulation (which bases only on the quasi-TEM mode approximation) is not very good for frequencies higher than 20 GHz. The resonant frequencies of the stub resonators are shifted by nearly 8 GHz in the simulation compared to the measurements.

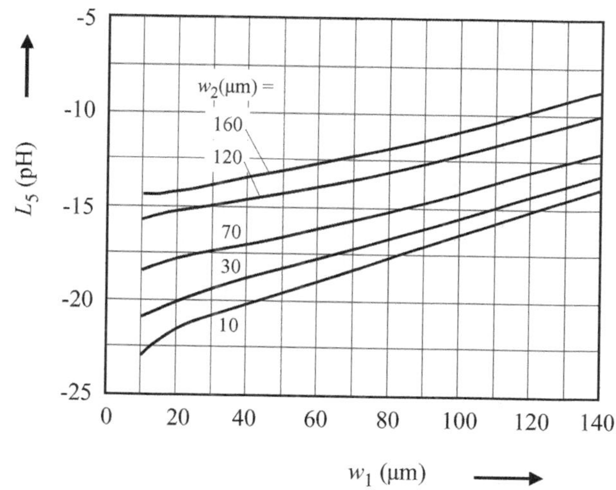

Fig. 3.5.89. Dependence of the equivalent inductance L_5 of a coplanar crossing on the geometric parameters w_1 and w_2. Other geometric parameters: $w_4 = w_1$, $w_3 = w_2$, $d_1 = d_2 = d_3 = d_4 = 170\,\mu m$, $t = 3\,\mu m$, air-bridge height $= 3\,\mu m$, substrate material GaAs, $\varepsilon_r = 12.9$, $h = 410\,\mu m$.

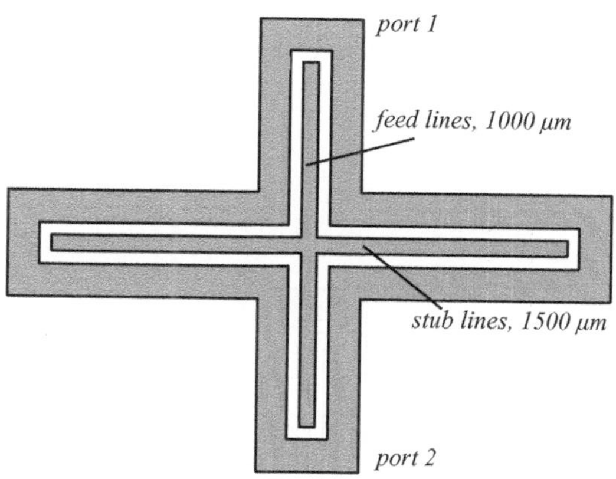

Fig. 3.5.90. Test structure including a coplanar crossing, two open-ended 50-Ω stub resonators and two 50-Ω feed lines on GaAs substrate to measure the resulting scattering parameters between port 1 and port 2. Test circuit without air bridges. Parameters of the 50-Ω lines: $w = 100\,\mu m$, $s = 75\,\mu m$, ground plane width $= 200\,\mu m$, substrate GaAs, $\varepsilon_r = 12.9$, $h = 410\,\mu m$.

A second test structure is similar to that shown in Fig. 3.5.90, but it contains four air bridges directly at the ports of the coplanar crossing to suppress the odd mode. It is shown in Fig. 3.5.92. The geometrical parameters are identical to those shown in Fig. 3.5.90. The used air bridges in this case are of type 2

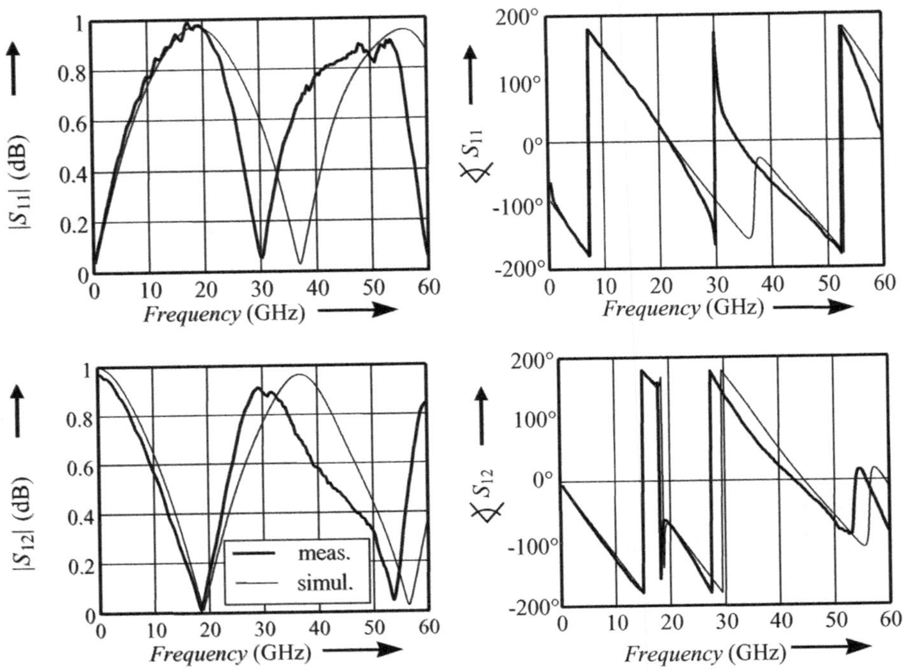

Fig. 3.5.91. Comparison between measured (thick lines) and simulated (thin lines) scattering parameters for the test structure shown in Fig. 3.5.90, plotted against the frequency.

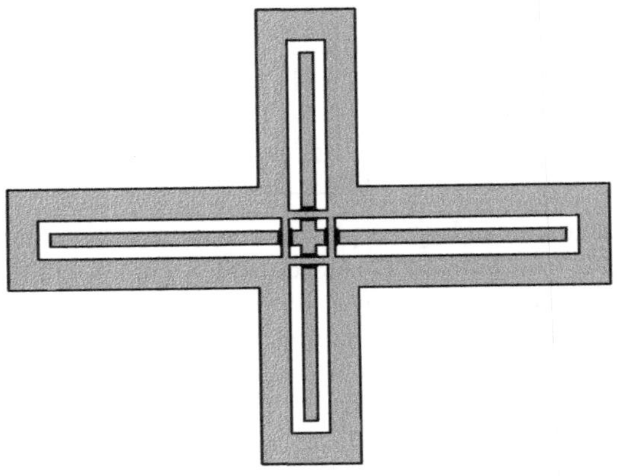

Fig. 3.5.92. Modified test structure that includes four air bridges at the ports of the coplanar crossing. All other line parameters as in Fig. 3.5.90. Air-bridge type 2: $bg = 8$ μm, $bw = 50$μm, $bs = 14$μm. For the details of the geometric parameters compare Fig. 5.6.6.

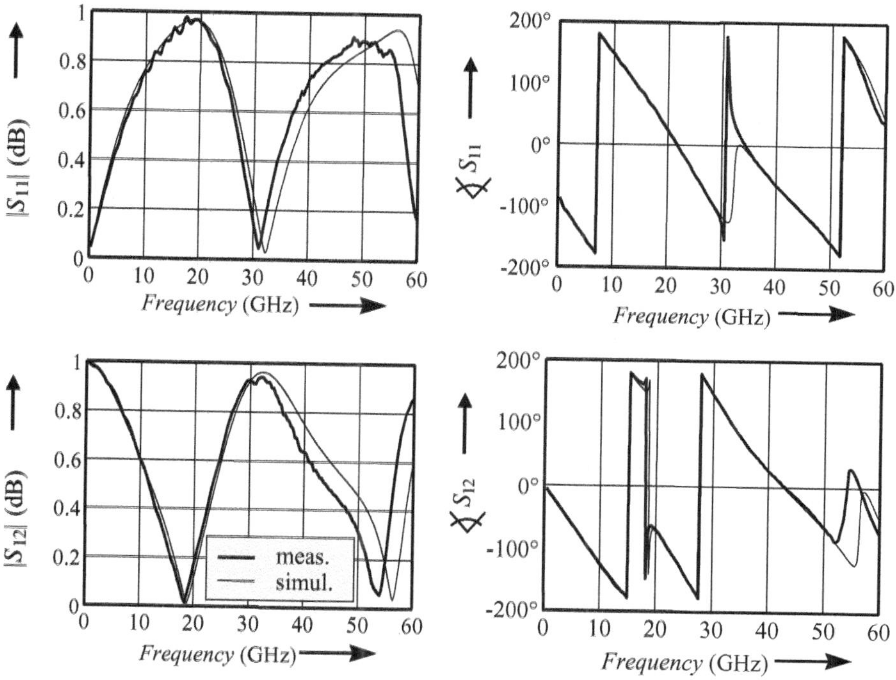

Fig. 3.5.93. Comparison between measured (thick lines) and simulated scattering parameters for the test structure shown in Fig. 3.5.92, plotted against the frequency.

(see Section 3.5.5). No big differences have been measured if other types of air bridges are used in this structure in the considered frequency range. Therefore these results are not discussed here additionally.

The comparison between measurement and simulation is shown in Fig. 3.5.93. It shows a much better agreement up to 60 GHz compared to the results shown in Fig. 3.5.91. Especially the first resonant frequency near 20 GHz is simulated fairly well. The still remaining shift of the resonant frequencies at higher frequencies results from the dispersion of the stub resonator lines which is not included in the simulation but which might easily be added.

BIBLIOGRAPHY AND REFERENCES

1. A. Farrar and A. T. Adams, Matrix methods for microstrip three-dimensional problems, *IEEE Trans. Microwave Theory Tech.*, vol. MTT-20, Aug. 1972, pp. 497–504.
2. P. Benedek and P. Silvester, Equivalent capacitances for microstrip gaps and steps, *IEEE Trans. Microwave Theory Tech.*, vol. MTT-20, Nov. 1972, pp. 729–733.
3. R. Horton, Equivalent representation of abrupt impedance steps in microstrip line, *IEEE Trans. Microwave Theory Tech.*, vol. MTT-21, Aug. 1972, pp. 562–564.

4. P. Silvester and P. Benedek, Equivalent capacitances of microstrip open circuits, *IEEE Trans. Microwave Theory Tech.*, vol. MTT-20, Aug. 1972, pp. 511–516.
5. P. Silvester and P. Benedek, Microstrip discontinuity capacitances for right-angled bends, T junctions and crossings, *IEEE Trans. Microwave Theory Tech.*, vol. MTT-21, May 1973, pp. 341–346.
6. R. Horton, The electrical characterization of a right-angled bend in microstrip line, *IEEE Trans. Microwave Theory Tech.*, vol. MTT-21, June 1973, pp. 427–429.
7. A. Gopinath and B. Easter, Moment method of calculating discontinuity inductance of microstrip right-angled bends, *IEEE Trans. Microwave Theory Tech.*, vol. MTT-22, Oct. 1974, pp. 880–883.
8. F. Thomson and A. Gopinath, Calculation of microstrip discontinuity inductances, *IEEE Trans. Microwave Theory Tech.*, vol. MTT-23, Aug. 1975, pp. 648–655.
9. B. Easter, The equivalent circuit of some microstrip discontinuities, *IEEE Trans. Microwave Theory Tech.*, vol. MTT-23, Aug. 1975, pp. 655–660.
10. A. Gopinath, A. F. Thomson, and I. M. Stephenson, Equivalent circuit parameters of microstrip step change in width and cross junctions, *IEEE Trans. Microwave Theory Tech.*, vol. MTT-24, March 1976, pp. 142–144.
11. C. Gupta and A. Gopinath, Equivalent circuit capacitance of microstrip step change in width, *IEEE Trans. Microwave Theory Tech.*, vol. MTT-25, Oct. 1977, pp. 819–822.
12. W. Menzel and I. Wolff, A method for calculating the frequency-dependent properties of microstrip discontinuities, *IEEE Trans. Microwave Theory Tech.*, vol. MTT-25, Feb. 1977, pp. 107–112.
13. J. A. G. Malherbe and A. F. Steyn, The compensation of step discontinuities in TEM-mode transmission lines, *IEEE Trans. Microwave Theory Tech.*, vol. MTT-26, Nov. 1978, pp. 883–885.
14. B. N. Naele and A. Gopinath, Microstrip discontinuity inductances, *IEEE Trans. Microwave Theory Tech.*, vol. MTT-26, Oct. 1978, pp. 203–207.
15. B. Easter, A. Gopinath, and I. M. Stephenson, Theoretical and experimental methods for evaluating discontinuities in microstrip, *The Radio and Electronic Eng.*, vol. 48, 1978, no. 1/2, pp. 74–84.
16. A. Gopinath and C. Gupta, Capacitance parameters of discontinuities in microstrip lines, *IEEE Trans. Microwave Theory Tech.*, vol. MTT-26, Oct. 1978, pp. 831–836.
17. R.-H. Jansen, Hybrid mode analysis of end effects of planar microwave and millimeterwave transmission lines, *IEE Proc., Part H (Microwaves, Optics and Antennas)*, vol. 128, 1981, no. 2, pp. 77–86.
18. M. Kirschning, R. H. Jansen, and N. H. L. Koster, Accurate model for open end effect of microstrip lines, *Electronic Lett.*, vol. 17, 1981, no. 3, pp. 123–124.
19. S. S. Bedair and M. I. Sobhy, Open-end discontinuity in shielded microstrip circuits, *IEEE Trans. Microwave Theory Tech.*, vol. MTT-29, Oct. 1981, pp. 1107–1109.
20. R. Mehran, *Grundelemente des rechnergestützten Entwurfs von Mikrostreifenleitungs-Schaltungen*, Aachen: Verlag H. Wolff, 1982.
21. M. Kirschning, R. H. Jansen, and N. H. L. Koster, Measurement and computer-aided modeling of microstrip discontinuities by an improved resonator method, in: *1983 IEEE MTT-S International Microwave Symposium Digest*, 1983, pp. 495–497.

22. R.-W. Jackson and D.-M. Pozar, Surface wave losses at discontinuities in millimeter wave integrated transmission lines, in: *1985 IEEE-MTT-S. International Microwave Symposium Digest*, St. Louis, MO, 4–6 June 1985, pp. 563–565.

23. A. Christ and H. L. Hartnagel, Three-dimensional finite-difference method for the analysis of microwave-device deembedding, *IEEE Trans. Microwave Theory Tech.*, vol. MTT-35, Aug. 1987, pp. 688–696.

24. T. Hirota, Y. Tarusawa, and H. Ogawa, Uniplanar MMIC hybrids—a proposed new MMIC structure," *IEEE Trans. Microwave Theory Tech.*, vol. MTT-35, June 1987, pp. 576–581.

25. H. Ogawa and A. Minagawa, Uniplanar MIC balanced multiplier—A proposed new structure for MIC's, *IEEE Trans. Microwave Theory Tech.*, vol. MTT-35, Dec. 1987, pp. 1363–1368.

26. X. Zhang and K. K. Mei, Time-domain finite difference method to the calculation of the frequency-dependent characteristics of microstrip discontinuities, *IEEE Trans. Microwave Theory Tech.*, vol. MTT-36, Dec. 1988, pp. 1775–1787.

27. R.-W. Jackson, Mode conversion due to discontinuities in modified grounded coplanar waveguide, in: *1988 IEEE MTT International Microwave Symposium Digest*, 25–27 May 1988, New York, vol. 1, 1988, pp. 203–206.

28. R.-N. Simons and G.-E. Ponchak, Modeling of some coplanar waveguide discontinuities, in: *1988 IEEE MTT International Microwave Symposium Digest*, 25–27 May 1988, New York, vol. 1, 1988, pp. 297–300.

29. R. N. Simons and G. E. Ponchak, Modeling of some coplanar waveguide discontinuities, *IEEE Trans. Microwave Theory Tech.*, vol. MTT-36, Dec. 1988, pp. 1796–1803.

30. N. M. Wigley, An efficient method for subtracting off singularities at corners for Laplace's equation, *J. Comput. Phys.*, vol. 78, 1988, pp. 369–377.

31. D. W. Kelly, R. J. Mills, and J. A. Reizes, A posteriori error estimates in finite difference techniques, *J. Comput. Phys.*, vol. 74, 1988, pp. 214–232.

32. N. H. L. Koster, S. Koßlowski, R. Bertenburg, S. Heinen, and I. Wolff, Investigation on air bridges used for MMICs in CPW technique, in: *Proceedings, 19th European Microwave Conference*, 1989, pp. 666–671.

33. X. Zhang and K. K. Mei, Time-domain calculation of microstrip components and the curve-fitting of numerical results, in: *1989 IEEE MTT-S International Microwave Symposium*, 1989, pp. 313–316.

34. R.-N. Simons, G.-E. Ponchak, K.-S. Martzaklis, and R.-R. Romanofsky, Channelized coplanar waveguide: discontinuities, junctions, and propagation characteristics, in: *IEEE 1989 MTT-S International Microwave Symposium Digest*, 13–15 June 1989, Long Beach, CA, vol. 3, 1989, pp. 915–918.

35. W. P. Harokopus and P. B. Katehi, An accurate characterization of open microstrip discontinuities including radiation losses, in: *1989 IEEE MTT-S Internat. Microwave Symp. Digest*, 1989, pp. 231–234.

36. C. J. Railton and J. P. McGeehan, Analysis of microstrip discontinuities using the finite difference time domain technique, in: *1989 IEEE MTT-S International Microwave Symposium Digest*, 1989, pp. 1009–1112.

37. C.-W. Kuo and T. Itoh, Characterization of coplanar waveguide step discontinuity using the transverse resonance method, in: *Proceedings, 19th European Microwave Conference*, 1989, pp. 662–665.

38. R.-W. Jackson, Mode conversion at discontinuities in finite-width conductor-backed coplanar waveguide, *IEEE Trans. Microwave Theory Tech.*, vol. 37, 1989, no. 10, pp. 1582–1589.

39. T. Tokumitsu, S. Hara, T. Takenaka, and M. Aikawa, Divider and combiner line-unified FETs as basic circuit function modules—part 1, *IEEE Trans. Microwave Theory Tech.*, vol. 38, Dec. 1990, pp. 1210–1217.

40. M. Naghed and I. Wolff, A three-dimensional finite-difference calculation of equivalent capacitances of coplanar waveguide discontinuities, in: *1990 IEEE MTT-S International Microwave Symposium Digest*, 8–10 May 1990, Dallas, TX, vol. 3, 1990, pp. 1143–1146.

41. M. Naghed and I. Wolff, Equivalent capacitances of coplanar waveguide discontinuities and interdigital capacitors using a three-dimensional finite-difference method, *IEEE Trans. Microwave Theory Tech.*, vol. MTT-26, Dec. 1990, pp. 1808–1815.

42. G. Bartolucci and J. Piotrowski, Full-wave analysis of shielded coplanar waveguide short-end, *Electronics Letters*, vol. 26, no. 19, 1990, pp. 1615–1616.

43. N.-I. Dib, P.-B. Katehi, G.-E. Ponchak, and R.-N. Simons, Coplanar waveguide discontinuities for p-i-n diode switches and filter applications, in: *1990 IEEE MTT-S International Microwave Symposium Digest*, 8–10 May 1990, Dallas, TX, vol. 1, 1990, pp. 399–402.

44. M. Naghed and I. Wolff, A three-dimensional finite-difference calculation of capacitances of coplanar waveguide discontinuities, in: *1990 IEEE MTT-S International Microwave Symposium Digest*, 1990, pp. 1143–1146.

45. M. Naghed, M. Rittweger, and I. Wolff, A new method for the calculation of the equivalent inductances of coplanar waveguide discontinuities, in: *1991 IEEE MTT-S International Microwave Symposium Digest*, 1991, pp. 747–750.

46. M. Naghed, M. Rittweger, and I. Wolff, A new method for the calculation of the equivalent inductances of coplanar waveguide discontinuities, in: *1991 IEEE MTT-S International Microwave Symposium Digest* (91CH2870-4), 10–14 June 1991, Boston, MA, vol. 2, 1991, pp. 747–750.

47. C.-W. Kuo, T. Kitazawa, and T. Itoh, Analysis of shielded coplanar waveguide step considering the finite metallization thickness effect, in: *1991 IEEE MTT-S International Microwave Symposium Digest*, 10–14 June 1991, Boston, vol. 2, 1991, pp. 473–475.

48. M. Guglielmi, Capacitive discontinuities: Rigorous multimode equivalent network representation, in: *1991 IEEE MTT-S International Microwave Symposium Digest*, 10–14 June 1991, Boston, vol. 2, 1991, pp. 477–480.

49. M. Rittweger, M. Abdo, and I. Wolff, Full-wave analysis of coplanar discontinuities considering three-dimensional bond wires, in: *1991 IEEE MTT-S International Microwave Symposium Digest*, 10–14 June 1991, Boston, vol. 2, 1991, pp. 465–468.

50. S. Alexandrou, R. Sobolewski, H. Nakano, B.-C. Tousley, and T.-Y. Hsiang, Picosecond characterization of bent coplanar waveguides, *IEEE Microwave and Guided Wave Lett.*, vol. 1, no. 9, 1991, pp. 236–238.

51. W.-P. Harokopus and P.-B. Katehi, Radiation loss from open coplanar waveguide discontinuities, in: *1991 IEEE MTT-S International Microwave Symposium Digest*, 10–14 June 1991, Boston, vol. 2, 1991, pp. 743–746.

52. D. S. Mirshekar and W.-K. Ofosu, Reflection coefficient of terminated coplanar strips, in: *IEE Colloquium on Computer Based Tools for Microwave Engineers*, 15 Oct. 1991, London, vol. no. 152, 1991, pp. 8/1–8/5.

53. M. Rittweger, N. H. L. Koster, S. Koßlowski, R. Bertenburg, S. Heinen, and I. Wolff, Full-wave analysis of a modified coplanar air bridge T-junction, in: *Proceedings, 21st European Microwave Conference*, Sept. 1991, pp. 993–998.

54. M. Rittweger, M. Abdo, and I. Wolff, Full-wave analysis of coplanar discontinuities considering three-dimensional bond wires, in: *1991 IEEE MTT-S International Microwave Symposium Digest*, June 1991, pp. 465–468.

55. N.-I. Dib, P.-B. Katehi, and G.-E. Ponchak, Analysis of shielded CPW discontinuities with air bridges, in: *1991 IEEE MTT-S International Microwave Symposium Digest*, 10–14 June 1991, Boston, vol. 2, 1991, pp. 469–472.

56. N.-I. Dib, L.-P.-B. Katehi, G.-E. Ponchak, and R.-N. Simons, Theoretical and experimental characterization of coplanar waveguide discontinuities for filter applications, *IEEE Trans. Microwave Theory Tech.*, vol. 39, no. 5, 1991, pp. 73–882.

57. M. Solaimani, E. van Lii, and A. van Capelle, Computation of static capacitance for discontinuities in grounded coplanar waveguide by moment method, *Microwave Eng. Eur.*, Oct. 1992, pp. 9, 52, 55.

58. S. Alexandrou, R. Sobolewski, and T.-Y. Hsiang, Time-domain characterization of bent coplanar waveguides, *IEEE J. Quantum Electronics*, vol. 28, no. 10, 1992, pp. 2325–2332.

59. T. Becks and I. Wolff, Analysis of 3*d* metallization structures by a fullwave spectral domain technique, *IEEE Trans. Microwave Theory Tech.*, vol. 40, Dec. 1992, pp. 2218–2227.

60. M. Rittweger, *Simulation transienter elektrodynamischer Ausbreitungsphänomene zur Analyse der Übertragungseigenschaften von Systemen der Mikro- und Millimeterwellentechnik*, Doctoral Thesis, Duisburg University, Duisburg, Germany, 1992.

61. M. Naghed, *Analyse koplanarer Mikrowellenstrukturen mit der Methode der quasistatischen Finiten Differenzen*, Doctoral Thesis, Duisburg University, Duisburg, Germany, 1992.

62. T. Becks and I. Wolff, Full-wave analysis of various coplanar bends and T-junctions with respect to different types of air bridges, in: *1993 IEEE MTT-S Internat. Microwave Symp. Digest*, June 1993, pp. 697–700.

63. F. Alessandri, G. Baini, M. Mongiardo, and R. Sorrentino, A 3-D mode matching technique for the efficient analysis of coplanar MMIC discontinuities with a finite metallization thickness, *IEEE Trans. Microwave Theory Tech.*, vol. 41, no. 9, 1993, pp. 1625–1629.

64. C.-W. Chiu, and R.-B. Wu, A moment method analysis for coplanar waveguide discontinuity inductances, *IEEE Trans. Microwave Theory Tech.*, vol. 41, no. 9, 1993, pp. 1511–1514.

65. N.-I. Dib, M. Gupta, G.-E. Ponchak, and L.-P.-B. Katehi, Characterization of asymmetric coplanar waveguide discontinuities, *IEEE Trans. Microwave Theory Tech.*, vol. 41, no. 9, 1993, pp. 1549–1558.

66. S. Sali, Coupling of electromagnetic fields to coplanar striplines with discontinuities, *IEE Proc., Part H (Microwaves, Antennas and Propagation)*, vol. 140, no. 6, 1993, pp. 481–487.
67. B.-N. Lyons, T.-E. O. Ciardha, P.-A.-F. Herbert, and W.-M. Kelly, Experimental evaluation of coplanar waveguide discontinuities, *Int. J. Infrared Millimeter Waves*, vol. 14, no. 10, 1993, pp. 2021–2053.
68. S.-J. Chung and T.-R. Chrang, Full-wave analysis of discontinuities in conductor-backed coplanar waveguides using the method of lines, *IEEE Trans. Microwave Theory Tech.*, vol. 41, no. 9, 1993, pp. 1601–1605.
69. A.-M. Tran and T. Itoh, Full-wave modeling of coplanar waveguide discontinuities with finite conductor thickness, *IEEE Trans. Microwave Theory Tech.*, vol. 41, no. 9, 1993, pp. 1611–1615.
70. K.-M. Rahman and C. Nguyen, On the analysis of single- and multiple-step discontinuities for a shielded three-layer coplanar waveguide, *IEEE Trans. Microwave Theory Tech.*, vol. 41, no. 9, 1993, pp. 1484–1488.
71. T.-W. Huang and T. Itoh, The influence of metallization thickness on the characteristics of cascaded junction discontinuities of shielded coplanar type transmission line, *IEEE Trans. Microwave Theory Tech.*, vol. 41, no. 4, 1993, pp. 693–697.
72. N.-I. Dib, G.-E. Ponchak, and L.-P.-B. Katehi, A theoretical and experimental study of coplanar waveguide shunt stubs, *IEEE Trans. Microwave Theory Tech.*, vol. 41, no. 1, 1993, pp. 38–44.
73. S.-D. Mirshekar, Computer evaluation of equivalent circuit of coplanar waveguide T-junctions, in: *Analysis, Design and Applications of Coplanar Waveguides*, London, GB, Oct. 19, 1993, IEE Colloquium, vol. 1993/186, pp. 1/1–1/7.
74. H. Jin and R. Vahldieck, Full-wave analysis of coplanar waveguide discontuities using the frequency domain TLM method, *IEEE Trans. Microwave Theory Tech.*, vol. 41, no. 9, 1993, pp. 1538–1542.
75. V. Radisic, D. R. Hjelme, A. Horrigan, Z. B. Popovic, and A. R. Mickelson, Experimentally verifiable modelling of coplanar waveguide discontinuities, *IEEE Trans. Microwave Theory Tech.*, vol. 41, no. 9, Sept. 1993, pp. 1524–1533.
76. M. Yu, R. Vahldieck, and K. Wu, Theoretical and experimental characterization of coplanar waveguide discontinuities for filter applications, *IEEE Trans. Microwave Theory Tech.*, vol. 41, no. 9, 1993, pp. 1638–1640.
77. A. A. Omar and Y. L. Chow, Coplanar waveguide with top and bottom shields in place of air bridges, *IEEE Trans. Microwave Theory Tech.*, vol. 41, no. 9, Sept. 1993, pp. 1559–1563.
78. K. Beilenhoff, H. Klingbeil, W. Heinrich, and H.-L. Hartnagel, Open and short circuits in coplanar MMIC's, *IEEE Trans. Microwave Theory Tech.*, vol. 41, no. 9, 1993, pp. 1534–1537.
79. M. Abdo Tuko, M. Naghed, and I. Wolff, Novel 18/36 GHz (M)MIC GaAs FET frequency doublers in CPW-techniques under the consideration of the effects of coplanar discontinuities, *IEEE Trans. Microwave Theory Tech.*, vol. 41, no. 8, 1993, pp. 1307–1315.
80. T. Becks, *Elektrodynamische Simulation von passiven, dreidimensionalen Komponenten in (M)MIC-Schaltungen mit dem Spektralbereichsverfahren*, Doctoral Thesis, Duisburg University, 1993.

81. M.-H. Mao, R.-B. Wu, C.-H. Chen, and C.-H. Lin, Characterization of coplanar waveguide open end capacitance—Theory and experiment, *IEEE Trans. Microwave Theory Tech.*, vol. 42, no. 6, 1994, pp. 1016–1024.

82. P. Sewell and T. Rozzi, Characterization of air-bridges in MM-wave coplanar waveguide using the complete mode spectrum of CPW, *IEEE Trans. Microwave Theory and Tech.*, vol. 42, no. 11, 1994, pp. 2078–2086.

83. R. Schmidt and P. Russer, Modeling of cascaded coplanar waveguide discontinuities by the mode-matching approach, in: *1995 IEEE MTT-S International Microwave Symposium*, 16–20 May 1995, Orlando, FL; and *IEEE Trans. Microwave Theory Tech.*, vol. 43, no. 12, pt. 2, 1995, pp. 2910–2917.

84. M.-D. Wu, S.-M. Deng, R.-B. Wu, and P. Hsu, Full-wave characterization of the mode conversion in a coplanar waveguide right-angled bend, *IEEE Trans. Microwave Theory Tech.*, vol. 43, no. 11, 1995, pp. 2532–2538.

85. T. Krems, W. Haydl, L. Verweyen, M. Schlechtweg, H. Maßler, and J. Rüdiger, Coplanar bond wire interconnections for millimeter-wave applications, in: *Electrical Performance of Electronic Packaging, IEEE 4th Topical Meeting*, Portland, OR, Oct. 2–4, 1995, pp. 178–180.

86. R.-N. Simons, N.-I. Dib, and L.-P.-B. Katehi, Modeling of coplanar stripline discontinuities, *IEEE Trans. Microwave Theory Tech.*, vol. 44, no. 5, 1996, pp. 711–716.

87. K. Beilenhoff, Simulation und Modellierung von Leitungs-Diskontinuitäten und Verzweigungen für monolithisch integrierte Millimeterwellenschaltungen," Report: *Fortschrittberichte VDI*, Reihe 9, 1996, pp. 1–156.

88. H. Klingbeil, K. Beilenhoff, and H.-L. Hartnagel, FDFD full-wave analysis and modeling of dielectric and metallic losses of CPW short circuits, *IEEE Trans. Microwave Theory Tech.*, vol. 44, no. 3, 1996, pp. 485–487.

89. L. Stephan, J.-P. Coupez, E. Rius, C. Person, and S. Toutain, Integration of various types of compensated dielectric bridges for mm coplanar applications, in: *1996 IEEE MTT-S International Microwave Symposium Digest*, vol. 1, San Francisco, CA, June 17–21, 1996, pp. 83–86.

90. D. Jaisson, Coplanar waveguide bend with radial compensation, *IEE Proc. (Microwaves, Antennas and Propagation)*, vol. 143, no. 5, 1996, pp. 447–450.

91. J.-C. Goswami and R. Mittra, An application of FDTD in studying the end effects of slotline and coplanar waveguide with anisotropic substrates, *IEEE Trans. Microwave Theory Tech.*, vol. 45, no. 9, 1997, pp. 1653–1657.

92. L.-L. Fang and B.-W. Ruey, Analysis of coplanar-waveguide discontinuities with finite-metallization thickness and nonrectangular edge profile, *IEEE Trans. Microwave Theory Tech.*, vol. 45, no. 12, pt. 1, 1997, pp. 2131–2138.

93. W.-C. Chien and B.-W. Ruey, Capacitance computation for CPW discontinuities with finite metallization thickness by hybrid finite-element method, *IEEE Trans. Microwave Theory Tech.*, vol. 45, no. 4, 1997, pp. 498–504.

94. K. Goverdhanam, R.-N. Simons, and L.-P.-B. Katehi, Coplanar stripline components for high-frequency applications, *IEEE Trans. Microwave Theory Tech.*, vol. 45, no. 10, pt. 1, 1997, pp. 1725–1729.

95. Y. Gao, *A near field measurement system for measuring electric and magnetic fields in planar high-frequency circuits.* Doctoral Thesis, Duisburg University, Duisburg, Germany, 1997.

96. C.-W. Chiu, Inductance computation for coplanar waveguide discontinuities with finite metallisation thickness, *IEE Proc. (Microwaves, Antennas and Propagation)*, vol. 145, no. 6, 1998, pp. 496–500.
97. C.-W. Chiu, Equivalent circuit parameters of coplanar stripline discontinuities, *IEE Proc. (Microwaves, Antennas and Propagation)*, vol. 149, no. 1, 2002, pp. 10–16.

4

COPLANAR LUMPED ELEMENTS

4.1 INTRODUCTION

One of the most essential goals of microwave integrated circuit design is the reduction of the needed space on the substrate material. This requirement restricts the application of distributed line components, because they are space-intensive. More and more lumped elements are used in microwave integrated circuit design [42]. Contrary to the distributed line elements that normally base on $\lambda/4$ or $\lambda/2$ long structures, the dimensions of the lumped elements are much smaller than the wavelength at the applied frequency. Therefore, the needed small space of such lumped element components is one of their great advantages. Another advantage of these components is that their properties, because of their small size, can quite accurately be described using simple equivalent circuit descriptions. Therefore their properties can be simulated in a straightforward manner, and their application leads to a high flexibility in circuit design.

On the other hand, there are also some disadvantages of lumped elements. The most essential disadvantage of these components is their small Q-factor. There are two reasons for the small Q-factors: (1) Because of the small size of the components, the electromagnetic field concentration inside these structures is much higher as compared to the distributed line elements, and (2) depending on the topology and size, the Q-factor is reduced to a large amount by the parasitic effects that influence the components properties heavily. The parasitic effects may also reduce the frequency range in which the lumped ele-

Coplanar Microwave Integrated Circuits, by Ingo Wolff.

ments can be used with respect to their primary properties. Another disadvantage of these components is that because of the complexity of their structures, an accurate simulation of their frequency-dependent properties is only possible using rigorous techniques [15, 16, 19, 24]. The computation time that is needed by these techniques makes it difficult to use them directly in computer-aided design techniques.

The most commonly used lumped elements of microwave integrated circuits are the interdigital capacitor, the MIM (metal–insulator–metal) capacitor, the spiral inductor, the spiral transformer, and the thin-film resistor. In the following sections, the realization of these components in coplanar technology will be discussed. The components in a first step will be modeled using simple equivalent circuit descriptions, and then the properties of the model parameters (i.e., the equivalent circuit elements) will be described using again the quasi-static electromagnetic analysis technique, as has already been described in Chapter 3 in connection with the coplanar waveguide discontinuities. Finally, some examples of lumped elements that have been fabricated on ceramic and GaAs substrate will be analyzed, and their properties will be discussed. The dependence of the equivalent circuit elements on the geometrical sizes and electrical parameters will be demonstrated, and possible techniques to reduce the influence of the parasitic elements will be discussed.

The above-mentioned lumped elements often are used as frequency defining elements in filter and oscillator structures [10, 15, 23]. But they are also used in matching and biasing circuits. Whereas the parasitic elements in oscillator applications are kept as small as possible, they can be used to realize the frequency-dependent transmission properties of filters or matching circuits strategically. This kind of application needs an accurate and broadband description of these second-order parasitic elements. Also, the structure-dependent feedback as well as the couplings inside the lumped element structure have to be considered in the analysis of the components. All these requirements can only be fulfilled by application of a three-dimensional field analysis technique. The three-dimensional quasi-static analysis technique that is used in this chapter for the analysis of the lumped elements fulfills all these requirements. Additionally, this method is fast enough to use it directly in a computer-aided circuit design technique. The results described in the following are taken mainly from the doctoral thesis of Naghed [27], who investigated the problems under the leadership of the author.

4.2 THE COPLANAR INTERDIGITAL CAPACITOR

4.2.1 The Lumped Element Modeling Approach

Interdigital capacitors in form of multifinger structures already have been used intensively in microstrip technology [4, 10, 16]. In Fig. 4.2.1 an interdigital capacitor in coplanar technology is shown. The capacitor itself is defined

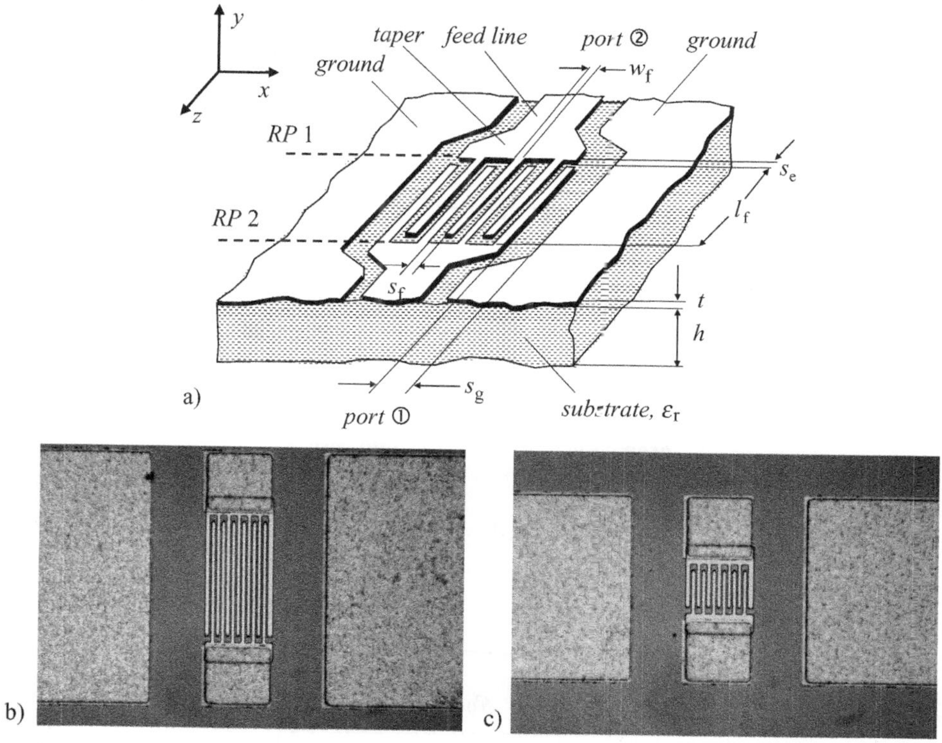

Fig. 4.2.1. Interdigital capacitor in coplanar technology. (a) General description of the structure. Geometrical parameters: finger width w_f, space width between the fingers s_f, finger length l_f, gap width to ground s_g, gap width at finger ends s_e, *RP* stands for reference plane. (b, c) Technological realization of two small capacitors of different lengths in coplanar environment on GaAs substrate.

between the two shown ports ① and ②, respectively. The feeding coplanar waveguides are connected to the capacitor using a tapered line structure.

For the following investigations it is assumed that the interdigital capacitor is a real lumped element; that is, it is assumed that the linear dimensions of the capacitor in any case are very much smaller than the wavelength of the used signal frequency inside the substrate material. It will be shown that by this approach the properties of the interdigital capacitor can be described with a high accuracy up to frequencies of about 30 GHz using static analysis techniques. For applications in the millimeter–wave range, the model has to be enhanced by taking the distributed properties of the structure into account (see Section 4.2.2).

The interdigital capacitor couples the two coplanar waveguides by the electromagnetic field in its region. The electric coupling can be represented by a coupling capacitor in the equivalent circuit as shown in Fig. 4.2.2. To simulate the magnetic coupling between the fingers, a transformer with the self-

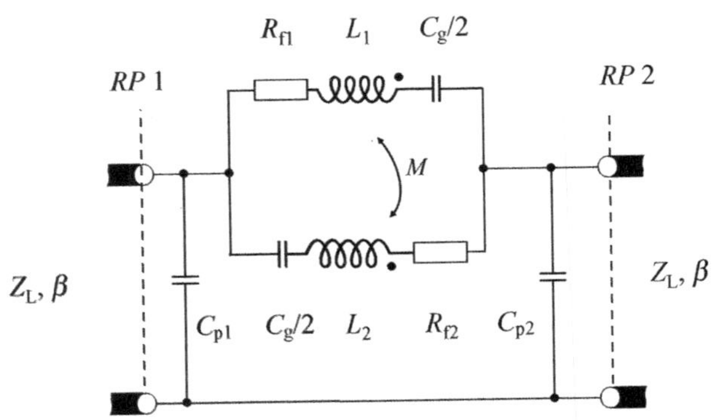

Fig. 4.2.2. The equivalent circuit model of the interdigital capacitor in coplanar technology. *RP* stands for reference plane.

even case **odd case**

$$\left.\begin{aligned}\varphi_1 &= +1\text{V}\\ \varphi_2 &= +1\text{V}\\ \varphi_0 &= 0\text{V}\end{aligned}\right\} \rightarrow C_{p1}, C_{p2} \qquad \left.\begin{aligned}\varphi_1 &= +1\text{V}\\ \varphi_2 &= -1\text{V}\\ \varphi_0 &= 0\text{V}\end{aligned}\right\} \rightarrow C_g$$

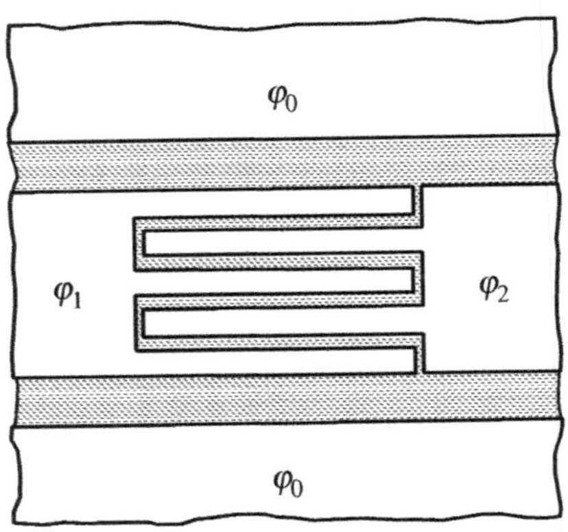

Fig. 4.2.3. Schematic description of the analysis method to determine the capacitive elements of the equivalent circuit shown in Fig. 4.2.2.

inductances L_1 and L_2 and the mutual inductance M is used. The ohmic losses that occur due to the current flow through the fingers can be described by two frequency dependent resistors R_{f1} and R_{f2}. All these elements form the equivalent circuit of the interdigital capacitor as is shown in Fig. 4.2.2.

The capacitances C_{p1} and C_{p2} represent the stray fields from the fingers to the ground plane. The capacitances C_{p1}, C_{p2}, and C_g are determined in a way that is analogous to the one already described in Section 3.5.3 for the gap in a coplanar waveguide using the quasi-static finite difference analysis of the electric field distribution and two different excitations as shown in Fig. 4.2.3.

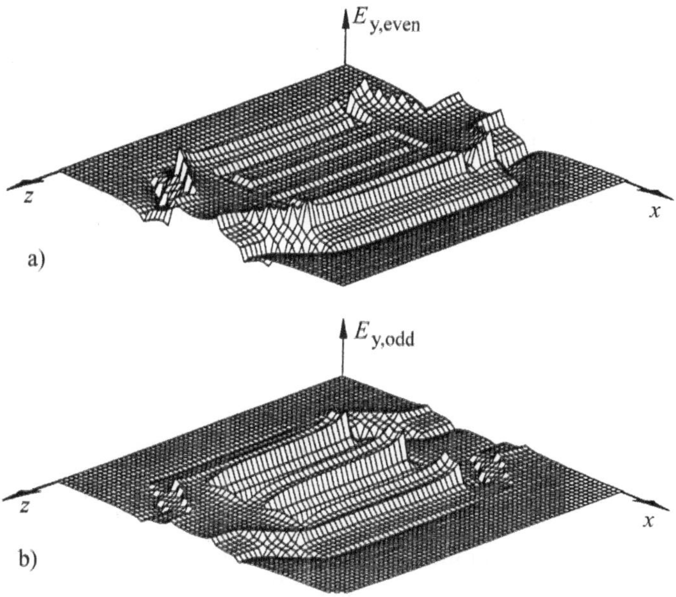

Fig. 4.2.4. The electric field component normal to the metalization plane of an interdigital capacitor in coplanar technology in the case of an even-mode (a) and an odd-mode (b) excitation at the input and output ports.

More detailed information of the method is given in the above-mentioned section of Chapter 3.

In Fig. 4.2.4 the analyzed distribution of the electric field strength in the metalization plane of an interdigital coplanar capacitor is shown for the case of the even-mode (a) and the odd-mode (b) excitation. The typical electric field distribution on the coplanar input and output line and the field in the finger structure can clearly be identified from these field plots.

For the determination of the magnetic coupling, only the section of the coupled fingers with lengths l_f is considered. This coupling is calculated assuming a pure TEM mode propagation on this structure. Under this assumption, the coupling section may be assumed to be homogeneous in longitudinal direction, so that the coupling can be analyzed directly from the cross section of the finger structure. Before going into a detailed description of how the parameters L_1, L_2, and M are determined, some explanations concerning the equivalent circuit in Fig. 4.2.2 must be given.

Figure 4.2.5 shows a lossless transformer. It is described by the following equations:

$$v_1 = j\omega L_1 i_1 + j\omega M i_2, \qquad (4.2.1)$$

$$v_2 = j\omega L_2 i_2 + j\omega M i_1, \qquad (4.2.2)$$

These equations can also be written in the form

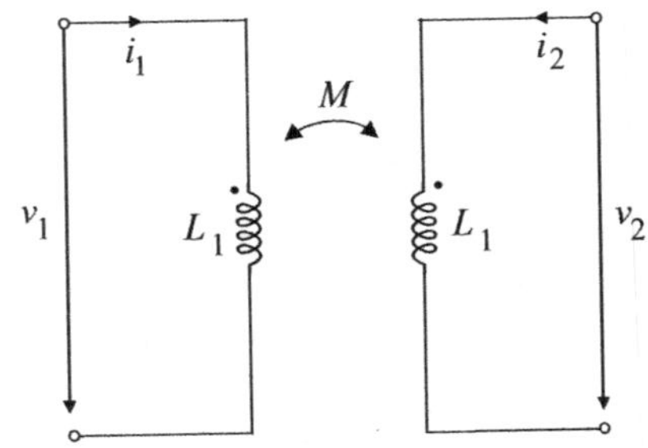

Fig. 4.2.5. Lossless transformer.

$$v_1 = j\omega[L_1 + (i_2/i_1)M]i_1, \tag{4.2.3}$$

$$v_2 = j\omega[L_2 + (i_1/i_2)M]i_2. \tag{4.2.4}$$

The expressions in rectangular brackets are, in principle, inductances (at least they have the units of inductances). Therefore, Eqs. (4.2.3) and (4.2.4) can also be written as

$$v_1 = j\omega L_a i_1, \tag{4.2.5}$$

$$v_2 = j\omega L_b i_2, \tag{4.2.6}$$

with

$$L_a = L_1 + (i_2/i_1)M, \tag{4.2.7}$$

$$L_b = L_2 + (i_1/i_2)M. \tag{4.2.8}$$

To determine the parameters L_1, L_2, and M, Eqs. (4.2.5) and (4.2.6) may be used in the following way: In a first step the electric potential of the fingers is chosen to be $\varphi_1 = \varphi_2 = 1\,\text{V}$ and the potential of the ground plane is chosen to be $\varphi_0 = 0\,\text{V}$. This is an excitation of the structure for the even case in that the voltages v_1 and v_2 are equal. Therefore from Eqs. (4.2.5) to (4.2.8) it follows that

$$L_{ae} = L_1 + (L_{ae}/L_{be})M, \tag{4.2.9}$$

$$L_{be} = L_2 + (L_{be}/L_{ae})M. \tag{4.2.10}$$

In the case of an odd excitation, if the potentials are chosen as $\varphi_1 = 1\,\text{V}$, $\varphi_2 = -1\,\text{V}$, and $\varphi_0 = 0\,\text{V}$, the equations for the inductances L_a and L_b are

$$L_{ao} = L_1 - (L_{ao}/L_{bo})M, \tag{4.2.11}$$

$$L_{bo} = L_2 - (L_{bo}/L_{ao})M. \tag{4.2.12}$$

Finally, from Eqs. (4.2.9)–(4.2.12), the parameters L_1, L_2, and M can be estimated. Details of the analysis technique are described in Section 3.4.2.

The parameters R_{f1} and R_{f2} may also be calculated from the computed current distribution on the fingers. The so-determined values are valid for a frequency of up to $f = 1\,\text{GHz}$. For other frequencies, the resistors, analogue to Eq. (2.2.27), can be given as follows:

$$R_{f1,2} = \begin{cases} R_{\text{dc}1,2} + \left\{R_{f1,2}(f = 1\text{GHz})\sqrt{f_{\text{lim}}/\text{GHz}} - R_{\text{dc}1,2}\right\}\dfrac{f}{f_{\text{lim}}} & \text{for } f \leq f_{\text{lim}}, \\ R_{f1,2}(f = 1\text{GHz})\sqrt{f/\text{GHz}} & \text{for } f \geq f_{\text{lim}}, \end{cases} \quad (4.2.13)$$

where $R_{\text{dc}1,2}$ is the dc resistance of the fingers at ports 1 and 2, respectively. f_{lim} is a limiting frequency as given in Eq. (2.2.25).

To evaluate the accuracy of this model for the interdigital coplanar capacitor, various capacitors have been fabricated on gallium arsenide substrate material ($\varepsilon_r = 12.9$, substrate height $h = 400\,\mu\text{m}$). Measured scattering parameters of these capacitors have been compared to computed values using the above described finite difference model. From the series of measurements that have been performed, only four capacitors with a finger number of 3, 5, and 7 and with capacitances between 50 fF and 300 fF are presented here. In all these structures, the distance s_g (see Fig. 4.2.1) to ground is chosen so that the coplanar waveguides behind the reference planes *RP*1 and *RP*2 have a characteristic line impedance $Z_L = 50\,\Omega$. The geometrical parameters of the capacitors as well as the computed model parameters are given in Table 4.2.1.

TABLE 4.2.1. Geometrical Parameters as Well as Computed Model Parameters of Four Interdigital Coplanar Capacitors on Gallium Arsenide Substrate Material. $\varepsilon_r = 12.9$

Parameter	Capacitor No. 1	2	3	4
Finger number, N	3	5	7	7
Finger width, w_f (μm)	40	40	40	40
Gap width, s_f (μm) = s_e (μm)	10	10	10	10
Finger length, l_f (μm)	200	200	300	500
Distance to ground, s_g (μm)	95	160	230	230
Metalization thickness, t (μm)	3	3	3	3
Coupling capacitance, C_g (fF)	47.7	94.4	189.6	310.9
Parallel capacitance, C_{p1} (fF)	28.6	25.8	35.8	56.9
Parallel capacitance, C_{p2} (fF)	9.4	11.5	18.7	27.3
Self-inductance, L_1 (pH)	89.4	87.8	133.5	217.7
Self-inductance, L_2 (pH)	127.3	103.2	149.5	240.2
Mutual inductance, M (pH)	75.7	78.5	25.6	200.9
Ohmic resistor, R_{f1} (mΩ)	13	8	9	15
Ohmic resistor, R_{f2} (mΩ)	21	10	10	17

R_{f1} and R_{f2} are the resistor values for a frequency of 1 Ghz.

The measured and calculated scattering parameters of the capacitors are shown in Fig. 4.2.6 and Fig. 4.2.7. The good agreement between the measured and simulated values up to frequencies of 30 GHz shows that the chosen model can be used over a large frequency range. Especially, the very accurate simulation of the phase angles (Fig. 4.2.7) proves the high quality of the model. The small deviations between measurement and simulation that can be observed from the magnitude comparison (Fig. 4.2.6) result, especially from neglecting possible radiation losses in the quasi-static finite difference model.

The main aim of the interdigital capacitor design is the realization of a coupling capacitance between two coplanar waveguides that to the highest possible degree is free of parasitic effects. This coupling capacitance is formed by the electric field strength in the gap region between the different fingers. In

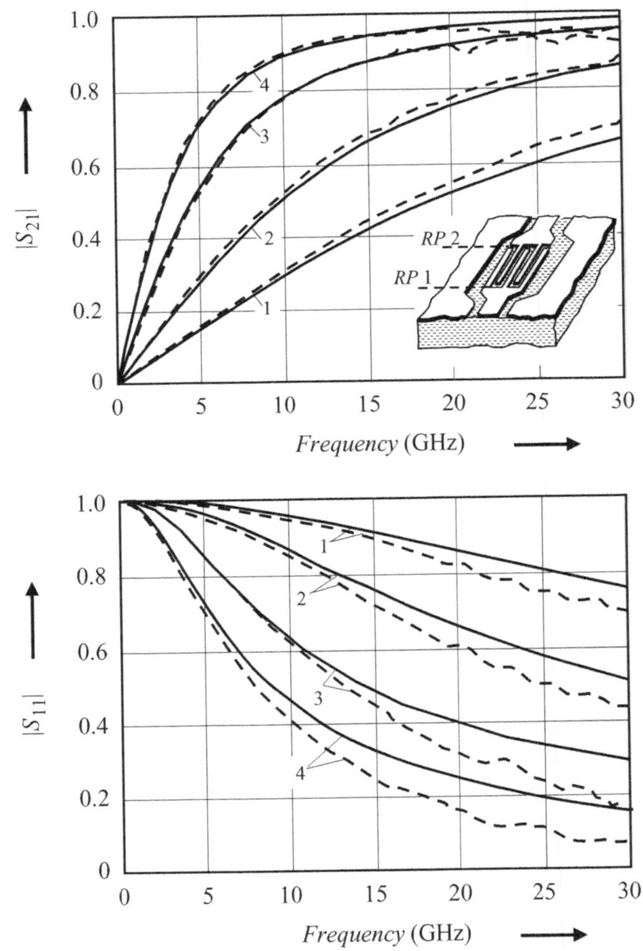

Fig. 4.2.6. Measured (– – –) and calculated (———) magnitudes of the scattering parameters for coplanar interdigital capacitors (for parameters see Table 4.2.1).

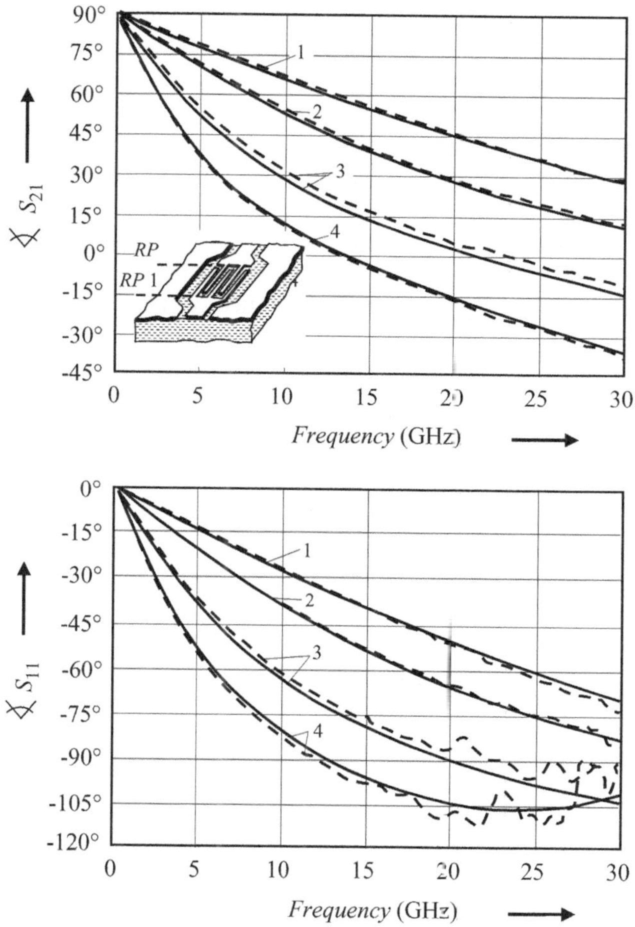

Fig. 4.2.7. Measured (– – –) and calculated (———) phases of the scattering parameters for coplanar interdigital capacitors (for parameters see Table 4.2.1).

order to estimate the influence of the parasitic effects of the structure on the transmission properties, a systematic investigation of the capacitors as a function of the geometrical and electrical parameters is needed. Only the dependencies on the main essential geometrical parameters will be discussed in the following. As mentioned earlier, capacitors on GaAs substrate material with a relative permittivity of $\varepsilon_r = 12.9$ and a substrate height of $h = 400\,\mu m$ are considered here. The metalization thickness, $t = 3\,\mu m$, is considered in the theoretical analysis. Due to symmetry aspects, only capacitors with an odd number of fingers are investigated since the excitation of odd modes on the coplanar structure can be avoided only under this assumption.

In a first step, the model parameters of the interdigital capacitors are analyzed as a function of the finger width w_f for different finger numbers N. To realize a coupling capacitance as high as possible, the distances s_f and s_e

(see Fig. 4.2.1) are chosen to be very small. The distances s_g of the fingers to ground and the finger lengths l_f are kept constant. The computed results for the equivalent circuit parameters are shown in Fig. 4.2.8 to Fig. 4.2.11.

As shown in Fig. 4.2.8, the value of the coupling capacitance C_p increases with increasing finger width w_f and increasing finger number N. The parallel capacitances C_{p1} and C_{p2} show, in principle, the same dependencies. C_{p1} is always larger than C_{p2} because the outside positioned fingers (see Fig. 4.2.1) have a larger capacitance to ground. This is also the reason for the stronger dependence of the capacitance C_{p1} on the finger width w_f. Furthermore, it can be observed from Fig. 4.2.8 that the dependence of the capacitance C_g on the

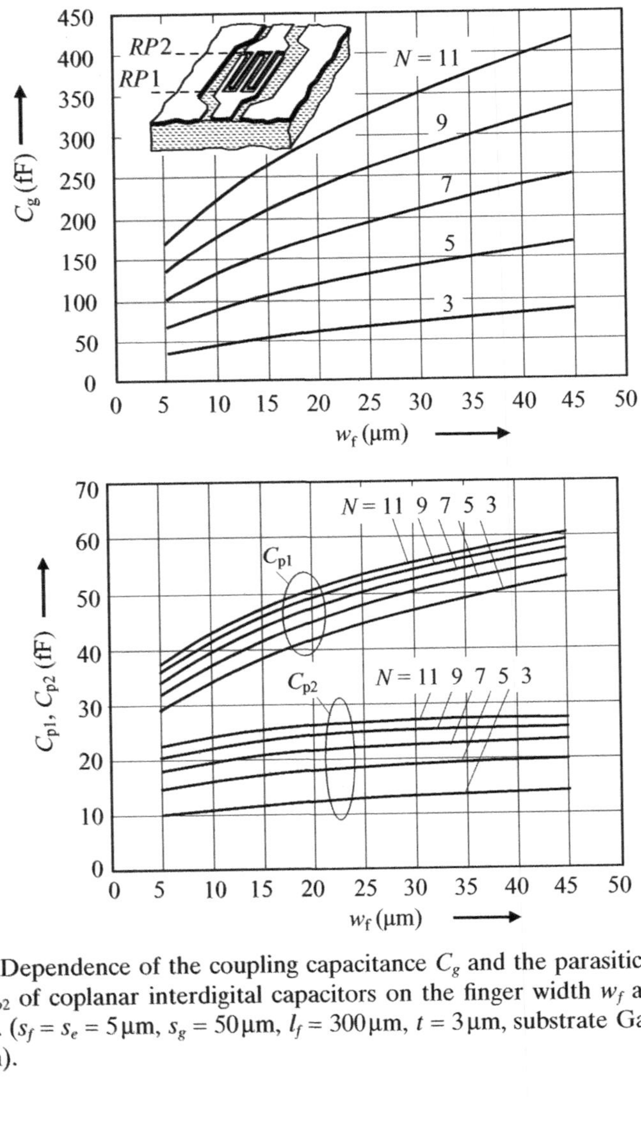

Fig. 4.2.8. Dependence of the coupling capacitance C_g and the parasitic capacitances C_{p1} and C_{p2} of coplanar interdigital capacitors on the finger width w_f and the finger number N. ($s_f = s_e = 5\,\mu m$, $s_g = 50\,\mu m$, $l_f = 300\,\mu m$, $t = 3\,\mu m$, substrate GaAs, $\varepsilon_r = 12.9$, $h = 400\,\mu m$).

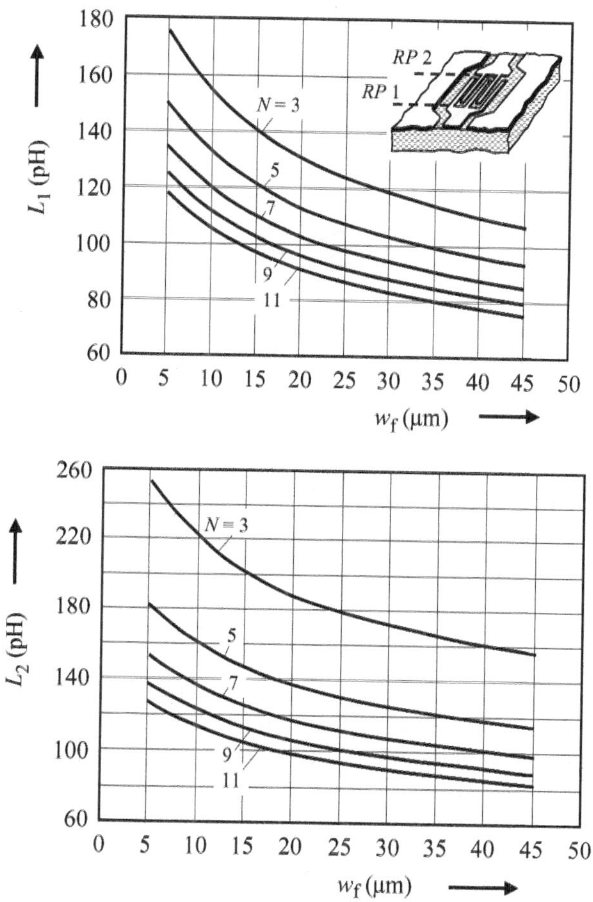

Fig. 4.2.9. Dependence of the self-inductances L_1 and L_2 of coplanar interdigital capacitors on the finger width w_f and the finger number N. Parameters: $s_f = s_e = 5\,\mu$m, $s_g = 50\,\mu$m, $l_f = 300\,\mu$m, $t = 3\,\mu$m, substrate GaAs, $\varepsilon_r = 12.9$, $h = 400\,\mu$m.

finger number N is stronger than its dependence on the finger width w_f. For the capacitance C_{p1}, these dependencies are vice versa. This fact may be used in such a way that when realizing a capacitor with a certain capacitance value, capacitors with a larger finger number and a smaller finger width are designed. In this way the effect of the parasitic capacitance can be kept small.

In Fig. 4.2.9 the influence of the finger width w_f on the inductances L_1 and L_2 is shown for various finger numbers N. In contrast to the capacitive elements, the inductive elements of the equivalent circuit (Fig. 4.2.2) decrease with increasing finger width w_f and finger number N. The inductance L_2 is always larger than L_1, but the difference between these two values becomes smaller and smaller with increasing finger number N. The capacitance C_g, which is a measure of the electric coupling, increases with increasing finger

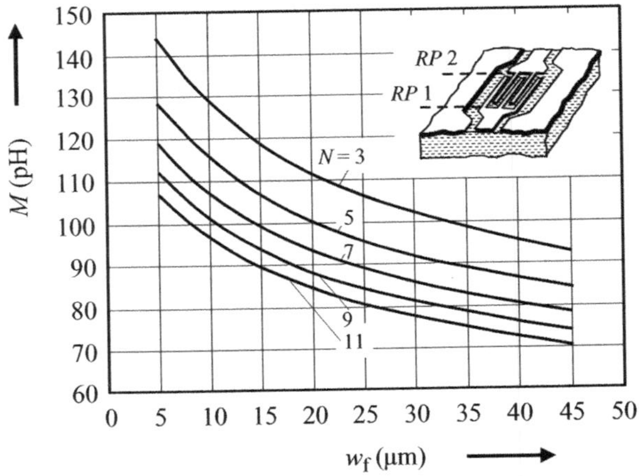

Fig. 4.2.10. Dependence of the mutual inductance M of a coplanar interdigital capacitor on the finger width w_f and the finger number N. Parameters $s_f = s_e = 5\,\mu m$, $s_g = 50\,\mu m$, $l_f = 300\,\mu m$, $t = 3\,\mu m$, substrate GaAs, $\varepsilon_r = 12.9$, $h = 400\,\mu m$.

number N and finger width w_f. The mutual inductance, which is a measure of the magnetic coupling between the two coplanar waveguides, however, decreases with increasing values of N and w_f.

The resistors R_{f1} and R_{f2} of the equivalent circuit (Fig. 4.2.2) represent the ohmic losses of the component. Their values depend on the skin effect and therefore are frequency-dependent. In Fig. 4.2.11 the values of these resistors for a frequency of 1 GHz are shown. Similar to the behavior of the self-inductances, L_1 and L_2, the recognizable difference between the two resistor values decreases with increasing finger number.

When interdigital capacitors are realized in microstrip technology, the essential problem is that the distances of the fingers to the ground plane are fixed by the given substrate height. Because of this fact, it is not possible to vary the value of the parasitic parallel capacitances C_{p1} and C_{p2} without simultaneously changing the value of the coupling capacitance C_g. In the case of the coplanar interdigital capacitor, however, it is very easy to change the parallel capacitances nearly without changing the other elements of the equivalent circuit (see Fig. 4.2.2) by simply changing the gap width s_g between the outside fingers and the ground plane.

Figure 4.2.12 shows the dependencies of the parallel capacitances C_{p1} and C_{p2} as well as of the coupling capacitance C_g on the gap width s_g for capacitors with different finger numbers. The parallel capacitances can be reduced to smaller values by increasing the distance of the outside finger to ground. At the same time the coupling capacitance increases only by some percent over the same gap width change.

A measure for the quality of the lumped elements is their Q-factor. Because the Q-factor can be defined uniquely only for a one-port, it cannot be given directly for the described coplanar interdigital capacitor. One possibility to

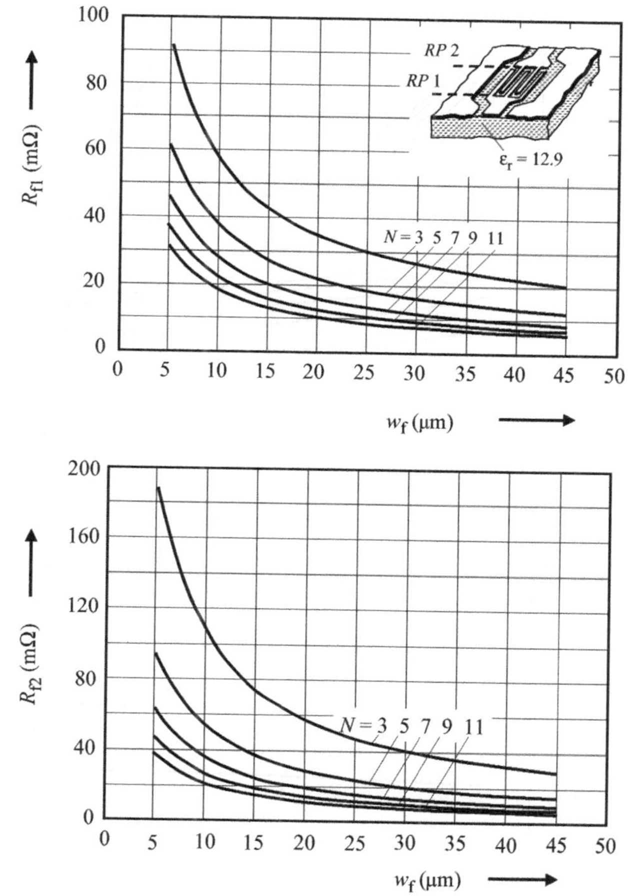

Fig. 4.2.11. Dependence of the ohmic resisor R_{f1} ($f = 1\,\text{GHz}$) (a) and R_{f2} ($f = 1\,\text{GHz}$) (b) of a coplanar interdigital capacitor on the finger width w_f and the finger number N. Parameters: $s_f = s_e = 5\,\mu\text{m}$, $s_g = 50\,\mu\text{m}$, $l_f = 300\,\mu\text{m}$, $t = 3\,\mu\text{m}$, substrate GaAs, $\varepsilon_r = 12.9$, $h = 400\,\mu\text{m}$.

nevertheless calculate a Q-factor of the capacitor is to short-circuit port 2 of the capacitor and, in this way, to define a one-port structure. This one-port then has an equivalent circuit that is shown in Fig. 4.2.13. If alternatively port 1 would have been short-circuited, the two outside-positioned fingers would not deliver a contribution to the coupling capacitance and the resulting capacitor, in principle, would have two fingers less compared to the original capacitor.

The Q-factor is defined as the quotient of the energy stored in the element and the power dissipated in the ohmic resistors multiplied by the cycling frequency ω. Because only the energy stored in the coupling capacitance C_g is of interest for the function of the interdigital capacitor in a circuit, the following definition for the Q-factor seems meaningful:

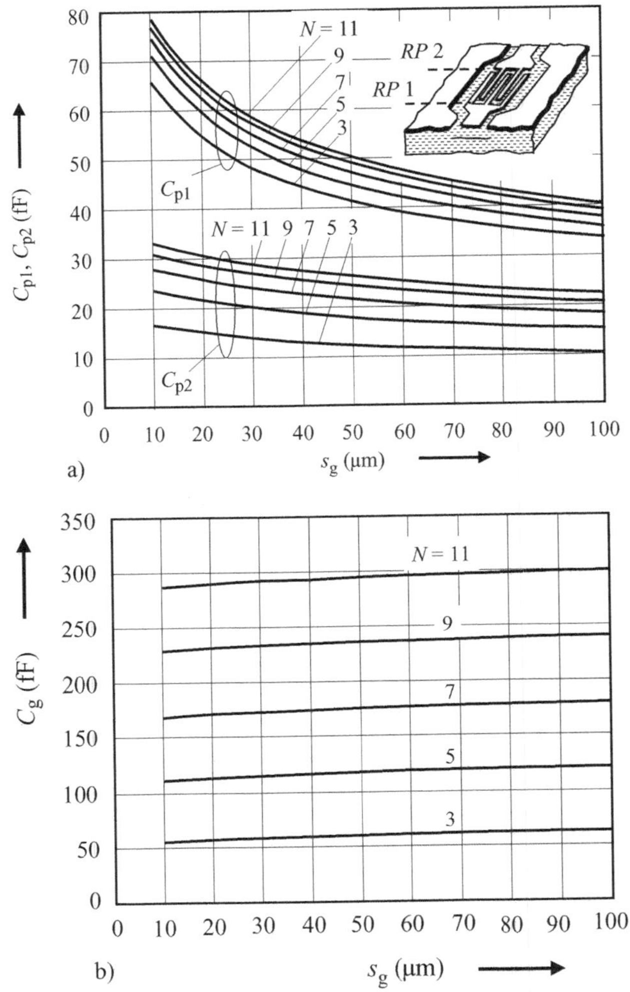

Fig. 4.2.12. Dependence of the parallel capacitances C_{p1} and C_{p2} (a) as well as of the coupling capacitance C_g (b) on the distance s_g of the outside positioned fingers to ground and the finger number N. Parameters: $w_f = 20\,\mu\text{m}$, $s_f = s_e = 5\,\mu\text{m}$, $l_f = 300\,\mu\text{m}$, $t = 3\,\mu\text{m}$, substrate GaAs, $\varepsilon_r = 12.9$, $h = 400\,\mu\text{m}$.

$$Q = \frac{\omega C_g}{\text{Re}\{Y_i\}}, \tag{4.2.14}$$

where Y_i is the input admittance of the equivalent circuit shown in Fig. 4.2.13. In Fig. 4.2.14 the so-calculated Q-factor is shown for capacitors listed in Table 4.2.1 and with a short-circuited port 2. As can be seen from the figure,

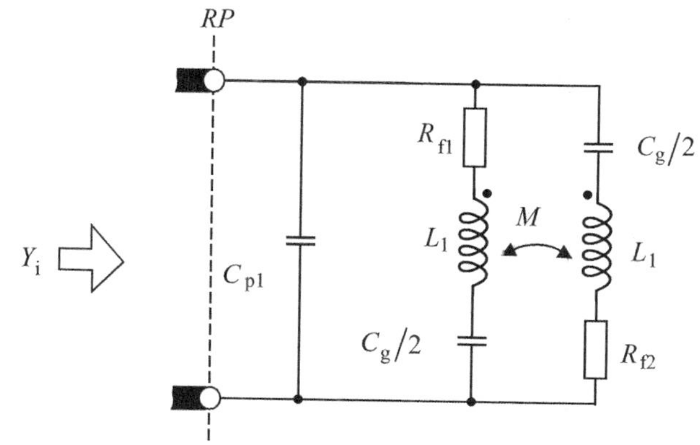

Fig. 4.2.13. The equivalent circuit of the coplanar interdigital capacitor with port ② short-circuited. *RP* stands for reference plane.

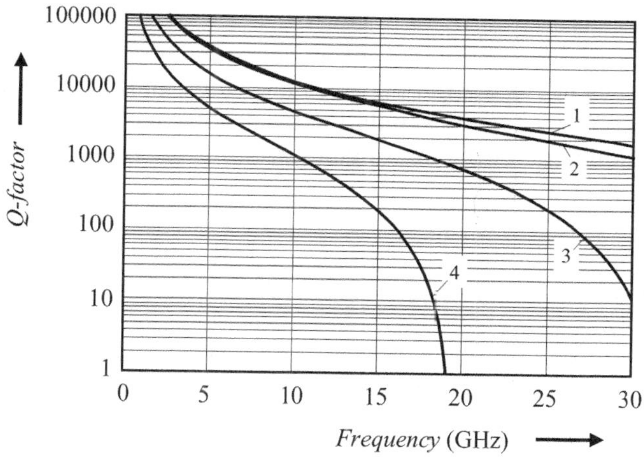

Fig. 4.2.14. The Q-factor of coplanar interdigital capacitors in dependence on the frequency. (For geometrical and other parameters see Table 4.2.1.)

the coplanar interdigital capacitors at low frequencies have a very large Q-factor. But with increasing frequency, this Q-factor is reduced heavily. This large reduction of the Q-factor is due to the influence of the parasitic elements in the equivalent circuit. These parasitic elements, depending on their values, may lead to a large reduction of the Q-factor even at low frequencies. This can be observed from the shown characteristic of capacitor no. 4, in which, at 19 GHz because of its large finger length, the parasitic elements are dominant and the Q-factor is reduced to the value $Q = 1$. Fortunately, as has been mentioned above, for the case of the coplanar lumped components, the parasitic elements may be reduced by an optimal choice of the geometrical parameters.

To test the simulation technique that has been described above over a large frequency range, various test structures (shown in Fig. 4.2.15a to Fig. 4.2.15d) have been built on GaAs substrate material with a height of $h = 400\,\mu m$ and a dielectric constant of $\varepsilon_r = 12.9$. The measurement results of four interdigital capacitors with $n = 2, 5, 10$, and 35 fingers are compared with the simulation results of the three-dimensional finite difference simulation technique described above over a frequency range from 45 MHz to 60 GHz.

It should be remembered here that the simulation approach discussed above for the interdigital capacitors considers the capacitors to be real lumped elements; that is, the three-dimensional electric field of the structure is analyzed using a quasi-static method. Finally, the equivalent circuit that simulates the component is derived from this field information. This means that it is assumed that the considered capacitors have linear dimensions that are very small compared to the wavelength of the applied signal. If such elements are used over a large frequency range (e.g., up to millimeter-wave frequencies), these conditions may be violated and the assumed basis for the simulation is no longer correct.

To find the limitations of the introduced simulation technique, the four capacitors shown in Fig. 4.2.15 have been measured and the measured results are critically compared to the simulation results. The measured and simulated scattering parameters for the interdigital capacitor of Fig. 4.2.15a that has two fingers are shown in Fig. 4.2.16.

The finger length is 300 μm. The capacitive elements of the equivalent circuit (see Fig. 4.2.2) are $C_{p1} = 49\,fF$, $C_{p2} = 51\,fF$, and $C_g = 27\,fF$; that is, the parasitic capacitances to ground (C_{p1}, C_{p2}) are the dominating elements of the capacitor.

Because the structure is symmetrical with respect to port 1 and port 2, the parasitic capacitances to ground are nearly equal. From Fig. 4.2.16 it can be observed that a quite good agreement between the measured and the simulated magnitude of the scattering parameters is given for frequencies up to 30 GHz. For higher frequencies the measurement results of the reflection coefficient are lower and the measurement results of the transmission coefficient are higher than the simulated results. The phases, however, are still in a quite good agreement up to frequencies of about 45 GHz.

For a frequency of about 35 GHz the wavelength of the signals on GaAs substrate with an effective dielectric constant between 6.5 and 7 is on the order of 3 mm. This means that at this frequency the finger length of the capacitors is on the order of $\lambda/10$, which is normally the upper limit for a lumped element approach.

Very similar results can be found for the capacitor with five fingers (see Fig. 4.2.15b), which are shown in Fig. 4.2.17. The elements of its equivalent circuit are given as $C_{p1} = 83\,fF$, $C_{p2} = 46\,fF$, and $C_g = 82\,fF$. Now the capacitance to ground C_{p1} is nearly of the same value as the coupling capacitance C_g.

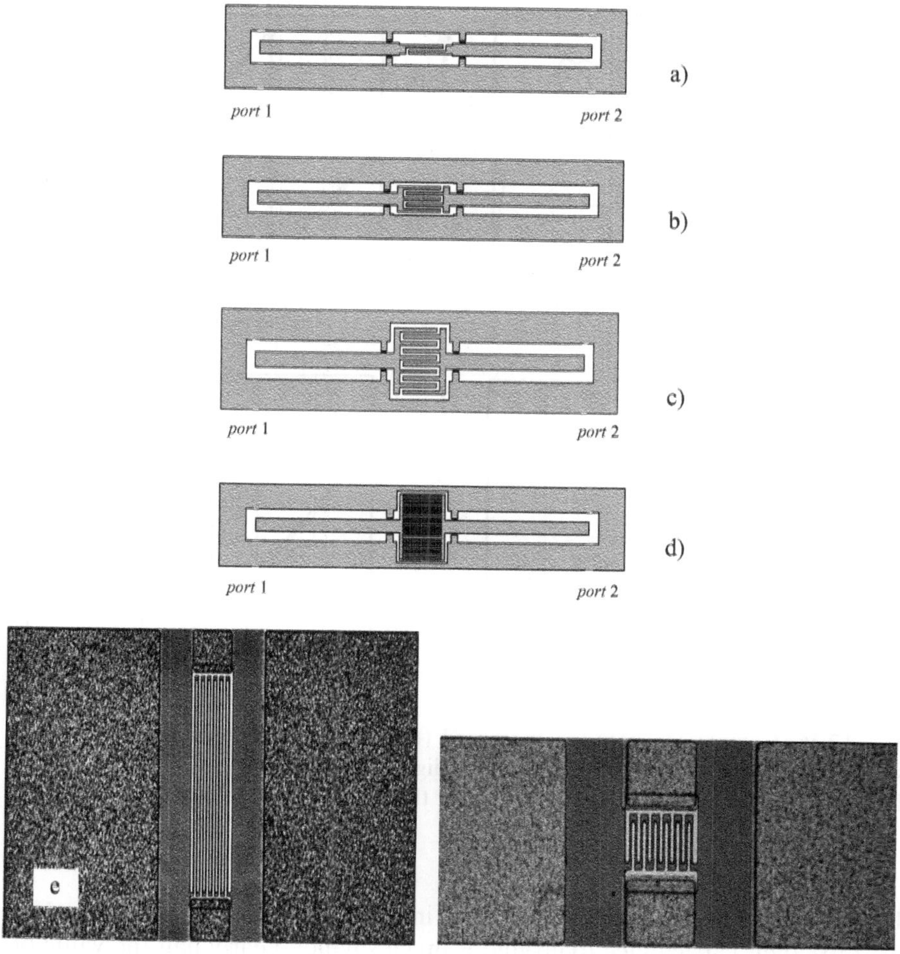

Fig. 4.2.15. Four interdigital coplanar capacitors as test structures for broadband measurements. The feed lines for the on-wafer measurements are 50-Ω coplanar waveguides of length $l = 1000\,\mu m$ ($w = 50\,\mu m$, $s = 20\,\mu m$ for cases a to c and $w = 20\,\mu m$, $s = 5\,\mu m$ for case d). Air bridges of type 1 (see Section 3.5.5) are introduced at the input port and the output port of the capacitors. Their parameters are $bw = 50\,\mu m$, $bs = 14\,\mu m$, and $bg = 8\,\mu m$; for details of the structure see Fig. V.6.6. Substrate GaAs, $\varepsilon_r = 12.9$, $h = 400\,\mu m$.

(a) $n = 2$, $w_f = 25\,\mu m$, $s_f = 30\,\mu m$, $l_f = 300\,\mu m$,
(b) $n = 5$, $w_f = 25\,\mu m$, $s_f = 30\,\mu m$, $l_f = 300\,\mu m$,
(c) $n = 10$, $w_f = 25\,\mu m$, $s_f = 30\,\mu m$, $l_f = 300\,\mu m$,
(d) $n = 35$, $w_f = 10\,\mu m$, $s_f = 5\,\mu m$, $l_f = 300\,\mu m$,
(e) technological realization of two coplanar interdigital capacitors.

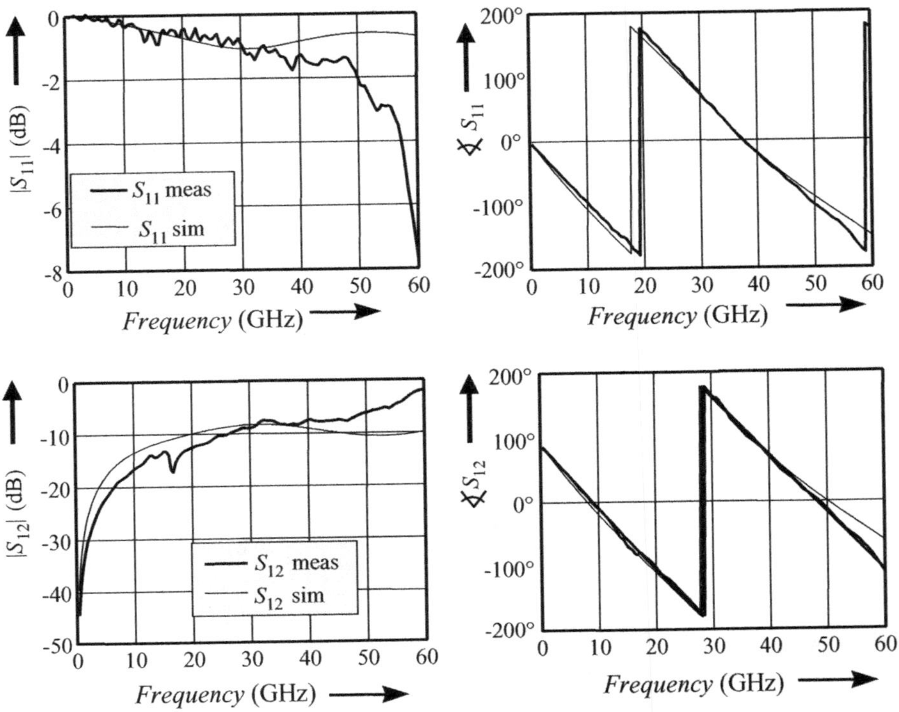

Fig. 4.2.16. Comparison between measured (thick lines) and simulated (thin lines) scattering parameters for a two-finger interdigital capacitor on GaAs substrate (ε_r = 12.9) as shown in Fig. 4.2.15a, plotted against the frequency.

Because the fingers at port 2 are the inner fingers of the capacitor (Fig. 4.2.15b), their capacitance to ground is much smaller, nearly half the value of C_{p1}. Again the frequency of 35 GHz is the upper limit for a good agreement between measurement and simulation.

The only difference compared to the results shown in Fig. 4.2.16 is that deviations between measurement and simulation are already observable for the phase at a frequency of about 35 GHz. From both results that are shown in Figs. 4.2.16 and 4.2.17, it may be concluded that for these two capacitors the finger length of 300 μm is the critical dimension that limits the application of the lumped element model at higher frequencies.

The measurement and simulation results for the 10-finger capacitor (Fig. 4.2.15c) are given in Fig. 4.2.18. The capacitances of the equivalent circuit now are C_{p1} = 89 fF, C_{p2} = 90 fF, and C_g = 190 fF. Now the coupling gap capacitance between the fingers is the dominating one. It may be observed that the 10-finger capacitor again is a symmetrical structure with respect to port 1 and port 2, and therefore the two parasitic capacitances to ground are nearly equal.

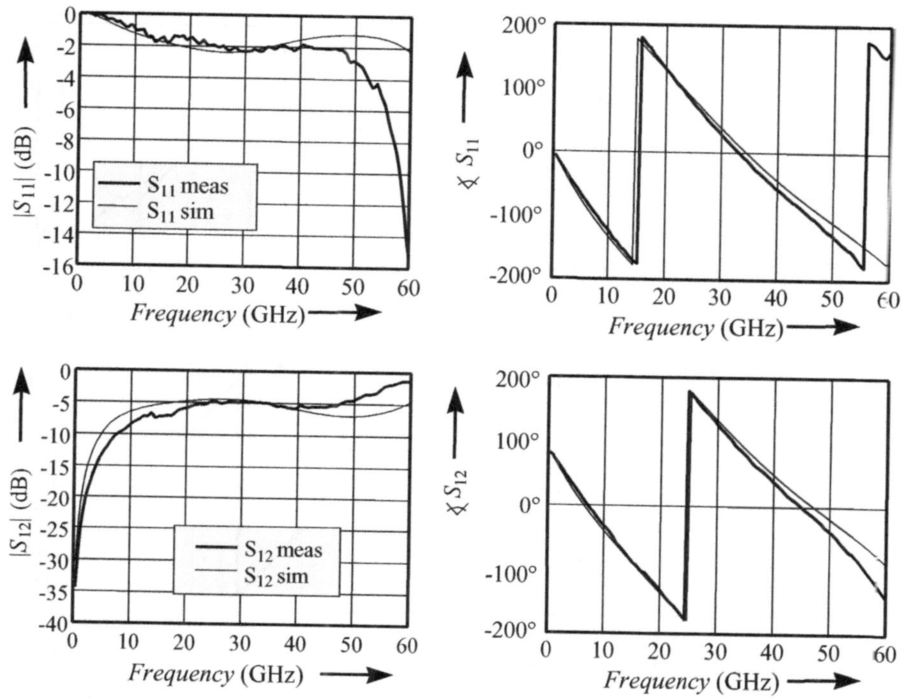

Fig. 4.2.17. Comparison between measured (thick lines) and simulated (thin lines) scattering parameters for a five-finger interdigital capacitor on GaAs substrate as shown in Fig. 4.2.15b in dependence on the frequency.

If the results for the 10-finger capacitor are considered, first resonant phenomena may be observed in the measurement results as shown in Fig. 4.2.18. The measurements show two resonant frequencies, the first one at about 42 GHz and the second one at about 53 GHz. The simulation predicts only one resonant frequency at about 55 GHz in the considered frequency range up to 60 GHz. Furthermore, a good agreement between measurement and simulation may only be found in this case for frequencies below 20 GHz. The reason for this is that now the dimensions of the capacitor perpendicular to the coplanar waveguide direction are larger than the finger length and therefore define the limiting frequency for the application of the lumped element model.

Finally, Fig. 4.2.19 shows the equivalent results for the 35-finger capacitor. Note that the dimensions of the finger width and the space between the fingers have been largely decreased for this capacitor (Fig. 4.2.15d). This test circuit has been simulated assuming two coplanar waveguide steps at the input and output port of the capacitor to consider the large discrepancies between the width of the feed line and the capacitor width correctly. The 35-finger capacitor again is an asymmetrical structure, so its parasitic capacitances

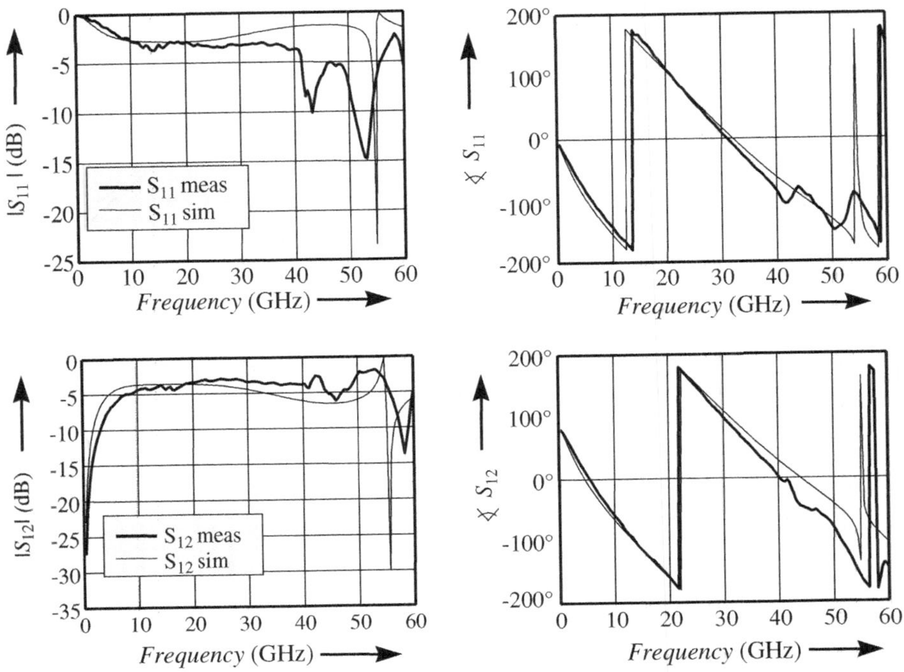

Fig. 4.2.18. Comparison between measured (thick lines) and simulated (thin lines) scattering parameters for a 10-finger interdigital capacitor on GaAs substrate as shown in Fig. 4.2.15c, plotted against the frequency.

$C_{p1} = 116\,\text{fF}$ and $C_{p2} = 87\,\text{fF}$ are of different values. The coupling capacitance $C_g = 790\,\text{fF}$ is nearly seven to nine times larger than the capacitances to ground.

The comparison between measurement and simulation shows that the simulated results are good enough for application only below a frequency of about 10 GHz. The first resonant frequency that can be measured at about 30 GHz is quite well predicted by the model. Even if it is considered that a capacitor of the dimensions given in Fig. 4.2.15d never would be used at frequencies above 10 GHz, the agreement between measured and simulated results is not satisfying in this case.

In conclusion, it may be observed that the lumped model approach for simulating the coplanar interdigital capacitors is a good approach as long as the capacitors are real "lumped," which means their largest linear dimension must be clearly below a tenth of the signal wavelength. If this condition is not fulfilled (i.e., if, for example, the finger length is needed to be longer than this limiting value), the propagation effects on the capacitor fingers must be considered in another simulation approach, as will be described in Section 4.2.2. To increase the transversal dimensions of the capacitors above, the previously mentioned limit is not advisable because in this case the resonating effects of

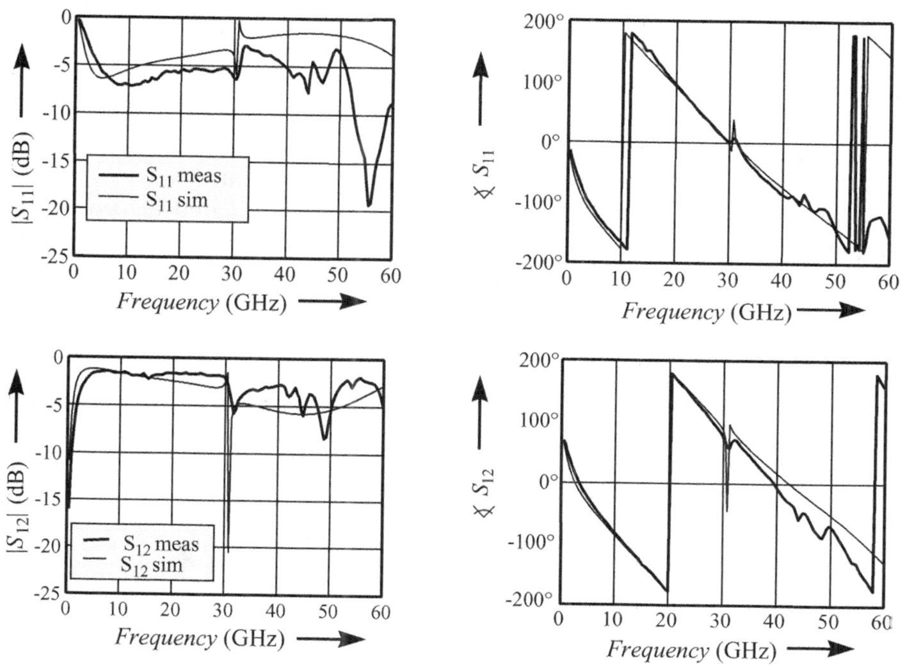

Fig. 4.2.19. Comparison between measured (thick lines) and simulated (thin lines) scattering parameters for a 35-finger interdigital capacitor on GaAs substrate as shown in Fig. 4.2.15d, plotted against the frequency.

the capacitor can only be controlled with the larger effort of a full-wave analysis. Also, the real capacitor in the circuit would no longer behave like a lumped element.

4.2.2 Enhancement of the Interdigital Capacitor Model for Application at Millimeter-Wave Frequencies

Utilizing the three-dimensional finite difference formulation as described above, the static electric field and the static magnetic field on the whole interdigital capacitor structure are calculated.

From these fields an *R*-*L*-*C* equivalent circuit is derived as described in Section 4.2.1. As has already been mentioned in the previous section, for this approach the overall dimensions of the capacitor have to be small in comparison to the wavelength on the structure. It means that the interdigital capacitor has to be a real "lumped" element. It can be observed, however, that the accuracy of this method decreases with increasing frequency and increasing length of the fingers. Therefore, for applications under these conditions an improvement of the model is necessary. This is found utilizing a coupled line model on the basis of the theory described in Section 2.2.10.

The basic module of the modified model consists of a section of n coupled lines in a coplanar environment. The $n \times n$-dimensional C'-, L'-, R'-, and G'-matrices in units per line length are calculated similar as described in Section 2.2.11.1 for the C'-matrix. An interconnection of the fingers is then made within the model. The feeding lines, the gaps between the ends of the fingers, and the spacing to the ground strips are taken into account.

A total of four interdigital capacitors have been evaluated to prove the validity of this enhanced model. All the interdigital capacitors are integrated in a 50-Ω coplanar environment with a line width w of 100 μm, a spacing to ground s of 75 μm, and a ground width of 200 μm. The number of fingers are 2, 5, 10 ($w_f = 25\,\mu m$, $s_f = s_e = 20\,\mu m$), and 35 ($w_f = 10\,\mu m$, $s_f = s_e = 10\,\mu m$) and the coupling length l_f is 300 μm.

Figure 4.2.20 depicts the magnitude of the S-parameters of the two-finger capacitor. The agreement of the old and the enhanced model with the meas-

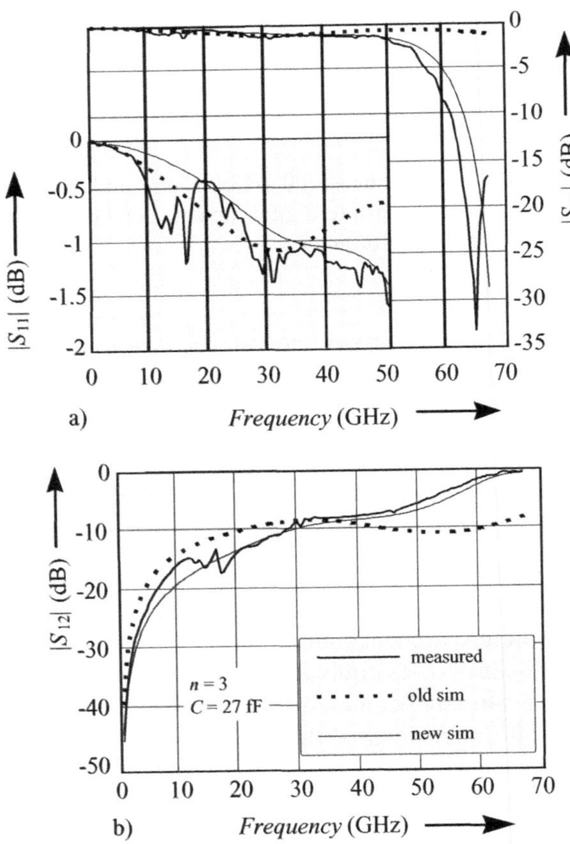

Fig. 4.2.20. Magnitude of the S-parameters of a coplanar interdigital capacitor with two fingers, coupling length = 300 μm, finger width = 25 μm, and spacing between the fingers = 20 μm.

urement results is excellent up to 30 GHz. Beyond 40 GHz, the advantage of the new model becomes evident: With the exception of a small deviation, the resonance at 65 GHz is very accurately simulated.

Another interdigital capacitor and its simulation and measurement results are depicted in Fig. 4.2.21. The number of fingers is $n = 5$. The test circuit is shown in Fig. 4.2.21c. Air bridges in front and behind the capacitor have been integrated to ensure that the coplanar odd mode will be suppressed at the discontinuity. Again, the simulation of the new and the enhanced model diverge beyond 40 GHz, while the enhanced method is accurate (compared to the measurements) up to 67 GHz.

Another interesting phenomenon that should be pointed out is that in the frequency range of 62–67 GHz, a constant coupling with a transmission value better than −1 dB can be achieved between port ① and ②. This can only be simulated with the enhanced model.

Nevertheless, the interdigital capacitor has some significant disadvantages. On the one hand, the parasitic elements have a strong influence on the electrical behavior of the element. As an example, the value of the parallel capacitance C_p (capacitance between the fingers and ground) for the capacitor in Fig. 4.2.15c is $C_{p1} = 83$ fF for the left and $C_{p2} = 46$ fF for the right electrode. On the

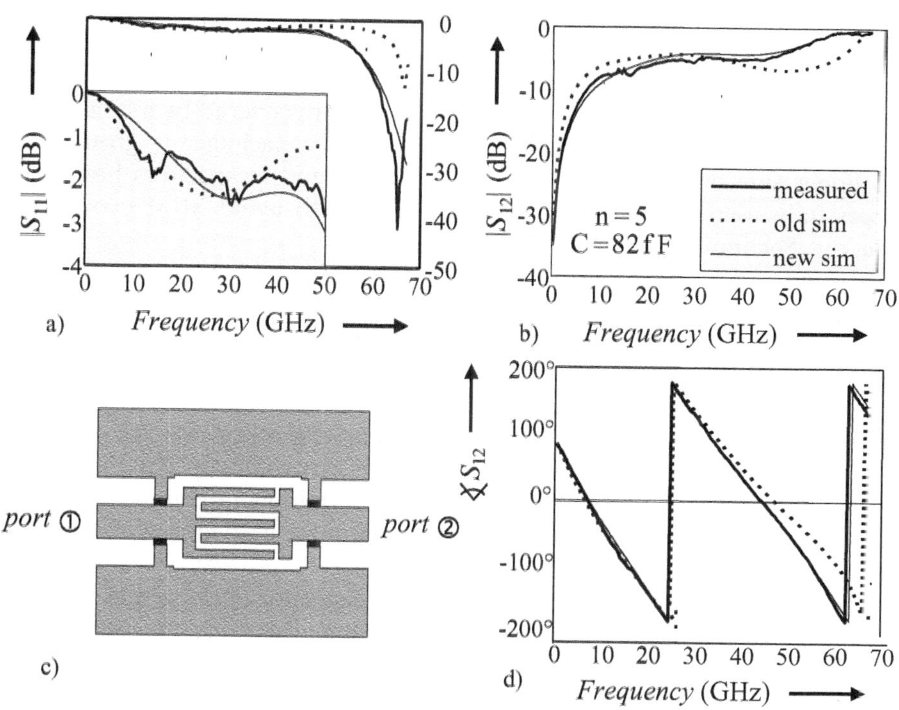

Fig. 4.2.21. *S*-parameters of a coplanar interdigital capacitor with five fingers, coupling length = 300 μm, finger width = 25 μm, and spacing between finger = 20 μm.

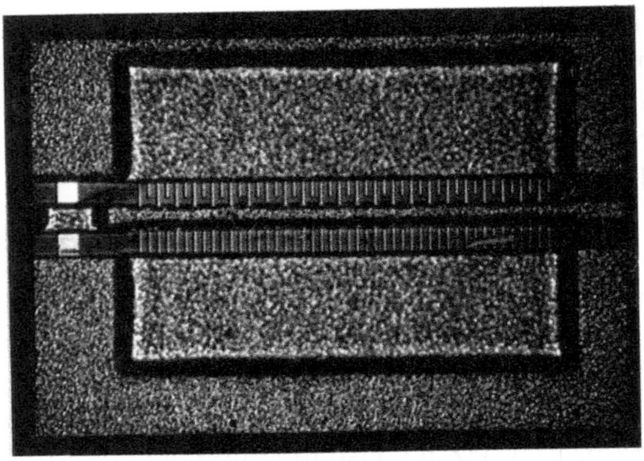

Fig. 4.2.22. A special form of the interdigital capacitor to ground with a metalization shielding produced in air-bridge technology.

other hand, the total size of the interdigital capacitor is large whereas the reachable capacitance value is small (e.g., for a capacitor with $n = 35$ fingers: $C_g = 0.8\,\text{pF}$).

An improvement may be found using an interdigital capacitor of a very special form shown in Fig. 4.2.22, which may be used in dc circuitry as an RF blocking capacitor. The fingers are designed between the center conductor of the coplanar waveguide and the ground planes so that a capacitance to ground is developed. To enlarge this capacity, the fingers are covered by a large "air-bridge" construction so that there is an additional capacitance to ground on the upper side of the finger construction. This kind of capacitors has been successfully used in microwave amplifier design to avoid MIM capacitors (see, e.g., Section 7.4.1) in the dc supply circuits.

4.3 THE COPLANAR METAL–INSULATOR–METAL (MIM) CAPACITOR

An alternative solution to the interdigital capacitor is the metal–insulator–metal (MIM) capacitor in a coplanar environment. This capacitor has a higher capacitance and a reduced space requirement compared to the interdigital one. Two versions of the MIM capacitor are used in coplanar circuit design. They are shown in Fig. 4.3.1 and Fig. 4.3.5.

The capacitor shown in Fig. 4.3.1 is a series capacitor in the center strip of a coplanar waveguide. It is often used as a coupling capacitor, for instance, at the input or output of a microwave circuit. It is built as shown in Fig. 4.3.2. In the technology process the bottom electrode is formed by the gate metalization layer. The top electrode is fabricated in the galvanic layer. The dielectric insulation between the electrodes has the permittivity ε_r and the thickness d_c.

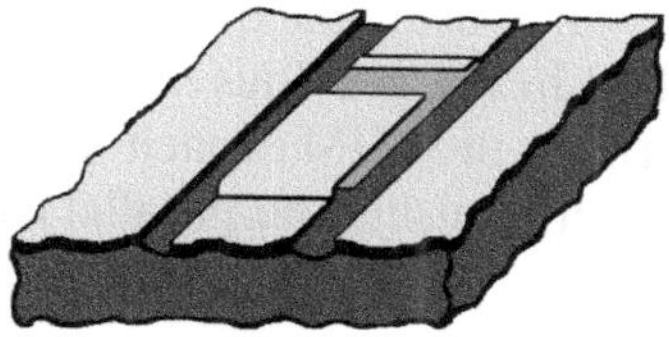

a)

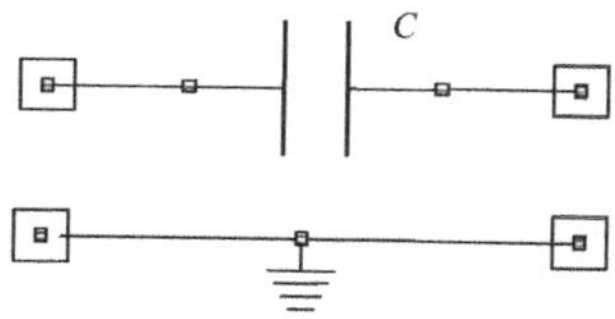

b)

c)

Fig. 4.3.1. Series MIM capacitor in a coplanar environment (a) and its fundamental equivalent circuit (b). Technological realization of a MIM capacitor in the center conductor of a coplanar waveguide as well as of two MIM capacitors of different size at the end of a coplanar waveguide to ground (c).

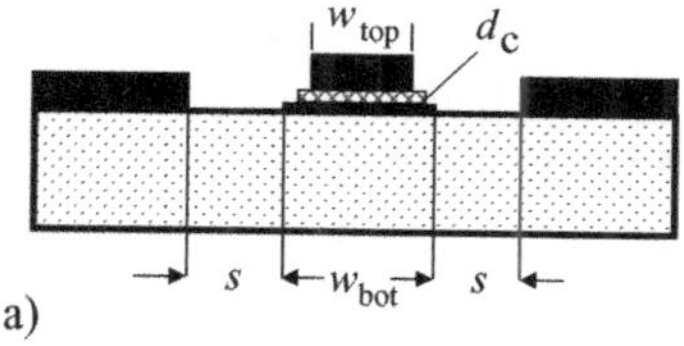

a)

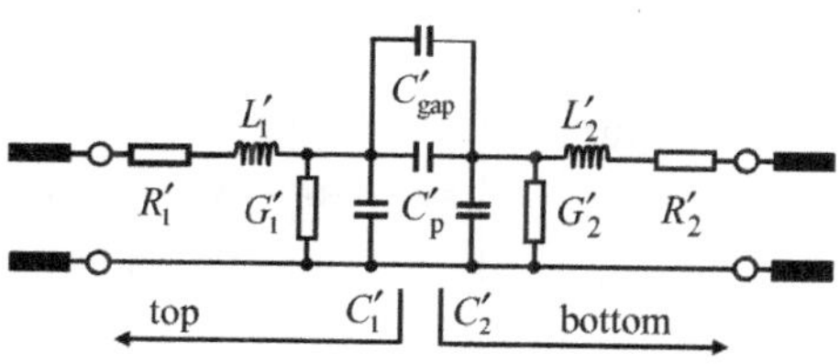

b)

Fig. 4.3.2. Cross section of the MIM capacitor as shown in Fig. 4.3.1. (a) and its equivalent circuit description (b).

As shown in Fig. 4.3.2, the width of the top electrode may be different from the width of the bottom electrode.

Similar to the modeling used in the enhanced model of the interdigital capacitor (Section 4.2.2), the 2×2 C'-, L'-, R'-, and G'-matrices for the line parameters per unit line length are determined using the quasi-static finite difference simulation technique as described in the previous sections. The coupling of the two coplanar waveguides forming the MIM capacitor is taken into account in a fundamental equivalent circuit by using the parallel capacitance formula:

$$C'_p = \frac{\varepsilon_0 \varepsilon_r w}{d_c}. \tag{4.3.1}$$

The total equivalent circuit of the capacitor is shown in Fig. 4.3.2b. The line parameters for the top and the bottom electrodes are determined in units per line length. The gap capacitance C_{gap} that is shunted parallel to C'_p describes the gap between the end of the top electrode and the connected bottom line. This element is independent of the coupling length and is calculated from the cross-sectional geometry of the MIM (metalization thickness and line width).

The bar chart in Fig. 4.3.3 depicts the range of available MIM capacitance values in comparison to the values of an interdigital capacitor (n = number of fingers). The bars in the case of the MIM capacitors start with a capacitor length of $l_c = w/2$ up to $l_c = 4w$, where w is the center strip width. The dielectric material thickness is 200 nm and its dielectric constant ε_r is 7.45. From this figure the advantages of the MIM configurations become evident: With less total size, higher capacitance values can be obtained. The geometry of all utilized 50-Ω coplanar line environments is listed in Table 4.3.1.

Interdigital capacitors have a maximum capacitance value of about 0.8 pF, while MIM capacitors can have capacitances higher than 10 pF. Another important advantage of MIM capacitors is that their parasitic elements are small (e.g., C_{p1} = 38 fF and C_{p1} = 35 fF for the capacitor shown in Fig. 4.3.4).

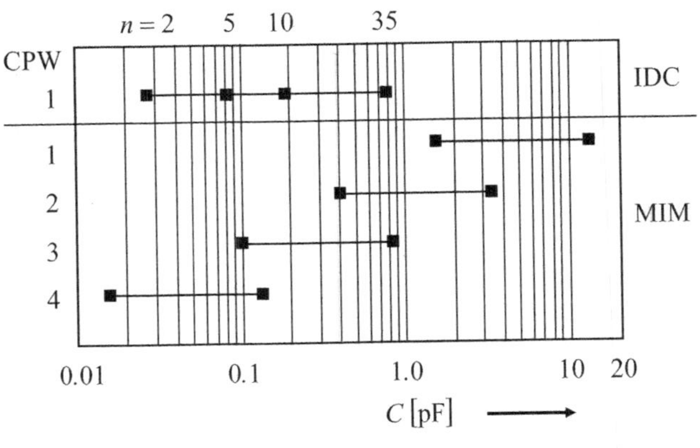

Fig. 4.3.3. A comparison of the capacitance values of MIM capacitors and interdigital capacitors (IDC) in coplanar environment. Parameters of the MIM: ε_r = 7.45 and d_c = 200 nm; see Table 4.3.1 for the different CPW structures.

TABLE 4.3.1. Geometry of the Four Investigated Coplanar Waveguide Configurations

Coplanar Waveguide Configuration	w (μm)	s (μm)
CPW1	100	75
CPW2	50	37
CPW3	25	20
CPW4	10	10

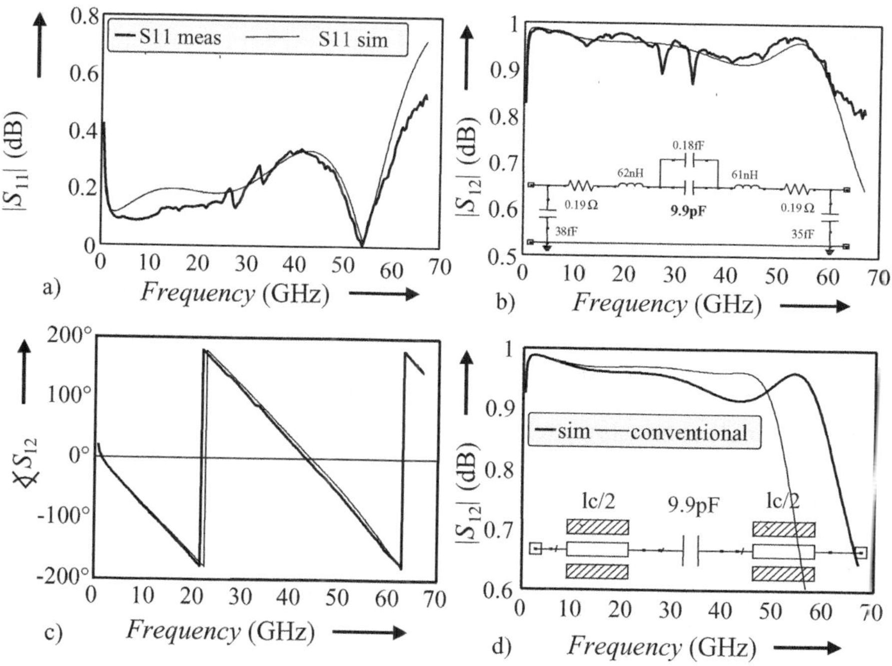

Fig. 4.3.4. S-parameters of a series MIM capacitor in CPW1 (Table 4.3.1) environment with a length $l_c = 300\,\mu m$. Comparison between simulation (thin lines) and measurement (thick lines) results for (a) magn (S_{11}), (b) magn (S_{12}), and (c) angle (S_{12}). (d) Comparison between simulation results from the model used here (see Fig. 4.3.2) and a simple line–capacitor–line model (see inset of part d).

Figure 4.3.4 depicts the measured and simulated S-parameters of a series MIM capacitor. This element is fabricated in a CPW1 (Table 4.3.1) environment. Top and bottom electrodes have the same line width. The capacitor length is 300 μm. The agreement between measurement and simulation is excellent up to the highest frequency of verification (67 GHz).

The comparison between the MIM capacitor simulation used here and a conventional line–capacitor–line model is depicted in Fig. 4.3.4d. This model

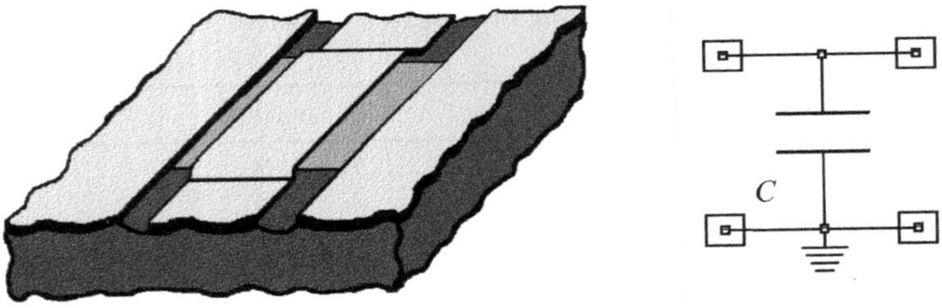

a) b)

Fig. 4.3.5. The parallel shunted MIM capacitor in a coplanar environment (a) and its fundamental equivalent circuit (b).

is built up with two transmission lines. Each one has half of the length of the coupling lines. The capacitance C_p is determined from the parallel plate capacitance equation (4.3.1). The problem with this type of model is that it is only accurate up to a maximum frequency of about 30 GHz. The advantage of the alternative model, shown in Fig. 4.3.2, is obvious.

Figure 4.3.5 shows the construction and the fundamental equivalent circuit of a parallel shunted MIM capacitor to ground in coplanar waveguide environment. The center line that is fabricated in the galvanic layer is led over a gate metalization area in a form of a type 1 air bridge (compare Section 3.5.5).

The rectangle conducting area in the gate level is connected with the ground strips of the coplanar waveguide. A thin-film dielectric block is located between the 2 electrodes, similar to the configuration of the series MIM capacitor. This element is very useful in circuits, where a capacitance to ground is needed (e.g., dc-bias circuits), but no junction is desired. Moreover, such a solution saves circuit size and reduces the parasitic effects.

Figure 4.3.6 shows the comparison between measured and simulated *S*-parameters of such a parallel MIM capacitor in coplanar line geometry CPW1 (Table 4.3.1) with a length of $l_c = 300$ mm. The agreement is excellent even at a frequency as high as 67 GHz. The simplified equivalent circuit diagram in the inset of the figure (Fig. 4.3.6b) demonstrates the small values of the parasitic elements.

4.4 THE COPLANAR SPIRAL INDUCTOR

The realization of inductances in microwave integrated circuits using distributed line structures has the disadvantage that a large space is needed. An alternative approach for realizing planar inductances in a coplanar environment is the spiral inductor [1, 5, 11, 17] that is used in circular and rectangular form. The method described in this section can be used to analyze both kinds of spiral inductors. Since the rectangular spiral inductors can be designed

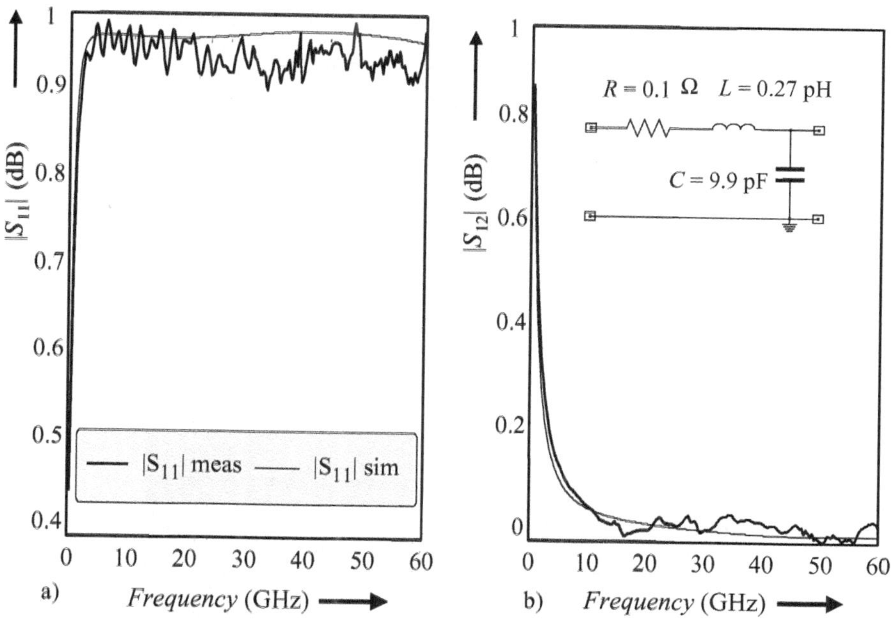

Fig. 4.3.6. Comparison between simulated (thin lines) and measured (thick lines) *S*-parameters of a parallel shunted MIM capacitor in a coplanar environment.

more effectively, saving space in coplanar technology, they are the only ones discussed here.

A typical layout of the coplanar spiral inductor is shown in Fig. 4.4.1a. The center conductor of the coplanar waveguide is coiled to a spiral inductor. The inner end of the inductor is connected to the output center conductor using an air-bridge construction. Depending on the number of turns used in the inductor—that is, depending on the size of the spiral inductor—the ground-plane dimensions must be changed. As will be discussed later, the design of the surrounding ground plane has a large influence on the frequency-dependent properties of the inductor.

Figures 4.4.1b to 4.4.1e show different technological realizations of spiral inductors in a coplanar environment. In Fig. 4.4.1b very small inductors with very narrow gaps between the different turns are shown. To produce the gaps (3 μm to 5 μm) with the needed accuracy, the inductors are produced using the lower (gate) metalization layer. The feeding coplanar waveguide and the ground planes are realized in the thicker second metalization layer. An air bridge, again produced in the second metalization layer, crosses the windings of the inductor and connects the center point of the spiral coil with the second feeding coplanar waveguide center strip.

Figures 4.4.1c and 4.4.1d show three examples of coplanar spiral inductors of larger size. They are produced in the second metalization layer. The center

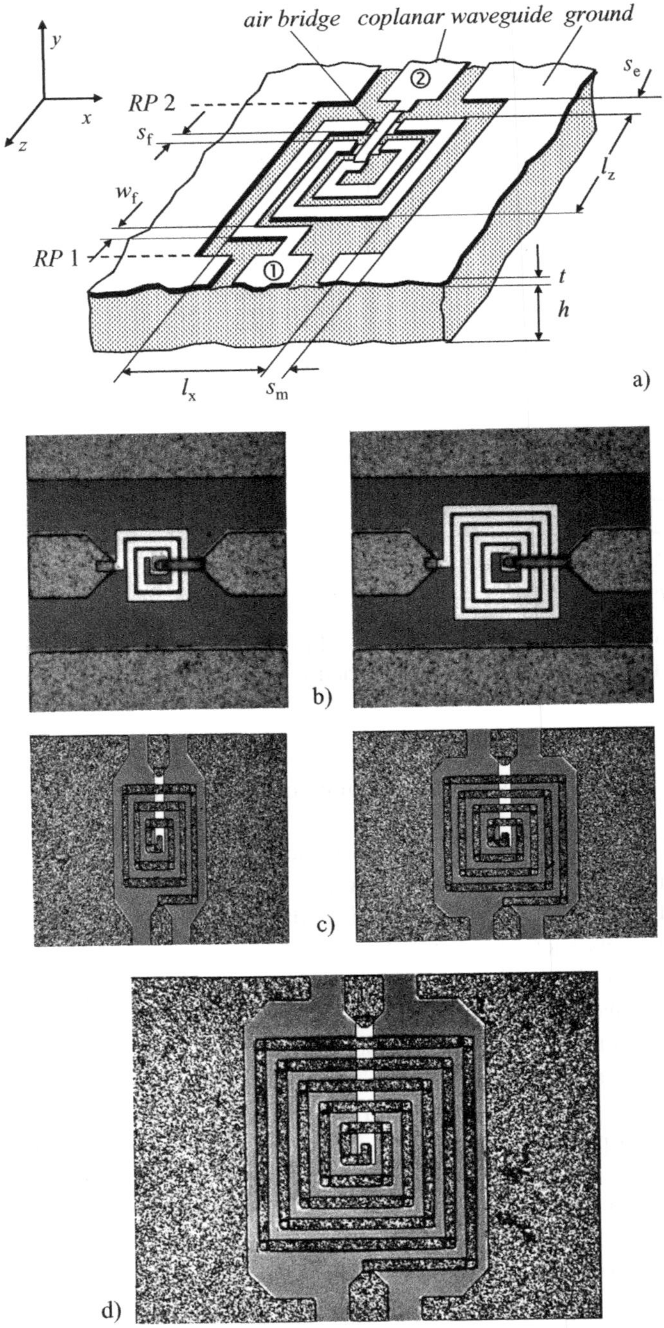

Fig. 4.4.1. The rectangular spiral inductor in coplanar waveguide technology. (a) Principal sketch (b–e) Technological realizations. *RP* stands for reference plane.

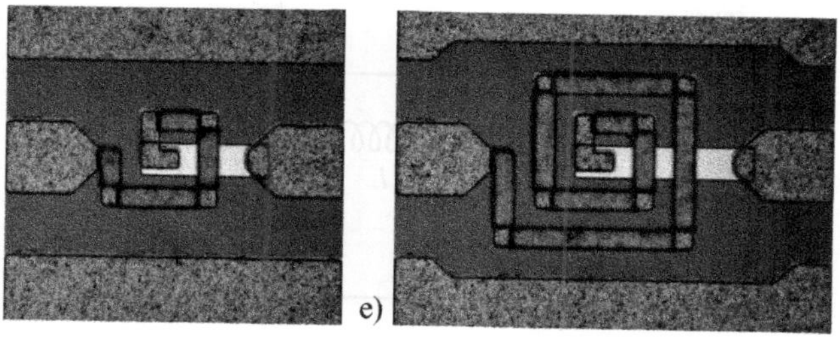

Fig. 4.4.1. (*Continued*).

points of the structures are connected to the coplanar waveguide using an underpass in the first (gate) metalization layer. Finally, Fig. 4.4.1e shows two spiral inductors where the windings of the coil are formed as an air-bridge structure at a height of about 2–3 μm above the substrate material.

This technique is used to reduce the fringing capacitances that heavily influence the frequency dependence of the inductance and to shift the first resonant frequency to higher frequencies. Additionally, using this construction, the Q-factor of the inductor can be improved because for this design form the electric stray field is mostly in air and not in the lossy substrate material.

The coiled structures shown in Fig. 4.4.1, in principle, are no longer a real coplanar waveguide structure. This means that analysis techniques like the segmentation method [14, 17, 20], which has been used for the analysis of microstrip spiral inductors, will not deliver accurate results in this case. An accurate description of the coplanar spiral inductors (as shown in Fig. 4.4.1) that is valid for a broad frequency range can only be found using a full-wave analysis technique as, for example, in reference 19. For most practical applications, the properties of the lumped elements are only of interest in a frequency range where the linear size of the components is much smaller than the wavelength of the adjungated working frequency. Under these conditions the properties of the components may be simulated using a simple equivalent circuit model. Such a model for the coplanar spiral inductor is shown in Fig. 4.4.2.

The parameter L of the equivalent circuit is the main inductance of the spiral inductor that has the wanted physical effect of the component. The parameters C_{p1} and C_{p2} are the parasitic capacitances of the inductor turns to ground whereas the parameter C_g describes the electric coupling between the different turns. A frequency-dependent resistor R_f represents the ohmic losses, which are analyzed on the basis of the skin effect.

For the analysis of the capacitances C_{p1}, C_{p2}, and C_g, the inductor is divided into two parts of equal lengths (Fig. 4.4.3). Both parts are set to two constant potentials φ_1 and φ_2. The ground-plane potential is φ_0. Using two different potential configurations of the conducting structure, the surrounding potential distribution can be analyzed using the finite difference technique as described

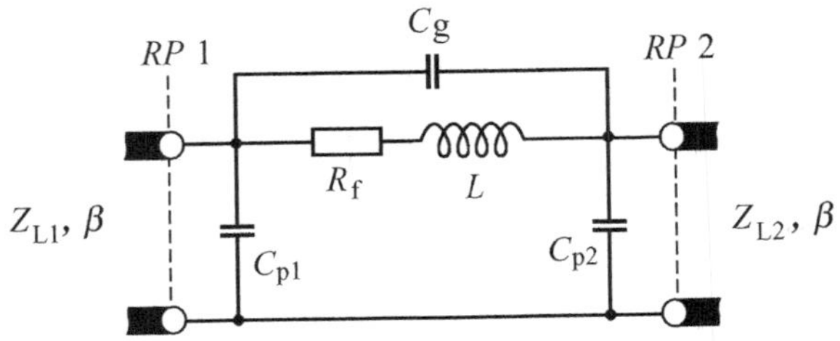

Fig. 4.4.2. The equivalent circuit model of the coplanar spiral inductor. *RP* stands for reference plane.

even case	**odd case**
$\left.\begin{matrix}\varphi_1 = +1\mathrm{V} \\ \varphi_2 = +1\mathrm{V} \\ \varphi_0 = 0\mathrm{V}\end{matrix}\right\} \rightarrow C_{p1}, C_{p2}$	$\left.\begin{matrix}\varphi_1 = +1\mathrm{V} \\ \varphi_2 = -1\mathrm{V} \\ \varphi_0 = 0\mathrm{V}\end{matrix}\right\} \rightarrow C_g$

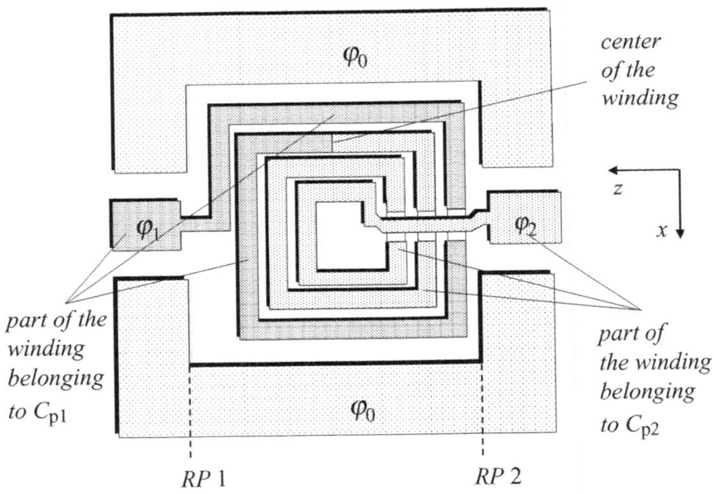

Fig. 4.4.3. Partitioning of the spiral inductor into two equal parts for the computation of the capacitances C_{p1}, C_{p2}, and C_g.

in Chapter 3. This analysis leads to a surface charge density distribution from which the capacitances can be computed as described in detail in Section 3.4.

Figure 4.4.4 shows the distribution of the normal electric field components perpendicular to the metalization layer along the turns of the spiral inductor. As can be observed from the figure, the electric field strength is concentrated near the outside turn. Therefore, the parasitic capacitance C_{p1} is always larger than C_{p2}.

To determine the coupling capacitance C_g using the above-mentioned method, the two parts of the spiral inductor are kept on two different poten-

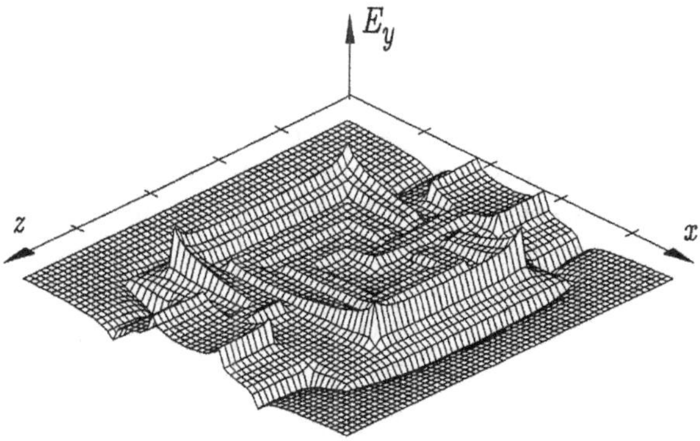

Fig. 4.4.4. Distribution of the normal electric field component perpendicular to the metalization layer of a coplanar spiral inductor.

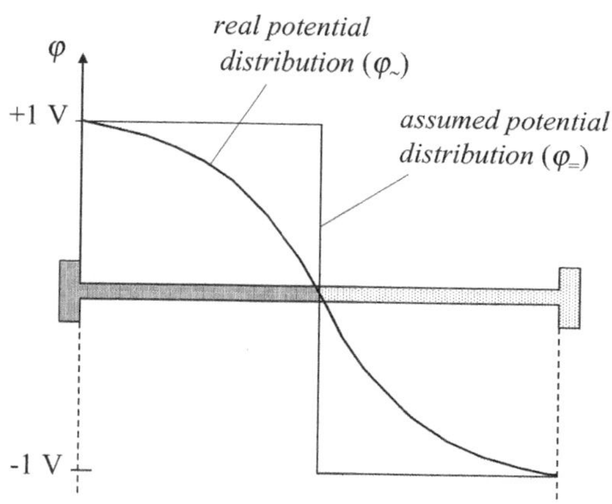

Fig. 4.4.5. The electric potential distribution along the turns of the spiral inductor for the case of a $\lambda/2$ resonance and the assumption of a constant phase velocity along the turns.

tials with opposite sign ($\varphi_1 = +1\,\text{V}$, $\varphi_2 = -1\,\text{V}$). By this method, in principle, a resonant situation is simulated on the spiral inductor where a half-wavelength signal is distributed along the turns. In this simulation it is assumed that the potential distribution along the strip forming the inductor is a rectangular function. Assuming a constant phase velocity of a wave along the turns of the spiral inductor, however, this function will in reality be a sinusoidal-like function as shown along the unwounded coil conductor between the two reference planes at port ① and port ② in Fig. 4.4.5. Also the rectangular distribution that has been assumed for the simulation is shown in this figure.

The assumption of a rectangular potential distribution ($\varphi_=$) instead of the real sinusoidal distribution ($\varphi_\sim$) leads to a capacitance value C_g that is too large. The reason for this is that under the assumption of the rectangular potential distribution, there would be more charge on the inductor strip as in the case of the real potential distribution. A first-order correction C_g^c can be found if the capacitance value C_g, calculated using the finite difference analysis technique (see Section 3.4) is corrected by weighting the two possible charge distributions with the ratio of the two potentials, that is,

$$C_g^c = C_g \frac{\int_0^{\pi/2} \varphi_{\sim} d\alpha}{\int_0^{\pi/2} \varphi_{=} d\alpha} = C_g \frac{2}{\pi}. \tag{4.4.1}$$

An accurate solution for the capacitance cannot be found using the quasi-static finite difference technique because in reality the different points on the turns do not have the same potentials as it has been assumed in the computation. Therefore, this applied analysis technique can only be a first-order approximation for determining the capacitance C_g. On the other hand, this method is much faster compared to a full-wave analysis that uses a moment method or a finite difference time domain analysis and therefore is of big advantage in a circuit analysis program.

The computation of the inductance follows the method that has been described in Sections 3.3 and 3.4. For the analysis of the magnetic field distribution, the spaces between the winding as well as the other slot spaces are set to different but constant magnetic potentials. Figure 4.4.6 shows such a choice of the magnetic potential for a spiral inductor of 2.5 turns.

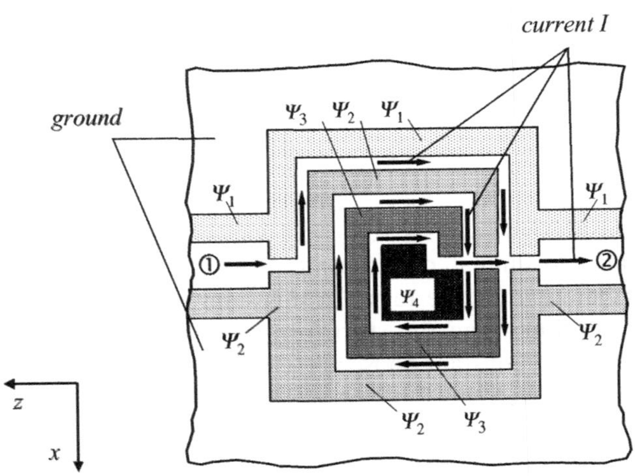

Fig. 4.4.6. Choice of the magnetic potential in the slot spaces of a spiral inductor for determining the magnetic field distribution. (Example: $\Psi_1 = +1\,\mathrm{A}$, $\Psi_2 = -1\,\mathrm{A}$, $\Psi_3 = -3\,\mathrm{A}$, $\Psi_4 = -5\,\mathrm{A}$, $I = 4\,\mathrm{A}$)

The solution of Laplace's differential equation for the magnetic potential delivers the magnetic potential distribution above the metalization plane. From this potential the normal component of the magnetic field strength follows from a derivation with respect to the y-coordinate (perpendicular to the xz-plane; see Fig. 4.4.1). The integration of this normal magnetic field component over the slot areas delivers the total magnetic flux Φ that is excited by the current I of the inductor. The inductance then can be computed from these values using Eq. (3.4.8).

The distribution of the magnetic field strength can be additionally used to determine the surface current density distribution in the surface of the conductors. For this, the tangential component of the magnetic field must be known at each point on the metalized areas. In Fig. 4.4.7 the magnitude of the surface current density $\boldsymbol{J}$ on a coplanar spiral inductor is shown.

This current density may then be used again to compute the ohmic losses considering the specific frequency-dependent surface resistance defined by the skin effect. As has been explained in detail in Section 2.2.4, the ohmic losses calculated in this way are only valid for frequencies higher than a limit frequency. This limit frequency f_{lim} is defined by the metalization thickness t and the skin depth $\delta = (\pi f \sigma \mu)^{1/2}$, where σ is the conductivity of the metalization and μ its permeability. The frequency f must at least be so high that the metalization thickness t is three times the skin depth δ or higher. Considering the dc resistor R_{dc} of the winding, the ohmic losses can be approximately calculated in the total frequency range using the equations (compare Eq. (2.2.24))

$$R_f = \begin{cases} R_{\text{dc}} + \left\{ R_f(f = 1\text{GHz})\sqrt{f_{\text{lim}}/\text{GHz}} - R_{\text{dc}} \right\} \dfrac{f}{f_{\text{lim}}} & \text{for } f \le f_{\text{lim}}, \\ R_f(f = 1\text{GHz})\sqrt{f/\text{GHz}} & \text{for } f \ge f_{\text{lim}}. \end{cases} \tag{4.4.2}$$

The frequency f_{lim} can be calculated using Eq. (2.2.25).

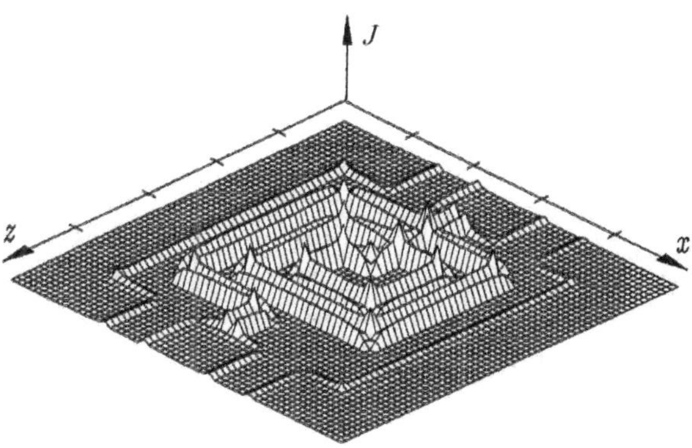

Fig. 4.4.7. Magnitude of the surface current density on the metalized areas of a coplanar spiral inductor.

To verify the above-described analysis technique, coplanar spiral inductors with 2.5, 3.5, and 4.5 turns have been fabricated on GaAs substrate material (substrate $\varepsilon_r = 12.9$, height $h = 400\,\mu\text{m}$) and their scattering parameters have been measured. The parameters of the equivalent circuit (Fig. 4.4.2) as well as the geometrical parameters of the spiral inductors are shown in Table 4.4.1. l_{strip} is the total length of the winding and $f(\lambda/4)$ is the frequency at which the phase angle of the transmission coefficient becomes −90°. This frequency is the upper limit for the application of the quasi-static analysis technique used here.

The calculated and measured scattering parameters of the three spiral inductors are shown in Figs. 4.4.8 and 4.4.9. The agreement between measurement and simulation is, especially for the phase angle, very good. It demonstrates the accuracy of the applied analysis technique in determining the reactive elements of the equivalent circuit. The reduced accuracy in the calculation of the ohmic losses (see discussion above) leads to some small discrepancies between measurement and simulation in the case of the scattering parameter magnitudes.

The parameters of the equivalent circuit (Fig. 4.4.2) are dependent on the geometrical parameters of the inductors in different ways. An investigation of the dependencies on all geometrical parameters shows that the inductance L, especially, is dependent on the total strip length of the inductor and on the number of turns N. The distances to ground s_m and s_e (Fig. 4.4.1), on the other hand, have their largest influences on the parallel capacitances C_{p1} and C_{p2}. The coupling capacitance is more dependent on the distance between the windings and on the number of turns.

TABLE 4.4.1. Geometrical Parameters as Well as Computed Model Parameters of Three Coplanar Spiral Inductors on GaAs Substrate ($\varepsilon_r = 12.9$, $t = 3\,\mu\text{m}$, $h = 400\,\mu\text{m}$)

	Inductor No.		
Parameter	1	2	3
Turn number, N	2.5	3.5	4.5
Turn width, w_f (μm)	25	25	25
Gap width, s_f (μm)	5	5	5
Size length ($l_x = l_z$), l_x, l_z (μm)	185	240	300
Distance to ground, s_m (μm)	50	50	50
Distance to ground, s_e (μm)	50	50	50
Length of winding, l_{strip} (μm)	1.245	2.055	3.155
Self-inductance, L (nH)	0.701	1.455	2.813
Parallel capacitance, C_{p1} (fF)	41.97	60.72	81.34
Parallel capacitance, C_{p2} (fF)	19.03	23.36	29.65
Coupling capacitance, C_g (fF)	21.38	22.16	32.67
Ohmic resistor, R_f (1 GHz)(Ω)	0.334	0.563	0.838
dc resistor, R_{dc} (Ω)	0.434	0.716	1.10
λ/4-frequency, f(λ/4) (GHz)	30	18	11

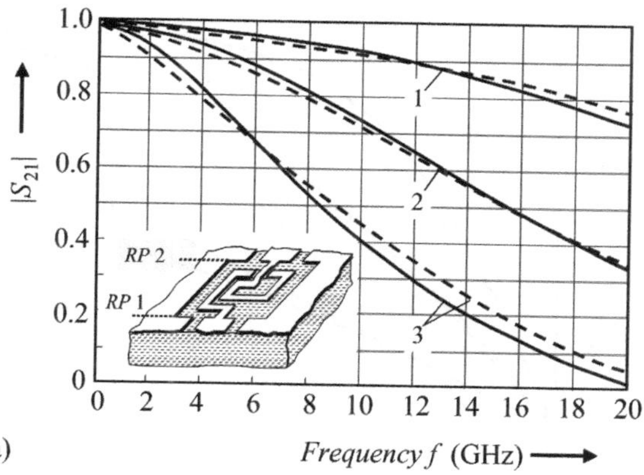

a)

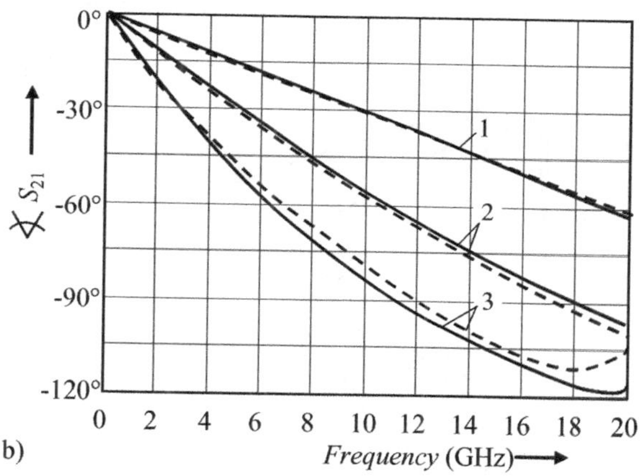

b)

Fig. 4.4.8. Measured (– – –) and simulated (———) magnitude (a) and phase angle (b) of the transmission coefficients of three spiral inductors on GaAs substrate materials, plotted against the frequency. For geometrical parameters see Table 4.4.1.

In Fig. 4.4.10 to Fig. 4.4.13 the dependencies of the equivalent circuit parameters on the lengths l_x and l_z of the winding are shown. The gap widths to the ground plane have been kept constant in these computations. As can be observed, the inductance L increases with increasing size of the spiral inductor and, therefore, with increasing total length l_{strip} of the strip line forming the spiral inductor.

It may also be shown that the inductance is dependent not only on the total strip length l_{strip} but also, assuming a constant length of the strip, on the number of turns the inductor has. A comparison of the depicted results for two inductors with the same line length but different turn numbers shows this.

An essential criterion for the applicability of coplanar spiral inductors in circuit design is the value of the Q-factor of the component. Analogous to the definition of the Q-factor of the interdigital capacitor (see Section 4.2), the Q-factor of the spiral inductor is defined by

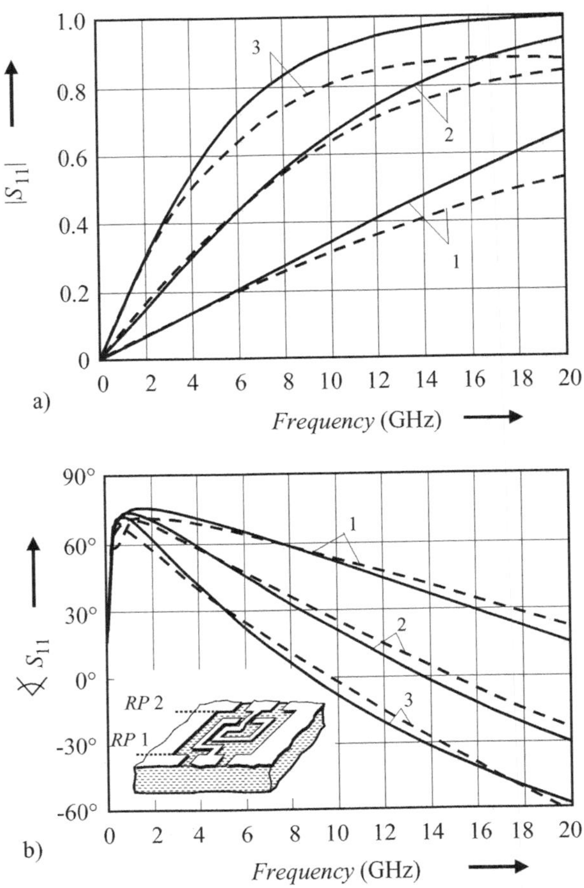

Fig. 4.4.9. Measured (– – –) and simulated (———) magnitude (a) and phase angle (b) of the reflection coefficient of three spiral inductors on GaAs substrate materials, plotted against the frequency. For geometrical parameters see Table 4.4.1.

$$Q = \frac{\omega L}{\mathrm{Re}\{Z_i\}}, \tag{4.4.3}$$

where Z_i is the input impedance of the equivalent circuit shown in Fig. 4.4.2 with port ② short-circuited. If port ① is short-circuited and the input impedance would be defined at port ②, then this leads to a higher value of the Q-factor because in this case the parasitic parallel capacitance C_{p1}, which is larger than C_{p2}, would also be short-circuited. The Q-factors of the spiral inductors listed in Table 4.4.1, calculated with Eq. (4.4.3), are shown in Fig. 4.4.14 as a function of frequency. As can be seen from this figure, the Q-factor has a maximum value at some certain frequency. The value of this frequency is dependent on the ratio of the parasitic reactance of the equivalent circuit and

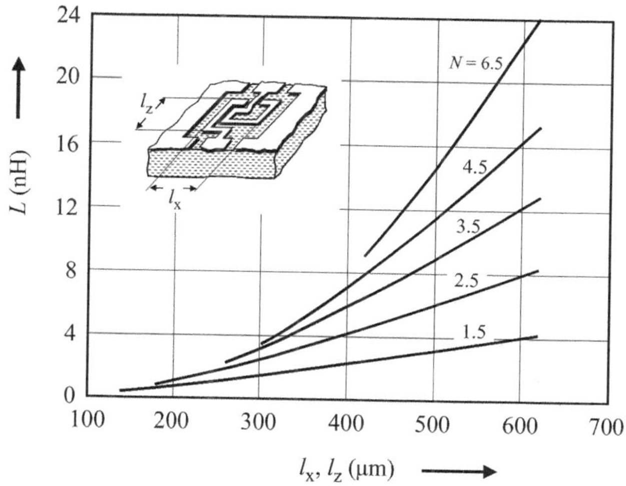

Fig. 4.4.10. Dependence of the self-inductance L of coplanar spiral inductors on the outside size $l_x = l_z$ (see inset) of the winding and on the number N of turns. $w_f = 20\,\mu m$, $s_f = 10\,\mu m$, $s_e = s_m = 50\,\mu m$, $h = 400\,\mu m$, $\varepsilon_r = 12.9$, $t = 3\,\mu m$.

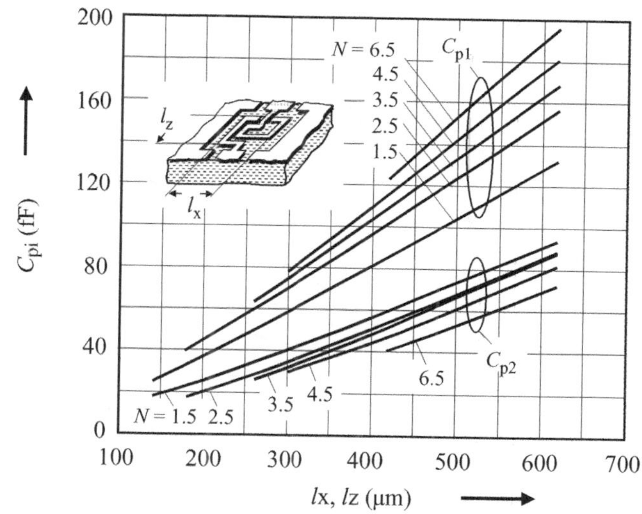

Fig. 4.4.11. Dependence of the parasitic capacitances C_{p1} and C_{p2} of coplanar spiral inductors on the outside size $l_x = l_z$ (see inset) of the winding and on the number N of turns. $w_f = 20\,\mu m$, $s_f = 10\,\mu m$, $s_e = s_m = 50\,\mu m$, $h = 400\,\mu m$, $\varepsilon_r = 12.9$, $t = 3\,\mu m$.

the self-inductance L of the inductor. For higher frequencies, the Q-factor is decreasing to a minimum if the parasitic reactance is dominant in the equivalent circuit. It can also be observed from Fig. 4.4.14 that coplanar spiral inductors with a small value of the self-inductance have a higher Q-factor.

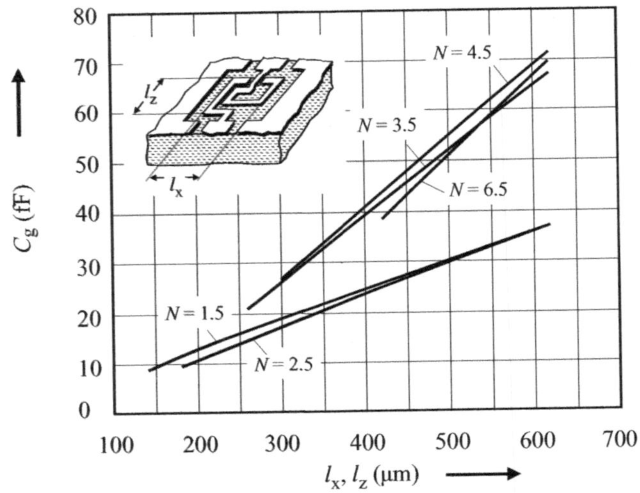

Fig. 4.4.12. Dependence of the parasitic capacitance C_g of coplanar spiral inductors on the outside size $l_x = l_z$ (see inset) of the winding and on the number N of turns. $w_f = 20\,\mu\text{m}$, $s_f = 10\,\mu\text{m}$, $s_e = s_m = 50\,\mu\text{m}$, $h = 400\,\mu\text{m}$, $\varepsilon_r = 12.9$, $t = 3\,\mu\text{m}$.

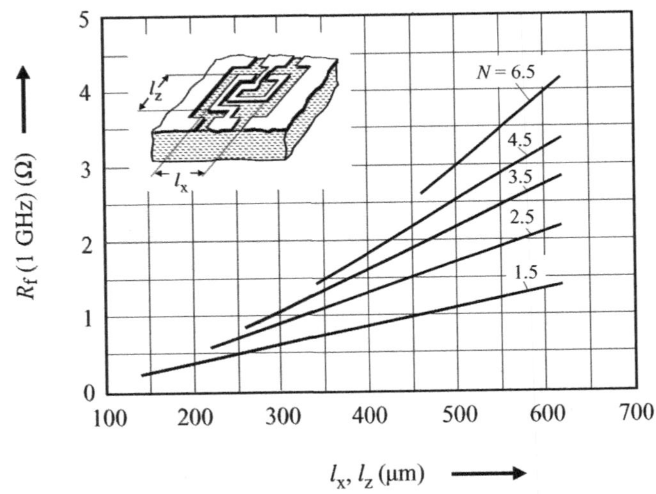

Fig. 4.4.13. Dependence of the ohmic resistor R_f (1 GHz) of coplanar spiral inductors on the outside size $l_x = l_z$ (see inset) of the winding and on the number N of turns. $w_f = 20\,\mu\text{m}$, $s_f = 10\,\mu\text{m}$, $s_e = s_m = 50\,\mu\text{m}$, $h = 400\,\mu\text{m}$, $\varepsilon_r = 12.9$, $t = 3\,\mu\text{m}$.

Even if the Q-factor may be optimized by a proper choice of the geometrical parameters, a Q-factor higher than 30 will not be realizable with common thin-film technology. One additional method to improve the Q-factor will be shortly discussed as an example in the next section: that is, the strips of the

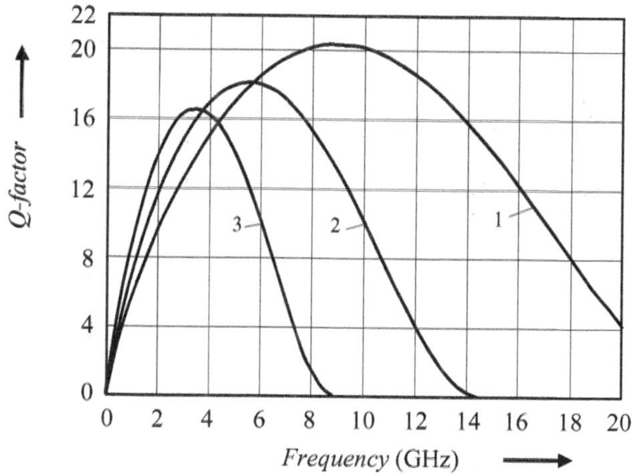

Fig. 4.4.14. Q-factors of three coplanar spiral inductors in dependence on the frequency. For parameters see Table 4.4.1.

inductor turns may be lifted above the substrate material using air-bridge technologies (see Section 4.5.5). This is a technique that does not require additional expense in monolithic microwave integrated circuit design and realization because air-bridge technologies are available in all MMIC technologies for producing transistors or other components. Nevertheless, the Q-factors of the spiral inductors have an upper limit. That means that the spiral inductors are, for example, not applicable in circuits requiring a high-frequency selection of small bandwidth (see also the discussion on lumped element filters in Chapter 6).

Spiral inductors with a high inductance often need large space in the circuit design. When space is a problem (which is normally the case), the application of different production technologies may help save space. In monolithic microwave integrated circuits, normally two different metalization layers are used: the gate layer with a thickness smaller than 1 μm (typical 0.8 μm) and the galvanic layer with a thickness of typically 3 μm. If the inductor is produced in the galvanic layer, space widths and line widths of smaller than 10 μm become difficult to be realized. In the gate layer, slot widths and line widths of about 4 μm to 5 μm may be produced.

Figure 4.4.15 shows how the application of the two different production techniques may reduce the size of the inductor. With the reduction of size, also the resonant frequency can be increased. However, because of the reduced metalization thickness, the smaller inductor has a much higher series resistor. But this, in special cases (for example, in the oscillator design), may even be an advantage. The first resonant frequency of the left inductor (Fig. 4.4.15) is about 20 GHz, and that of the larger inductor is about 8 GHz. The needed space has been reduced from 250 μm^2 to 80 μm^2.

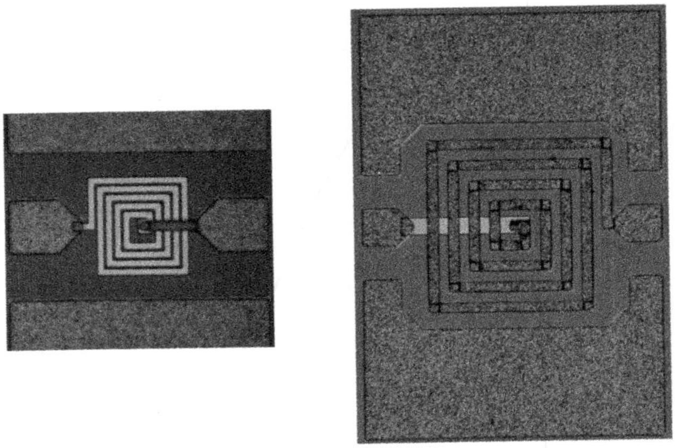

Fig. 4.4.15. Two spiral inductors of 5.5 turns, produced in different metalization layers: Left figure, an inductor produced in the gate layer (0.8 μm metalization thickness). Right figure, inductor produced in the galvanic layer (3-μm metalization thickness).

4.4.1 Enhancement of the Inductor Model for Millimeter-Wave Frequencies

If the above-described model for the spiral inductor in coplanar environment is used for frequencies higher than 20–30 GHz (depending on the size of the inductor), its accuracy is not good enough because at millimeter-wave frequencies the effects of the distributed structure have to be considered. To describe the spiral inductor for higher frequencies more accurately, the equivalent circuit shown in Fig. 4.4.16 can be used.

The model that has been described in the previous section is changed in such a way that each turn of the inductor is segmented into quarter turns in order to handle the distributed effects that become obvious at higher frequencies or for a large turn number. For each quarter turn a basic equivalent circuit, BEC (see Fig. 4.4.16), is computed utilizing a quasi-static three-dimensional finite difference approach as described earlier in this chapter.

Simulation, measurement, and layout of a 1.5-turn inductor are given in Fig. 4.4.17. This small inductor with a track of 25 μm, a spacing of 20 μm, and a first S_{11}-resonance at 14.5 GHz was designed for integration into a millimeter-wave system. Comparing the simulation results of the standard model (as described in the previous section) and the enhanced model, it is clearly visible that there are discrepancies between the measured and the simulated S-parameters when using the standard model. In the case of the enhanced model, the agreement between measurement and simulation is very good up to the frequency of 60 GHz.

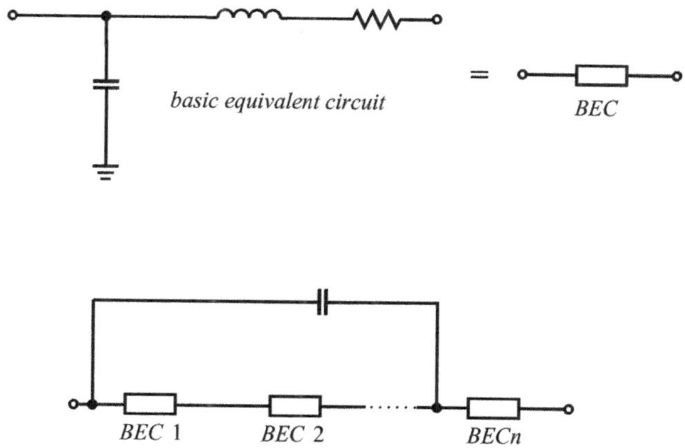

Fig. 4.4.16. Enhanced equivalent circuit for the spiral inductor in a coplanar environment.

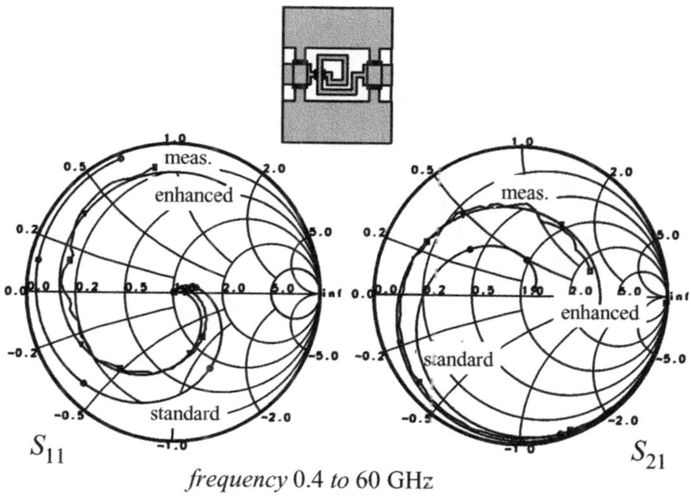

Fig. 4.4.17. Comparison of the simulation quality of the standard model and the enhanced model for a 1.5-turn planar spiral inductor in a coplanar environment. Left diagram: S_{11}, right digram: S_{21}, the layout of the inductor is shown in centre of the figure.

4.4.2 Coupled Coplanar Rectangular Inductors

Planar spiral inductors often are used in dc bias networks of microwave integrated circuits to save space. If there is more than one bias network, the spiral inductors possibly are placed near each other as shown in Fig. 4.4.18, for example. In the construction shown, the two spiral inductors may couple to

Fig. 4.4.18. Two closely spaced spiral inductors in a coplanar environment.

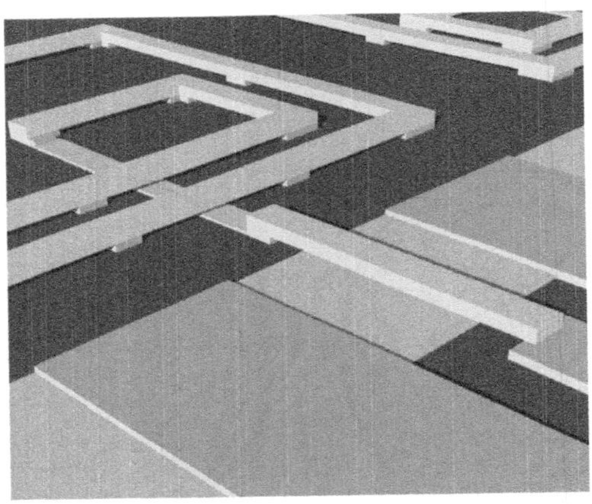

Fig. 4.4.19. Details of the planar spiral inductor in a coplanar environment.

each other via their magnetic field and the circuit may possibly not work as it was designed for.

In Fig. 4.4.19 some details of the spiral inductors are shown. To improve the Q-factor of the inductors, they are placed about 3 μm above the substrate material as has already been discussed at the end of the last section. To form the air bridges that are needed to realize the contact to the inner winding of the inductor, two different metalization layers (the thin gate layer and the galvano-technically enhanced metalization layer) are used.

To study the effect of the different possible constructions when designing two closely spaced spiral inductors, the structures shown in Figs. 4.4.20a–4.4.20d have been analyzed and realized.

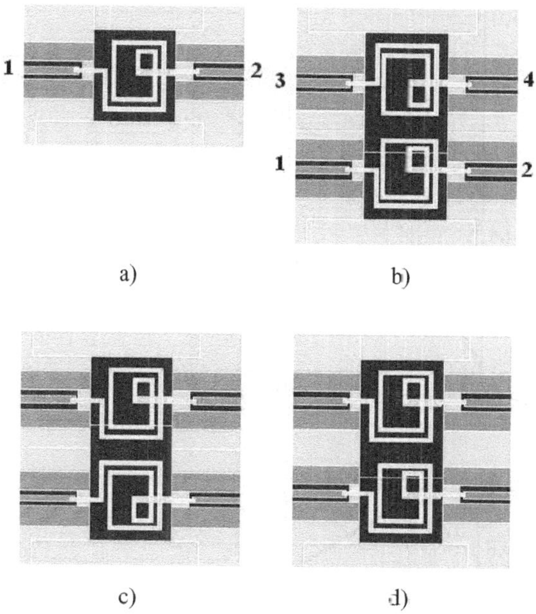

Fig. 4.4.20. A single planar spiral inductor in a coplanar environment (a) and three possible construction of two coupled planar spiral inductors that are placed near each other (b–d).

Figure 4.4.20a shows a single spiral inductor in a coplanar environment that works as a reference for the other three considered cases. Figures 4.4.20b to 4.4.20d show three possible ways of placing two planar inductors near each other. Because the construction of the spiral inductor is not symmetrical, each of these three circuits should have different properties. How much these properties are really different has been studied using a full-wave three-dimensional finite difference time domain (FDTD) analysis technique (see Section 2.1).

For the constructions shown in Fig. 6.4.20, the scattering parameters have been analyzed and measurements have been performed for the structure shown in Fig. 4.4.20b. The results of the simulations and the measurements for the single spiral inductor and the two coupled inductors shown in Fig. 4.4.20b are given in Fig. 4.4.21 for comparison. For the eigenreflection coefficient $|S_{11}|$ it can be observed that the simulation results for the single spiral inductor (Fig. 4.4.20a) and the two coupled inductors (Fig. 4.4.20b) are nearly identical. It means that the eigenreflection coefficient of the single inductor is nearly not changed if a second spiral inductor is placed nearby (as shown in Fig. 4.4.20b).

The measured results for case b show some deviations from the simulation but in principle still are in quite good agreement with the simulated results.

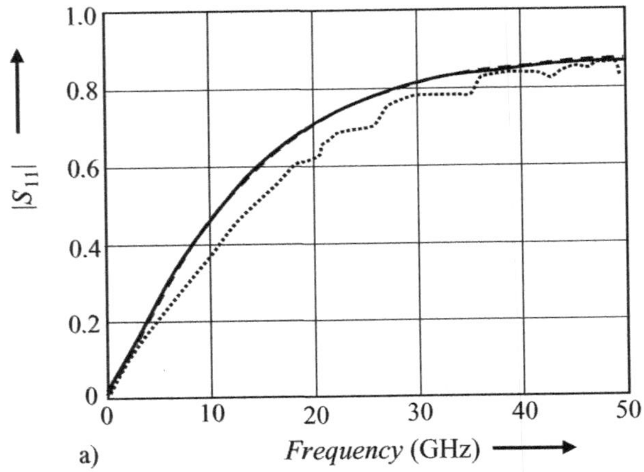

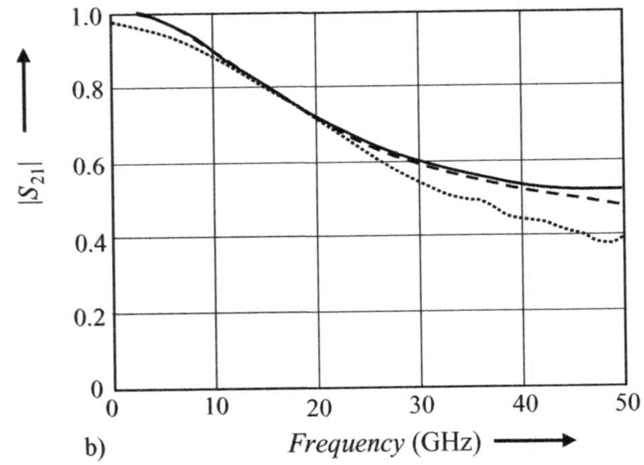

Fig. 4.4.21. The scattering parameters $|S_{11}|$ and $|S_{21}|$ for a single spiral inductor and two coupled spiral inductors in a coplanar environment. For the port numbers see Fig. 6.4.20. Solid line, case a; dashed line, case b; dotted line, measurements case b.

The simulated transmission coefficients $|S_{21}|$ of the single spiral inductor is also nearly not changed by the second inductor; only at very high frequencies the simulations show some deviations from the measurements. These deviations that can be more clearly seen in the comparison between simulation and measurements at high frequencies give a hint that the spiral inductor models should not be used at very high frequencies (compare also the discussion in Section 4.4.1).

As can be noted from the discussion above, the properties of the single inductors are not changed very much by the nearby placed second inductor. But there is an essential coupling between the two structures that may be of disadvantage in microwave circuit design. Additionally, this coupling is a function of the structure chosen as can be seen from Fig. 4.4.22. The coupling coef-

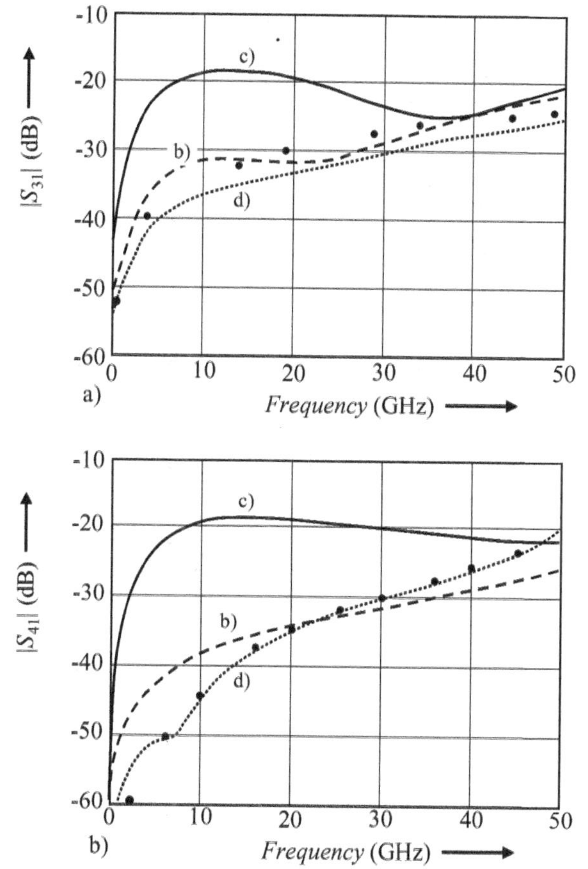

Fig. 4.4.22. The coupling coefficients $|S_{31}|$ and $|S_{41}|$ describing the coupling effect between the two adjacent spiral inductors. Solid line, simulation case c; dashed line, simulation case b; dotted line, simulation case d; Dots, measurement results for case b.

ficients $|S_{31}|$ and $|S_{41}|$ (for the port numbering see Fig. 4.4.20) may differ by more than 10 dB, depending on the considered structure (case b, c, or d). Case c is especially critical because it shows a coupling coefficient higher than −20 dB between both ports of the coupled inductors, which can be too high for many applications. This construction therefore should be avoided in microwave circuit design.

4.5 THE COPLANAR RECTANGULAR SPIRAL TRANSFORMER

Two spiral inductors as described in Section 4.4 may be wound into each other to form a planar transformer. They are used in microwave integrated circuits such as couplers in broadband phase shifters or impedance transformers in amplifier circuits. Figure 4.5.1 shows such a spiral transformer in coplanar technology. The position of the four ports is chosen in such a way that the transformer is symmetrical and has two windings of the same length.

There are very few publications in the open literature that deal with the application of planar spiral inductors in microwave integrated circuits, [9, 12]. Also there are only few publications describing analysis techniques for these complex components [13, 17, 18, 25]. The moment method analysis technique described in reference 13 is a computer-time-intensive method that can be used in CAD techniques only in connection with a look-up table strategy. Moreover, using this analysis approach, the structure must be divided into partial structures and the effect of coupled bends and air bridges is not generally and only approximately considered. Also the method described in reference 18, which is a simple approximate method, does not consider the above-mentioned effects. The analysis technique described here, on the other hand, allows the analysis of the transformer as a three-dimensional, total structure including the air bridges and the four connected coplanar waveguides. However, since an equivalent circuit describes the properties of the spiral transformers, the results are only valid in a frequency range where the linear size dimensions of the structure are smaller than a quarter wavelength at the operating frequency.

The equivalent circuit of the coplanar spiral transformer is shown in Fig. 4.5.2. C_g is a capacitance that describes the electrical coupling between the two windings of the transformer. The two lossy transformers simulate the magnetic coupling. As can be seen from Fig. 4.5.2, the electrical coupling is assumed to be concentrated in the center of the windings. The capacitances C_{p1} and C_{p2} represent the parasitic electric stray fields from the windings to the ground

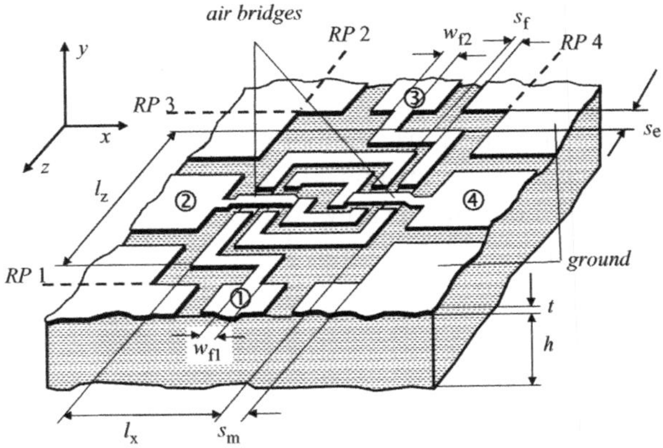

Fig. 4.5.1. Coplanar rectangular spiral transformer. Parameters: w_{f1}, w_{f2} = width of the winding strips, s_f = space between the windings, l_x, l_z = outside size of the transformer, s_m = space between transformer output and ports ② and ④, s_e = space between transformer output and ports ① and ③ (reference planes: *RP* 1 to *RP* 4).

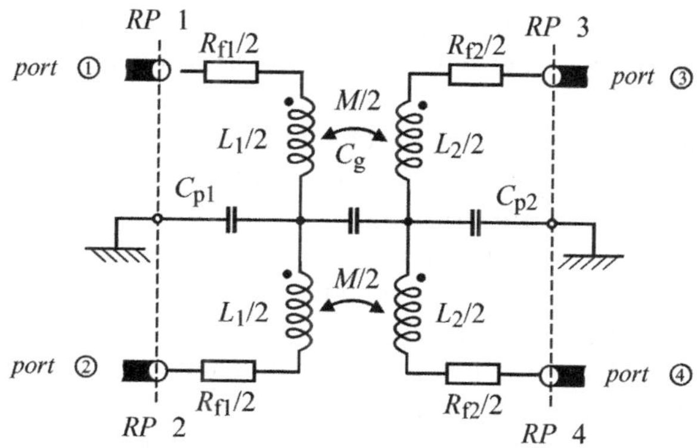

Fig. 4.5.2. Equivalent circuit model of the rectangular coplanar spiral transformer.

even case **odd case**

$$\left.\begin{array}{l}\varphi_1 = +1\text{V}\\ \varphi_2 = +1\text{V}\\ \varphi_0 = 0\text{V}\end{array}\right\} \rightarrow C_{p1}, C_{p2} \qquad \left.\begin{array}{l}\varphi_1 = +1\text{V}\\ \varphi_2 = -1\text{V}\\ \varphi_0 = 0\text{V}\end{array}\right\} \rightarrow C_g$$

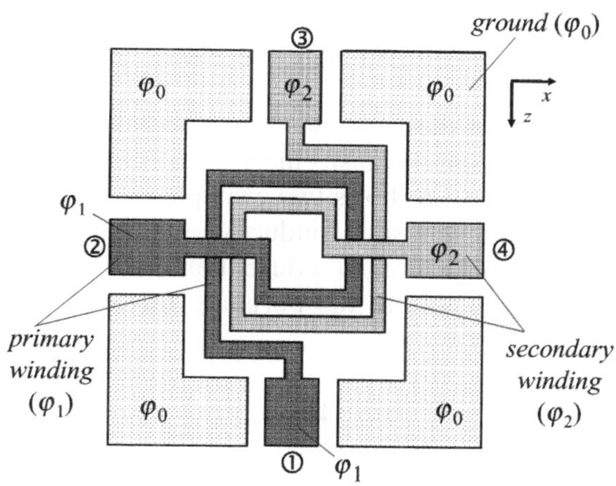

Fig. 4.5.3. Demonstration of the schematic way for determining the capacitive elements of the equivalent circuit model for the coplanar spiral inductor.

plane. As in the case of the spiral inductor, the frequency dependence of the resistances R_{f1} and R_{f2} are considered using the simulating equations given in Section 4.4 [see Eq. (4.4.2)].

The capacitive elements can be determined in an analogous way as described in Section 4.4 for the coplanar spiral inductor. Two different potential distributions are used. The method is schematically demonstrated in Fig. 4.5.3. More details can be found in Section 4.4. As may be seen from the electric field distribution on the metalized areas of the transformer (Fig. 4.5.4), the effects of the air bridges as well as of the line width steps are fully considered in the analysis of the capacitances.

The determination of the inductive elements of the equivalent circuit is performed in an analogous way as in the case of the spiral inductor (see Section 4.4). For the case of the transformer, however, three computation steps are needed to determine the self-inductances L_1 and L_2 and the mutual inductance M. For the calculation of the self-inductances L_1 and L_2, a current flow in one winding is simulated and the current flow in the other winding is set to zero. The analysis of the magnetic field distribution using this current distribution and the quasi-static finite difference analysis technique (see Section 3.3) then leads to the values of the self-inductances. For the determination of the mutual inductance M, a current flow in both windings is simulated. In Fig. 4.5.5 a schematic way to determine the three inductive parameters is demonstrated. Also shown in the figures are the adjoined values of the magnetic potentials that are assumed to be constant in the slot spaces between the windings and between the turns of the windings.

The magnetic field distribution that is used for the analysis of the inductances can also be used for the computation of the ohmic resistors R_{f1} and R_{f2} of the windings. As in the case of the spiral inductor (see Section 4.4), also for the spiral transformer, the dc resistance of the windings must be additionally considered (see Eq. (4.4.2)) to simulate the losses also at low frequencies accurately.

To verify the validity of the equivalent circuit model for the spiral transformer, various coplanar spiral transformers have been fabricated on GaAs substrate material. Figure 4.5.6 shows two examples in the form of the classically side-coupled rectangular spiral inductors. They are fabricated in the second, galvanic metalization layer to reduce the losses. The connections from the inner part of the inductors to the ports are fabricated using a classical

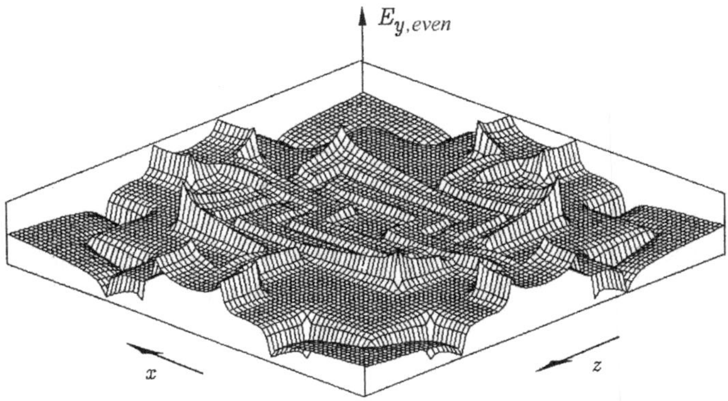

Fig. 4.5.4. The normal component of the electric field strength on the metalized areas of a coplanar spiral transformer. Even case (see text).

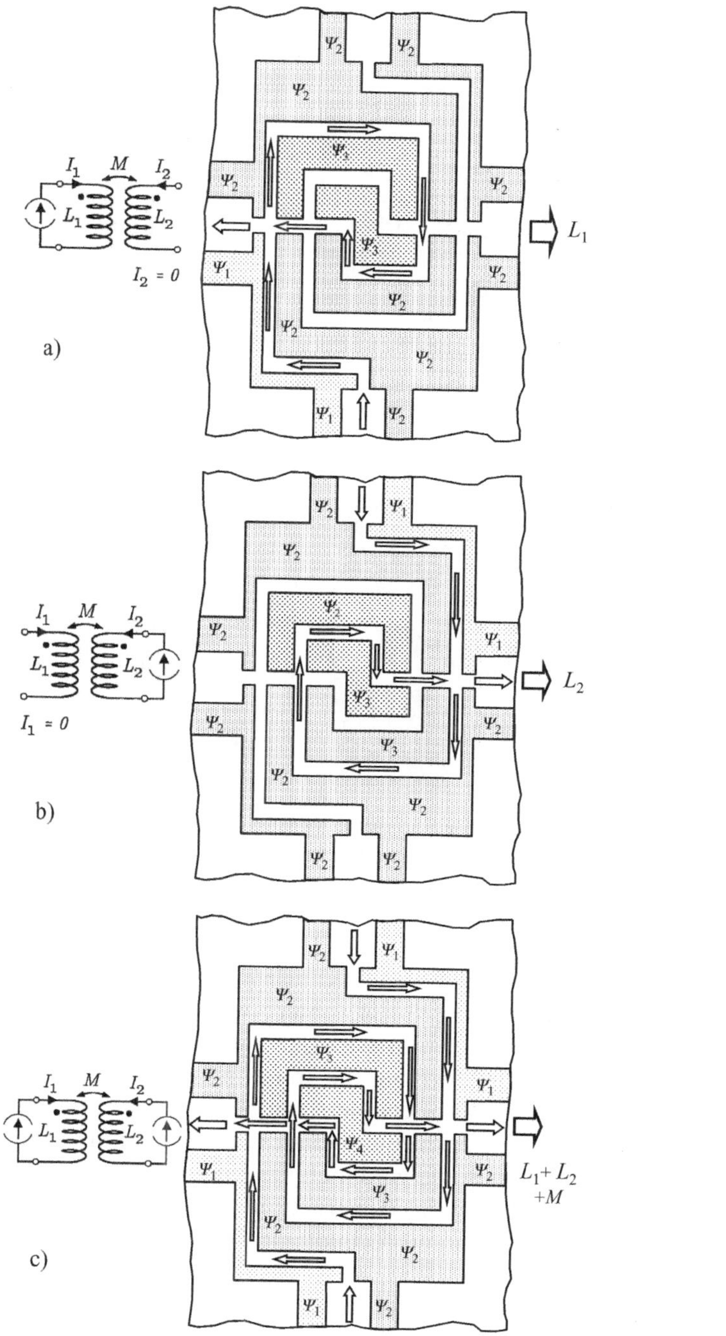

Fig. 4.5.5. Schematic way to determine the self-inductances and the mutual inductance of the coplanar spiral transformer using the quasi-static finite difference technique. (a) excitation of a current $I = I_1 = 4A$ between port ① and port ② to determine L_1 ($\Psi_1 = 1A$, $\Psi_2 = -1A$, $\Psi_3 = -3A$), (b) excitation of a current $I = I_2 = 4A$ between port ③ and port ④ to determine L_2 ($\Psi_1 = 1A$, $\Psi_2 = -1A$, $\Psi_3 = -3A$), (c) excitation of a current $I = I_1 = 4A$ between port ① and port ② and of a current $I = I_2 = 4A$ between port ③ and port ④ to determine $L_1 + L_2 + M$ ($\Psi_1 = 1A$, $\Psi_2 = -1A$, $\Psi_3 = -3A$, $\Psi_4 = -5A$).

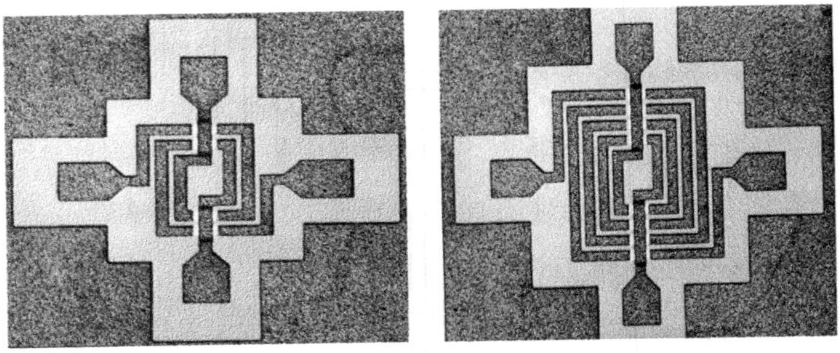

Fig. 4.5.6. Two coplanar rectangular spiral transformers with a winding of 1.5 turns and 2.5 turns, respectively, fabricated on GaAs substrate ($\varepsilon_r = 12.9$, $h = 400\,\mu\text{m}$).

Fig. 4.5.7. Two coplanar circular spiral transformers with a winding of 1.5 turns and 2.5 turns, respectively, fabricated on GaAs substrate ($\varepsilon_r = 12.9$, $h = 400\,\mu\text{m}$).

air-bridge technology of type 1 (see Chapter 3.5.5). As Fig. 4.5.7 shows, circular transformers can also be realized using the same technology.

If a high capacitive coupling is wanted, two broadside coupled spiral inductors in a technology shown in Fig. 4.5.8 can be used. A thin dielectric layer between the two strips forming the spiral inductors separates the two parts. The analysis of this kind of couplers can be done using the design equations given in Section 2.3.2.2 or the quasi-static field simulation technique. The lower inductor is fabricated in the gate metalization level, the upper in the galvanic metalization layer. Using this construction, a reduction in size can be realized. The figure also shows some technological details at the input ports of the structure.

The scattering parameters of these transformers have been measured using an on-wafer measurement technique. The transformers have been simulated using the above-mentioned analysis technique and the simulated and measured results have been compared.

Figure 4.5.9 and 4.5.10 show the comparison for two transformers with windings of 1.75 and 2.25 turns, respectively. They are designed in the classical form shown in Fig. 4.5.6. As can be seen from the figures, the measured

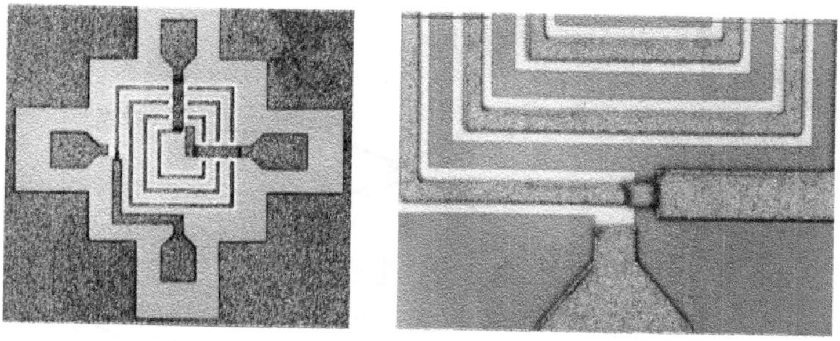

Fig. 4.5.8. A broadside-coupled rectangular spiral transformers with a winding of 3.5 turns fabricated on GaAs substrate ($\varepsilon_r = 12.9$, $h = 400\,\mu m$).

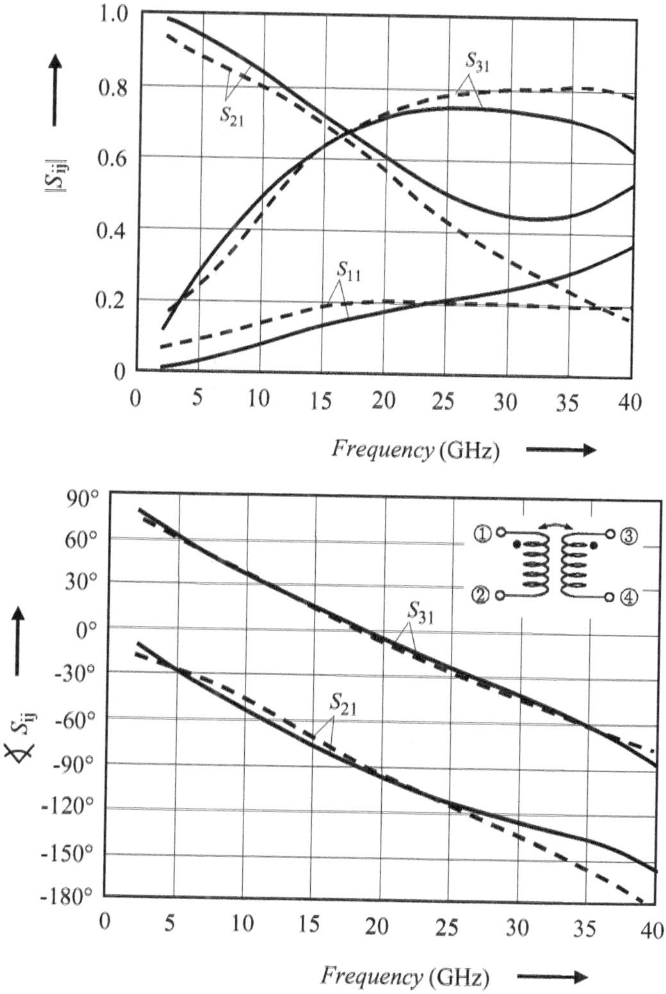

Fig. 4.5.9. Measured (---) and simulated (———) scattering parameters of a coplanar spiral transformer with 1.75 windings on a GaAs substrate material as a function of the frequency. Geometrical parameters: $w_{f1} = w_{f2} = 20\,\mu m$, $s_f = 5\,\mu m$, $s_m = 50\,\mu m$, $s_e = 50\,\mu m$, $h = 400\,\mu m$, $t = 3\,\mu m$. Equivalent circuit parameters: $L_1 = L_2 = 0.942\,nH$, $M = 0.53\,nH$, $C_{p1} = C_{p2} = 35\,fF$, $C_g = 175\,fF$, $R_{f1} = R_{f2} = 0.594\,\Omega$ at $f = 1\,GHz$.

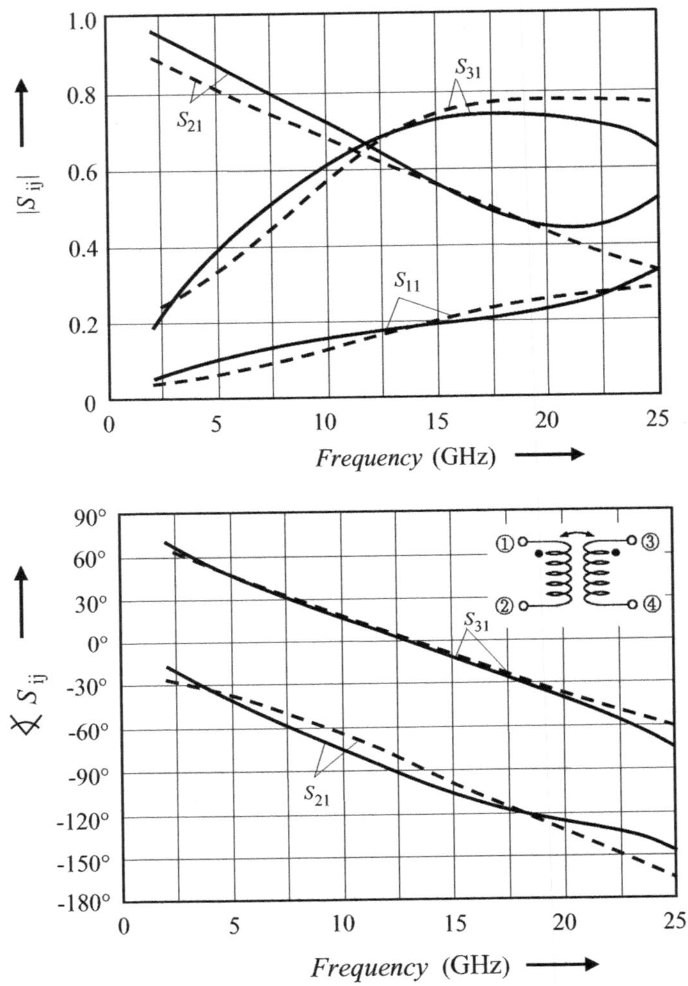

Fig. 4.5.10. Measured (– – –) and simulated (———) scattering parameters of a coplanar spiral transformer with 2.25 windings on a GaAs substrate material in dependence on the frequency. Geometrical parameters: $w_{f1} = w_{f2} = 20\,\mu\text{m}$, $s_f = 10\,\mu\text{m}$, $s_m = 50\,\mu\text{m}$, $s_e = 50\,\mu\text{m}$, $h = 400\,\mu\text{m}$, $t = 3\,\mu\text{m}$. Equivalent circuit parameters: $L_1 = L_2 = 1.69\,\text{nH}$, $M = 0.977\,\text{nH}$, $C_{p1} = C_{p2} = 49\,\text{fF}$, $C_g = 222\,\text{fF}$, $R_{f1} = R_{f2} = 0.914\,\Omega$ at $f = 1\,\text{GHz}$.

and simulated scattering parameters are in good agreement up to the first ($\lambda/4$-) resonance. But also for higher frequencies a quite good agreement, especially for the phase angle, may be observed.

It can also be observed from the figures that the fabricated coplanar transformers have nearly constant phase differences between the two transmission coefficients, S_{21} and S_{31}. They are, therefore, well-suited for an application as couplers in a microwave integrated circuit. It should be mentioned here that the measurements presented in Figs. 4.5.9 and 4.4.10 have been performed using a time domain measurement technique. Therefore, at very high and very low frequencies a certain measurement error may occur.

As has already been discussed above, the windings of the spiral transformer are coupled electrically as well as magnetically. The ratio of these two kinds of coupling and their values and, therefore, the effect of the parasitic elements of the equivalent circuit largely depend on the geometrical design of the transformer. Thus, for example, the electrical coupling can be reduced by the choice of small line widths w_{f1} and w_{f2} and a large gap space s_f between the windings, or it can be increased using the broadside coupled inductors as shown in Fig. 4.5.8.

The parasitic capacitances C_{p1} and C_{p2} may be reduced by increasing the distances s_m and s_e of the transformer windings to the ground plane. If the transformer is to be used as an impedance transformer, the best way to do this is to choose a transformer with different line widths for the two windings. Such a transformer has different properties with respect to the primary and the secondary winding, which can well be used for an impedance transforming application.

4.6 THE COPLANAR THIN-FILM RESISTOR

Thin-film resistors are used in many applications for microwave integrated circuit design. Normally, these resistors, which can be produced in the gate level layer, are considered to be ideal resistors, and special modeling (such as dependence on frequency) is not usually done. In most circuits, transmission line effects of lumped elements are not taken into account during the design phase, and these elements are assumed to be "ideal lumped" up to the highest frequencies of interest. But it is well known that there are discrepancies between measured and simulated S-parameters at high frequencies for some geometrical dimensions. For example, a resistor with a nominal high resistance shows a significant dispersion, and at 40 GHz the resistance is only a few percent of its dc value. This effect is also visible for microstrip thin-film resistors.

In this section a brief investigation of the electrical behavior of coplanar thin-film resistors is given. The thin-film resistor is described by a per unit-length equivalent circuit (lossy transmission line). The elements of this equivalent circuit are derived from the very fast 3D–FD quasi-static approach for the calculation of coplanar waveguide structures as has been described in Chapters 2 and 3. For the measurements, thin-film resistors placed between the end of a coplanar waveguide and its ground, as shown in Fig. 4.6.1, have been used.

It was observed from measurements and simulations that the impedance of resistors with low resistance (e.g., 50 Ω) is nearly constant for frequencies from dc up to 67 GHz. A thin-film resistor with such a low resistance can be handled ideally during the design phase.

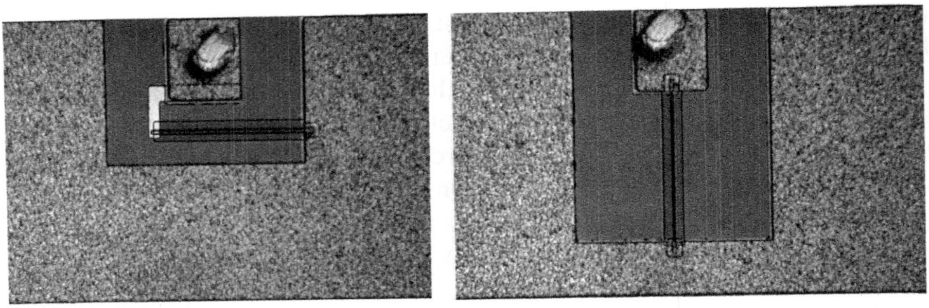

Fig. 4.6.1. Two thin-film resistors at the end of a coplanar waveguide used for the measurements.

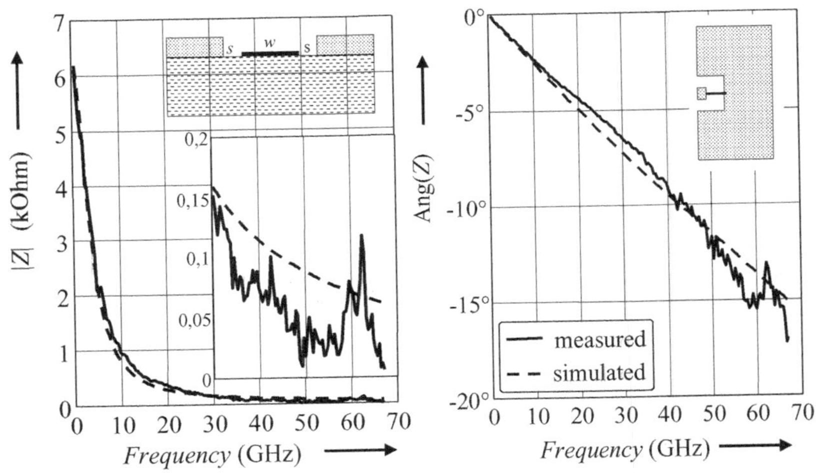

Fig. 4.6.2. Impedance of a coplanar 6.2-kΩ thin-film resistor ($w = 5\,\mu m$, $l = 64\,\mu m$, and $R_s = 490\,\Omega$/square).

In contrast to this, the case of the impedance of a nominal 6 kΩ resistor is shown in Fig. 4.6.2. In this case the thin film resistor shows a significant dispersion even for frequencies below 10 GHz. Using this element for example in a gate bias structure, the circuit function would fail if the dispersion is neglected during the design. The agreement between simulation and measurement is excellent for all thin film resistors considered and the simulation method described above.

BIBLIOGRAPHY AND REFERENCES

1. D. A. Daly, S. P. Knight, M. Caulton, and R. Ekholdt, Lumped elements in microwave integrated circuits, *IEEE Trans. Microwave Theory Tech.*, vol. MTT-15, Dec. 1967, pp. 713–721.

2. M. Caulton and W. E. Poole, Designing lumped elements into microwave amplifiers, *Electronics*, no. 4, 1969, pp. 100–110.
3. H. N. Toussaint and R. Hoffmann, Integrierte Mikrowellenschaltungen—Stand und Tendenzen der Entwicklung, *Frequenz*, vol. 25, no. 4, 1971, pp. 100–109.
4. C. S. Aitchison, Konzentrierte Elemente für Mikrowellenfrequenzen, *Philips techn. Rdsch.*, vol. 32, no. 9, 1971/72, pp. 326–335.
5. H. M. Greenhouse, Design of planar rectangular microelectronic inductors, *IEEE Trans. Parts, Hybrids, and Packaging*, vol. PHP-30, no. 2, 1974, pp. 101–109.
6. P.-A-R. Holder, Some examples of slotline and coplanar waveguide circuits, in: *Colloquium on Microwave Integrated Circuits*, London, 24 May 1974, pp. 8/1–8/2.
7. M. Houdart, C. Aury, and Frederic.-A. Jean, Coplanar lines: Application to lumped and semi-lumped microwave integrated circuits, in: *Proceedings, 7th European Microwave Conference*, Bella Center, Copenhagen, Denmark, 5–8 Sept. 1977, pp. 450–454.
8. Y.-K. Jain and S.-K. Varshney, Properties of coplanar type MIS-SIM structure chip capacitor, *Electrocomponent Sci. Technol.*, vol. 7, no. 4, 1981, pp. 227–228.
9. S. A. Jamison, A. Podell, M. Helix, P. Ng, and C. Dhao, Inductively coupled push–pull amplifiers for low cost monolithic microwave ICs, in: *IEEE GaAs IC Symposium Digest*, 1982, pp. 91–93.
10. R. Esfandiari, D. W. Maki, and M. Siracusa, Design of interdigitated capacitors and their application to Gallium Asenide monolithic filters, *IEEE Trans. Microwave Theory Tech.*, vol. MTT-31, no. 1, 1983, pp. 57–64.
11. R. H. Jansen, L. Wiemer, H. J. Finlay, J. R. Suffolk, B. D. Roberts, and R. S. Pengelly, Theoretical and experimental broadband characterisation of multiturn square spiral inductors in sandwich type GaAs MMICs, in: *Proc. 14th European Microwave Conf.* 1984, pp. 946–951.
12. D. Ferguson, P. Bauhahn, J. Keuper, R. Lokken, J. Culp, C. Chao, and A. Podell, Transformer coupled high-density circuit technique for MMIC, in: *1984 IEEE MTT-S International Microwave Symposium Digest*, 1984, pp. 34–36.
13. L. Wiemer, R. H. Jansen, I. D. Robertson, and J. B. Swift, Computer simulation and experimental investigation of spiral transformers for MMIC applications, in: *IEE Colloquium on Computer Aided Design of Microwave Circuits Digest*, no. 99, 1985, pp. 2/1–2/5.
14. E. Pettenpaul, H. Kapusta, A. Weisgerber, H. Mampe, J. Luginsland, and I. Wolff, CAD models of lumped elements on GaAs up to 18 GHz, *IEEE Trans. Microwave Theory Tech.*, vol. 36, Feb. 1988, pp. 294–304.
15. G. Kibuuka, R. Bertenburg, M. Naghed, and I. Wolff, Coplanar lumped elements and their application in filters on ceramic and Gallium Arsenide substrates, in: *Proceedings, 19th European Microwave Conference*, 1989, pp. 656–661.
16. W. Wertgen, *Elektrodynamische Analyse geometrisch komplexer (M)MIC-Strukturen mit effizienten numerischen Strategien*, Doctoral Thesis, Duisburg University, 1989.
17. L. Wiemer, *Interdigitated and Spiral Components in Planar Technology in Monolithic Integrated Microwave Circuits*, Doctoral Thesis, Duisburg University, Duisburg, Germany, 1989.

18. E. Frlan, S. Meszaros, M. Cuhaci, and J. S. Wight, Computer aided design of square spiral transformers and inductors, in: *1989 IEEE MTT-S International Microwave Symposium Digest*, 1989, pp. 661–664.
19. M. Rittweger and I. Wolff, Analysis of complex passive (M)MIC-components using the finite difference time-domain approach, in: *1990 IEEE MTT-S International Microwave Symposium Digest*, 1990, pp. 1147–1150.
20. G. Kibuuka, *Computation of Lumped* and *Semi-Lumped Elements in Microstrip and Coplanar Technique Based on Spectral Domain Analysis of Planar Lines*, Doctoral Thesis, Duisburg University, Duisburg, Germany, 1990.
21. A. R. Djordjevic, C. K. Allen, T. K. Sarkar, and Z. A. Maricvic, Inductance of perfectly conducting foils including spiral inductors, *IEEE Trans. Microwave Theory Tech.*, vol. 38, no. 10, 1990, pp. 1407–1414.
22. M. Naghed and I. Wolff, A three-dimensional finite-difference calculation of equivalent capacitances of coplanar waveguide discontinuities, in: *1990 IEEE MTT-S International Microwave Symposium Digest*, pp. 1143–1145.
23. B. Roth, R. Tempel, W. Hui, and A. Beyer, A MMIC for the construction of nearly arbitrary microwave oscillators, in: *Proceedings, 21st European Microwave Conference*, 1991, pp. 172–177.
24. T. Becks and I. Wolff, Calculation of three-dimensional passive structures including bond-wires, via-holes and air-bridges using the spectral domain analysis method, in: *Proceedings, 21st European Microwave Conference*, 1991, pp. 571–576.
25. J. Borkes, M. Naghed, and I. Wolff, Measurement and analysis of coplanar MMIC four-port spiral transformers, in: *Proceedings, 21st European Microwave Conference*, 1991, pp. 1023–1028.
26. N. I. Dib, L. P. Katehi, G. E. Ponchak, and R. N. Simons, Theoretical and experimental characterisation of coplanar waveguide discontinuities for filter applications, *IEEE Trans. Microwave Theory Tech.*, vol. 39, no. 5, 1991, pp. 873–882.
27. M. Naghed, *Analyse koplanarer Mikrowellenstrukturen mit der Methode der quasistatischen Finiten Differenzen*, Doctoral Thesis, Duisburg University, Duisburg, Germany, 1992.
28. C.-W Chiu and R.-B. Wu, A moment method analysis for coplanar waveguide discontinuity inductances. *IEEE Trans. Microwave Theory Tech.*, vol. 41, 1993, no. 9, pp. 1511–1514.
29. K. Beilenhoff, H. Klingbeil, W. Heinrich, and H.-L. Hartnagel, Open and short circuits in coplanar MMIC's, *IEEE Trans. Microwave Theory Tech.*, vol. 41, no. 9, 1993, pp. 1534–1537.
30. H. Jin and R. Vahldieck, Full-wave analysis of coplanar waveguide discontinuities using the frequency domain TLM method, *IEEE Trans. Microwave Theory Tech.*, vol. 41, no. 9, 1993, pp. 1538–1542.
31. N.-I. Dib, M. Gupta, G.-E. Ponchak, and L.-P-B. Katehi, Characterization of asymmetric coplanar waveguide discontinuities, *IEEE Trans. Microwave Theory Tech.*, vol. 41, no. 9, 1993, pp. 1549–1558.
32. A.-A. Omar and Y.-L. Chow, Coplanar waveguide with top and bottom shields in place of air-bridges, *IEEE Trans. Microwave Theory Tech.*, vol. 41, no. 9, 1993, pp. 1559–1563.

33. P. Pogatzki, R. Kulke, T. Sporkmann, D. Köther, R. Tempel, and I. Wolff, A comprehensive evaluation of quasi-static 3D-FD calculation for more than 14 CPW structures—Lines, discontinuities and lumped elements, in: *1994 IEEE MTT-S International Microwave Symp. Digest*, vol. 2, pp. 1289–1292.

34. P. Pogatzki and O. Kramer, A coplanar element library for the accurate CAD of (M)MICs, *Microwave Engineering Europe*, 1994, pp. 41–46.

35. P. Pogatzki, D. Köther, R. Kulke, S. Sporkmann, and I. Wolff, Coplanar hybrids based on an enhanced inductor model for mixer applications up to 67 GHz, in: *Proceedings, European Microwave Conference*, Cannes, 5–8. Sept. 1994, vol. 1, pp. 254–257.

36. Y. Kalayci, R. Tempel, B. Hopf, J.-J. Borkes, R. Gründler, and I. Wolff, Miniaturizing of K-band coplanar MMIC-amplifiers by using lumped elements, in: *Proceedings, 24th European Microwave Conference*, vol 1, Cannes, Sept. 5–8, 1994, vol. 1, pp. 343–348.

37. A.-C. Reyes, Ghazaly.-S.-M. El, S.-J. Dorn, M. Dydyk, D.-K. Schroder, and E. Patterson, Coplanar waveguides and microwave inductors on silicon substrates, *IEEE Trans. Microwave Theory Tech.*, vol. 43, no. 9, 1995, Part I, pp. 2016–2022.

38. K. Beilenhoff, W. Heinrich, and H.-L. Hartnagel, Analysis of MIM series capacitances for coplanar MMICs, in: *MIOP 95, Mikrowellen und Optronik, 8. Kongreßmesse für Hochfrequenztechnik*, Sindelfingen, Germany, 30 May–1 June, 1995, pp. 124–128.

39. C.-Y. Chi and G.-M. Rebeiz, Planar microwave and millimeter-wave lumped elements and coupled-line filters using micro-machining techniques, *IEEE Trans. Microwave Theory Tech.*, vol. 43, no. 4, Part I, 1995, pp. 730–738.

40. G.-M. Shau, K.-C. Hwann, and H.-C. Chun, Modeling of lumped-element coplanar-stripline low-pass filter, *IEEE Microwave Guided Wave Lett.*, vol. 8, no. 3, 1998, pp. 141–143.

41. A. Bessemoulin, M. Sedler, H. Massler, W.-H. Haydl, D. Geiger, H. Brugger, P. Quentin, and M. Schlechtweg, A complete coplanar element library in commercially available foundry process for millimeter-wave integrated circuit design, in: *European Microwave Week 2000, 30th European Microwave Conf., GAAS 2000, European Conference on Wireless Technol. 2000*, Conference Proceedings, Paris, F, Oct. 2–6, 2000 (in CD-ROM).

42. I. Bahl, *Lumped Elements for Microwave Circuits*, Boston: Artech House, 2003.

5

COPLANAR ELEMENT LIBRARY AND CIRCUIT DESIGN PROGRAM

5.1 INTRODUCTION

On the basis of the quasi-static analysis technique (as described in Section 2.2 and Chapters 3 and 4) a coplanar circuit design tool *Coplan for ADS*™ that forms a state-of-the-art software for the simulation and the design of CPW circuits has been developed. This library offers the complete set of coplanar elements such as transmission lines, discontinuities, junctions, and lumped elements for the design and fabrication of (M)MICs that have been described in the previous chapters. A CD with a test version of the program has been attached at the end of the book.

Although there are many encouraging activities in CPW design, this technique has not yet achieved the real breakthrough. One of the reasons for this has been the lack of a complete, accurate CAD-oriented CPW element library. With *Coplan for ADS*™ such a library is now available in the environment *Agilent ADS*™. At present, there are four possible techniques for the design of coplanar circuits. Figure 5.1.1 shows the comparison between these techniques due to the cost and time consumption as well as accuracy and validity aspects.

Applying the measurement-based modeling, a large number of test structures have to be realized and measured. The measured data are then used for the modeling of these structures. Realization and measurement of the test structures takes a long time and leads to high costs. This technique, however,

Coplanar Microwave Integrated Circuits, by Ingo Wolff.

	Measurement-Based Modeling	Analytical Approximation	Full-wave EM Analysts	*COPLAN* 3D-FDM Analysis
Costs	High	Low	Low	Low
Time	High	Low	High	Low
Accuracy	High	Low	High	High
Limitations	High	High	Low	Low

Fig. 5.1.1. Coplanar design techniques.

has the advantage of high accuracy, but this accuracy is limited to those dimensions and substrate parameters, used in the measurements. Additionally, the results are valid only for the frequency range in which the measurements have been performed.

A second method that is often used for modeling CPW circuits is the application of models based on analytical approximating formulas. However, compared to microstrip line techniques, there are only a small number of such models available for coplanar structures and components.

An alternative method is the application of EM-field solvers, which overcome the problem of cost and limitation of validity range. But, the application of such tools is very complicated and computer time-consuming. As a result, the design of large circuits and the optimization of circuit performances normally is not possible.

Coplan for ADS™ enables the designer to analyze and optimize the coplanar circuits in a relatively short time. There is no limit due to the structure dimensions, and the results are very accurate for a wide frequency range. Figure 5.1.2 shows the features of *Coplan for ADS*™ at a glance.

A broad spectrum of coplanar elements as they have been discussed in the previous chapters is implemented. Elements from single and multiple coupled coplanar lines over discontinuities like a waveguide step, gap and bend up to the three and four port junction elements with and without air bridges are available. To complete this library, also lumped elements such as rectangular spiral inductors, interdigital capacitors, MIM capacitors and thin film resistors are also included in the library. Besides the accurate simulation of such elements, the schematic circuit entry and a multilevel layout generation are completely supported.

The kernel of this state-of-the-art library is based on the field-theoretical modeling of coplanar structures using the quasi-static finite difference technique that is described in detail in Section 2.2 and Chapters 3 and 4. The

- ***Complete Library of Elements***
- ***Fast, Accurate 3D-FD Field Simulation***
- ***Fully Integrated in Agilents ADS™***
- ***Automatic CPW-Layout Generation***
- ***Verified up to 67 GHz (to 205 GHz recently by CSIRO, Australia)***
- ***Tuning/Optimization***
- ***Suitable for Many Applications***

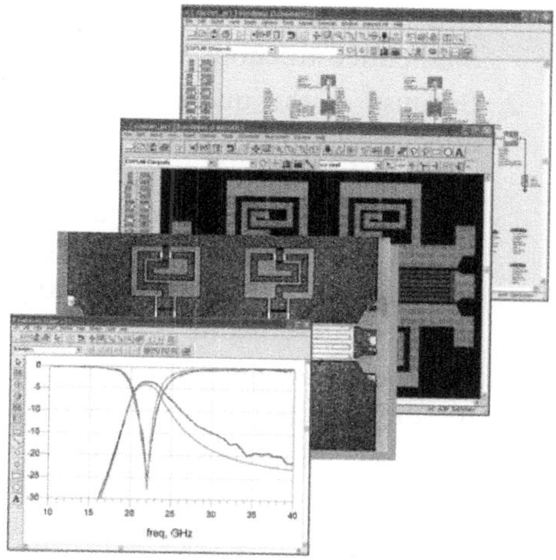

Fig. 5.1.2. Features of *Coplan for ADS*™ at a glance.

method is applied to each of the elements in such a way that a parametric description (equivalent circuit) of the elements results. Thus, the actual implementation of the numerical calculations also allows circuit optimization, for instance.

To make the utilization of *COPLAN for ADS*™ more efficient, a smart cache memory speeds up optimizations and statistical analysis. The implemented CACHE works as follows: The first numerical analysis of a CPW element completely applies the finite difference field analysis algorithms. In the second run, the CACHE management identifies already analyzed structures and makes available these previously calculated results. Therefore, optimization of CPW circuits, including numerical CPW elements, becomes possible in a matter of seconds.

It can be concluded that the CPW elements of *Coplan for ADS*™ can be used equivalently to the well-known microstrip elements in *Agilent ADS*™ software. The complete layout generation of CPW designs is supported up to the level of foundry requirements. Thus, using *Coplan for ADS*™, it becomes possible to accurately design CPW circuits with only a small manual effort.

The following instructions help to apply the software as efficiently as possible.

1. *Coplan for ADS*™ utilizes a numerical field calculation. Therefore, the structures to be calculated are discretized using an automatically generated mesh. The structure dimensions have to fit into the mesh; otherwise the simulator changes incompatible dimensions (see C_GRID, Section 5.6.2.5) in order to adapt them to the mesh. As a result, if the changes of the structure dimensions are smaller than 0.25DL (with DL being the

grid size), these changes are not taken into account. It is therefore recommended to choose grid compatible dimensions for the structure.

2. Because of the quasi-static nature of the applied method, dispersion effects are not taken into account. On the other hand, these effects are negligible in case of coplanar structures provided that the structure dimensions are small compared to the substrate height (see also Section 2.2). As an example, the dispersion of the effective dielectric constant of a coplanar line with a ground-to-ground spacing of 625 μm on a 250-μm-thick GaAs substrate is less than 3% from 1 to 40 GHz.
3. The presented method assumes a pure TEM-wave propagation as a basis for the calculations. This means that longitudinal field and transversal current components are not considered. As a result, higher-order modes are not taken into account. In case of coplanar lines, there are at least two modes that can be excited: the even mode and the odd mode (see also Chapter 1 and Section 2.1). The even mode is normally the basis for coplanar microwave circuits, and therefore only this mode will be taken into account in the simulation. The excitation of other modes must be avoided. The coplanar odd mode may be excited at discontinuities or other asymmetrical structures. It can be suppressed using air-bridge structures (compare Section 3.5.5).
4. If the backside of the substrate is metalized (conductor-backed coplanar line), the coplanar even mode is partially mixed with a so-called parasitic microstrip mode (compare also Sections 2.2.8 and 2.3.2.4). The influence of this microstrip mode on the structure characteristics is considered by *COPLAN*. However, the dispersion effect due to this mode is not taken into account. However, the parasitic microstrip mode has no considerable effect if the ground-to-ground space is smaller than the substrate height.
5. Since each element is simulated separately, only the coupling effects within the elements themselves are considered. The interaction between the elements that are closely connected together cannot be taken into account. The user should take care of a proper distance between the elements in order to avoid undesired couplings.

5.2 MODELING, CONVERGENCE, AND ACCURACY

The integration of the coplanar library into the *Agilent ADS*™ software and the intensive use of these programs makes it necessary to investigate some properties of the simulator, such as the numerical stability, accuracy setting, and convergence behavior. On the one hand, this will help the user to minimize computation time and to optimize the numerical results. On the other hand, series of simulations will be used for an automatic determination of some model parameters, like the discretization or the iteration boundary. The following calculations are made for a GaAs substrate ($\varepsilon_r = 12.9$) with a thickness of 450 μm. To explain the investigations, the model of the coplanar line that has the advantage of a simple geometry was used. In addition, the C_LIN model was implemented with ground strips that are not in contact with the electrical sidewalls. This gives the opportunity to simulate coplanar lines with finite ground strips (see Section 5.6.2.2, C_LIN, Notes) and to investigate the influence of the electrical sidewalls (see Section 5.6.2.5, C_GRID, Notes) while increasing the distance to the CPW configuration.

Figure 5.2.1a depicts the geometrical configuration of the C_LIN model. Three dielectric layers are put in a box with electrical walls on the left and right side and at the top and the bottom. The bounded region is divided into

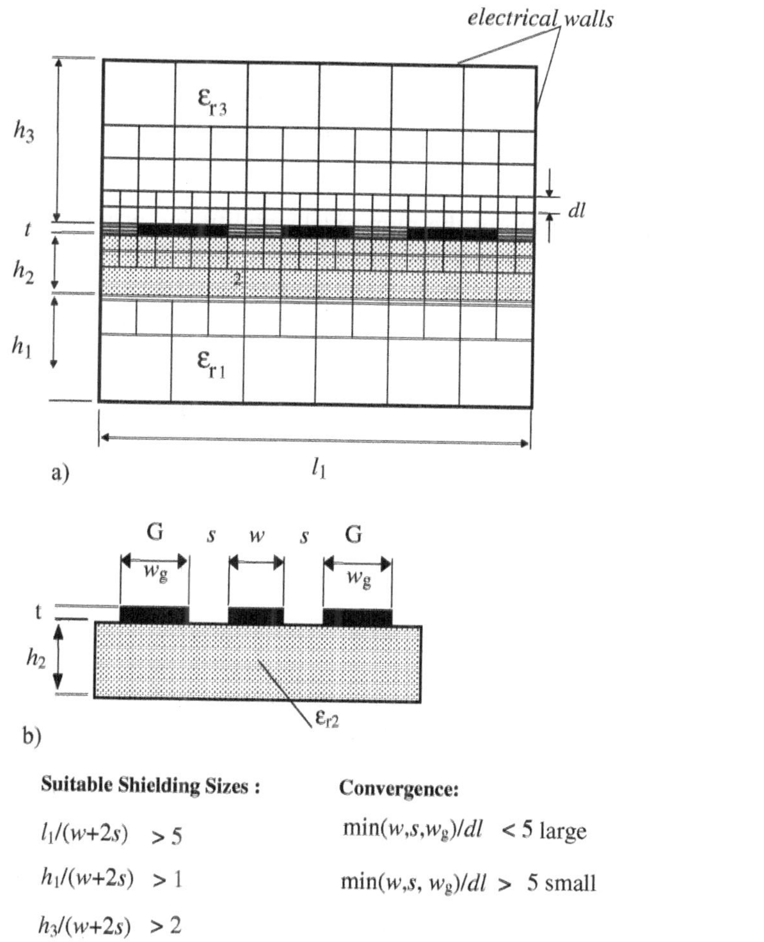

Fig. 5.2.1. (a) Applied discretization (two-dimensional grid). (b) Parameter setting for the calculation of coplanar lines.

elementary cells using a nonequidistant Cartesian grid (smallest cell size = DL). The user has to choose the grid dimension so that the grid crossing points are positioned on the edges of the metallization. Otherwise, the simulator adapts the line dimensions to the grid by changing the line dimensions (warnings are given!).

The calculations of very narrow and wide coplanar lines have been evaluated to study the convergence behavior of the simulator. A relative error that is always lower than 1% can be obtained if the predefined iteration error BOUND is 1×10^{-5} and the cell size DL is smaller than the smallest line dimension divided by 10. For a fast simulation run or a large number of calculations (e.g., during an optimization), a coarse discretization and a higher iteration error (1×10^{-3} or 1×10^{-4}) should be used. The following two statements describe the relation of the discretization size and the iteration error in a basic formulation:

1. A reduction of the iteration error does not lead to a better simulation result if a coarse discretization size is used.
2. A large iteration error leads to a poor calculation result even if a very small discretization size is used.

For an investigation of the numerical stability and accuracy of the simulator, several combinations of discretization sizes and iteration errors are chosen for the calculation of a CPW line. The relationship between discretization size DL and the iteration error BOUND is given in Table 5.2.1.

A coplanar line configuration with the following parameters has been chosen to ensure that the grid crossings are always on the edges of the metalization: $h_1 = 0\,\mu\text{m}$, $h_2 = 450\,\mu\text{m}$, $h_3 = 900\,\mu\text{m}$, $\varepsilon_{r1} = 1$, $\varepsilon_{r2} = 12.9$, $\varepsilon_{r3} = 1$, $w = 45\,\mu\text{m}$, $s = 30\,\mu\text{m}$, $l_1 = 1000\,\mu\text{m}$ with a metalization of gold having a thickness of $t = 3\,\mu\text{m}$. The results are plotted in Fig. 5.2.2. The discretization setting $dl = 15\,\mu\text{m}$ (= variable DL in the program) results from the smallest geometrical dimension (here: $s = 30\,\mu\text{m}$) divided by 2. With increasing accuracy DL decreases with a factor of 2 and the iteration error with a factor of 10 (see Table 5.2.1).

Figure 5.2.2 depicts the calculation of the characteristic impedance and its relative error as a function of the accuracy. The analysis error decreases from about 4% to 0.1% if it is assumed that the best performance for the combination no. 6 (Table 5.2.1) is the exact solution. A similar behavior can be seen

TABLE 5.2.1. Examples for the Combination of the Iteration Error BOUND and the Cell Size *dl* for Analyzing Coplanar Waveguides

Example Number:	1	2	3	4	5	6
dl (μm):	15	7.5	3.75	1.875	0.9375	0.46875
BOUND:	1×10^{-3}	1×10^{-4}	1×10^{-5}	1×10^{-6}	1×10^{-7}	1×10^{-8}

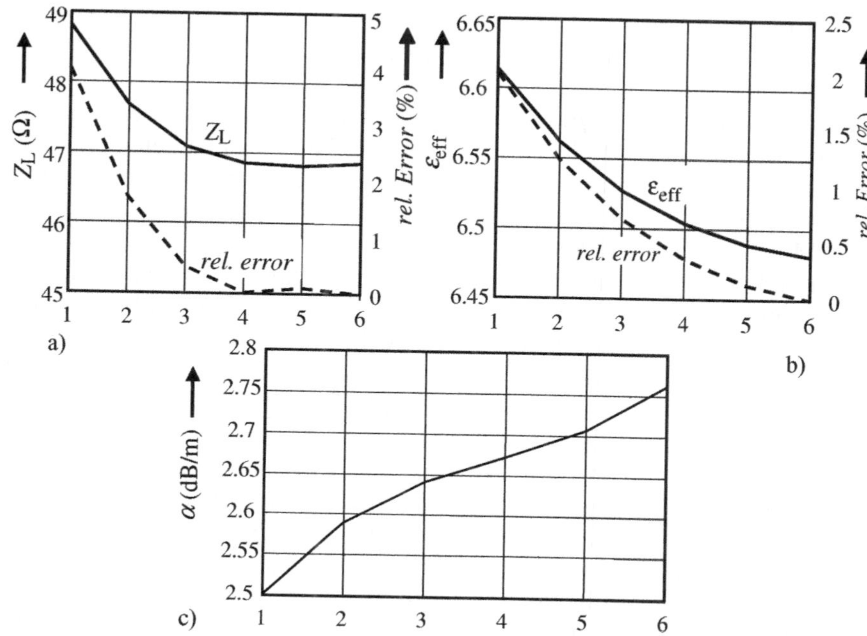

Fig. 5.2.2. (a) Characteristic impedance Z_L. (b) Effective dielectric constant ε_{eff}. (c) Conductor losses α as a function of the accuracy settings for examples 1 to 6 given in Table 5.2.1.

in Fig. 5.2.2b for the effective permittivity. The relative error decreases from about 2% down to 0.1%.

These results prove that even for example no. 6 the simulator is numerically stable, since the calculation time rises from a few seconds up to hours for the accuracy setting no. 6. Only accuracy settings from no. 1 to no. 4 are useful in circuit design. Figure 5.2.2c depicts the behavior of the conductor losses α as a function of an increasing accuracy requirement. No convergence can be observed. This results from the current density distribution that is infinite at the edges of the metalization strips.

5.3 OVERVIEW ON *COPLAN FOR ADS*™

As described in Section 5.1, *Coplan for ADS*™ is a numerical simulation tool that is implemented as a library of coplanar elements into the environment of *Agilent ADS*™ software. All elements are calculated as real 3D-elements using field-theoretical calculations, and the numerical results are used for the extraction of an equivalent circuit.

Since the numerical computations are performed in the background, the user does not need to take care for the field calculations and can handle the library similar to other libraries (like, for example, Microstrip Elements) within *ADS*™. The implementation of *Coplan for ADS*™ in *Agilent ADS*™ is shown in Fig. 5.3.1.

After installation of *Coplan for ADS*™, circuits can be designed within *schematic window* of *ADS*™ using the coplanar data items and elements that are available in the library palette. During a frequency-independent pre-analysis the equivalent circuit parameters for each element are calculated by the FDM-solver (compare Section 2.2 and Chapters 3 and 4).

The number of iterations, the residual error, and the name of the model are displayed inside the *ADS*™ *Status Window*. Input parameters and the results (equivalent circuit parameters) are stored by the cache management in a data base file. During the simulator start-up, this file is loaded into memory and works like a cache. Based on these results, a frequency-dependent analysis is started and the *S*-parameters are calculated.

Coplan for ADS™ consists of two palettes of coplanar elements and a palette of special data items. Each element represents a specific coplanar structure like T-junction, bend, spiral inductor, and so on. Special data items are introduced in order to reduce the number of input parameters for each model. These data items have a function similar to that of the well-known data items such as the MSUB item of the microstrip library. The list of implemented data items and coplanar elements as well as their use will be described below.

The *COPLAN for ADS*™ software can be accessed by selecting the palette "*Coplanar Elements*" in the schematic and layout windows, as shown in Fig. 5.3.2.

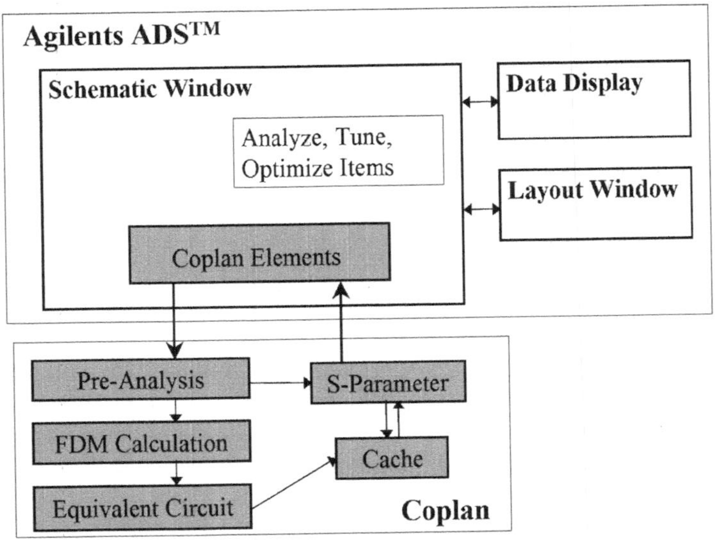

Fig. 5.3.1. Implementation of *Coplan for ADS*™ in *Agilent ADS*™.

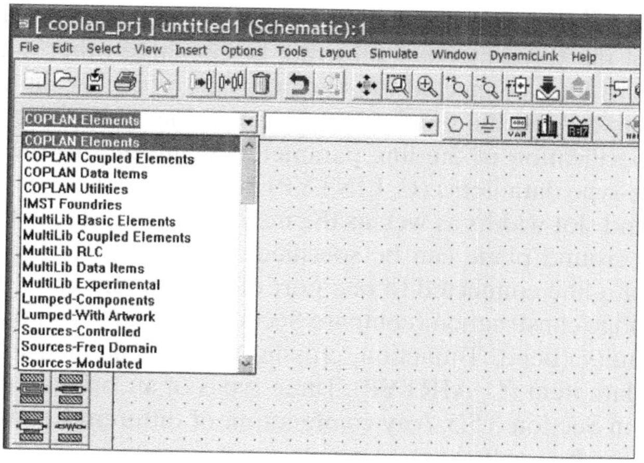

Fig. 5.3.2. Selection of *COPLAN for ADS*™ in *Schematic Window*.

TABLE 5.3.1. Coplanar Data Items Used in *Coplan for ADS*™

Data Items	Description
C_SUB	Substrate definition for coplanar structures
C_GRID	Definition of grid and shielding sizes for the EM finite difference approach
C_LINTYP	Definition of cross-sectional dimensions of coplanar lines
C_NL_TYP	Definition of cross-sectional dimensions of coplanar coupled lines (up to 10)
C_AIRTYP	Definition of air-bridge parameters
C_PROCES	Foundry selection for layout generation and process-related simulation
C_TECH	Definition of technological data for a selected foundry
C_LAYER	Definition of layer data for a selected foundry

5.3.1 Data Items

Coplan for ADS™ has eight special data items (see Table 5.3.1) that are available in the palette "*Coplanar Data Items.*" These items are described in detail in Section 5.6 (Coplanar Data Items).

One of the problems during the implementation of the CPW-Library into *Agilent ADS*™ environment is that a very large number of material and geometrical parameters is necessary for the definition of the coplanar elements such as junctions, lumped elements, and single or coupled lines. Furthermore, many of the necessary input parameters are common for a lot of components in a given design. To overcome this problem, data items have been defined. Common parameters are stored in data items and can be referred to by all used modules.

Besides the substrate data item (C-SUB), which specifies the material parameters of the three possible substrate layers, three other data items for the definition of lines connected to the component ports, for the definition of air bridges, and for specific simulation control parameters are defined.

For the specification of the line parameters connected to the component ports, the line-type data items (C_LINTYP) are used. In these data items, line parameters and slot widths as well as the metal level used for both the center line and the ground plane can be specified. Each data item represents the parameters of a line connected to one port of a circuit element.

The air-bridge construction (compare Section 3.5.5) at each port of a coplanar discontinuity (bend, T-junction, crossing) can be defined using the air-bridge-type-data item (C_AIRTYP). Three types of air bridges are available, as described in Section 3.5.5. Any combination of different line types and air bridge types can be applied.

The parameters for controlling the numerical simulation process are stored in the grid data item (C_GRID). The box and grid sizes as well as the error boundary for the iteration procedure can be predefined using this data item.

Additionally, the library supports a process-related layout generation. All necessary information (oversize, layer configurations, etc.) is stored in two additional data items. Technological data such as the material parameters of the applied layers (dielectric constants, loss factors, and resistivity) and the layer height are defined in the C_TECH data item. A second data item C_LAYER is intended for layer data such as layer number and the specific oversize of each layer. In case of an available DEFAULT-foundry, these data items are identified as C_TECH_DEFAULT and C_LAYER_DEFAULT. The DEFAULT-foundry is a standard feature of *Coplan for ADS*™ with predefined parameters. Several of these parameters can be readjusted by the user to the actually used process. Additionally, other foundry parameters can be implemented into *Coplan for ADS*™ as a special service.

Coplan for ADS™ supports the use of various foundries and processes in parallel. In order to specify the actually used foundry, the name of the foundry (for example: DEFAULT) has to be selected in C_PROCES data item.

Using the data items, the entry of the component and process parameters is considerably simplified. Another advantage of the data items is that connected components always have compatible line parameters by selecting the correct line type at their ports.

Figure 5.3.3 shows how the simulator and the layout tool use the coplanar data items. As can be observed from the figure, some of the data items like C_SUB or C_LAYER have an effect only on the simulation area or only on the layout generation, while some items have an effect on both the simulation process and the layout generation. The access to the palette of "Coplanar Data Items" is shown in Fig. 5.3.4.

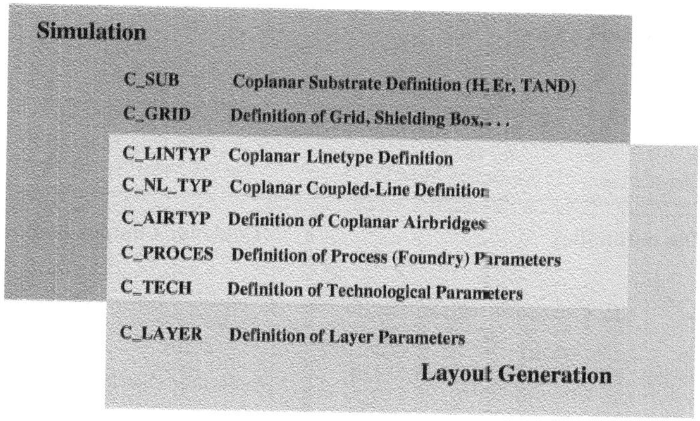

Fig. 5.3.3. Coplanar Data Items as used during the simulation and layout generation.

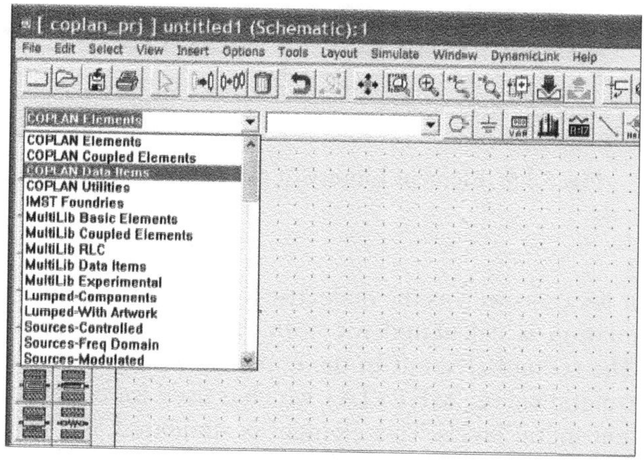

Fig. 5.3.4. Selection of palette "Coplanar Data Items" in *Schematic Window.*

5.3.2 Library Elements

The coplanar elements within *Coplan for ADS*™ are separated into two palettes: "*Coplanar Elements*" and "*Coplanar Coupled Elements.*" The complete list of elements available in these two palettes is shown in Fig. 5.3.5. The library of *Coplan for ADS*™ contains three types of elements: coplanar transmission lines, coplanar discontinuities and coplanar lumped elements. In the group of transmission lines, there is a coplanar transmission line definition (C_LIN) with arbitrary geometrical dimensions and substrate parameters. The effect of finite ground planes as well as of backside metalization is considered.

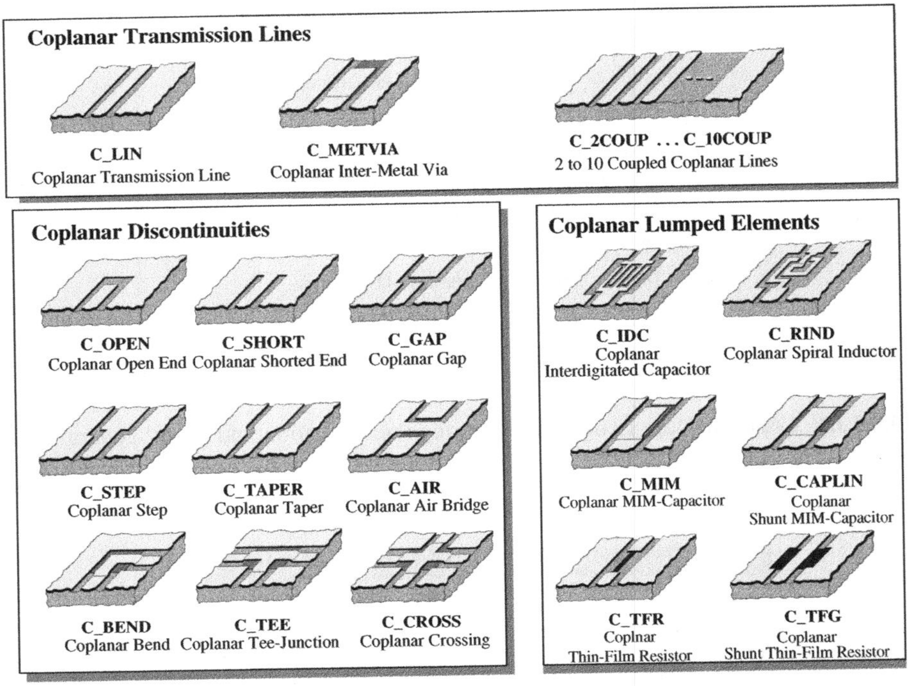

Fig. 5.3.5. Element Library of *Coplan for ADS*™.

The conductor losses as well as the dielectric losses are taken into account and the line attenuation is calculated for both DC and RF currents (compare Section 2.2). The second element in the coplanar waveguide spectrum is the coplanar inter-metal via (C_METVIA), which is used for connecting coplanar lines in different metal levels. This element is simulated as a coplanar transmission line with a conductor thickness that is the sum of the metalization thickness of the two utilized metal levels. C_LINE and C_METVIA are available in the palette "*Coplanar Elements*."

A set of coupled coplanar lines with 2–10 coupled center lines (C_2COUP ... C_10COUP) and ground planes of finite width are also implemented into *Coplan for ADS*™ and can be used for the realization of couplers or filters (compare also Chapter 6). The matrices of the coupling parameters are calculated using the quasi-static finite difference method, and the scattering parameters are calculated for the specified lengths of the coupled lines (for the analysis method compare Section 2.2.11). These elements are available in the palette "*Coplanar Coupled Elements*."

The group of discontinuities that can be simulated using COPLAN consists of discontinuities that are needed for the design of modern (M)MICs. From one-port coplanar open lines (C_OPEN) and short-circuited lines (C_SHORT) over two-port elements like coplanar gaps (C_GAP), coplanar steps (C_STEP), coplanar tapers (C_TAPER), and coplanar air bridges (C_AIR) to more complicated junction like components as coplanar bends (C_BEND), coplanar T-junctions (C_TEE), and coplanar cross-junctions (C_CROSS) are all elements needed for an exact circuit design available. All junctions can be combined in an arbitrary arrangement and combination with feed lines and the available air bridge types (compare Section 3.5.5). These elements are available in the palette "*Coplanar Elements.*"

The group of lumped elements which can be simulated with *COPLAN* contains the following elements: the coplanar interdigital capacitor (C_IDC), the coplanar rectangular spiral inductor (C_RIND), two types of coplanar thin-film resistors (C_TFR and C_TFG), and two types of coplanar MIM-capacitors (C_MIM and C_CAPLIN). These elements are also available in the palette "*Coplanar Elements.*" A detailed description of all elements and their application rules is given in Section 5.7.

5.4 CACHE MANAGEMENT

A smart cache memory management is implemented in order to speed up the statistical analysis and optimization process. The calculated parameters of coplanar elements are stored in cache during the first analysis run and will then be actualized if the structure data are changed. The calculated parameters are stored in look-up tables in binary format.

The look-up table files are named "cpw_element.cdb," where "*element*" is the name of the corresponding coplanar element (for example C_LIN, C_TEE). Each file contains the input parameters and the output equivalent circuit parameters of all coplanar elements of equal kind in the circuit as well as the characteristic line parameters of the coplanar lines connected at the ports of these elements. The user can define an environment variable in order to store the look-up table files. If this variable is not set, the look-up table files will be written into the TEMP directory.

In order to enable the user to have an access to the calculated equivalent circuit parameters, a C_DEBUG element is implemented. This element activates flags that force the simulator to create ASCII-Files containing the input and output parameters of all elements used in the circuit. The use of this element and the interpretation of its output are described in Section 5.6. This element is available in a separate palette named "COPLAN-Utilities." Figure 5.4.1 shows how to access this palette.

5.5 LAYOUT

Coplan for ADS™ runs an automatic layout generation feature. The layout of a designed coplanar circuit is generated using the circuit element information and taking into account the process-related information (oversizes, minimum

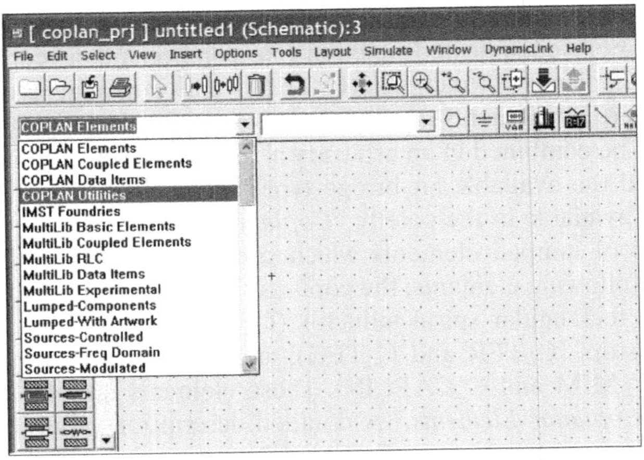

Fig. 5.4.1 Selection of palette "COPLAN-Utilities" in *Schematic Window*.

layer dimensions, etc.) of the selected foundry. A DEFAULT foundry is implemented in the standard packet that can be used for demonstration of the layout feature. Other customized foundries can be implemented additionally.

Detailed information about layout handling within *ADS*™ can be found in the corresponding *ADS*™ handbooks. The layout description of each coplanar element is given in Section 5.7.

Because the width of the ground planes of all coplanar elements can be selected separately, an overlapping of some ground sections may result. To avoid this problem, a default ground width (GW_DEF) can be defined in the C_PROCES data item. A negative value for GW_DEF activates the ground width selected by each element.

5.6 COPLANAR DATA ITEMS

5.6.1 Overview

One of the problems in the direct parameterization of CPW elements is the large number of input parameters. For some elements like the coplanar T-junctions, more than 30 parameters are necessary to define all the geometrical dimensions. A solution to reduce the problem is the use of special data items (see Fig. 5.6.1).

Coplan for ADS™ uses several data items to set up geometry data (C_LINTYP, C_NL_TYP and C_AIRTYP), substrate data (C_SUB), and process data (C_PROCES) as well as simulation control parameters (C_GRID). All data item names defined by *Coplan for ADS*™ start with "C_."

Using *Agilent ADS*™ software, some data items such as the MSUB item (definition of a microstrip substrate) are well known to the user. The C_SUB

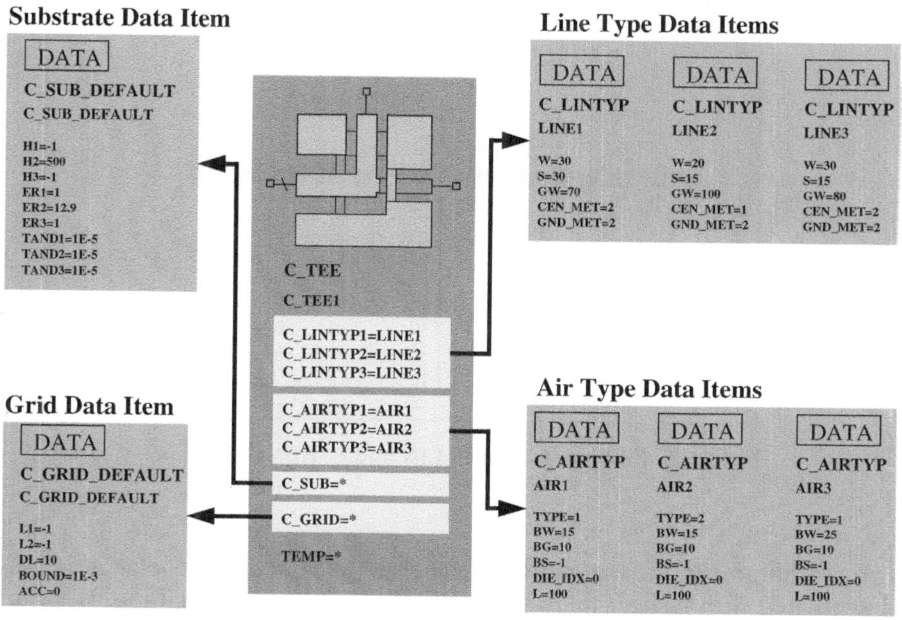

Fig. 5.6.1 Definition of parameters for a coplanar T-junction using coplanar data items.

item defines the substrate parameters of the coplanar circuit and corresponds to the MSUB data item.

A simple microstrip line is described only by its width. In case of a coplanar line, the cross section is described by the widths of the center line, the slot, and the ground plane. In order to reduce the number of input parameters for the elements, the cross section of the coplanar lines used in connection with the elements are defined in a special data item, called C_LINTYP.

Since *Coplan for ADS*™ uses an optimized numerical field solver for simulating the components, some simulation control information has to be passed to the simulator. The grid and control information is also stored in a special data item, called C_GRID. Because the grid parameters depend on the dimensions of the structure, they have to be defined for each component that is to be analyzed. However, in most cases only a small number of different C_GRID parameters are sufficient for a complete circuit design.

The same situation is given in connection with the air bridges. Most of the bridges have an equal structure. In *Coplan for ADS*™, components are defined with up to four air bridges to avoid an odd mode excitation. The introduction of the C_AIRTYP data item reduces the number of input parameters that are needed for describing the air bridges drastically.

An important feature is the process-related simulation and layout generation. During simulation, *Coplan for ADS*™ takes into account process-related information like minimal lengths, permittivity, loss of the dielectric layers, and

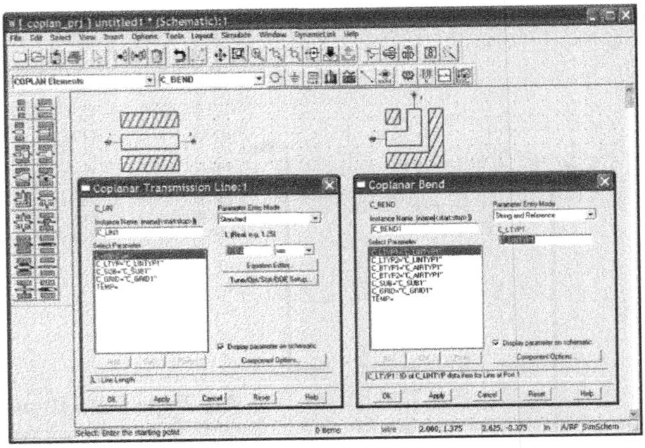

Fig. 5.6.2. Edit box for the C_LIN and C_BEND element.

so on. Additionally, the automatically generated layout of the coplanar circuit is generated with respect to a selected foundry. This means that the user can define oversize parameters, layer numbers and configurations etc. They are then used for the layout generation.

In order to have a very flexible user interface in defining the foundry data, three data items are introduced: the C_PROCES item, the C_TECH item and the C_LAYER item. *Coplan for ADS*™ is potentially able to support more than one foundry process. The name of the foundry being used for simulation and layout processing is specified in C_PROCES. The process data itself (layer configuration, permittivity etc.) are defined in the C_TECH and C_LAYER data items. The user can edit them in a similar way as for the other data items. In the standard package of *COPLAN*, a so-called *Default Foundry* is implemented. The user of a standard package cannot define a foundry (process) other than the default foundry. The user can edit the foundry data (layer configuration, permittivities etc.) however in a similar way as for other data items. A special service to implement an additional customized foundry into the *Coplan for ADS*™ software is available on request by IMST GmbH.[1]

5.6.2 Description of the Data Items

The coplanar data items used in *Coplan for ADS*™ are shown below.

In the following, the items briefly discussed above shall be described a little more in detail. It will be shown which parameters are defined, what the parameters mean, and what additional conditions must be obeyed.

[1] IMST GmbH, Carl-Friedrich-Gauß-Str. 2, D-47475 Kamp-Lintfort, Germany. contact@imst.de.

TABLE 5.6.2. Coplanar Data Items Used in *Coplan for ADS*™

Data Items	Description
C_SUB	Substrate definition for coplanar structures
C_GRID	Definition of grid and shielding sizes for finite difference approach
C_LINTYP	Definition of cross-sectional dimensions of coplanar line
C_NL_TYP	Definition of cross-sectional dimensions of coplanar coupled lines (up to 10)
C_AIRTYP	Definition of air-bridge parameters
C_PROCES	Foundry selection for layout generation and process-related simulation
C_TECH	Definition of technological data for selected foundry
C_LAYER	Definition of layer data for selected foundry

5.6.2.1 Coplanar Substrate Data Definitions C_SUB

Symbol:

C_DATA

Illustration:

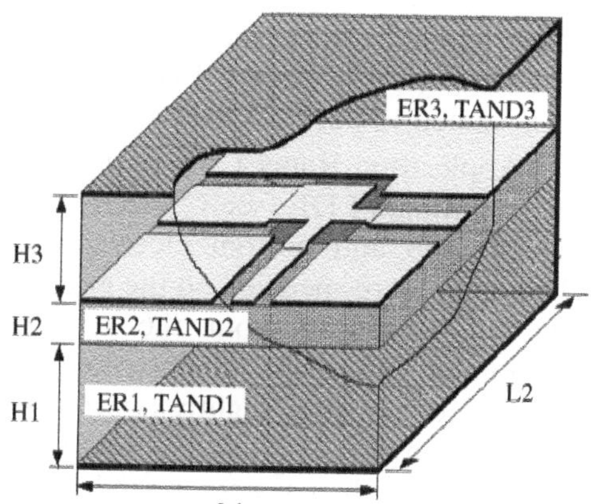

Fig. 5.6.3. The three-layer dielectric substrate structure.

Parameters:

L1	Width of the shielding box (L1 = –1, –2 → auto-sizing)
L2	Depth of the shielding box (L2 = –1, –2 → auto-sizing)
H1	Height of substrate layer 1 (H1 = –1, –2 → auto-sizing)
H2	Height of substrate layer 2

H3	Height of substrate layer 3 (H3 = –1, –2 → auto-sizing)
ER1	Relative permittivity of substrate layer 1
ER2	Relative permittivity of substrate layer 2
ER3	Relative permittivity of substrate layer 3
TAND1	Dielectric loss factor of substrate layer 1 (set to zero in the standard version)
TAND2	Dielectric loss factor of substrate layer 2
TAND3	Dielectric loss factor of substrate layer 3 (set to zero in the standard version)

Application Range[2]**:**

$$H1 = 4 \times n \times DL, \qquad n = 0, 1, 2, \ldots$$

$$H2 = n \times DL, \qquad n = 4, 5, 6, \ldots$$

$$H3 = (4 \times n + 3) \times DL, \quad n = 2, 3, \ldots$$

$$ER1 \geq 1$$

$$ER2 \geq 1$$

$$ER3 \geq 1$$

Notes:

1. C_SUB contains the parameters for a three-layer substrate. For each layer, the substrate height and dielectric constant as well as the loss tangent can be specified (in the standard version of the program, layer 1 and layer 3 are assumed to have TAND1 = 0 and TAND3 = 0).
2. As shown above (Application Range), the parameters H1, H2, and H3 must be an integer multiple of DL (see C_GRID for more information on DL), so that the top and bottom electric walls as well as the dielectric layer boundaries are specified on the grid. However, if the user specifies a maximum deviation from the grid (ACC in selected C_GRID), the simulator can adapt incompatible dimensions into the grid by changing these dimensions slightly (warnings will be reported). For ACC = 0 (no deviation from grid) the simulator reports an error and the next compatible value of the parameter will be suggested in the error message.
3. In case of coplanar lines and in order to keep the effects of the upper and the lower cover as small as possible, H1 and H3 should be as large as possible. *The simulator automatically adapts values for* H1 *and* H3, *if these parameters are set to –1 or –2 (auto-sizing).* A value of –1 for H1 and H3 is recommended.

[2] See C_GRID for more information on DL.

4. If the user likes to investigate the effects of the upper or lower cover on the structure characteristics, H1 and H3 can be specified manually. For such applications, the minimum values for H1 and H3 and their compatibility to the grid (Application Range) have to be considered.
5. For the simulation of a conductor-backed coplanar line (or to simulate the situation when the substrate is directly positioned on a metal—for example, during an on-wafer measurement), H1 = 0 has to be specified. The parameter H3 can be applied to simulate packaging effects.
6. ER1 and ER2 may be arbitrary values larger than or equal to 1. ER1 and ER3 are chosen to be 1 (air) for the simulation of simple coplanar lines.

5.6.2.2 Coplanar Line-Type Data Definition C_LINTYP

Symbol:

C_DATA

Illustration:

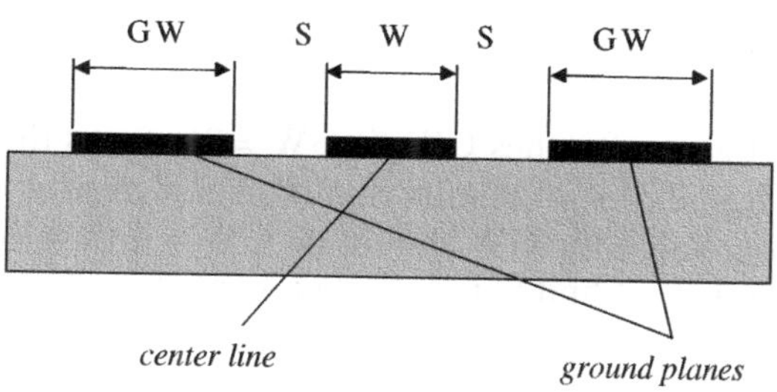

Fig. 5.6.4. Definition of the coplanar waveguide parameters.

Parameters:

W	Center conductor width
S	Slot width between center conductor and ground planes
GW	Width of ground planes (GW = 0 → infinite ground planes)
CEN_MET	Identification of metal level for the center line
GND_MET	Identification of metal level for the ground planes (used for layout only)

Application Range[3]**:**

$$W = 0.5 \times n \times DL, \qquad n = 2, 3, 4, \ldots$$

$$S = 0.5 \times n \times DL, \qquad n = 1, 2, 3, \ldots$$

[3] See C_GRID for more information on DL.

$$\text{GW} = 0.5 \times n \times \text{DL}, \quad n = 1, 2, 3, \ldots$$

$$\text{CEN\_MET} = 1 \text{ or } 2 \text{ (DEFAULT-Foundry)}$$

$$\text{GND\_MET} = 1 \text{ or } 2 \text{ (DEFAULT-Foundry)}$$

Notes:

1. The geometrical dimensions of a coplanar transmission line and the metal levels of the center line and ground planes are defined in the C_LINTYP data item.
2. The line parameters (W, S, and GW) must be chosen in such a way that the metalization corners lie on the grid (see Application Range). However, if the user specifies a maximum deviation from the grid size ACC (see C_GRID for details), the simulator can adapt the incompatible structure parameters into the grid by changing these parameters slightly (warnings will be reported). For ACC = 0 (no deviation from grid) the simulator reports an error and the next compatible value will be recommended in the error message. *Note that these changes are not considered in the automatic layout generation.*
3. If the parameter GW is set to 0, infinite ground planes are assumed. This means that the ground planes are connected to the electric walls at the right and left side of the shielding box, and they have the same electric potential as the shielding.
4. The width of ground planes GW (if GW $\neq$ 0) affects the line characteristics (see Fig. 2.2.24 for the effect on the characteristic impedance Z_L). The minimum value for GW should be GW = d (d = W + 2S) in order to keep the effect of finite ground planes small.
5. *Coplan for ADS*™ is able to handle more than one metal level. The number of metal levels depends on the currently selected foundry. Two different metal levels can be used in the case of the DEFAULT foundry.
6. If DEFAULT foundry is chosen, C_TECH_DEFAULT is selected. The integer value 1 for CEN_MET (or GND_MET) then indicates the metal layer 1 in C_TECH data item (parameters t1 and roh1) while the value 2 is used for the metal layer 2 in C_TECH data item (parameters t2 and roh2). Both levels may have different thickness and different resistivity that again are defined by the parameters t and roh in C_TECH.
7. In the present version of *Coplan for ADS*™, GND_MET is used for layout generation *only*. For simulation, GND_MET is set always equal to CEN_MET.

5.6.2.3 Coplanar Coupled Lines Data Definition C_NL_TYP

Symbol:

C_DATA

Illustration:

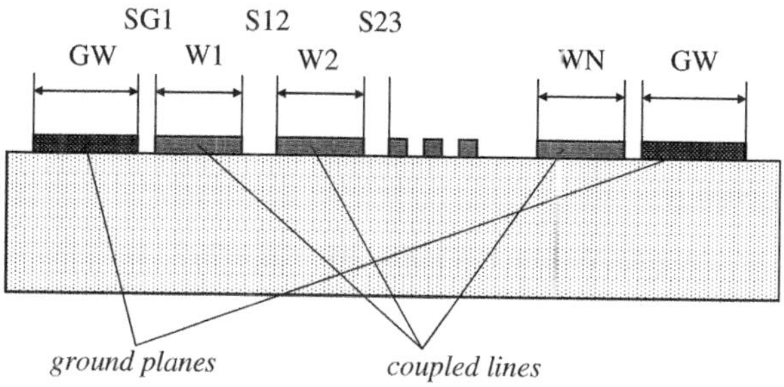

Fig. 5.6.5. Definition of the coupled coplanar waveguide parameters.

Parameters: n coupled coplanar lines ($2 \leq n \leq 10$)

GW	Width of ground planes
GND_MET	Identification of metal level for the ground planes (used for layout only)
SG1	Slot between first line and left ground plane
SG2	Slot between nth line and right ground plane
W1	Width of line 1
LEVEL1	Identification of metal level for line 1
S1_2	Slot between line 1 and line 2
W2	Width of line 2
LEVEL2	Identification of metal level for line 2
S2_3	Slot between line 2 and line 3
W3	Width of line 3
LEVEL3	Identification of metal level for line 3
S3_4	Slot between line 3 and line 4
W4	Width of line 4
LEVEL4	Identification of metal level for line 4
S4_5	Slot between line 4 and line 5
W5	Width of line 5
LEVEL5	Identification of metal level for line 5
S5_6	Slot between line 5 and line 6
W6	Width of line 6
LEVEL6	Identification of metal level for line 6
S6_7	Slot between line 6 and line 7
W7	Width of line 7
LEVEL7	Identification of metal level for line 7
S7_8	Slot between line 7 and line 8
W8	Width of line 8
LEVEL8	Identification of metal level for line 8

S8_9	Slot between line 8 and line 9
W9	Width of line 9
LEVEL9	Identification of metal level for line 9
S9_10	Slot between line 9 and line 10
W10	Width of line 10
LEVEL10	Identification of metal level for line 10

Application Range[4]:

$$\text{W1 to W10} = 0.5 \times n \times \text{DL} \quad n = 2, 2, 4, \ldots$$

$$\text{S1\_2 to S9\_10} = 0.5 \times n \times \text{DL} \quad n = 1, 2, 3, \ldots$$

$$\text{SG1, SG2} = 0.5 \times n \times \text{DL} \quad n = 1, 2, 3, \ldots$$

$$\text{GW} = 0.5 \times n \times \text{DL} \quad n = 2, 3, 4, \ldots$$

$$\text{LEVEL1 to LEVEL10} = 1 \text{ or } 2 \text{ (DEFAULT-Foundary)}$$

$$\text{GND\_MET} = 1 \text{ or } 2 \text{ (DEFAULT-Foundry)}$$

Notes:

1. The geometrical dimensions of coupled coplanar lines are defined in the C_NL_TYP data item.
2. The coupled line parameters (W1 to W10, S1_2 to S9_10, SG1, SG2, and GW) must be chosen in a way that metalization corners lie on the grid (see Application Range). However, if the user specifies a maximum deviation from the grid size ACC (see C_GRID for details), the simulator can adapt the incompatible structure parameters into the grid by changing these parameters slightly (warnings will be reported). For ACC = 0 (no deviation from grid) the simulator reports an error and the next compatible value will be recommended in the error message. *Note that these changes are not considered by the automatic layout generation.*
3. Depending on the number of lines of the considered element, some parameters in C_NL_TYP are ignored. For example, for five coupled coplanar lines (element C_5COUP), only the parameters with an index less than 5 are used (it means W1 to W5, LEVEL1 to LEVEL5 and S1_2 to S4_5). The parameters GW, SG1, SG2 and GND_MET are always used.
4. In the current version of *Coplan for ADS*™ all center lines must have the same metal level LEVEL.
5. In the current version, GND_MET is used for layout generation *only*. For the simulation, GND_MET is set equal to LEVEL1.

[4] See C_GRID for more information on DL.

5.6.2.4 *Coplanar Bridge-Type Data Definition C_AIRTYP*

Symbol:

C_DATA

Illustration:

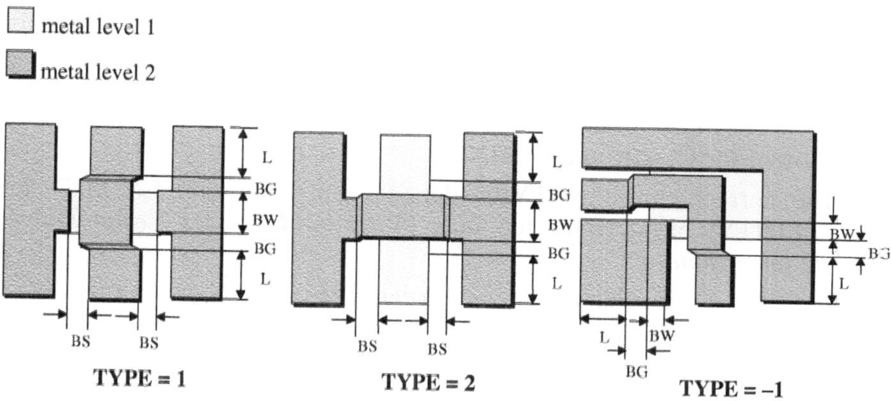

Fig. 5.6.6. Three types of air bridges and their parameters.

Parameters:

TYPE	Identifies the metal level of bridge (TYPE = 0 → no bridge)
BW	Bridge width
BG	Bridge gap
BS	Bridge spacing (BS = −1 → BS = S of connected Linetype)
DIE_IDX	Dielectric layer definition index for bridges in the selected Foundry
L	Length of feed line, connected to the bridge

Application Range[5]:

$$\text{TYPE} = 0, 1, 2, -1 \quad (\text{see illustration})$$

$$\text{BW} = 0.5 \times n \times \text{DL}, \quad n = 1, 2, 3, \ldots$$

$$\text{BG} = 0.5 \times n \times \text{DL}, \quad n = 0, 1, 2, 3, \ldots$$

$$\text{BS} = 0.5 \times n \times \text{DL}, \quad n = 1, 2, 3, \ldots (\text{or} -1)$$

$$\text{DIE\_IDX} = (0, 1 \text{ or } 2 \text{ in DEFAULT foundry})$$

[5] See C_GRID for more information on DL.

Notes:

1. The C_AIRTYP data item describes the construction of air bridges used in bends, junctions, and spiral inductors.
2. The value of TYPE defines the metal level of the bridge (the connection of the two ground planes). Two different options (TYPE = 1 and TYPE = 2) are available for the air-bridge element (C_AIR) and all other structures that contain bridges (C_BEND, C_TEE, C_CROSS, and C_RIND). In addition, for the junction elements C_BEND, C_TEE, and C_CROSS a special bridge type, the air bridge TYPE = –1, is also available (air-bridge bend, air-bridge T-junction, and air-bridge crossing; see illustrations in Table 5.7.3 and compare to Sections 3.5.5, 3.5.6, 3.5.7, and 3.5.9). TYPE = 0 switches off the bridge definition. TYPE = 0 is not allowed for C_AIR and C_RIND.
3. The correct choice of TYPE depends on value of CEN_MET in C_LINTYP of the connected lines. *TYPE and CEN_MET cannot have the same value.*
4. If TYPE = 0 is selected, *the parameters BW and BG are not set automatically to zero!* In this case, the bridge is removed but a piece of coplanar line with the length of (BW + 2BG + 2L) will remain in the circuit. This ensures that the reference planes of the element are not changed if the air bridge is switched off.
5. If TYPE = 0 or BS = –1 or BS > S, BS will be set automatically equal to S. (S is the slot width of the line type C_LINTYP connected to the bridge.) In the case of C_RIND, BS is always set equal to the slot width between the turns of the inductor (ST in C_RIND).
6. The dielectric material used under the bridge is defined in the foundry data item C-TECH (parameter die_ern, die_hn, and die_td$n$ with n = 0, 1, 2). DIE_IDX indicates the corresponding index n of the dielectric parameter in C_TECH data item. In the case of DEFAULT, foundry DIE_IDX can be 0, 1, or 2. For example, if DIE_IDX = 0, die_h0 will be the bridge height, die_er0 the dielectric constant of the dielectric material under the bridge, and die_td0 the loss tangent of this dielectric material. See C_TECH for more information on dielectric layer parameters.
7. L is the length of the coplanar line connected to the bridge at the ports of the considered element (see illustration for TYPE = –1). The configuration of the feed line connected to the bridge is given in the C_LINTYP data item of the corresponding element. For example, in Fig. 5.6.6, bridges are used in a C_BEND element. Each air bridge is connected to the ports of C_BEND over a piece of coplanar line of length L. In the case of C_AIR, the same line length L is connected to both ports (as shown in the illustration). *L is ignored in the case of C_RIND.*

8. A zero or negative value of L can be meaningful during the simulation. However, for $L < 0$, the automatic layout generation works incorrectly and the layout has to be edited manually.

5.6.2.5 *Coplanar Grid Data Definition C_GRID*

Symbol:

C_DATA

Illustration:

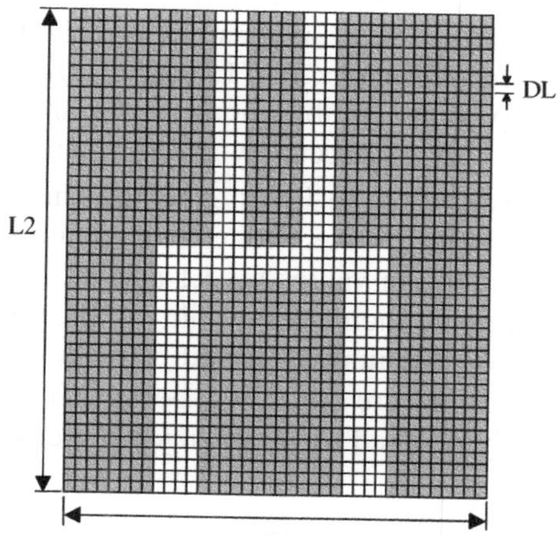

Fig. 5.6.7. Definition of the grid parameters.

Parameters:

BOUND	Maximum iteration error used by the iterative solver
L1	Width of shielding box (L1 = – 1, –2 → auto-sizing)
L2	Length of shielding box (L2 = –1, –2 → auto-sizing)
DL	Size of smallest grid element
ACC	Maximum allowed deviation from grid in %

Application Range:

BOUND	$1 \times 10^{-6} \leq \text{BOUND} \leq 1 \times 10^{-2}$
L1, L2	See notes of related element
DL	>0
ACC	$0 \leq \text{ACC} \leq 100$

Notes/Equations/References:

1. In this data item, grid and simulation control data are defined for the implemented finite difference solver.
2. The maximum allowed iteration error for the applied "Overrelaxation Method" (see Section 2.2, Eq. (2.2.14)) can be selected using the parameter BOUND. Table 5.6.3.1 shows the influence of this parameter on the simulation time and the accuracy of the results for a coplanar cross-junction. The relative time and error values are given with respect to the results for BOUND = 10^{-6}. A value of 10^{-3} is recommended for a first-run simulation. For the final analysis, a BOUND = 10^{-4} is always sufficient.
3. The shielding box sizes have an effect on the accuracy of the results. In order to keep these effects as small as possible, L1 (box width) and L2 (box depth) should be as large as possible. *The simulator adapts the box sizes on the structure automatically if* L1 *and* L2 *are set to* −1 *or* −2. Table 5.6.3.2 shows the possible combinations for L1 and L2 (for BOUND = 10^{-4}) and their effect on the accuracy and simulation time.
4. In cases where users want to investigate the shielding effects on the structure characteristics, the box sizes can be specified manually. For such applications, the minimum values for L1 and L2 (given in the notes of the related element) and their compatibility to the grid (see notes below) are to be considered. Manually selected box sizes can lead to errors if the structure dimensions are optimized, because in this case the box size is not automatically adapted to the changed dimensions.

TABLE 5.6.3.1. Effect of Iteration Error BOUND on Simulation Time and Accuracy

No.	BOUND	Simulation Time Compared to No. 5	Error Compared to No. 5
1	10^{-2}	9%	60%
2	10^{-3}	28%	2.5%
3	10^{-4}	54%	0.3%
4	10^{-5}	77%	0.03%
5	10^{-6}	100%	0%

TABLE 5.6.3.2. Effect of Box Sizes L1 and L2 on Simulation Time and Accuracy

No.	L1	L2	Simulation Time Compared to No. 4	Error Compared to No. 4
1	−2	−2	14%	8%
2	−1	−2	36%	7%
3	−2	−1	36%	6%
4	−1	−1	100%	0%

5. The parameter L2 is ignored for the elements: C_LIN, C_METVIA, C_TFG, C_TFR, C_MIM, C_CAPLIN.
6. DL is the external grid size, whereas the internal grid size is DL/2. The metalization edges of the structure have to lie on grid points. This means that the structure dimensions have to be an integer multiple of DL/2. However, if the user specifies a maximum allowed deviation ACC from the grid size, the simulator can adapt the incompatible structure dimensions into the grid by changing the dimensions slightly (warnings will be reported). For ACC = 0 (no deviation from grid is allowed) the simulator reports an error and the next compatible value will be suggested in the error message. These changes are not considered in the automatic layout generation. In order to avoid any possible deviations between simulation and layout, a value of ACC = 0 is recommended.
7. The value of the metalization thickness tn ($n = 1, 2$) and height of the dielectric layers, die_hn ($n = 0, 1, 2$) can be chosen independently from the grid size DL. However, for these two parameters, the ratios t/DL and die_h/DL should be between 0.2 and 5.0. Otherwise, the accuracy of the simulation cannot be guaranteed.

5.6.2.6 Process (Foundry) Used for Fabrication C_PROCES

Symbol:

C_DATA

Parameters:

FOUNDRY	Foundry identifier (= DEFAULT for DEFAULT foundry)
GW_DEF	Default ground width definition (for layout generation only)

Notes:

1. *Coplan for ADS*™ can handle any arbitrary foundry. For this purpose, the foundry data have to be implemented using data items. The name of a used foundry can be specified in C_PROCES. In the standard packet of *COPLAN*, a so-called DEFAULT foundry is implemented.
2. *The only name of the* C_PROCES *data item that is allowed in the standard version of Coplan for ADS*™ *is* "C_PROCES_DEFAULT". Use of other names causes the simulator to abort! The C_PROCES data item has always to be defined.
3. Special foundries are available as a special service. If no special foundry is implemented, the value only allowed for FOUNDRY is DEFAULT.

4. The ground-plane widths of all coplanar elements can be selected separately. Therefore an overlapping of some ground sections can result during the layout generation. To avoid this problem, a global default ground-plane width can be defined using the parameter GW_DEF. If GW > 0, this value will be used as the ground plane width for all elements. A negative or zero value for GW_DEF activates the ground-plane width definition of each element.
5. *The value of* GW_DEF *has no effect on the simulation results,* since for the simulation, the defined ground plane width for each element is always used.

5.6.2.7 Technological Data Definition (Default Foundry) C_TECH

Symbol:

C_TECH

Parameters:

die_h0	Height of dielectric layer for air bridges (for simulation only)
die_er0	Relative permittivity of dielectric layer for air bridges (for simulation only)
die_td0	Dielectric loss tangent of dielectric layer for air bridges (for simulation only)
cap_os1	Oversize of bottom metal plate of MIM capacitors over the dielectric layer dimensions
cap_l1	Length of the connection region of MIM capacitors with dielectric layer 1
die_h1	Height of dielectric layer 1 (for simulation only)
die_er1	Relative permittivity of dielectric layer 1 (for simulation only)
die_td1	Dielectric loss tangent of dielectric layer 1 (for simulation only)
cap_os2	Oversize of bottom metal plate of MIM capacitors over dielectric layer 2 dimensions
cap_l2	Length of the connection region of MIM capacitors with dielectric layer 2
die_h2	Height of dielectric layer 2 (for simulation only)
die_er2	Relative permittivity of dielectric layer 2 (for simulation only)
die_td2	Dielectric loss tangent of dielectric layer 2 (for simulation only)
res_l	Length of the connecting region for resistors
res_rs	Sheet resistivity of resistive layer (ohm/square) (for simulation only)
t1	Thickness of metal level 1 (cond) (for simulation only)
rho1	Resistivity of metal level 1 (cond) (for simulation only)
t2	Thickness of metal level 2 (cond2) (for simulation only)
rho2	Resistivity of metal level 2 (cond2) (for simulation only)

Application Range[6]**:**

$$\text{die\_h}n \quad 5\text{DL} < \text{die\_h}n < 0.2\ \text{DL} \quad (\text{for } n = 0, 1, 2)$$

$$\text{die\_er}n \quad \geq 1 \quad (\text{for } n = 0, 1, 2)$$

$$\text{cap\_os}n \quad \geq 0 \quad (\text{for } n = 1, 2)$$

$$\text{cap\_l}n \quad \geq 0 \quad (\text{for } n = 1, 2)$$

$$\text{res\_1} \quad \geq 0$$

$$\text{t}n \quad 5\text{DL} < \text{t}n < 0.2\ \text{DL} \quad (\text{for } n = 1, 2)$$

Notes:

1. In this data item, the process data of a DEFAULT foundry are stored. This is an arbitrary foundry and the user can modify the parameters.
2. In the DEFAULT foundry, two metal layers, three dielectric layers, and a resistive layer are available.
3. If only the DEFAULT foundry is available (standard packet of *Coplan for ADS*™), *the only name for* C_TECH *allowed is* "C_TECH_DEFAULT." The use of another name causes the simulator to abort!
4. The parameters die_h0, die_er0 and die_td0 represent the definition of the dielectric layer that is used in air bridges (C_AIRTYP) addressed by the parameter DIE_IDX. The application of this layer for C_MIM and C_CAPLIN is not allowed.
5. The parameters die_h*n*, die_er*n*, die_td*n*, cap_os*n* and cpa_l*n* (n = 1, 2) represent the definition of dielectric layers used in capacitors (C_CAPLIN and C_MIM) addressed by the parameter DIE_IDX. For example if DIE_IDX in C-MIM is set to 1, a dielectric layer of thickness die_h1 and dielectric constant of die_er1 and loss tangent of die_td1 is used for the capacitor. In this case, cap_os1 is the oversize of the bottom plate and cap_l1 is the length of connection region. The two dielectric layers can also be used for air bridges; however, in this case the parameter cap_os*n* and cap_l*n* (n = 1, 2) are ignored.
6. The parameters res_l and res_rs represent the definition of resistive layers for the resistor elements C_TFR and C_TFG addressed by the parameter RES_ IDX.
7. The parameters t1, rho1 and t2, rho2 represent the definition of metal layers addressed by CEN_MET, GDN_MET or LEVEL (see C_LINTYP, C_NL_TYP, C_IDC, C_RIND).
8. roh1 and roh2 are the specific resistivity of the metal layers 1 and 2 related to gold. For perfect gold, this value is 1.

[6] See C_GRID for more information on DL.

9. The value for the metalization thickness tn ($n = 1, 2$) and the dielectric height die_hn ($n = 0, 1, 2, \ldots$) does not depend on the grid size. However, for these two parameters, *the ratio* tn/*DL and die_hn/DL should be between* 0.2 *and* 5.0. *Otherwise, the accuracy of the simulation cannot be guaranteed.*

5.6.2.8 Layer Data Definition (Default Foundry) C_LAYER

Symbol:

C_TECH

Parameter:

via_ol	Overlap for inter metal via
cond_os	Oversize of metal layer 1 (cond)
cond_wm	Minimum width of metal layer 1 (cond)
cond_sm	Minimum slot width of metal layer 1 (cond)
cond2_os	Oversize of metal layer 2 (cond2)
cond2_wm	Minimum width of metal layer 2 (cond2)
cond2_sm	Minimum slot width of metal layer 2 (cond2)
resi_os	Oversize of resistive layer
diel_os	Oversize of dielectric layer 1
diel2_os	Oversize of dielectric layer 2

Notes:

1. In this data item, the layer information of the default foundry is stored. These data are used only for the layout generation and *have therefore no effect on the simulation results.*
2. If only the DEFAULT foundry is available (standard packet of *Coplan for ADS*™), *the only allowed name for* C_LAYER *is* "C_LAYER_DEFAULT." Use of another name causes the simulator to abort!
3. via_ol is the minimum length of the overlapping area for the connection of metal level 1 with metal level 2.
4. cond_os and cond2_os as well as diel_os and diel2_os are used for the generation of an oversized layout.
5. cond_wm and cond2_wm can be used for the definition of a minimum conductor size in metal level 1 and metal level 2, respectively.
6. cond_sm and cond2_sm can be used for the definition of a minimum slot size in metal level 1 and metal level 2 respectively.
7. The following values can lead to an incorrect layout generation:

$$\text{via\_ol} > \text{cap\_1 (see C\_TECH)}$$

$$\text{via\_ol} > \text{res\_1 (see C\_TECH)}$$

$$\text{cond\_wm} < 2 \times \text{cond\_os} \quad \text{if cond\_os} < 0$$

$$\text{cond\_sm} < 2 \times \text{cond\_os} \quad \text{if cond\_os} > 0$$

$$\text{cond\_wm} < 2 \times \text{cond\_os} \quad \text{if cond\_os} < 0$$

$$\text{cond2\_sm} < 2 \times \text{cond\_os} \quad \text{if cond\_os} > 0$$

5.7 THE COPLANAR COMPONENTS AND THEIR MODELS

The element library of *COPLAN for ADS*™ consists of three groups of elements: coplanar transmission lines, coplanar discontinuities, and coplanar lumped elements. In the group of transmission lines, there are two elements: the symmetrical coplanar waveguide (C_LIN) and the coplanar inter metal via (C_METVIA) as shown in Table. 5.7.1. For these elements an equivalent circuit of distributed parameters is calculated (see Table 5.7.1). As a result, the simulation of these elements is valid even for frequencies beyond the $\lambda/4$ resonant frequency.

The group of discontinuities that can be simulated by *COPLAN for ADS*™ consists of all usually needed coplanar discontinuities with and without air bridges (see Tables 5.7.2 and 5.7.3). Nine elements are available in this group, five in the group of discontinuities without air bridges: the coplanar open line (C_OPEN) with and without connected ground planes, the coplanar short circuited line (C_SHORT), the coplanar gap (C_GAP), the coplanar waveguide step (C_STEP), and the coplanar taper (C_TAPER).

Four discontinuities with air bridges are available: the air bridge itself (C_AIR, in three different versions; see Fig. 5.6.6), the coplanar bend (C_BEND), the coplanar Tee-junction (C_TEE), and the coplanar crossing (C_CROSS). All junctions can have an arbitrary arrangement and a different combination of feed lines and air-bridge types at their ports. The equivalent

TABLE 5.7.1. Coplanar Transmission Lines in *COPLAN for ADS*™ and Their Equivalent Circuits

Name	Physical	Equivalent Circuit
C_LIN Coplanar transmission line	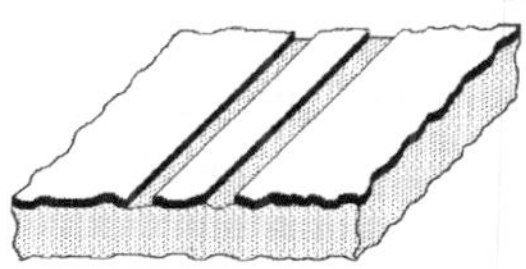	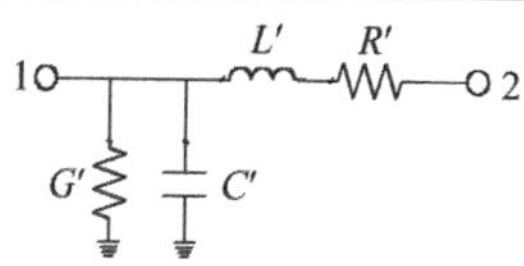
C_METVIA Coplanar inter-metal via (no step)	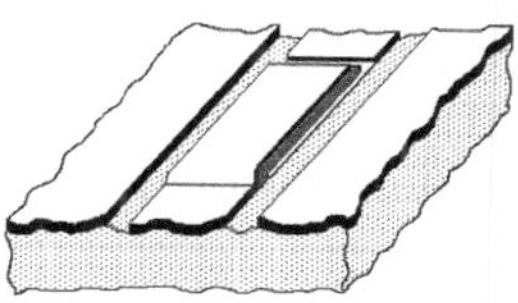	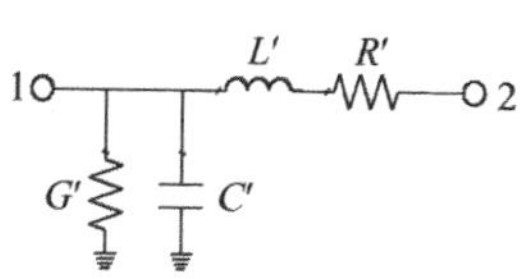

TABLE 5.7.2. Coplanar Discontinuities in *COPLAN for ADS*™ and Their Equivalent Circuits

Name	Physical	Equivalent Circuit
C_OPEN Coplanar open	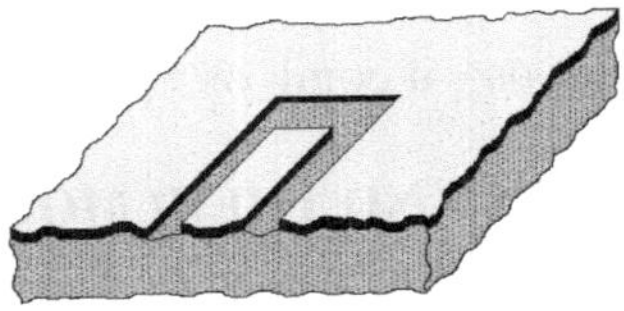	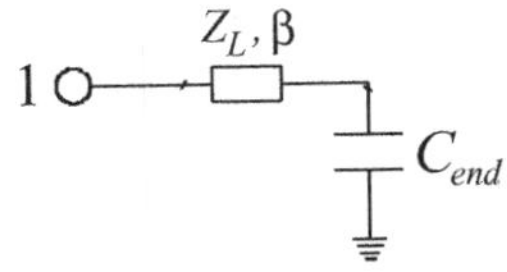
C_SHORT Coplanar short	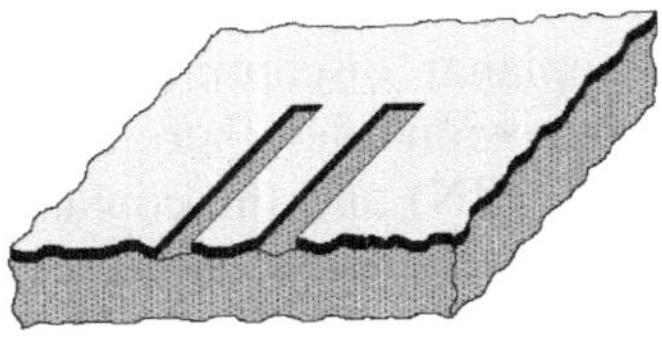	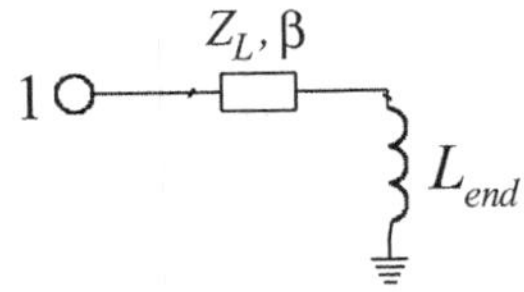
C_GAP Coplanar gap		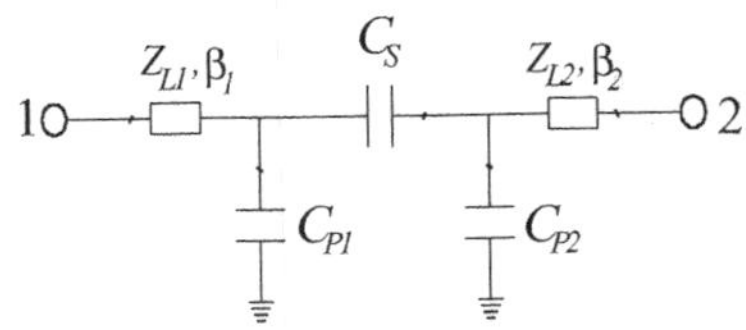
C_STEP Coplanar step		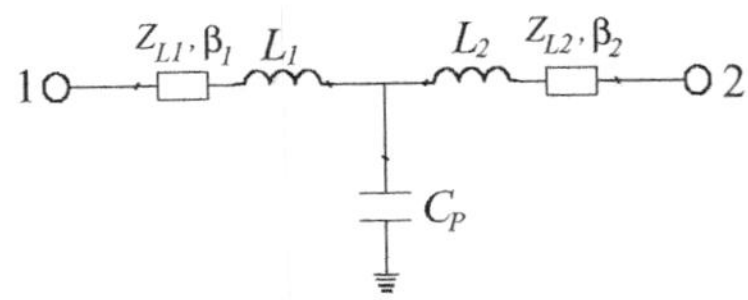
C_TAPER Coplanar taper	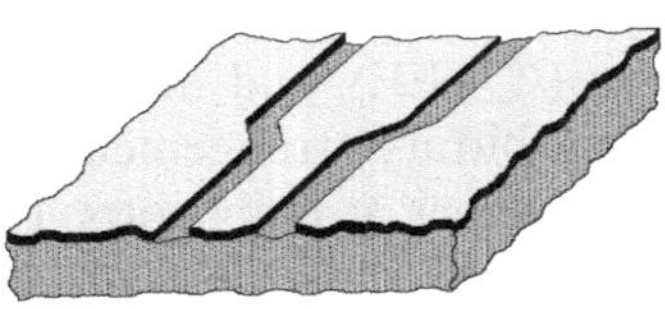	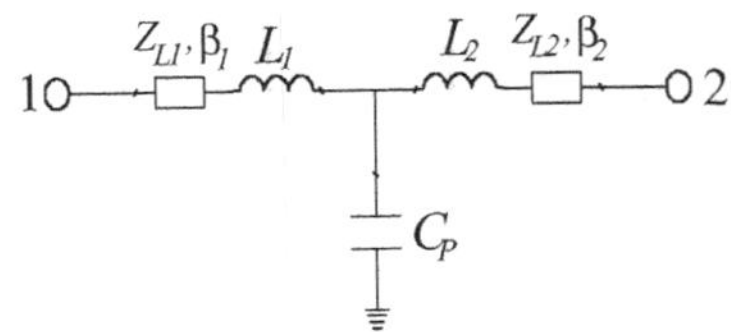

circuits describing the electromagnetic behavior of these elements are also shown in Tables 5.7.2 and 5.7.3.

The group of lumped elements that can be simulated by *COPLAN for ADS*™ contains the mainly used components. There are six elements available in this group: the coplanar interdigital capacitor (C_IDC), the coplanar rectangular spiral inductor (C_RIND), the coplanar thin-film resistor as a series (C_TFR) or a shunt resistor (C_TFG), and the coplanar MIM-capacitor as a series (C_MIM) or shunt a capacitor (C_CAPLIN). Table 5.7.4 shows the available coplanar lumped element models and their equivalent circuits.

TABLE 5.7.3. Coplanar Discontinuities (with Air Bridges) Available in *COPLAN for ADS*™ and Their Equivalent Circuits

Name	Physical	Equivalent Circuit
C_AIR Coplanar air bridge	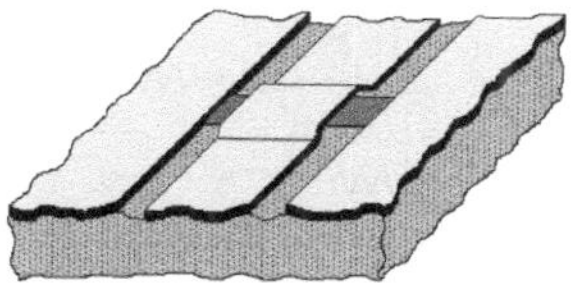	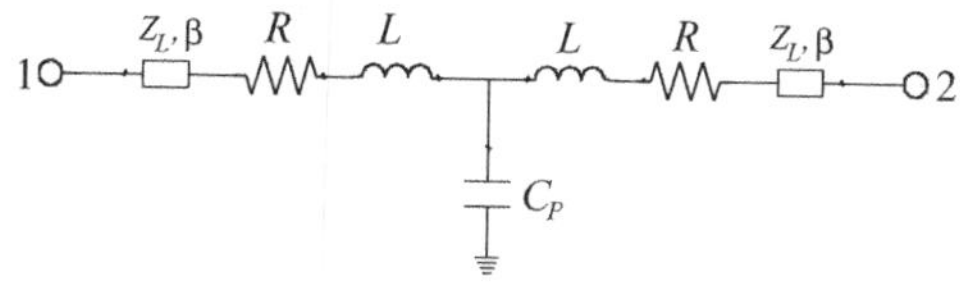
C_BEND Coplanar bend	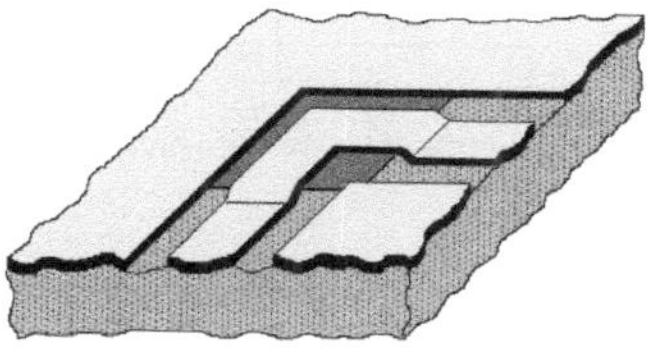	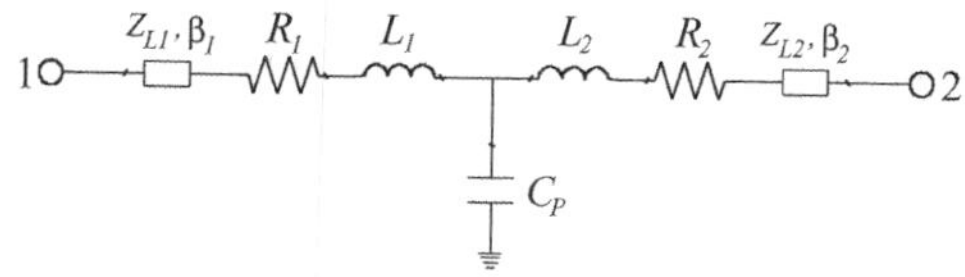
C_TEE Coplanar T-junction	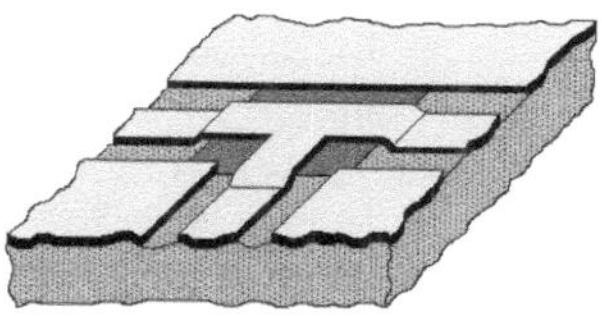	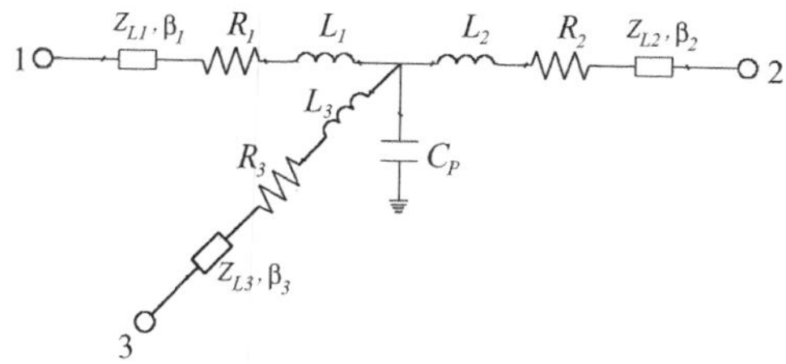
C_CROSS Coplanar cross-junction	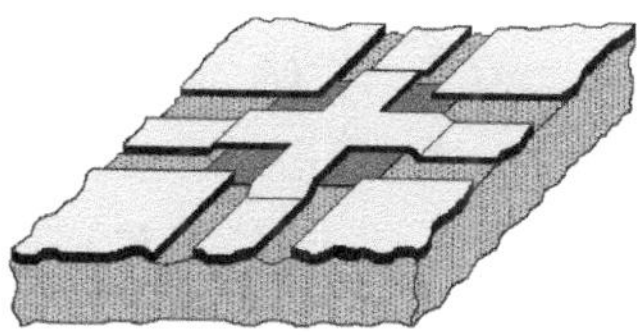	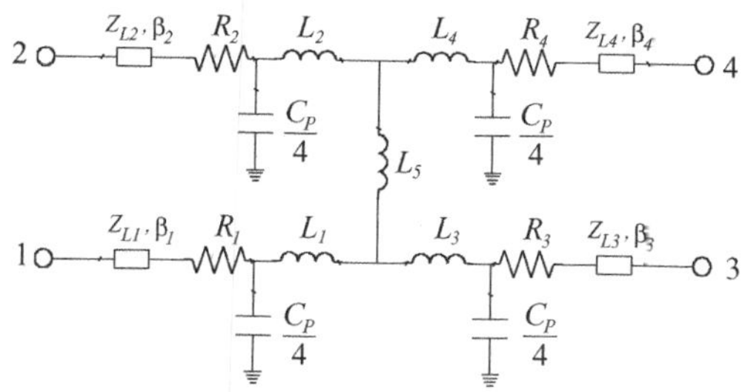

In order to have a compatible coplanar port, a C_PORT (see below) element is also implemented into the coplanar palette. This element is used for layout generation only and has no effect on simulation results.

In the following, a short description of all components and their parameters that can be addressed in the *COPLAN for ADS*™ software will be given.

TABLE 5.7.4. Coplanar Lumped Elements in *COPLAN for ADS*™ and Their Equivalent Circuits

Name	Physical	Equivalent Circuit
C_IDC Coplanar interdigital capacitor	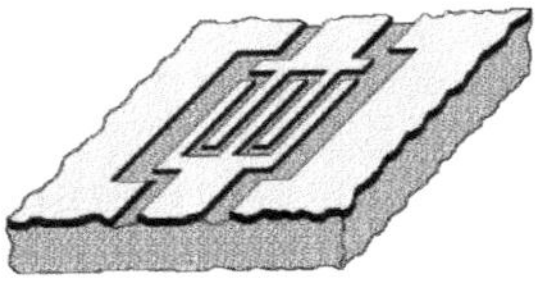	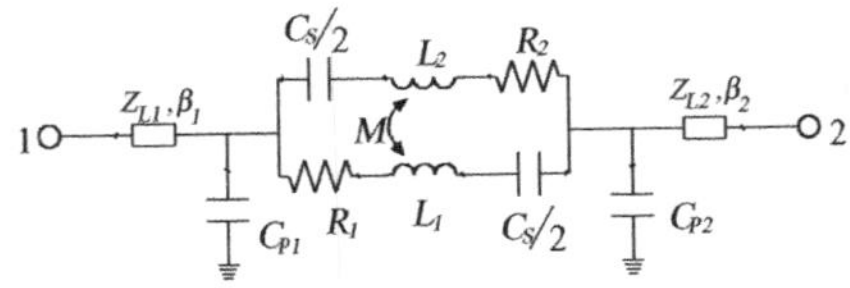
C_RIND Coplanar rectangular inductor	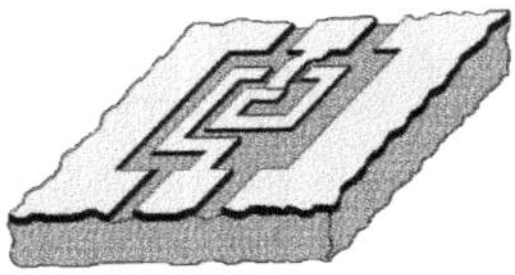	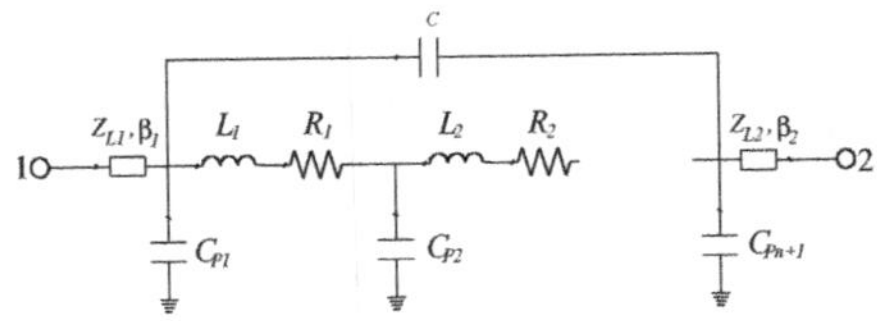
C_MIM Coplanar MIM-capacitor	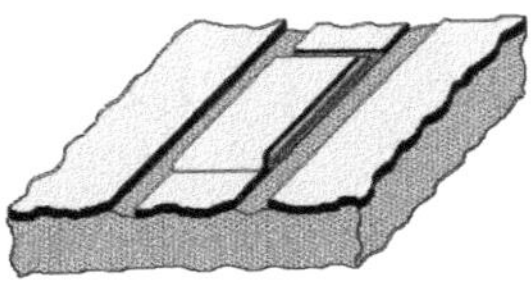	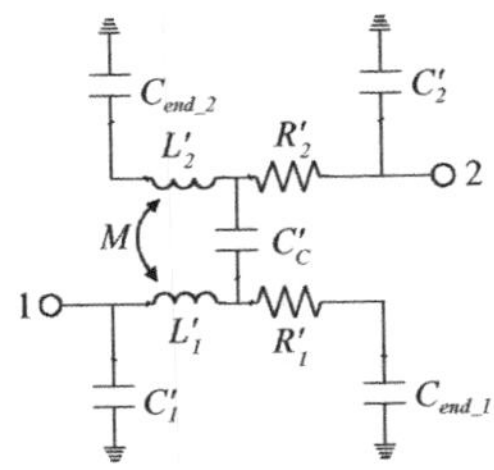
C_CAPLIN Coplanar MIM-capacitor to ground	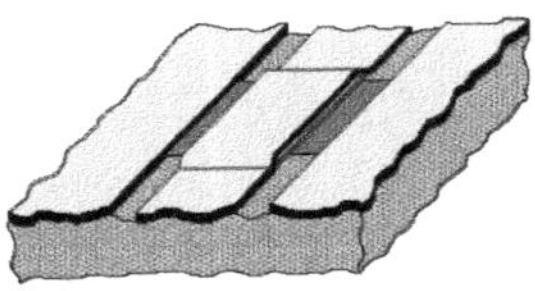	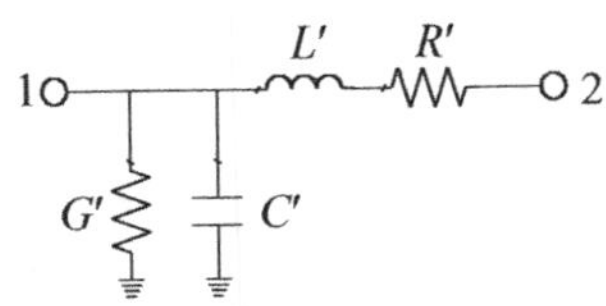
C_TFG Resistively loaded transmission line	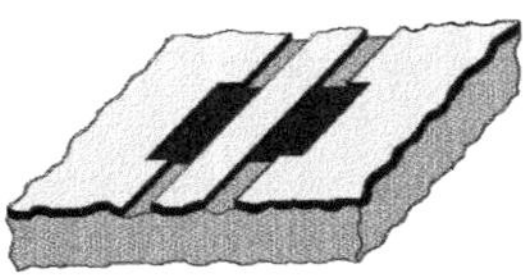	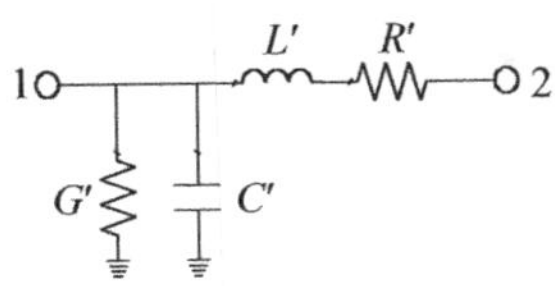
C_TFR Coplanar thin-film resistor	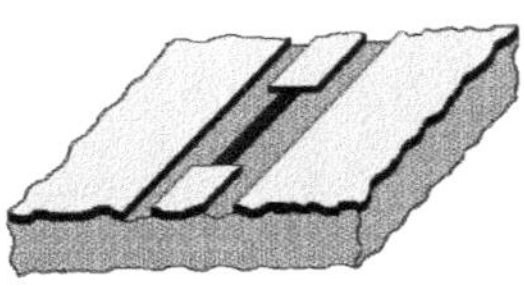	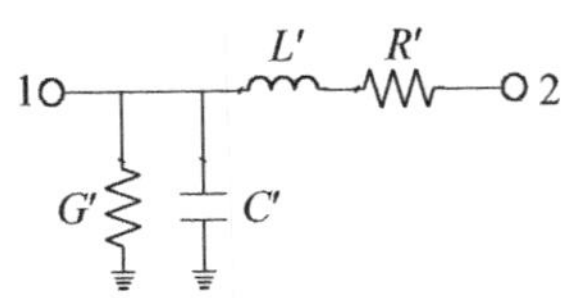

5.7.1. Coplanar Waveguide RF-Port C_PORT

Symbol:

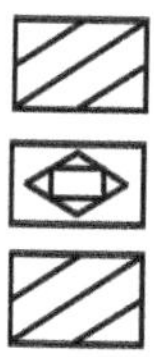

Illustration:

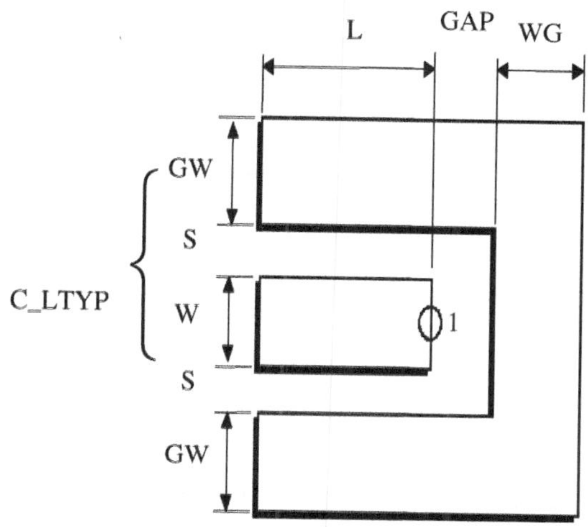

Fig. 5.7.1. The coplanar waveguide RF port.

Parameters:

C_PORT	Port number
L	Length of port section
GAP	Gap (spacing) between the end of the center conductor and ground plane (only for TYPE = *Ground_Connection*)
WG	Width of ground plane at the end of the center conductor (only for TYPE = *Ground_Connection*)
TYPE	Type of RF port (see Notes)
C_LTYP	ID of coplanar transmission Line applied at port 1

Application Range:

$$\text{TYPE} = Ground\_Connection$$
$$No\_Ground\_Connection$$

(see Notes for definition)

Notes:

1. The C_PORT module is introduced <u>*only*</u> for layout generation (a schematic representation is also available). This module has no effect on the electrical behaviour of the circuit to be analyzed.
2. The parameter L is ignored in the available version of the software.
3. Figure 5.7.2 shows the two port types available.

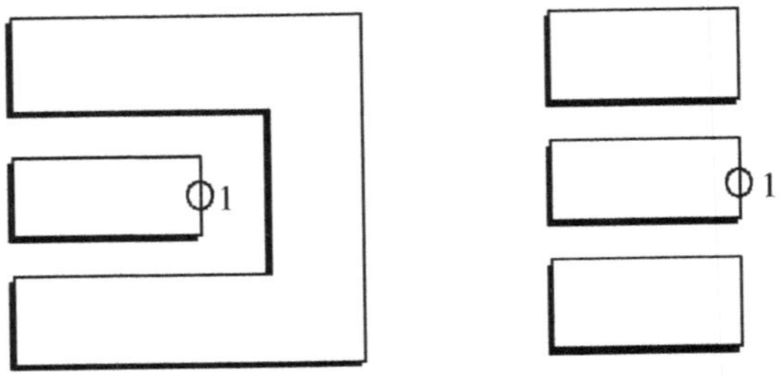

TYPE = *Ground_Connection* TYPE = *No_Ground_Connection*

Fig. 5.7.2. The two possible coplanar ports.

4. If TYPE = *No_Ground_Connection* is selected, the parameter WG and GAP are ignored.
5. See also the notations for the correct selection of the C_LTYP in Section 5.6.2.2.

5.7.2 Coplanar Transmission Line C_LIN

Symbol:

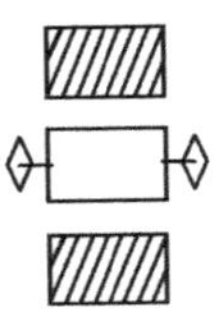

Illustration:

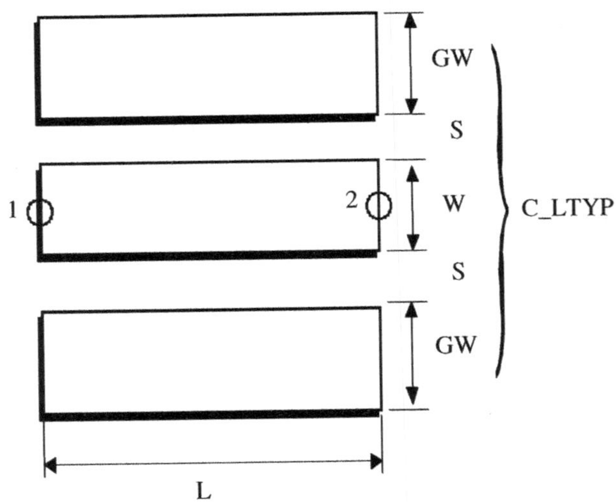

Fig. 5.7.3. The coplanar transmission lines.

Parameters:

L	Line length, stretchable
C_LTYP	IDs of the coplanar transmission lines applied at port 1 and at port 2
C_SUB	ID of coplanar substrate definition
C_GRID	Definition of variables needed for the equivalent circuit generation
TEMP	ID of the element temperature definition used for noise computation

Notes:

1. A zero or negative value for L can be meaningful during the simulation. In such cases, the automatic layout generation works incorrectly and the generated layout has to be edited manually.
2. Note that the finite width of ground planes as well as the metalization thickness affect the line characteristics. As can be seen from Fig. 2.2.24, the effect of finite ground planes can be neglected if GW > $2(W + 2S)$.
3. See also the notations for the correct selection of the C_LTYP in Section 5.6.2.2.

Equivalent Circuit:

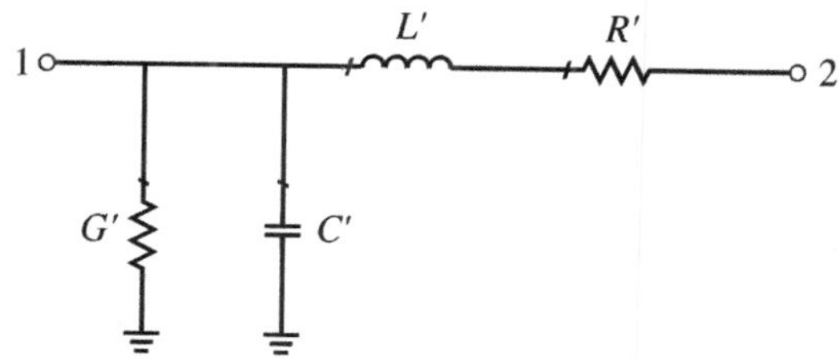

Fig. 5.7.4. The equivalent circuit of the coplanar transmission line.

5.7.3 Coplanar Inter-metal Via (No Step) Connection C_METVIA

Symbol:

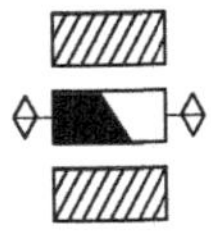

Illustration:

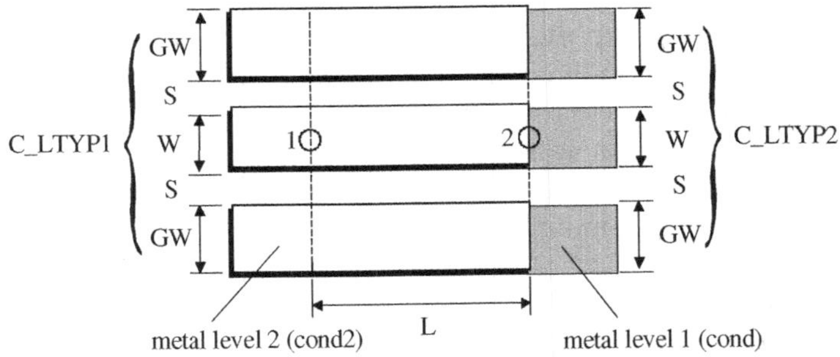

Fig. 5.7.5. The coplanar inter-metal via connection.

Parameters:

L	Length of via-section
C_LTYP1	ID of the coplanar transmission line applied at port 1
C_LTYP2	ID of the coplanar transmission line applied at port 2
C_SUB	ID of the coplanar substrate definition
C_GRID	Definition of the variables needed for the equivalent circuit generation
TEMP	ID of the element temperature definition used for noise computation

Notes:

1. The C_METVIA module is introduced to connect two coplanar transmission lines fabricated in different metal levels. The C_METVIA module is not able to calculate a step. Therefore, both transmission lines must have the same line and slot widths.
2. The minimum allowed value for L depends on the overlap defined in the currently selected foundry. In the case of C_LAYER_DEFAULT, L should be larger than or equal to via_ol (see Section 5.6.2.8).
3. If L is set to zero, the automatic layout generation works incorrectly if a via_ol is defined for the inter-metal via. In this case the generated layout has to be edited manually.
4. In the case of interactive layout generation using ACADEMY, selecting the C_METVIA module causes ACADEMY to generate an error message because both LINTYP's have the same LEVEL by default. In order to get a correct layout, the user has to attach a valid pair of LINTYPs to the C_METVIA module using the dialog box.
5. See also the notations for the correct selection of the C_LTYP1 and C_LTYP2 in Section 5.6.2.2.

Equivalent Circuit:

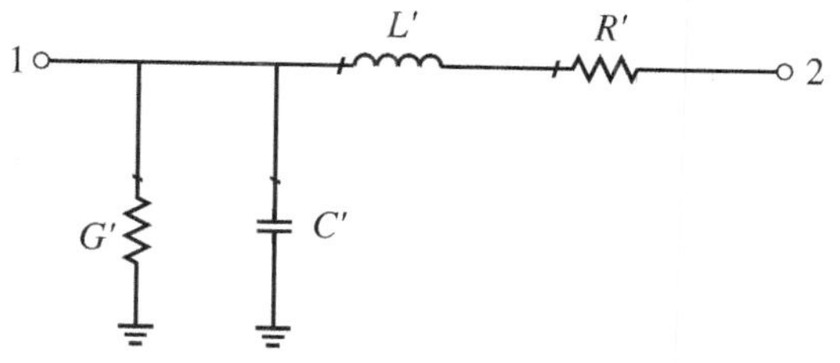

Fig. 5.7.6. The equivalent circuit of the inter metal via connection.

5.7.4 Coplanar Resistively Loaded Transmission Line C_TFG

Symbol:

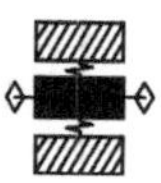

Illustration:

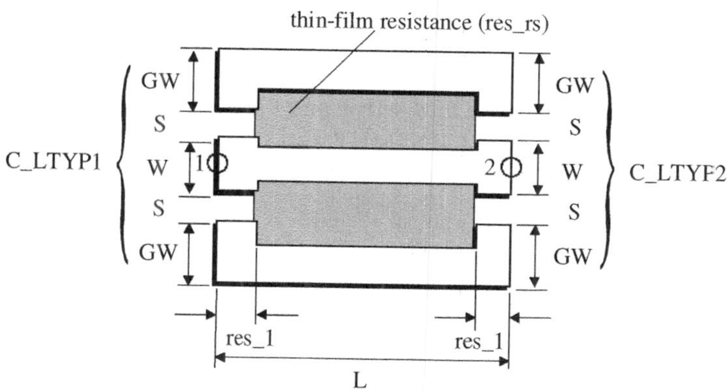

Fig. 5.7.7. The resistively loaded coplanar transmission line.

Parameters:

L	Total length of the resistor
C_LTYPR	ID of coplanar transmission line applied to the resistive section
C_LTYP1	ID of the coplanar transmission line applied at port 1
C_LTYP2	ID of the coplanar transmission line applied at port 2
RES_IDX	Index to sheet resistivity (ohm/square) definition in the currently selected foundry (res_rs)

C_SUB	ID of coplanar substrate definition
C_GRID	Definition of variables needed for equivalent circuit generation
TEMP	ID of element temperature definition used for noise computation

Notes:

1. The C_TFG module is introduced to have a resistive connection between the center conductor and the ground planes (resistor to the ground).
2. The C_TFG module is *not* able to calculate a step. Therefore, if a step configuration is used for layout reasons, the effect of such a discontinuity is neglected in the simulation.
3. In case of the default foundry, RES_IDX is always 1.
4. The minimum value for L depends on the value of res_l and the minimum resistor length res_mim_w (10 μm) in the currently selected foundry. Minimum value for res_l is res_min_w (10 μm). The effective length of the resistive section is given by L − 2 × res_l.
5. If L is set to zero, the automatic layout generation works incorrectly if resi_l > 0 or resi_os > 0. In this case the generated layout has to be edited manually.
6. The parameters CEN_MET and GND_MET in C_LTYPR are ignored. The parameters CEN_MET and GND_MET in C_LTYP1 are used to specify the line within the resistive section.

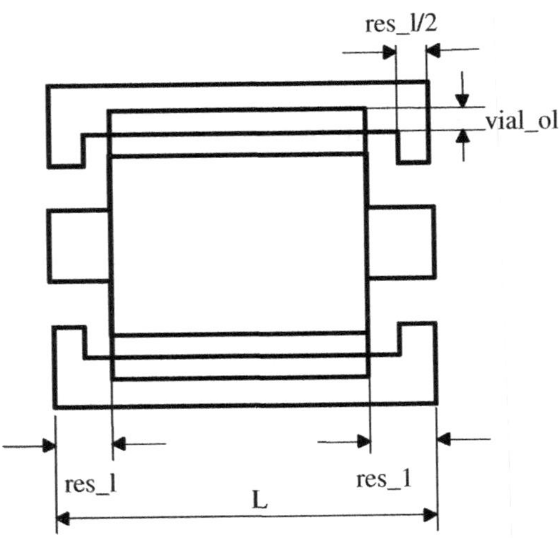

Fig. 5.7.8. Structure of the resistively loaded coplanar transmission line.

7. See also the notations for the correct selection of the C_LTYP1, C_LTYP2 and C_LTYPR in Section 5.6.2.2.

Equivalent Circuit:

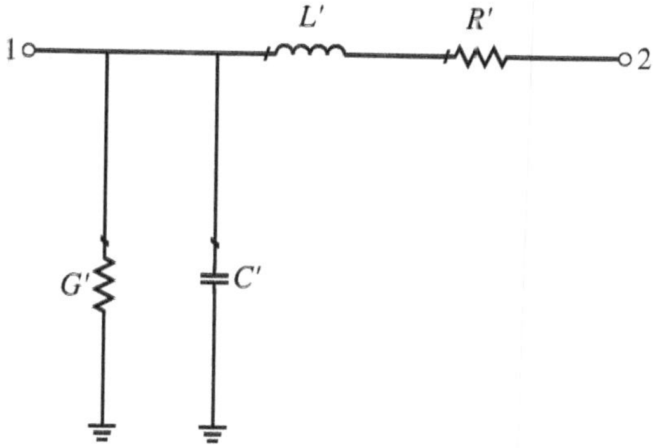

Fig. 5.7.9. The equivalent circuit of the resistively loaded transmission line.

5.7.5 Coplanar MIM-Capacitor to Ground C_CAPLIN

Symbol:

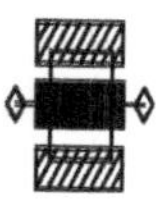

Illustration:

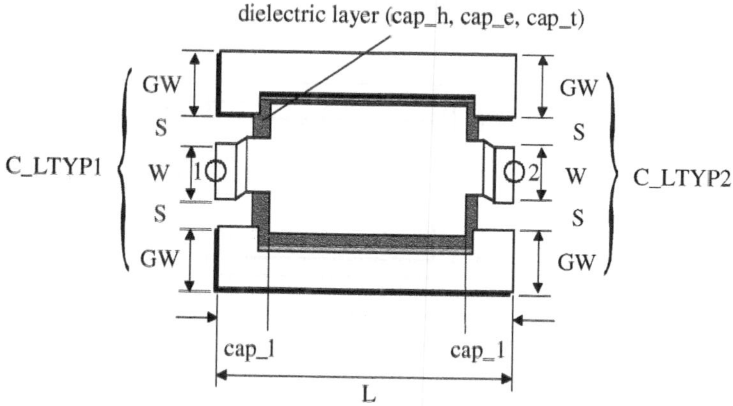

Fig. 5.7.10. The coplanar MIM capacitor to ground.

Parameters:

L	Total length of the capacitor
C_LTYPC	ID of coplanar transmission line connected to the capacitive section
C_LTYP1	ID of coplanar transmission line connected to port 1
C_LTYP2	ID of coplanar transmission line connected to port 2
DIE_IDX	Index to define the dielectric layer of capacitors in the currently selected foundry (die_h, die_er, die_td)
C_SUB	ID of coplanar substrate definition
C_GRID	Definition of variables needed for equivalent circuit generation
TEMP	ID of element temperature definition used for noise computation

Notes:

1. The C_CAPLIN module is introduced to generate a capacitive connection between the center conductor and the ground planes.
2. In the current version of *Coplan for ADS*™, the C_CAPLIN module is <u>not</u> able to calculate a step. Therefore, if a step configuration is used for layout reasons, the effect of such a discontinuity is neglected in the simulation.
3. In case of DEFAULT foundry, two different dielectric layers are available. The index DIE_IDX is used for the selection of the desired layer. DIE_IDX can be 1 or 2. If DIE_IDX = 1, the parameter die_h1, die_er1 and die_td1 are used.
4. The minimum value for L depends on the value cap_l in the currently selected foundry. Minimum value for cap_l is cap_min (10 μm). Note that the effective length of the capacitive section is given by L − 2 × cap_l. The oversize of lower metal plate is defined by the keyword "cap_os" (see Figure 5.7.11).

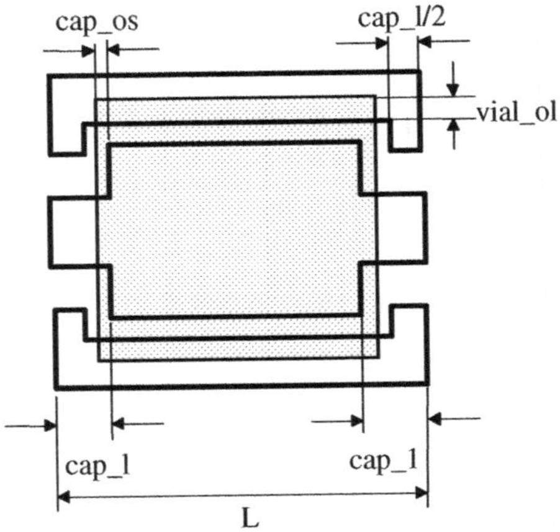

Fig. 5.7.11. The structure of the coplanar MIM capacitor to ground.

5. If L is set to zero, the automatic layout generation works incorrectly if cap_1 > 0 is defined. In this case the generated layout has to be edited manually.
6. For the layout generation, the parameter CEN_MET and GND_MET in C_LTYPR are ignored. The metalization layer GND_MET in C_LTYP1 is used for the ground plane in the capacitive section.
7. See also the notations for the correct selection of the C_LTYP1, C_LTYP2 and C_LTYPC in Section 5.6.2.2.

Equivalent Circuit:

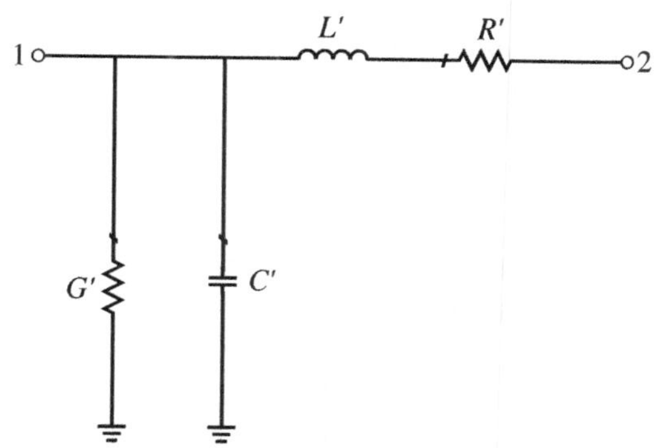

Fig. 5.7.12. The equivalent circuit of the coplanar MIM capacitor to ground.

5.7.6 Coplanar Open-Ended Transmission Line C_OPEN

Symbol: **Illustration:**

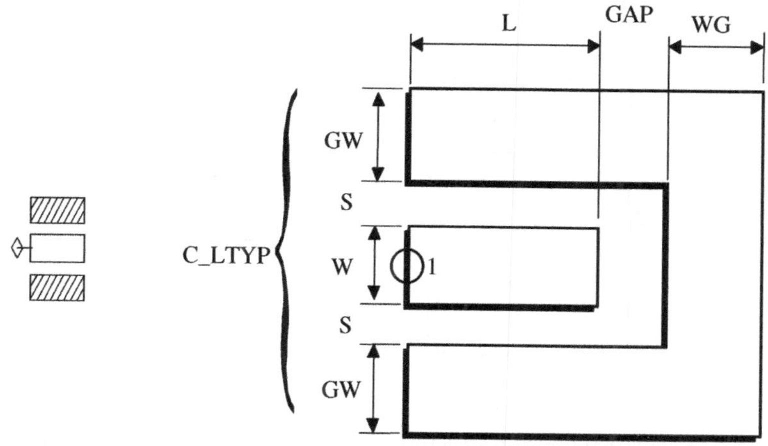

Fig. 5.7.13. The coplanar open transmission line.

Parameters:

L	Length of transmission line connected to the coplanar open end.
GAP	Gap (spacing) between the end of the centre conductor and the ground planes.
WG	Width of the ground plane at the end of the center conductor.
C_LTYP	ID of coplanar transmission line connected at port 1
C_SUB	ID of coplanar substrate definition
C_GRID	Definition of variables needed for equivalent circuit generation
TEMP	ID of element temperature definition used for noise computation
TYPE	*Ground_Connection* *No_Ground_Connection* (see Notes and Figure 5.7.14 for definition)

Notes:

1. The minimum value for L depends on the metal layer oversize in the currently selected foundry. In the case of C_LAYER_DEFAULT, L should be larger than or equal to max(|cond_os|, |cond2_os|).
2. If L is set to zero, the automatic layout generation works incorrectly if an oversize is defined for the cond or for the cond2 layer. In this case the generated layout has to be edited manually.
3. Fig. 5.7.14 below depicts the open-ended transmission line types that are available in *Coplan for ADS*™.
4. If TYPE = *No_Ground_Connection* is selected, the parameters WG and GAP are ignored.
5. See also the notations for the correct selection of the C_LTYP in Section 5.6.2.2.

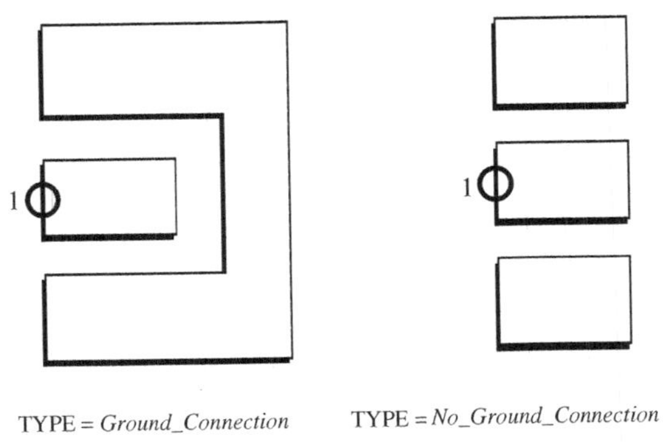

Fig. 5.7.14. The two possible structures of a coplanar open transmission line.

Equivalent Circuit:

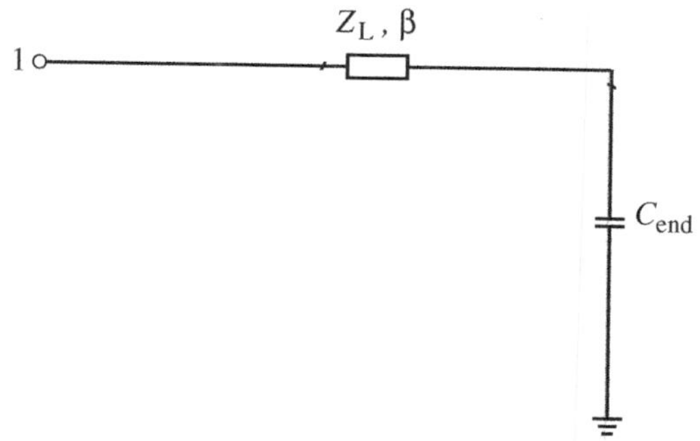

Fig. 5.7.15. The equivalent circuit of the coplanar open transmission line.

5.7.7 Coplanar Short-Circuited Transmission Line C_SHORT

Symbol: **Illustration:**

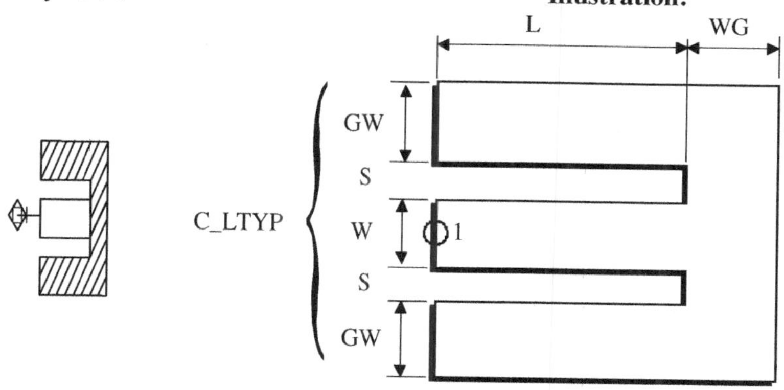

Fig. 5.7.16. The coplanar short-circuited transmission line.

Parameters:

L	Length of the transmission line connected to the coplanar short-circuited coplanar waveguide
WG	Width of the ground plane at the end of the center conductor
C_LTYP	ID of the coplanar transmission line connected to port 1
C_SUB	ID of coplanar substrate definition
C_GRID	Definition of variables needed for equivalent circuit generation
TEMP	ID of element temperature definition used for noise computation

Notes:

1. The minimum value for L depends on the metal layer oversize in the currently selected foundry. In case of C_LAYER_DEFAULT, L should be larger than or equal to max(|cond_os|, |cond2_os|, see Section 5.6.2.8).
2. If L is set to zero, the automatic layout generation works incorrectly if an oversize is defined for the cond or for the cond2 layer. In this case the generated layout has to be edited manually.
3. If CEN_MET and GND_MET in the corresponding C_LTYP are at different levels, a positive nonzero via_ol (see Section 5.6.2.8) should be defined.
4. See also the notations for the correct selection of the C_LTYP in Section 5.6.2.2.

Equivalent Circuit:

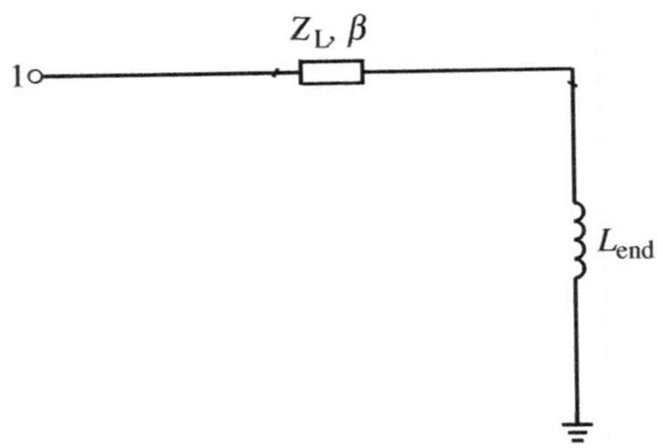

Fig. 5.7.17. The equivalent circuit of the coplanar shorted transmission line.

5.7.8 Gap in a Coplanar Transmission Line C_GAP

Symbol:

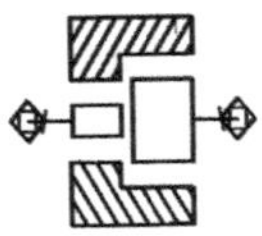

Illustration:

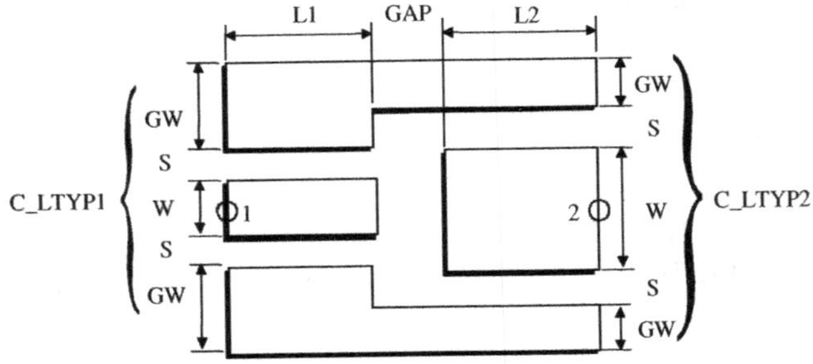

Fig. 5.7.18. The gap in a coplanar transmission line.

Parameters:

L1	Length of the transmission line connected to port 1
L2	Length of the transmission line connected to port 2
GAP	Gap spacing
C_LTYP1	ID of coplanar transmission line connected at port 1
C_LTYP2	ID of coplanar transmission line connected at port 2
C_SUB	ID of coplanar substrate definition
C_GRID	Definition of variables needed for equivalent circuit generation
TEMP	ID of element temperature definition used for noise computation

Notes:

1. The minimum allowed value for L1 and L2 depends on the metal layer oversize in the currently selected foundry. In case of C_LAYER_DEFAULT, L1 and L2 should be larger than or equal to max(|cond_os|, |cond2_os|) (see Section 5.6.2.8).
2. If L1 or L2 is set to zero, the automatic layout generation works incorrectly if an undersize is defined for the cond or the cond2 layer. In this case the generated layout must be edited manually.
3. The minimum value of GAP depends on cond_sm (or cond2_sm; see Section 5.6.2.8) in the currently selected foundry.
4. C_LTYP of the connected lines should have the same CEN_MET and the same GND_MET (no integrated inter-metal via).
5. See notations for the correct selection of the C_LTYP1, C_LTYP2 in Section 5.6.2.2.

Equivalent Circuit:

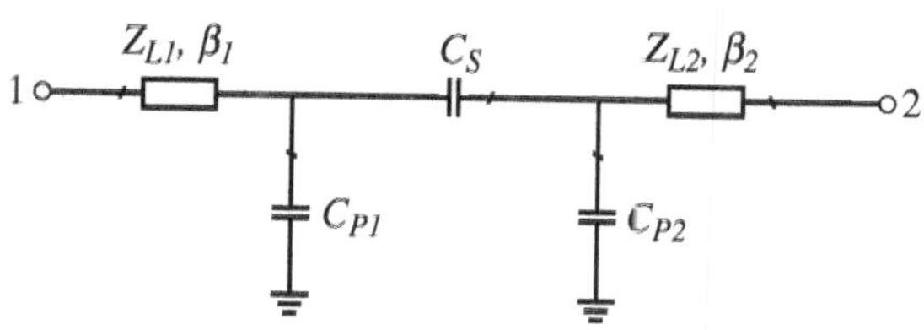

Fig. 5.7.19. The equivalent circuit of the gap in a coplanar transmission line.

5.7.9 Step in a Coplanar Transmission Line C_STEP

Symbol:

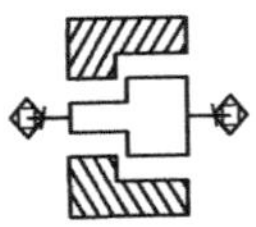

Illustration:

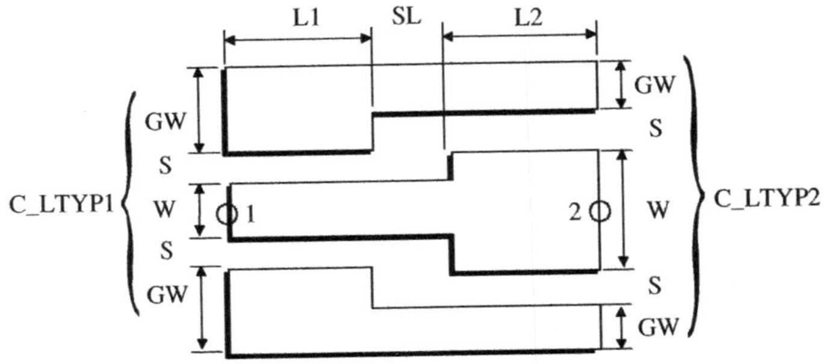

Fig. 5.7.20. The step in a coplanar transmission line.

Parameters:

L1	Length of the transmission line connected to port 1
L2	Length of the transmission line connected to port 2
SL	Step length
C_LTYP1	ID of coplanar transmission line connected at port 1
C_LTYP2	ID of coplanar transmission line connected at port 2
C_SUB	ID of coplanar substrate definition
C_GRID	Definition of variables needed for equivalent circuit generation
TEMP	ID of element temperature definition used for noise computation

Notes:

1. The minimum value for L1 and L2 depends on the metal layer oversize in the currently selected foundry. In the case of C_LAYER_DEFAULT, L1 and L2 should be larger than or equal to max(|cond_os|, |cond2_os|) (see Section 5.6.2.8).
2. If L1 or L2 is set to zero, the automatic layout generation works incorrectly if an undersize is defined for the cond or for the cond2 layer. In this case the generated layout must be edited manually.
3. The minimum value of SL depends on cond_sm (or cond2_sm) (see Section 5.6.2.8) in the currently selected foundry.
4. C_LTYP of the connected lines should have the same CEN_MET and the same GND_MET (no integrated inter metal via).
5. See notations for the correct selection of the C_LTYP1, C_LTYP2 (in Section 5.6.2.2).

Equivalent Circuit:

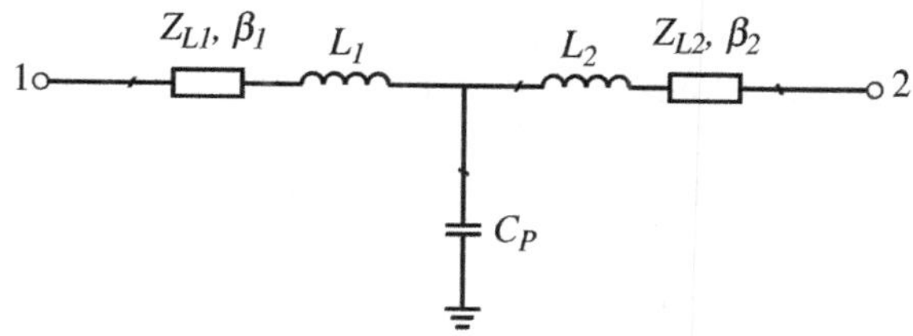

Fig. 5.7.21. The equivalent circuit of the step in a coplanar transmission line.

5.7.10 Coplanar Waveguide Taper C_TAPER

Symbol:

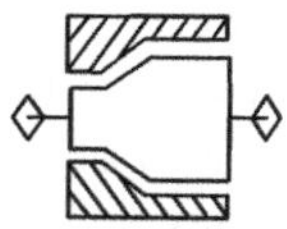

Illustration:

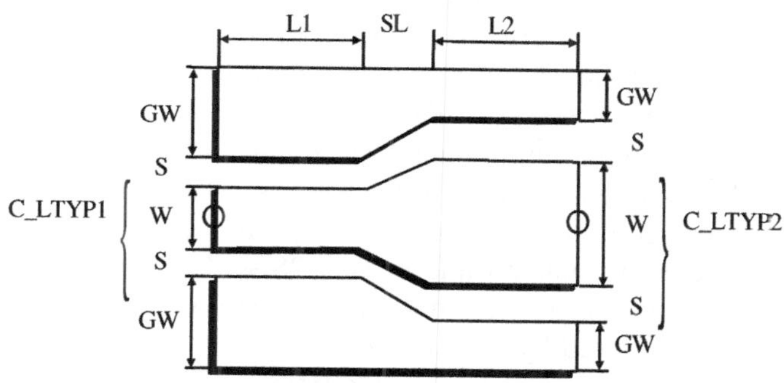

Fig. 5.7.22. The coplanar waveguide taper.

Parameters:

L1	Length of transmission line connected to the port 1
L2	Length of transmission line connected to the port 2
TL	Taper length
C_LTYP1	ID of coplanar transmission line applied at port 1
C_LTYP2	ID of coplanar transmission line applied at port 2
C_SUB	ID of coplanar substrate definition
C_GRID	ID of simulation control data
TEMP	ID of element temperature definition used for noise computation

Application Range[7]:

$$\mathrm{TL} = 0.5 \times n \times \mathrm{DL}^7, \qquad n = 1, 2, 3, \ldots.$$

Notes/Equations/References:

1. The C_TAPER module represents a piece of a tapered coplanar line. The dimensions of lines at both sides of a taper are defined using data items addressed by C_ LTYP1 and C_TYP2. TL is the step length. L1 and L2 are the lengths of connected coplanar lines at ports.
2. A zero or negative value for L1 or L2 can be meaningful during the simulation. In such cases, the automated layout generation works incorrectly and the generated layout has to be edited manually. The minimum values for L1 or L2 depend on the metal layer over sizes in the currently selected foundry. In case of C_LAYER_DEFAULT, L1 and L2 should be greater than or equal to max(|cond_os|, |cond2_os|).
3. C_LTYP1 and C_LTYP2 must have the same CEN_MET and GND_MET.
4. The difference between the center line widths (W2 – W1) has to be an integer multiple of DL. Otherwise the simulator will shift the line at port 1 by a half of grid size (DL/2). In this case the step will become asymmetrical and the user gets the message: "(W2 – W1)/DL no integer → asymmetrical taper." Please note that this change is not considered by layout generation. *The layout of a taper is always symmetrical.*
5. If W1 + 2S1 + 2GW1 ≠ W2 + 2S2 + 2GW2, there could be a step in the outer contour of the ground strips. In this case, the ground width of the line with smaller value of W + 2S + 2GW is increased (see layout), so that there is no step in the outer contour of ground planes. This change is considered by the simulation.
6. If L1 and L2 in selected C_GRID are not set to –1 or –2 (auto-sizing), the minimum values for L1 and L2 are $L1_{min} = \max(W1 + 2*S1 + 2*GW1, W2 + 2*S2 + 2*GW2)$, $L2_{min} = 8*DL\{C\_GRID\} + TL$.

   See notations in Section 5.6.2.2 for the correct selection of the C_LTYP1, C_LTYP2.

Equivalent Circuit:

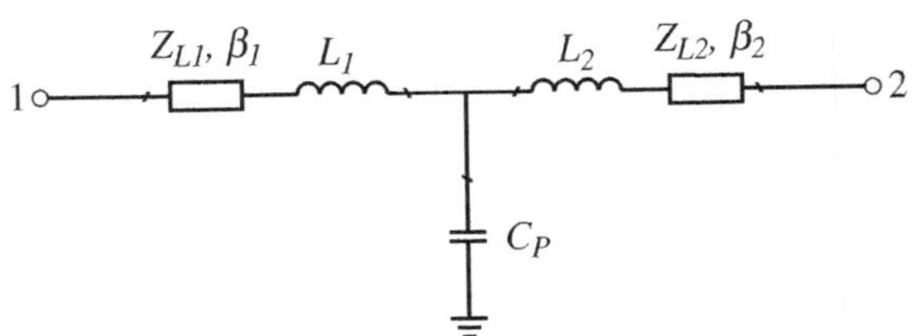

Fig. 5.7.23. Equivalent circuit of the coplanar waveguide taper.

[7] See C_GRID for information on DL.

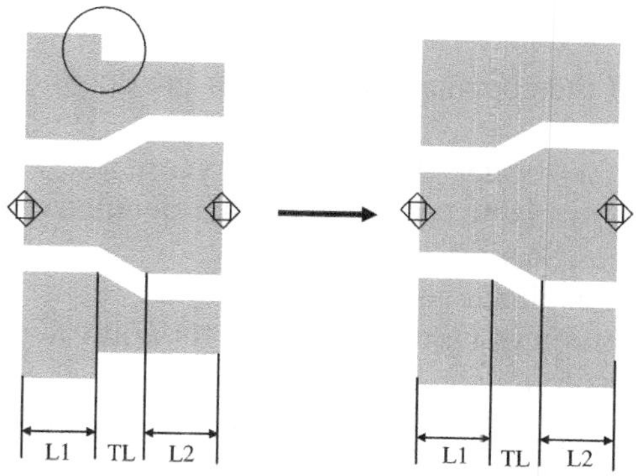

Fig. 5.7.24. Layout of the coplanar waveguide taper.

Layout:

1. If W1 + 2S1 + 2GW1 ≠ W2 + 2S2 + 2GW2, the ground width of the line with smaller value of W + 2S + 2GW is changed (no step in outer contour of ground planes).
2. If the parameter GW_DEF in C_PROCES is set to a positive nonzero value, this value is used as ground width for layout generation.

5.7.11 Coplanar Air Bridges C_AIR

Symbol:

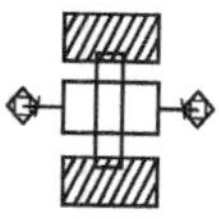

Illustration:

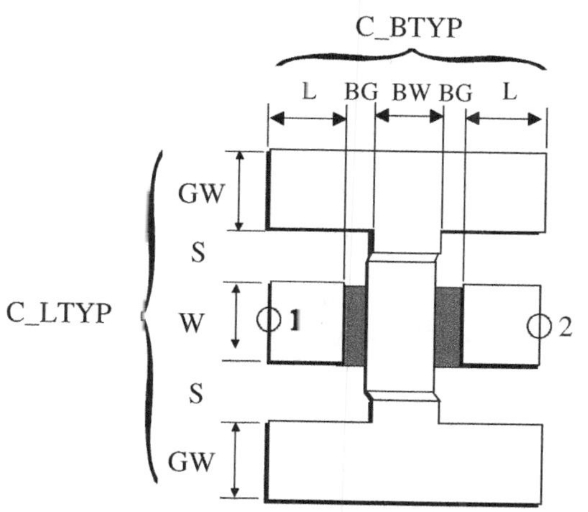

Fig. 5.7.25. The coplanar air bridge, type 2 as an example. Compare Fig. 5.6.6 for other types.

Parameters:

C_LTYP	ID of the coplanar transmission lines connected at port 1 and at port 2
C_BTYP	ID of air-bridge-type definitions (see also Fig. 5.6.6 for the three available types and further detailed parameters of air bridges, Section 5.6.2.4)
C_SUB	ID of coplanar substrate definition
C_GRID	Definition of variables needed for equivalent circuit generation
TEMP	ID of element temperature definition used for noise computation

Notes:

1. The minimum value for L in C_BTYP depends on the metal layer oversize in the currently selected foundry. In the case of C_LAYER_DEFAULT, L1 and L2 should be larger than or equal to max(|cond_os|, |cond2_os|) (see Section 5.6.2.8).
2. If L in C_BTYP is set to zero, the automatic layout generation works incorrectly if an undersize is defined for the cond or for the cond2 layer. In this case, the generated layout has to be edited manually.
3. For this element, TYPE = 0 and TYPE = −1 in C_BTYP are not allowed (see also Section 5.6.2.4, C_AIRTYP, TYPE).
4. The total sizes of the element are:

$$X = \text{C\_BTYP}\{2 \times L + 2 \times \text{BG} + \text{BW}\}$$

$$Y = \text{C\_LTYP}\{W + 2 \times S + 2 \times \text{GW}\}$$

5. See notations for the correct selection of the C_LTYP and C_BTYP in Section 5.6.2.2.

Equivalent Circuit:

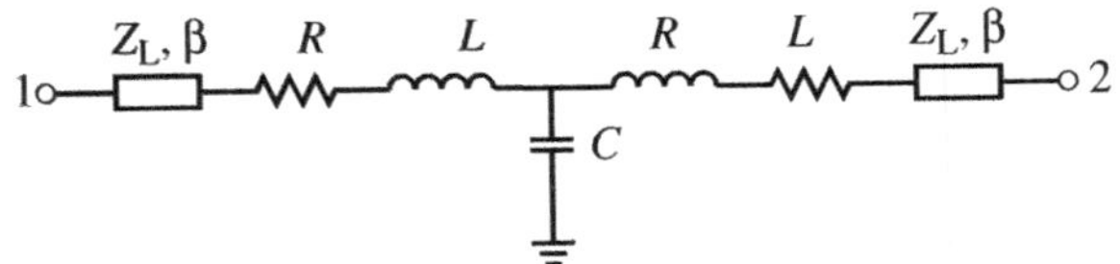

Fig. 5.7.26. The equivalent circuit of the coplanar air bridge.

5.7.12 Bend in a Coplanar Transmission Line C_BEND

Symbol:

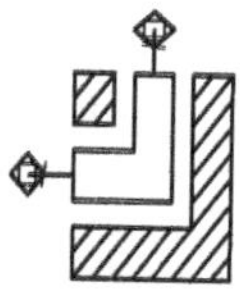

Illustration:

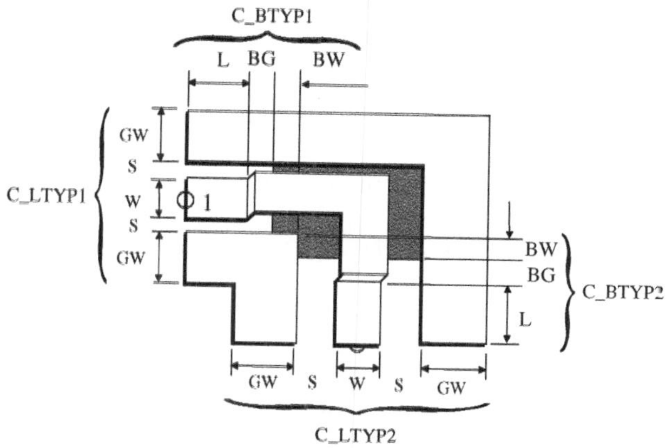

Fig. 5.7.27. The bend in a coplanar transmission line (example: air-bridge bend).

Parameters:

C_LTYP1	ID of coplanar transmission line connected at port 1
C_LTYP2	ID of coplanar transmission line connected at port 2
C_BTYP1	ID of the air bridge used at port 1
C_BTYP2	ID of air bridge used at port 2
C_SUB	ID of coplanar substrate definition
C_GRID	Definition of variables needed for equivalent circuit generation
TEMP	ID of element temperature definition used for noise computation

Notes/Equations/References:

1. The C_LINTYP and C_AIRTYP data items control the bend configuration.
2. If the feed lines defined by C_LTYP1 and C_LTYP2 have different levels (see CEN_MET and GND_MET keyword of C_LINTYP), an inter-metal via is automatically included in the bend structure.
3. The keyword TYPE of the respective C_AIRTYP data item controls whether the structure contains air bridges or not. Normally, air bridges are necessary to suppress the odd-mode excitation. Due to the applied air bridges, additional feed lines with a length defined in the respective C_AIRTYP data item are connected to the bend structure. These additional feed lines connect the reference planes with bridges. There is no additional line between the bridges and the bend structure itself.

4. The bend module is able to handle an integrated step, which means that the two feed lines may have different line widths and slot widths.
5. If the keyword TYPE in one of the C_AIRTYP data items is set to −1 (air-bridge bend; see Fig. 5.7.27), the TYPE of the other C_AIRTYP data item has also to be −1.
6. Both C_AIRTYP data items should have the same value for the keyword DIE_IDX.
7. The total sizes of the discontinuity are

$$X = C\_BTYP1\{L + BG + BW\} + C\_LTYP2\{W + 2 \times S + GW\}$$
$$Y = C\_BTYP2\{L + BG + BW\} + C\_LTYP1\{W + 2 \times S + GW\}$$

8. If the keyword TYPE in one of the C_AIRTYP data items is set to zero and the parameters BG and BW are nonzero values, these parameters will *not* be set automatically to zero. This means that the total size of the discontinuity is not changed if only TYPE is set to zero.
9. See also the notations for the correct selection of the C_LTYP and C_BTYP in Section 5.6.2.4.

Equivalent Circuit:

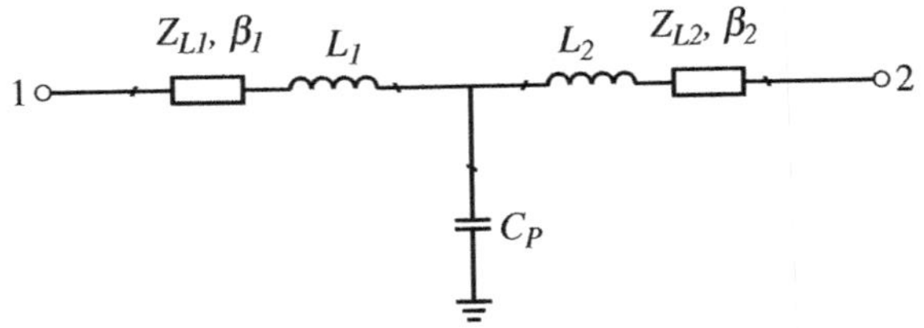

Fig. 5.7.28. The equivalent circuit of the bend in a coplanar transmission line.

5.7.13 T-Junction in Coplanar Transmission Lines C_TEE

Symbol:

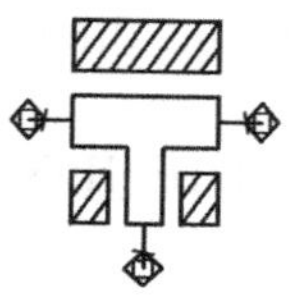

Illustration:

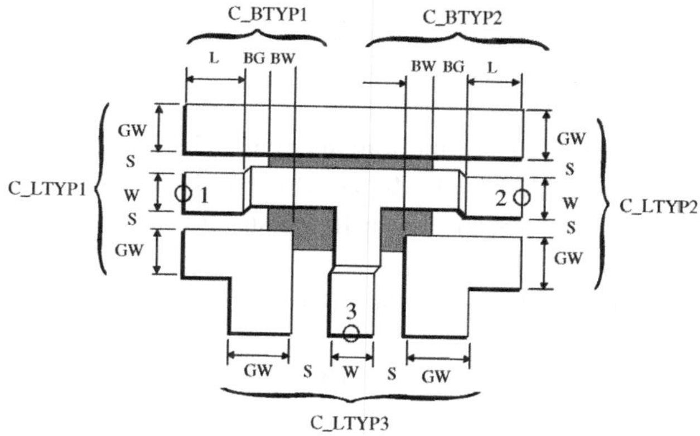

Fig. 5.7.29. The T-junction in coplanar transmission lines (example: air-bridge T-junction).

Parameters:

C_LTYP1	ID of coplanar transmission line connected at port 1
C_LTYP2	ID of coplanar transmission line connected at port 2
C_LTYP3	ID of coplanar transmission line connected at port 3
C_BTYP1	ID of the air bridge used at port 1
C_BTYP2	ID of the air bridge used at port 2
C_BTYP3	ID of the air bridge used at port 3
C_SUB	ID of coplanar substrate definition
C_GRID	Definition of variables needed for equivalent circuit generation
TEMP	ID of element temperature definition used for noise computation

Notes:

1. The C_LINTYP and C_AIRTYP data items control the T-junction configuration.
2. If the feed lines defined by C_LTYP1, C_LTYP2 or C_LTYP3 have different levels (see CEN_MET and GND_MET keyword of C_LINTYP), an inter-metal via is automatically included in the tee structure.
3. The keyword TYPE of the respective C_AIRTYP data item controls whether the structure contains bridges or not. Normally, bridges are necessary to suppress the odd-mode excitation. Please note that due to the applied air bridges, additional feed lines with a length defined in the respective C_AIRTYP data item are connected to the T-junction

structure. These additional feed lines connect the reference planes with the air bridges. There are no lines between the bridges and the T-junction itself.

4. The C_TEE module is able to handle integrated steps, which means that the feed lines may have different line widths and slot widths.
5. If the keyword TYPE in one of the C_AIRTYP data items is set to −1 (air-bridge T-junction; see Fig. 5.7.29), the TYPE values of the other C_AIRTYP data items have also to be −1.
6. All C_AIRTYP data items should have the same keyword DIE_IDX value.
7. The total sizes of the discontinuity are

$$X = \text{C\_BTYP1}\{L + BG + BW\} + \text{C\_LTYP3}\{W + 2 \times S\} + \text{C\_BTYP2}\{L + BG + BW\}$$

$$Y = \text{C\_BTYP3}\{L + BG + BW\} + \text{C\_LTYP1}\{W + 2 \times S + GW\}$$

8. If the keyword TYPE in one of the C_AIRTYP data items is set to zero and the parameters BG and BW are nonzero values, these parameters will *not* be set automatically to zero. This means that the total size of the discontinuity is not changed if only the TYPE value is set to zero.
9. See also the notations for the correct selection of the C_LTYP and C_BTYP in Section 5.6.2.2.

Equivalent Circuit:

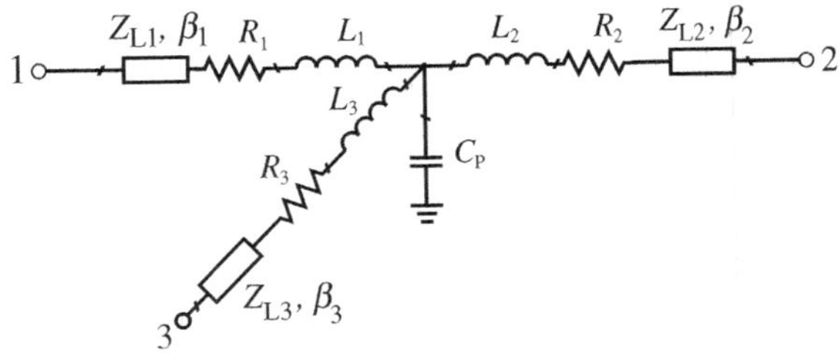

Fig. 5.7.30. The equivalent circuit of the T-junction in coplanar transmission lines.

5.7.14 Crossing of Coplanar Transmission Lines C_CROSS

Symbol:

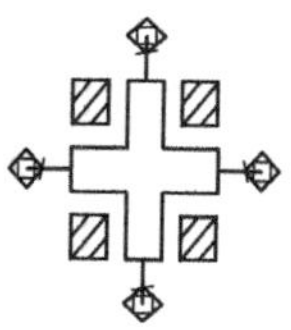

Illustration:

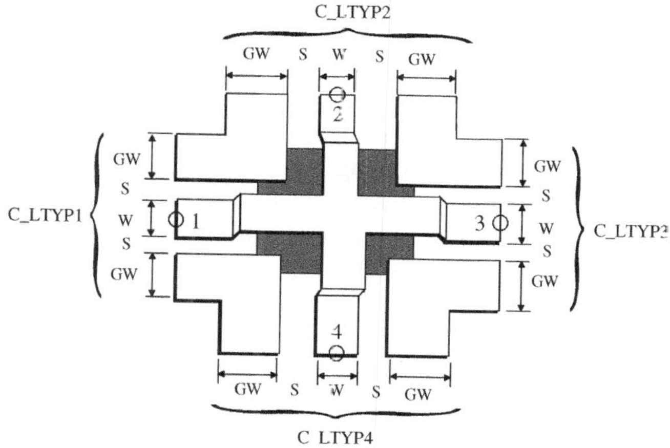

Fig. 5.7.31. The crossing of coplanar transmission lines (example: air-bridge crossing).

Parameters:

C_LTYP1	ID of coplanar transmission line connected to port 1
C_LTYP2	ID of coplanar transmission line connected to port 2
C_LTYP3	ID of coplanar transmission line connected to port 3
C_LTYP4	ID of coplanar transmission line connected to port 4
C_BTYP1	ID of the air bridge connected to port 1
C_BTYP2	ID of the air bridge connected to port 2
C_BTYP3	ID of the air bridge connected to port 3
C_BTYP4	ID of the air bridge connected to port 4
C_SUB	ID of coplanar substrate definition
C_GRID	Definition of variables needed for equivalent circuit generation
TEMP	ID of element temperature definition used for noise computation

Notes:

1. The C_LINTYP and C_AIRTYP data items control the cross-junction configuration.
2. If the feed lines defined by C_LTYP1, C_LTYP2 C_LTYP3 or C_LTYP4 have different metalization levels (see keywords CEN_MET and GND_MET of C_LINTYP) an inter-metal via is automatically included into the cross-structure.
3. The keyword TYPE of the respective C_AIRTYP data item controls whether the structure contains air bridges or not. Normally, air bridges are necessary to suppress the odd mode excitation. Please note that due

to the applied air bridges, additional feed lines with a length defined in the respective C_AIRTYP data item are connected to the crossing. These additional feed lines connect the reference planes with the bridges. There are no lines between the air bridges and the cross-junction itself.

4. The C_CROSS module is able to handle integrated steps, which means that the feed lines may have different line widths and slot widths.
5. If the keyword TYPE in one of the C_AIRTYP data items is set to –1 (air-bridge crossing, see Fig. 5.7.31), the TYPE value of the other C_AIRTYP data items has also to be –1.
6. All C_AIRTYP data items should have the same DIE_IDX keyword value (see Section 5.6.2.4).
7. The total sizes of the discontinuity are

$$\begin{aligned} X &= \mathrm{C\_BTYP1}\{L+BG+BW\}+\mathrm{C\_BTYP3}\{L+BG+BW\} \\ &\quad + \max(\mathrm{C\_LTYP2}\{W+2\times S\}, \mathrm{C\_LTYP4}\{W+2\times S\}) \\ Y &= \mathrm{C\_BTYP2}\{L+BG+BW\}+\mathrm{C\_BTYP4}\{L+BG+BW\} \\ &\quad + \max(\mathrm{C\_LTYP1}\{W+2\times S\}, \mathrm{C\_LTYP3}\{W+2\times S\}) \end{aligned}$$

Equivalent Circuit:

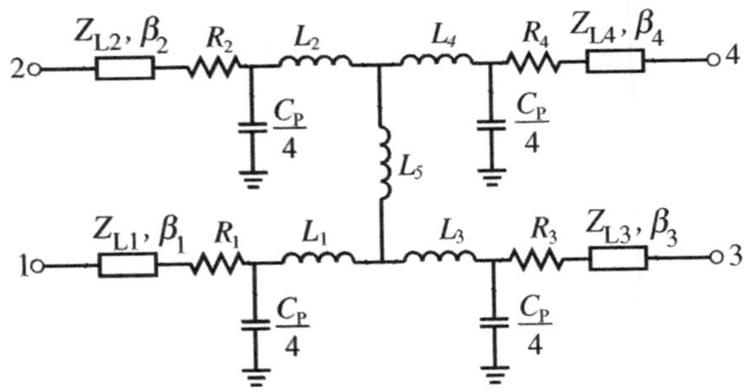

Fig. 5.7.32. The equivalent circuit of the cross junction in coplanar transmission lines.

8. If the keyword TYPE in one of the C_AIRTYP data items is set to zero and the parameters BG and BW are nonzero values, these parameters will *not* be set automatically to zero. This means that the total size of the discontinuity is not changed if only the TYPE value is set to zero.
9. See also the notations for the correct selection of the C_LTYP and C_BTYP in Section 5.6.2.2.

5.7.15 Coplanar Interdigital Capacitor C_IDC

Symbol:

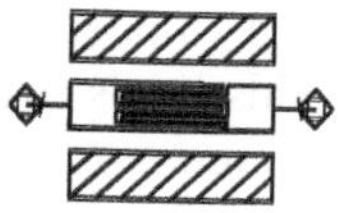

Illustration:

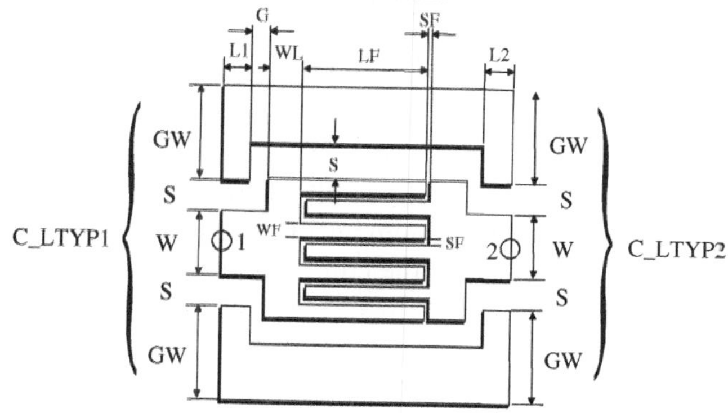

Fig. 5.7.33. The coplanar interdigital capacitor.

Parameters:

L1	Length of transmission line connected to the coplanar capacitor at port 1
L2	Length of transmission line connected to the coplanar capacitor at port 2
WF	Finger width
SF	Slot width between fingers
LF	Finger length
NF	Number of fingers (>1)
S	Slot width between ground and fingers
G	Gap width between ground and feed line
WL	Width of feed line
IDCLEVEL	Identification of metal level of the interdigital section
C_LTYP1	ID of coplanar transmission line connected at port 1
C_LTYP2	ID of coplanar transmission line connected at port 2
C_SUB	ID of coplanar substrate definition
C_GRID	Definition of variables needed for equivalent circuit generation
TEMP	ID of element temperature definition used for noise computation

Notes:

1. The minimum values of L1 and L2 depend on the metal layer oversize in the currently selected foundry. In the case of C_LAYER_DEFAULT, L1 and L2 should be larger than or equal to max(|cond_os|, |cond2_os|) (see Section 5.6.2.8).
2. If L1 or L2 is set to zero, the automatic layout generation works incorrectly if an undersize is defined for the cond or for the cond2 layer

(see Section 5.6.2.8). In this case, the generated layout has to be edited manually.

3. The minimum value for WF, WL, SF, G, and S depends on the parameter cond_wm (or cond2_wm) and cond_sm (or cond2_sm) in C_LAYER data item of the selected foundry (see Section 5.6.2.8).
4. LF should always be larger than SF.
5. NF is an integer number. The minimum value for NF is 2. Please note that for odd numbers of NF, the number of fingers connected to port 1 is larger than the number of fingers connected to port 2. This leads to different values of the equivalent capacitances CP1 and CP2 (see equivalent circuit; compare also Section 4.2).
6. See also the notations for the correct selection of the C_LTYP1 and C_LTYP2 in Section 5.6.2.2.

Equivalent Circuit:

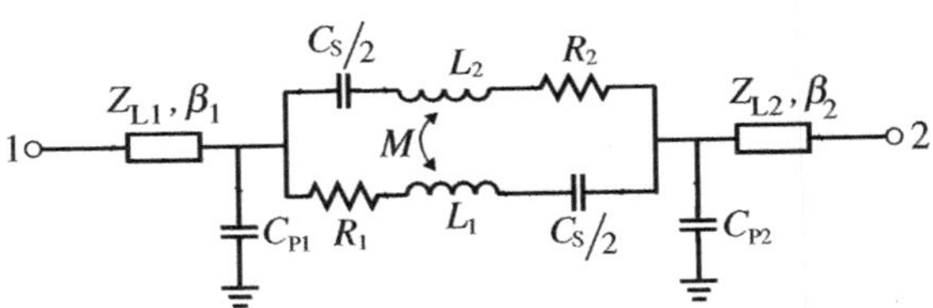

Fig. 5.7.34. The equivalent circuit of the coplanar interdigital capacitor.

5.7.16 Coplanar Rectangular Inductor C_RIND

Symbol:

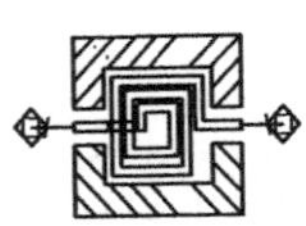

Illustration:

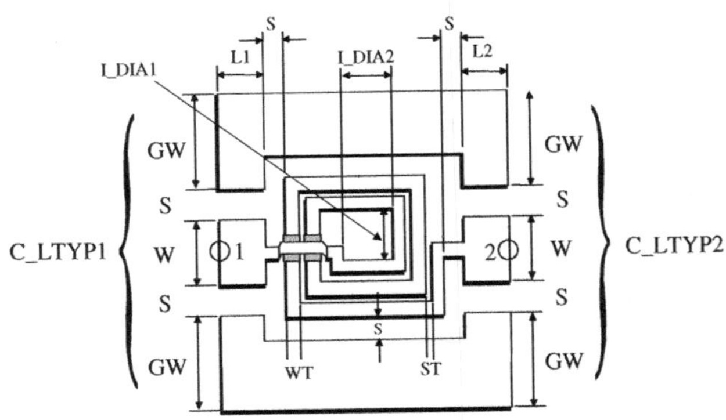

Fig. 5.7.35. The coplanar rectangular inductor.

Parameters:

L1	Length of transmission line connected to inductor port 1
L2	Length of transmission line connected to inductor port 2
N	Number of turns (1.5, 2.5, 3.5, . . .)
WT	Track width
ST	Slot width between tracks
I_DIA1	Inner diameter 1
I_DIA2	Inner diameter 2
S	Slot width between track and ground
INDLEVEL	Identification of the metal level of inductor windings
C_LTYP1	ID of coplanar transmission line connected to port 1 (backtrack)
C_LTYP2	ID of coplanar transmission line connected to port 2
C_BTR	ID of backtrack bridge
C_SUB	ID of coplanar substrate definition
C_GRID	Definition of variables needed for equivalent circuit generation
TEMP	ID of element temperature definition used for noise computation

Notes:

1. The minimum value of L1 and L2 depends on the metal layer oversize in the currently selected foundry. In the case of C_LAYER_DEFAULT, L1 and L2 should be larger than or equal to max(|cond_os|, |cond2_os|) (see Section 5.6.2.8).
2. If L1 or L2 is set to zero, the automatic layout generation works incorrectly if an undersize is defined for the cond or for the cond2 layer (see Section 5.6.2.8). In this case, the generated layout has to be edited manually.
3. The minimum value of WT, ST, and S depends on the parameter cond_wm (or cond2_wm) and cond_sm (or cond2_sm) in the C_LAYER data item of the selected foundry (see Section 5.6.2.8).
4. I_DIA1 and I_DIA2 should always be larger than WT + ST.
5. The minimum value for N is 1.5.
6. The parameter TYPE in the data item addressed by C_BTR should have values between 1 and 2. The value for INDLEVE and the center metal level of the line at port 1 (CEN_MET in C_LTYP1) have to be equal.
7. See also the notations for the correct selection of the C_LTYP1 and C_LTYP2 in Section 5.6.2.2.

Equivalent Circuit:

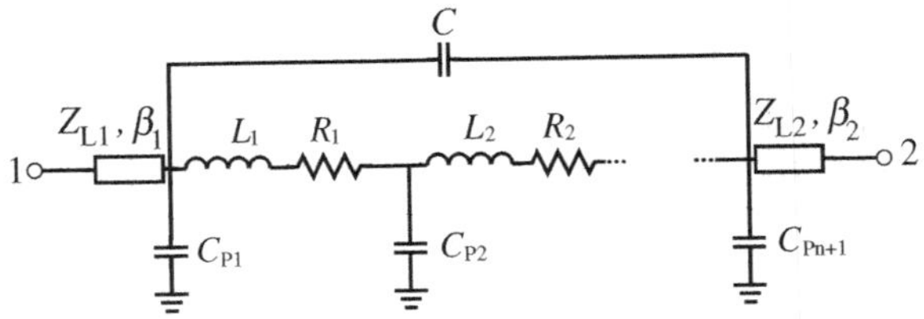

Fig. 5.7.36. The equivalent circuit of the coplanar rectangular inductor.

5.7.17 Coplanar Thin-Film Resistor C_TFR

Symbol:

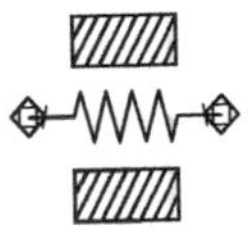

Illustration:

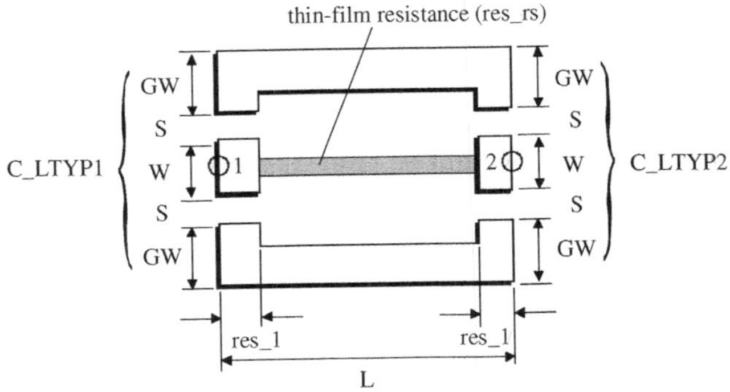

Fig. 5.7.37. The coplanar thin film resistor.

Parameters:

L	Total length of the resistor
C_LTYPR	ID of the coplanar transmission line in the resistive section
C_LTYP1	ID of coplanar transmission line connected to port 1
C_LTYP2	ID of coplanar transmission line connected to port 2
RES_IDX	Index of the sheet resistivity (ohm/square) definition in the currently selected foundry (res_rs; see Section 5.6.2.8)
C_SUB	ID of coplanar substrate definition

C_GRID	Definition of variables needed for equivalent circuit generation
TEMP	ID of element temperature definition used for noise computation

Notes:

1. The C_TFR module represents a series thin-film resistor in coplanar line technique.
2. The C_TFR module is *not* able to additionally calculate a step (i.e. the same C_LTYP ID for both ports and the resistive section are needed). Therefore, if a step configuration is used for layout reasons, the effect of such a discontinuity on the component properties is neglected.
3. In the case of the default foundry, RES_IDX is always 1 (compare Section 5.6.2.4).
4. The minimum value for res_l is res_min (10 μm). The effective length of the resistive section is given by L – 2 × res_l.
5. If L is set to zero, the automatic layout generation works incorrectly if res_l > 0. In this case, the generated layout has to be edited manually.
6. The parameters CEN_MET and GND_MET in C_LTYPR are ignored. GND_MET in C_LTYP1 is used for the ground planes within the resistive section.
7. See also the notations for the correct selection of the C_LTYP1. C_LTYP2, and C_LTYPR in Section 5.6.2.2.

Equivalent Circuit:

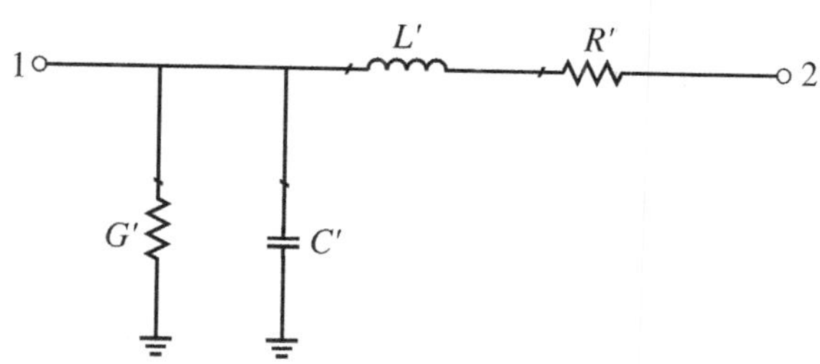

Fig. 5.7.38. The equivalent circuit of the coplanar thin film resistor.

5.7.18 Coplanar Metal–Insulator–Metal Capacitor C_MIM

Symbol:

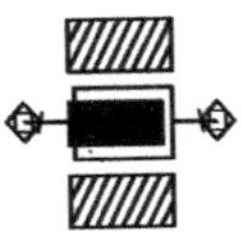

Illustration:

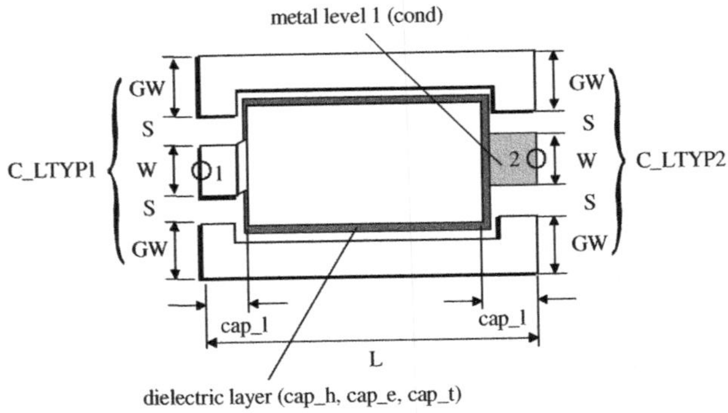

Fig. 5.7.39. The coplanar metal–insulator–metal (MIM) capacitor.

Parameters:

L	Total length of the capacitor
C_LTYPC	ID of the coplanar transmission line in the capacitive section
C_LTYP1	ID of coplanar transmission line connected to port 1
C_LTYP2	ID of coplanar transmission line connected to port 2
DIE_IDX	Index for dielectric layer definition of capacitors in the currently selected foundry (die_h, die_er, die_td; see Section 5.6.2.7)
C_SUB	ID of coplanar substrate definition
C_GRID	Definition of variables needed for equivalent circuit generation
TEMP	ID of element temperature definition used for noise computation

Notes:

1. The C_MIM module represents a series metal–insulator–metal capacitor (MIM capacitor) in coplanar line technique.
2. The C_MIM module is *not* able to additionally calculate a step. Therefore, if a step configuration is used for layout reasons, the effect of such a discontinuity is neglected in the component simulation.
3. In case of the default foundry, two different dielectric layers are available. The index DIE_IDX is used for the selection of the desired layers. DIE_IDX can be 1 or 2. If DIE_IDX = 1, the parameter die_h1, die_er1, and die_td1 are used (see Section 5.6.2.7).
4. The minimum value for L depends on the value of cap_l and the minimum length of the dielectric layer in the currently selected foundry.

The effective length of the capacitive section is given by L – 2 × cap_l. The oversize of the lower metal plate is defined by the keyword "cap_os" of the actual process.

5. If L is set to zero, the automatic layout generation works incorrectly if cap_l > 0. In this case the generated layout has to be edited manually.
6. See notations for the correct selection of the C_LTYP1, C_LTYP2 and C_LTYPC in Section 5.6.2.2.

Equivalent Circuit:

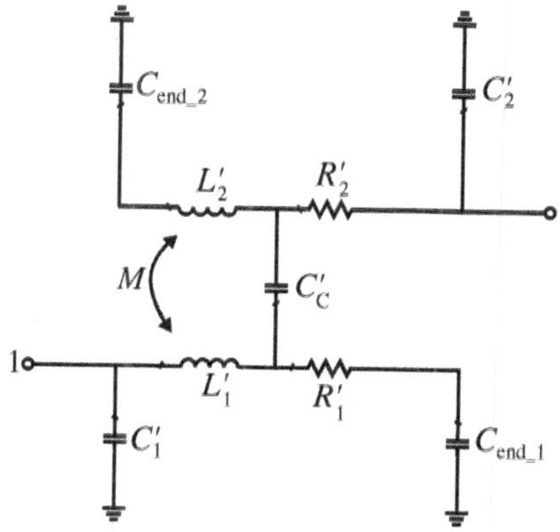

Fig. 5.7.40. Equivalent circuit of the coplanar metal–insulator–metal (MIM) capacitor.

BIBLIOGRAPHY

1. N. H. L. Koster, S. Koßlowski, R. Bertenburg, S. Heinen, and I. Wolff, Investigation on air bridges used for MMICs in CPW technique, in: *Proceedings, 19th European Microwave Conference*, 1989, pp. 666–671.
2. G. Kibuuka, R. Bertenburg, M. Naghed, and I. Wolff, Coplanar lumped elements and their application in filters on ceramic and gallium arsenide substrates, in: *Proceedings, 19th European Microwave Conference*, 1989, pp. 656–661.
3. M. Naghed and I. Wolff, Equivalent capacitances of coplanar waveguide discontinuities and interdigitated capacitors using a three-dimensional finite difference method, *IEEE Trans. Microwave Theory Tech.*, vol. MTT-38, no. 12, Dec. 1990, pp. 1808–1815.
4. M. Naghed and I. Wolff, A three-dimensional finite-difference calculation of equivalent capacitances of coplanar waveguide discontinuities, in: *IEEE MTT-S International Microwave Symposium Digest*, 1990, pp. 1143–1145.
5. M. Naghed and I. Wolff, Multiple coupled asymmetrical coplanar waveguides and their application in interdigital filters, in: *Proceedings, 20th European Microwave Conference*, 1990, pp. 913–918.
6. M. Naghed, M. Rittweger, and I. Wolff, A new method for the calculation of the equivalent inductances of coplanar waveguide discontinuities, in: *IEEE MTT-S International Microwave Symposium Digest*, 1991, pp. 747–750.

7. M. Rittweger, M. Abdo, and I. Wolff, Three-dimensional finite difference time-domain analysis of complex coplanar discontinuities, in: *PIERS Proceedings*, Boston, July 1991, p. 308.

8. M. Rittweger, N. H. L. Koster, S. Koßlowski, R. Bertenburg, S. Heinen, and I. Wolff, Full wave analysis of a modified coplanar air bridge T-junction, in: *Proceedins, 21st European Microwave Conference*, Stuttgart, Sept. 1991, pp. 993–998.

9. J. Borkes, M. Naghed, and I. Wolff, Measurement and analysis of coplanar MMIC four-port spiral transformer, in: *Proceedings, 21th European Microwave Conference*, Sept. 1991, Stuttgart, pp. 1023–1028.

10. U. Mueller, M. Rittweger, and A. Beyer, Coplanar short considered by the TLM-method with symmetrical condensed nodes, in: *Proceedings, 21th European Microwave Conference*, Stuttgart, Sept. 1991, pp. 999–1003.

11. T. Becks and I. Wolff, Investigations of various air-bridge structures within coplanar bends and T-junctions by a full-wave method, in: *PIERS Proceedings*, Pasadena, 1993, p. 562.

12. T. Becks and I. Wolff, Investigation on the reduction of package densities in coplanar circuits, in: *IEEE MTT-S International Microwave Symposium Digest*, 1993, pp. 869–872.

13. T. Becks and I. Wolff, Full-wave analysis of various coplanar bends and T-junctions with respect to different types of airbridges, in: *IEEE MTT-S International Microwave Symposium*, Atlanta, 1993, pp. 697–700.

14. R. Kulke and T. Sporkmann, Coplanar waveguide elements for a European CAD environment, in: *Proceedings, 23rd European Microwave Conference*, 1993, pp. 209–211.

15. P. Pogatzki and O. Kramer, A coplanar element library for the accurate CAD of (M)MICs, *Microwave Eng. Eur.*, Dec./Jan. 1993/1994, pp. 41–46.

16. M. Rittweger, M. Werthen, R. Kulke, B. Hopf, P. Pogatzki, and I. Wolff, Miniaturization of MMIC inductors using a 3D FDTD approach with a SI method, in: *IEEE MTT-S International Microwave Symposium Digest*, San Diego, May 1994, pp. 1297–1300.

17. P. Pogatzki, R. Kulke, T. Sporkmann, D. Köther, R. Tempel, and I. Wolff, A comprehensive evaluation of quasi-static 3D-FD calculations for more than 14 CPW structures—Lines, discontinuities and lumped elements, in: *IEEE MTT-S International Microwave Symposium Digest*, vol. 2, San Diego, May 1994, pp. 1289–1292.

18. B. Hopf, I. Wolff, and M. Guglielmi, Coplanar MMIC bandpass filters using negative resistance circuits, in: *Proc. Microwave and Millimeter-Wave Monolithic Circuits Symposium*, San Diego, May 1994, pp. 229–231.

19. P. Pogatzki, D. Köther, R. Kulke, T. Sporkmann, and I. Wolff, Coplanar hybrids based on an enhanced inductor model for mixer applications up to 67 GHz, in: *Proceedings, 24th European Microwave Conference*, Cannes, 1994, pp. 254–257.

20. R. Kulke, P. Pogatzki, D. Köther, T. Sporkmann, and I. Wolff, Enhancement of coplanar capacitor models and verification up to 67 GHz for (M)MIC circuit design, in: *Proceedings, 24th European Microwave Conference*, Cannes, 1994, pp. 258–262.

21. D. Köther, B. Hopf, S. Koßlowski, T. Sporkmann, and I. Wolff, Active CPW MMIC circulators for the 40 GHz band, in: *Proceedings, 24th European Microwave Conference*, Cannes, 1994, pp. 542–547.

22. R. Kulke, T. Sporkmann, M. Werthen, and I. Wolff, Investigations on various coupling effects in coplanar circuits, in: *PIERS Proceedings*, Session 5A4, Noordwijk (NL), July 1994, p. 500.

23. R. Kulke, T. Sporkmann, D. Köther, I. Wolff, and P. Pogatzki, Coplanar elements support circuit designs to 67 GHz, *Microwaves & RF*, vol. 33, no. 13, December 1994, pp. 103–104, 106, 108–109, 112, 114, 116.

24. R. Kulke, T. Sporkmann, D. Köther, I. Wolff, and P. Pogatzki, Modeling and analysis aid coplanar designs, *Microwaves & RF*, vol. 34, no. 1, January 1995, pp. 89–90, 92, 95–96.

25. R. Kulke, T. Sporkmann, D. Köther, I. Wolff, and P. Pogatzki, Examine the applications of coplanar circuits, *Microwaves & RF*, vol. 34, no. 2, February 1995, pp. 112, 114, 116–117.

6

COPLANAR FILTERS AND COUPLERS

This chapter does not deal with general filter or coupler theory and design. Only very special aspects of filters designed in coplanar technology shall be discussed here. For this, several fundamental investigations on the applicability of coplanar lumped elements as filter elements or even as one element filters as well as various coplanar distributed waveguide filters will be presented here. The chapter deals more with the aspect of how to design successfully filters in coplanar technology than how to design filters in general. The same aspect applies to the design of special coplanar coupler structures that will be described in this chapter. Again no general overview on couplers will be given, but only design aspects in coplanar technology will be discussed.

6.1 COPLANAR LUMPED ELEMENT FILTERS

6.1.1 The Coplanar Spiral Inductor as a Filter

In this first section on coplanar filter structures, a brief discussion on the possibility of designing a microwave filter using only one or more coplanar spiral inductors will be given. It will be shown that this idea, in principle, leads only to the realization of band-stop and low-pass filters of low quality. But in special applications such simple filters may be an interesting solution to the circuit designer.

Coplanar Microwave Integrated Circuits, by Ingo Wolff.

The spiral inductor, shown in Fig. 6.1.1, has been simulated using the quasi-static finite difference method (see Section 4.4) and the finite difference time domain (FDTD) technique (see Section 2.1 and references 18 and 23). As has been discussed in Section 4.4, the quasi-static finite difference simulator calculates the parameters of an equivalent circuit of the inductor (Fig. 6.1.2) as a function of the geometrical dimensions, whereby the line width w_f and the slot spacing s_f are kept constant over all turns.

As can be seen from the equivalent circuit, the resulting filter is always described by longitudinal inductive elements and vertical shunt capacitances. This means that only a filter of low-pass character can be realized on this basis if the inductors are used in the frequency range below their first resonant frequency. In particular, it is not possible to realize a band-pass filter on this basis.

For the simulation using the FDTD technique, the structure of the inductor has been divided into boxes using a three-dimensional nonequidistant Cartesian grid, whereby the grid is equidistant in the horizontal level. The mesh size depends on the smallest size of the inductor, which is the size of the line width w_f or the slot spacing s_f. Moreover, the FDTD simulation needs a fictitious air bridge (bond wire) from one ground plane to the other ground plane at each port to suppress the odd mode on the coplanar structure, because

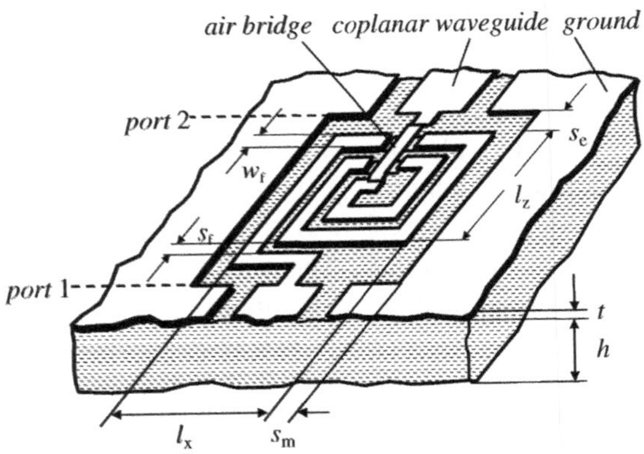

Fig. 6.1.1. Perspective view of the spiral inductor.

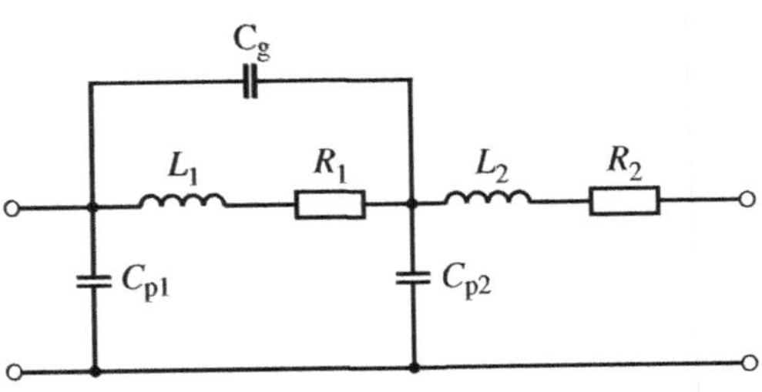

Fig. 6.1.2. Equivalent circuit of the spiral inductor with $n + 0.5$ turns.

of the asymmetrical inductor layout. The FDTD simulator calculates the scattering parameters of the spiral inductor.

6.1.2 Design and Realization

To check up the approximation limits, a series of continually expanding quadratic spiral inductors ($l_x = l_z$, see Fig. 6.1.1) with different geometrical sizes has been fabricated on an alumina substrate. The following geometrical dimensions have been used: $w_f = 25\,\mu m$, $s_f = 25\,\mu m$, $s_m = 25\,\mu m$, $s_e = 50\,\mu m$.

Table 6.1.1 shows a list of the investigated inductors. Both the quasi-static finite difference and the FDTD simulation technique have been employed for analyzing the inductors. In addition some simulated (quasi-static) low-pass filters have been produced and tested.

Spiral inductors fabricated on alumina substrate utilize bond wires instead of air bridges. Using the quasi-static approximation the bond wire inductance, L_B has been added to the inductance of the inductor L_2 of the equivalent circuit shown in Fig. 6.1.2. The inductance L_B of the bond wires is calculated using the following equation [3]:

$$L_B = 0.2 l_B [\ln(l_B/d_B) + 0.5 + 0.44\, d_B/l_B] \quad \text{nH/mm}, \qquad (6.1.1)$$

where l_B is the wire length and d_B is the wire diameter in mm. The inductance is given in nH. On the other hand, the spiral inductors fabricated on GaAs substrate have air bridges and the quasi-static analysis includes the effect of the air bridges as described in Section 3.5.5.

TABLE 6.1.1. Investigated Spiral Inductors for Filter Applications

Spiral Inductor No.	$l_x = l_z$ (μm)	Number of Windings, n
001	200	1.5
002	300	1.5
003	400	1.5
004	500	1.5
005	600	1.5
006	300	2.5
007	400	2.5
008	500	2.5
009	600	2.5
010	400	3.5
011	500	3.5
012	600	3.5
013	500	4.5
014	600	4.5
015	600	5.5

Employing the FDTD technique, the effect of the bond wires has been simulated by setting the field components of the corresponding grid points to zero. The air bridges have been simulated in the same way.

Also, the interconnection of a grounded and an open spiral inductor with a coplanar line, as depicted in Fig. 6.1.3 and Fig. 6.1.4, respectively, has been studied. These circuits have a filter characteristic that is comparable to the filter characteristic of a coiled stub coupled to a coplanar line.

To simulate the interconnection of the coplanar line and the spiral inductor accurately, the air-bridge T-junction (see Section 3.5.7) and the spiral inductor (see Section 4.4) have been described by their complete equivalent circuits. Because of the reflection asymmetry of the spiral inductor and the two possible terminations (open/short) of the second inductor port, the following four possible circuits have been studied:

1. Inductor port 1 is connected to the coplanar line and port two is open.
2. Inductor port 1 is connected to the coplanar line, and port 2 is grounded.

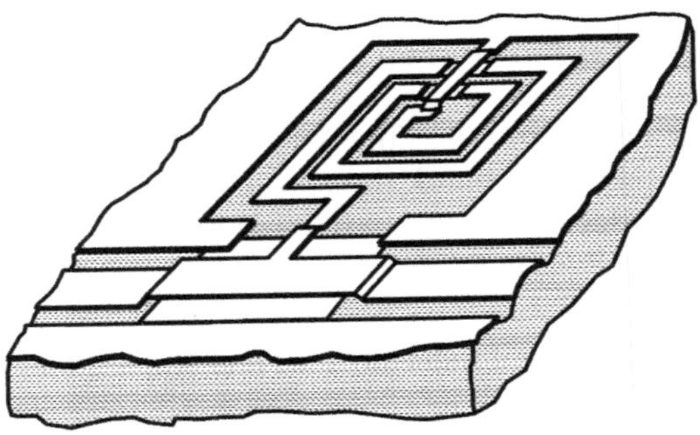

Fig. 6.1.3. Coplanar line with a spiral inductor short-circuited at one port and coupled to the coplanar waveguide by an air-bridge T-junction [36].

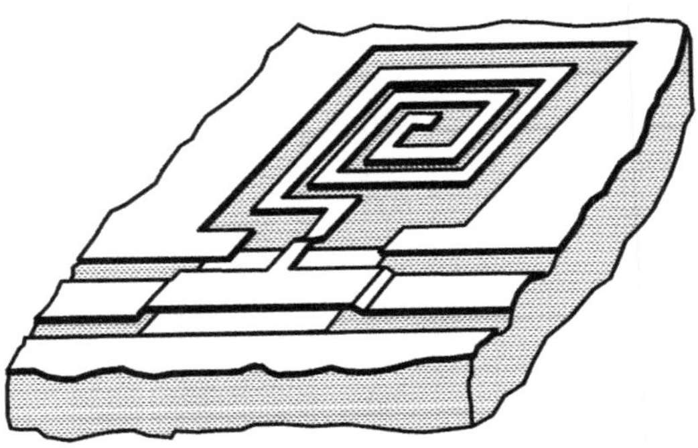

Fig. 6.1.4. Coplanar line with a spiral inductor open ended at one port and coupled to the coplanar waveguide by an air-bridge T-junction [36].

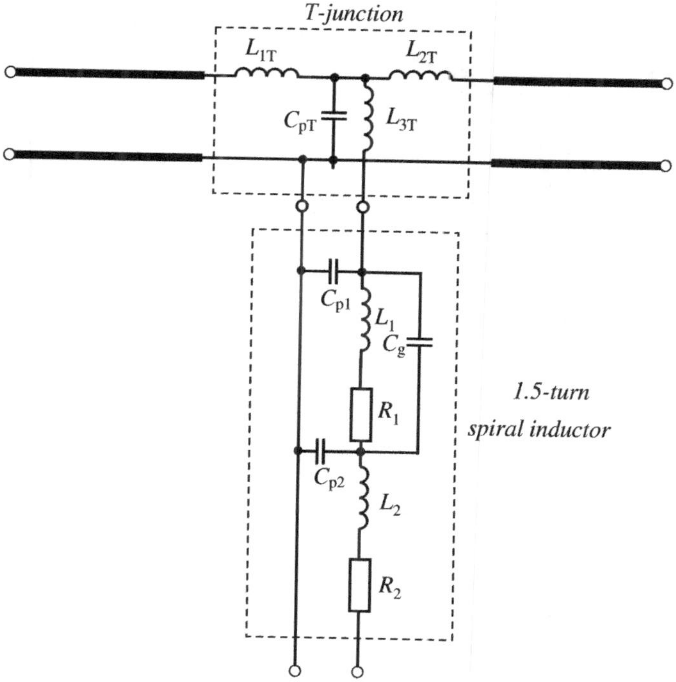

Fig. 6.1.5. Complete equivalent circuit of the interconnection of a 1.5-turn open-ended spiral inductor and a coplanar line.

3. Inductor port 2 is connected to the coplanar line, and port 1 is open.
4. Inductor port 2 is connected to the coplanar line, and port 1 is grounded.

As an example the equivalent circuit of case 1 is shown in Fig. 6.1.5.

6.1.3 Results

Comparison of both simulation methods (the quasi-static finite difference method and the FDTD technique) for the structure shown in Fig. 4.1.1 is represented in Figs. 6.1.6a and 6.1.6b for the reflection coefficient and in Fig. 6.1.7 for the transmission coefficient. The $\lambda/4$ resonant frequency of the used inductor 002 (see Table 6.1.1) is calculated to 35.3 GHz.

Up to this frequency, the quasi-static finite difference approximation is in good agreement with the measurements. On the other hand, the FDTD simulation is in a good agreement with the measurements over the whole measured frequency interval. But this simulation technique needs a larger CPU time. The results show that it is not possible to design a low-pass or a bandstop filter with good characteristics using only a single spiral inductor (real-

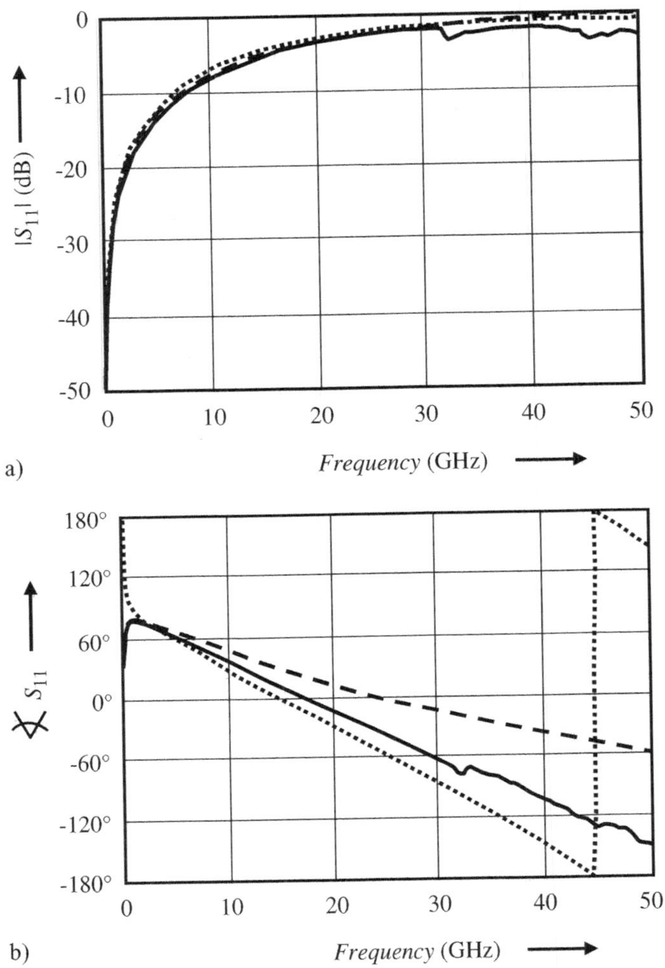

Fig. 6.1.6. Reflection coefficient. (a) Magnitude and (b) phase angle of the spiral inductor 002 (see Table 6.1.1). (———) Measured, (– – –) simulated using the quasi-static finite difference technique, (•••) simulated using the finite difference time domain (FDTD) technique.

ized either on GaAs or on an alumina substrate), because its fringing capacitances are too small to form a filter characteristic.

For example, a spiral inductor whose quasi-static simulation results (Fig. 6.1.8) promise an excellent 2.4-GHz low-pass filter has been fabricated on alumina substrate. The line length of this inductor is 24.85 mm, and the $f_{\lambda/4}$ resonant frequency is simulated to 1.3 GHz. The bad agreement between simulation and measurement above the frequency $f_{\lambda/4}$ is shown in Fig. 6.1.8. But also the FDTD simulation of various spiral inductors with turn-dependent line

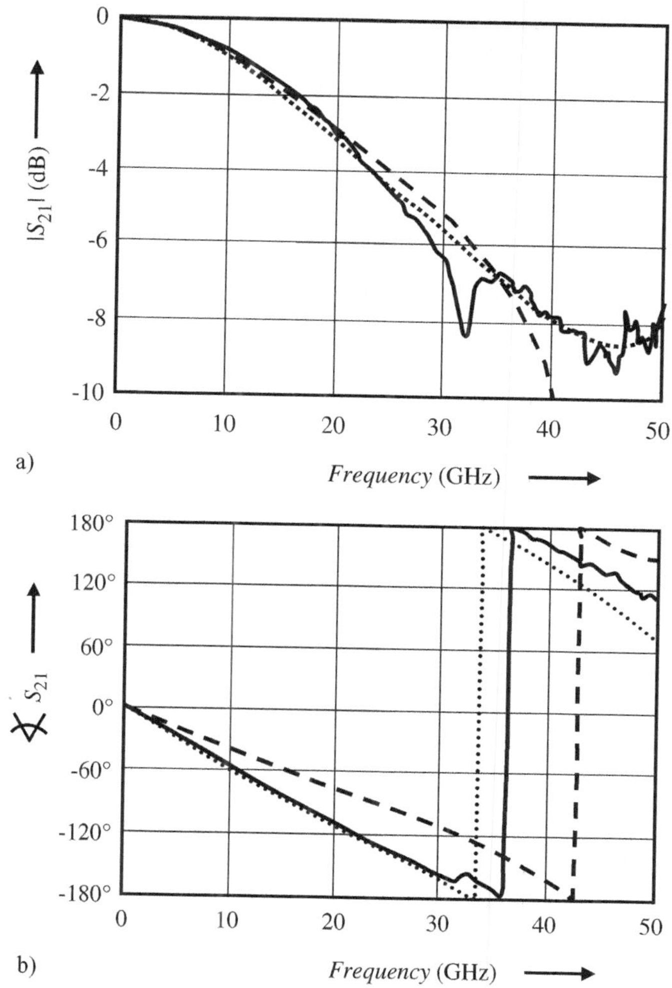

Fig. 6.1.7. Transmission coefficient. (a) Magnitude and (b) phase angle of the spiral inductor 002 (see Table 6.1.1). (———) Measured, (– – –) simulated using the quasi-static finite difference technique, (••••) simulated using the FDTD technique.

width w_f and slot spacing s_f has not led to a sufficient single spiral inductor filter with good filter properties.

Only low-pass filters or band-stop filters with bad filter properties can be realized on the basis of a single spiral inductor. For example, the transmission coefficient of inductor 005 and 011 (see Table 6.1.1) is plotted in Figs. 6.1.9 and 6.1.10. The spiral inductor 005 can be interpreted as a bad low-pass filter and inductor 011 as a bad band stop filter.

Band stop filters have been realized by connecting the spiral inductor to a coplanar line via a T-junction as shown in Fig. 6.1.3 and 6.1.4. Four different

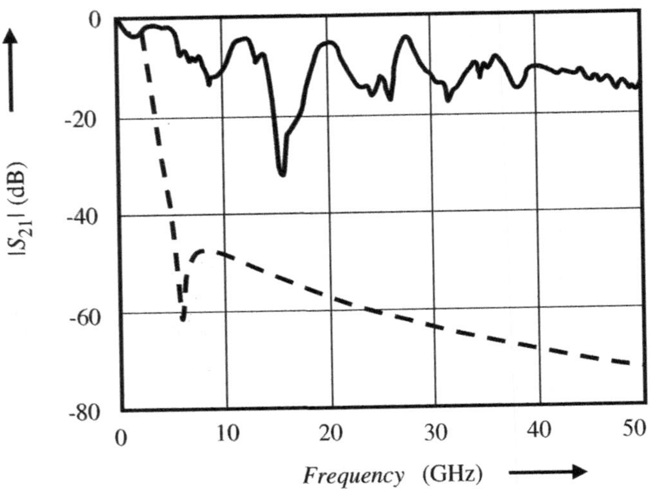

Fig. 6.1.8. Transmission coefficient of a 1.5-turn spiral inductor. $f_{\lambda/4}$ = 1.3 GHz. Line width w = 50 μm, slot width s = 300 μm, line length l = 24.85 mm. (———) Measured, (– – –) simulated.

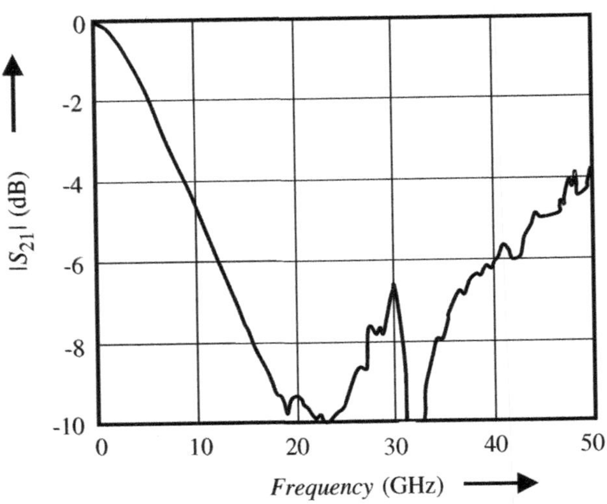

Fig. 6.1.9. Measured transmission coefficient of spiral inductor 005 (see Table 6.1.1).

spiral inductors of the mentioned interconnection type (Figs. 6.1.3 and 6.1.4) have been fabricated on alumina substrate.

The comparison of the measured and simulated transmission properties is shown in Figs. 6.1.11 and 6.1.12. The sample inductor has a line length of 6.375 mm and its simulated $f_{\lambda/4}$ resonant frequency is 5.3 GHz, and the simu-

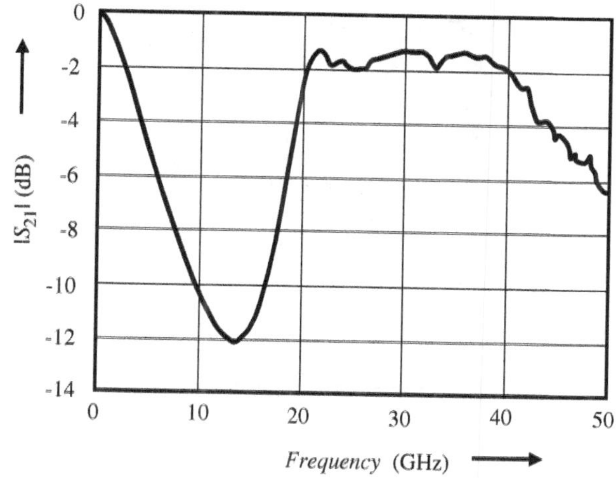

Fig. 6.1.10. Measured transmission coefficient of spiral inductor 011 (see Table 6.1.1).

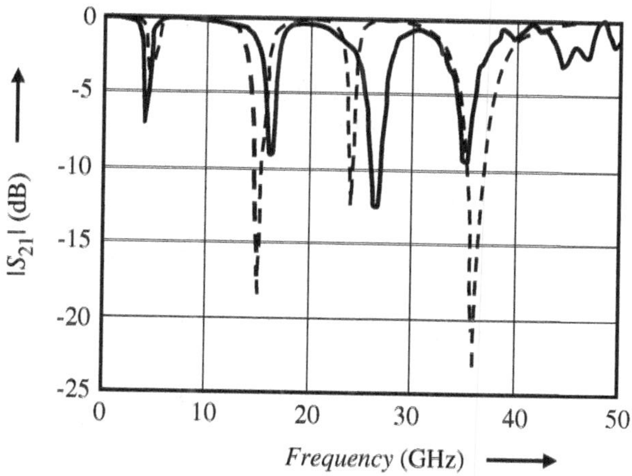

Fig. 6.1.11. Transmission coefficient of a spiral inductor coupled to a coplanar waveguide via a T-junction, as shown in Fig. 6.1.4. Inductor port ① is connected to the coplanar waveguide, port ② is open. (———) Measured, (– – –) simulated.

lated $f_{\lambda/2}$ resonant frequency is 12.2 GHz. The agreement between simulation and measurement is satisfactory up to the $f_{\lambda/2}$ frequency. As a first result, it may be concluded that the application of a single spiral inductor as a microwave filter is not very promising also for these structures.

For applications below the first resonant frequency of the inductor the quasi-static approximation supplies good results with a short CPU time. Above the first resonant frequency the FDTD simulation has to be employed. It was not possible to simulate and fabricate a filter with good filter characteristics

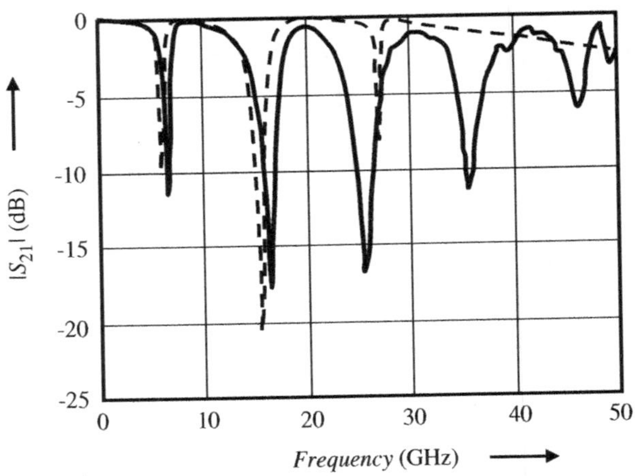

Fig. 6.1.12. Transmission coefficient of a spiral inductor coupled to a coplanar waveguide via a T-junction, as shown in Fig. 6.1.3. Inductor port ① is connected to the coplanar waveguide, port ② is shorted. (———) Measured, (– – –) simulated.

using only one spiral inductor because the fringing capacitances of the spiral inductors to ground are too small. Only the interconnection of a single open-ended or short-circuited spiral inductor to a coplanar line via a T-junction shows some promising band-stop filter characteristics. The use of two spiral inductors connected symmetrically to a coplanar waveguide using a cross junction seems to be promising to build up a band-stop filter with acceptable properties. Also, as will be shown in the next section, the series connection of more than one spiral inductor may lead to interesting solutions for filters with some special properties.

6.1.4 Phase-Shifting Filter Circuits

A phase-shifting filter circuit for the design of a balanced frequency doubler (see Section 7.6) shall be considered in this section. For the 18- to 36-GHz balanced doubler design, a 180° phase-shifting circuit is required between the gates of the two transistors in order to obtain the cancellation of the fundamental signal at the output of the transistors [38]. Such a circuit can be realized using properly designed spiral inductors. A combination of three spiral inductors results in a circuit that has a perfect 180° phase-shifting property at 18 GHz, with a minimal insertion loss. The structure has a low-pass filter characteristic. The photograph of such a circuit is shown in Fig. 6.1.13.

The circuit is realized both on ceramic and on gallium arsenide substrates, and the respective simulated and measured performance of the monolithic version is given in Fig. 6.1.14. As can be seen from the depicted dependencies, the realization of the filter on a GaAs substrate has a good performance. The

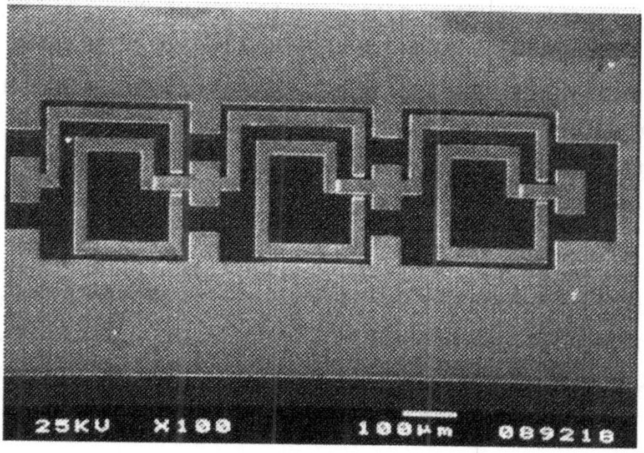

Fig. 6.1.13. Photograph of a 180° phase-shifting circuit using three spiral inductors in series. In reference to Section 4.4, Fig. 4.4.1a, each spiral inductor has the following dimensions: $w_f = 20\,\mu m$, $s_f = 30\,\mu m$, $s_e = s_m = 15\,\mu m$, $l_x = 250\,\mu m$, $l_z = 240\,\mu m$, $t = 3\,\mu m$, substrate GaAs, $\varepsilon_r = 12.9$, $h = 500\,\mu m$.

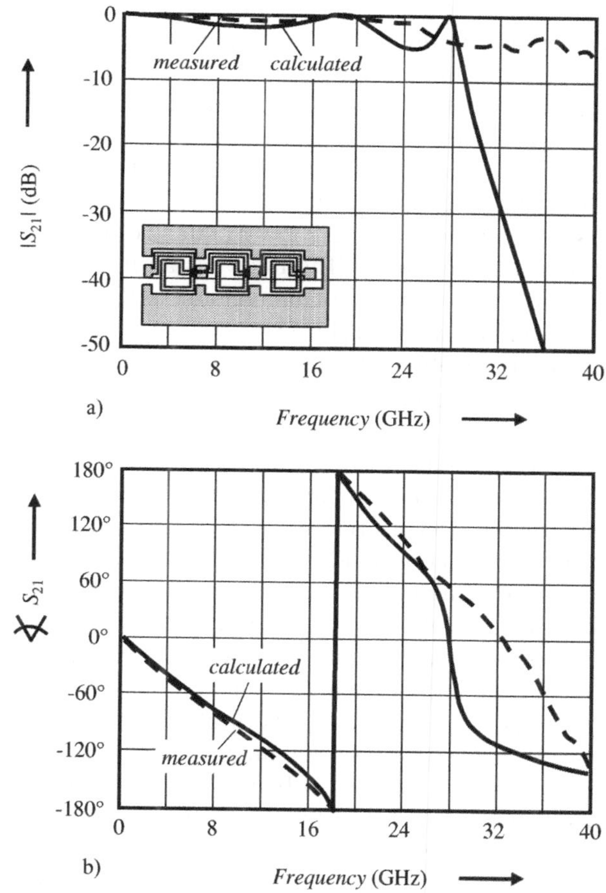

Fig. 6.1.14. Measured and calculated magnitude (a) and phase angle (b) of the transmission coefficient of a 180° phase-shifting circuit on GaAs substrate. (– – –) Measured, simulated (———).

results of repeated realizations of this circuit proved the reproducibility of the design.

In Fig. 6.1.14 discrepancies between measurement and simulation results at frequencies above 30 GHz can be observed. The reason for this is, among possible technology inaccuracies, the fact that during the simulation each spiral inductor is treated as an independent component and the interaction or coupling between the spiral inductors is not taken into consideration (see also Section 4.4.2 for the coupling effects of nearby positioned spiral inductors). Obviously, such inter-circuit interactions and loading effects are significant and unavoidable in the high-frequency region. However, in the frequency range in which the circuit is required to operate (18 GHz), the agreement between measurement and simulation is very good and the circuit fulfils the desired application.

6.2 COPLANAR PASSIVE LUMPED-ELEMENT BAND-PASS FILTERS

The aim of this section is to discuss passive lumped-element band-pass filters in coplanar waveguide technique. As examples, filters for a center frequency of 16 GHz and 29 GHz will be discussed. Lumped-element filters are reported in several publications, but there are only few papers concerning a complete analysis of such complicated structures based on electromagnetic field calculations [41]. The lumped element microstrip band-pass filter reported by Esfandiary [8] is designed for relatively low frequencies (center frequency of 12 GHz) and has insufficient slopes. The filter has an insertion loss of 1.5 dB that is relatively low. A layout of a coplanar version of this filter has been presented by Kibuuka and co-workers [12, 13].

Lumped-element filters offer several advantages such as small size, flexible design, and narrow bandwidth. On the other hand, the low-quality factor of the lumped elements leads to a relatively high insertion loss of the filters. Another problem is the parasitic effects of the lumped elements that have a negative influence on the filter characteristics. In the case of coplanar lumped elements, the parasitic effects are relatively small and can be varied, thereby changing the distances to the ground planes. This property can be used with advantage to design the lumped elements in such a way that the parasitic effects are utilized as part of the capacitive or inductive components of the filter. This presupposes, however, that (a) the lumped elements are described by broadband models taking the higher resonant effects into account and (b) the model parameters can be calculated accurately and under consideration of all geometrical effects. The analysis of lumped elements using simple models is often not accurate enough for narrow-band filter applications. On the other hand, a full-wave analysis [12, 13] of the whole filter (or even every lumped element involved) is a time-consuming and memory-storage-consuming technique and gives no information on the influence of the various elements of the filter. The segmentation method offers an alternative, but in this technique, some effects such as the influence of the air bridges and connected transmission lines on the ports as well as the interaction between the segmented parts are not taken into account. In the presented technique, the filter is divided into a number of subsections (lumped elements, discontinuities, and short lines) with no or a negligible coupling between them. A three-dimensional quasi-static analysis of each subsection considering all geometrical effects delivers the field distribution. The equivalent circuit parameters of the subsections are determined using the field information and a suitable broadband model. The characteristics of the filter are then calculated by matching the circuits of all subsections.

6.2.1 Theoretical Background

The filters that will be considered here are based on a second-order Chebyschev band-pass filter as shown in Fig. 6.2.1. The advantages of this kind of filters, generally, are the narrow bandwidth and their simple realization.

To analyze the filters, each element in the circuit shown in Fig. 6.2.1 is replaced by its lumped-element equivalent. Such a replacement is especially valid if the parasitic elements of the component have a negligibly small influence on the transmission properties of the circuit. In the case of planar lumped elements such as interdigital capacitors or spiral inductors, the parasitics have a large influence on the characteristics of the elements and they, therefore, cannot be neglected. But there is a possibility to use these effects for the

filter topology

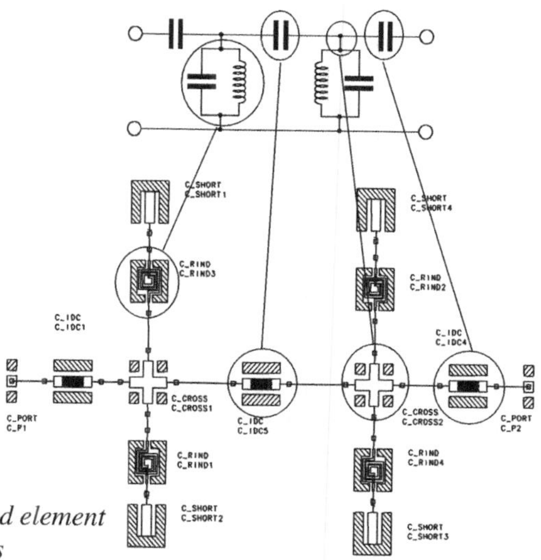

coplanar-lumped element filter schematics

Fig. 6.2.1. The lumped-element Chebychev band-pass filter and its schematic layout using the COPLAN software (see Chapter 5).

desired purpose if the broadband properties of the lumped elements can be modeled by accurate equivalent circuits. Moreover, a reliable analysis method to calculate the mentioned equivalent circuit elements [18, 23, 35] must be available.

For the realization of the Chebyschev band-pass filter shown in Fig. 6.2.1 in coplanar waveguide technique, two different lumped-element components are used: interdigital capacitors and spiral inductors. For the spiral inductors, the equivalent circuit already given in Section 4.4 (see Fig. 4.4.2) is valid. The interdigital capacitors are modeled using the equivalent circuit already discussed in Section 4.2 (compare with Fig. 4.2.2). The circuit consists of the coupling gap capacitance C_g and two parasitic capacitances C_{p1} and C_{p2} as well as a transformer representing the magnetic coupling between the fingers. The metalization losses are modeled by the frequency-dependent resistances R_{f1} and R_{f2}.

Also shown in Fig. 6.2.1 is a schematic circuit description of the filter, as realized in coplanar waveguide technology. In addition to the lumped elements that have been mentioned above, coplanar waveguide sections, coplanar crossings (see also Section 3.5.9) and coplanar short circuits (see Section 3.5.2) are used in the filter design. All these elements are described by their equivalent circuits that have been derived from the electromagnetic field distribution using the quasi-static finite difference technique, as has been described in detail in Chapters 3 and 4. The schematic description of the filters is the basis of the circuit description used in the developed circuit simulator, COPLAN, which has been described in Chapter 5.

The cross-junction component is the coplanar waveguide discontinuity that connects the above-mentioned lumped elements. It has a large influence on the characteristics of the filter and must therefore be modeled accurately. For the suppression of undesired higher modes, a full metalization underpass is used when the filter is produced on GaAs substrate (refer Section 3.5.9 and especially Fig. 3.5.84). This underpass keeps the coplanar ground planes at the same potential, which is the condition for the propagation of the coplanar even mode. If a hybrid technology is applied for fabricating the filters, bond wires substitute the air bridges across the feeding coplanar waveguide sections.

6.2.2 Properties of the Coplanar Hybrid Band-Pass Filters

A first version of the lumped-element filter for the center frequency of 17 GHz is designed and fabricated on a ceramic substrate.

The layout of this filter is shown in Fig. 6.2.2. The measured and calculated results are shown in Fig. 6.2.3. The good agreement between the calculated and measured results indicates the accuracy of the calculation method even for such relatively high frequencies.

The filter has a 3-dB insertion loss. Optimizing the components of the filter can further reduce this loss. However, an insertion loss lower than 1.5 dB cannot be reached by this kind of pure passive filter. This is because of the low

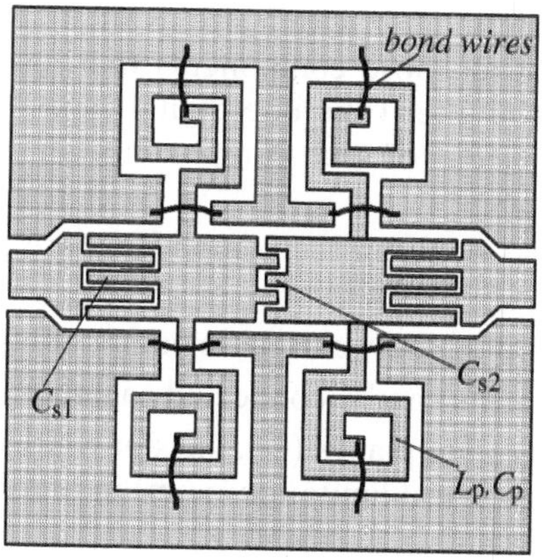

Fig. 6.2.2. Layout of the coplanar lumped-element Chebychev band-pass filter for a center frequency of 17 GHz fabricated on ceramic substrate.

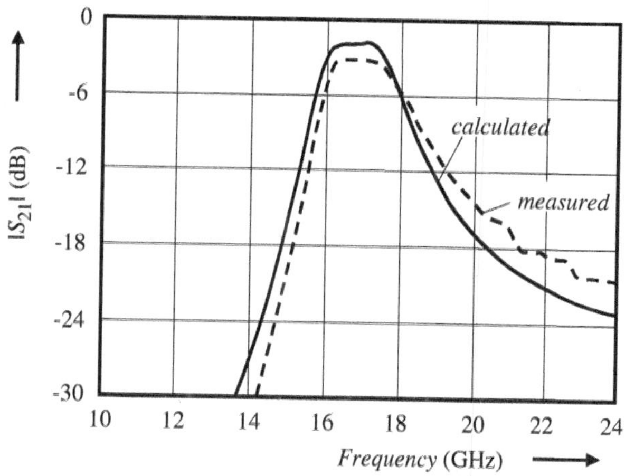

Fig. 6.2.3. Magnitude of the scattering parameters of the hybrid coplanar lumped-element Chebychev band-pass filter.

Q-factor of the planar lumped elements, especially that of the spiral inductors. An alternative solution is to use superconducting materials for the filter design or additional active circuit elements (see Section 7.3). The insufficient slope of the filter is due to the small parallel capacitance C_p that comprises of parasitic capacitances of the lumped elements to the ground planes.

As a next step, two band-pass filters for center frequencies of 5.4 GHz and 22 GHz have been designed and realized on GaAs-substrate (substrate height = 500 μm) in monolithic technology. For the MMIC design of the filter layout, several aspects are taken into account:

1. In order to realize the filter with a size as small as possible, no electroplating is used for all those components that have little effect on the insertion loss of the filter (like interdigital capacitors). As a result, the geometrical dimensions of such components can be chosen to be very small.
2. Optimum distances between the components are chosen in such a way that
   (a) undesired coupling between these components is very small and
   (b) these components can be connected together via very short lines.
3. Distances to the ground planes (i.e., the slot widths) are chosen as small as possible (technological limitation: 5 μm) in order to have large capacitances to ground.

Figure 6.2.4a shows a photograph of the 22-GHz filter. Details of the airbridge construction in the area of the cross-junction are shown in Fig. 6.2.4b. A small number of turns with large turn width are used for the spiral inductors in order to keep the losses as low as possible and to get a better quality factor for the inductor. All parts of the circuit are produced in the galvanic layer (thickness 3 μm) except the interdigital capacitors that are realized in the gate metallization layer (thickness 0.4 μm). This was necessary because the minimum slot width of 5 μm between the fingers (which was needed to have large enough capacitances) could only be produced in this layer.

The measured and calculated insertion loss of the filter is plotted in Fig. 6.2.5. The agreement between measured and calculated data is good. The center frequency is slightly shifted to higher frequencies (from 22 GHz to 22.5 GHz).

The results of the 5.4-GHz band-pass filter are plotted in Figure 6.2.6. As can be seen from this figure, the measured center frequency of two identical filters on the same wafer are between 5.3 GHz and 5.5 GHz, which is ±2% of the desired center frequency. This shows that an accurate simulation of such filters is meaningful only when the technological tolerances of the production process are very small. The measured insertion loss is 2 dB higher than the calculated value, indicating the reduced accuracy of the applied method for calculating the losses.

6.3 SPECIAL COPLANAR WAVEGUIDE FILTERS

Some special single coplanar waveguide filters will be described in this section. Again it is not the aim of this section to discuss the theory of filter design, but only to show the possibilities and the problems that are encountered when designing filters in coplanar technology. The filters described here are for use in the design of a single-device frequency doubler from 18 GHz to 36 GHz (see

a)

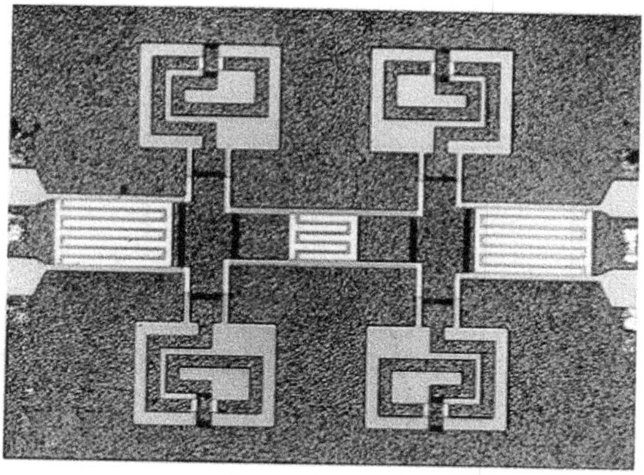

b)

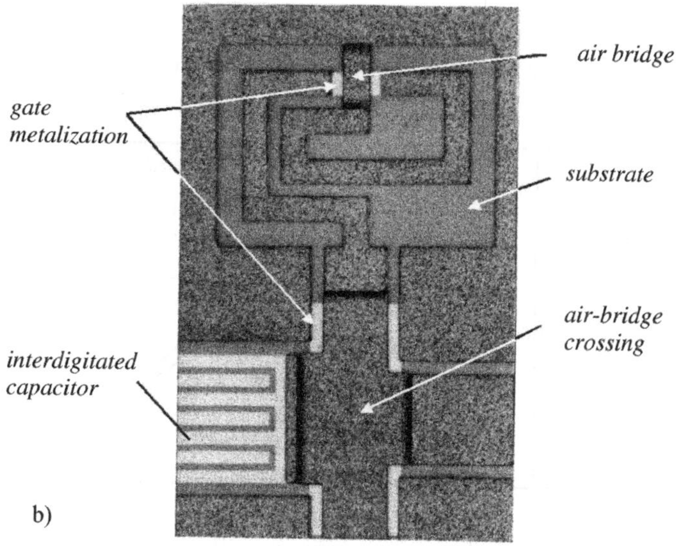

Fig. 6.2.4. Photograph of the 22-GHz lumped-element filter on GaAs (a) and a detail enlargement (b).

Section 7.6) [38]. The filters are designed for the output network in such a way that they suppress the fundamental frequency (18 GHz) and at the same time select the desired output frequency (36 GHz) optimally. In practice, this means they must have a rejection level of at least 30 dB at the fundamental frequency and a minimum insertion loss at the desired output frequency. The output filters can be of band-reject type or of band-pass type. The design of each of these filters, which are good examples for the design of coplanar distributed filter structures, will be discussed in the following sections.

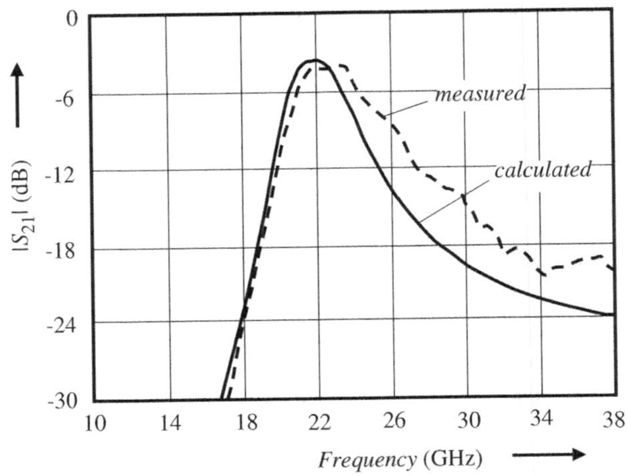

Fig. 6.2.5. Transmission parameter of the 22-GHz lumped-element coplanar filter.

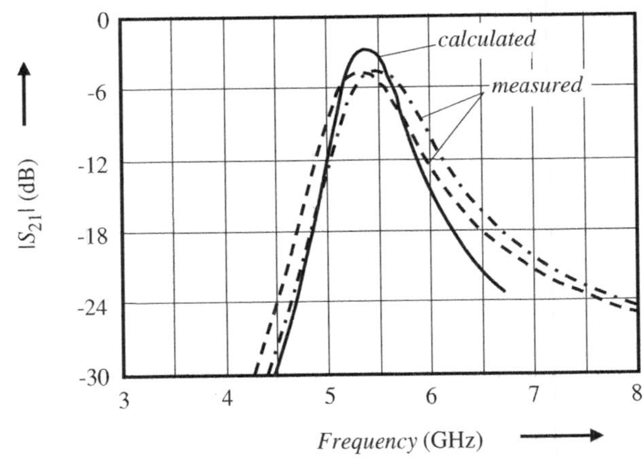

Fig. 6.2.6. Transmission parameter of the 5.4-GHz lumped-element coplanar filter.

6.3.1 The Coplanar Band-Reject Filter

6.3.1.1 The Hybrid Band-Reject Filter. Different open-ended coplanar structures (stubs) have been investigated for possible application in the realization of an output band-reject filter that will then be used in the design of frequency doublers (see Section 7.6). Most of the structures investigated have shortcomings in that either their rejection level is too small or, due to the asymmetry, the generation of higher-order modes is excessive, or both. On the other hand, a symmetrical and bent stub has the best performance regarding the fundamental rejection level and the suppression of higher-order modes. The layout of one of these structures is given in Fig. 6.3.1, and its measured scattering parameters are shown in Fig. 6.3.2.

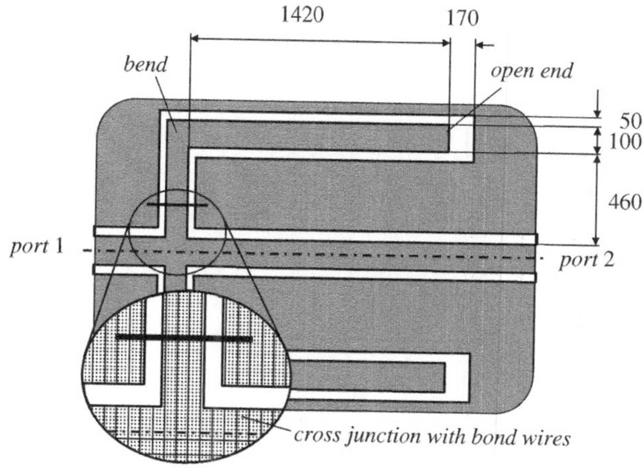

Fig. 6.3.1. Symmetrical and bent open-ended stub band reject filter on ceramic substrate. Geometrical parameters are measured in micrometers.

The design of this structure (to be called the band-reject filter from now on) is then optimized with respect to the number, length, and location of the bond-wires used, and realized on a 635-μm Al_2O_3-substrate with a permittivity of $\varepsilon_r = 9.8$. Moreover, the filter has been realized using a symmetrical stub in order to suppress the generation of odd modes without necessitating the use of additional bond wires. The band-reject filter is designed to resonate at 18 GHz and is supposed to have an effective length of 1780 μm [38]. However, since it is bent, the size it occupies is effectively reduced. The measured and calculated S-parameters of the optimized filter are compared in Fig. 6.3.2.

The essential difference between the measurement and calculation is observed around 26 GHz, where a sharp resonance phenomenon is observed in the calculated results. The measurement results also show this phenomenon but to a lesser extent, and this can be explained with the interaction of the connected discontinuities together with higher-order modes, different from the symmetric quasi-TEM mode on the stubs [38].

Another important consideration is the length and location of the bond wires. They must be positioned as close to the cross-junction as possible. Several measurement and simulation results show that if the bond wires are placed far away from the cross-junction, the resonant frequency of the filter increases. The use of more than one bond wire assures a good potential balance on both sides of the ground planes.

6.3.1.2 The Monolithic Band-Reject Filter. The problem with a hybrid realization of the band-reject filter is that the resonant frequency is influenced by the length and location of the bond wires used [38], and hence there is a difficulty reproducing the circuit accurately. To overcome this problem, the design

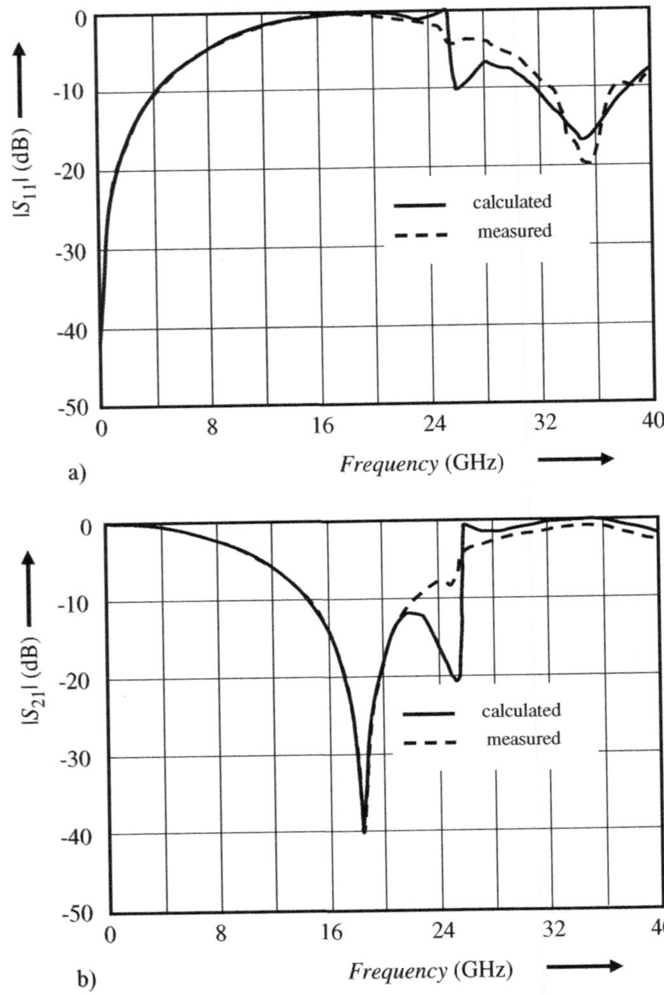

Fig. 6.3.2. The measured and calculated magnitude of the reflection coefficient (a) and the transmission coefficient (b) for the 18-GHz hybrid band-reject filter, plotted against the frequency *f*.

of the filter in the preceding section is repeated using monolithic integrated circuit technique. One of the advantages of the monolithic realization is that there is no need for bond wires. Instead, use is being made of the coplanar air bridges discussed in Section 3.5.5. Another advantage of the MMIC realization is that the circuit is accurately reproducible without any difficulty. The monolithic band-reject filter in question is realized on a 500-μm-thick GaAs substrate with a permittivity of $\varepsilon_r = 12.9$.

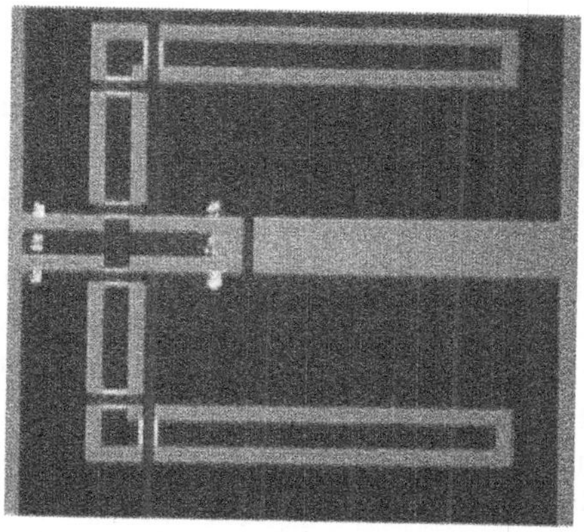

Fig. 6.3.3. Photograph of the band-reject filter using coplanar bends with air bridges passing over the inner conductor (type 2 air bridges; see Section 3.5.5).

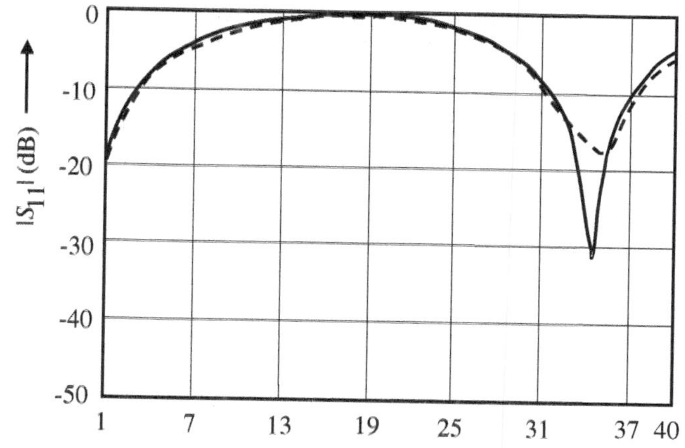

a)

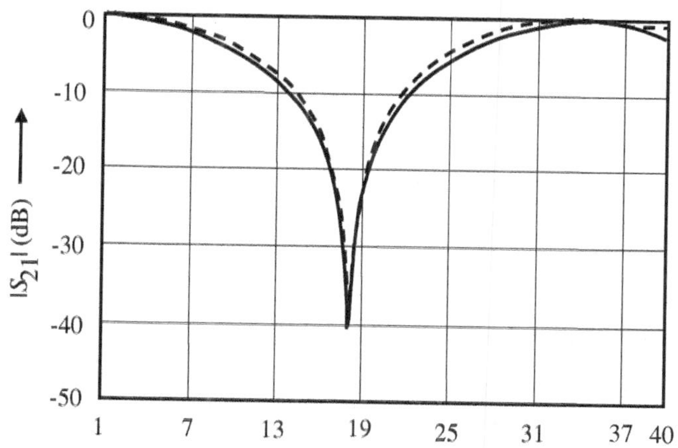

b)

Fig. 6.3.4. Measured (- - -) and calculated (———) magnitude of (a) the reflection coefficient and (b) the transmission coefficient for the monolithic band-reject filter using bends with air bridges passing over the inner conductor, plotted against the frequency.

Two versions of the monolithic band-reject filter are realized using the air-bridge bends described in Section 3.5.6. The photographs of the realized filters are shown in Figs. 6.3.3 and 6.3.5a, while the measured and calculated scattering parameters of the filters are given in Figs. 6.3.4 and 6.3.5b.

As can be seen from these results, the filter that uses air bridges of type 2 (see Section 3.5.5)—namely, the type with air bridges passing over the inner conductor—has a better performance regarding the insertion loss at 36 GHz. However, because of the superiority of the air-bridge bend in suppressing the generation of odd modes (compare also Chapter 3.5.7), the filter with air-bridge bend type 2 has been used for the final design of the frequency doublers (see Section 7.6). The discrepancy between measurement and simulation results can be attributed, among other things, to the fact that metallic losses of the discontinuities are not taken into consideration in the simulation. Furthermore, radiation loss at the open ends that is also not considered in the simulation can be another cause for the discrepancy.

This section will be concluded by presenting the results that show the performance of the overall output circuit when the band-reject filter is incorporated into the output matching network. The photograph of the complete circuit is shown in Fig. 6.3.6a.

As can be seen from the picture, the output-matching network is composed of two spiral inductors that are connected in series. The desired impedance matching can be achieved with these two spiral inductors and their associated parasitic capacitances, if the geometric dimensions and the number of windings of the spiral inductors are properly designed, which is the case here. The spiral inductors used consist of 1.5 turns. It is possible, in principle, to use only one spiral inductor of appropriate turns (say 2.5 turns) instead of cascading two spiral inductors of fewer turns. However, this causes a problem in that the first resonant frequency of the spiral inductor with several turns is lower than the one with fewer turns; therefore, there is a danger that this resonant frequency is in the frequency range at which the circuit is desired to operate. Moreover, the spiral inductor with fewer turns has more capacitance to ground, and this is exactly the property that one likes in order to obtain the desired impedance matching. In fact, this is a property that is true only for coplanar components and not for microstrip components. The measured transmission coefficients of the designed circuit with and without the output-matching network are given in Fig. 6.3.6b.

6.3.2 Coplanar Millimeter-Wave Filters

Two passive coplanar filters on GaAs substrate for V-band applications have been investigated. The first one shows a Chebyshev characteristic and utilizes coplanar waveguide segments, while the second one is a Cauer filter, where the elements of the prototype filter parameters have directly been transformed into coplanar lumped elements.

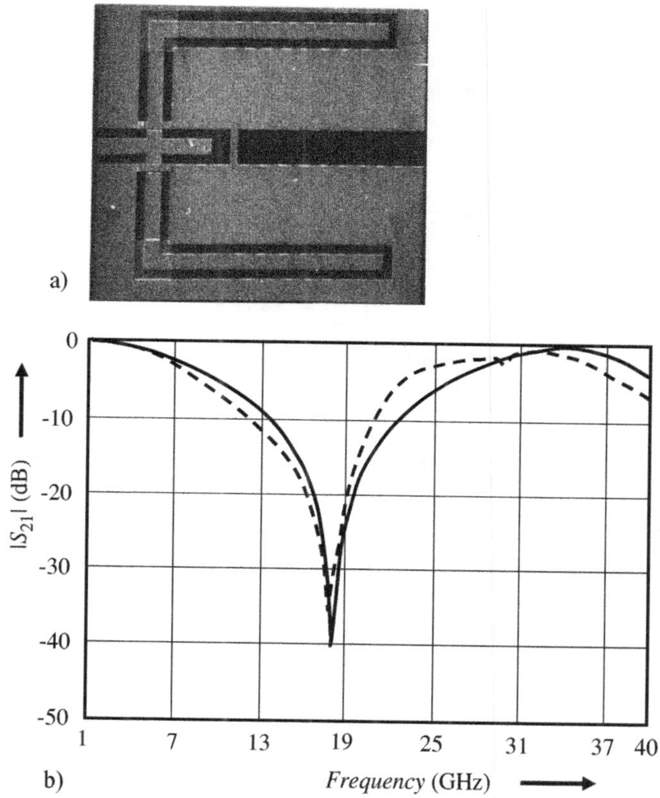

Fig. 6.3.5. (a) Photograph of the monolithic band-reject filter using coplanar air-bridge bends with metallic underpass. (b) Measured (– – –) and simulated (———) transmission coefficients in dependence on the frequency.

The specifications for the following filter synthesis come from a mixer project, where the up-converter has the LO frequency at 56.8 GHz, the IF at 5.2–9.2 GHz, and the RF at 62–66 GHz. The undesired lower side band of the mixer is at 47.6–51.6 GHz, which should be suppressed by a passive filter. The method that has been chosen for the synthesis is a very conventional one. In a first step the filter requirements for the prototype have been defined and parameters were taken from reference 6: For stop band, 48–52 GHz and $|S_{12}| < -35$ dB; for pass band, 62–66 GHz and $|S_{12}| > -0.0988$ dB.

Two filters have been developed in parallel: One consists of line-stub elements, and the other utilizes lumped elements of the coplanar library (see Chapter 5). For the first case, the requirements will be fulfilled if a Chebyshev type filter with an order of $n = 11$ is used. The second prototype filter has been

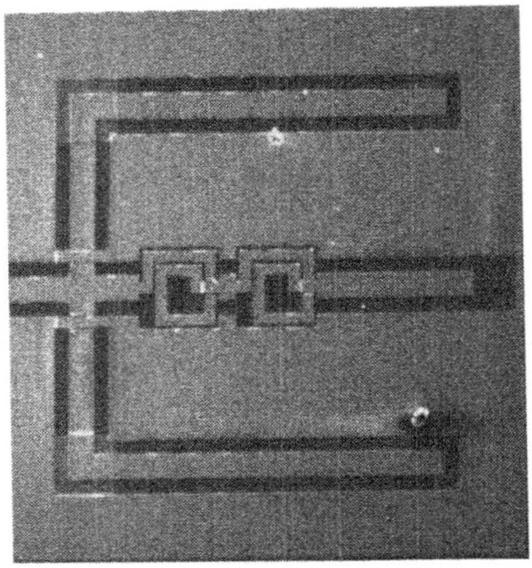

a)

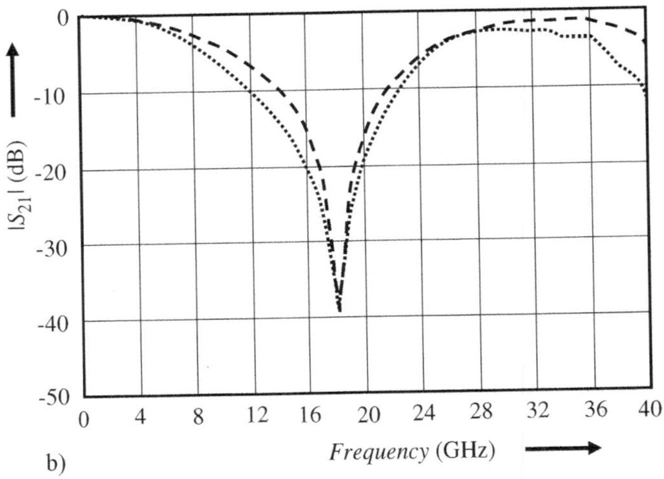

b)

Fig. 6.3.6. (a) Photograph of the complete output circuit. (b) Measured transmission coefficients with (dotted line) and without (dashed line) the spiral inductors to show the effect of the matching network on the output filter.

chosen to be a Cauer filter with an order of $n = 6$. Figure 6.3.7 shows the filter characteristics of both prototype filters.

A very conventional method of filter synthesis that leads very fast to a sufficient filter characteristic by utilizing commercially available design tools is used. The synthesis starts with the specification of a prototype filter with a stop band at 48–52 GHz and an out-of-band insertion loss of >35 dB and a pass band between 62 GHz and 66 GHz with an insertion loss of <0.0988 dB. The filter parameters were taken from the tables given in reference 6.

In the second design step, the coefficients of the prototype filters have been transformed into coplanar waveguide sections or lumped elements. For the

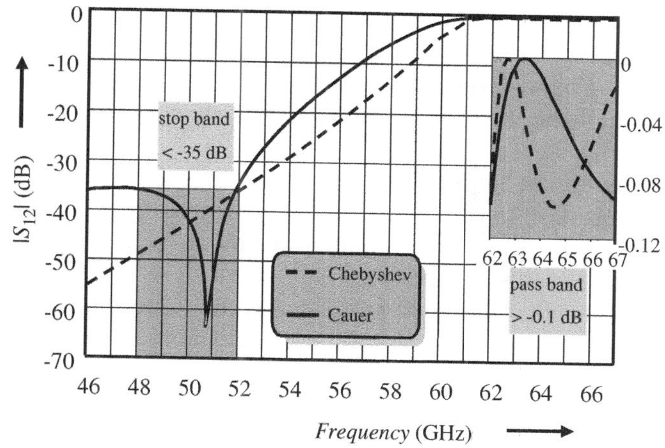

Fig. 6.3.7. Transmission coefficient $|S_{12}|$ of the Chebyshev and the Cauer prototype filters.

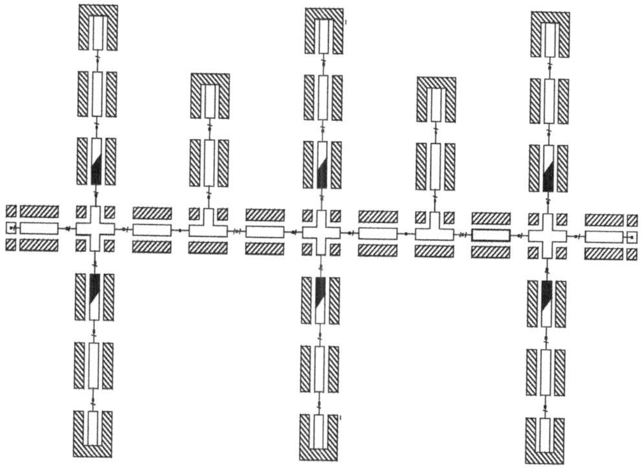

Fig. 6.3.8. Schematic circuit of the Chebyshev filter, built up with CPW elements.

Chebyshev type a line filter has been determined, where each line is described by its characteristic impedance and electrical length. A simple optimization routine determines the corresponding coplanar lines. The final circuit is built up in the schematic editor of Agilent ADS™ software. This is illustrated in Fig. 6.3.8, which shows the schematic layout. The filter characteristic changes if coplanar T- and cross-junctions are used instead of ideal nodes. The best performance of the coplanar filter has been achieved after the optimization of $|S_{12}|_{\text{CPW-filter}}$ to $|S_{12}|_{\text{Prototype-filter}}$. It is evident that the coplanar filter cannot reach the same performance as the prototype filter. Nevertheless, the simulated transmission coefficient in the pass band is better than –3 dB and lower than –25 dB in the stop band. Even more important is the good agreement between the predicted and the measured filter, which is shown in Fig. 6.3.11 for the Chebyshev filter.

When all the design steps are processed, the circuit will be ready for the layout synchronization in the circuit design program (Fig. 6.3.9). All relevant foundry parameters such as layer numbering, undersize, and oversize parameters are included in this layout.

The second filter that has been fabricated is the Cauer filter with an order of $n = 6$. In contrast to the Chebyshev filter, the elements of the prototype filter have been directly transformed into coplanar MIM capacitors and spiral inductors. The layout of this structure is illustrated in Fig. 6.3.9, too. The major advantage of this filter is its small size (8 times smaller than the Chebyshev filter).

Fig. 6.3.10 shows the realized filters on GaAs substrate material and Fig. 6.3.11 a comparison between the measured and simulated filter characteristics

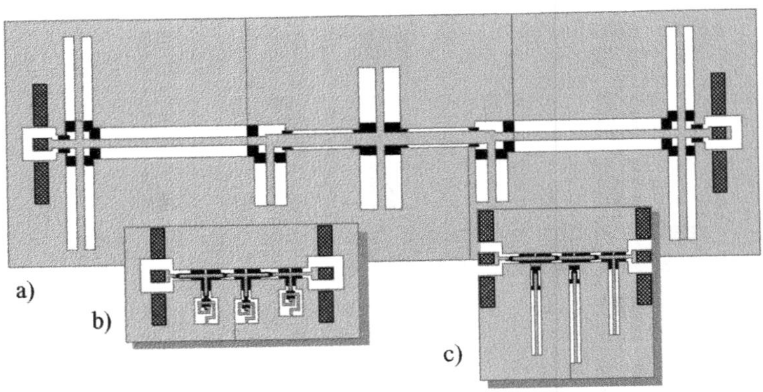

Fig. 6.3.9. (a) Chebyshev filter with line stub elements (size = 1.45 mm^2, length 1960 µm). (b) Cauer filter with lumped elements (size = 0.11 mm^2). (c) Cauer filter with lumped elements and line stubs (size = 0.19 mm^2, length 380 µm).

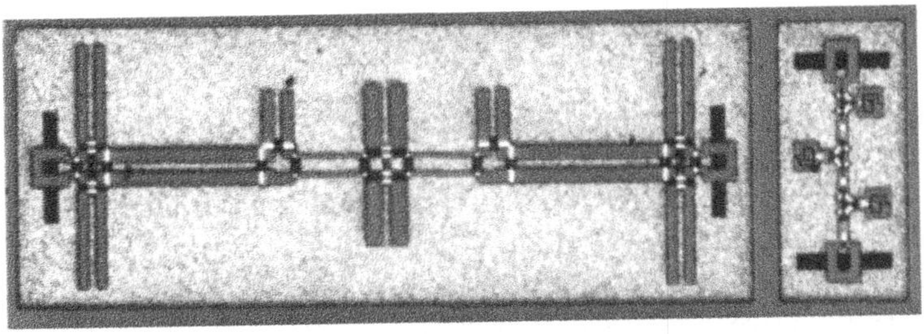

Fig. 6.3.10. Realized Chebyshev filter (left) and Cauer filter (right) for application in V-band.

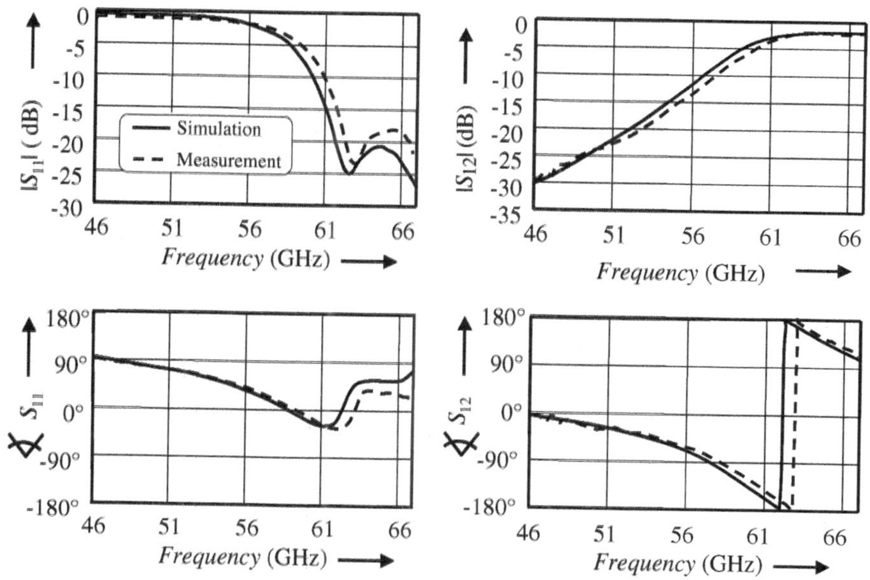

Fig. 6.3.11. Chebyshev filter: comparison between measurement and simulation.

for the case of the Chebyshev filter. The agreement between simulation and measurement is quite good which shows that the quasi-static models for the coplanar waveguide and the discontinuities like crossings and T-junctions can be used with an acceptable accuracy up to frequencies of 50–60 GHz. The realized Cauer filter, however, behaves not as expected, because the accuracy of the spiral inductor simulation was too low at these high frequencies. The coplanar inductor model with 1.5 turns is valid only up to 40 GHz (see also Chapter 6.4).

To overcome this problem, a modified Cauer filter with inductive lines instead of spiral inductors has been designed. The layout is shown in Fig. 6.3.12c. The size is slightly larger than that of the first design shown in Fig. 6.3.9. The simulated characteristic in comparison with the first, ideal prototype is presented in Fig. 6.3.12.

By the technique described here, a method that allows a fast and accurate design of passive filters for millimeter-wave applications in coplanar line technology has been developed. The parameters for the prototype filters have been taken from standard literature, while the optimization has been made with commonly available CAD software. Additional libraries for the analysis and the automatic design synchronization have been integrated into this program (see Chapter 5) to enable the design of coplanar circuits. This has been demonstrated with the examples of two passive coplanar filters.

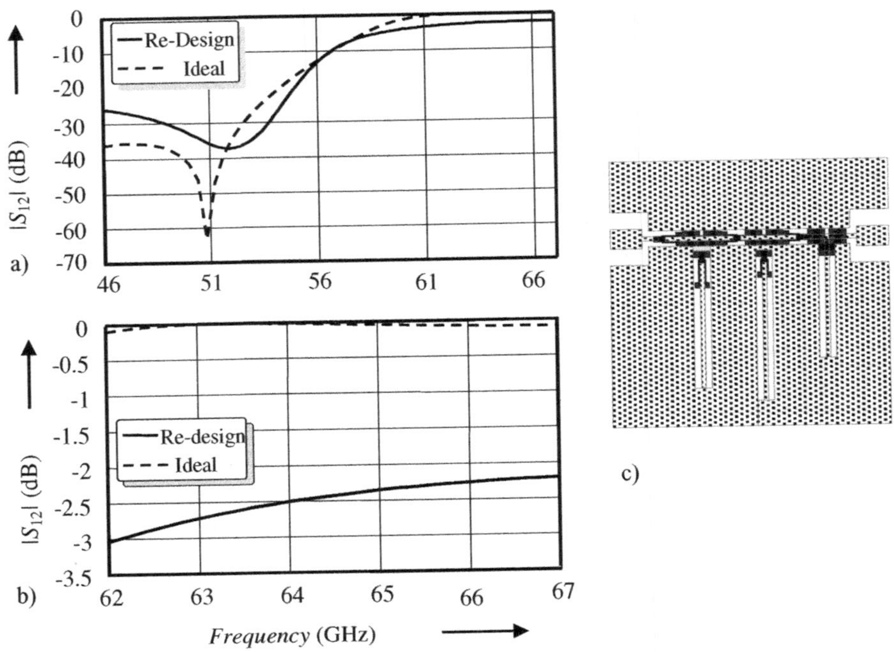

Fig. 6.3.12. The redesigned Cauer filter, replacing lumped elements by coplanar waveguide segments. (a) Comparison between the ideal filter properties and the coplanar line Cauer filter. (b) Enlarged characteristic for the pass band. (c) New filter layout.

6.4 COPLANAR EDGE-COUPLED LINE STRUCTURES

In this section, the application of coupled coplanar waveguides, as described on the quasi-static basis in Section 2.2, shall be used to realize interdigital filter structures with $\lambda/2$- or $\lambda/4$-resonators. Figure 6.4.1 shows several examples of this kind of filters in microstrip technology. The end and parallel-coupled filters (Figs. 6.4.1a and 6.4.1b) can be easily realized. The interdigital filters (Fig. 6.4.1c) and the comb-line filters (Fig. 6.4.1d), however, need a higher technological effort, because via holes have to be applied for the ground connections within these filters. Furthermore, the application of such via holes reduces the Q-factor of the filters because they are no real short connections. Nevertheless, such interdigital filters are often used in microstrip circuit design because of their small insertion loss and especially because of their small sizes.

If these kind of filters are to be designed in coplanar technology, the problem with realizing the short connections do not exist because the ground plane is on the same side of the substrate material as the filter structure. No via-hole technology is needed.

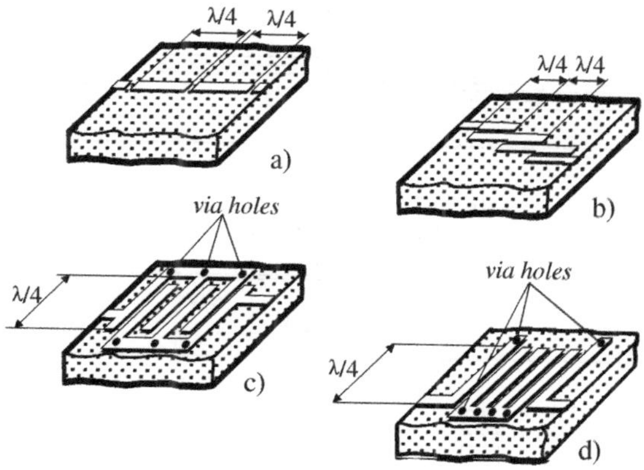

Fig. 6.4.1. Microstrip line filters. (a) End-coupled resonator filter. (b) Parallel-coupled resonator filter. (c) Interdigital filter. (d) Comb-line filter.

6.4.1 Verification of Coupling Between Coupled Coplanar Waveguides

Edge coupled coplanar line structures are frequently used in microwave circuits. In microwave filters, the coupling effect between two or more lines is a needed effect.

On the other hand, in designing dense packaged circuits, the coupling effect may be a nonwanted effect. In any case, it is most important to accurately predict the coupling effects and their influence on the circuit performance. Therefore, several structures built of coupled line segments have been tested by measurements, and the results have been compared to results from simulations on the basis of the analysis technique described in Section 2.2.11.

Figure 6.4.2a depicts a typical geometry for such a structure. The investigated structures have a coupling length of 926 μm, and the spacing between the lines ranges from 37 μm up to 185 μm. All measured structures have been realized on 450-μm-thick GaAs substrate material. The measurements are conducted utilizing a 67-GHz on-wafer measurement system in combination with the well-established LRM calibration method.

A ground strip is introduced between the two coupled lines in the structure shown in Fig. 6.4.2, but also structures without a ground between the two coupled line sections have been investigated.

In Fig. 6.4.2b a schematic description of the structure shows how the different elements of the structure are simulated in the applied *COPLAN* software that was developed on the basis of the finite difference analysis technique (for more details see Chapter 5). In the center of the schematic picture the coupled line section is shown. Both lines end in an open end (C_OPEN; see also Section 3.5.1). The ground between the two lines is simulated using a third strip of adequate width that ends in a coplanar short (C_SHORT; see also Section 3.5.2). At the input and output of the coupled line section, an air bridge (C_AIR; see Section 3.5.5) prevents the excitation of the odd mode on the

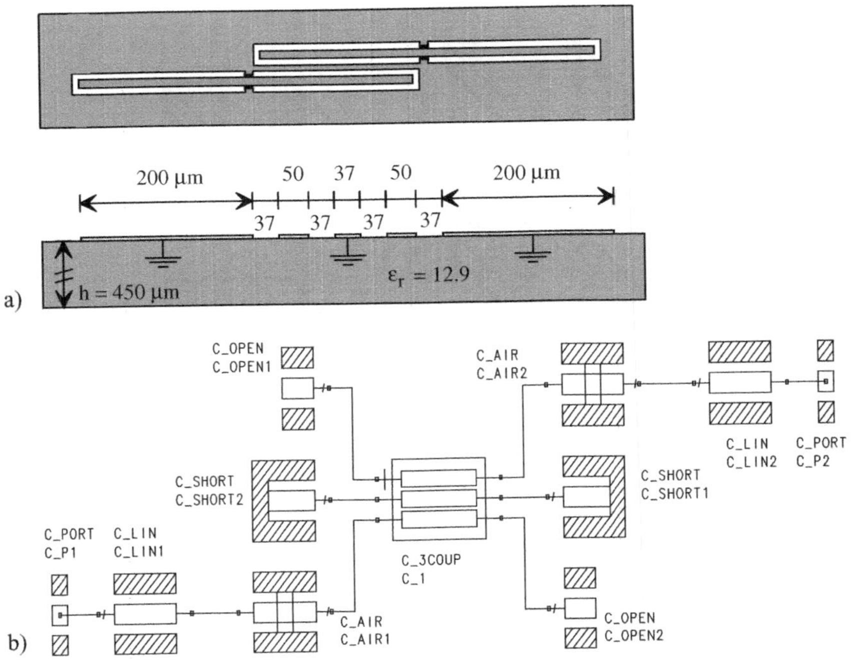

Fig. 6.4.2. Typical edge-coupled line structure: (a) Layout for an edge-coupled line structure with ground between the lines. Coupled line length 926 μm, feed line length 1000 μm. (b) Schematic layout for simulating edge coupled lines with ground strips.

coplanar feed lines (C_LIN) that lead to the measurement ports for on-wafer measurements (C_PORT).

The total spacing between the lines is 111 μm, while the ground width between the lines is 37 μm. The line width and the outside ground width are 50 μm and 200 μm, respectively, and the gap width is 37 μm. For this structure, there is a maximum coupling of −17 dB from the input port to the output port. All the structures discussed in this section are simulated utilizing the two-dimensional finite difference analysis for the coupled line structure, as described in Section 2.2.10. The discontinuities are analyzed using a three-dimensional version of the method as described in detail in Chapter 3. The agreement between measurement and simulation is depicted in Fig. 6.4.3. In this figure, the solid lines represent the simulation and the dotted lines are measured results. It is clear from the good match of the curves that for this structure the coupling effect and all other effects can sufficiently be taken into account up to very high frequencies by segmenting the geometry as shown in Fig. 6.4.2b.

Figure 6.4.4 shows how the coupling situation is changed if the ground strip between the lines is removed. The total gap between the two coupled coplanar line structures is again 111 μm, identical to the distance of the lines shown

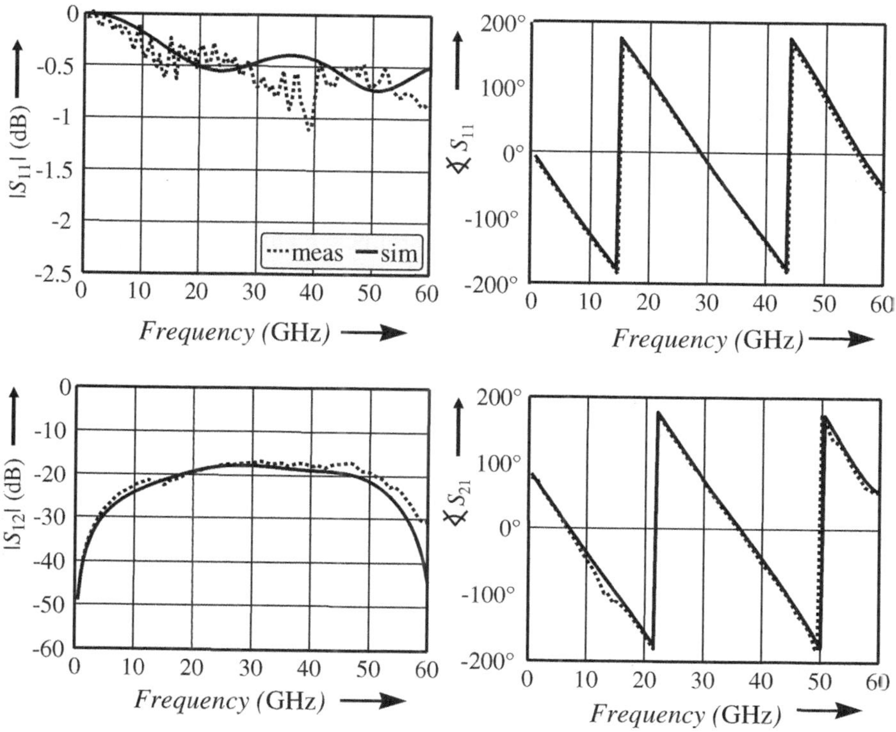

Fig. 6.4.3. Measured and simulated scattering parameters of the structure depicted in Fig. 6.4.2a.

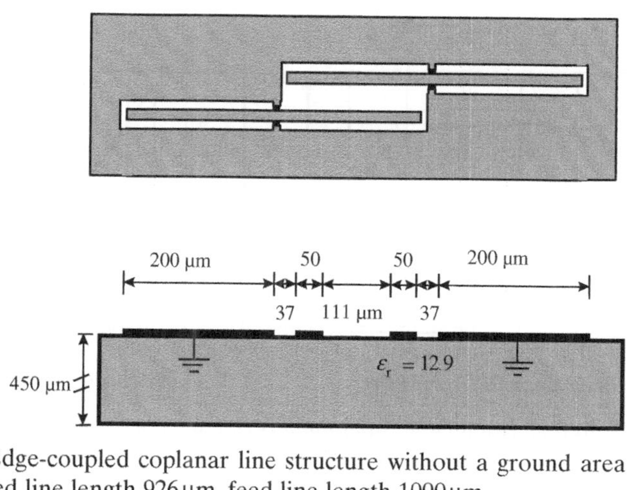

Fig. 6.4.4. Edge-coupled coplanar line structure without a ground area between the lines. Coupled line length 926 µm, feed line length 1000 µm.

in Fig. 6.4.2a, but no metal strip is between them now. The measured S-parameters $|S_{11}|$ and $|S_{12}|$ in magnitude and phase are shown in Fig. 6.4.5 in comparison to the simulation results. From these figures it may be determined that the coupling of the structure without ground strip between the lines is increased by about 10 dB compared to the structure shown in Fig. 6.4.2 and is now about −7 dB (measured) compared to −10 dB (simulated).

The remaining question now is to determine the coupling in both cases as a function of the total spacing between the lines. Figure 6.4.6 depicts $|S_{21}|$ as function of the spacing for both edge-coupled lines with and without a ground area between the strips at a frequency of 30 GHz. It can be observed that the coupling is between −2 dB and −12 dB for the coupled coplanar waveguide without ground between the strips. With ground between the strips, however, the coupling ranges from −12 dB to −25 dB. It can also be seen that the coupling is decreasing more rapidly (steeper slope) in the case when a ground is introduced between the edge-coupled lines. The simulation shown here is for a frequency of 30 GHz, which is also the highest coupling frequency for this geometry (refer to Fig. 6.4.5). The length for the coupled line sections again is 926 μm.

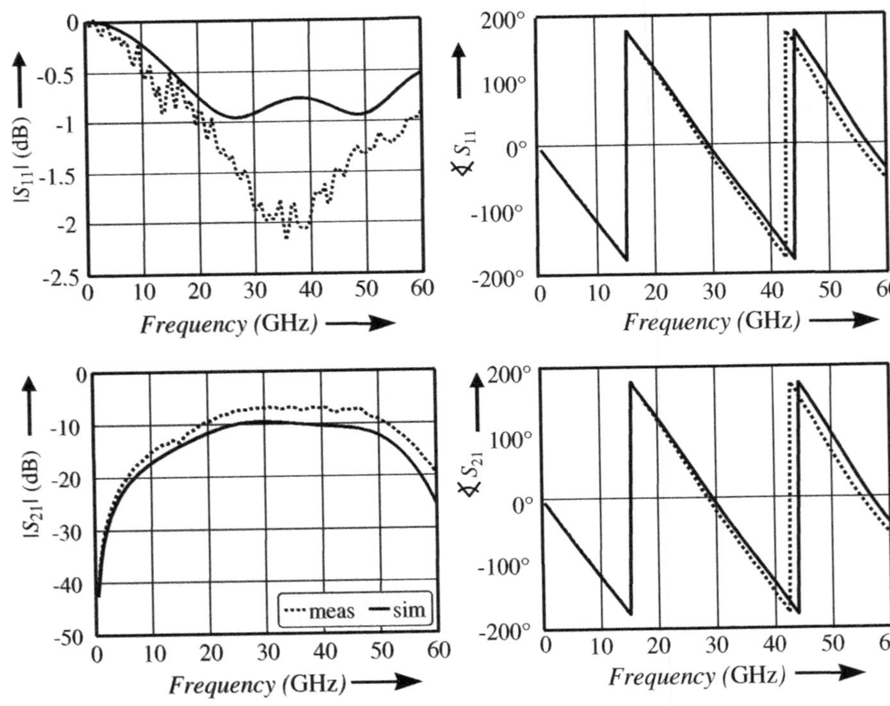

Fig. 6.4.5. Measured and simulated S-parameters of edge-coupled line structures (w = 50 μm, s = 37 μm, ground width = 200 μm), without ground strip between the lines (total spacing between the lines 111 μm).

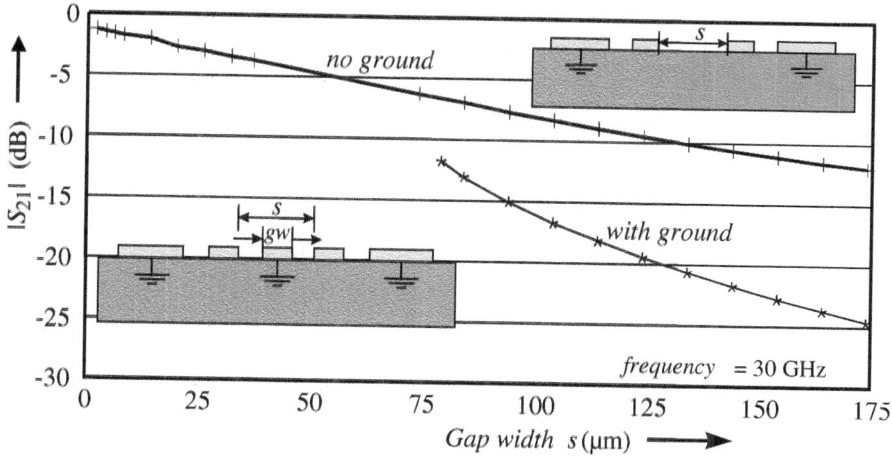

Fig. 6.4.6. $|S_{21}|$ as function of the spacing between the strips.

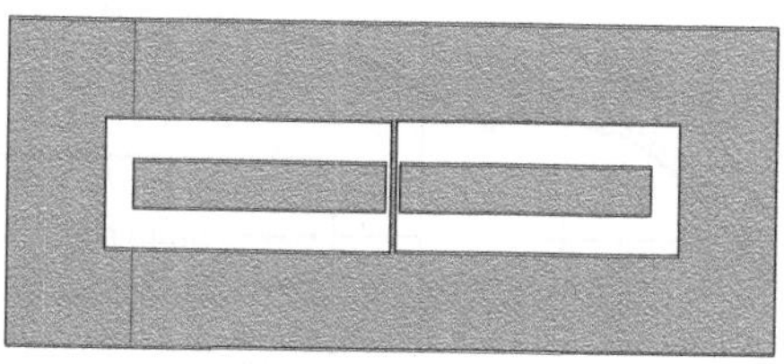

Fig. 6.4.7. Top view of a typical geometry for an end-coupled line structure with a grounded strip in the gap.

6.4.2 End-Coupled Coplanar Line Structures

Various types of end-coupled line structures have been investigated. A typical structure of this kind is depicted in Fig. 6.4.7. The figure shows two end-coupled coplanar waveguides with (or alternatively without) a grounded strip in the gap between the two waveguides. The total gap widths for these structures are varied from 20 μm to 50 μm. In these cases the spacing to ground at the end of the lines is fixed to 5 μm, while the ground width in the gap is varied.

Structures without ground in the gap can accurately be simulated with the above-mentioned quasi-static finite difference FD method. The coplanar element library, however, does not cover the simulation of the structures with ground in the gap. Thus, only the simulation utilizing (for example) the simulation tool *em*™ or a FDTD technique is possible. Simulations utilizing the *em*™ software have been conducted. Figure 6.4.8a depicts measured results of $|S_{12}|$ as function of the frequency for three different grounded strip widths (10 μm, 20 μm and 40 μm).

As expected, the coupling decreases with increasing gap or ground width. Figure 6.4.8b, on the other hand, compares simulated and measured results for

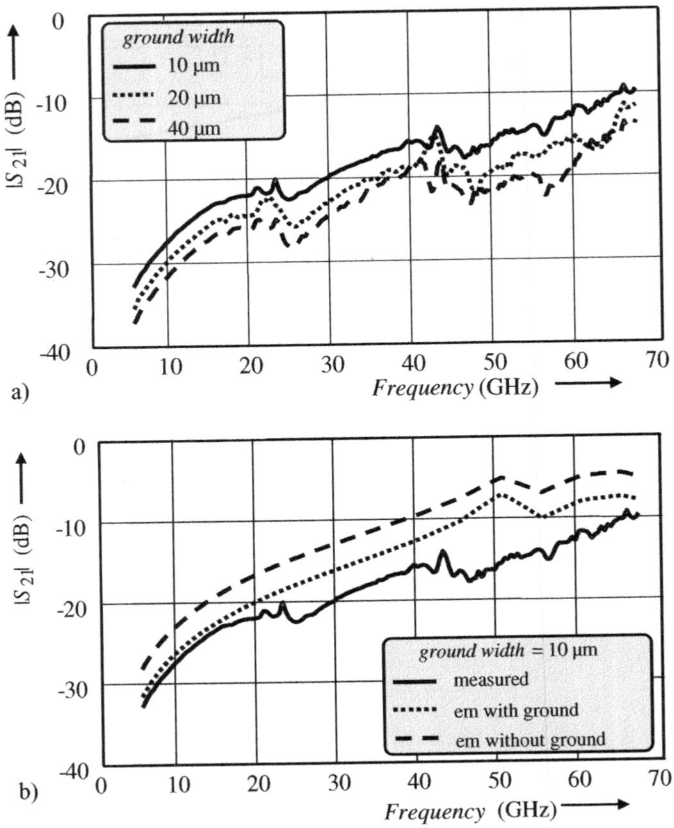

Fig. 6.4.8. (a) Measured $|S_{12}|$ for the structure shown in Fig. 6.4.7 with a grounded strip in the gap of width 10 μm, 20 μm, and 40 μm. (b) Measured and simulated (field solver *em*™) results of $|S_{12}|$ for the structure shown in Fig. 6.4.7 with grounded area of width 10 μm in the gap between the strips.

the structure with a grounded strip width of 10 μm. At frequencies above 20 GHz, the discrepancy between measurements and simulation is in the order of 3–5 dB. The simulated coupling is higher than the measured one. These poor results are due to the neglecting of the metalization thickness within *em*™ software. Since the distance to ground is 5 μm while the metalization is 3 μm thick, a lower coupling due to the increased "shielding" of the lines can be expected. In addition, it can be observed that there are two resonances simulated by the *em*™ software at frequencies around 50 GHz. These resonances are box resonances due to the definition of the structure within the *em*™ software. It should be pointed out, however, that these resonances also occur if absorbing boundary conditions are used.

6.4.3 Coplanar Waveguide End-Coupled to an Orthogonal Coplanar Waveguide

In this chapter, the structure shown in Fig. 6.4.9a will be discussed. This structure consists of two coplanar waveguides. One is used as a feeding line, the other as a waveguide resonator. It is measured as a one-port structure; the wide coplanar line is used to excite the resonator. Since a good coupling is wanted for measuring this structure, the dimensions of the resonating waveguide are very small ($w = 20\,\mu m$, $s = 5\,\mu m$). The impedance of the resonator waveguide is about $33\,\Omega$. The length of the measured resonator is $1130\,\mu m$. Figure 6.4.9b depicts the schematic layout for simulating the structure given in Fig. 6.4.9a utilizing the FD method.

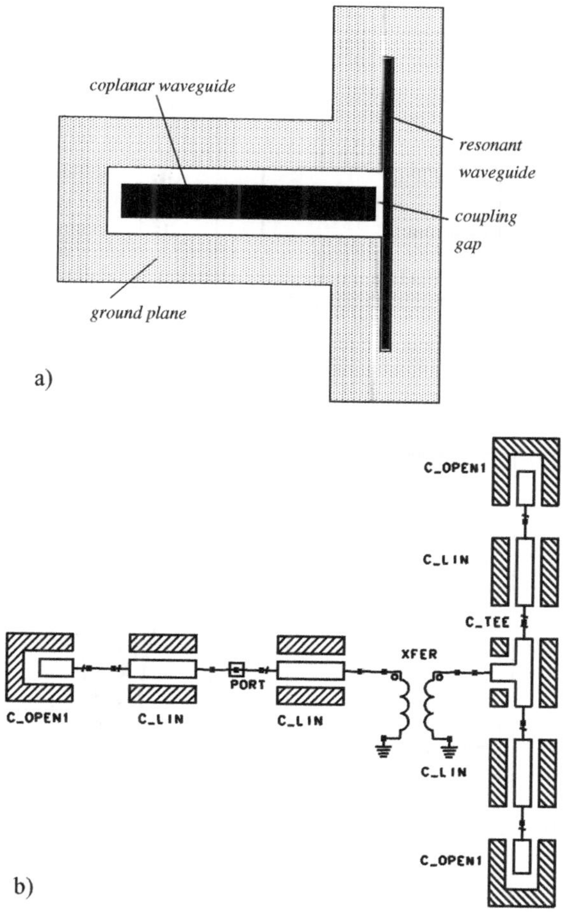

Fig. 6.4.9. Structure for coupling to an orthogonal coplanar waveguide. (a) Layout of the coupling structure. (b) Schematic layout for simulation.

Since such a coupling structure to an orthogonal waveguide is not available as a standard element in the FD software (see Chapter 5), the coupling effect was emulated by utilizing an ideal transformer. Additionally, full-wave simulations using a finite difference time domain (FDTD) technique and the *em*™ field solver have been performed. In Fig. 6.4.10a it can be observed that the simulated results from the *em*™ software and the FDTD simulator agree well. This is true for all simulated results of this structure. However, the resonant frequency is about 10% lower than the measured value that is depicted in Fig. 6.4.10b. Also the bandwidth of the resonant curve is quite different for the simulation and the measurements.

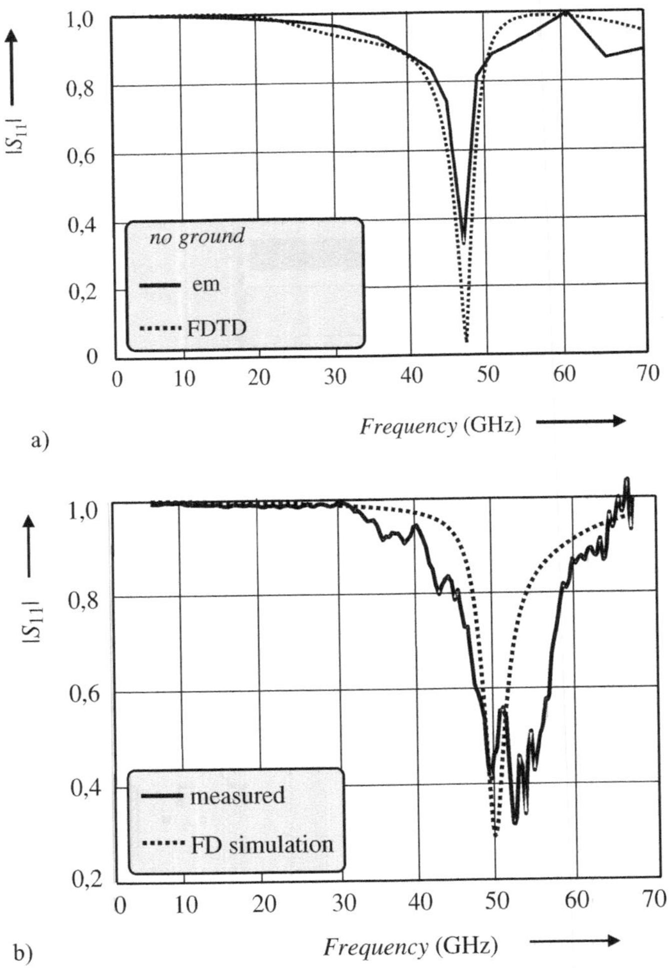

Fig. 6.4.10. Simulated and measured results for the structure in Fig. 6.4.9: (a) **em** and FDTD simulation with $t = 0$. (b) Measured results and FD simulation with finite metalization thickness ($t = 3\,\mu$m).

This disagreement is due to the fact that the metalization thickness is neglected in both simulation techniques. Because of the close proximity of ground and signal line in the resonator, however, this effect cannot be neglected without loosing accuracy. It also seems that loss effects in the resonator are not taken into account well.

Due to the large dynamic range in the geometry of this structure (small gaps and long resonator) the needed computation time for both programs are quite large. It should also be mentioned that the number of frequency points (resolution = 750 MHz in this case) utilized in full-wave simulators may sometimes not be sufficient for accurately simulating sharp resonances.

Like in the case of the end-coupled waveguides, a resonance can be observed in the *em*™ simulation at about 60 GHz. This resonance does not occur in the measurements of the realized structure. Figure 6.4.10b shows measured results and simulated results using the FD simulation technique described in earlier chapters, considering finite metalization thickness. These results are in a good agreement with the measured *S*-parameters. Only the bandwidth of the resonant curve is somewhat larger in the measured case compared to the simulations (as already mentioned above).

Various coupling structures have been discussed in this section. For some structures it was found that simulations with zero metalization thickness lead to an error > 10%, especially in predicting the resonant frequencies.

Table 6.4.1 depicts the transformation ratios for the three structures as function of the coupling space. These values are determined by introducing an ideal transformer at the location of the coupling area. An example for such a configuration is shown in Fig. 6.4.9b.

TABLE 6.4.1. Overview of Transformation Ratio (1/Coupling) for the Three Investigated Structures

	No Ground			With Ground Strip		
	20 μm	30 μm	50 μm	20 μm	30 μm	50 μm
	2.1	2.36	3.15	7.6	11.4	19.6
	12.7	15.6	21.5	18.5	26	31
	7.0	—	—	7.0	5.8	4.5

If the coupling space is reduced for the third structure, the resonance of the resonator is more and more affected, because the current flow in the ground strip becomes more and more disturbed. In all cases it is found, however, that the static 3D-FD simulations agree well with measurements.

6.5 COUPLED COPLANAR WAVEGUIDE FILTERS AND COUPLERS

6.5.1 Interdigital Filter Design

Analogous to the microstrip structure shown in Fig. 6.4.1c, several band pass filters have been designed and realized in coplanar technology [19, 24]. To check the applicability of this kind of filters, the insertion loss and the slope of the filter curve have been investigated as a function of the main geometrical parameters. Band-pass filters with three to six coupled coplanar strip lines have been designed for a center frequency of 5 GHz, 10 GHz, and 20 GHz. Additionally, they have been fabricated on ceramic substrate (Al_2O_3) and GaAs substrate. To approximate the requirements of a quasi-TEM wave propagation on the filter structure, the linear size of the cross section was chosen to be smaller than the filter length. Therefore the networks and components at the ends of the filter structure may be assumed to be quasi-lumped.

The following design rules are valid for the interdigital filters:

- The center band frequency is dependent on the line lengths of the filters and may be changed over large frequency ranges.
- The bandwidths of the filters may only be changed between 40% and 60%. Filters with a large number of lines and smaller line widths have a large bandwidth and a larger slope of the filter curve. The application of lines with smaller strip width, however, leads to higher conductor losses and thereby to a higher insertion loss of the filter.
- The slope of the filter curves becomes larger with increasing bandwidth. The application of lines with small line widths (which is needed for such filters), however, leads to higher losses of the line elements and, as a result, to higher insertion loss.
- The insertion loss of the filters is a function of the line lengths on the one hand and of the skin effect on the other hand. The resistance of a line is directly proportional to the line length of the filter. But because the skin effect resistance increases only with the square root of the frequency, the insertion losses of filters at higher center band frequency are smaller than those at lower frequencies.
- The distances between the different line elements of the filters influence the ripple of the filter curve in the pass band. Smaller space between the lines leads to a higher ripple, but at the same time the slope of the filter curve is increased.

Two interdigital filters in coplanar technology will be discussed in some detail in the following. The above-mentioned design rules will be investigated using these filters as examples. The first filter that shall be investigated is an interdigital filter built from four coupled coplanar strips in a coplanar environment. The coplanar strips are alternately grounded or open at their ends. The layout and the geometrical dimensions of this first filter are shown in Fig. 6.5.1. For measurement reasons the input and output lines of the filter are connected to 50-Ω lines using coplanar tapers (Fig. 6.5.1a; see also Section 3.5.4).

For the analysis of the filter, it is divided into a coupling section and the network that describes the end effects at the ends of the coupled lines, as shown in Fig. 6.5.1b. Two of the totally six coupling coplanar strips are assumed as the two ground planes. The electric shielding that is used in the analysis (see Section 2.2.11) is taken as a reference line. Its distance from the filter

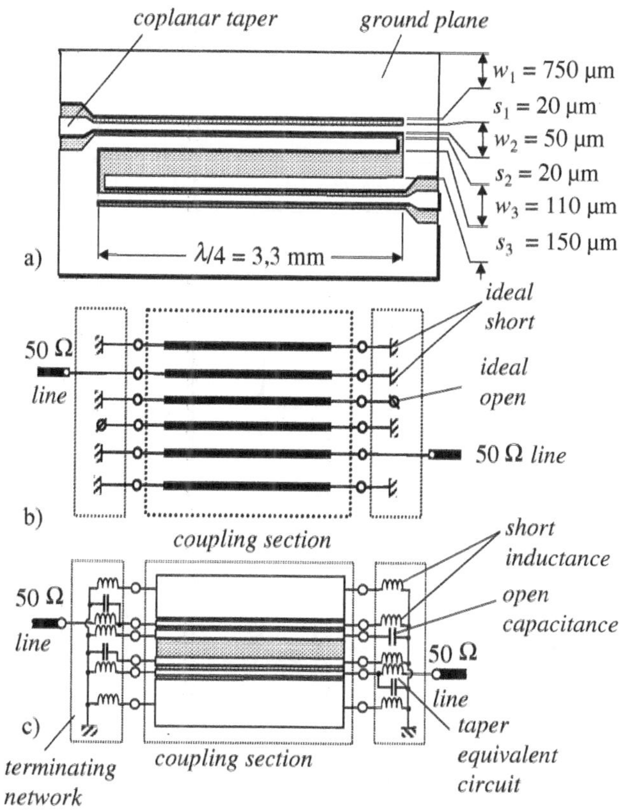

Fig. 6.5.1. (a) Interdigital band-pass filter with four coupled coplanar waveguides in a coplanar environment (conductor thickness t = 5 μm, substrate Al_2O_3, ε_r = 9.8, h = 625 μm). (b) Coupling section with ideal open ends and grounded ends. (c) Coupling section with modeled open ends and grounded ends.

structure is chosen in such a way that it does not have a significant effect on the filter properties.

A detailed modeling of the networks that describe the effects at the end of the lines is only needed, if the line number is large. The inductances of the shorts and the capacitances of the open ends may be considered as discussed in Chapter 3 and as depicted in Fig. 6.5.1c. For the filter investigated here (which has only four coupled coplanar lines), ideal grounded and open ends have been assumed in the analysis.

In the first step of the analysis the inductance, capacitance, and resistance per unit line length matrices of the coupled line sections are computed using the method described in detail in Section 2.2.11. The results are four symmetrical 6×6 matrices for the capacitances, inductances, and resistances per unit line length. The solution of the eigenvalue problem defined in Eq. (2.2.36) for the special line configuration of the filter shown in Fig. 6.5.1 delivers the propagation coefficients for the six quasi-TEM modes that may propagate on the different strips of the filter. Using these propagation coefficients, the effective dielectric constants and the attenuation coefficients of the modes may be calculated. Figure 6.5.2 shows the current distribution on the strips in the cross

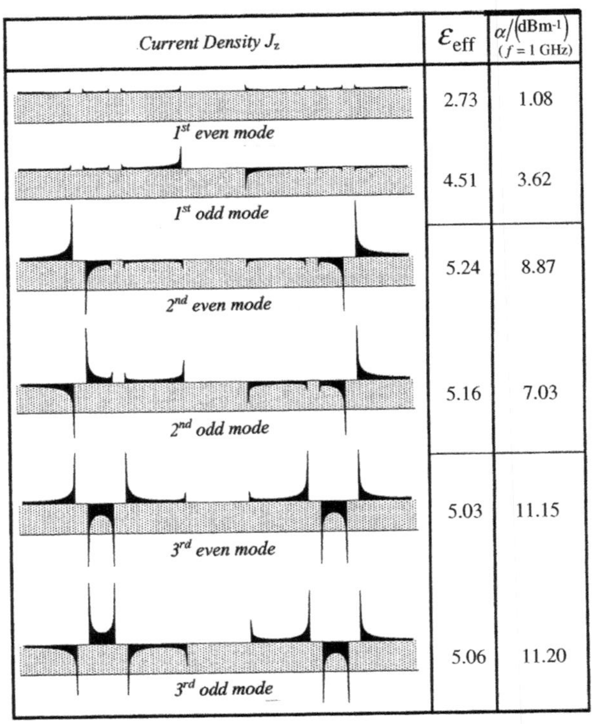

Fig. 6.5.2. Current distribution on the strips of the filter as shown in Fig. 6.5.1 for the six quasi-TEM modes that may propagate on the strips.

section of the filter structure together with the computed effective dielectric constants and the attenuation coefficients.

From the figure it may be observed that the current distribution of the different modes are axial symmetric (even modes) or point symmetric (odd modes). Therefore, these modes may be divided into three types of fundamental modes, each of which occurs in an even mode and an odd mode form. The scattering matrix of the coupling section is then calculated using these six propagating modes. If the ports of this structure are connected to an open or short as shown in Fig. 6.5.1, the scattering matrix of the filter describing the interconnection between the input and the output signal can be finally derived.

The comparison of the filter scattering matrix (calculated in the above-described way) with measurements as shown in Fig. 6.5.3 demonstrates a very good agreement over a large frequency range. The small discrepancies between the calculated and measured magnitude of the transmission

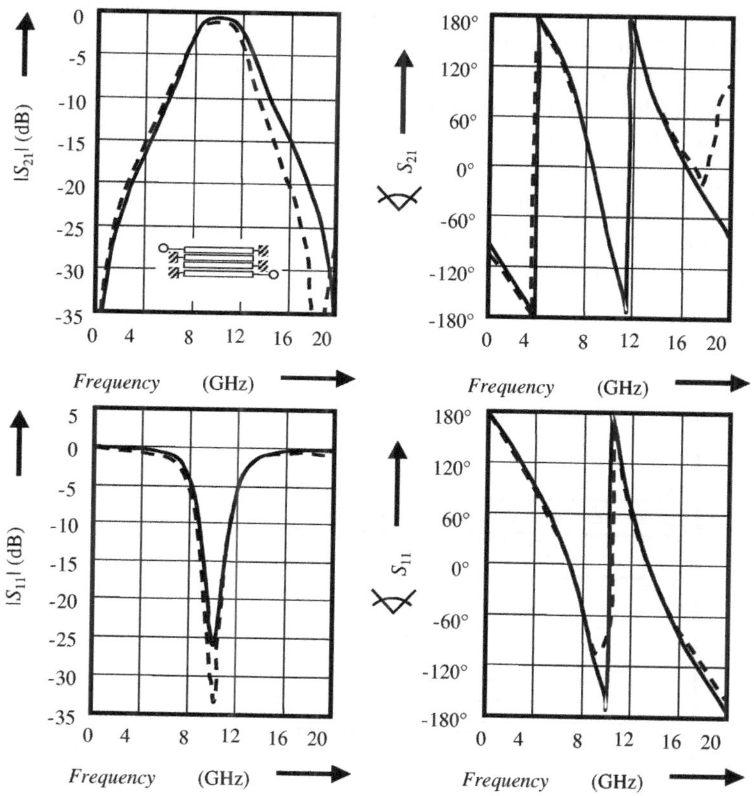

Fig. 6.5.3. The measured (– – –) and calculated (———) scattering parameters of an interdigital coplanar band-pass filter as a function of frequency (b). For geometrical parameters see Fig. 6.5.1.

coefficient $|S_{21}|$ at higher frequencies can be explained by radiation losses at the open coplanar waveguide ends. They are not considered in the quasi-static analysis used for the design here.

Also, it can be observed that the $\lambda/2$-resonant frequency has been measured at 18 GHz instead of at 20 GHz, which was the theoretical result. Again the quasi-static analysis technique is the reason for this deviation. The analysis does not take into account the (small) dispersion of the coplanar waveguide properties. It must be mentioned that the consideration of the metalization thickness of $t = 5\,\mu m$ has a considerable effect on the analysis results because its effect may not be neglected in the case of the closely coupled lines in the filter structure (Fig. 6.5.1).

The measured insertion loss of 1.15 dB at a center band frequency of 10 GHz is a value that is relatively small for a planar microwave filter. It cannot be reduced further without increasing the size of the filter drastically. An increase of the strip line widths would reduce the insertion loss by only a small amount. At the same time the slope of the filter curve would become smaller. The 4.5 GHz (45%) bandwidth of the filter is much too large for many applications. It can be reduced only if a higher insertion loss is tolerated.

The slope of the filter curve may also be increased by an increase of the used strip line number. This shall be demonstrated with the design of a coplanar interdigital filter using six coupled coplanar waveguides. The layout and the geometrical parameters of this filter are shown in Fig. 6.5.4. The center band frequency, again, is 10 GHz. Therefore, the length of the filter is unchanged compared to the filter shown in Fig. 6.5.1. The cross-section size and especially the spaces between the single strips have been increased to have a ripple, in the pass band, as small as possible.

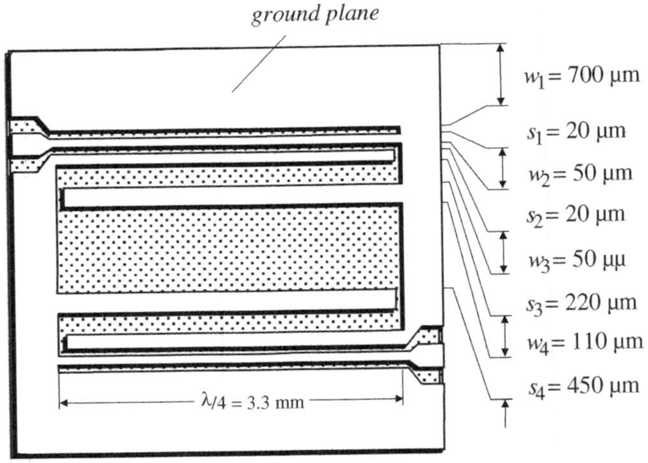

Fig. 6.5.4. Coplanar interdigital band-pass filter with six coupled coplanar waveguides (metalization thickness $t = 5\,\mu m$, substrate height $h = 625\,\mu m$, substrate Al_2O_3, $\varepsilon_r = 9.8$).

Figure 6.5.5 shows the comparison between the calculated and the measured scattering parameters of this second filter. A small shift of the center band frequency to lower frequencies may be explained by the fact that the cross-section size of the filter no longer fulfills the requirement of being small as compared to the longitudinal dimensions. This leads to the conclusion that the assumption of ideal network elements at the ends of the strip lines leads to a larger error in the analysis of this filter. Also, the relatively large s_4/h value (450 µm/625 µm) results in a larger dispersion of the filter characteristics. Therefore, the quasi-static analysis is no longer as accurate as it is in the case of smaller frequencies or smaller circuit element dimensions.

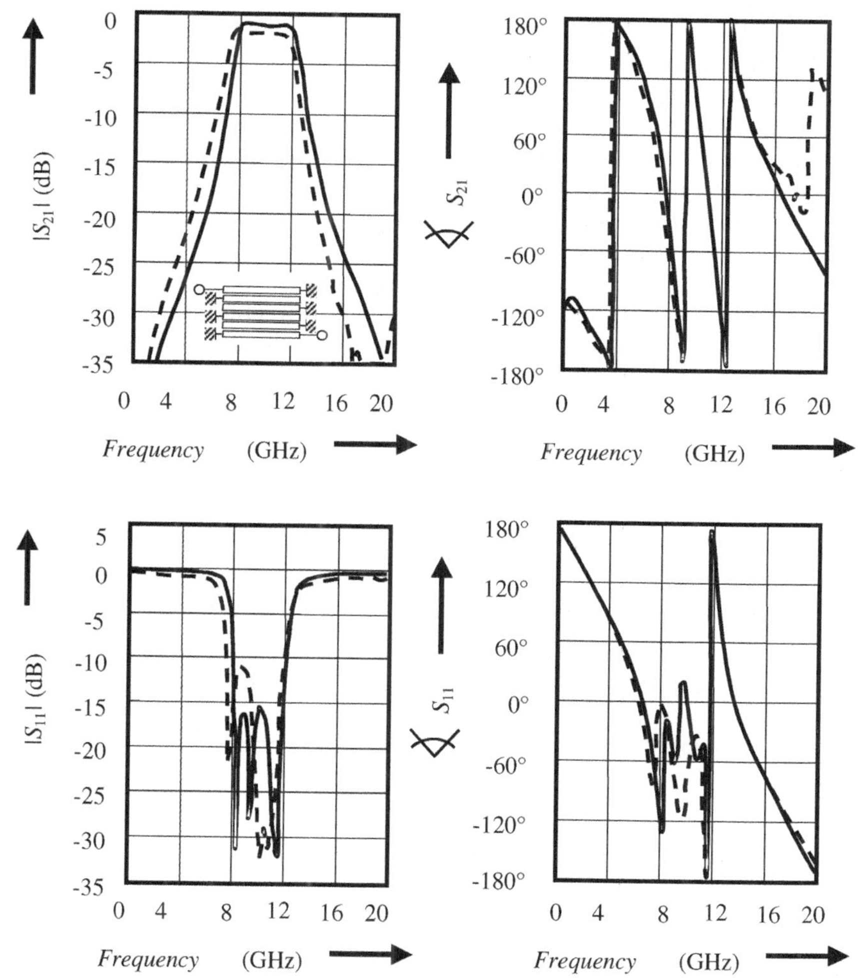

Fig. 6.5.5. Measured (– – –) and calculated (———) scattering parameters of the interdigital coplanar band pass filter with six coupled coplanar waveguides on GaAs substrate, plotted against the frequency. For geometrical parameters see Fig. 6.5.4.

The improved slope of the filter characteristic has been successfully reached. At the same time the insertion loss has increased to about 2 dB. Also, the bandwidth of the filter is a little bit large (5.2 GHz) compared to the filter with four coupled coplanar waveguides (Fig. 6.5.1). Moreover, the reflection coefficient in the pass band (–15 dB) is somewhat larger than that of the first filter (<–20 dB).

6.5.2 Coplanar Waveguide Couplers

Coupler structures, especially waveguide couplers, play an important role in microwave integrated circuits. Often, they are used as 3-dB couplers in various amplifier circuits and phase shifters as well as in mixers and adjustable attenuators. Also, they are used in power divider structures. Figure 6.5.1 shows three examples of commonly used microstrip couplers for application in planar microwave integrated circuits.

The realization of the branch line and the ring coupler is not easy in the coplanar technique because large ground-plane areas are enclosed by the line sections and it is not very easy to keep these isolated areas on the common ground potential. The coplanar interdigital coupler, however, can be built up without large problems. The construction of the coplanar interdigital coupler is very similar to that in microstrip technology, as shown in Fig. 6.5.6c. In this section, a solution for the realization of coplanar directional couplers with a coupling of 3–10 dB using conventional (M)MIC technique will be discussed [24]. In the case of a realization in monolithic integrated circuit technology, air bridges replace the unavoidable bond wires. Moreover, air bridges (or bond

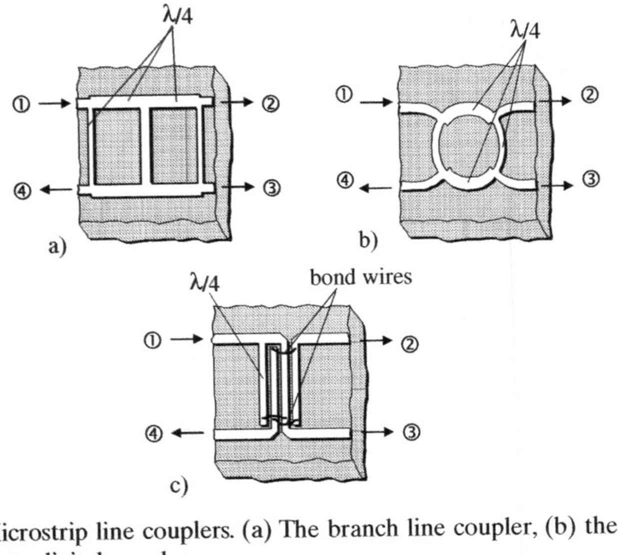

Fig. 6.5.6. Microstrip line couplers. (a) The branch line coupler, (b) the ring coupler, and (c) the interdigital coupler.

wires in the case of a hybrid construction) are needed at the input and the output ports of the coupler to hold the ground planes at the same potential and thereby suppress the odd coplanar waveguide mode.

The analysis method for the couplers described in the following, again, is based on the analysis of the coupled coplanar waveguides as presented in Section 2.2.11. The method allows the consideration of finite substrate and metalization thickness as well as the influence of air bridges (see Section 3.5.5) used for the suppression of higher-order modes. In order to demonstrate the accuracy and efficiency of the applied method, two coplanar directional couplers for different applications (power detection at 1 GHz and power splitting/combining at 20 GHz) are designed and realized in coplanar hybrid technology and MMIC technology.

Figure 6.5.7 shows the schematic description of the used analysis method based on the theory of multi conductor systems. The quasi-static finite difference method as developed in Section 2.2.2 is applied to the system of multiply coupled coplanar waveguides that forms the coupler cross section. The application of a smart nonequidistant discretization as well as the utilization of the successive overrelaxation (SOR) method for the solution of Laplace's finite difference equation leads to a very fast numerical determination of the

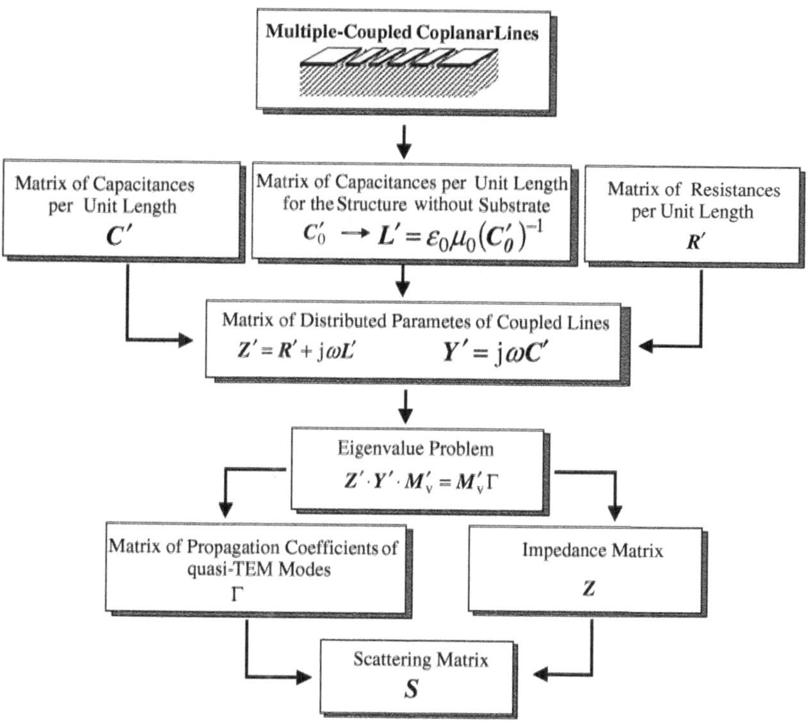

Fig. 6.5.7. Quasi-TEM mode analysis of coupled lines as applied to the coupler analysis.

charge and surface current distribution on all conductors. From these quantities, the distributed parameters of the coupled lines, meaning the capacitance, inductance, and resistance per unit length ($\boldsymbol{C}'$, $\boldsymbol{L}'$, and $\boldsymbol{R}'$) matrices, can be calculated.

The matrix of the series impedances as well as the matrix of the shunt admittances per unit length of the coupled line section is used for the solution of the eigenvalue problem that delivers the propagation coefficients of the propagating quasi-TEM modes and the impedance matrix of the coupled line system (compare Section 2.2.11).

Figure 6.5.8, as an example, shows the propagation characteristics (effective dielectric constant ε_{eff}, attenuation coefficient α, and surface current density distribution) calculated for the six possible quasi-TEM modes of the 3-dB MMIC coupler shown in Fig. 6.5.11. Using normalization impedances at the ports, the scattering matrix of the coupled lines can be calculated. The parasitic effects of the bridge construction between the ports of the coupler (see Fig. 6.5.11b) and the coupled line section cannot be neglected at high frequencies. These effects are considered in a network of coplanar lines and air bridges, which is analyzed using the 3D finite difference method as described in Chapter 3 and using the CPW-Library software as described in Chapter 5.

The first structure that is investigated is a 10-dB coplanar coupler for the center frequency of 1 GHz used for power detection. Figure 6.5.9 shows the layout of the coupler. The coupler is fabricated in a hybrid technology on ceramic substrate using two parallel center conductors in a coplanar environment.

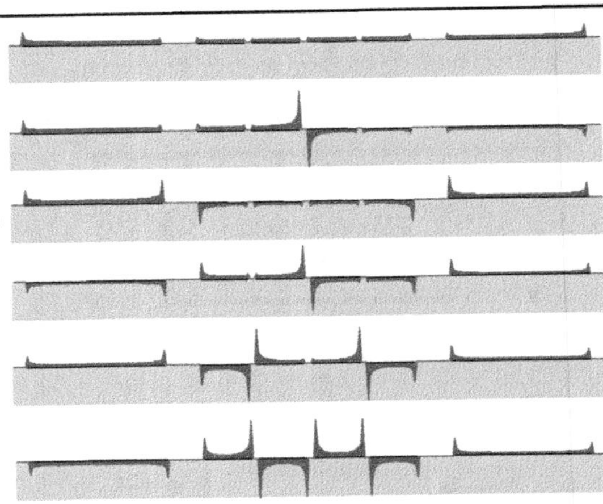

Quasi-TEM Modes	Surface Current Distribution	ε_{eff}	α @1GHz dB/m
1st Even Mode		7.263	4.042
1st Odd Mode		6.772	8.118
2nd Even Mode		6.649	15.129
2nd Odd Mode		6.088	24.967
3rd Even Mode		6.388	27.830
3rd Odd Mode		6.190	29.758

Fig. 6.5.8. Propagation characteristics (ε_{eff}, α) of the quasi-TEM modes on six coupled strips (four coupled center conductors and two ground planes of finite width) used in a 3-dB coplanar coupler structure on GaAs substrate.

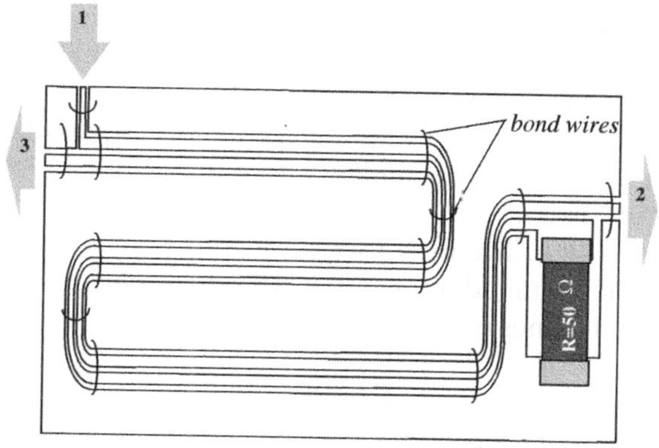

Fig. 6.5.9. Layout of a 10-dB coplanar coupler on ceramic substrate. Line width = 140 μm, slot space between the coupling strips = 60 μm, space to the ground = 60 μm, length of the coupling section = 30 mm. Substrate: alumina ceramic, Al_2O_3.

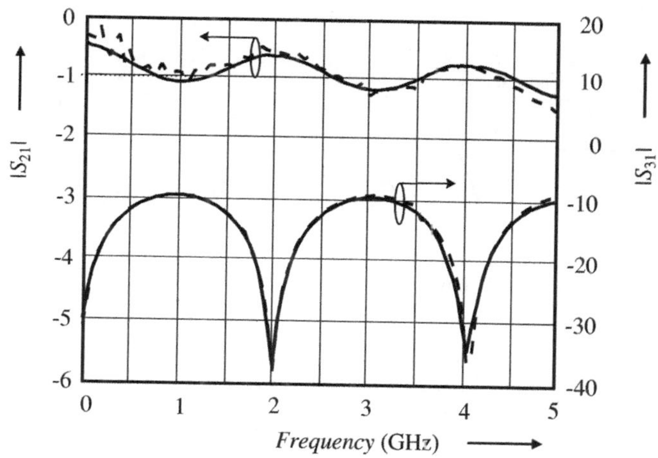

Fig. 6.5.10. Measured (- - -) and calculated (———) scattering parameters of the 10-dB coplanar coupler on ceramic substrate.

In order to minimize the size of the coupler, the coupled coplanar waveguides are folded several times. Bond wires are used to suppress the parasitic higher order modes that could be excited at the bends and at the input and output ports. Port 4 is terminated using a 50-Ω SMD resistor.

The scattering parameters plotted in Fig. 6.5.10 show a very good agreement between the calculated and the measured results indicating the accuracy of the applied method even beyond the first and the second resonant frequency. The coupling of 10 dB is achieved. The input matching is better than −20 dB and the isolation is higher than 30 dB.

In the next step a 3-dB coplanar coupler for a center frequency of 20 GHz is designed and realized on GaAs substrate (substrate height = 500 μm). The main goal was to realize a small size coupler for an application as power divider/combiner. In order to reach as much coupling as possible, four coupled center strips in a coplanar environment with a line width of 35 μm and a line spacing of 5 μm are used. Four coupled strips in a coplanar environment are used for this application. The use of more than four lines leads to losses that are too high for the assumed application. The distance between the coupled lines and the ground planes is 25 μm, and the total length of the coupled section is 1.45 mm.

Figure 6.5.11 shows a photograph of the realized coupler. Air bridges at the ports suppress the excitation of higher-order modes. Two further air bridges

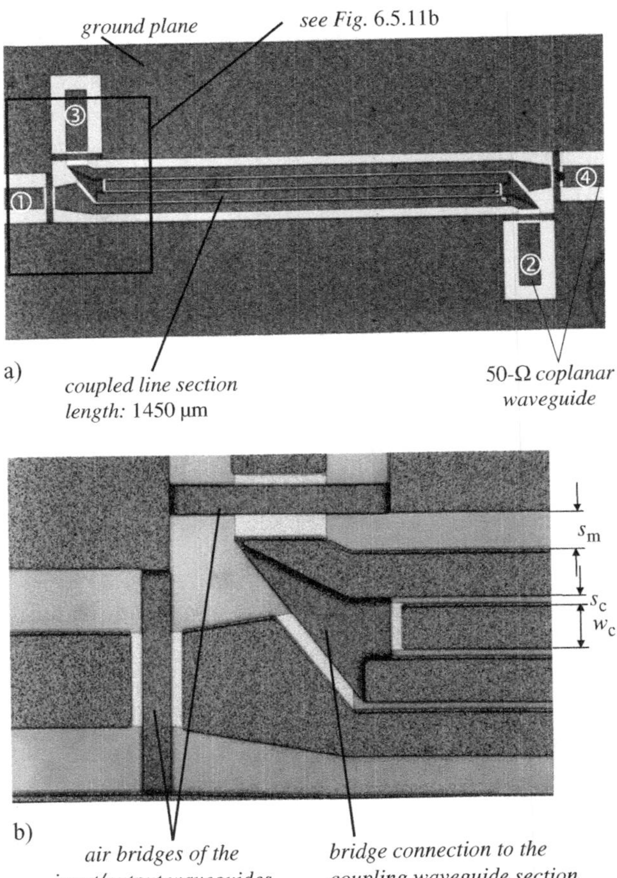

Fig. 6.5.11. The coplanar interdigital 3-dB coupler in monolithic technology on GaAs substrate. (a) The total coupler structure. (b) Details of the air bridge and connection technique at the end of the coupling strips. Parameters: $w_c = 35\,\mu m$, $s_c = 5\,\mu m$, $s_m = 25\,\mu m$.

are necessary for the connection of the coupled lines at both ends of the coupling section. In Fig. 6.5.11b the used air-bridge construction can be seen in detail. The air bridges at the input and the output coplanar waveguide are of conventional type 2 (see Section 3.5.5). In the vicinity of the air bridges the metalization height of the coplanar waveguide center conductor has been reduced to the height of the fundamental metalization layer (gate layer). The strips of the coupling waveguide section are connected to port ③ (see Fig 6.5.11a) using a more complicated air-bridge structure as can be seen from Fig. 6.5.11b. The complete coupler is simulated using the COPLAN software described in Chapter 5. The achieved results are plotted in Fig. 6.5.12.

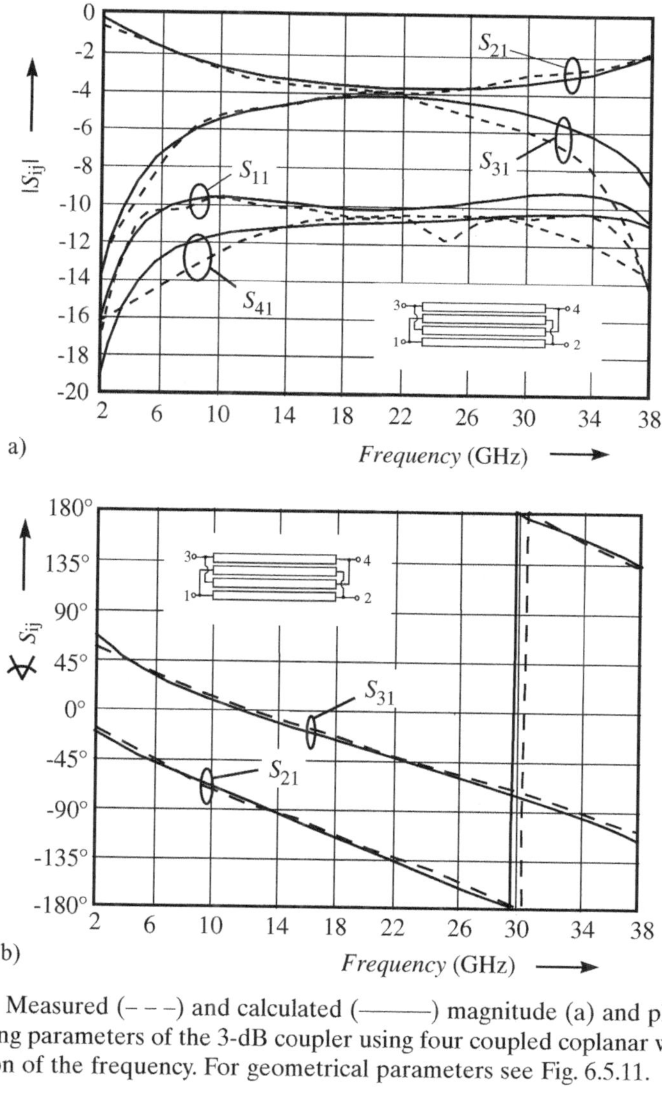

Fig. 6.5.12. Measured (– – –) and calculated (———) magnitude (a) and phase (b) of the scattering parameters of the 3-dB coupler using four coupled coplanar waveguides as a function of the frequency. For geometrical parameters see Fig. 6.5.11.

The maximum achievable coupling is 4.18 dB (directivity = 4.06 dB). The input matching and isolation of –10.6 dB is not satisfactory. The main reasons for this are the bridge construction at the ports and the connection between the lines at the end of the coupled line section. The phase balance is within ±5° from 15 GHz to 25 GHz. The excellent agreement between simulation and measurement, especially also in the case of the S-parameter phases, indicates the validity of the quasi-static analysis up to relatively high frequencies.

In a summary, coplanar directional couplers of relatively small sizes and good performances can be realized using conventional (M)MIC technology. Relatively high coupling is achieved utilizing multiply coupled coplanar lines. However, the input reflection coefficient and the isolation are not at all sufficient in this case. An efficient and powerful simulation tool using two- and three-dimensional FDM is used for the fast and accurate simulation of the coupler characteristics. The accuracy of the applied method has been verified in a wide range of frequencies using measurements on two coplanar couplers for different applications on ceramic and GaAs substrates.

6.6 COPLANAR MMIC WILKINSON COUPLERS

Standard Wilkinson couplers are required for many applications such as power amplifiers, mixers, or active circulators. In reference 1, such n-port power dividers were first presented. For MMICs, most designers avoid using these couplers on-chip since normally the space requirement is quite large due to the $\lambda/4$ length of the coupling section. In designing MMICs, the Wilkinson coupler is typically utilized at medium and high frequencies (>10 GHz). Most papers on coupler types are still considering conventional dimensions for the coupling section. A paper by Merneyei et al. [37], for instance, explains the design of a broadside-coupled line structure where the two coupled lines are on two different metalization layers. Similar couplers have been described before. The drawback in all these cases is that such couplers become quite large at low frequencies. First attempts in miniaturizing couplers were shown in reference 20. Also, the idea of applying lumped elements for couplers is not new [31]. In addition, there are clearly a lot of applications at lower and higher frequencies where compact couplers are needed. For low frequencies, for instance, the phase shift of 90° can hardly be realized on-chip by using transmission lines. At high frequencies, on the other hand, the applied discontinuities (such as T-junctions) lead to an undesired response that in turn must be considered by a modified structure. Therefore, alternative types of Wilkinson couplers are required. In this section, practical design guides for two types of coplanar Wilkinson couplers will be given. The simulations shown in the following are based on the 3D-FD algorithm for simulating discontinuities and lumped elements that was verified in large extent up to about 67 GHz (see Chapters 3 and 4). Finally, coupler structures realized in monolithic technology on GaAs substrate material will be presented, and their properties will be compared to the simulated results.

6.6.1 Conventional Wilkinson Couplers

A standard Wilkinson coupler consists of two transmission lines with a characteristic impedance of $\sqrt{2}Z_L$ (Z_L = port reference impedance) and an extension of a quarter wavelength. Both branches are connected by a resistor with a value of $R = 2Z_L$ (Fig. 6.6.1). This fact makes it impossible to realize such a coupler at a frequency of 1.8 GHz or lower in a small MMIC design. On GaAs substrate, for instance, the length of such a coupler will be around 12 mm. The scattering parameters of this ideal structure are shown in Fig. 6.6.2. It can be clearly seen that the isolation bandwidth is limited by the quarter wavelength requirement.

6.6.2 Wilkinson Couplers with Discrete Elements

At low frequencies, a Wilkinson coupler can also be designed on the basis of discrete elements. The basic idea in this case is that the 90° phase shift may be realized by using inductors and capacitors (Fig. 6.6.3).

For better matching and for increasing the bandwidth, some additional elements are also included in the circuit shown in Fig. 6.6.3. In Fig. 6.6.4 the theoretically predicted properties of this structure using ideal discrete components are depicted. This design shows a wider bandwidth than the ideal Wilkinson coupler (see Fig. 6.6.2) because only discrete elements are utilized. This approach has the big advantage that the frequencies for such a coupler can also be much lower than 1.8 GHz, for instance. Of course, such a design

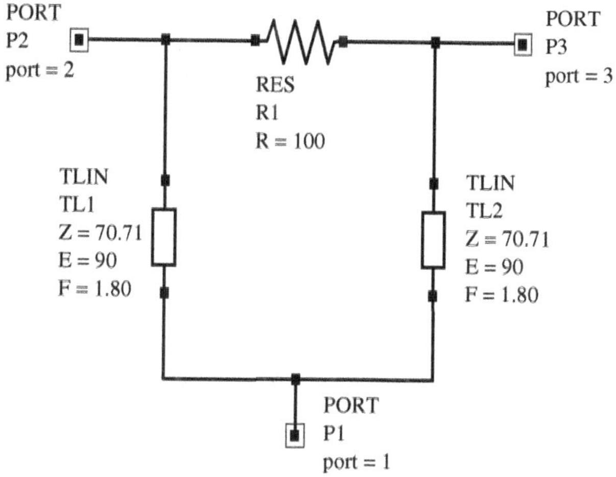

Fig. 6.6.1. Schematic of a standard Wilkinson coupler at 1.8 GHz.

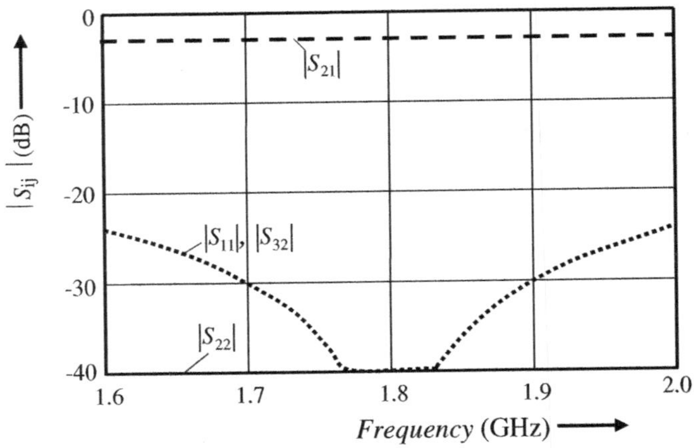

Fig. 6.6.2. *S*-parameters of an ideal standard Wilkinson coupler for 1.8 GHz.

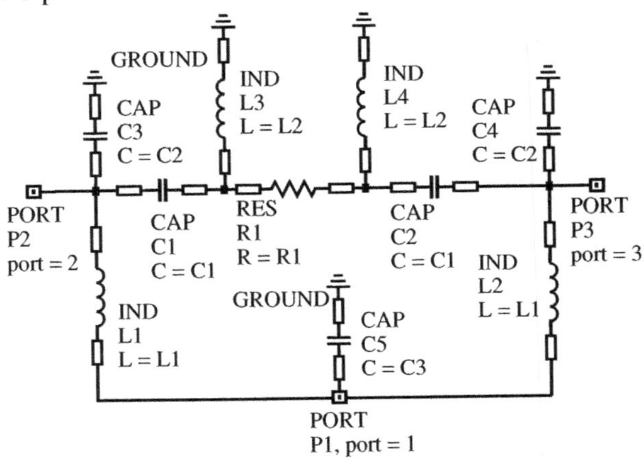

Fig. 6.6.3. Schematic of a Wilkinson coupler with discrete elements for 1.8 GHz. Parameters: C_1 = 2.92 pF, C_2 = 1.19 pF, C_3 = 1.85 pF, L_1 = 5.02 nH, L_2 = 11.70 nH, R_1 = 111.98 Ω.

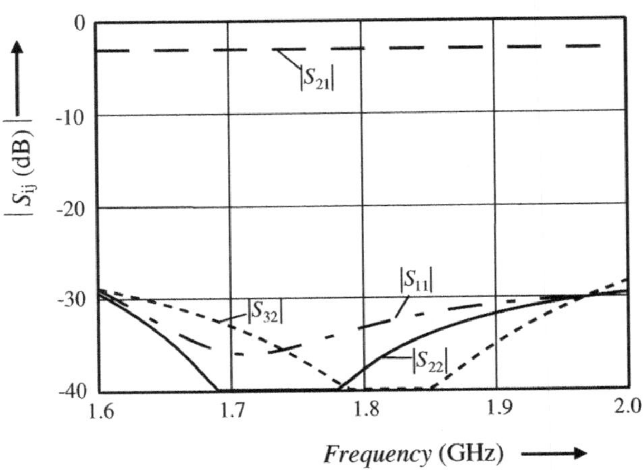

Fig. 6.6.4. Theoretically predicted *S*-parameters of a Wilkinson coupler with discrete elements for 1.8 GHz.

with discrete elements makes only sense, if the discrete elements are finally realized by lumped elements in monolithic integrated circuit technology—for example, on the basis of a coplanar circuit technology. How this can be done will be shown in the next section.

6.6.3 MMIC Applicable Wilkinson Couplers with Coplanar Lumped Elements

For the realization of a Wilkinson coupler in monolithic integrated technology, a modified version of the above-described discrete element coupler is possible. Such a coupler may utilize coplanar lumped elements as shown in Fig. 6.6.5. In this case, all connecting transmission lines have a length of only 10 µm. These lines are required only for separating the different lumped elements.

The complete layout and a photograph of this circuit are shown in Fig. 6.6.6. The coupler consists of two inductors, a resistor, three T-junctions, six capacitors, and 17 coplanar waveguide sections. The size of this circuit, as shown in Fig. 6.6.6, is only 1400 µm × 900 µm and could even be reduced further. Compared to the size of a classical Wilkinson coupler, which at a frequency of 1.8 GHz is defined by the $\lambda/4$ length of about 12 mm, this size reduction is a big improvement. Due to the small size, the S-parameters (see Fig. 6.6.7) are comparable to those of an ideal Wilkinson coupler.

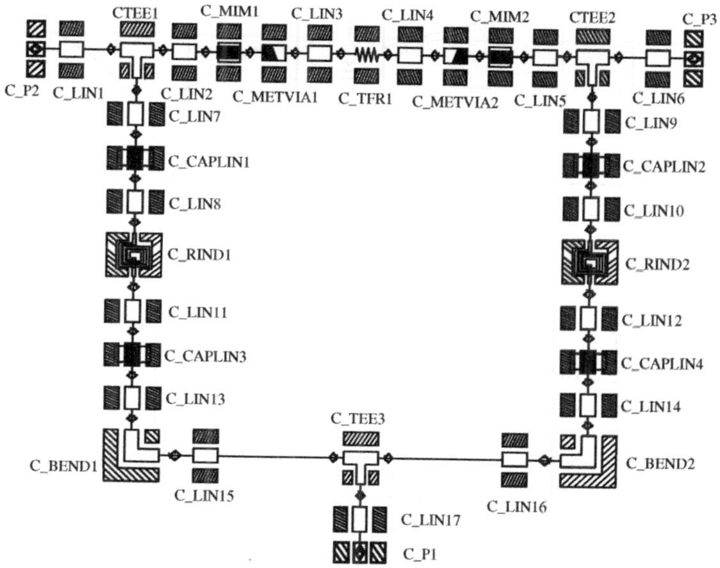

Fig. 6.6.5. Schematic layout of a 1.8-GHz Wilkinson coupler with coplanar-lumped elements (for the schematic layout elements see Chapter 5).

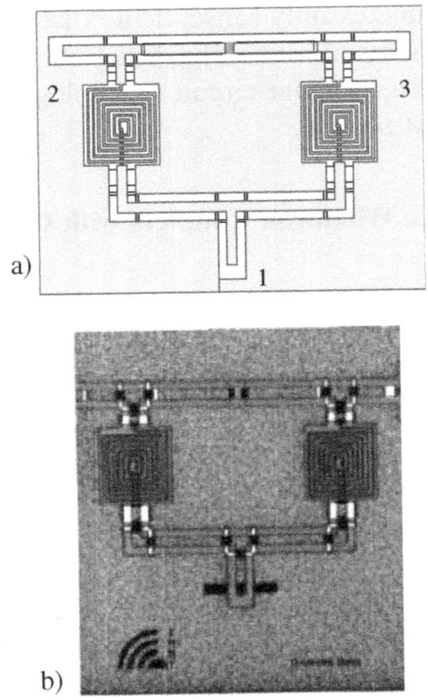

Fig. 6.6.6. Layout (a) and realization (b) of a Wilkinson coupler with coplanar-lumped elements for 1.8 GHz (1400 μm × 900 μm) in monolithic GaAs technology.

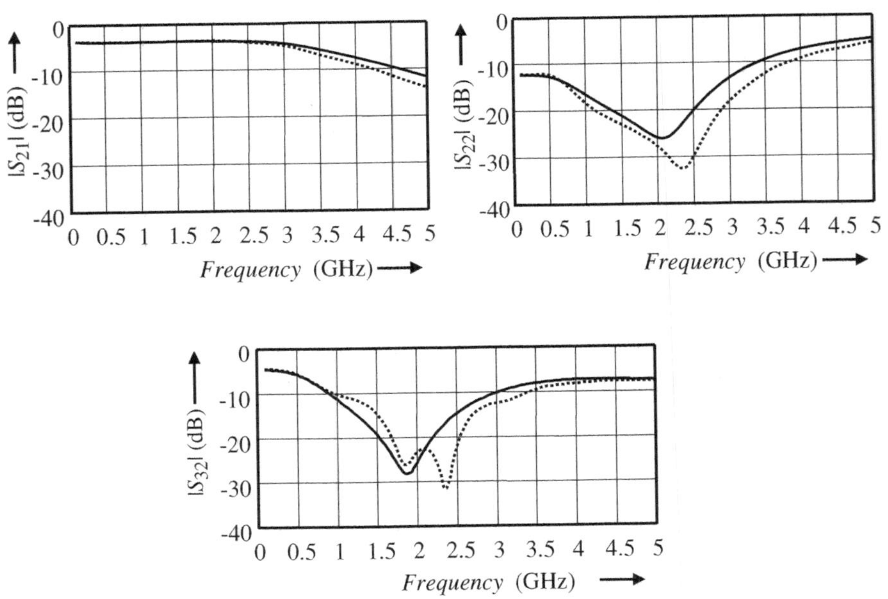

Fig. 6.6.7. Simulated (———) and measured (•••) *S*-parameters of a Wilkinson coupler with CPW-lumped elements for a frequency of 1.8 GHz.

Figure 6.6.7 shows the magnitude of the scattering parameters for this 1.8-GHz Wilkinson coupler. The agreement between simulation and measurement is very good. It may be observed that the bandwidth of this coupler (input reflection coefficient $|S_{22}| < -20\,\text{dB}$ and transmission coefficient $|S_{32}| < -20\,\text{dB}$) at a center band frequency of 1.8 GHz is 1 GHz, which is a relative bandwidth of about 50%! These results could even be improved by a careful redesign of the coupler. The insertion loss between input and output of the coupler is about 3.1 dB. The isolation of the coupler is around 25 dB. In some publications, couplers with about 3.5-dB insertion loss have been reported. In no case, however, could values of about 3.1 dB be achieved.

It should also be pointed out here that the losses of the components are realistically taken into account for the simulation. Such a coupler is ideally suited for low frequencies, where the geometrical equivalent of $\lambda/4$ dimensions are large and where a low insertion loss is required.

6.6.4 Wilkinson Coupler in Coplanar Waveguide Technique for Millimeter-Wave Frequencies

For high frequencies, "conventional" Wilkinson couplers in the sense that line sections are used for the needed phase shifts may be designed. Figure 6.6.8 shows the schematic layout of such a Wilkinson coupler in coplanar technology for application in the millimeter-wave range. Because of the high

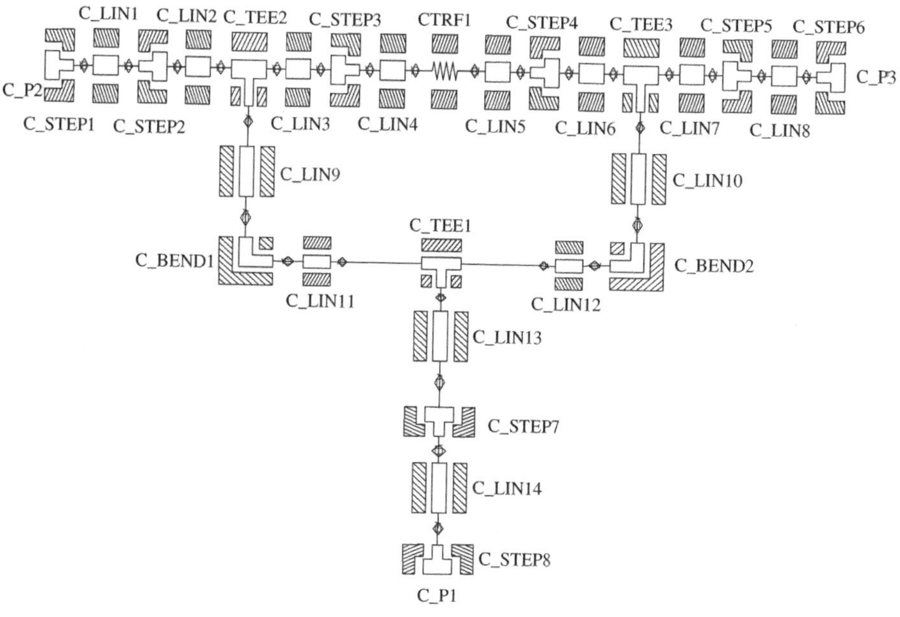

Fig. 6.6.8. Schematic layout of a Wilkinson coupler in CPW technique for 80 GHz.

frequencies, however, all effects due to discontinuities have to be taken into account in the design phase. It is no longer possible to apply inductors for realizing the 90° phase shift. But the phase shift can now be realized using transmission lines as in the standard Wilkinson coupler. The size of the nominal $2 \times Z_L$ resistor must be appropriately designed with respect to the frequencies (waveguide effects; see also Section 4.6) and the technological limitations. In order to compensate for the nonideal T-junctions, some additional matching elements have been introduced.

The layout for such a coupler is shown in Fig. 6.6.9a. In Fig. 6.6.9b two realized couplers for frequencies of 40 GHz and 80 GHz are shown as a photograph.

Figure 6.6.10 shows a comparison between the simulated and the measured scattering parameters. Considering the reduced measurement accuracy (due

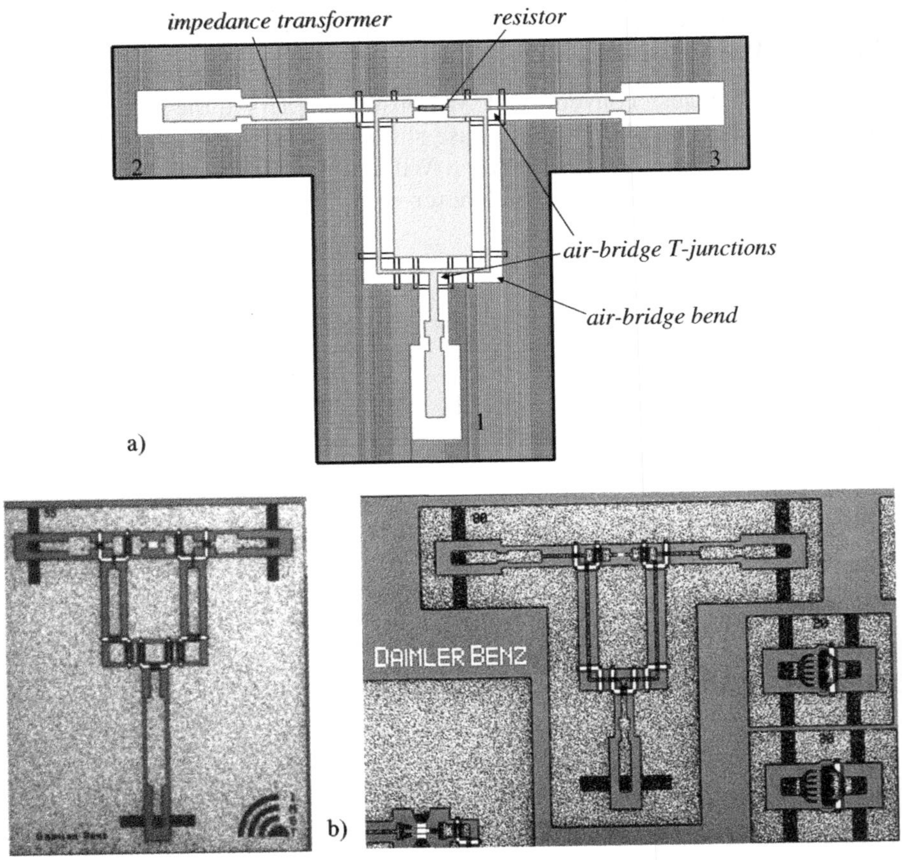

Fig. 6.6.9. (a) Layout of a Wilkinson coupler in CPW line technique. (b) Photograph of two Wilkinson couplers in CPW line technique for 40-GHz left side and 80-GHz right side (design by IMST GmbH, foundry: Daimler Benz AG) [46].

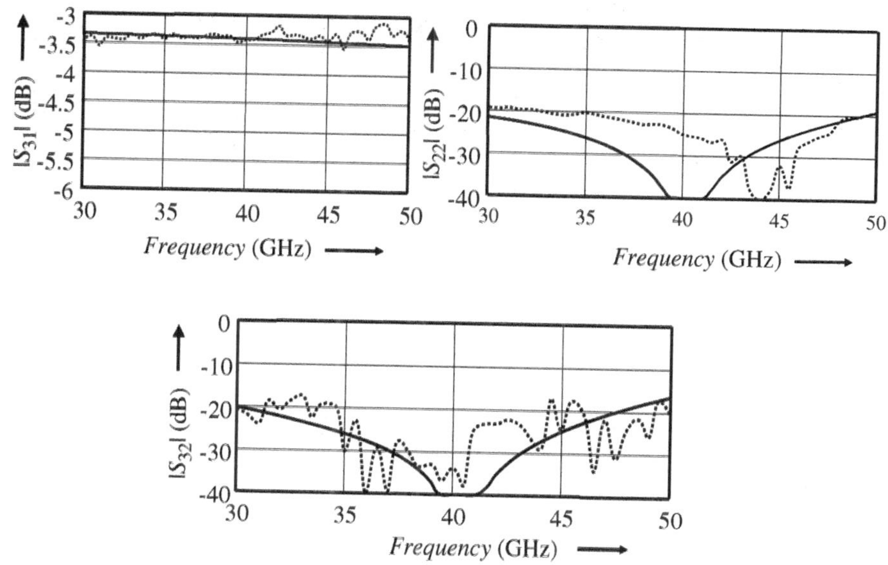

Fig. 6.6.10. Simulated (———) and measured (•••) *S*-parameters of a 40-GHz Wilkinson coupler in CPW technique.

to not available optimum calibration standards at millimeter-wave frequencies), the agreement between measurement and simulation is good. Using the same criteria as in the case of the 1.8-GHz coupler, the measured bandwidth is about 20 GHz. This means again a relative bandwidth of 50% can be realized for this coupler. It is also interesting that the insertion loss of the 40 GHz coupler only increases to a value of 3.3 dB.

The comparison between measured and simulated scattering parameters (Fig. 6.6.11) for the case of the 80 GHz coupler is not so good. But this mostly is due to the reduced measurement accuracy. Because no real three-port measurement equipment was available at this frequency, the measurement results were extracted from a two-port measurement where the third port was matched by a resistor. An accurate determination of the bandwidth under these condition is very difficult, but it should be somewhere on the order of 15–20 GHz. The additional insertion loss is on the order of 0.7 dB. Input reflection coefficient and isolation again are on the order of –20 dB and 20 dB, respectively. Generally, it can be stated that the properties of this 80-GHz Wilkinson coupler are comparable to those of the coupler at lower frequencies.

In conclusion, Wilkinson couplers for coplanar MMIC application at 1.8 GHz, 40 GHz, and 80 GHz have been demonstrated in this section. The investigations show that such couplers no longer need to be avoided in designing coplanar MMICs. All interesting frequencies can be covered by the

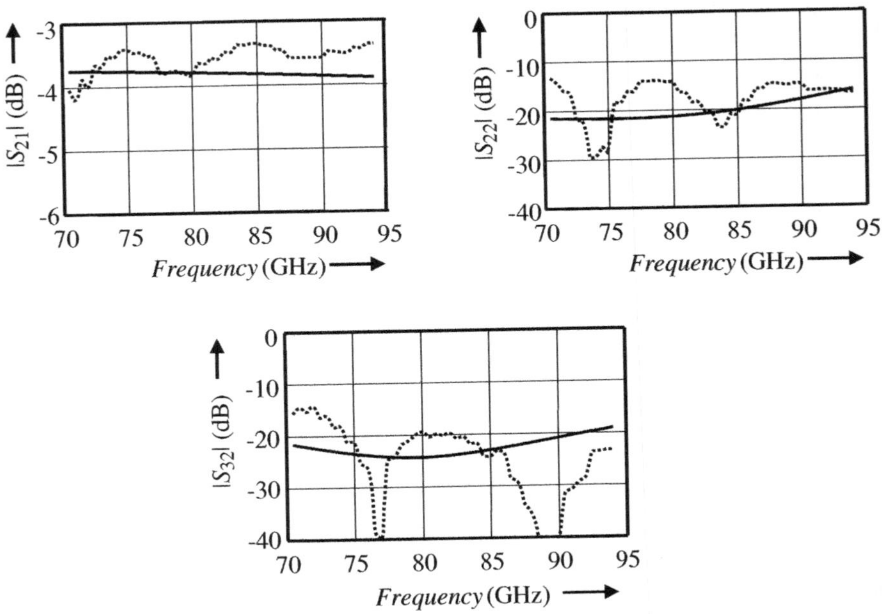

Fig. 6.6.11. *S*-parameters of an 80-GHz Wilkinson coupler in CPW technique with a size of 1200 µm × 800 µm.

described coupler designs. Lumped coplanar waveguide couplers are especially suitable at the low-frequency end.

It is evident that such couplers cannot only be realized in coplanar waveguide technique but also in conventional microstrip line technique. All designs feature a very low insertion loss around 3.1 dB (3 dB for the ideal coupler) and isolation as well as input reflection below –25 dB with a bandwidth of over 20%.

BIBLIOGRAPHY AND REFERENCES

1. J. Wilkinson, An *n*-way hybrid power divider, *IRE Trans. Microwave Theory Tech.*, vol. MTT-8, 1960, pp. 116–118.
2. M. Houdart, Coplanar lines: application to broadband microwave integrated circuits, in: *Proceedings, 6th European Microwave Conference* Rome, 14–17 Sept., 1976, pp. 49–53.
3. G.-J. Laughlin, A new impedance-matched wide-band balun and magic tee, *IEEE Trans. Microwave Theory Tech.*, vol. MTT-24, no. 3, 1976, pp. 135–141.
4. M. Houdart, C. Aury, Frederic, and A. Jean, Coplanar lines: Application to lumped and semi-lumped microwave integrated circuits, in: *Proceedings, 7th European Microwave Conference*, Copenhagen, Denmark, 5–8 Sept., 1977, pp. 450–454.

5. M. Houdart and C. Aury, Various excitation of coplanar waveguide, in: 1979 *IEEE MTT-S International Microwave Symposium Digest*, April 30–May 2, 1979, Orlando, FL, pp. 116–118.

6. R. Saal, *Handbuch zum Filterentwurf* (*Handbook of Filter Design*), Allgemeine Elektrizitäts-Gesellschaft AEG-Telefunken, 1979.

7. D.-F. Williams and S.-E. Schwarz, Design and performance of coplanar waveguide bandpass filters, *IEEE Trans. Microwave Theory Tech.*, vol. MTT-31, no. 7, 1983, pp. 558–566.

8. R. Esfandiari et al., Design of interdigitated capacitors and their application to gallium arsenide filters, *IEEE Trans. Microwave Theory Tech.*, vol. 31, no. 1, 1983, pp. 57–64.

9. G. Ghione, C. Naldi, and R. Lich, *Q*-factor evaluation for coplanar resonators, *Alta Frequenza*, English Edition, vol. 52, no. 3, 1983, pp. 191–193.

10. B.-J. Janiczak, Multiconductor planar transmission-line structures for high-directivity coupler applications, in: *1985 IEEE-MTT-S. Int. Microwave Symposium Digest*, St. Louis, USA, 4–6 June 1985, pp. 215–218.

11. H. Baundrand, M. Kaddour, and M. Ahmadpanah, An active directional coupler on semiconductor substrate, in: *MIOP '86, Fachmesse und Konferenz für Hoechstfrequenztechnologie*, Conference Proceedings, Mikrowellentechnologie und Optologie, Wiesbaden, Germany, 10–12 June 1986, vol. 2, pp. 6.1–6.12.

12. G. Kibuuka, R. Bertenburg, M. Naghed, and I. Wolff, Coplanar lumped elements and their application in filters on ceramic and Gallium Arsenide substrates, *Proceedings, 19th European Microwave Conference*, 1989, pp. 656–661.

13. G. Kibuuka, *Computation of lumped and semi-lumped elements in microstrip and coplanar technique based on spectral domain analysis of planar lines*, Doctoral Thesis, Duisburg University, Germany, 1990, 111 pages.

14. D. Leistner, W. Schmid, and G. Eggers, Application of coplanar waveguide microwave integrated circuits at C- and Ku-band frequencies, in: *Proceedings, 20th European Microwave Conference*, Budapest, Hungary, 10–13 Sept. 1990, vol. 2, pp. 1021–1026.

15. V.-K. Tripathi and A. Biswas, Coplanar and broadside coupled finline six port hybrids, in: *Proceedings, 20th European Microwave Conference*, Budapest, Hungary, 10–13 Sept. 1990, vol. 2, pp. 1033–1038.

16. C.-K.-C. Tzuang, C.-C. Yit, and S. Shyang, Design of a quasi-planar broadside end-coupled bandpass filter, in: *1990 IEEE MTT-S International Microwave Symposium Digest*, 8–10 May 1990, Dallas, TX, vol. 1, pp. 407–410.

17. T. Hirota, A. Minakawa, and M. Muraguchi, Reduced-size branch-line and rat-race hybrids for uniplanar MMIC's, *IEEE Trans. Microwave Theory Tech.*, vol. 38, no. 3, 1990, pp. 270–275.

18. M. Naghed and I. Wolff, Equivalent capacitances of coplanar waveguide discontinuities and interdigitated capacitors using a three-dimensional finite difference method, *IEEE Trans. Microwave Theory Tech.*, vol. 38, no. 12, Dec. 1990, pp. 1808–1815.

19. M. Naghed and I. Wolff, Multiple coupled asymmetrical coplanar waveguides and their application in interdigitated filters, in: *Proceedings, 20th European Microwave Conference*, 1990, pp. 913–918.

20. T. Hirota, A. Minakawa, and M. Muraguchi, Reduced-size branch-line and rat-race hybrids for uniplanar MMIC's, *IEEE Trans. Microwave Theory Tech.*, vol. 38, no. 3, March 1990, pp. 270–275.
21. N.-I. Dib, L.-P-B. Katehi, G.-E. Ponchak, and R.-N. Simons, Theoretical and experimental characterization of coplanar waveguide discontinuities for filter applications, *IEEE Trans. Microwave Theory Tech.*, vol. 39, no. 5, 1991, pp. 873–882.
22. Y.-H. Shu, J.-A. Navarro, and K. Chang, Electronically switchable and tunable coplanar waveguide-slotline band-pass filters, *IEEE Trans. Microwave Theory Tech.*, vol. 39, no. 3, 1991, pp. 548–554.
23. M. Naghed, M. Rittweger, and I. Wolff, A new method for the calculation of the equivalent inductances of coplanar waveguide discontinuities, in: *IEEE MTT-S International Microwave Symposium Digest*, 1991, pp. 747–750.
24. M. Naghed, *Analyse koplanarer Mikrowellenstrukturen mit der Methode der quasistatischen Finiten-Differenzen,* Doctoral Thesis, Duisburg University, Duisburg, Germany, 1992.
25. A.-K. Rayit and N.-J. McEwan, Coplanar waveguide filters: New design data and case studies, in: *Analysis, Design and Applications of Coplanar Waveguides*, London, GB, Oct. 19, 1993, IEE Colloquium, vol. 1993/186, 1993, pp. 11/1–11/11.
26. S. Sattler, T. Felgentreff, and P. Russer, A coplanar millimeterwave resonator on silicon, in: *1993 IEEE Microwave and Millimeter-Wave Monolithic Circuits Symposium*, Atlanta, June 14–15, 1993, pp. 57–60.
27. J.-K-A. Everard and K.-K-M. Cheng, High performance direct coupled bandpass filters on coplanar waveguide, *IEEE Trans. Microwave Theory Tech.*, vol. 41, no. 9, 1993, pp. 1568–1573.
28. A.-K. Rayit and N.-J. McEwan, Coplanar waveguide filters, in: *1993 IEEE MTT-S International Microwave Symposium Digest*, vol. 2, Atlanta, June 14–18, vol. 2, 1993, pp. 1317–1320.
29. R.-N. Simons and S.-R. Taub, Coplanar waveguide radial line double stub and application to filter circuits, *Electronics Lett.*, vol. 29, no. 17, 1993, pp. 1584–1586.
30. A. K. Rayit and N. J. McEwan, Coplanar waveguide filters, in: *1993 IEEE MTT-S International Microwave Symposium Digest*, 1993, pp. 1317–1320.
31. A. Maas, *Microwave Mixers*, Boston: Artech House, 1993, pp. 253–256.
32. S. Uysal, K.-S. Ang, and P.-S. Kooi, Coplanar waveguide active bandpass filter, *Electronics Lett.*, vol. 30, no. 19, 1994, pp. 1605–1606.
33. W. Schwab, F. Boegelsack, and W. Menzel, Multilayer suspended stripline and coplanar line filters, *IEEE Trans. Microwave Theory Tech.*, vol. 42, no. 7, part 2, 1994, pp. 1403–1407.
34. M. Gillick, I.-D. Robertson, and J.-S. Joshi, Coplanar waveguide two-stage balanced MMIC amplifier using impedance-transforming lumped-distributed branchline couplers, *IEE Proc. (Microwaves, Antennas and Propagation)*, vol. 141, no. 4, 1994, pp. 241–245.
35. P. Pogatzki, R. Kulke, T. Sporkmann, D. Köther, R. Tempel, and I. Wolff, A comprehensive evaluation of quasi-static 3D-FD calculations for more than 14 CPW structures—Lines, discontinuities and lumped elements, in: *IEEE MTT-S International Microwave Symposium Digest*, San Diego, May 1994, vol. 2, pp. 1289–1292.

36. B. Hopf, I. Wolff, and M. Guglielmi, Coplanar MMIC bandpass filters using negative resistance circuits, in: *IEEE Microwave and Millimetre-Wave Monolithic Circuits Symposium*, San Diego, May 1994, pp. 229–231.

37. Mernyei, I. Aoki, and H. Matsuura, A novel MMIC coupler—Measured and simulated data, in: *IEEE MTT-S International Microwave Symposium Digest*, 1994, pp. 229–232.

38. M. Abdo Tuko, *Theoretical and experimental investigation of the design and realization of microwave frequency multipliers in coplanar (M)MIC techniques*, Doctoral Thesis, Duisburg University, Duisburg, Germany, 1994.

39. Pogatzki, D. Köther, R. Kulke, B. Hopf, T. Sporkmann, and I. Wolff, Coplanar hybrids based on an enhanced inductor model for mixer applications up to mm-wave frequencies, in: *Proceedings, 24th European Microwave Conference*, Cannes, 1994, pp. 258–262.

40. K. Wada, Y. Noguchi, E. Higashino, and J. Ishii, Tapped-feed combline type coplanar waveguide resonator bandpass filters, in: *1995 IEEE MTT-S Symposium on Technology for Wireless Applications Digest*, Vancouver, Canada, Feb. 20–22, 1995, vol. 1, pp. 141–146.

41. R. Stephan and J. Möhring, Monolithisch integrierte Mikrowellenfilter in GaAs-Koplanartechnologie, in: *40th IWK/1995, 40th International Scientific Colloquium of the Technical University Ilmenau*, Ilmenau, Germany, 18–21 Sept., 1995, vol. 3, pp. 169–174.

42. F.-L. Lin, C.-W. Chiu, and R.-B. Wu, Coplanar waveguide bandpass filter—a ribbon-of-brick-wall design, *IEEE Trans. Microwave Theory Tech.*, vol. 43, no. 7, part I, 1995, pp. 1589–1596.

43. C.-Y. Chi and G.-M. Rebeiz, Planar microwave and millimeter-wave lumped elements and coupled-line filters using micro-machining techniques, *IEEE Trans. Microwave Theory Tech*, vol. 43, no. 4, part I, 1995, pp. 730–738.

44. U. Karacaoglu, R. Khatri, M. Gillick, N. Azefor, I. D. Robertson, and M. Guglielmi, An investigation of CPW bandpass filters using end-coupled resonators and square dual-mode rings, in: *Proceedings, 25th European Microwave Conference*, Bologna, 1995, pp. 519–523.

45. W. Menzel, W. Schwab, and G. Strauss, Investigation of coupling structures for coplanar bandpass filters, in: *IEEE MTT-S International Microwave Symposium Digest*, 1995, pp. 1407–1410.

46. D. Köther, B. Hopf, and T. Sporkmann, MMIC Wilkinson couplers for frequencies up to 110 GHz, in: *1995 IEEE MTT-S International Microwave Symposium Digest*, Orlando, FL, May 1995, pp. 663–666.

47. M.-R. Lyons and C.-A. Balanis, Transient coupling reduction and design considerations in edge-coupled coplanar waveguide couplers, *IEEE Trans. Microwave Theory Tech.*, vol. 44, no. 5, 1996, pp. 778–783.

48. A.-K. Rastogi, Design and performance of coplanar waveguide slotline bandpass filter, in: *Int. J. Infrared Millimeter Waves*, vol. 17, no. 5, 1996, pp. 915–920.

49. F. Tefiku, E. Yamashita, and J. Funada, Novel directional couplers using broadside-coupled coplanar waveguides for double-sided printed antennas, *IEEE Trans. Microwave Theory Tech.*, vol. 44, no. 2, 1996, pp. 275–282.

50. L. Verweyen, W. H. Haydl, A. Tessmann, H. Massler, T. Krems, and J. Schneider, Coplanar branch-line and rate-race couplers for W-band applications, in: *Proceedings, 26th European Microwave Conference*, 1996, pp. 602–606.
51. T. Gokdemir, I. D. Robertson, Q. H. Wang, and A. A. Rezazadeh, K/Ka-band coplanar waveguide directional couplers using a three-metal-level MMIC process, *IEEE Microwave and Guided Wave Lett.*, vol. 6, no. 2, 1996, pp. 76–78.
52. K. Wada, I. Awai, and Y. Yamashita, Characteristics of lambda /4 CPW resonators with tap-excitation and their application to bandpass filters, in: *1997 IEEE MTT-S International Microwave Symposium Digest*, 8–13 June 1997, Denver, CO, vol. 2, 1997, pp. 717–720.
53. T.-M. Weller, K.-J. Herrick, and L.-P.-B. Katehi, Quasi-static design technique for MM-wave micromachined filters with lumped elements and series stubs, *IEEE Trans. Microwave Theory Tech.*, vol. 45, no. 6, 1997, pp. 931–938.
54. IMST GmbH, *COPLAN for ADS*™, Kamp-Lintfort, Germany.
55. G.-M. Shau, K.-C. Hwann, and H.-Chen Chun, Modeling of lumped-element coplanar-stripline low-pass filter, *IEEE Microwave and Guided Wave Lett.*, vol. 8, no. 3, 1998, pp. 141–143.
56. O. Wohlgemuth, T. Krems, R. Reuter, M.-J.-W. Rodwell, W. Haydl, and M. Schlechtweg, Integrated directional coupler for 90 and 180 GHz, *IEEE Microwave and Guided Wave Lett.*, vol. 9, no. 8, 1999, pp. 308–310.
57. B. Hopf, *Aktive monolithisch integrierte Mikrowellen-Kopplungsfilter in Koplanarleitungstechnik*, Doctoral Thesis, Duisburg University, Duisburg, Germany, 2000.
58. J.-H. Park, H.-T. Kim, Y. Kwon, and Y.-K. Kim, Tunable millimeter-wave filters using a coplanar waveguide and micromachined variable capacitors, *J. Micromech. Microeng.*, vol. 11, no. 6, 2001, pp. 706–712.
59. C.-L. Liao and C.-H. Chen, A novel coplanar-waveguide directional coupler with finite-extent backed conductor, *IEEE Trans. Microwave Theory Tech.*, vol. 51, no. 1, 2003, pp. 200–206.

7

COPLANAR MICROWAVE INTEGRATED CIRCUITS

7.1 INTRODUCTION

Whereas the previous chapters have been dealing with coplanar waveguides and components as well as coplanar technique in general, the following sections are dedicated to the corresponding applications in circuit design. This chapter starts with a short description of the used active elements and their simulation background. Then tunable resonant circuits that will later be used in the oscillator design will be described. Afterwards, active filters in CPW technique are introduced since their proper design offers the opportunity to improve losses and to reduce required chip size compared to conventional approaches based on passive components. Design examples for coplanar switches, coplanar active filters, and coplanar amplifiers are given in the following sections. Section 7.5 describes a coplanar electronic circulator and Section 7.6 a special frequency doubler in coplanar technology. Sections 7.7 and 7.8 deal with oscillators, and it is worth noting that the intention is not simply to present various designs in coplanar technology. Considerations on decreased chip area and discussions on how to include the parasitics of lumped elements in the circuit design will be the central design guidelines.

During the last 30 years, microwave integrated circuits have been produced mainly (more than 98%) in microstrip technology. Therefore, before starting the discussion of coplanar microwave integrated circuits in detail, a comparison between these two technologies will be given in the beginning of this

Coplanar Microwave Integrated Circuits, by Ingo Wolff.

chapter. This intensive comparison of microstrip and coplanar waveguide structures leads to the conclusion that coplanar technique has more advantages than disadvantages compared to the microstrip circuit design. This is especially true for frequencies beyond 30 GHz, where monolithic microwave integrated circuits (MMICs) have been established and the fabrication of air bridges, which are necessary in coplanar circuits, is standard. A major goal of this introduction is to overcome the widespread prejudices that coplanar waveguides have higher line losses, are more dispersive, and are complicated to handle for the circuit design. It will be demonstrated that the opposite is true and that coplanar waveguide circuit design is as easy (or even easier) as microstrip circuit design, delivering even a higher modeling accuracy.

7.1.1 The Effect of the Shielding on Modeling

The Finite Difference (FD) technique for the analysis of coplanar waveguides as described in Chapter 2 is suitable to calculate the microstrip line properties, too. These properties have been exploited to investigate the effect of the electrical sidewalls, cover plane and ground metalization on the simulation results. This is important because the FD model calculates the structure in a box with electric or magnetic walls. Nevertheless, *on-wafer configurations* with no housing effect can be simulated if some simple rules are observed. These rules are summarized in Fig. 7.1.1 for the coplanar as well as for the microstrip line.

Above and below the carrier substrate, two further dielectric layers with the heights h_3 (cover height) and h_1 (ground plane distance) are given in the model configuration (see also Fig. 2.2.1). The effect of the electric or magnetic planes covering the structure can be neglected if the bottom substrate has the same height and the top plane height is 2 times the thickness of the carrier substrate. In the case of the microstrip line this factor is 10. This large difference results from the field distribution that is concentrated in the slots between the center strip and the ground strips in the case of the coplanar waveguide. On the other hand, the microstrip field has a strong concentration in the air

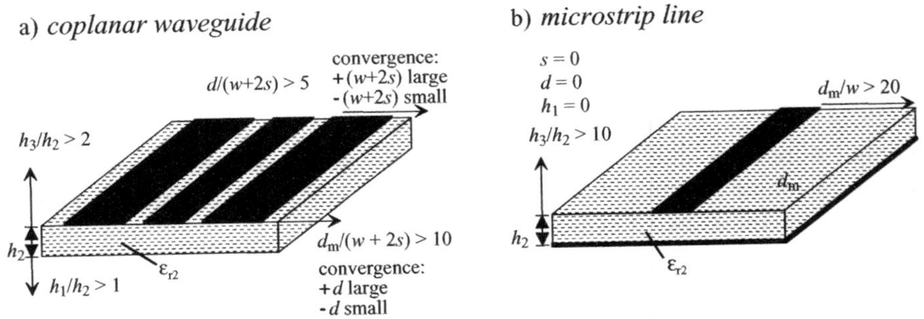

Fig. 7.1.1. Rules for simulating the coplanar waveguide and the microstrip line.

region above the substrate material and therefcre can be easily influenced in its properties by an assumed shielding.

7.1.2 The Waveguide Properties

The most popular prejudice against coplanar waveguides is that the line losses are higher than in microstrip lines. Results shown in Fig. 7.1.2 illustrate that this assumption is, in general, wrong.

The losses of four different 50-Ω coplanar lines are compared with that of a 50-Ω microstrip line. Since there is one more design parameter available for the design engineer in the case of the coplanar waveguide (center strip width w and slot width s) compared to the microstrip case (where only the strip width w can be changed if the substrate height is fixed), different designs for a coplanar waveguide of 50-Ω impedance are possible. As Fig. 7.1.2 shows, the transmission coefficients $|S_{12}|$ of coplanar and microstrip line sections are very similar in both technologies if the total line width of the coplanar waveguide ($w + 2s$) is 125 μm or more. Coplanar lines with small center strip widths have higher losses, but with increasing frequencies both technologies have again losses of the same order as can be seen from Fig. 7.1.2.

A clear advantage of the coplanar waveguide is that dispersion effects are lower than in the case of the microstrip line. This fact is well known in the literature and has also been demonstrated in Chapters 1 and 2. The example given in Fig. 7.1.3 makes this very clear for the case of waveguide components or discontinuities. Two T-junctions, one in microstrip technology and one in coplanar waveguide technology (even including an air bridge), are compared.

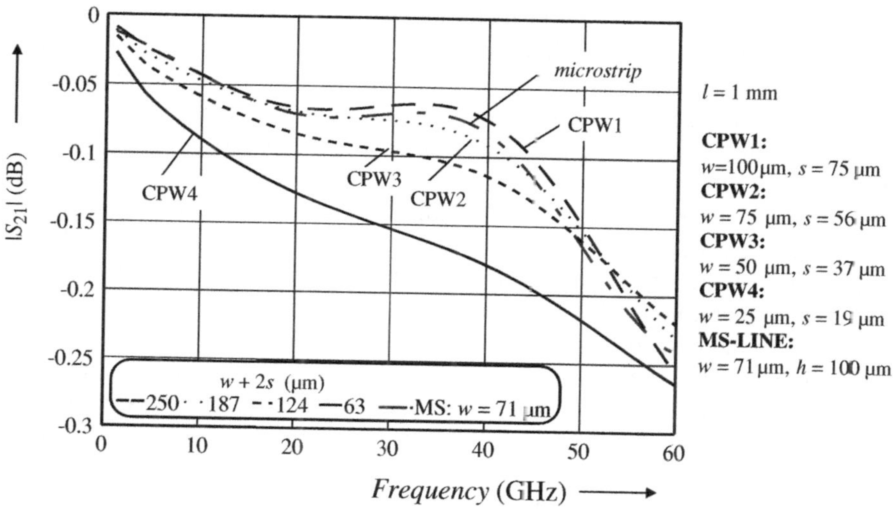

Fig. 7.1.2. Comparison of the transmission properties of various 50-Ω coplanar waveguides and a 50-Ω microstrip line.

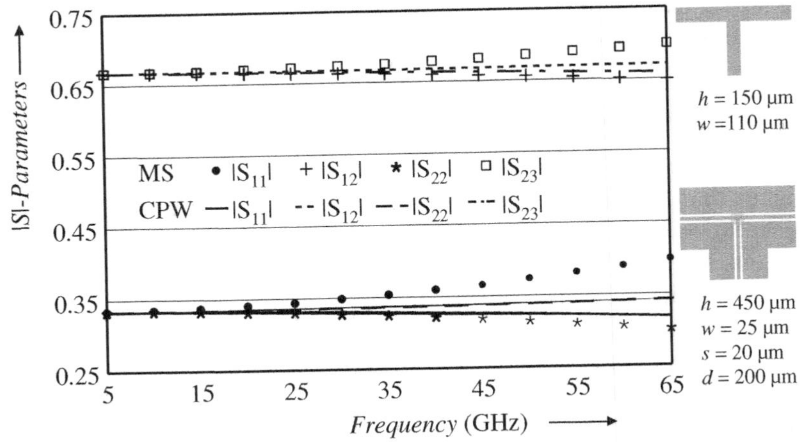

Fig. 7.1.3. Comparison of the dispersion of a microstrip T-junction and a coplanar T-junction (including the necessary air bridges).

Due to the strong field concentration in the slot of the coplanar waveguide (the concentration even increases with the frequency), the curves of the *S*-parameters are more flat over the frequency, for the coplanar case.

Various examples for coupled lines in microstrip technique and coplanar waveguide configurations have been investigated (see also Section 4.4). The main result of these investigations shows that the coupling of two neighbored microstrip lines is much stronger (more than 30 dB) than in CPW configurations with an extra shielding ground strip between the center lines. This advantage of the coplanar waveguide technology is very useful in circuit design, where a coupling of two neighbored circuit parts can be avoided, if needed. Because of this lower coupling of neighbored waveguides and components, the coplanar waveguide circuits can normally be spaced more closely and about 30% of the substrate space can be saved compared to an equivalent microstrip circuit layout.

In the production techniques for coplanar waveguide circuits, three expensive technology steps are not needed and can be avoided: (1) the backside preparation of the substrate material, (2) the backside metalization, and (3) the via-hole technology. This reduces the costs of coplanar waveguide circuits compared to microstrip circuits and increases the production yield.

Of course, there are disadvantages associated with the coplanar technology. At each discontinuity the coplanar odd mode will be excited. This mode has a zero cutoff frequency, which makes air bridges in each coplanar circuit necessary. These air bridges are a short circuit from the left to the right ground strip (to ensure an equal potential on both grounds). They have been extensively described in Section 3.5.5. They have to be located at each input and output port of a waveguide discontinuity.

TABLE 7.1.1. Comparison of the Advantages and Disadvantages of the Microstrip and the Coplanar Waveguide Technology

Parameter	Microstrip	Coplanar
Technology		
Thinning	Yes	No
Backside metal	Yes	No
Via etching	Yes	No
Yield	Lower than CPW	Higher than MS
Circuit size	Fair	Compact
Technology compliance	Excellent	Excellent
Parasitics		
Device source inductance	Yes	No
Ground capacitance	Higher than in CPW	Lower than in MS
Line coupling	Higher than in CPW	Lower than in MS
Loss	Typically lower than CPW loss	For narrow lines higher than MS loss
Dispersion	Higher than in CPW	Lower than in MS
Performance		
Gain	Lower than for CPW	Higher than for MS
Frequency range	Low–high	Low–very high
General		
Handling	Difficult for thin wafers	Easy due to thick wafers
Acceptance	Very high	Very low
CAD tools	Good	Very good
Production costs	Higher than for CPW	Lower than for MS

In MMICs the fabrication of air bridges is no extra effort. In hybrid circuits, bond vias have to be placed, which increases the manufacturing expense and cost. Their production is also not of high reproducibility. For MMIC applications, three different types of air bridges can be used (see Section 3.5.5) and their models are fully integrated into the coplanar library (see Chapter 5), so all parasitics can be considered in the circuit design.

The second disadvantage of CPW technology is the worse thermal dissipation in power applications because of the thicker substrate, the unavailable (because normally not needed) via technology, and the missing backside metalization. The flip-chip technology is one solution that can be used to overcome this problem.

The design of a coplanar MMIC utilizing available software tools, especially the COPLAN software (described in Chapter 5) and also the available general-purpose electromagnetic field solvers, leads in many cases to a successful design in the first or second through-run. Research labs are continuously demonstrating coplanar circuits up to now, but only a few companies show serious interest in coplanar circuit design and production. At this time, there are various possibilities to design coplanar circuits. These design solutions can be divided as follows:

- Analytical models (various)
- Measured data base models
- EM solvers (various)
- COPLAN software together with a circuit design program as described in Chapter 5

Various attempts of coplanar circuit designs were not successful in the past because of missing design tools, but in recent years, more and more coplanar applications have been demonstrated. Thus, the interest in this technique is steadily increasing and there is a fair chance that coplanar circuit design will replace some of the microstrip designs in the coming years simply because coplanar circuits are cheaper and easier to design.

In this chapter, a short overview of the evolution in coplanar circuit design has been given. As has been seen in the previous chapters, one of the main advantages of the circuit design in coplanar waveguide technique is that the electric characteristics of the circuit are determined mainly by the layout of the metallic conductor on the surface of the wafer. Compared to typical microstrip designs, neither a thinning of the monolithic substrate nor any via-hole techniques are necessary. Moreover, because of the low dispersion of coplanar waveguides, circuit simulations based on frequency-independent characteristic impedances as well as effective permittivities are valid over a wide frequency range. Therefore, the simulation of the circuit behavior is simplified. On the other hand, the designer has to ensure that there is always a proper grounding even if the transmission lines and the biasing lines are cutting up the metallic surface in numerous separate parts. Multimode propagation along the RF signal paths has to be suppressed using an adequate number of air bridges.

7.2 COPLANAR TRANSISTORS AND COPLANAR SWITCHES

7.2.1 Active Power Dividers and Combiners and Switches

7.2.1.1 Power Dividers and Combiners. In this section, coplanar active three-ports based on conventional foundry devices are presented. Utilizing these devices, active power combiners, dividers, and switches can be realized. The modeling of these devices is simply done by networking individual FETs or HEMTs [207]. Measurement and simulation results are in excellent agreement. Three different devices will be compared, and their scaling results are demonstrated. Additionally, the results of a SPDT switch are shown. Depending on the device periphery, the application frequency can easily go beyond 40 GHz.

Power dividers and combiners are utilized in almost all microwave circuits. Passive structures not only have the disadvantage of being lossy, but also have a severe drawback in that there is normally low isolation between the ports. In some cases, such as mixers and active isolators or circulators [49, 180], good isolation is mandatory. There are various ways to reduce the insertion loss and to increase the isolation for such structures.

Conventional HEMT (or FET) cells are combined to create three structures. The device technology used for demonstration is a 0.25-μm PM-HFET technology [260]. Figure 7.2.1 depicts the layout and the schematic representation of these three devices. The three ports are constructed by networking two foundry HEMT cells with coplanar transmission lines. Important for this design is that all three ports must conform to 50-Ω coplanar waveguides. Due to the cell-oriented design strategy, the size of these devices is about 0.15 mm^2. As shown in the section on scaling (see below), the size of such three-ports may even be reduced by about 50%. From the functionality point of view, it is clear that a dual gate common source (DGCS) device works as a power combiner whereas the dual drain common source (DDCS) device can be identified as a power divider. Five peripheries ranging from 6 × 10-μm up to 6 × 50-μm gate width have been designed, fabricated, and measured for each of the devices depicted in Fig. 7.2.1.

The scattering parameter behavior as function of the frequency for the three devices is depicted in Fig. 7.2.2. The DGCS and DSCG devices behave like signal combiners while the DDCS device is a signal divider. The isolation of the DDCS device is around −25 dB, and for the other devices it is on the order of −15 dB. For the DGCS and DDCS devices the isolation between port 1 and port 2 is around −12 dB due to the isolation-amplification chain of the two devices.

While the inherent match of the DDCS is good ($-12\,\text{dB} < |S_{11}| < -6\,\text{dB}$) up to 20 GHz, the insertion loss of the other devices corresponds to conventional

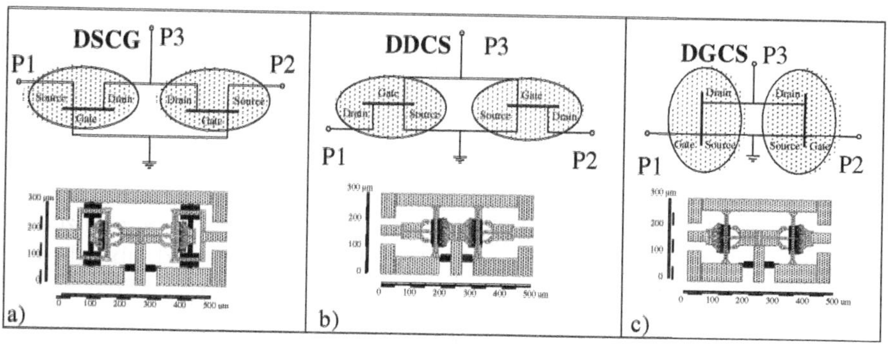

Fig. 7.2.1. Layout and schematic representation of three coplanar port structures. (a) Dual source common gate device (DSCG). (b) Dual drain common source device (DDCS). (c) Dual gate common source device (DGCS).

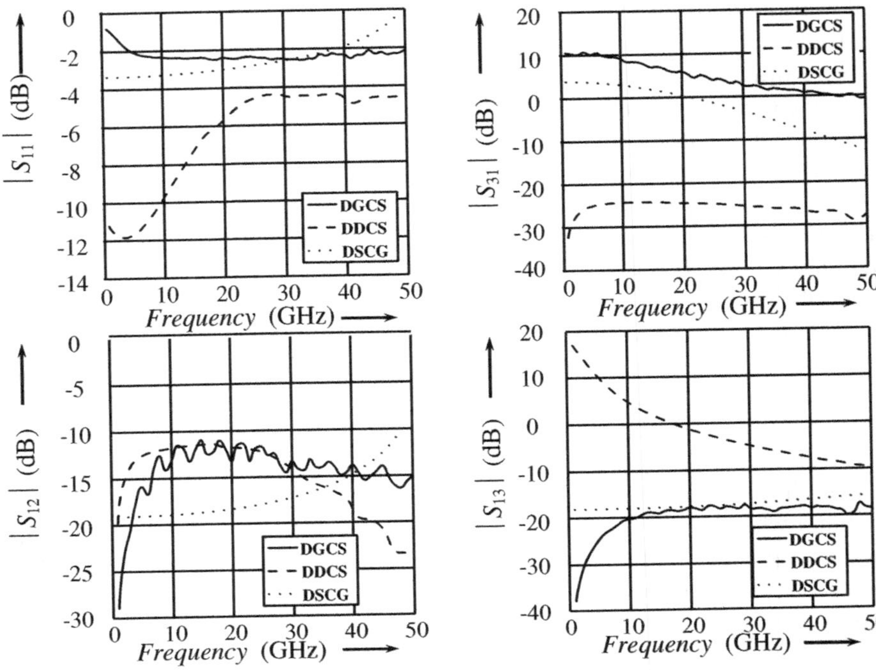

Fig. 7.2.2. Comparison of the three different three-ports (DDCS, DGCS, DSCG) with $6 \times 10\,\mu m$ total gate width.

FETs. It should be pointed out, however, that the scattering parameters depicted in Fig. 7.2.2 are for a 50-Ω environment; that is, the transistors have a very large gate width (see parameters given above). In addition, it may be found that matching of these devices is quite difficult since the devices tend to be unstable over a broad frequency band. This typically results in lossy matching networks for any application.

7.2.1.2 Fundamental Coplanar Switch Circuits. Another interesting application for the devices described above are microwave switches. MMIC switches may be realized with diodes or utilizing FET/HEMT circuits in T- or π-configuration. However, the three-ports discussed in the previous section are capable of replacing such circuits. This type of switch features gain and a good isolation. As in the combiner and divider case, switches can also be realized in two ways. In the following, the example of a SPDT (single-pole double throw) switch is given. In the convention used here, this device could be a DDCS (see Fig. 7.2.1) device, for instance. Figure 7.2.3 depicts some of the S-parameters for such a DDCS-switch as a function of frequency. To determine these results, the device between port 1 and port 3 was in the ON condition while the device between port 2 and 3 was switched OFF. The bias points were set to 0 V at the

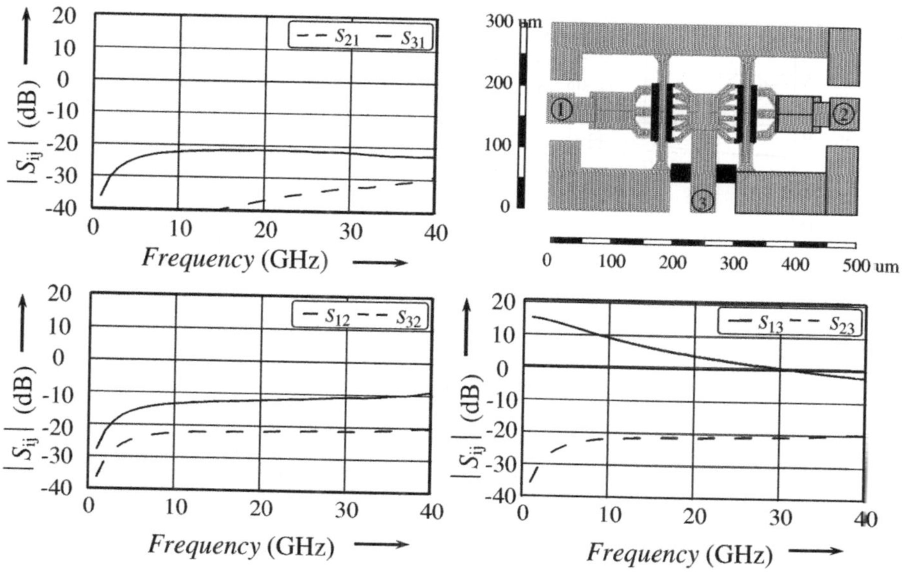

Fig. 7.2.3. *S*-parameters and layout for a SPDT switch (DDCS-switch) in coplanar technology.

gates and to 3V and 0V at the two drains. It can be seen that only a small amount ($|S_{23}| < -20$ dB) of the incoming signal goes from port 3 to port 2, in this case, while an amplified part of the signal ($|S_{13}|$) goes to port 1 of the switch. The isolation between ports 1 and 2 is about 15 dB ($|S_{12}|$) and 20 dB ($|S_{21}|$), respectively. Even though the raw performance of such switches is good, they clearly need matching elements for stabilization and for broadband applications. The advantages of these switches are that they feature gain, have compact size, and can be integrated into MMICs.

7.2.1.3 Results and Measurements. Scattering parameter measurements of three-ports are difficult. Therefore, three two-port measurements were conducted in order to determine the three-port *S*-parameters. The nonideal termination was considered by applying a time gate at the third port. For some measurements, however, this error correction was not applied. In Fig. 7.2.2, for instance, it can be observed that small measurement errors occurred due to inaccuracies of the nonideal load at the third port (ripples in the measurements). The results shown in Figs. 7.2.3 and 7.2.4, however, were achieved utilizing the time gate function of the network analyser. Thus, the curves are much smoother and have no ripples. Above 45 GHz, additional problems, such as higher mode effects due to the bend calibration structure (specified up to 40 GHz only), can be observed. However, the results of the measurements are quite reasonable for all cases.

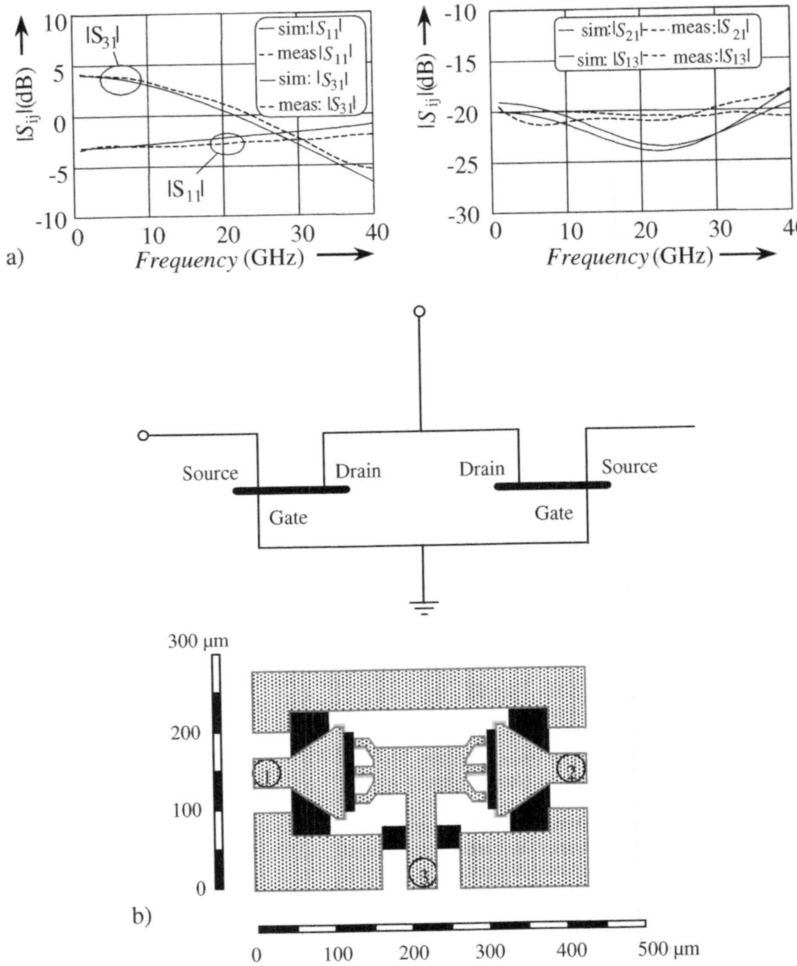

Fig. 7.2.4. (a) Simulated and measured scattering parameters for a 300-µm gate width DSCG device as shown in part b.

The simulation of all three-port structures is based only on one device model that is discussed in the next section. To determine this model, only common source devices with various gate widths were measured. Then, the scaleable model was determined. The results shown in Figs. 7.2.4 and 7.2.5 are derived simply from networking this model in the appropriate configurations. With this background, it is interesting to see how well the simulation of a DSCG device agrees with measurements. The devices are rotated in comparison with standard HEMTs for this simulation. Figure 7.2.4 depicts the magnitudes of S_{11}, S_{31}, S_{21}, and S_{13} for such a 300-µm DSCG device. The port numbering is depicted in the layout of the device in Fig. 7.2.4b.

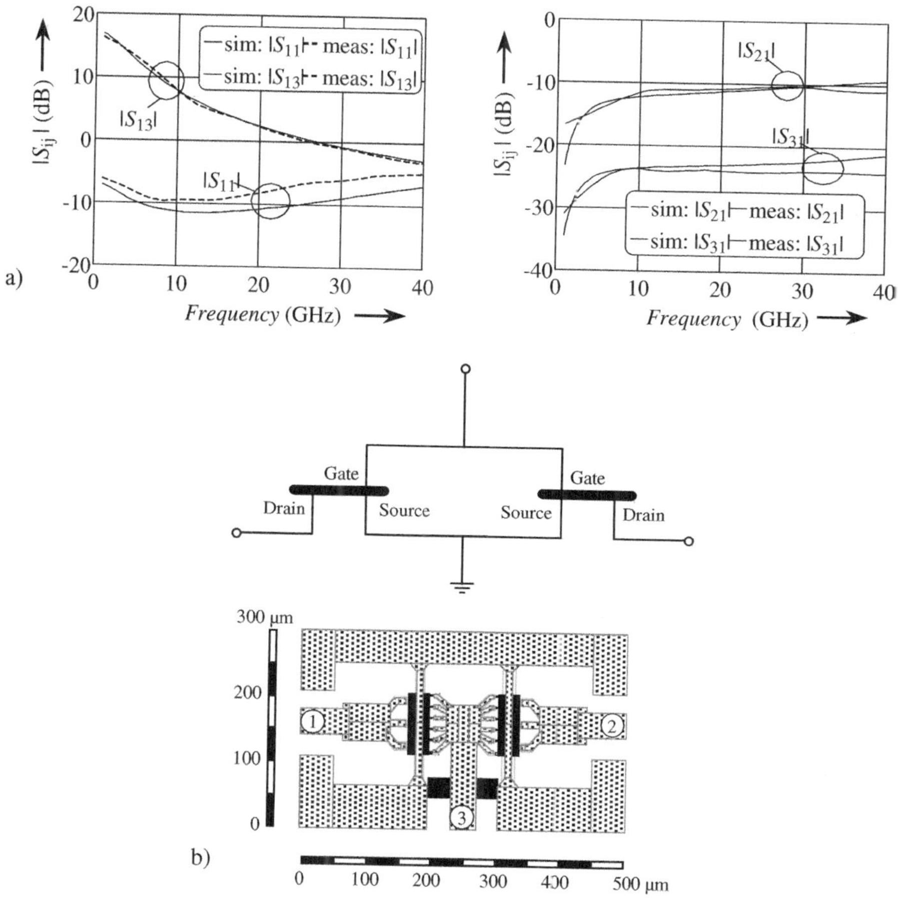

Fig. 7.2.5. (a) Simulated and measured scattering parameters for a 180-μm DDCS device as shown in part b.

The agreement between simulation and measurement for all S-parameters is good up to the highest measurable frequency. In other words, a common source model was successfully applied to a device with common gate transistors. Compared to conventional gain curves, the achieved forward gain is quite low for this device. One reason for this is that the S-parameters are measured and presented for a 50-Ω environment. On the other hand, the isolation between port 1 and port 2 as well as between port 1 and port 3 is about 20 dB for this device.

Figure 7.2.5 depicts the scattering parameters for a 180-μm DDCS-HEMT. This device behaves like a power divider (active tee), where port 3 is the input and port 1 and port 2 are the output ports. The modeling in this case is in an even better agreement with measurements than in the case of the DSCG structure. In this case, $|S_{13}|$ that is the forward gain from port 3 to port 1 looks identical to the corresponding $|S_{21}|$ of the 180 µm HEMT. For this device an isolation of about 25 dB can be observed between port 1 and port 3.

7.2.1.4 Device Scaling. A scalable model for MESFETs has been developed [185]. The model itself and the relations for the equivalent circuits are depicted in Fig. 7.2.6. It is obvious that an identical procedure can be applied to HEMT devices.

It is not directly obvious, however, if such a model can be applied to the design of three-port structures as shown in Fig. 7.2.1. For this reason, five geometries for each of the three ports have been realized. The layouts with the corresponding dimensions for the DDCS-scaling are given in Fig. 7.2.7. These depicted DDCS devices have a gate width of 60 µm, 120 µm, 180 µm, 240 µm, and 300 µm, respectively.

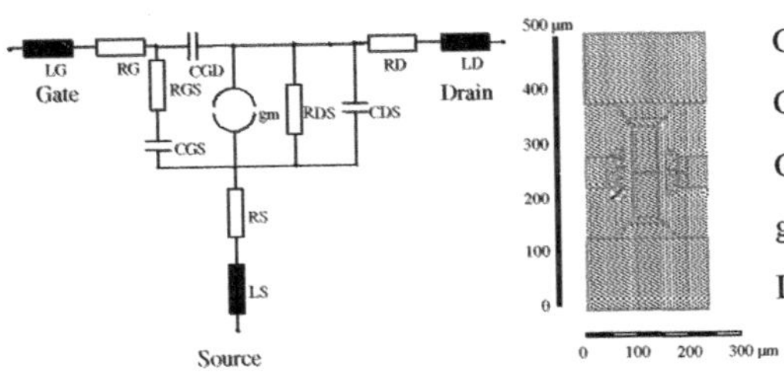

$C_{gs} = C_{gs0} * Z$ $\quad$ $R_g = R_{g0} * Z / (M*M)$

$C_{gd} = C_{gd0} * Z$ $\quad$ $R_d = R_{d0} / Z$

$C_{ds} = C_{ds0} * Z$ $\quad$ $R_s = R_{s0} / Z$

$g_m = g_{m0} * Z$ $\quad$ $R_{gs} = R_{gs0} / Z$

$L_g = L_{g0} * Z$ $\quad$ $R_{ds} = R_{ds0} / Z$

$Z = \text{gatewidth} / 100\ \mu m$

$M = \text{number of fingers}/6$

Fig. 7.2.6. Scaleable HEMT model.

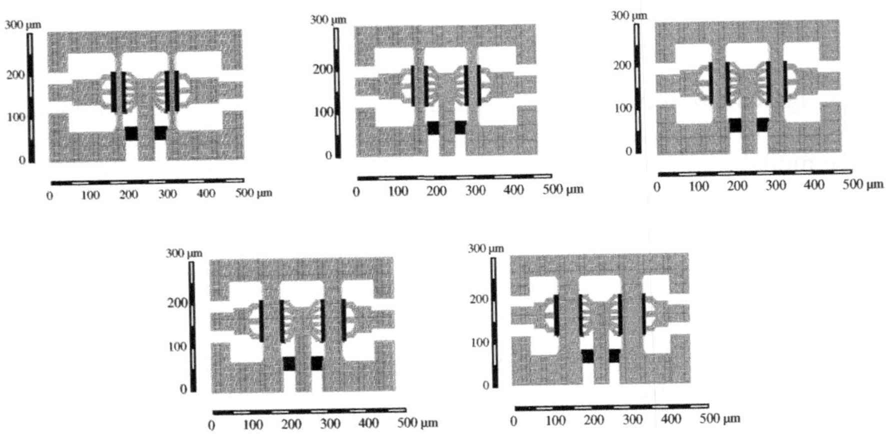

Fig. 7.2.7. Layout and geometry for DDCS-scaling (gate widths = 60/120/180/240/300 µm).

The layout in Fig. 7.2.6 depicts the basic cell structure that is utilized for determining the values of C_{gs0}, C_{gd}, C_{gd0}, C_{ds0}, g_{m0}, L_{g0}, R_{g0}, R_{d0}, R_{s0}, R_{gs0}, and R_{ds0}. This basic cell has a gate width of 100 μm and six fingers. The equivalent circuit elements can, for instance, be determined from measurements. In Fig. 7.2.7, it can be seen that devices with a periphery of 60 μm or 300 μm have almost the same outer dimensions. It should also be pointed out that the devices have not been optimized for small size. This size constraint depends on to the foundry cell orientation of the devices.

A size reduction of about 50% is realistic if only the intrinsic foundry device is used and the extrinsic environment is optimized for size. One possible example of such a size reduction is shown in Fig. 7.2.8. For both devices in this figure, identical intrinsic device peripheries have been used while the extrinsic environment was optimized for size. Both structures in Fig. 7.1.8 represent a DSCG device with HEMTs of 60-μm gate width.

Based on the scaleable HEMT model depicted in Fig. 7.2.6, results for various DDCS devices are analyzed. The results are shown in Fig. 7.2.9, where the magnitudes of S_{21}, S_{31}, and S_{13} are plotted as a function of the frequency and with the gate width of the single device ranging from 50 μm to 500 μm in steps of 50 μm as a parameter. The isolation between port 1 and port 2 is better than 10 dB for all devices. However, small devices with 50-μm gate width have isolation close to 20 dB. The scattering parameter S_{13} behaves similar to the S_{21} counterpart of a normal HEMT. Large devices show high gain with a steep slope, while small devices have lower gain at low frequencies but a much less steep gain-slope.

Thus, it is clear that small devices are especially suitable for high frequencies. These relations are depicted in Fig. 7.2.9. The isolation between port 1 and port 3 is better than 20 dB for all 10 devices that are simulated. For this parameter (S_{31}), there is almost no variation as a function of the gate width. Based on this model, it is quite simple to choose the appropriate gate width for a given frequency. In contrast, it is also clear that, for smaller devices, lower input power is required. Information on maximum incident power, however, is not available from the linear model utilized here.

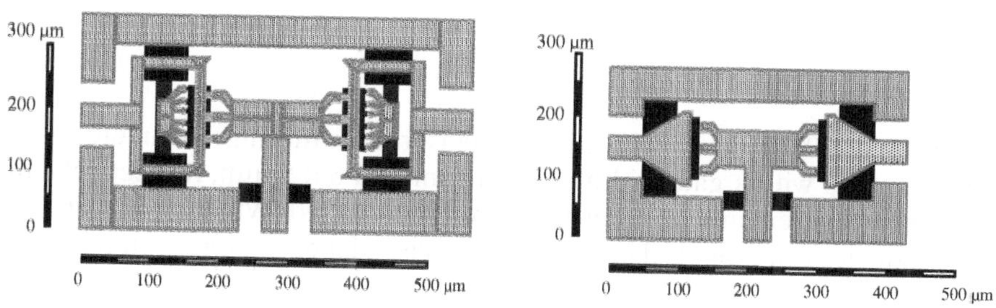

Fig. 7.2.8. Two types of DSCG structures with 60-μm gate width devices (0.192 mm^2 and 0.113 mm^2).

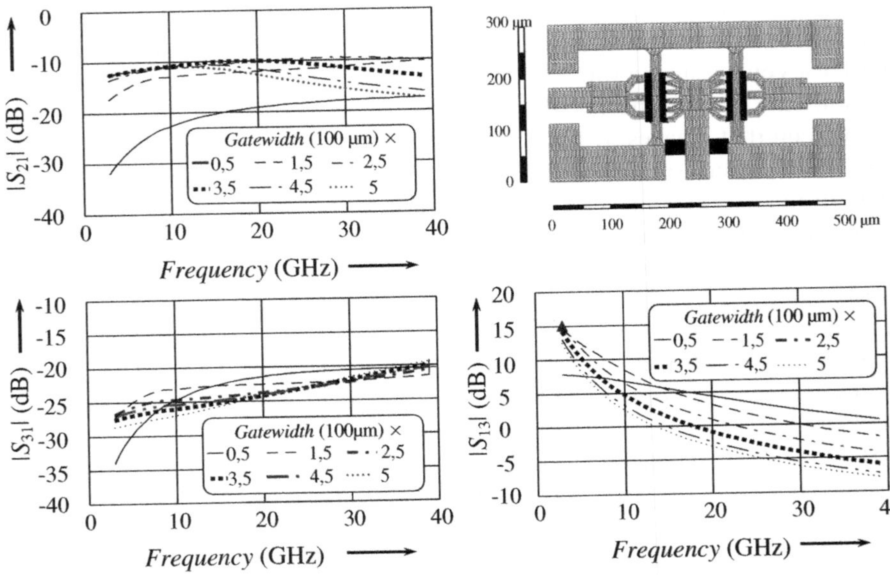

Fig. 7.2.9. Scaling of DDCS devices (gate width = 50–500 µm).

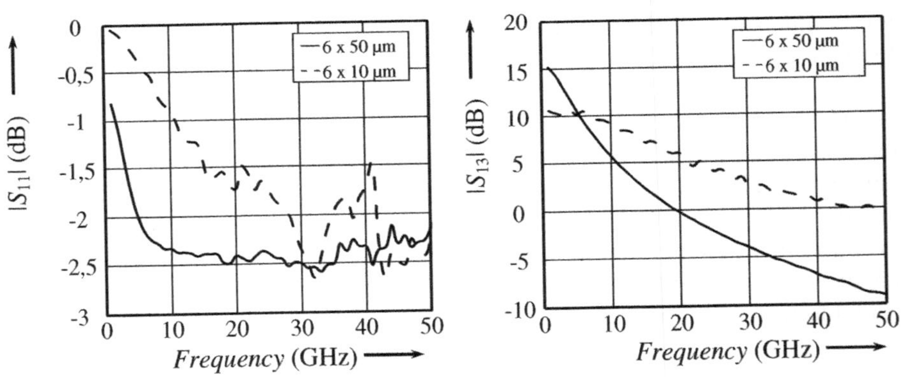

Fig. 7.2.10. Scattering parameters of DDCS devices (gate width = 6 × 10 µm and 6 × 50 µm).

Figure 7.2.10 depicts some of the measured scattering parameters for DDCS devices with a gate width of 6 × 10 µm and 6 × 50 µm, respectively. In this case, measurements up to 50 GHz were conducted. While $|S_{11}|$ is similar for both devices, the forward gain $|S_{13}|$ in this case depends largely on the total gate width. This effect can also be observed in Fig. 7.2.8. It is obvious that the large device has a higher $|S_{13}|$ at frequencies up to 5 GHz. On the other hand, the small device shows gain at 40 GHz, while the $|S_{13}|$ of the large device crosses

0 dB at 20 GHz. From this investigation, it is clear that the total gate width has to be chosen carefully based on the frequency requirements. Both devices have more than 15 dB isolation. The large device has isolation even better than 22 dB up to 50 GHz.

It should be noted that the scattering parameters in Fig. 7.2.9 have been derived from simulations while the results in Fig. 7.2.10 are measured for 60-μm and 300-μm gate width devices, respectively. The 60-μm device is the smallest and the 300-μm device is the largest device that was realized in this investigation.

7.2.1.5 Design and Realization of Coplanar RF Switches. Switches are needed in many circuits (e.g., as R/T-switch, level shifter, or signal selector). Usually electronic switches in microwave applications are realized by diodes because of their good RF performances. Drawbacks are the poor possibilities of monolithic integration with other active elements and the high dc power consumption. To overcome these poor properties, the above-described transistor circuits have been used to build up microwave switches in coplanar environment [207].

Figure 7.2.11 shows the fundamental circuit diagram of a π-switch using three transistors: two in a shunt connection and one in a series connection. This kind of transistor switch is suitable for a large range of applications in monolithic microwave integrated circuits. The switches allow small and wide-band applications with insertion losses around 1 dB and an isolation better than 40 dB. By using the coplanar waveguide technology, the layout requires only small space. Also the power supply for the switching voltage can be integrated, as will be shown below. Because the power consumption of the switches is nearly zero, the space requirement for the power supply is also small.

The developed switches are built up in a π-configuration as shown in Fig. 7.2.12. This configuration allows a 50-Ω matching in the ON as well as in the OFF state. To meet this requirement, each gate width of the applied MESFET's is optimized. As has been shown above, the circuit diagram of the

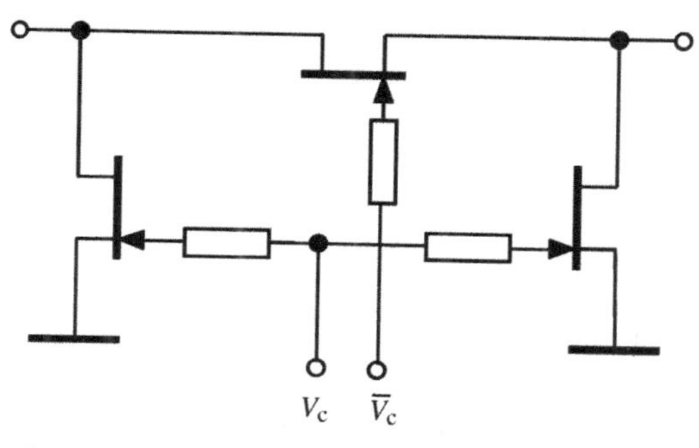

Fig. 7.2.11. Circuit diagram of a single π-switch.

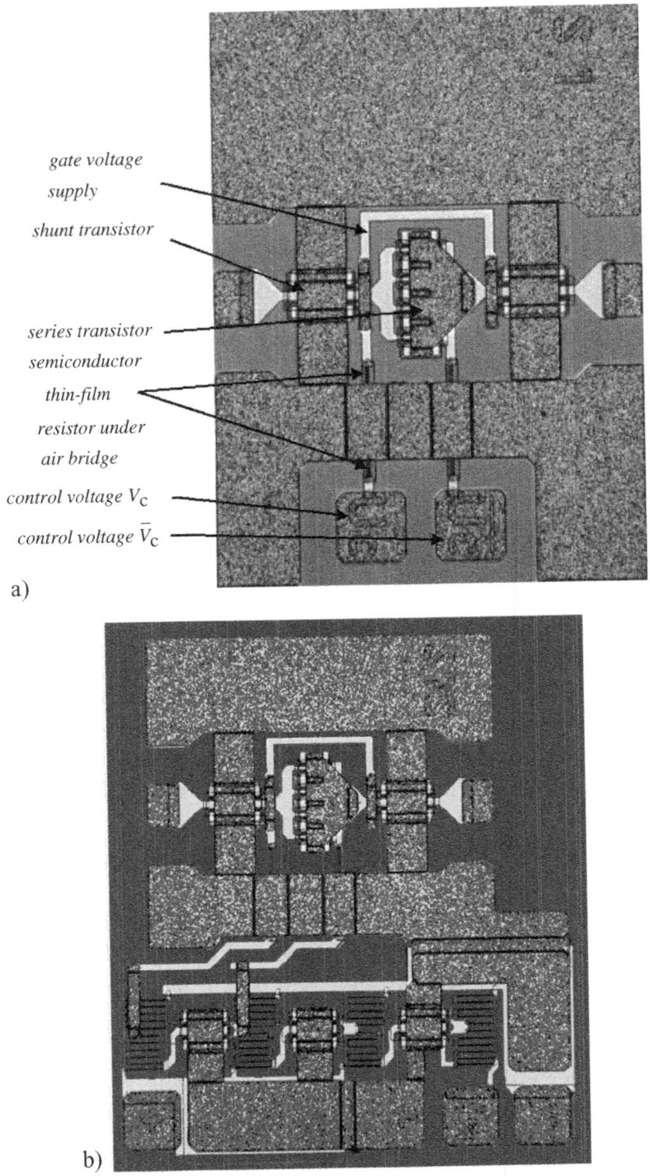

Fig. 7.2.12. Coplanar SPDT switch for application at 1–15 GHz (a) and the switch with a power supply (b).

MESFET's is scalable and gate widths of 40 μm, 80 μm, 160 μm, 320 μm, and 640 μm have been used.

Figure 7.2.12a shows an example of a realized broadband SPDT switch in coplanar technology. Figure 7.2.12.b shows the same switch with a TTL-compatible driver. The input and output of this coplanar switch is matched to 50 Ω, and the operating frequency band ranges from about 1 GHz to 15 GHz. Over the whole frequency band the return loss is better than 10 dB and the isolation of this switch is better than 25 dB (>40 dB @ 2 GHz) with about 1.5-dB insertion loss (Fig. 7.2.13).

This 0.2-mm^2 circuit was realized utilizing a 0.5-μm MESFET process and the *COPLAN* software (Chapter 5) for simulation and prediction. Figure 7.2.14 shows the measured insertion loss of the switch versus the input power.

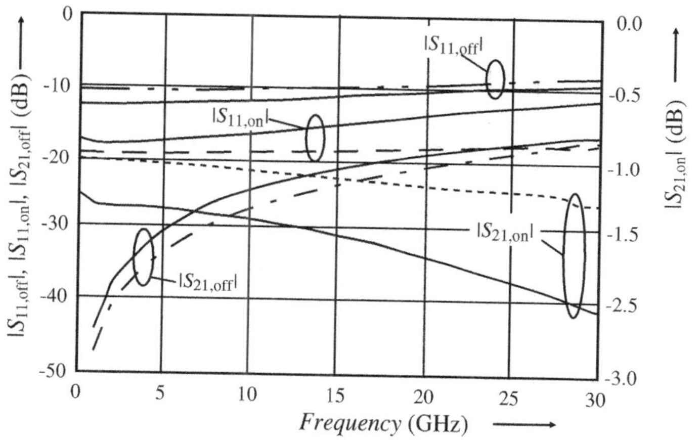

Fig. 7.2.13. Simulated (dashed lines) and measured (solid lines) scattering parameters of the single π-switch.

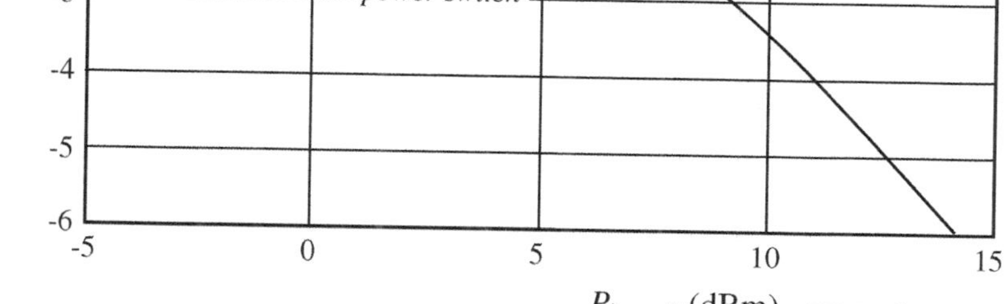

Fig. 7.2.14. Measured insertion loss for the ON state of a simple and a power p-switch at a frequency of 1 GHz.

It may be observed that the above-mentioned simple switch has a saturation power of about 16 dBm.

A switch with improved performance is shown in Fig. 7.2.15 in the form of a double π-switch. The layout also includes the power supply for the switch circuit. Its small signal scattering parameters are shown in Fig. 7.2.16. Again the return loss is better than 10 dB over the interesting frequency range in the ON and the OFF state. The power consumption of the circuit is less than 1 dBm.

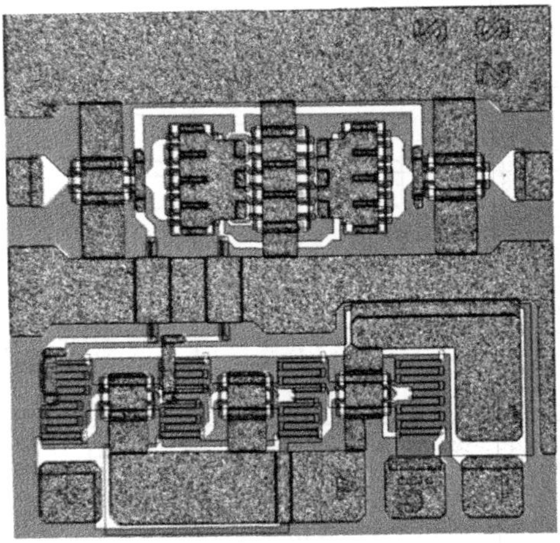

Fig. 7.2.15. Layout of a double π-switch and the voltage supply.

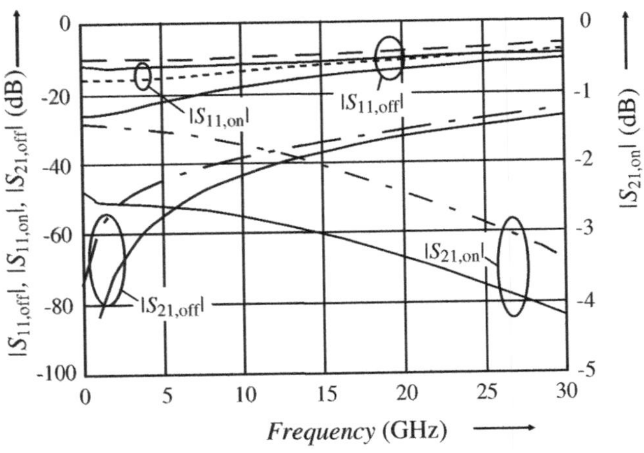

Fig. 7.2.16. Simulated (dashed lines) and measured (solid lines) scattering parameters of a double π-switch in dependence on the frequency.

7.3 COPLANAR MICROWAVE ACTIVE FILTERS

7.3.1 Introduction

An interesting alternative to classical passive filters that have been discussed in Chapter 6 is the use of active filters whereby transistors are utilized to reduce the losses and eventually varactor diodes are applied to tune the center band frequency or the bandwidth of the filter. Most of the active microwave filters can be classified in the following four groups: transversal-recursive active filters [30, 56, 78, 104], active filters using simulated active inductors (gyrators) [50, 52, 60–63, 95–97], active filters consisting of the cascade connection of passive filters and amplifiers [65, 81, 123], and active filters based on the negative resistance technique [2–4, 57, 75, 100, 101]. In this chapter the use of negative resistance circuits in coplanar line technique for the application in MMIC band-pass filter circuits will be discussed [253].

Negative resistance technique means use of active devices that generate a negative resistance to compensate the passive filter losses of the inductors (inductors are those passive components that have the highest loss in passive filter circuits). A common-source capacitive feedback circuit of a FET (Fig. 7.3.1a) or a common-gate inductive feedback circuit of a FET (Fig. 7.3.1b) generates a negative resistance, whereby the common-source capacitive feedback circuit can be described by a series circuit of this negative resistor and a capacitor. The common-gate inductive feedback circuit can be described by a parallel circuit of the negative resistor and an inductor.

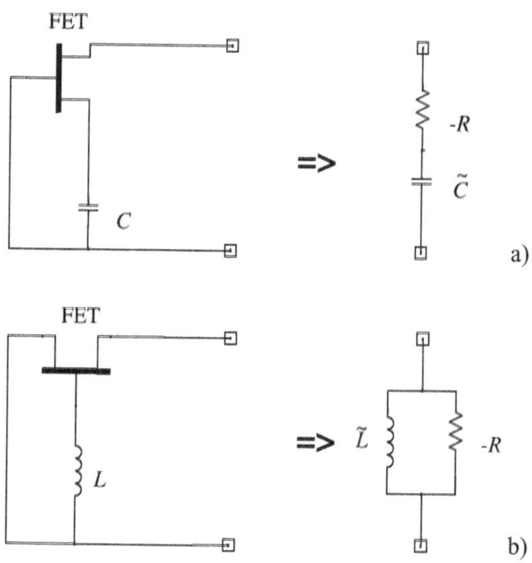

Fig. 7.3.1. Negative resistance circuits, (a) the common-source capacitive feedback circuit of a FET, (b) the common-gate inductive feedback circuit of a FET and their equivalent circuits.

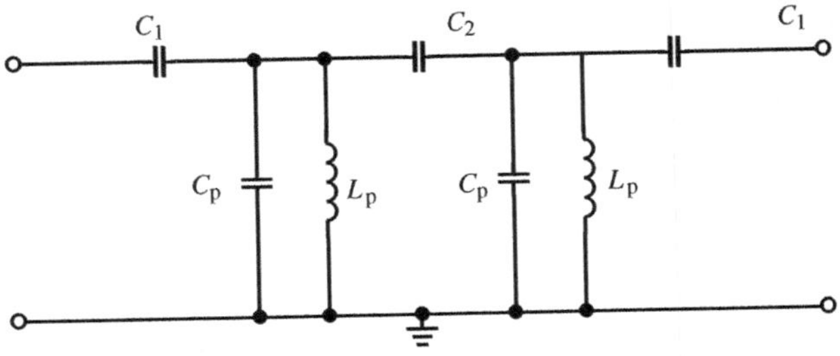

Fig. 7.3.2. Circuit diagram of the second-order band-pass filter.

As a first example, a second-order band-pass filter topology that has been used to design an active band-pass filter in coplanar waveguide technology is shown in Fig. 7.3.2. The equivalent passive filter has a low Q value because of the coplanar spiral inductor (L_p) losses. Therefore, these inductors have been substituted by the active inductors as shown in Fig. 7.3.1b. At UHF frequencies, these active inductors have been built using an inverted common-collector circuit of a bipolar transistor [2–4], which is the analogous circuit to the common-gate inductive feedback circuit of the FET. To bias the FET of the active inductor (Fig. 7.3.1b), some additional elements have to be added (see Section 7.3.4).

For the technological realization of the active filters, multifinger FETs have been used (see Section 7.1). Each of the gate fingers has a length of 0.5 μm and a width of 40 μm. At 1.8 GHz, eight-gate finger FETs with a total gate width of 320 μm have been used; at 5.5 GHz, four gate finger FETs with a total gate width of 160 μm have been employed. All coplanar waveguides, spiral inductors, junctions, and MIM capacitors have been calculated using the quasi-static finite difference method as described in the Chapters 2 to 5.

7.3.2 The Coplanar Active Inductor

To generate an active inductor at 1.8 GHz and at 5.5 GHz by using the above mentioned FETs, very large feedback spiral inductors are required. To save space, these inductors have been realized in the gate metalization (see also Section 4.4, Fig. 4.4.1) with a strip width of 3 μm, a slot spacing of 5 μm, and 14.5 turns or 7.5 turns (corresponding to an inductance of 28 nH or 5.16 nH, respectively). The following figures show a photo of the 1.8-GHz active inductor and the comparison between the simulated and measured resistance and inductance values of these two active inductors (Fig. 7.3.3). The expression "active inductor" means that the circuit behaves inductive and has a negative resistance. Both values, the inductance and the resistance, are frequency-dependent. Therefore, these active inductors are only applicable in narrow band applications, like in the filters presented here. One disadvantage of the active filters is that because of a possible foundry-parameter drift, especially

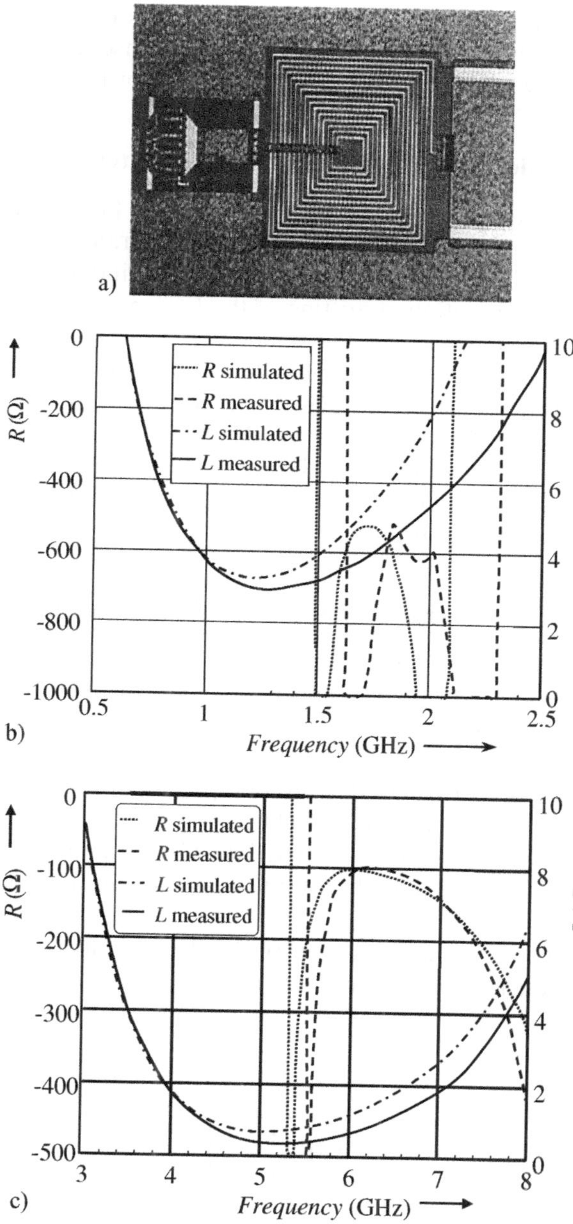

Fig. 7.3.3. (a) Photograph of the 1.8-GHz active inductance in coplanar environment and simulated and measured resistance and inductance of the active inductors at (b) 1.8 GHz and (c) 5.5 GHz.

in the case of the FET parameters, the values of the inductances can possibly be changed, as may be recognized from Fig. 7.3.3. Consequently, in this case the center frequency of the filters is shifted to higher or lower frequencies.

7.3.3 The First-Order Active Coplanar Band-Pass Filter

As a first step a first-order 5.5-GHz active band-pass filter without internal bias networks has been realized by substituting the parallel inductor L_p (Fig. 7.3.2) by an active inductor. Figure 7.3.4 shows the circuit diagram of this filter. Because of the above-mentioned drifting of the FET-parameters, the realized filter works at 5.74 GHz instead of the simulated 5.5 GHz.

The comparison between measurement and simulation using the measured FET parameters is demonstrated in Fig. 7.3.5. This filter has an insertion gain of +2.9 dB at the center frequency of 5.74 GHz and a bandwidth of 1.14 GHz, which is 20% of the center frequency. The out-of-band insertion loss is −8 dB at 4.2 GHz and 9.7 GHz, and the return loss is −14 dB at 5.6 GHz. The filter occupies an area of $0.6 \times 0.38\,\text{mm}^2$.

7.3.4 The Fixed Center Frequency Second-Order Active Filter

The cascade connection of two first-order band-pass filters leads to a second-order active band-pass filter for a fixed center frequency, as shown in Fig. 7.3.6. The results of a 1.8-GHz second-order active band-pass filter are demonstrated as an example. This filter has been fabricated in monolithic technology including all biasing networks. In the photograph shown in Fig. 7.3.6, the filter structure can be clearly identified. A coplanar waveguide in the middle of the structure forms the π-section of the filter. Two MIM (metal–isolation–metal) capacitors C_2 at the input and the output separate the dc voltages to the outside.

The capacitor C_1 between the two air-bridge T-junctions is part of the filter structure. Two T-junctions couple the two active inductors to the main copla-

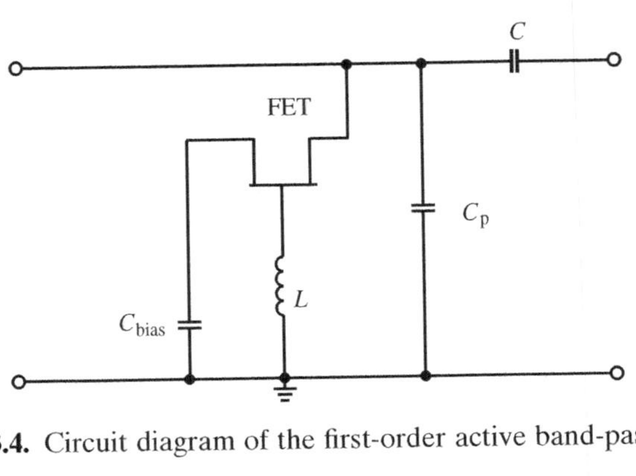

Fig. 7.3.4. Circuit diagram of the first-order active band-pass filter.

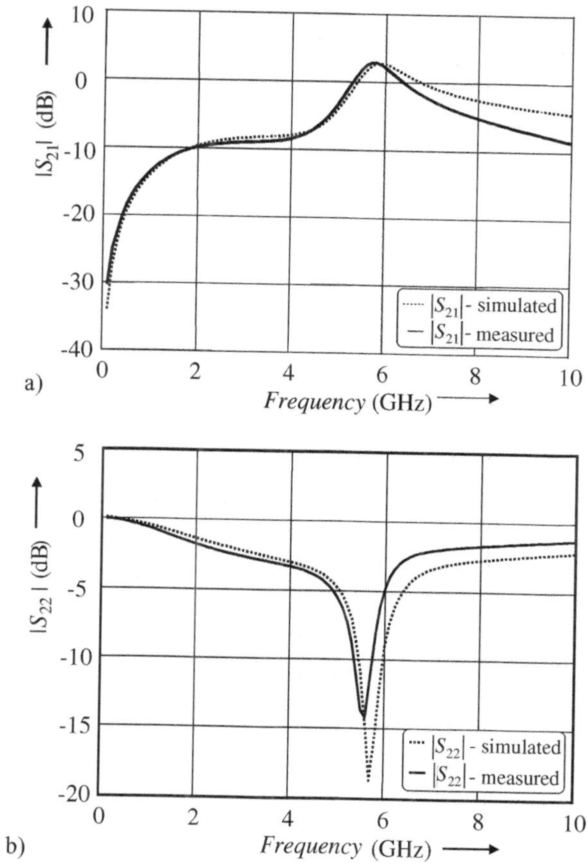

Fig. 7.3.5. Performance of the first-order active band-pass filter.

nar waveguide. They are connected to ground via two bias capacitors C_{bias}. The two spiral inductors on the upper side of the filter structure separate the RF signal from the dc bias circuit. The comparison between simulation and measurement of this filter is shown in Fig. 7.3.7 for a dc power supply of $V_{ds} = 4\,V$ and $I_d = 176\,mA$.

The center frequency of the measured filter is shifted by 20 MHz (~1% of the designed center frequency) from 1.79 GHz to 1.81 GHz. Because of the center frequency shifting, the filter curve is asymmetric. A maximum insertion gain of 0.17 dB and a ripple of 2 dB have been measured. The 3-dB bandwidth of this filter is 110 MHz (~6% of the center frequency) instead of the simulated 90 MHz (5% of center frequency). An out-of-band rejection of –20 dB at 1.68 GHz and at 2 GHz is achieved. The inband return loss of the filter is less than 5 dB. It is also possible to reduce the ripple and the return loss by increasing the order of the filter.

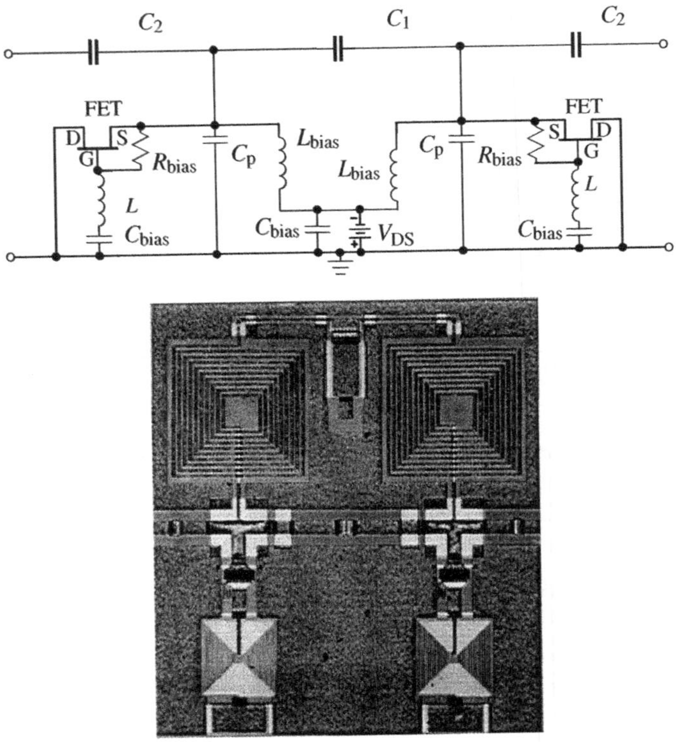

Fig. 7.3.6. Circuit diagram and photograph of the second-order active band-pass filter for a fixed center frequency.

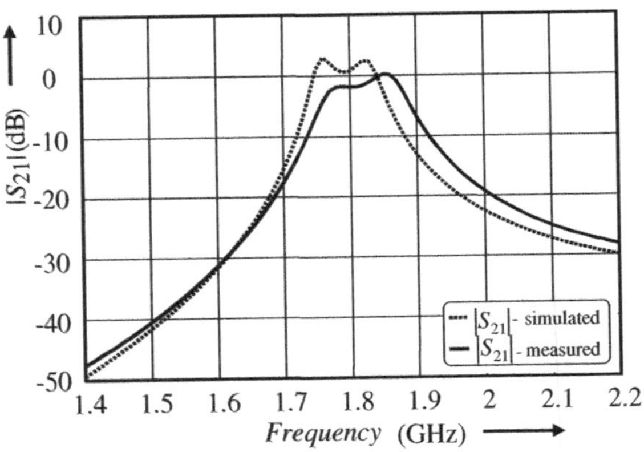

Fig. 7.3.7. Performances of the second-order active band-pass filter for a fixed center frequency.

To determine the dynamic range of this active filter, the in-band compression output power curves are measured. For three different frequencies, these curves are illustrated in the following figures. At the center frequency of 1.81 GHz (Fig. 7.3.8.b) the 1-dB compression is –2.6 dBm. But at the pole frequencies the 1 dB compression is only –4.8 dBm at 1.77 GHz (Fig. 7.3.8a) and –7.2 dBm at 1.85 GHz (Fig. 7.3.8c). Therefore, the 1-dB compression of this filter has been specified with –7.2 dBm.

7.3.5 The Coplanar Active Tunable Filter

Figure 7.3.9 shows the circuit diagram of a tunable second-order active bandpass filter, and Fig. 7.3.10 shows the photograph of the filter fabricated in coplanar monolithic integrated circuit technology. The parallel capacitors C_p of the fixed center frequency filter (Fig. 7.3.6) are substituted by varactor diodes tc realize the voltage-controlled tuning of the filter.

For several tuning voltages (0–5 V) the measured scattering parameters are plotted in Fig. 7.3.11, whereby the dc power supply is kept constant. The center frequency can be changed over a range of about 200 MHz from 1.7 GHz to 1.9 GHz. By controlling the biasing and the tuning voltage, a varying of the center frequency with a nearly constant insertion gain is possible. The other small-signal performances are almost the same as those of the fixed center frequency filter as discussed in the previous section.

Each of the 1.8-GHz filters occupies an area of $1.5 \times 1.6\,mm^2$ on GaAs substrate, whereby one-half of the area is used for the biasing elements as can be seen from the photograph shown in Fig. 7.3.10.

The value of the negative resistance, realized using the approach described above, depends on the bias voltage of the FETs. To stabilize the filter at different bias points, a cold FET is additionally used as a variable series resistor as shown in the next layout. Additionally, as in the previous example, the parallel capacitors C_p have been replaced by diodes to tune the center frequency of this filter. Figure 7.3.12 shows the circuit diagram and Fig. 7.3.13 the layout and a photograph of this filter.

A comparison between simulation and measurement results is shown in Fig. 7.3.14 for a filter with a center frequency of 2 GHz. Again, the realized FETs have slightly different parameters from the ones that have been used for simulation. Therefore the filter has been measured at a slightly different bias point than that assumed in the simulation, to guarantee the stability cf the filter.

The filter has been measured at the operating point of $V_{ds} = 3.4\,V$ and $I_d = 180\,mA$. With the exception of the in-band insertion loss of 4 dB instead of a simulated in-band gain of 1 dB, the simulation agrees well with the measurement. The measured 3-dB bandwidth of this filter is 24 MHz, which is 1.2% of the center frequency. The out-of-band rejection is –40 dB at 1.9 GHz and –35 dB at 2.1 GHz. An in-band returns loss better than –25 dB is achieved.

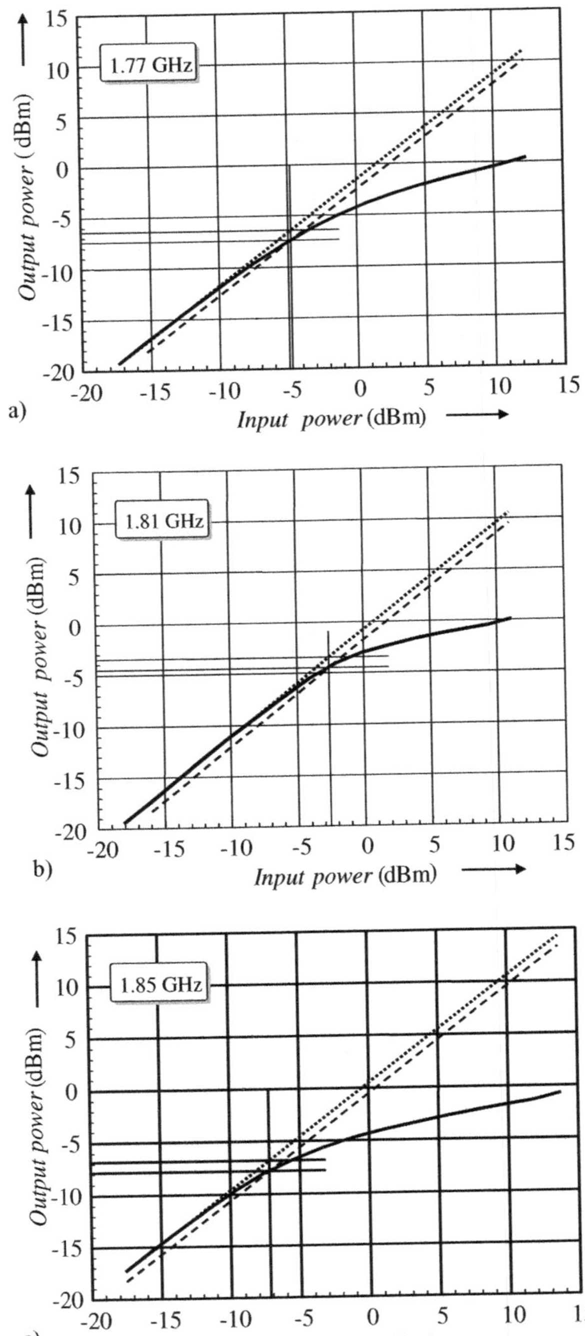

Fig. 7.3.8. Compression output power curves of the active fixed center frequency band-pass filter at different center band frequencies: (a) 1.77 GHz, (b) 1.81 GHz, (c) 1.85 GHz.

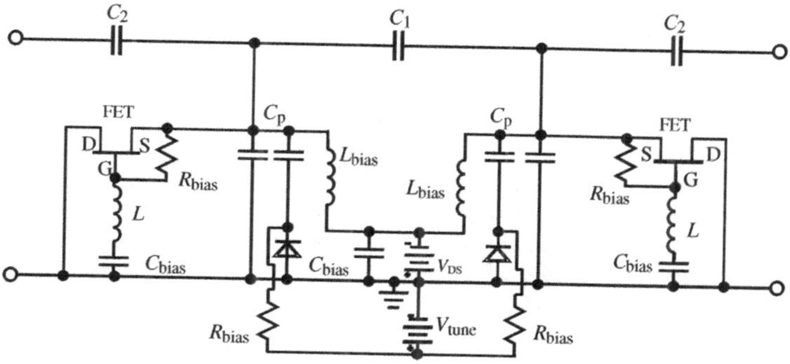

Fig. 7.3.9. Circuit diagram of the tunable second-order active band-pass filter.

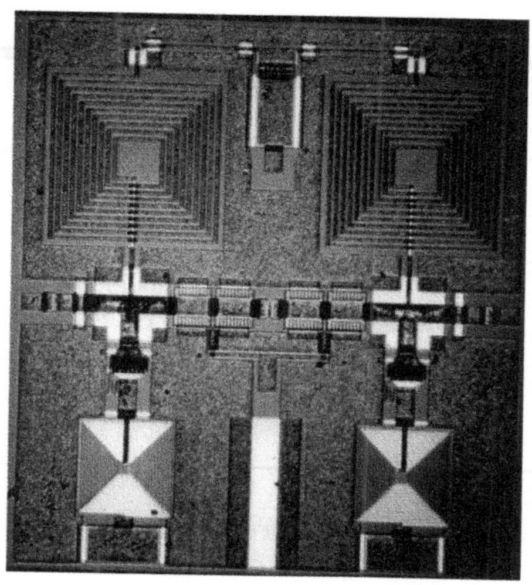

Fig. 7.3.10. Photo of the tunable second order active band-pass filter.

Figure 7.3.15 shows the measured tuning range of the center frequency from 1.92 GHz to 2.0 GHz for a tuning voltage from 0–3 V. In this tuning range, the in-band attenuation and bandwidth is nearly constant.

Broadband measurements up to 65 GHz of this filter for the above-mentioned operating point are illustrated in Fig. 7.3.16. With the exception of the frequency range from 24 GHz to 38 GHz the out-of-band rejection is better than –35 dB and the out-of-band rejection is better than –23 dB in mentioned frequency range.

As has already been mentioned above, process-induced variations of the FET parameters lead to slightly different inductance values if the bias point has been chosen for loss compensation. Consequently, the resonant frequency of each resonator is shifted. For narrow-band application it is necessary to

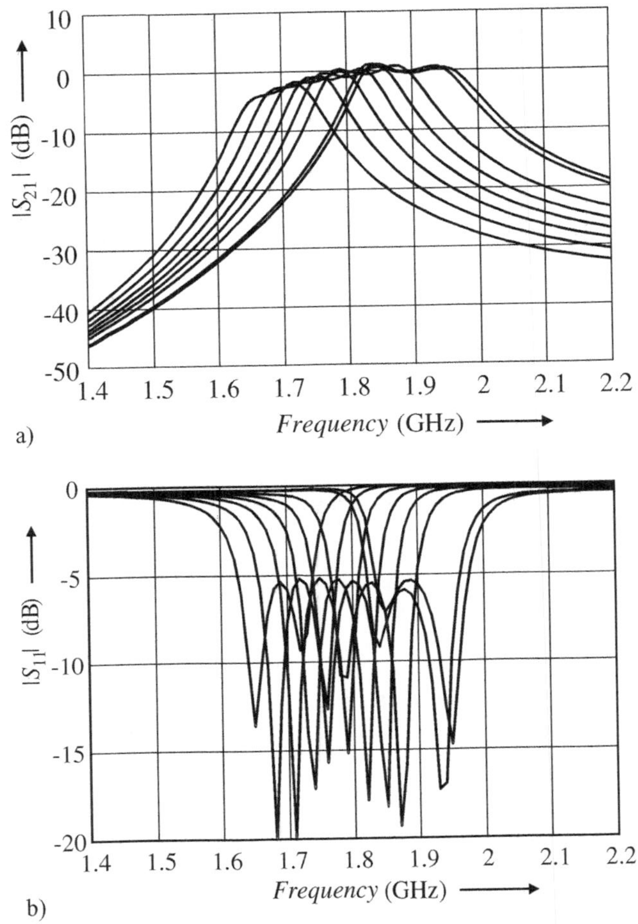

Fig. 7.3.11. Measured scattering parameters of the tunable second-order active band-pass filter for several tuning voltages.

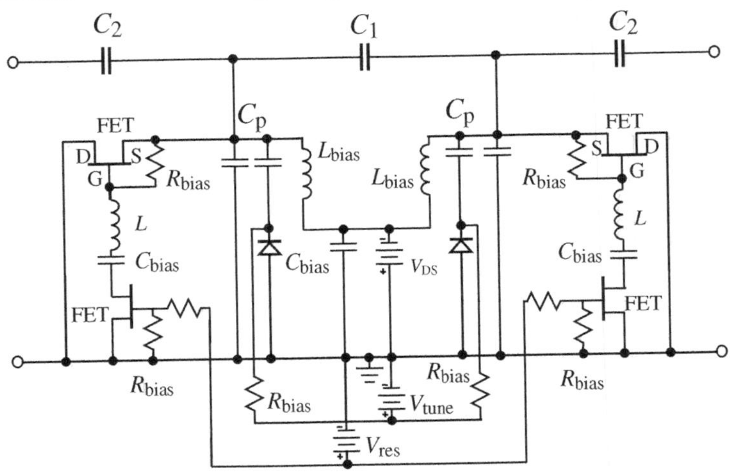

Fig. 7.3.12. Circuit diagram of the active band-pass filter including a cold FET for stabilization.

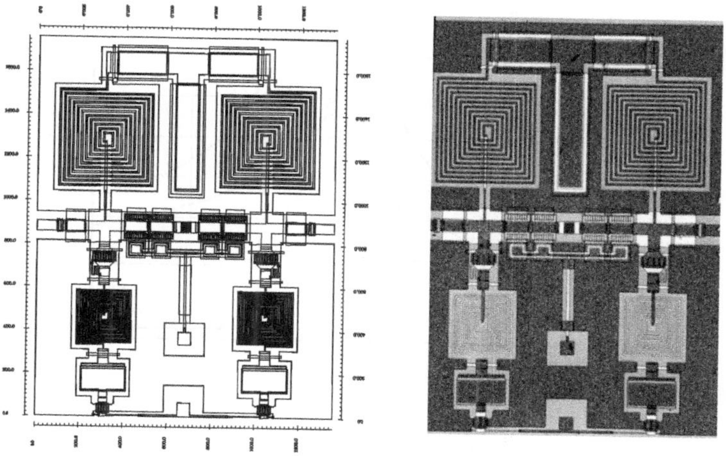

Fig. 7.3.13. Layout and photograph of the active band-pass filter including a cold FET for stabilization.

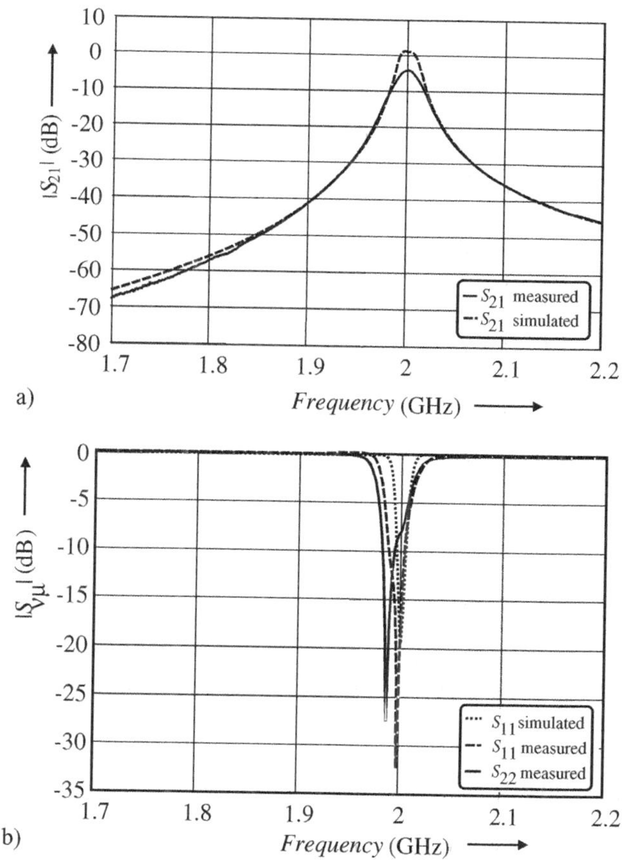

Fig. 7.3.14. Comparison between simulated and measured (a) transmission coefficients and (b) reflection coefficients.

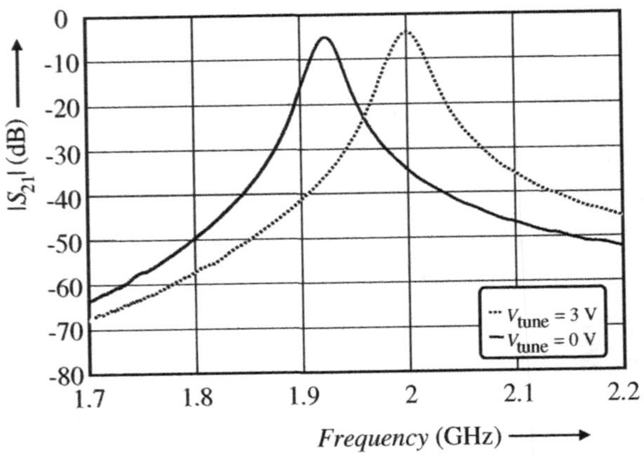

Fig. 7.3.15. Measured tuning range of the filter.

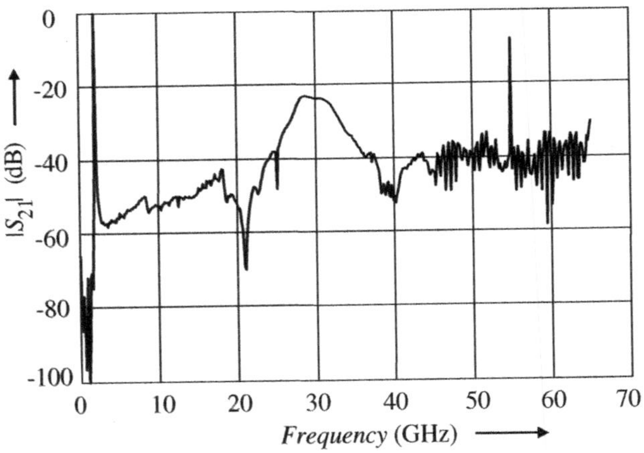

Fig. 7.3.16. Measured transmission coefficient of the filter.

compensate this frequency shift. One possibility is to substitute the parallel capacitors C_p by diodes to tune the capacitance and consequently to tune the resonant frequency, as has already been discussed. Additionally, the coupling capacitor C_2 can be substituted by a diode to tune the bandwidth of the filter. The circuit diagram and a photograph of this filter are shown in Figs. 7.3.17 and 7.3.18, respectively.

Each active inductor has been realized using an eight-gate finger FET with a gate length of 0.5 μm and a total gate width of 320 μm. The feedback inductor is a 14.5-turn spiral inductor with a strip and slot width of 4 μm realized in

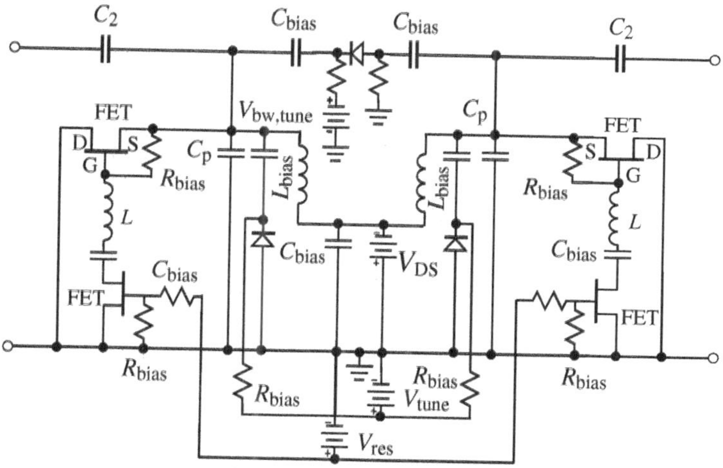

Fig. 7.3.17. Circuit diagram of the tunable second-order active band-pass filter.

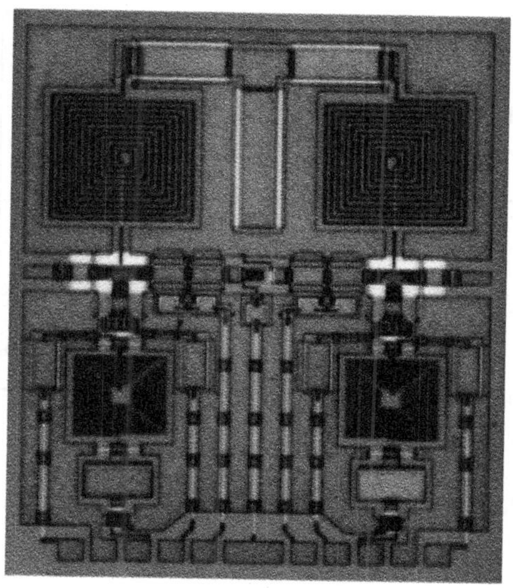

Fig. 7.3.18. Photograph of the tunable second order active band-pass filter.

the gate metalization. All used coplanar elements have been calculated with the coplanar library as described in Chapter 5. This filter occupies an area of 1.4 mm × 1.9 mm, where nearly half of the space is used for the biasing networks.

The measurements in Figs. 7.3.19 and 7.3.20 show the tuning possibilities of this filter. The tuning voltage for each diode is 0–5 V and the dc power con-

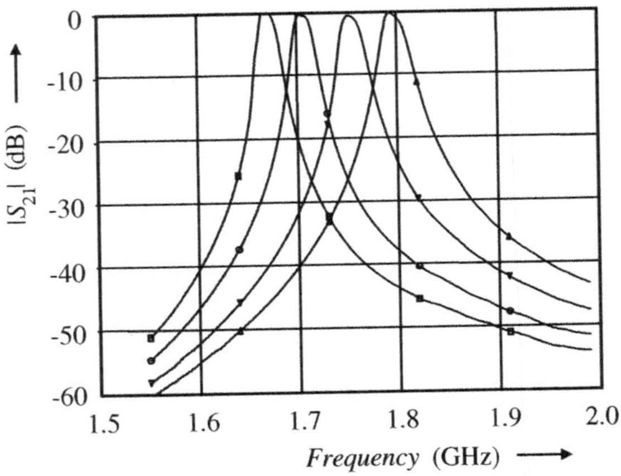

Fig. 7.3.19. Measured tuning range of the center frequency of the filter.

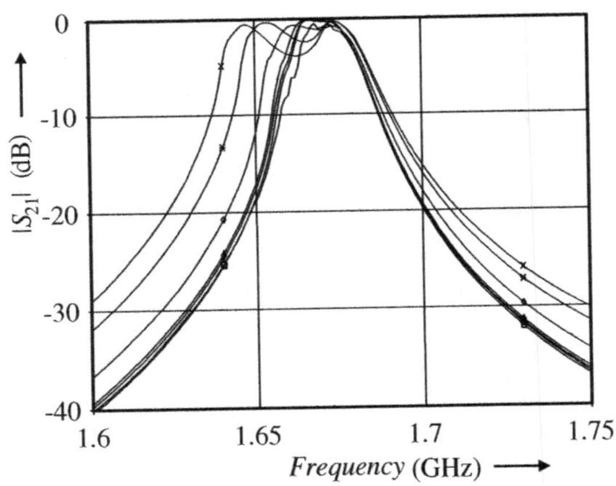

Fig. 7.3.20. Measured tuning range of the 3-dB bandwidth of the filter.

sumption is nearly 0.45 W, which depends on the used FETs. Figure 7.3.19 shows the tuning range of the center frequency for the smallest 3-dB bandwidth of 18 MHz—that is, 1%. The center frequency can be tuned over the range of 130 MHz from 1670 MHz to 1800 MHz. The insertion loss of the filter is nearly zero.

For a fixed resonator tuning voltage the bandwidth tuning is shown in Fig. 7.3.20. The 3-dB bandwidth is tunable from 18 MHz to 40 MHz, but the ripple of the filter increases with increasing bandwidth. For wide-band applications, increasing the number of resonators can decrease the ripple.

7.4 COPLANAR MICROWAVE AMPLIFIERS

7.4.1 Coplanar Microwave Amplifiers in Waveguide Design

7.4.1.1 Introduction. This chapter demonstrates the design of two three-stage Ka-band monolithic amplifiers in coplanar waveguide technique. The first MMIC is based on AlGaAs/GaAs HEMT devices with a gate length of 0.25 μm. The second MMIC is realized with 0.25-μm InAlAs/InGaAs HEMT devices on InP. Both of the applied types of transistors, the GaAs and the InP HEMT, are characterized by a similar lateral layout [132–134]. Therefore, an identical circuit layout has been investigated in both applications: the GaAs and the InP MMIC. Measurements demonstrate a gain of 18 dB and 29 dB for the GaAs and the InP amplifier, respectively. This 11-dB increase in gain confirms the benefit of the idea to just replace GaAs HEMT devices by InP-based HEMTs in the same circuit layout. In particular, in the case of MMICs in coplanar waveguide technique the advantage of increased transconductance and reduced feedback capacitance of the InP HEMTs compared to their GaAs counterparts can be achieved without any restriction concerning the circuit realization and chip handling.

InP-based HEMTs demonstrate high cutoff frequencies, power gain, and low noise figures at millimeter-wave frequencies [105, 119, 121]. The distributed amplifier early presented by Majidi-Ahy et al. [73, 74] already pointed out the excellent opportunities resulting from InP MMIC design in coplanar waveguide technique. From the technological point of view, the MMIC technology required for a coplanar waveguide circuit realization is nearly identical to that of single HEMT fabrication. With respect to the desired coplanar waveguide design, it can be stated here that the expense incurred in the realization of a MMIC on GaAs and InP substrate are comparable. Consequently, the motivation for such a direct comparison of a GaAs amplifier and an InP MMIC based on the identical circuit layout is obvious. Assuming the same lateral topology for each of the HEMT devices, the observed results are related directly to the applied material system. It should be emphasized here that the permittivity of GaAs and InP are quite similar, which means that almost identical phase velocity for coplanar waveguides realized on both substrates can be assumed. This aspect is important for our comparison since the matching networks of the three-stage Ka-band amplifiers presented here mainly consist of transmission lines as will be described later.

The most important result of this investigation is the achievement of an 11-dB increase in gain for the InP-based Ka-band amplifier compared to the GaAs MMIC. The parameters of the equivalent circuit of the HEMT devices show that this remarkable improvement is achieved by the drastically increased transconductance and reduced feedback capacitance of the InP devices. These results demonstrate the validity of the idea to improve, significantly, the characteristics of a Ka-band amplifier in coplanar technique by just replacing GaAs HEMTs by InP-based devices in the same MMIC layout.

7.4.1.2 Circuit Design and Technological Aspects. Originally, the three-stage Ka-band amplifier in coplanar waveguide technique presented here has been developed as a MMIC on GaAs substrate. Based on AlGaAs/GaAs-HEMTs with a gate length of 0.25 μm and a gate width of 120 μm, the design objective has been to achieve an approximately 20-dB gain from 26 GHz to 30 GHz with reasonable input and output return losses and unconditional stability ($K > 1$) not only within the pass band but also over the whole range of frequency. The matching networks have been designed to consist of mainly coplanar transmission lines. The characteristic impedances and line lengths of the coplanar waveguides used in the matching networks are obtained by optimizing the circuit with respect to the design goals described above. From field theoretical investigations (see Sections 2.1 and 2.2) it is well known that the characteristic impedances of coplanar waveguides are mainly determined by the ratio of strip width to slot width (w/s). For this particular application, values of the characteristic impedance Z_L from 40 Ω up to 60 Ω were required. This range for the characteristic impedance can be covered by a ratio of s/w varying from 0.3 up to 1.2. Assuming minimum lateral geometry, $s_{min} = w_{min} =$ 10 μm, the total width of such coplanar waveguides is much smaller than the substrate height h, which is typically 450 μm for GaAs as well as InP. Therefore the characteristic impedance of such coplanar structures can be considered to be independent of the substrate height.

With respect to the required separation of the different stages, interdigital capacitors are introduced in the waveguide structures. Regarding the simulation of the capacitors, the circuit development models presented in Chapter 4 have been applied. The number of coupling fingers, the finger width, and the finger separation are adjusted to assure the total device geometry to be much smaller than the substrate height. The main reason of using these interdigital capacitors in the circuit design has been to, additionally, investigate in how far conventional MIM capacitors might be avoided with respect to future demands and applications such as a possible reduction of the number of technological processes in passive circuits or the elimination of dielectric losses in capacitors at higher frequencies. These details of the MMIC described above are obvious in Fig. 7.4.1.1.

Although this photograph depicts the three-stage Ka-band amplifier in coplanar waveguide technique on InP substrate, there is no difference in the layout when compared to the corresponding GaAs design as already explained. Therefore all details of the applied passive components for the GaAs MMIC as well as for the InP MMIC can be demonstrated by this figure.

Concerning the on-chip biasing of the HEMT devices, a common gate bias and a common drain bias solution has been chosen. For DC-blocking, a new type of capacitor has been introduced in the circuit design. This element is shown in Fig. 7.4.1.2 and is named "covered interdigital capacitor (CIC)." It is designed similar to a conventional interdigital capacitor between the center conductor of the coplanar waveguide and the ground plane except that at the final step the fingers (thickness 0.3 μm) are completely covered by using an

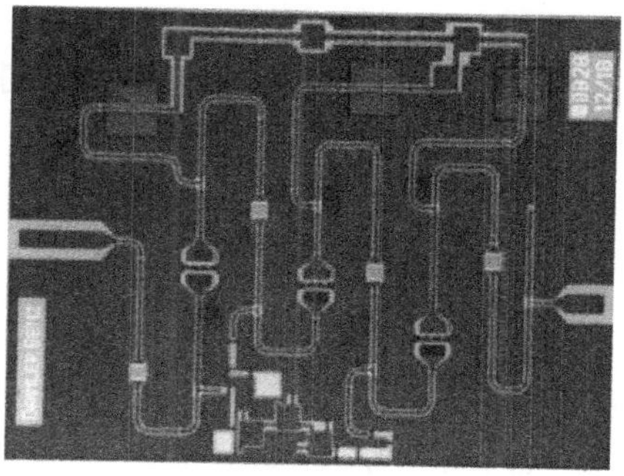

Fig. 7.4.1.1. Photograph of the three-stage Ka-band amplifier in coplanar waveguide technique on InP substrate. The MMIC is based on 0.25-μm InAlAs/InGaAs HEMTs with 120-μm gate width (chip size: 1.9 mm × 2.4 mm) [134].

Fig. 7.4.1.2. Photograph of the applied covered interdigital capacitor (CIC) for dc-blocking.

air-bridge-like structure. However, the center conductor of this CIC is electroplated with respect to the drain bias current. In comparison to a conventional interdigital capacitor, the main advantages of this structure are the increased capacitance values as well as the improved shielding characteristics.

For the coplanar waveguide amplifier on GaAs substrate a 0.25-μm AlGaAs/GaAs-HEMT device with a gate width of 120 μm has been used. The element values of the equivalent circuit (Fig. 7.4.1.3) of this device are given in Table 7.4.1.1.

The parameters of a corresponding InAlAs/InGaAs HEMT (derived from measurements on a prototype FET) are given in the same table. These transistors of course differ in the layer sequence and the material system, but their lateral layout is identical, which is important for the described investigations regarding similar parasitics for both devices.

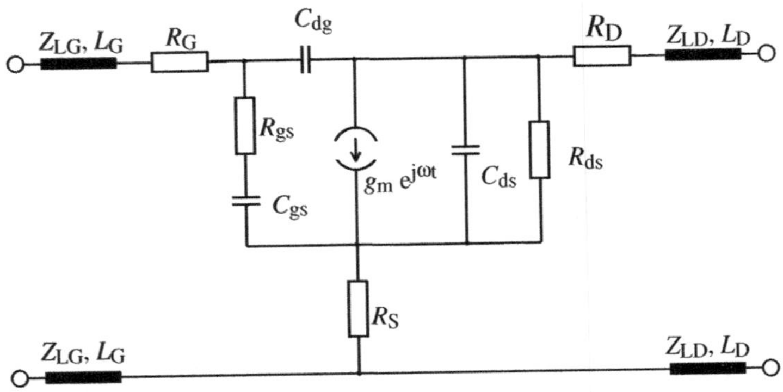

Fig. 7.4.1.3. Equivalent circuit for small signal description of the applied HEMT devices.

TABLE 7.4.1.1. Parameters of the Small-Signal Equivalent Circuit for the AlGaAs/GaAs- and the InAlAs/InGaAs-HEMT Devices

Parameter	AlGaAs/GaAs-HEMT Device $w_G = 120\,\mu m$ $L_G = 0.25\,\mu m$	InAlAs/InGaAs-HEMT Device $w_G = 120\,\mu m$ $L_G = 0.25\,\mu m$
R_{gs} (Ω)	2.2	3.3
C_{gs} (fF)	122.2	110.3
C_{dg} (fF)	24.2	12.2
g_m (mS)	53.6	68.6
τ (ps)	0.49	0.53
R_{ds} (Ω)	308	641
C_{ds} (fF)	45.9	22.7
R_D (Ω)	7.7	11.0
R_S (Ω)	3.9	0.9
R_G (Ω)	4.8	5.0
Z_{LG} (Ω)	50	50
L_G (μm)	64.1	116.2
Z_{LD} (Ω)	50	50
L_D (μm)	13.6	62.0

The results of the comparison are strongly correlated to the material system. Obviously, the main advantages of the InP device are the increased transconductance g_m, the significantly increased drain-source resistance R_{ds}, and also a drastically decreased feedback capacitance C_{dg}. These improvements in parameters for the InP-based device compared to its GaAs counterpart will result in an improved power gain performance. From the circuit designers point of view, it should be emphasized that all other parameters of the different devices are nearly identical, as is to be expected. The conse-

50 nm GaAs	ND = 5×10^{18} cm^{-3}
40 nm $Al_{0.25}Ga_{0.75}As$	ND = 2×10^{18} cm^{-3}
2 nm $Al_{0.25}Ga_{0.75}As$	spacer layer
800 nm GaAs i	
AlGaAs/GaAs i	superlattice
100 nm GaAs	buffer
GaAs i	substrate

InGaAs	surface depleted cap layer
25 nm InAlAs	undoped
10 nm InAlAs	ND = 5×10^{18} cm^{-3}
2nm InAlAs i	spacer layer
32 nm InGaAs i	
InAlAs/InGaAs i	superlattice
40 nm InAlAs	buffer
InAlAs/InGaAs i	superlattice
InP s.i.	substrate

Fig. 7.4.1.4. Layer sequence for the AlGaAs/GaAs- and the InAlAs/InGaAs-HEMT devices.

quences of these aspects concerning the device characteristics are discussed in the following section after the description of the principal layer structures.

Figure 7.4.1.4 demonstrates the layer sequence for the two types of HEMT devices. The AlGaAs/GaAs layer structure is quite conventional, using a homogeneously doped donor layer and a highly doped thick cap layer to reduce parasitic resistances. In order to improve the carrier confinement, an AlGaAs/GaAs superlattice followed by a thick GaAs layer is used as a buffer.

For the lattice-matched InAlAs/InGaAs HEMT, the surface-depleted cap layer approach in combination with a planar-doped donor layer is used to allow for high C_{gs}/C_{gd} and g_m/g_d ratios [115]. Another advantage of this approach is that now a high drain bias operation ($V_{DS} > 2.0\,V$) is possible for the InP-based HEMTs that yields improved gain performance, too. To characterize the quality of the layer structure, Hall measurements were performed at a temperature of 77 K. For the GaAs-based structure a sheet carrier concentration of $n_s = 1.2 \times 10^{12}\,cm^{-2}$ and a mobility of $\mu_0 = 25{,}000\,cm^2/V \cdot s$ was measured after the cap layer was removed. For the InP-based layer structure values of $n_s = 2.7 \times 10^{12}\,cm^{-2}$ and a mobility of $\mu_0 = 33{,}000\,cm^2/V \cdot s$ were measured.

The sequence of the processing steps for circuit fabrication are: mesa etching, mesa sidewall etching in the case of InAlAs/InGaAs HEMTs, Ge/Ni/Au-ohmic contact formation, e-beam lithography for the gate definition, wet chemical recessing, Ti/Pt/Au-gate metal evaporation, PECVD-Si_3N_4 passivation, and air-bridge formation.

7.4.1.3 Results and Comparison with Measurements. The circuit optimization was carried out for the originally designed AlGaAs/GaAs-HEMT device MMIC amplifier. Figure 7.4.1.5a demonstrates the simulated gain and return

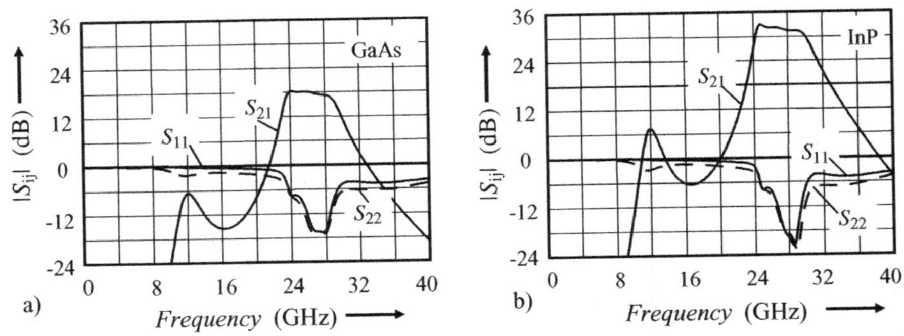

Fig. 7.4.1.5. Simulated results for the Ka-band amplifiers in coplanar waveguide technique. (a) Gain and return losses versus frequency for the design based on AlGaAs/GaAs-HEMT devices. (b) Results for the corresponding MMIC based on InAlAs/InGaAs devices.

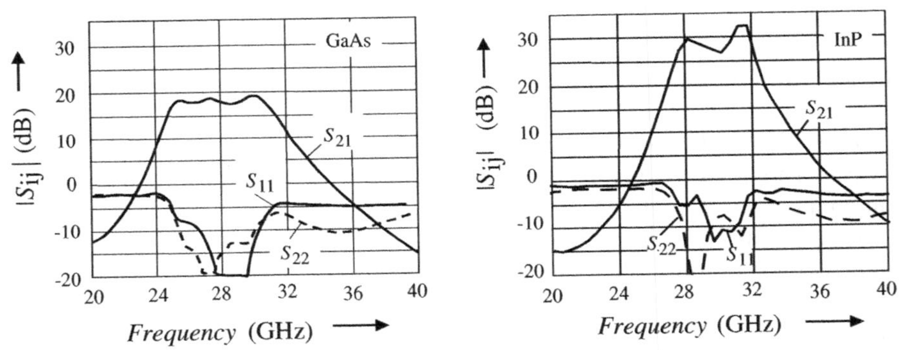

Fig. 7.4.1.6. Measured results for the Ka-band amplifiers in coplanar waveguide technique. (a) Gain and return losses for the design based on AlGaAs/GaAs-HEMT devices. (b) Results for the corresponding MMIC based on InAlAs/InGaAs devices.

losses for the coplanar amplifier on GaAs substrate. Obviously, a gain of 18 dB and return loss better than −10 dB is expected for this amplifier. The measured results for this MMIC are shown in Fig. 7.4.1.6a. For frequencies from 26 GHz up to 30 GHz an average gain of 18 dB has been measured, which is in an excellent agreement with the predicted results. Regarding the return losses the measured results have also been predicted correctly. These results confirm the validity and applicability of the coplanar elements that have been introduced in the design such as air-bridge T-junctions and covered interdigital capacitors.

The same circuit layout has been applied to realize the equivalent Ka-band amplifier on InP substrate. No reoptimization of the passive components has been carried out for this MMIC design based on the InAlAs/InGaAs-HEMT devices. It means that the same mask set as before was used. The simulated results for the amplifier on InP substrate are presented in Fig. 7.4.1.5b. A comparison with the GaAs version of the circuit demonstrates a significant increase in gain. That is, the average calculated gain could be improved from 18 dB to almost 32 dB. This is mainly achieved by the increased transconductance and output resistance as well as the reduced feedback capacitance of the InAlAs/InGaAs HEMT device. As predicted, the pass band and the insertion loss for the InP-based MMIC is very close to the simulated results for the amplifier based on AlGaAs/GaAs HEMTs. This is due to the similar parasitics of the two types of HEMT devices, which results from their identical lateral layout. It is worth noting that the criterion $K > 1$, for unconditional stability, is satisfied for the InP based design, too.

A photograph of the realized InP Ka-band amplifier in coplanar waveguide technique is shown in Fig. 7.4.1.1. The measured results for this MMIC are given in Fig. 7.4.1.6b. Compared to the predicted results, a shift of the pass band is observed (28–32 GHz instead of 26–30 GHz, which has been calculated based on the data of an InAlAs/InGaAs prototype). This shift in frequency is accompanied by a slight degradation of the input and output return losses compared to the calculated values. The reasons for this deviation are due to the differences between the gate recess of the used InAlAs/InGaAs HEMT and the prototype device. Nevertheless, the most significant characteristics of the simulated gain and return losses can be verified by measurement. In particular, the measured average gain of 29 dB for the InP-MMIC demonstrates that the theoretically expected improvement as compared to the GaAs counterpart of this Ka-band circuit can be utilized in practical application. Although the gain is slightly reduced due to the explained decreased input and output return losses, the average gain of the InP-based amplifier is about 11 dB higher than for the GaAs-based amplifier. These results of comparing two coplanar amplifiers realized with an identical lateral layout (for the active as well as the passive components) on different material systems demonstrate the advantages of increased g_m and R_{ds} and decreased C_{dg} of the InAlAs/InGaAs HEMT in circuit applications. Moreover, it should be mentioned again that for circuits in the coplanar waveguide technique, no thinning of the substrate is necessary and therefore no problems in handling the GaAs and InP wafers have occurred.

7.4.2 Coplanar Lumped-Element MMIC Amplifiers

7.4.2.1 Introduction. As MMICs became more widespread in commercial applications, the cost of production gained increased attention. For the MMIC designer it is important to reduce the chip size while maintaining the electrical characteristics. To demonstrate what size reduction is possible using coplanar technology and lumped elements, this chapter demonstrates the comparison of a distributed and a lumped-element MMIC K-band amplifier in coplanar line technique. The distributed element amplifier is a two-stage design and needs a size of $3\,mm^2$. In the frequency range from 18 to 20 GHz the gain is more than 12dB. The lumped-element amplifier is a three-stage design, which has a size of $1\,mm^2$. For the same frequency range the gain is more than 23dB [164, 165].

The objective here is to show what size reduction is possible by using lumped elements in MMIC coplanar design. Therefore, two different amplifiers have been fabricated on a GaAs substrate in coplanar waveguide technique. The coplanar technology can be realized without via-hole technique and no backside preparation or metalization as already discussed earlier. Therefore, low-cost production of monolithic integrated circuits is possible. The coplanar lines, interdigital capacitors, MIM (metal–insulator–metal) capacitors, rectangular spiral inductors, and discontinuities (such as air bridges, T-junctions, bends, and stubs) are calculated with a quasi-static finite difference method (see Chapters 3, 4, and 5). Both circuits have been realized with multifinger FETs. Each gate finger has a length of 0.5 μm and a width of 40 μm.

7.4.2.2 MMIC Design and Results. The first MMIC amplifier has been realized with distributed elements, using two FETs. Each of the FETs has a total gate width of 80 μm. A simplified schematic circuit topology of this two-stage amplifier is shown in Fig. 7.4.2.1 and the corresponding photograph is illustrated in Fig. 7.4.2.2.

The chip size of the fabricated two-stage distributed element amplifier is $1.48 \times 2.05\,mm^2$, including bias networks. In the two-stage amplifier, coplanar lines and interdigital capacitors have been used as matching networks, MIM capacitors, and thin-film resistors in the bias networks.

The comparison of the simulated and measured performance of the two-stage distributed element amplifier is shown in Fig. 7.4.2.3. The simulated and measured scattering parameters agree well. The amplifier chip has been biased

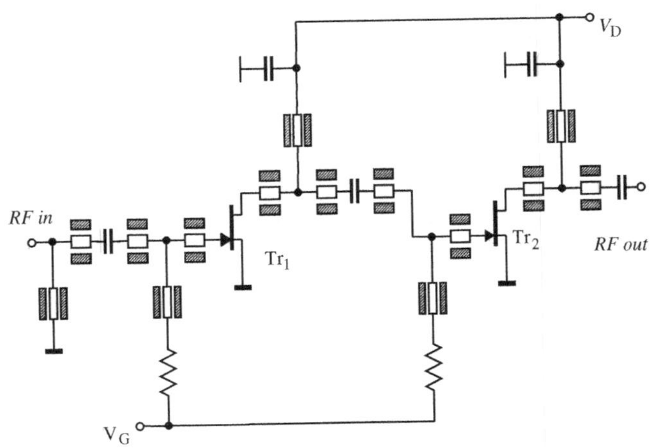

Fig. 7.4.2.1. Electrical schematic diagram of the two-stage distributed element amplifier.

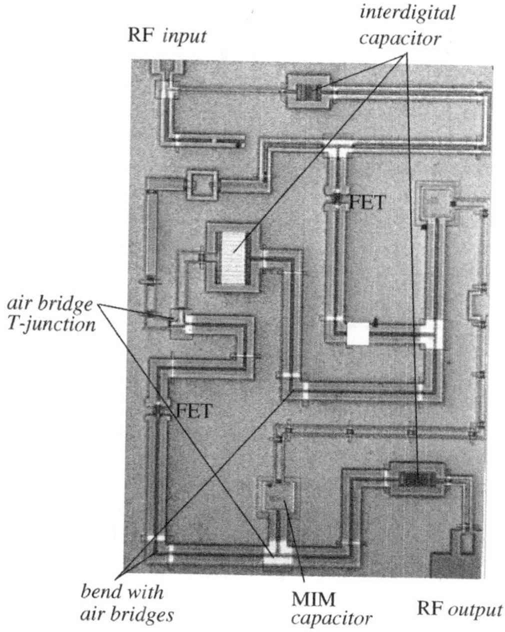

Fig. 7.4.2.2. Photograph of the two-stage amplifier in CPW-technique (size: $1.48 \times 2.05\,\text{mm}^2$).

at a drain voltage of 3 V and a drain current of 21.2 mA. A gain of 12.5 dB and a ripple of 1 dB have been measured in the frequency band 17.8 to 19.4 GHz. The voltage standing wave ratio VSWR is less than 1.7:1, and the circuit is unconditionally stable ($K > 1$) over the whole frequency range. Reverse isolation of the two-stage amplifier is more than 20 dB in the entire measurement frequency range.

The second MMIC amplifier is designed with lumped elements, using three FETs. In the matching networks of this three-stage amplifier the coplanar lines have been replaced by rectangular spiral inductors and the interdigital capacitors by MIM capacitors. Bias networks are realized with MIM capacitors, thin-film resistors, and rectangular spiral inductors. For a minimum quadratic chip size of the circuit, some transmission lines must be used to connect the lumped elements and the FETs. Using three 80-μm FETs the amplifier occupies an area of $1.0 \times 1.2\,\text{mm}^2$. Replacing the 80-μm FETs with 160-μm FETs further reduced the chip size area. This is made possible due to the size reduction of the matching structures.

The results shown are for the three-stage amplifier utilizing 160-μm FETs. A simplified schematic circuit topology of the three-stage amplifier is illustrated in Fig. 7.4.2.4. The fabricated circuit shown in Fig. 7.4.2.5 has a size of $0.94 \times 1.05\,\text{mm}^2$, including the bias networks. The technology of reference 261 has been used as a foundry.

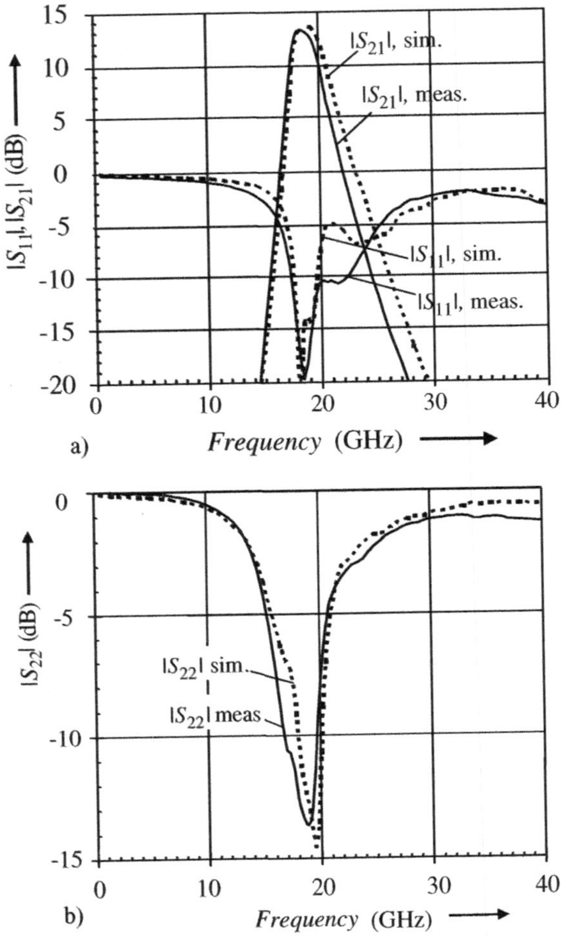

Fig. 7.4.2.3. Performance of the two-stage 18- to 20-GHz distributed element amplifier.

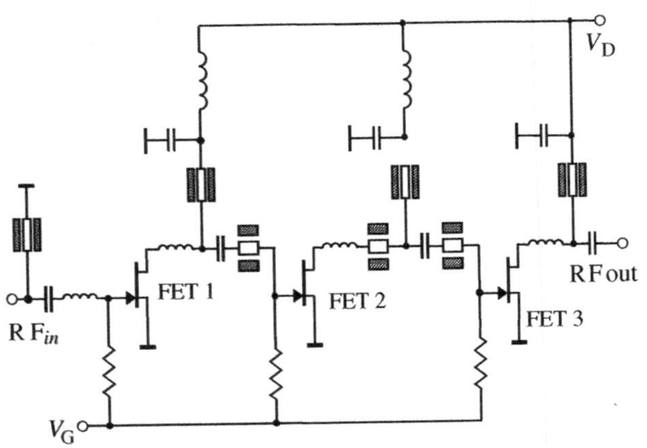

Fig. 7.4.2.4. Electrical schematic diagram of the three-stage lumped element amplifier.

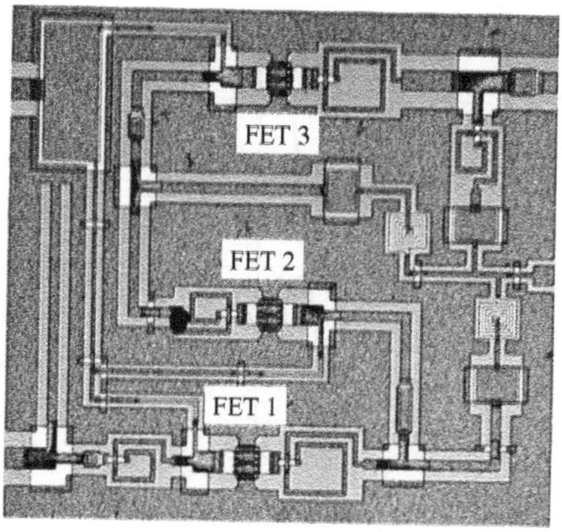

Fig. 7.4.2.5. Photograph of the three-stage amplifier in CPW-technique (size: $0.94 \times 1.05\,mm^2$).

For a biased drain voltage of 3 V and a drain current of 55.2 mA the comparison between the simulated and the measured scattering parameters of the three-stage lumped-element amplifier is shown in Fig. 7.4.2.6. The simulated and measured performances agree well. A gain of 23.5 dB and a ripple of 1 dB have been measured in the frequency band of 18 to 19.9 GHz. The VSWR is less than 2:1 and the circuit is unconditionally stable ($K > 1$) over the whole frequency range. Reverse isolation of the three-stage amplifier is more than 30 dB in the entire measured frequency range.

Figures 7.4.2.7 and 7.4.2.8 illustrate the comparison of the chip size and the measured gain of the two coplanar MMIC K-band amplifiers. The three-stage lumped-element amplifier has nearly 10 dB more gain than the two-stage distributed amplifier while requiring only about 30% of the chip size of the two-stage distributed amplifier.

7.4.3 Influence of the Backside Metalization on the Design of a Coplanar Low-Noise Amplifier

7.4.3.1 Modeling the Transistor and Its Noise Properties. In this chapter a possible influence of a backside metalization on a coplanar X-band low-noise amplifier will be described. After selecting the transistors for the amplifier, simulation results and the layout neglecting the backside metalization will be presented in a first design step. Using the measured results for the low-noise amplifier fabricated on a backside-metalized semiconductor (GaAs), the influence of the backside metalization will be discussed [259].

Transistor devices used in low-noise amplifiers must fulfill one important criterion: Using a corresponding circuitry, the transistor device must have both a good noise match and a good power match. Keeping this in mind, a transistor was chosen for the LNA specifications: $f = 8$–$12\,GHz$, gain = 18 dB,

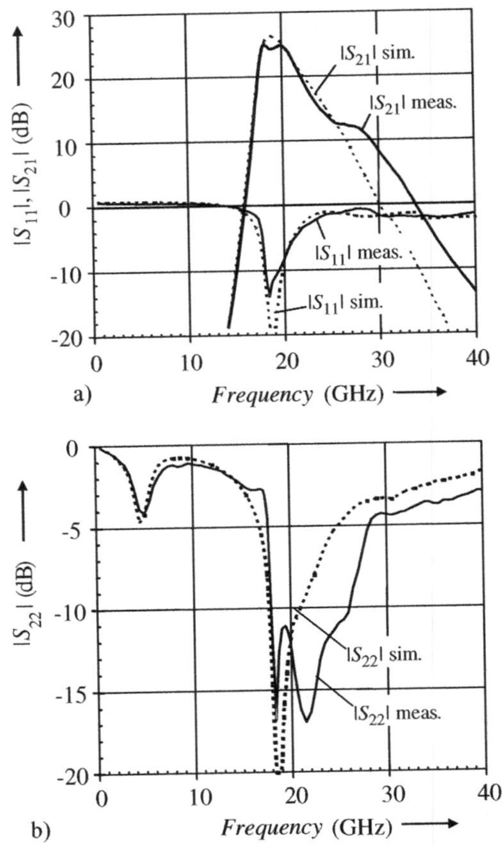

Fig. 7.4.2.6. Performance of the three-stage 18- to 20-GHz lumped-element amplifier: (a) Input reflection and transmission coefficient. (b) Output reflection coefficient.

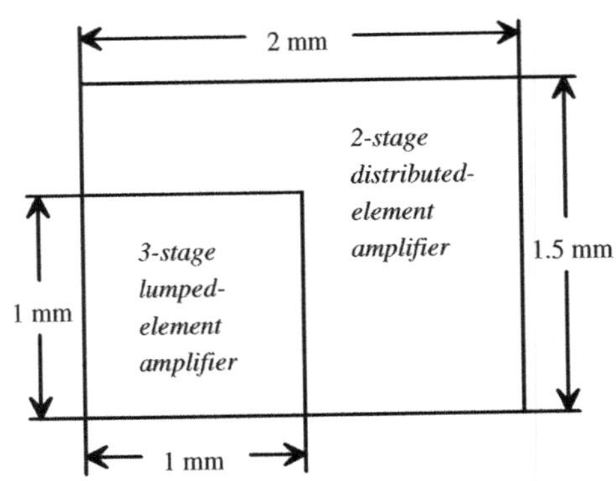

Fig. 7.4.2.7. Comparison of the chip size of the two coplanar MMIC amplifiers.

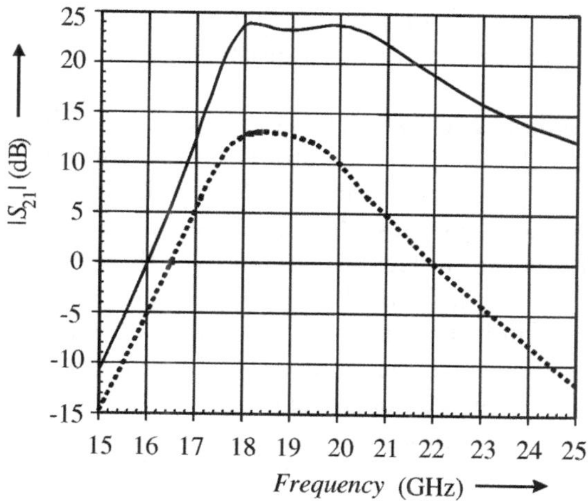

Fig. 7.4.2.8. Comparison of the measured gain of the two coplanar MMIC amplifiers. (· · ·) Two-stage distributed-element amplifier. (———) Three-stage lumped-element amplifier.

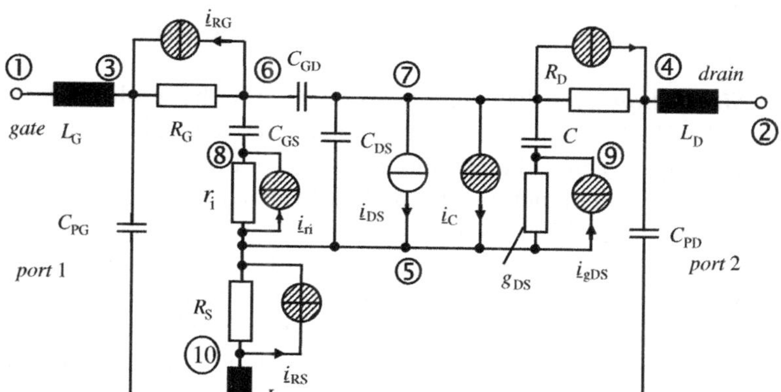

Fig. 7.4.3.1. Noise equivalent circuit for the FET.

$N_f \leq 1.4\,\text{dB}$, $P_{1\,\text{dB}} = 12.0\,\text{dBm}$, and $TOI \geq 22.0\,\text{dBm}$. For noise parameter extraction and simulation, the TOPAS equivalent circuit [246] as shown in Fig. 7.4.3.1 was used. The 10 nodes of the equivalent circuit are numbered in order to determine the 10×10 Y-matrix. In correspondence with the Y-matrix, a 10×10 noise matrix is defined. The position of the noise sources in the matrix is given by the node numbers at the output nodes of the noise sources (Fig. 7.4.3.1). Using the method given in reference 246, the noise figure of the amplifier can be calculated [247]. For the extraction of the noise parameters, the calculated noise figure can be compared to the measured one to determine the required RF noise parameters of the transistor device.

7.4.3.2 The Coplanar LNA Design. For the design of the LNA, the UMS T624 transistor was chosen. This transistor is a 6 × 40-μm HEMT device delivering excellent noise performance and satisfying gain. Figure 7.4.3.2 shows the principal set up of the two stages LNA. The first stage features a serial inductive feedback. This causes a decreasing gain over the frequency that is compensated by the second stage. Both stages are matched at input and output using short-circuited inductors. These inductors are also used to supply the dc power. Capacitors are used at the input and the output port as well as between the stages to decouple the dc power supply from the RF circuitry. The parallel circuitry consisting of a capacitor and a resistor at the output of the second stage takes care of a constant efficiency and delivers more stability. This element in the layout of the LNA is a little bit critical in design and simulation of the amplifier because there is no such element in the *COPLAN* library (see Chapter 5) available. Therefore it has been simulated using the model of a MIM capacitor in parallel with an ideal resistor. This delivers an acceptable accuracy in the simulation for X-band frequencies. The LNA was produced at UMS using the PH25 process on a thinned wafer with a thickness of 100 μm. A backside metalization was applied for mounting and stability reasons.

7.4.3.3 Simulation Results. Figure 7.4.3.3 shows the realized low-noise amplifier in MMIC technique. The mentioned inductive feedback circuit at the input of the amplifier and the RC circuit at the output are indicated in Fig. 7.4.3.3. Also, the spiral inductors and the MIM capacitors for dc power supply can be well-recognized. The MMIC size is 1.8 × 0.89 mm^2 (1.60 mm^2).

The LNA operates under the bias conditions: $V_{G1} = V_{G2} = -0.3$ V and $V_{D1} = V_{D2} = 3.0$ V. All simulations have been carried out for a temperature of 27°C. The current consumption is 73 mA. The simulation results for the scattering parameters S_{11}, S_{21}, and S_{22} using the mentioned transistor model and the *COPLAN* library for the passive components are depicted in Fig. 7.4.3.4. Input and output matching is better than −14 dB, and the gain is 18.5 dB.

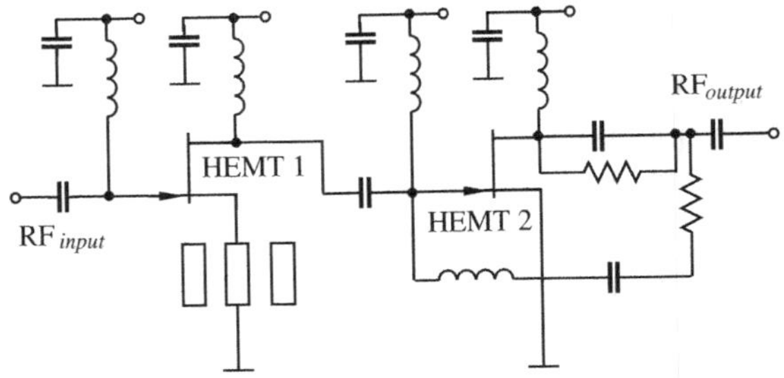

Fig. 7.4.3.2. Schematic layout of the two-stage low-noise amplifier.

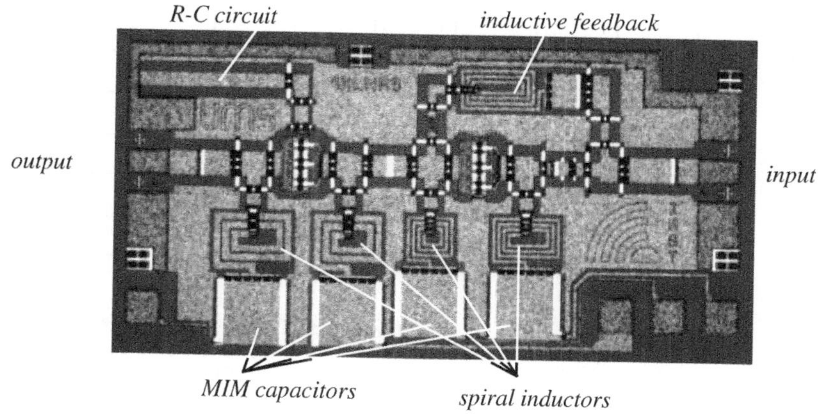

Fig. 7.4.3.3. Technological realization of the low-noise amplifier.

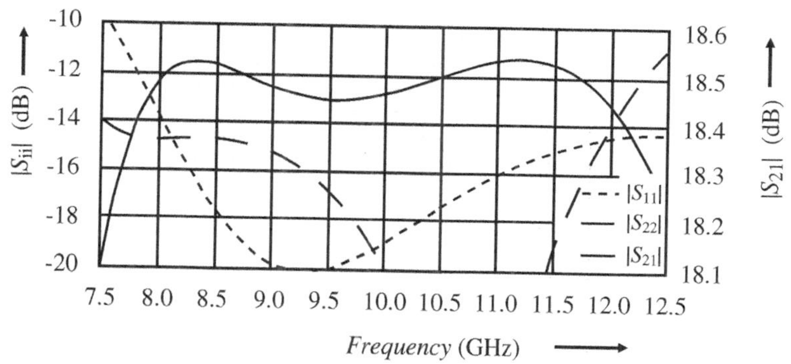

Fig. 7.4.3.4. Simulated scattering parameters of the low-noise amplifier.

From the *K*-factor (Fig. 7.4.3.5) it may be suggested, that the second stage may become instable. A closer look at the stability of the amplifier shows that the LNA is conditionally unstable, but stable in the frequency range of interest.

The simulation of the noise figure and the minimum noise figure as shown in Fig. 7.4.3.6 demonstrates that the noise matching is of high performance.

7.4.3.4 Measurement Results. To exclude external influences and disturbances on the measurements, the MMIC is mounted on a ceramic substrate. All dc bond pads have been connected to bypass capacitors (1000 pF and 47 μF). The gate voltage is connected through a 500-Ω resistor. Figure 7.4.3.7 shows the measured scattering parameters of the low-noise amplifier versus frequency in a frequency range from 0 to 20 GHz. Despite the precautions mentioned above, the measured scattering parameters of the amplifier show

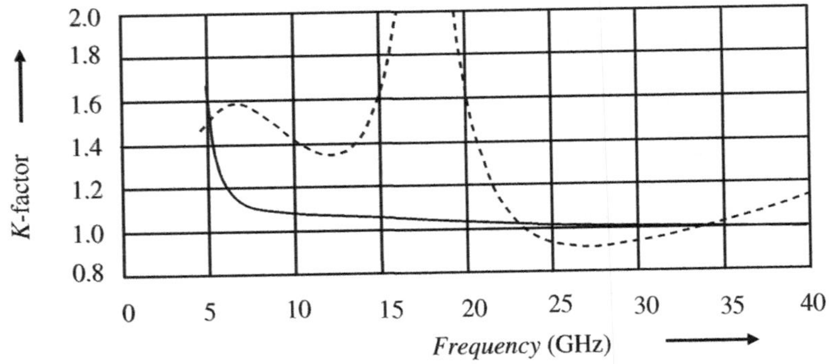

Fig. 7.4.3.5. *K*-factor of the low noise amplifier.

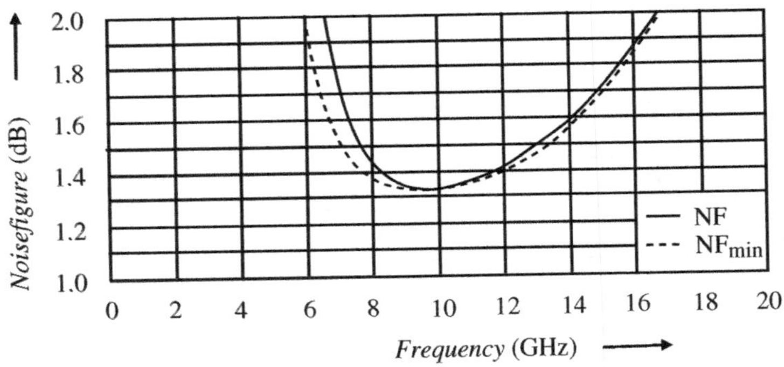

Fig. 7.4.3.6. Noise figure and minimum noise figure of the low-noise amplifier.

resonant effects at a frequency of approximately 13 GHz (see Fig. 7.4.3.7, dashed curves). Examinations of the frequency spectrum with a spectrum analyzer, however, show that no oscillations occur in the amplifier circuit. The reason for the shown behavior of the scattering parameters are substrate modes between the backside metalization and the metalization on top of the carrier material as they have been also described in reference 235 in a similar form. To prove this suggestion, the backside metal of the MMIC was removed using a gold etchant and hydrofluoric acid. To protect the MMIC topside from this etch solution, it was covered with wax before starting the etch process. In this way the backside metalization could be removed successfully.

After this treatment, new measurements have been carried out; their results are also shown in Fig. 7.4.3.7 (dotted curves). As can be observed from the figures, no more resonant effects have been measured under these conditions and the measurements agree very well with the simulations (solid curves).

Finally, Fig. 7.4.3.8 shows the measured noise figure that again agrees well with the simulation results. The same is true for the simulated and measured output power that is depicted in Fig. 7.3.4.9.

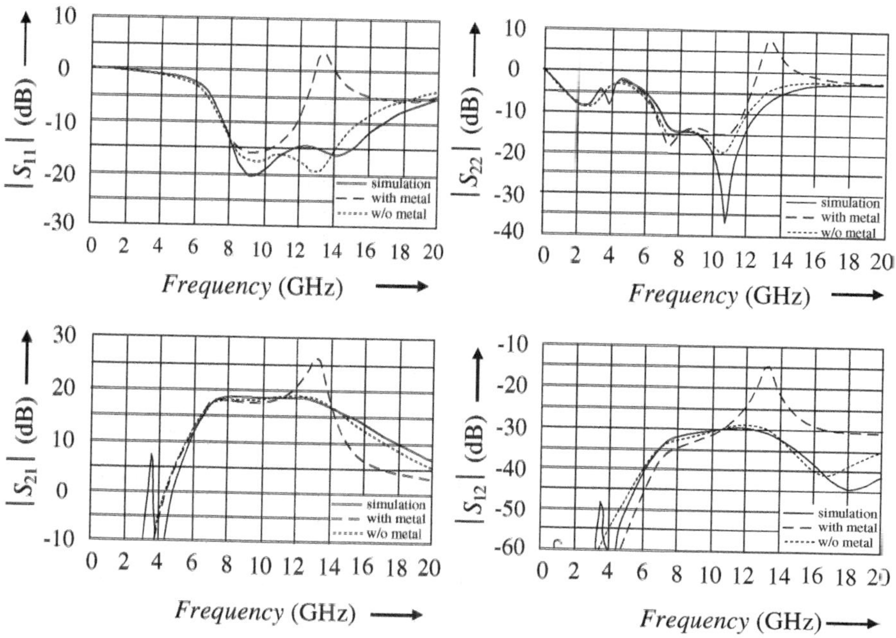

Fig. 7.4.3.7. Simulated and measured scattering parameters of the low-noise amplifier versus frequency. Comparison of the measurement results for the circuit with and without backside metalization.

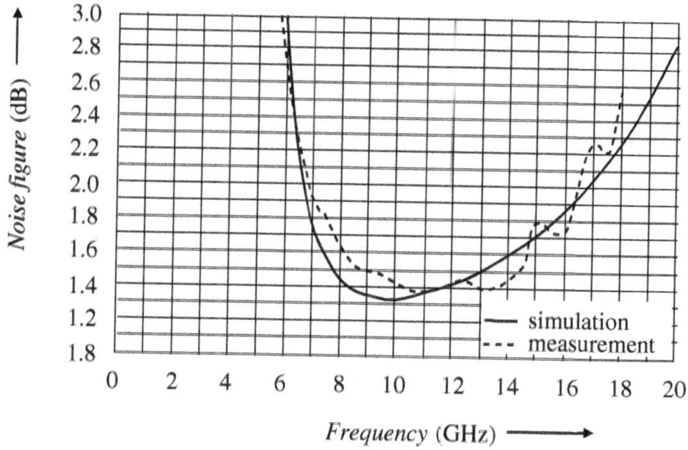

Fig. 7.4.3.8. Simulated and measured noise figure of the low-noise amplifier versus frequency.

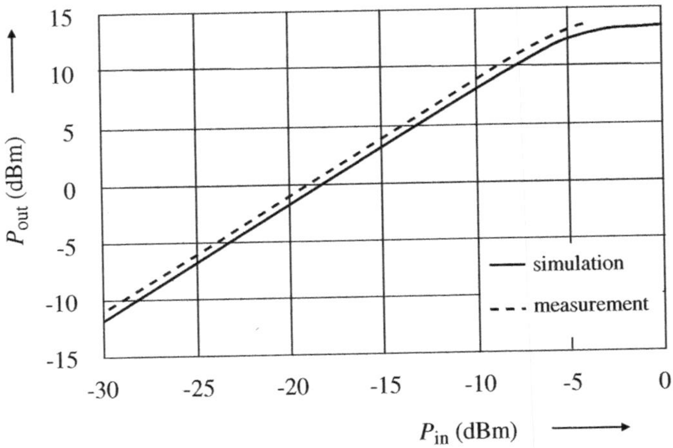

Fig. 7.4.3.9. Simulated and measured output power of the low-noise amplifier.

7.4.4 Miniaturized Ka-band MMIC High-Gain Medium-Power Amplifier in Coplanar Waveguide Technique

7.4.4.1 Introduction. The objective of this section is to investigate a small-sized low-cost Ka-band MMIC power amplifier in coplanar line technique by using a conventional 0.5-μm MESFET technology. The miniaturized MESFET amplifier is fabricated on a GaAs substrate using the CPW technique. Using this technology and by miniaturization of the chip size, a low-cost and large-volume production of MMICs is possible [148, 164, 165, 186]. The coplanar elements such as transmission lines, MIM capacitors, steps, and T-junctions, which are used in the amplifier, are calculated with the quasi-static finite difference method as described in Section 2.2 and Chapter 5. The amplifier design has been performed with the *Agilent ADS*™ software including the coplanar element library and by using the CURTICE cubic model for the MESFET [29, 173].

7.4.4.2 MMIC Design and Results. Figure 7.4.4.1 shows a simplified circuit diagram of the amplifier, and the corresponding photograph is illustrated in Fig. 7.4.4.2. The technology of reference 261 has been used as foundry. The amplifier was designed and optimized for high-gain, good input and output return losses, medium output power, and small chip size.

The amplifier is realized using a 16-gate finger MESFET. Each of the gate fingers has a width of 40 μm. Coplanar transmission lines and MIM capacitors were utilized for the matching networks. The one-stage amplifier occupies a chip area of 0.295 mm^2, including all elements of the bias network. The two versions shown in Fig. 7.4.4.2 differ only in the design of the gate bias circuitry.

The Ka-band amplifier was measured with a combined on-wafer linear and nonlinear measurement system up to 60 GHz. A comparison of the simulated

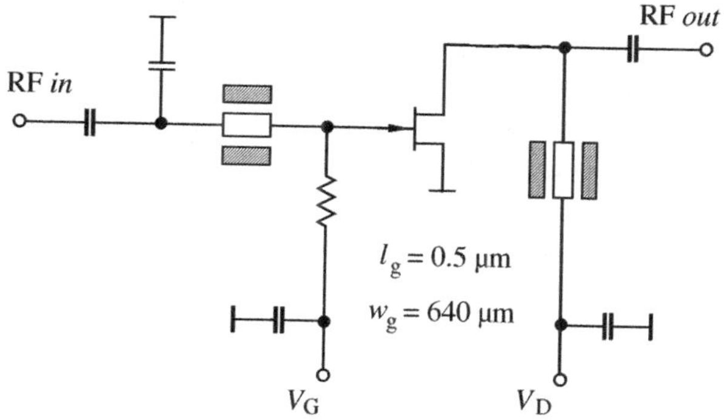

Fig. 7.4.4.1. Simplified circuit diagram of the one-stage 0.5-µm MMIC MESFET amplifier.

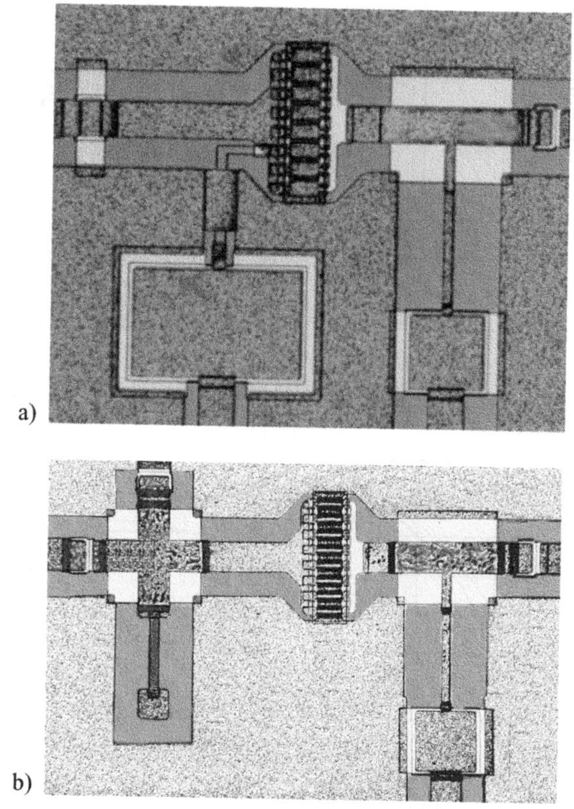

Fig. 7.4.4.2. Photograph of the coplanar Ka-band MESFET amplifier (chip size: $0.46 \times 0.64\,mm^2$) in two realized versions (a) and (b) with different bias circuitry.

and measured small-signal parameters of the Ka-band amplifier is shown in Fig. 7.4.4.3 for two different bias points.

The amplifier circuit has been biased at a drain voltage of 3 V and a drain current of 126 mA in the first case and a drain voltage of 5 V and a drain current of 109 mA in the second case (Figs. 7.4.4.3a and 7.4.4.3b, respectively). A small shift of the center band frequency can be observed due to the changed properties of the transistor in the different bias points, but in both cases the agreement between simulated and measured results is excellent. In the frequency band from 28 GHz to 30 GHz (in the first case) and from 26.3 GHz to 28.3 GHz (in the second case), the input and output reflection coefficients are less than –10 dB and the gain is better than 7 dB with a ripple of 1 dB. The circuit is unconditionally stable ($K > 1$) over the whole frequency range. Reverse isolation of the amplifier is better than 15 dB in the entire measured frequency range.

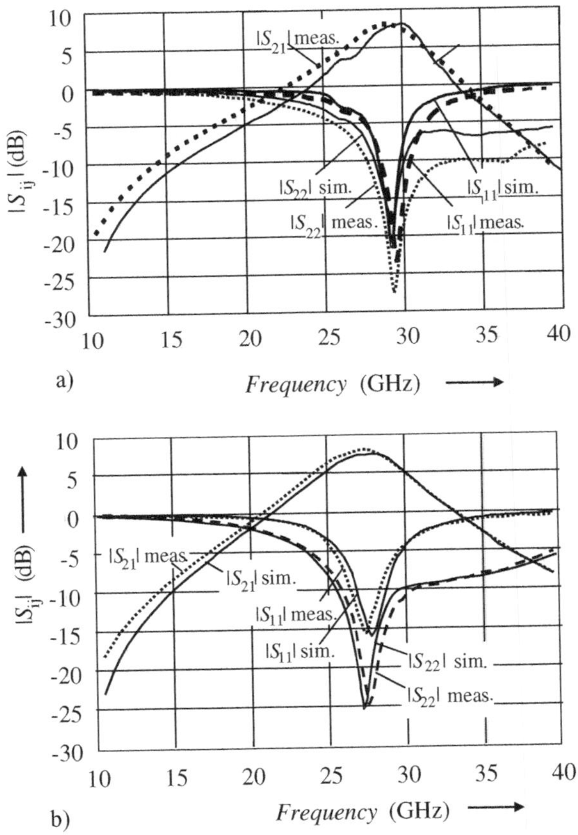

Fig. 7.4.4.3. Comparison of the simulated and measured small signal parameters of the 0.5 μm MESFET Ka-band amplifier for two different power supplies: (a) $V_D = 3$ V, $I_D = 126$ mA. (b) $V_D = 5$ V, $I_D = 109$ mA.

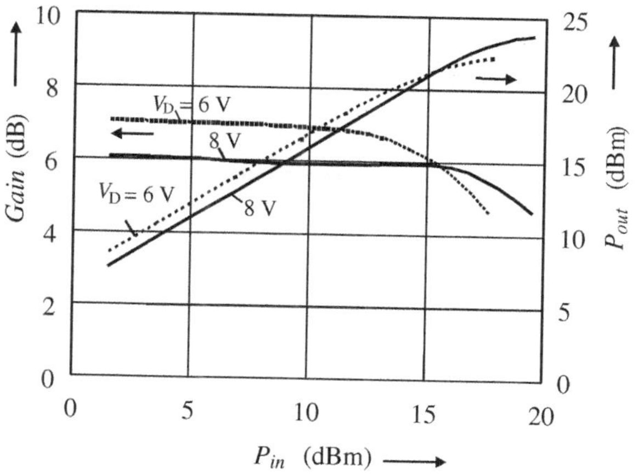

Fig. 7.4.4.4. Power gain and output power versus input power at 26.5 GHz for the drain voltages $V_D = 6\,V$ and $V_D = 8\,V$ and for a fixed gate voltage of $V_G = -0.5\,V$.

Higher drain voltage will increase the output power. Therefore, Fig. 7.4.4.4 shows the measured output power and power gain versus input power for the drain voltages of $V_D = 6\,V$ and $V_D = 8\,V$ and for a fixed gate voltage $V_G = -0.5\,V$.

7.5 COPLANAR ELECTRONIC CIRCULATORS

Circulators are essential components that are needed, for example, at a transmitter/receiver input/output for connecting the antenna. On the other side, realizing them with ferrite material always leads to a large-volume solution that is not compatible with microwave integrated circuits. Using coplanar technology and making use of the principles described in references 49, 88, and 89, electronic circuits that behave like circulators may be realized. They possibly can replace ferrite circulators. In the following it shall be investigated how such quasi-circulators can be designed using coplanar technology and what their electronic properties look like [162, 208].

The electronic circuit layout of the quasi-circulator is shown in Fig. 7.5.1. This circuit is only a quasi-circulator in the sense, that a signal can be transported from port 1 to port 2 and from port 2 to port 3, but there is no signal path back from port 3 to port 1. Figure 7.5.2 depicts the geometrical layout of two types of the realized circulators. While in Fig. 7.5.2a conventional matching strategies using coplanar transmission lines are applied, the second design utilizes rectangular lumped-element inductors for matching and dc biasing. The space reduction from the first design (type 1) to the second design (type 2) is about 70%. In real numbers, the circulator in Fig. 7.5.2b has a total chip size of $4.5\,mm^2$ whereas the space needed for type 1 is about $15\,mm^2$. In both

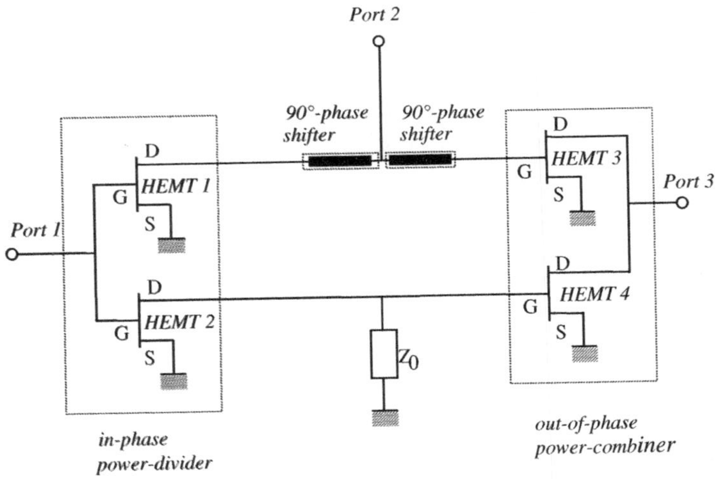

Fig. 7.5.1. Possible design of an active quasi circulator using an in-phase power divider and an out-of-phase power combiner on the basis of conventional foundry HEMT structures.

cases the design is such that the gate bias is set to zero and the drain bias can be applied in the center of the circuit.

The HFET device used in the presented design is a PM-HFET fabricated at Daimler Benz in Ulm, Germany [260] with 0.25-μm gate length and 300-μm gate width. It should be pointed out that the device periphery has not been optimized for the application described here. It is expected, for instance, that for a HEMT with 120-μm gate width an even better performance will be achieved.

Several active coplanar circulators have been investigated, designed, and fabricated for the 40-GHz band. The introduced designs combine new FET structures and standard foundry HEMT cells. The return loss of the realized quasi-circulators is better than −15 dB, and the insertion loss is typically 3 dB for a bandwidth of about 6 GHz at a center frequency of 40 GHz. The isolation between the ports is better than 20 dB in this frequency band.

The measured and simulated results of the structure shown in Fig. 7.5.2a are depicted in Fig. 7.5.3. In this figure, matching at port 1 and port 2 as well as the insertion loss and the isolation between these ports is shown. For comparison of simulation and measurement as shown in Fig. 7.5.3 a single HEMT device from the wafer accommodating the circulators has been measured after production. Then, the circuit with this HEMT data and the passive circuit design was simulated to avoid deviations of the HEMT parameters in the production process. Matching at all ports of this circuit is better than −10 dB while the insertion loss between the channels is around 3 dB at 40 GHz. Isolation between the ports is better than 20 dB. For a simple testing procedure, only one bias port (at the center of the structure) was used.

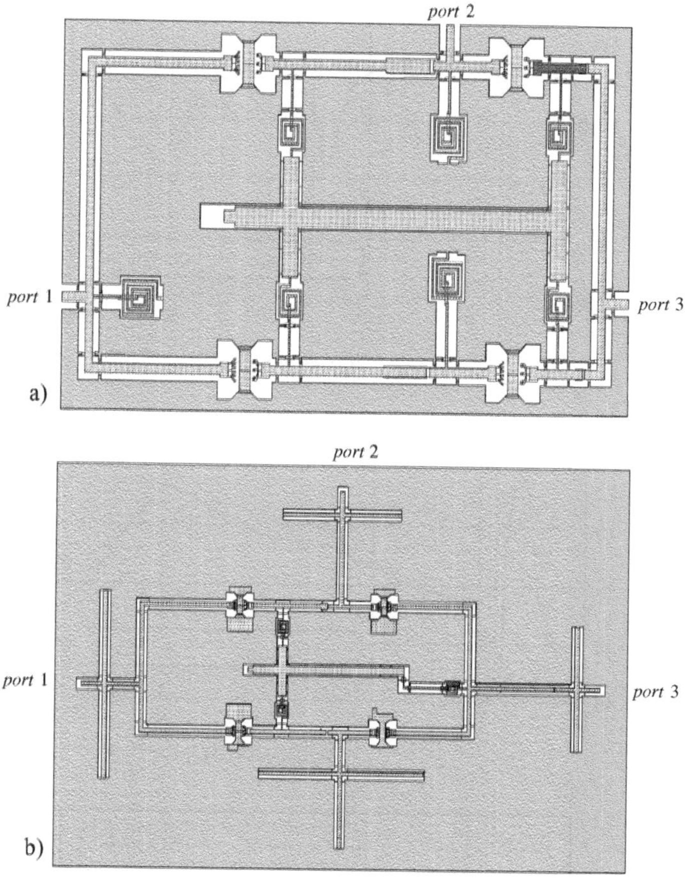

Fig. 7.5.2. Two types of active circulators in coplanar waveguide technology: (a) Active circulator of type 1, matching circuit with transmission lines. (b) Active circulator of type 2, matching circuit with lumped elements.

In design for real application the gate voltage of the devices should be controlled so that the isolation may be increased. All circuits are simulated and optimized utilizing the coplanar waveguide, discontinuity, and lumped-element models as described in Chapters 2 to 5.

With the described design, active coplanar quasi-circulators for 40 GHz (and lower frequencies) applications can be realized. Good performance can be achieved for the return loss, the isolation, and the insertion loss. In addition, excellent agreement between simulation and measurements is observed. The demonstrated circuits are the result of a first shot design that indicates the good accuracy of the applied linear models of the passive elements. The bandwidth of the quasi-circulator is about 15%.

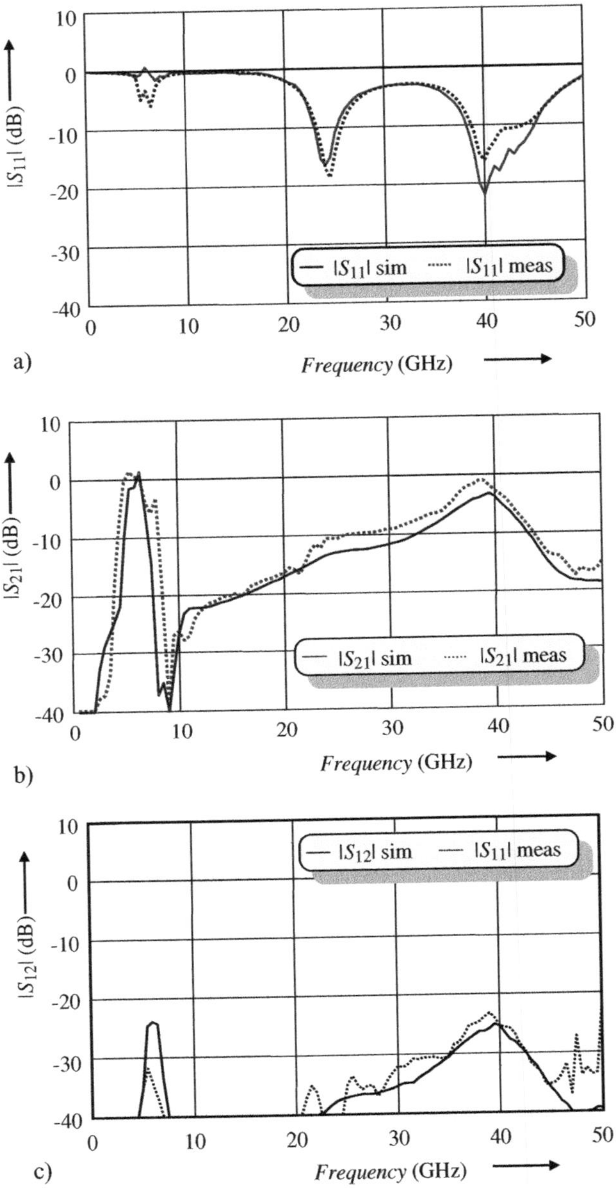

Fig. 7.5.3. Measured and simulated scattering parameters for the type 1 circulator.

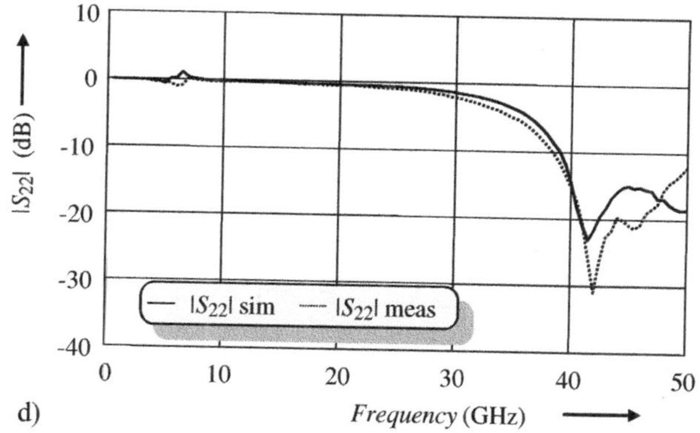

Fig. 7.5.3. (*Continued*)

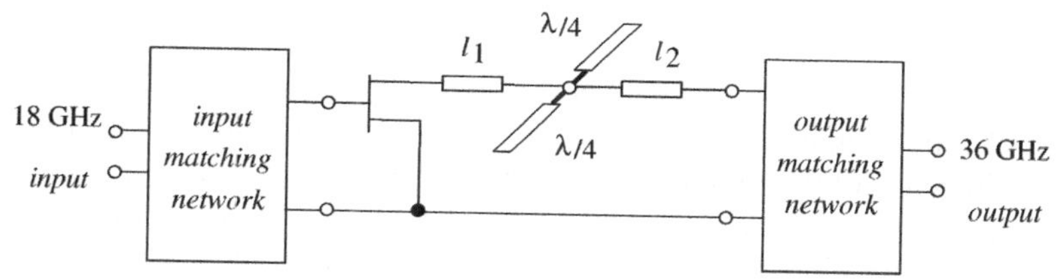

Fig. 7.6.1. Block diagram of the single-device frequency doubler.

7.6 COPLANAR FREQUENCY DOUBLERS

7.6.1 Different Realization Concepts of FET Frequency Doublers

There are at least three different circuit configurations for realizing FET frequency doublers. These are: the single-device configuration, the balanced or push–push configuration, and the wideband circuit configuration. Each of these will be explained briefly in general terms in the following sections [153, 135, 149].

7.6.1.1 The Single-Device FET Frequency Doubler. Figure 7.6.1 shows the block diagram of a single-device FET frequency doubler. It consists of a single FET, input and output impedance matching networks, and an output filter. The output filter can be a band-reject filter centered at the fundamental frequency or a band-pass filter centered at the second-harmonic output frequency. The design of each of the circuit components needed for the design is already explained in detail in the previous chapters of this book (see Section 6.3).

7.6.1.2 The Balanced (Push–Push) FET Frequency Doubler. The block diagram of a balanced FET frequency doubler is shown in Fig. 7.6.2. It con-

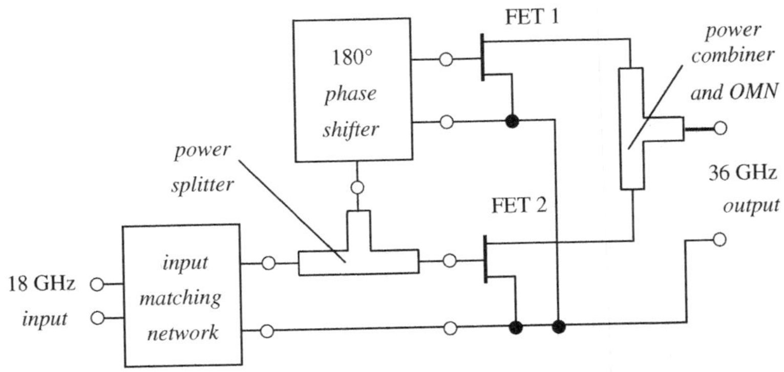

Fig. 7.6.2. Block diagram of the balanced (push–push) frequency doubler (OMN stands for output matching network).

sists of two identical FETs, a 180° hybrid (coupler) for phase-shifting purpose, input and output matching networks, a power splitter, and power combiner circuits. The gates of the two transistors are driven by signals having 180° phase difference, so the fundamental and the odd-order harmonic components of the drain currents are out of phase. This results in a situation in which each of the two FETs effectively short-circuits the other one at the fundamental and odd-harmonic frequencies and creates a virtual ground at the drain.

Since the second-harmonic drain currents have no phase difference, they combine and add up. In this way the fundamental and odd-harmonic frequencies at the drain will be suppressed and the second-harmonic signal is enhanced. This will result in a balanced doubler having 3dB more output power than an equivalent single-device circuit. Usually, the fourth-harmonic frequency signal of a well-designed FET frequency doubler is almost nonexistent. This can be observed from the fact that the zero of current $\hat{i}_4$ of the fourth-harmonic frequency lies in the vicinity of the peak of the current $\hat{i}_2$ of the second-harmonic signal. Therefore, the fourth and higher harmonics are too small to be of any concern to the designer.

In his book, Mass [54] has listed some advantages of the balanced doubler circuit over its single-device circuit counterpart. These include the fact that since the output filter is not necessary here, the output matching circuit can be located near the drain. The absence of the filter means absence of parasitic effects due to this filter, and this allows the balanced doubler to have a wider bandwidth than the single-device doubler circuit. Moreover, the load impedance of a balanced doubler is easier to realize than that of a single-device doubler. This is true, because the load impedance of the balanced doubler needs only be half of that required by the single-device doubler, and this fact eases the task of matching the output circuit at the second-harmonic frequency [54].

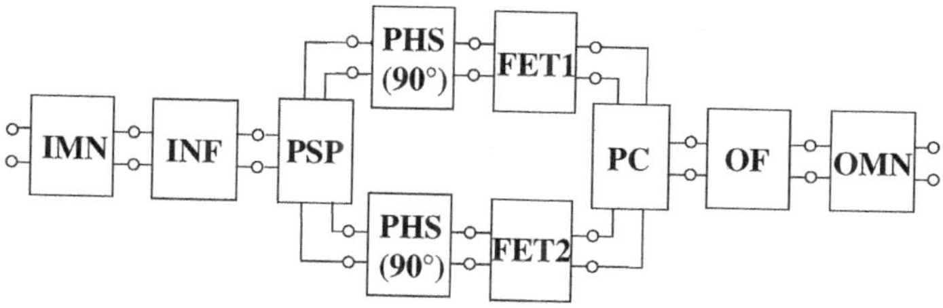

Fig. 7.6.3. Block diagram of the wideband frequency doubler. IMN, input matching network; INF, input filter; PSP, power splitter; PHS, phase shifter; PC, power combiner; OF, output filter; OMN, output matching network.

7.6.1.3 The Wideband FET Frequency Doubler. Figure 7.6.3 shows the block diagram of a wideband balanced frequency doubler. It consists of an input band-pass filter covering the input frequency range, an output band-pass filter covering the output frequency range, power splitter and power combiner circuits, a phase shifting circuit and input/output impedance matching networks. The power splitter and power combiner circuits can be realized with passive or active components and the band-pass filters may be realized using interdigital coplanar lines (see Section 6.5). Unlike the conventional single-frequency doublers, the wideband input and output-matching circuits must have a low Q-factor in order to maximize the bandwidth. One difficulty in designing a single-device doubler covering an octave or greater bandwidth is that the highest frequency in the band to be doubled will overlap with the desired lowest second-harmonic frequency in the output band. Consequently, tuning the FET input to the fundamental frequencies and tuning the output to the second-harmonic frequencies will result in a compromised performance at the lower edge of the output band. To overcome this problem, the balanced doubler concept is usually used for the design of wideband doublers.

In this way the anti-symmetrical configuration of the balanced doubler is used to effectively cancel the fundamental and odd-harmonic frequency signals at the output, as already explained in Section 7.6.1.2. Fundamental frequency signal rejection can also be obtained by using an additional balanced amplifier stage. Moreover, the amplifier provides gain at the second-harmonic.

7.6.2 Realization of Coplanar Frequency Doublers

As has been discussed above, some of the problems that one encounters in the design of the single-device MIC doubler can be overcome by making use of the balanced or push–push doubler. The block diagram of such a doubler has already been shown in Fig. 7.6.3. As mentioned in Section 7.6.1.2, it consists of a pair of identical transistors, a 180° phase-shifting circuit, power splitter/combiner circuits, and the input/output matching networks, which can also be treated as being part of the power splitter/combiner circuits. This doubler is designed and realized in both the hybrid and monolithic coplanar MIC techniques as described below.

7.6.2.1 The Coplanar Balanced Hybrid MIC Frequency Doubler. The layout of the balanced hybrid coplanar MIC doubler is shown in Fig. 7.6.4 [153]. Contrary to the case of the single-device doubler, there is no need of utilizing filters in the balanced doubler since the filtering action for odd harmonics can be achieved by the identical transistors that are connected in push–push operation. However, for a successful operation of this circuit the 180° phase shifter must be accurately designed. That means that if the phase-shifter circuit is properly designed, then the gates of the two FETs will be driven by signals having 180° phase difference, and therefore the fundamental frequency (which is taken to be 18 GHz in this design example) components of the drain currents are out of phase, resulting in their self-elimination. On the other hand, the desired second harmonic current components are in phase and therefore they add up, thus enhancing the generation of the second-harmonic power (here at 36 GHz). For a well-designed doubler, the fourth harmonic is often nearly nonexistent, and therefore there is no need to be concerned about the fourth and higher harmonics [54] as already discussed above.

As can be seen from the layout, the balanced doubler proposed here does not utilize the conventionally used and space occupying 180° coupler [7, 54, 71]. Instead, simple and small-size spiral inductors in coplanar technology and their parasitic capacitances are used to obtain the desired 180° phase shift

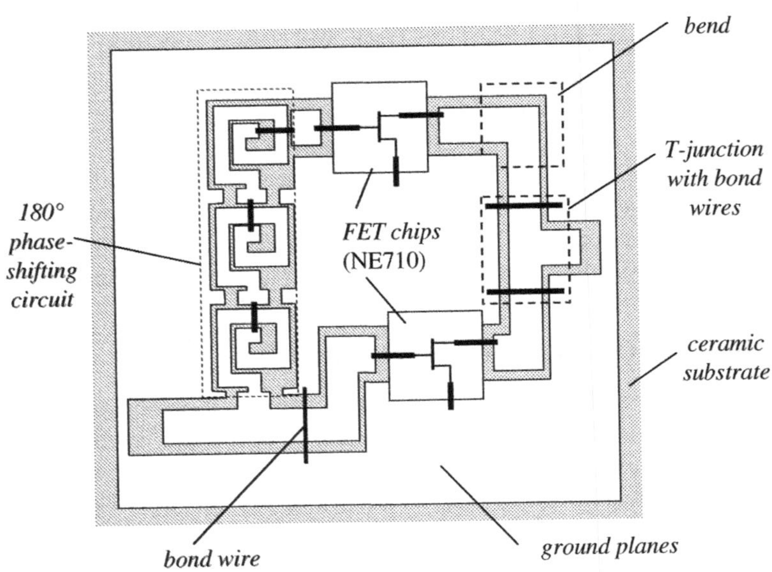

Fig. 7.6.4. Layout of the balanced hybrid MIC doubler.

between the gates of the two transistors. The necessary number of spiral inductors and the corresponding number of windings for the phase-shift purpose are optimized by use of the design method described for spiral inductors (see Sections 4.4 and 6.1.4).

The measured output power of the balanced hybrid MIC doubler in coplanar technology is given in Fig. 7.6.5. As can be seen from these results the doubler has a conversion loss of about 7 dB, a clear improvement of 3 dB when compared with a single-device doubler that has not been presented here. Moreover, there is a good agreement between measurement and calculation results. Fig. 7.6.6 shows the output signal of the frequency doubler. In Fig. 7.6.7 it is depicted that the fundamental frequency signal at the output of the

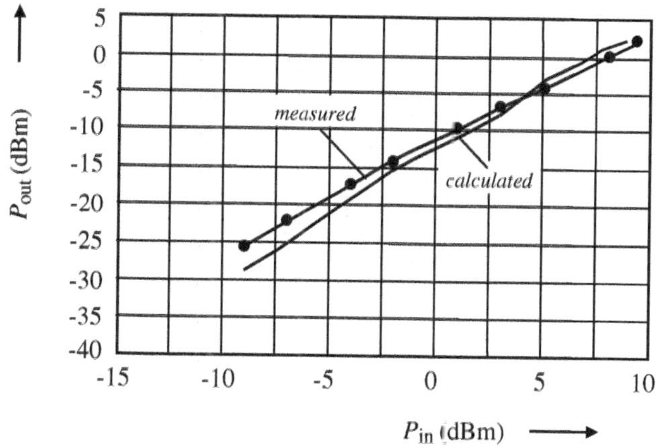

Fig. 7.6.5. Measured and calculated output power of the 36-GHz balanced hybrid MIC frequency doubler.

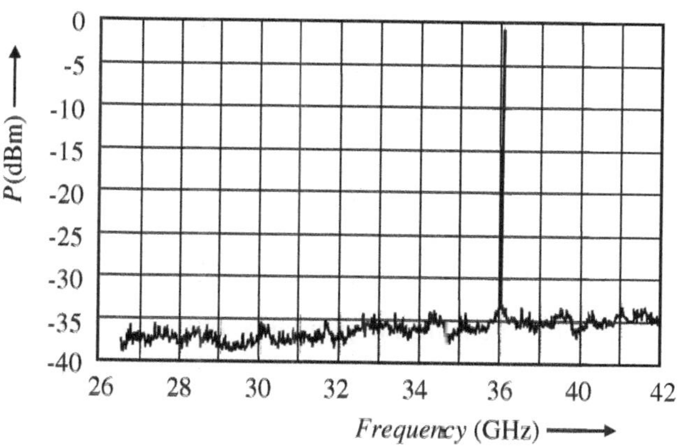

Fig. 7.6.6. The output power spectrum of the balanced hybrid MIC doubler corresponding to an input power of 6 dBm (measurement).

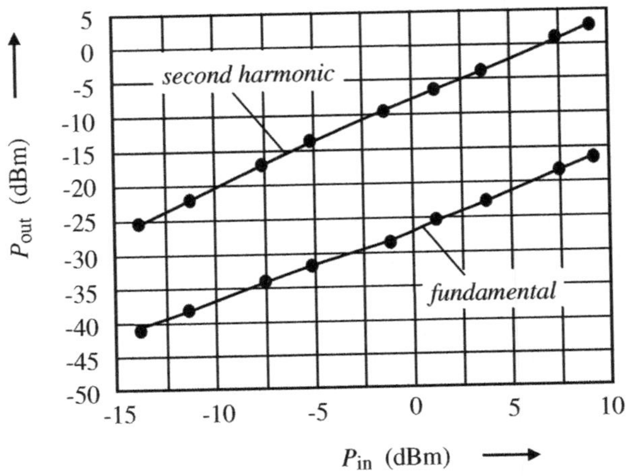

Fig. 7.6.7. Comparison of the fundamental frequency and the second-harmonic frequency power levels at the output of the doubler (measurement).

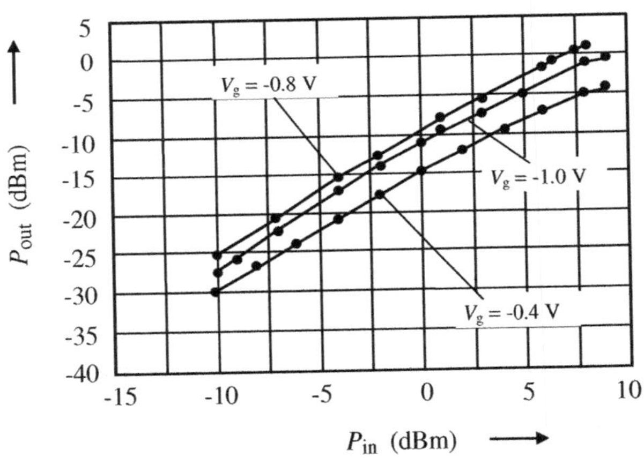

Fig. 7.6.8. Dependence of the second-harmonic output power on the input power of the balanced hybrid MIC doubler, with the bias voltage taken as a parameter (measurement).

doubler circuit is about 15–20 dB lower than the second order frequency signal.

In Fig. 7.6.8 finally the measured dependence of the output power is shown for different values of the gate voltage. For a gate voltage of –0.8 V a maximum output power at the second-harmonic frequency can be found.

7.6.2.2 The Coplanar Balanced Monolithic MIC Frequency Doubler. The layout of the monolithic version of the balanced (push–push) coplanar MIC frequency doubler is shown in Fig. 7.6.9. It is very similar to the hybrid version layout. As in the previous case of the doubler in hybrid technology, the doubler is optimized, taking into account the effects of the coplanar discontinuities, which in fact are exploited to serve as a part of the phase shifting and the impedance matching network.

The phase-shifting circuit is designed by making use of two or three spiral inductors of proper dimensions. Based on this, two versions of the balanced MMIC doubler have been realized: The first version uses three spiral inductors (of 1.5 windings each) connected in series for the purpose of phase shifting, whereas the other version uses only two spiral inductors, also connected in series, having 2.5 windings each. The input-matching network is also designed in the same way.

The simulated output power of both versions as a function of the input power is given in Figs. 7.6.10 and 7.6.11, respectively. One observes that both doublers have identical performance, and the results predict a conversion gain of about 6dB in both cases. Obviously, the doubler with two spiral inductors has the advantage of being more compact and is therefore more preferable for fabrication.

It has been discussed above that (a) the enhancement of second-harmonic generation depends upon how the device is biased and (b) optimum doubler operation is achieved when the transistor is biased either in the vicinity of pinch-off, or in the vicinity of forward direction. To verify this fact, investiga-

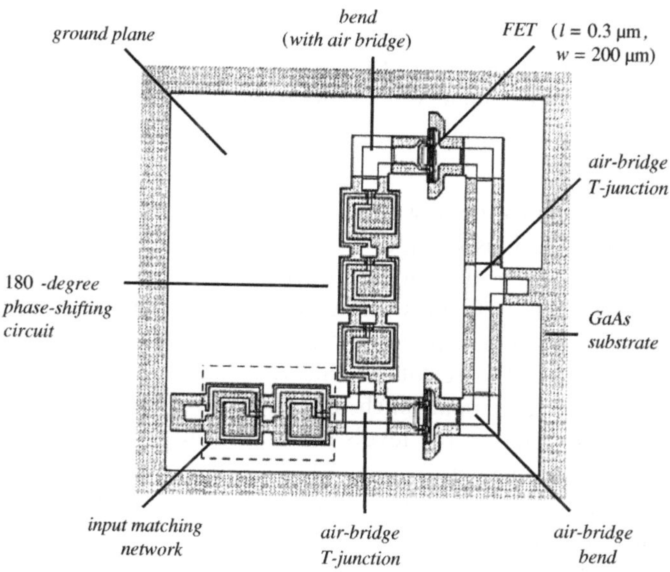

Fig. 7.6.9. Layout of the balanced 18- to 36-GHz monolithic MIC frequency doubler with three spiral inductors in the phase shifting circuit.

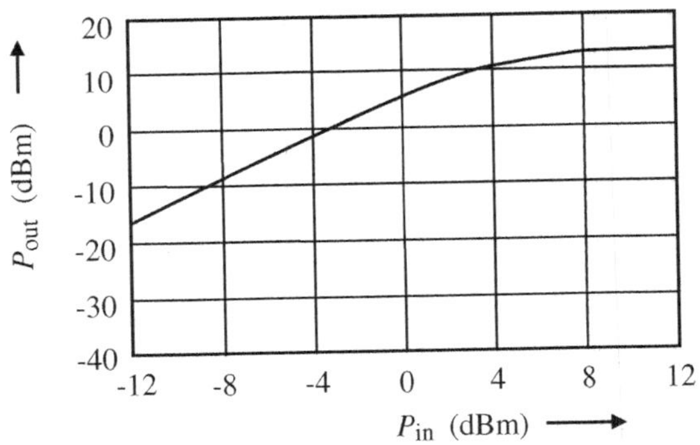

Fig. 7.6.10. Simulated output power of the 18- to 36-GHz balanced monolithic MIC doubler as a function of the input power; with three spiral inductors connected in series forming the phase-shifting circuit.

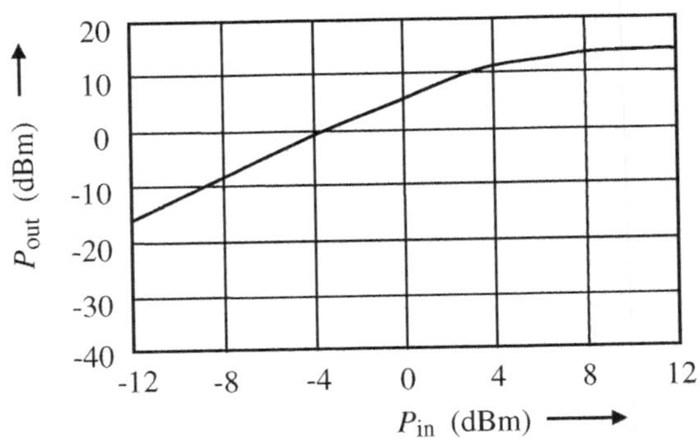

Fig. 7.6.11. Simulated output power of the 18- to 36-GHz balanced monolithic MIC doubler as a function of the input power, with two spiral inductors connected in series forming the phase-shifting circuit.

tions have been made to see how the circuit performance is really influenced by the variation of the bias voltage.

Figure 7.6.12 shows the dependence of the second-harmonic output power on the input power when the bias voltage is taken as a varying parameter (for the hybrid doubler the corresponding results have already been given in Fig. 7.6.8). It is clear from the results in Fig. 7.6.12 that the output power is a maximum for bias voltage near pinch-off ($V_g = -1.1\,\text{V}$) and that it drops when the bias voltage is varied to the levels far below and above pinch-off (i.e., to $V_g = -1.5\,\text{V}$ and $V_g = -0.7\,\text{V}$, respectively).

Finally, the photographs of both versions of the balanced monolithic MIC doubler circuits that were fabricated on a gallium arsenide substrate are shown in Figs. 7.6.13 and 7.6.14.

The measured and simulated output power of the realized doublers is compared in Fig. 7.6.15 as a function of the input power. The simulated result in

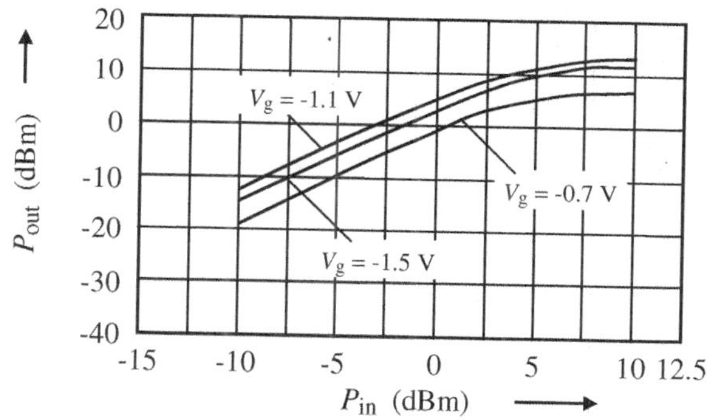

Fig. 7.6.12. Dependence of the second-harmonic output power on the input power of the balanced monolithic MIC doubler, with the bias voltage taken as a parameter (simulation).

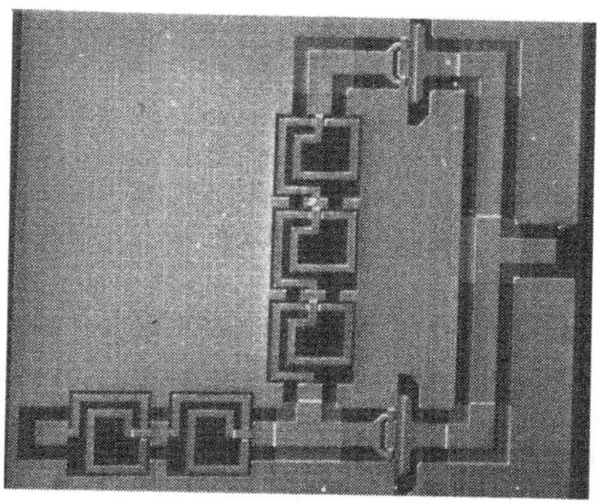

Fig. 7.6.13. Photograph of the balanced monolithic MIC doubler, with three spiral inductors connected in series forming the phase-shifting circuit.

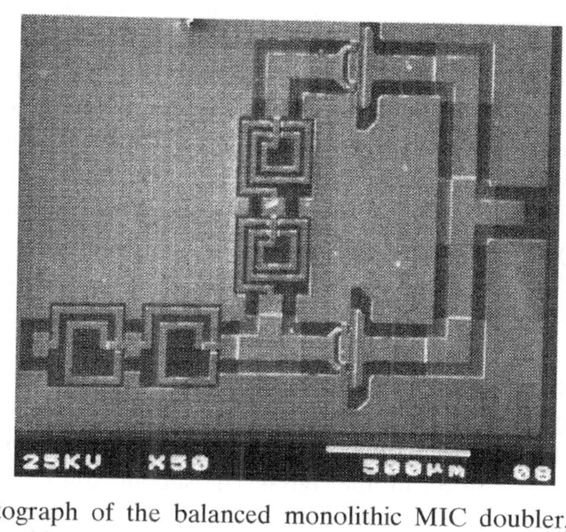

Fig. 7.6.14. Photograph of the balanced monolithic MIC doubler, with two spiral inductors connected in series forming the phase-shifting circuit.

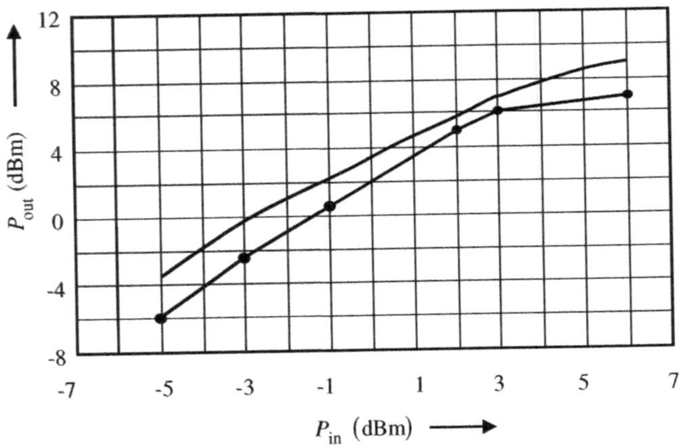

Fig. 7.6.15. Measured (· · ·) and simulated (———) output power of the 18- to 36-GHz balanced MMIC frequency doubler as a function of the input power.

this figure corresponds to the measured data of the transistor that is found on the wafer on which the circuit is fabricated. The measured output power versus input power shows that a conversion gain of 2–3 dB can be realized with the demonstrated monolithic coplanar doubler circuit.

7.6.3 A Coplanar Times Five Frequency Multiplier

In microwave systems it is often required that oscillator frequencies should be stabilized with the aid of a connection to a lower-frequency source. As a special example, a sensor signal at 24.8 GHz was needed and should be stabilized using a stable clock signal at low frequencies [212]. Since well-performing frequency dividers for the 20- to 30-GHz frequency region cannot be designed easily, a times five frequency multiplier, which transfers the oscillator signal of about 5-GHz frequency (see Section 7.7.2) to 25 GHz and which can more easily be realized, was developed. The frequency multiplier has been fabricated in a 0.25-μm HFET technology [260]. Besides the use of the coplanar library that has been described in Chapter 5, nonlinear models for the field effect transistors were mandatory for this circuit design. As an example, the result for the output power from such a model is shown in Fig. 7.6.16 for the case of a HFET used in the times five multiplier. While the accuracy of the modeled fundamental frequency output power is good, the fifth harmonic is reproduced with a sufficient accuracy only for an input power of around 10 dBm. This is due to the limited ability of nonlinear transistor models to correctly describe higher harmonics.

The times five multiplier is a two-stage device (Fig. 7.6.17). The first stage multiplies the input signal. It is optimized using a nonlinear FET model and the circuit design program *Agilent ADS*™ to produce the highest possible output power at the fifth-harmonic signal. The second stage serves as a buffer

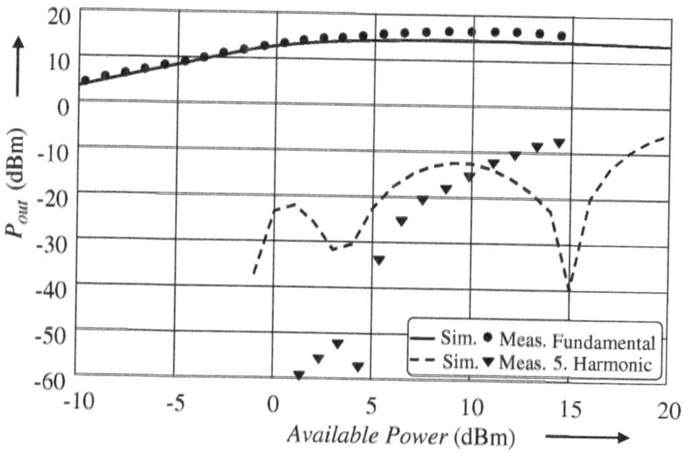

Fig. 7.6.16. Measured and modeled output power of an HFET at $f = 5.3\,\text{GHz}$, $V_{DS} = 3\,\text{V}$, and $V_{GS} = 0\,\text{V}$.

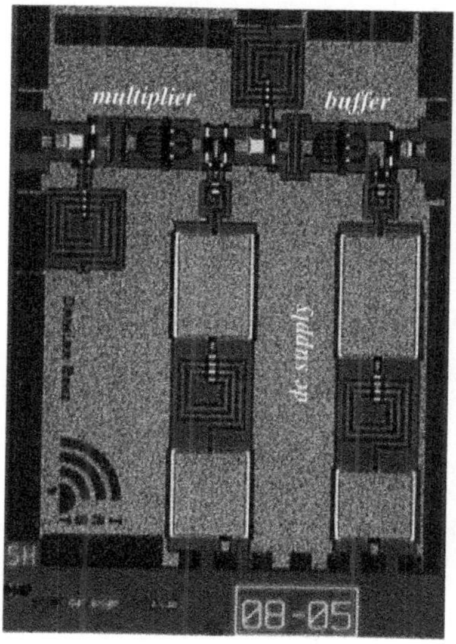

Fig. 7.6.17. The times five (5–25 GHz) coplanar FET multiplier (left side of the figure) and the buffer amplifier (right side of the figure).

amplifier. Most of the space on the chip is devoted to the dc-bias circuitry that is built using rectangular spiral inductors and MIM capacitors (lower part of the figure).

The performance of the times five multiplier is depicted in Fig. 6.6.18. The output power of the fifth harmonic versus the available input power of the fundamental frequency for the frequencies 4.7, 5.0, and 5.3 GHz is shown.

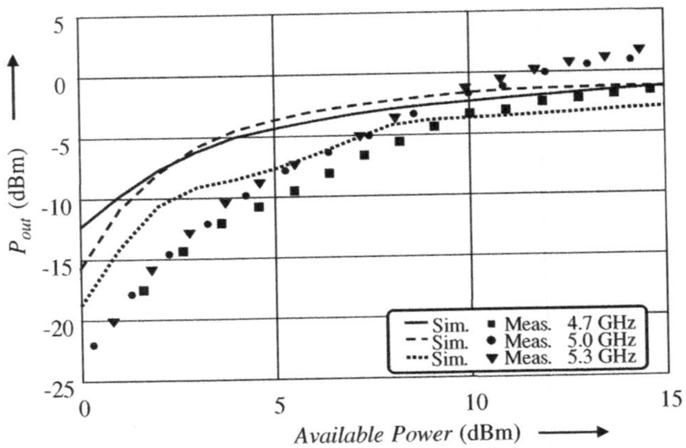

Fig. 7.6.18. Measured and simulated output power of the fifth harmonic versus input power.

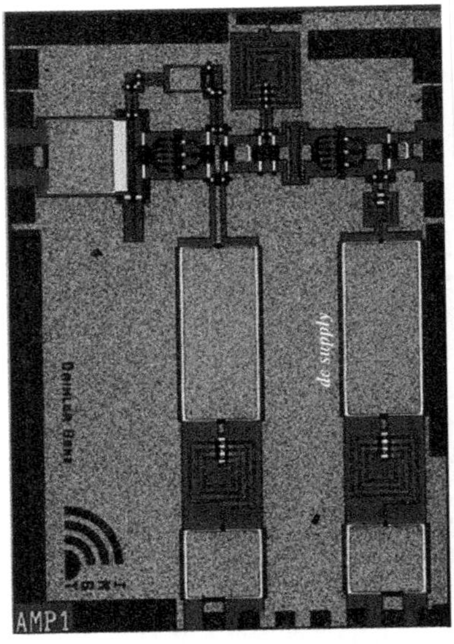

Fig. 7.6.19. The 25-GHz coplanar amplifier for a sensor system at the output of the times five multiplier.

There is a good agreement between the measurement results and the simulated values around 10-dBm input power. As has already been mentioned above, it is very difficult to model a transistor including the fifth-harmonic generation! The conversion loss at 10 dBm input power (which is the oscillator output power) is around 12 dB. The input and output return loss is lower than −5 dB and −15 dB, respectively.

In the sensor application, a 25-GHz amplifier, shown in Fig. 7.6.19, follows the multiplier. The large signal behavior of the amplifier can be seen in Fig.

7.6.20. It exhibits a gain of approximately 15 dB. For an input power of −2 dBm (from the multiplier) an output power of 13 dBm was achieved. This is in a very good agreement with the simulated data. The simulated value of the saturated output power is 15 dBm, while the measured value is around 17 dBm, both for 3-V bias voltage. This difference occurs because the saturation current of the HFET has been higher than the expected value. The input as well as the output return loss is lower than −15 dB. The frequency conversion behavior of the complete system is shown in Fig. 7.6.21. There is a suppression of more than 60 dB of the fundamental LO frequency and the overall conversion

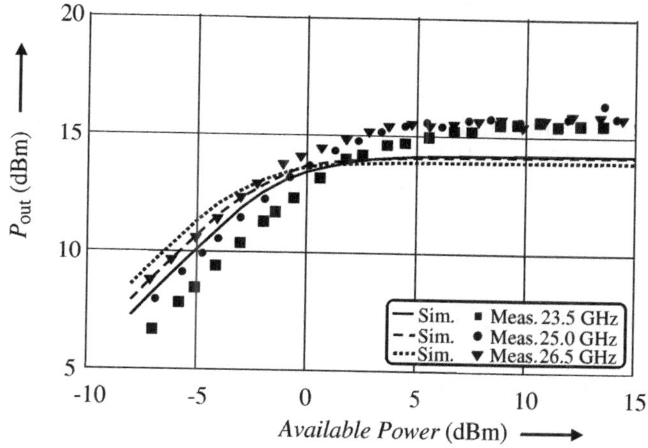

Fig. 7.6.20. Measured and simulated output power of the amplifier.

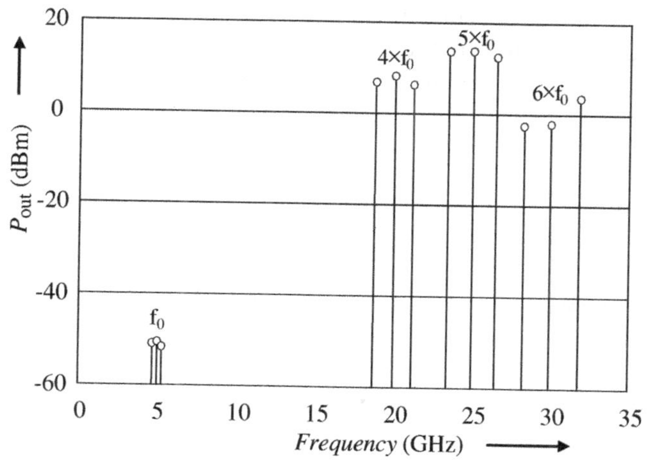

Fig. 7.6.21. Measured power spectrum of the complete sensor system. The utilized LO power is $P_{LO} = 12$ dBm.

gain is around +1 dB. The minimum sideband suppression is 6 dB in the case of the fourth harmonic and at least 10 dB in the case of the sixth harmonic.

7.7 MICROWAVE AND MILLIMETER-WAVE OSCILLATORS IN COPLANAR TECHNOLOGY

7.7.1 Coplanar Microwave Oscillators

In microwave integrated circuits, oscillators in over 90% of all cases base on a transistor circuit. Only one transistor may be used for generating the microwave oscillation; however, in many cases it is better to provide a second transistor for building up an additional buffer stage. The buffer stage minimizes the coupling from the load to the resonating circuit and provides a better and stable termination to the oscillator stage. Figure 7.7.1 shows a general oscillator circuit of this kind, which was intensively discussed in reference 150 and in references 176 and 184.

The first stage (T_1, Y_1, Y_2) is the basic oscillator stage. The admittances Y_1 and Y_2 serve as resonating circuit or feedback element, respectively. The coupling network Y_c mostly consists of waveguide elements or filters, but in some cases it also can be replaced by a direct through feed from the oscillator to the buffer stage. The output load of the circuit is represented by the admittance Y_3.

Two cases have to be considered concerning the admittances Y_1 and Y_2 (compare also Section 7.3):

(a) Y_1 works as an inductive feedback element. Then a negative conductance can be observed between the source terminal T_1 and the circuit ground (without considering Y_2). This circuit can be treated as a voltage-controlled current source. This means that the resonating circuit represented by the admittance Y_2 with preference should be a parallel resonant circuit. In this case its parasitic conductance can be

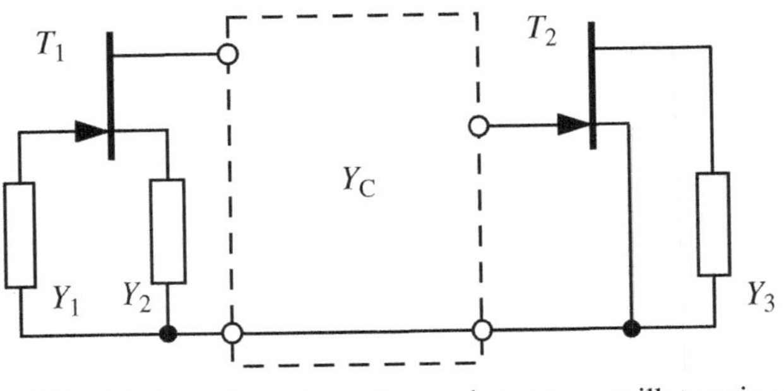

Fig. 7.7.1. Universal topology for a microwave oscillator circuit.

canceled by the mentioned negative conductance of the oscillator circuit.

(b) If Y_2 is constructed as a capacitive feedback element, a negative resistance between the gate of transistor T_1 and the circuit ground can be observed (not considering the admittance Y_1). This circuit behaves like a current-controlled voltage source. In this case the admittance Y_1 with advantage is built up as a series resonating circuit to receive a proper oscillating of the oscillator circuit.

Of course there are various other possible concepts for designing an oscillator circuit. For example, a two-port feedback circuit can be used between gate and drain of the field effect transistor. However, most requirements on a microwave oscillator can be well-fulfilled using the circuit given in Fig. 7.7.1. The two mentioned operation modes lead to two classes of MMIC oscillators. If it is assumed that first the circuit is designed without the resonating circuit (which is represented by the admittance Y_2 in case a and by the admittance Y_1 in case b), two fundamental circuits are derived which can be used to form various oscillators by just adding a suitable resonant circuit.

In a first step the realization of the circuit described as case a above shall be discussed. Figure 7.7.2 shows the circuit topology and Fig. 7.7.3 the realized circuit in coplanar monolithic microwave integrated technique on GaAs substrate using a 0.5-μm FET technology. The topology shown in Fig. 7.7.2 can be compared to the topology shown in Fig. 7.7.1: The admittance Y_1 in Fig. 7.7.1 is realized by a planar spiral inductor (1) that is modeled by a *RLC* circuit. The parallel resonant circuit Y_2 has to be externally connected at the input port *I–I′*. A second spiral inductor (2) and two MIM capacitors C_1 and C_2 allow routing the drain bias from pad ① to the oscillating transistor T_1. This construction can be interpreted as a bias tee. The resonator at the input port *I–I′* should be a short circuit for dc current to enable the drain current of transis-

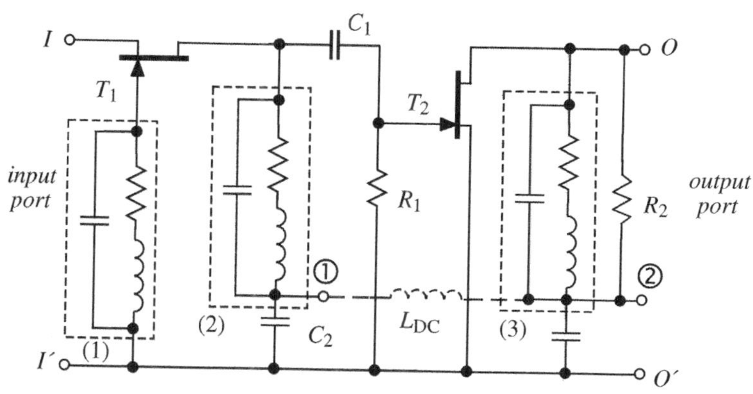

Fig. 7.7.2. Circuit topology of a negative conductance circuit of type a.

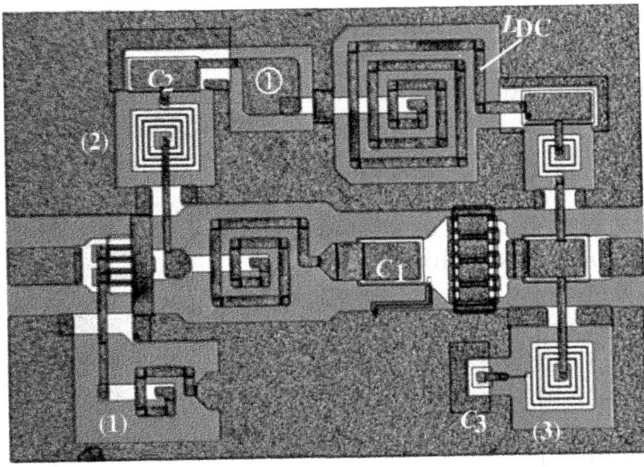

Fig. 7.7.3. Realized negative conductance circuit as a coplanar waveguide circuit on a GaAs substrate, case a.

tor T_1 to flow. The active devices are optimized to give a maximum gain at zero gate bias that is realized by the spiral inductor (1), too.

The second stage is formed mainly by the transistor T_2. The resistor R_1 realizes its gate bias. The network consisting of the two capacitors C_1 and C_2 (2) is needed not only for bias purposes but also to optimize the RF performance of the circuit. With their help the magnitude of the reflection coefficient visible at the input port I–I', which is larger than 1, can be enhanced significantly. The output network, consisting of the spiral inductor (3), MIM capacitor C_3, and resistor R_2, provides the drain bias from pad ② to transistor T_2 and optimizes the output matching at the output terminal O–O'.

The drain voltage can be fed into the circuit from only one pad ① if an additional spiral inductor L_{dc} is used for connecting pad ① and pad ② decoupling the two circuit parts. A photograph of the circuit is shown in Fig. 7.7.3. It contains the above-mentioned additional inductor L_{dc}. At the input of the circuit, a parallel resonant circuit that provides a dc connection between port I and I' must be applied (Fig. 7.7.2). The measured and simulated scattering parameters are shown in Fig. 7.7.4.

Figure 7.7.4 shows very clearly how the circuit that was optimized for a frequency range of about 30 GHz works. It shows a reflection coefficient larger than 1 at the input port I–I', transmission gain from port I–I' to port O–O', reverse transmission reduction, and a good output matching. This proves that it is possible to cancel and overcome the loss of a resonator connected to the input port I–I' and to deliver RF power from the output port O–O' to a connected load.

Figure 7.7.5 shows two parallel resonant circuits (in coplanar integrated technology) that can be connected to the ports I–I' of the above-discussed circuits to form a microwave oscillator. In Fig. 7.7.5a a fixed-frequency resonator

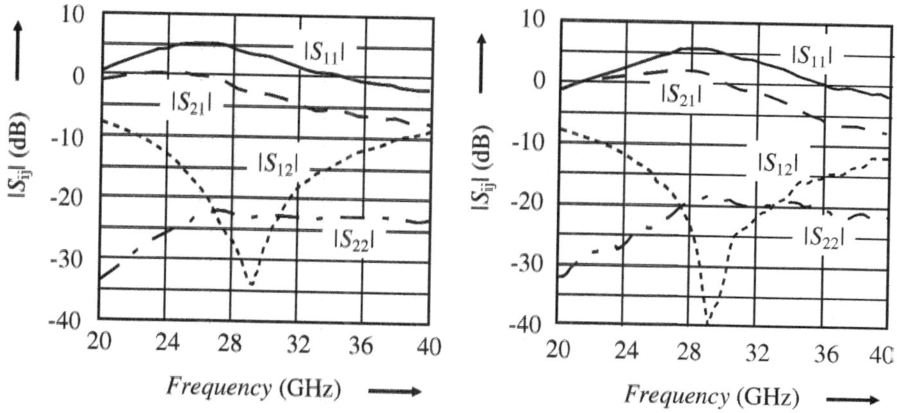

Fig. 7.7.4. Simulated (left) and measured (right) scattering parameters of the negative conductance circuit shown in Figs. 7.7.2 and Fig. 7.7.3.

that consists of a rectangular spiral inductor (produced in the gate metalization layer) and a MIM capacitor is shown. In Figs. 7.7.5b and 7.7.5c, tunable resonators are shown. The capacitor of the resonant circuit is formed by a varactor diode that has been realized in common FET technology. It is, in principle, a field effect transistor without the drain structure. Applying a negative voltage to the gate, a tunable capacitance can be realized. The tuning voltage is applied to the varactor diode via a 1-kΩ resistor that can be seen in Fig. 7.7.5b under the air bridge near the varactor diode. A big MIM capacitor takes care of the separation of the transistor and diode bias circuits. The difference between the two circuit versions shown in Figs. 7.7.5b and 7.7.5c is that in the case of Fig. 7.7.5b the spiral conductor is built in the galvanic layer to produce an inductor with a *Q*-factor as high as possible. The spiral inductor is placed inside the ground area. The version in Fig. 7.7.5c shows a small planar inductor located in the gate metalization layer. In this way, small spiral inductors that can be placed in the slot area of the coplanar waveguide can be produced.

Figure 7.7.6 shows a tunable oscillator created on the basis of the circuit type a as discussed above and using the tunable parallel resonant circuit. Pad *A* serves as the dc connection for the varactor diode.

The topology that meets the requirements of the above-described case b circuit is shown in Fig. 7.7.7. As mentioned above, a series resonator (Y_1 in Fig. 7.7.1) should be connected to the input port *I–I′* of the circuit given in Fig. 7.7.7. The capacitive feedback element is formed by an interdigital capacitor C_4 and the parasitic capacitance of the spiral inductor (1). This inductor also works as a bias shunt. The circuit works in a similar manner as that discussed above for the case a.

Figure 7.7.8 depicts the simulated and measured scattering parameters of the circuit that again shows clearly its ability to work as an oscillator if a res-

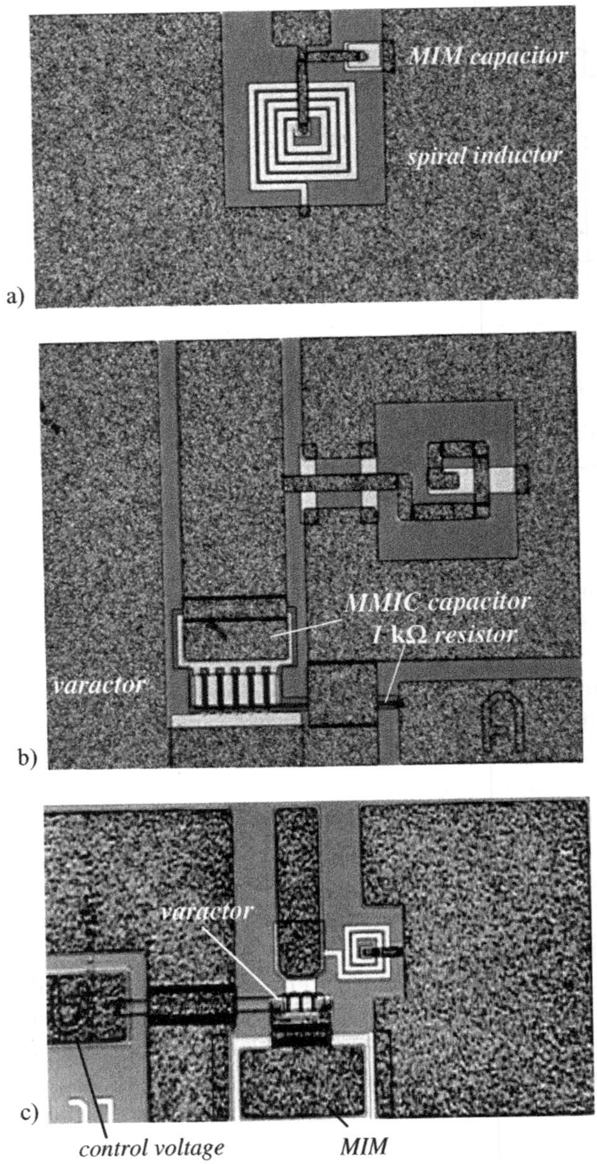

Fig. 7.7.5. A fixed-frequency integrated parallel resonant circuit (a) and two tunable parallel integrated resonant circuits (b, c).

onant circuit is applied to the input port *I–I′* and a load to the output port *O–O′*.

As already mentioned above, when the two different circuits of type a and type b are compared to each other, the type b circuit needs a series resonant circuit to be connected to the input port *I–I′* if an oscillator is to be realized.

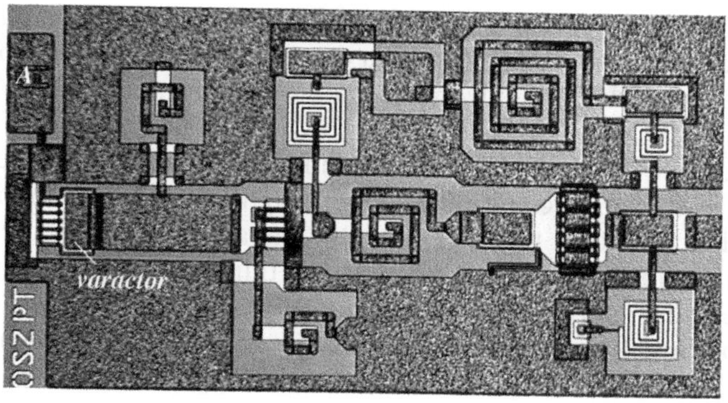

Fig. 7.7.6. A coplanar tunable oscillator on the basis of the fundamental circuit type a as described above. On the left side the added (compare with Fig. 7.7.5b) parallel resonant circuit with a tuning varactor capacitor can be observed. Pad A is for supplying the tuning voltage.

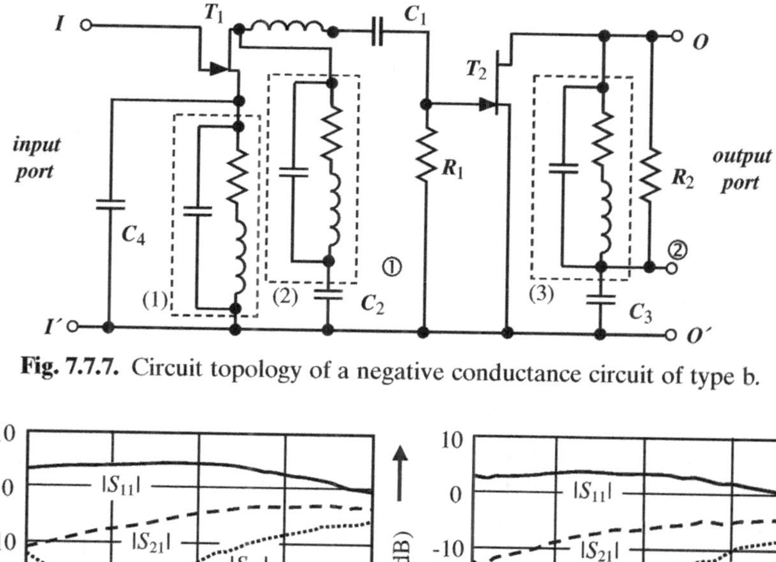

Fig. 7.7.7. Circuit topology of a negative conductance circuit of type b.

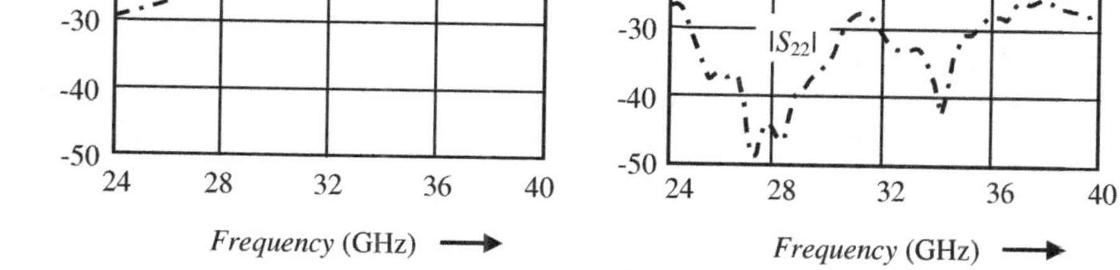

Fig. 7.7.8. Simulated (left) and measured (right) scattering parameters of the negative conductance circuit shown in Fig. 7.7.7.

These resonant circuits can be circuits outside the integrated circuit or they can be directly integrated on the chip in the same coplanar technology as that used for the circuit design. Only this kind of resonant circuits will be discussed here.

In Fig. 7.7.9a a MIM capacitor to ground that is designed as a series resonant circuit is shown. Because of the large length of the capacitor, it also has a finite inductance that together with the capacitance forms the resonant circuit. At the input of this resonant circuit a thin-film resistor to ground has been connected. This resistor is needed for the dc gate bias of the oscillator. However, because it is not a real resistance (compare Section 4.6), its properties must be considered in the resonant circuit design.

In Figs. 7.7.9b and 7.7.9c, frequency tunable series resonant circuits are shown. They consist of a rectangular spiral inductor in the center strip of a coplanar waveguide, a large MIM capacitor, and a varactor diode in series. The large MIM capacitor is needed to separate the dc bias circuits of the oscillator and the varactor diode. Again a dc bias resistor for the gate bias is part of the circuit. The control voltage for the varactor diode is fed in from a pad over a thin-film resistor (to be seen, for example, on the lower right side of Fig. 7.7.9b, partly under the air bridge in this area).

Figure 7.7.10 shows the input admittance of the thin-film resistor, which consists of a nearly constant real part and a capacitive imaginary part. The resistor considered here is a 10-kΩ resistor of 125-μm length and 2-μm width. The measured data of this resistor show that a parallel capacitance of about 8.86 fF must be considered if the resonant circuit is to be properly designed.

Using the discussed resonant circuit constructions, two coplanar integrated oscillators have been built. They are shown in Fig. 7.7.11a (fixed frequency oscillator) and Fig. 7.7.11b (tunable oscillator).

7.7.2 A 5-GHz Coplanar Voltage-Controlled Oscillator

Using the above-described fundamental concepts for the design of coplanar integrated microwave oscillators, a tunable oscillator for the frequency range from 4.7 GHz to 5.3 GHz was designed and fabricated. This oscillator is the central circuit of a 24.8-GHz sensor system used in the coplanar technique, and it is shown in Fig. 7.7.12 [212].

The oscillator is basically a three-stage device. While the first stage is the oscillator stage itself, the second and the third stage serve as a buffer. The oscillator is designed using a serial feedback as described in the discussion given above. The three-stage design was chosen for low-frequency pulling due to load changes. Each stage has an active-load circuit as a bias supply. This reduces the possibility of low-frequency oscillations. The bias pulling is suppressed by an internal bias supply for the first stage. For a monolithic integration including the tuning element, two Schottky diodes are connected in series as can be seen on the upper left side of the figure. The chip area is $2.6 \times 1.3\,\text{mm}^2$.

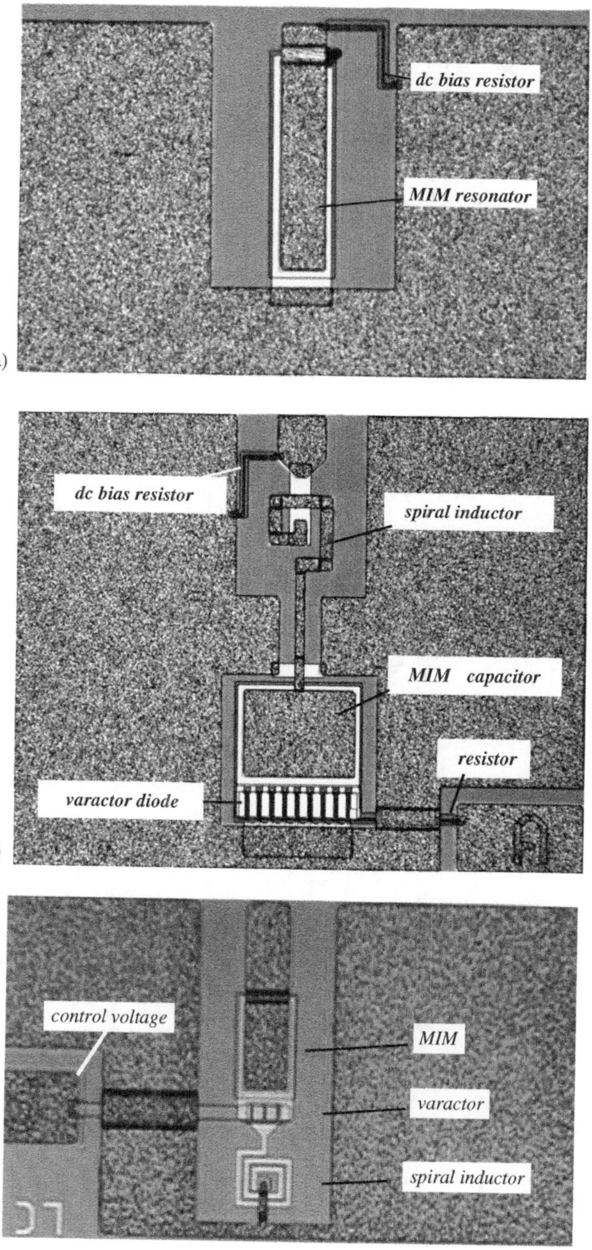

Fig. 7.7.9. Three series resonant circuits in coplanar technology for application in integrated oscillator circuits of type b. (a) Fixed frequency resonator. (b, c) Tunable resonators.

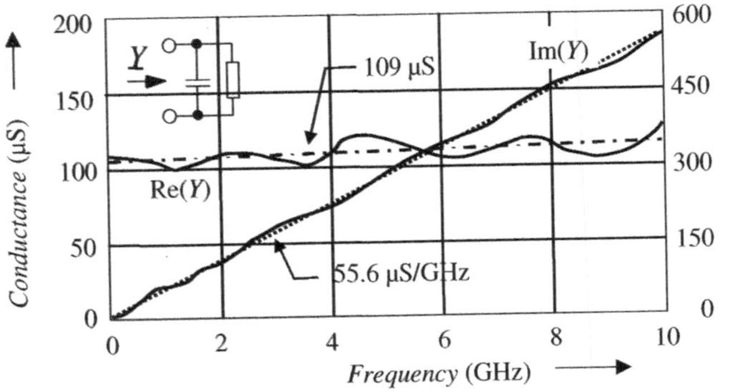

Fig. 7.7.10. The frequency-dependent input impedance of a thin-film resistor.

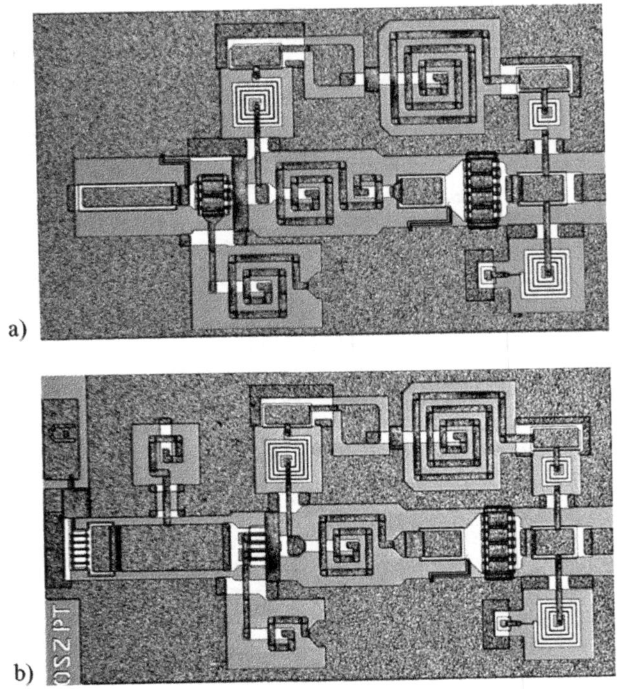

Fig. 7.7.11. A fixed frequency and a tunable frequency coplanar integrated oscillator on the basis of the type b circuit as discussed in this chapter.

Measurements have been carried out using a universal measurement stand. This equipment allows for on-wafer S-parameter, nonlinear, and load pull measurements and includes the characterization of frequency-converting circuits [6]. The oscillator performance is shown in Fig. 7.7.13. It works at 5-V bias supply. The output power is $11.6 \pm 0.5\,\text{dBm}$ over the required frequency

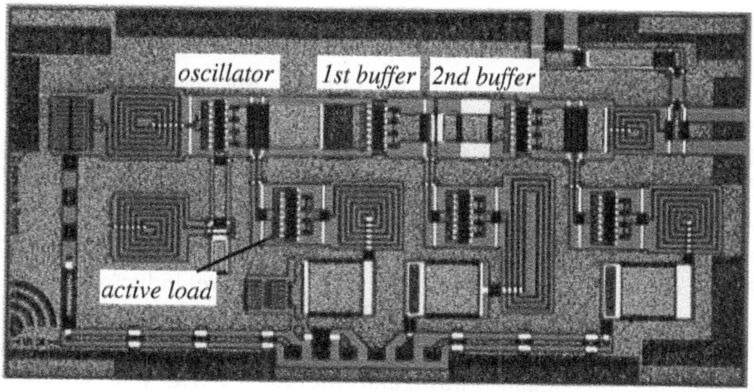

Fig. 7.7.12. A three-stage 5-GHz voltage-controlled oscillator in coplanar technology using an active load in the bias circuit.

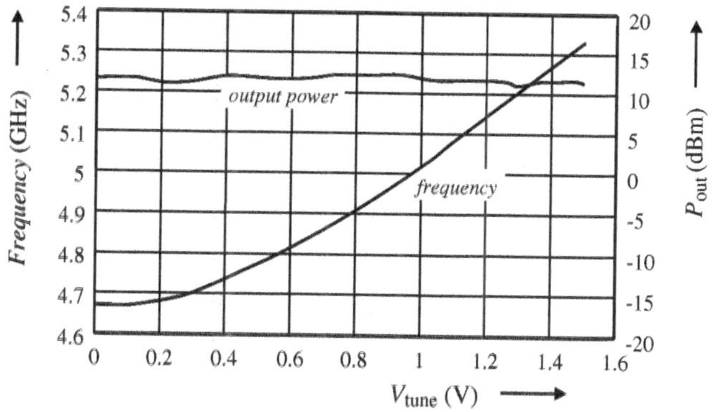

Fig. 7.7.13. Frequency tuning characteristic and dependence of the oscillator output power on the tuning voltage.

range from 4.7 GHz to 5.3 GHz. The suppression of the second and third harmonics is better than 20 dB and 30 dB, respectively. The output return loss is lower than −15 dB.

The oscillator has a second output with an output power of around −10 dBm that is needed in connection with a frequency divider. The frequency pulling caused by a load mismatching of 6-dB return loss at the output port is lower than 1 MHz. To reduce the bias pulling, an internal voltage source for the first stage is applied resulting in a frequency change of only 1 MHz/V. Replacing the integrated tuning diode by an external varactor, the frequency dependence on the tuning voltage becomes even more linear.

BIBLIOGRAPHY AND REFERENCES

1. G. L. Matthaei, L. Young, and E. M. T. Jones, *Microwave Filters, Impedance-Matching Networks, and Coupling Structures*, New York: McGraw-Hill, 1964.
2. D. K. Adams and R. Y. C. Ho, Active filters for UHF and microwave frequencies, *IEEE Trans. Microwave Theory Tech.*, vol. MTT-17, Sept. 1969, pp. 662–670.
3. R. V. Snyder and D. L. Bozarth, Analysis and design of a microwave transistor active filter, *IEEE Trans. Microwave Theory Tech.*, vol. MTT-18, Jan. 1970, pp. 2–9.
4. E. Fliegler, Operating criteria for active microwave inductors, *IEEE Trans. Microwave Theory Tech.*, vol. MTT-19, Jan. 1971, pp. 89–91.
5. J.-C. Vokes, W.-P. Barr, J.-R. Dawsey, B.-T. Hughes, and S.-J-W. Shrubb, A low noise FET amplifier in coplanar waveguide, in: *1977 IEEE MTT-S International Microwave Symposium Digest*, San Diego, 21–23. June 1977, pp. 185–186.
6. P.-A.-R. Holder, X-band microwave integrated circuits using slotline and coplanar waveguide, *Radio Electronics Eng.*, vol. 48, no. 1/2, 1978, pp. 8–42.
7. A. Presser, Interdigitated microstrip coupler design, *IEEE Trans. Microwave Theory Tech.*, vol. MTT-26, no. 10, 1978, pp. 801–805.
8. R.-E. Stegens, Coplanar waveguide FET amplifiers for satellite communications systems, *Comsat Tech. Rev.*, vol. 9, no. 1, 1979, pp. 255–267.
9. M. Aikawa and H. Ogawa, 2 Gb double-balanced PSK modulator using coplanar waveguides, in: *1979 IEEE International Solid-States Circuits Conference Digest of Technical Papers*, Feb. 14–16, 1979, Philadelphia, PA, 1979, pp. 172–173.
10. R.-A. Davidheiser, The coplanar-slot transition and a digital phase modulator, in: *1979 IEEE MTT-S International Microwave Symposium. Digest*, April 30–May 2, 1979, Orlando, FL, 1979, pp. 112–115.
11. P.-L. Fleming, T. Smith, H.-E. Carlson, and W.-A. Cox, GaAs SAMP device for Ku-band switching, *IEEE Trans. Microwave Theory Tech.*, vol. 27, no. 12, 1979, pp. 1032–1035.
12. M. S. Gupta, R. W. Laton, and T. T. Lee, Frequency multiplication with high-power microwave field-effect transistors, in: *IEEE MTT-S International Microwave Symposium. Digest*, June 1979, pp. 498–500.
13. C. Rauscher and H. A. Willing, Simulation of nonlinear microwave FET performance using a quasi-static model, *IEEE Trans. Microwave Theory Tech.*, vol. MTT-27, no. 10, Oct. 1979, pp. 834–840.
14. H. Abe and Y. Aono, 11-GHz GaAs power MESFET load-pull measurements utilizing a new method of determining tuner Y parameters, *IEEE Trans. Microwave Theory Tech.*, vol. MTT-27, no. 5, May 1979, pp. 394–399.
15. W. R. Curtice, A MESFET model for use in the design of GaAs integrated circuits, *IEEE Trans. Microwave Theory Tech.*, vol. MTT-28, no. 5, May 1980, pp. 448–456.
16. B.-J. Clifton, G.-D. Alley, R.-A. Murphy, and I.-H. Mroczkowski, High-performance quasi-optical GaAs monolithic mixer at 110 GHz, *IEEE Trans. Electron Devices*, vol. 28, no. 2, 1981, pp. 155–157.
17. E. Habekotte, B. Hoefflinger, W. Renker, and G. Zimmer, A coplanar CMOS power switch, *IEEE Journal of Solid-State Circuits*, vol. 16, no. 3, 1981, pp. 212–226.

18. H. Ogawa, M. Akaike, M. Aikawa, T. Karaki, and J. Watanabe, A 26-GHz band integrated circuit of a double-balanced mixer and circulators, *IEEE Trans. Microwave Theory Tech.*, vol. MTT-30, 1982, no. 1, pp. 34–41.
19. A. Gopinath and J. B. Rankin, Single-gate MESFET frequency doublers, *IEEE Trans. Microwave Theory Tech.*, vol. MTT-30, no. 6, June 1982, pp. 869–875.
20. P.-S. Cross, R.-A. Baumgartner, and B.-H. Kolner, A travelling wave, microwave optical modulator, in: *Proceedings, of SPIE. Picosecond Optoelectronics*, San Diego, CA, Aug. 24–26, 1983, vol. 439, pp. 153–158.
21. C. Rauscher, High-frequency doubler operation of GaAs field-effect transistors, *IEEE Trans. Microwave Theory Tech.*, vol. MTT-31, no. 6, June 1983, pp. 462–473.
22. G. S. Dow and L. S. Rosenheck, A new approach for mm-wave generation, *Microwave J.*, Sept. 1983, pp. 147–162.
23. E. Camargo, R. Soares, R. A. Perichon, and M. Goloubkoff, Sources of nonlinearity in GaAs MESFET frequency multipliers, in: *IEEE MTT-S International Microwave Symosium. Digest*, 1983, pp. 343–345.
24. R. Mehran, J. Ludewig, and L. Szabo, Integrated Ka-band finline mixer/modulator, *Electronics Lett.*, vol. 20, no. 22, 1984, pp. 934–935.
25. P.-S. Cross, R.-A. Baumgartner, and B.-H. Kolner, Microwave integrated optical modulator, *Appl. Phys. Lett.*, vol. 44, no. 5, 1984, pp. 486–488.
26. D. Neuf and S. Spohrer, Ultrasmall MIC mixer designed for ECM applications, *Microwave Syst. News, MSN*, vol. 15, no. 11, 1985, pp. 70–82.
27. K. Kawamoto, K. Hirota, N. Niizaki, Y. Fujiwara, and K. Ueki, Small size VCO module for 900 MHz band using coupled microstrip-coplanar lines, in: *1985 IEEE-MTT-S. International Microwave Symposium. Digest*, St. Louis, MO, 4–6 June 1985, pp. 689–692.
28. E. Allamando, N. E. Radhy, and E. Constant, Broadband high-order frequency multipliers by using dual-gate MESFET's, in: *Proceedings, 15th European Microwave Conference*, 1985, pp. 617–622.
29. W. R. Curtice and M. Ettenberg, A nonlinear GaAs FET model for use in the design of output circuits for power amplifiers, *IEEE Trans. Microwave Theory Tech.*, vol. MTT-33, December 1985, no. 12, pp. 1383–1393.
30. C. Rauscher, Microwave active filters based on transversal and recursive principles, *IEEE Trans. Microwave Theory Tech.*, vol. MTT-33, Dec. 1985, pp. 1350–1360.
31. M. Riaziat, I. Zubeck, S. Voly, and G. Zdasiuk, Coplanar waveguides used in 2–18 GHz distributed amplifier, in: *IEEE-MTT-S Internat. Microwave Symposium Digest*, Baltimore, MD, June 2–4, 1986, pp. 337–338.
32. Y. Tarusawa, H. Ogawa, and K. Ohwada, 26 GHz band 'Planar MMIC' hybrid circuit, *Trans. Inst. Electronics Commun. Engi. Japan*, Part E, vol. E69, no. 4, 1986, pp. 329–330.
33. M. Izutsu, H. Nakanishi, and T. Sueta, Design consideration for guided-wave light modulators with coplanar waveguide type electrodes, *Trans. Inst. Electronics Commun. Eng. Japan*, Part E, vol. E69, no. 4, 1986, pp. 415–417.
34. D.-F. Williams and S.-F. Schwarz, A planar subharmonically-pumped 71 GHz receiver with integral feed antenna, *Int. J. Infrared Millimeter Waves*, vol. 7, 1986, no. 2, pp. 183–195.

35. R. Gilmore, Design of a novel FET frequency doubler using a harmonic balance algorithm, in: *IEEE MTT-S International Microwave Symposium Digest*, 1986, pp. 585–588.
36. R. Gilmore, Nonlinear circuit design using the modified harmonic balance algorithm, *IEEE Trans. Microwave Theory Tech.*, vol. MTT-34, no. 12, December 1986, pp. 1294–1307.
37. T. Hirota, Y. Tarusawa, and H. Ogawa, Uniplanar MMIC hybrids—A proposed new MMIC structure, *IEEE Trans. Microwave Theory Tech.*, vol. 35, no. 6, 1987, pp. 576–581.
38. R.-E. Stegens and D.-N. Alliss, Coplanar microwave integrated circuit for integrated subsystems, *Microwave Systems News and Communications Technology, MSN*, vol. 17, no. 11, 1987, pp. 4, 86, 89, 91, 93, 96.
39. M.-M. Stallard, Optimisation of travelling wave integrated optic modulators in GaAs; a study, *Optics Communications*, vol. 63, no. 6, 1987, pp. 385–389.
40. J. Browne, Broadband amps sport coplanar waveguide, *Microwaves & RF*, vol. 26, no. 2, Feb. 1987, pp. 131–134.
41. J. Browne, Coplanar MIC amplifier bridges 0.5 to 18.0 GHz, *Microwaves & RF*, vol. 26, no. 6, June 1987, pp. 194–195.
42. R. Gilmore, Concepts in the design of frequency multipliers, *Microwave J.*, March 1987, pp. 129–139.
43. T. Hirota, Y. Tarusawa, and H. Ogawa, Uniplanar MMIC hybrids. A proposed new MMIC structure, *IEEE Trans. Microwave Theory Tech.*, vol. 35, no. 6, June 1987, pp. 576–581.
44. T. Ohira, T. Tokumitsu, T. Hiraoka, Y. Kihata, K. Akaki, and H. Kato, Development of key monolithic circuits to Ka-band full MMIC receivers, in: *IEEE Microwave and Millimeter-Wave Monolithic Circuits Symposium Digest*, 1987, pp. 69–74.
45. J.-K.-A. Everard, Novel coplanar three-wave mixer for coherent and heterodyne detection of optical signals, *Electronics Lett.*, vol. 24, no. 14, 1988, pp. 883–885.
46. M. Muraguchi, T. Hirota, A. Minakawa, K. Ohwada, and T. Sugeta, Uniplanar MMICs and their applications, in: *IEEE 1988 Microwave and Millimeter-Wave Monolithic Circuits Symposium Digest*, 24–25 May 1988, New York, pp. 75–78.
47. T. Hirota, M. Muraguchi, A. Minakawa, and K. Osafune, A uni-planar MMIC 26-GHz-band receiver, in: *10th IEEE Gallium Arsenide Integrated Circuit Symposium Technical Digest*, 6–9 Nov. 1988, Nashville, TN, pp. 185–188.
48. K.-M. Strohm, J. Buechler, E. Kasper, J.-F. Luy, and P. Russer, Millimeter wave transmitter and receiver circuits on high resistivity silicon, in: *IEE Colloquium on "Microwave and Millimetre Wave Monolithic Integrated Circuits,"* 11 Nov. 1988, London, UK, vol. 117, 1988, pp. 11/1–11/4.
49. M. Muraguchi, T. Hirota, A. Minakawa, K. Ohwada, and T. Sugeta, Uniplanar MMIC's and their application, *IEEE Trans. Microwave Theory Tech.*, vol. 36, no. 12, Dec. 1988, 1896–1901.
50. H.-C. Chien and J. Frey, Monolithically integrable active microwave inductor, *Electronics Lett.*, vol. 24, no. 14, July 1988, pp. 905–906.
51. S. Hara, T. Tokumitsu, T. Tanaka, and M. Aikawa, Broad band monolithic microwave active inductor and application to a miniaturized wide band

amplifiers, *IEEE MTT-S International Microwave Symposium Digest*, 1988, pp. 117–120.

52. S. Hara, T. Tokumitsu, T. Tanaka, and M. Aikawa, Broad-band monolithic microwave active inductor and its application to miniaturized wide-band amplifiers, *IEEE Trans. Microwave Theory Tech.*, vol. 36, Dec. 1988, pp. 1920–1924.
53. I. Bahl and P. Bhartia, *Microwave Solid State Circuit Design*, New York: John Wiley & Sons, 1988.
54. S. A. Maas, *Nonlinear Microwave Circuits*, Norwood, MA: Artech House, 1988.
55. G.-S. Dow, T.-N. Ton, and K. Nakano, Q-band coplanar waveguide amplifier, in: *IEEE MTT-S International Microwave Symposium Digest*, 13–15 June 1989, Long Beach, CA, vol. 2, 1989, pp. 809–812.
56. M. J. Schindler and Y. Tajima, A novel MMIC active filter with lumped and transversal elements, *IEEE Trans. Microwave Theory Tech.*, vol. 37, Dec. 1989, pp. 2148–2153.
57. C. Y. Chang and T. Itoh, Narrowband planar microwave active filter, *Electronics Lett.*, vol. 25, no. 18, Aug. 1989, pp. 1228–1229.
58. L.-J. Ecuyer, W.-J. Hoefer, and G.-B. Gajda, Coplanar bias stabilizes gain in finline amps, *Microwaves & RF*, vol. 28, no. 4, 1989, pp. 73–74, 76, 78–79.
59. S. Hara, T. Tokumitsu, T. Tanaka, and M. Aikawa, Correction to "Broad-band monolithic microwave active inductor and its application to a miniturized wide-band amplifiers," *IEEE Trans. Microwave Theory Tech.*, vol. MTT-37, July 1989, p. 1155.
60. S. Hara, T. Tokumitsu, and M. Aikawa, Lossless, broadband monolithic microwave active inductor, in: *IEEE MTT-S International Microwave Symposium Digest*, 1989, pp. 955–958.
61. S. Hara, T. Tokumitsu, and M. Aikawa, Lossless, broad-band monolithic microwave active inductor, *IEEE Trans. Microwave Theory Tech.*, vol. 37, Dec. 1989, pp. 1979–1984.
62. L. E. Larson, C.-S. Chou, and M. J. Delaney, An ultrahigh-speed GaAs MESFET operational amplifier, *IEEE J. Solid-State Circuits*, vol. 24, no. 6, June 1989, pp. 1523–1528.
63. E. M. Bastida, G. P. Donzelli, and L. Scopelliti, GaAs monolithic microwave integrated circuits using broadband tuneable active inductors, in: *Proceedings, 19th European Microwave Conference*, London, 1989, pp. 1282–1287.
64. J. Browne, Coplanar limiters boast 90-dB gain from 2 to 6 GHz, *Microwaves & RF*, vol. 28, 1989, no. 5, pp. 234–235.
65. S. E. Sussman-Fort, Design concepts for microwave GaAs FET active filters, *IEEE Trans. Microwave Theory Tech.*, MTT-37, Jan. 1989, pp. 1418–1424.
66. D. Cathana, A new coplanar waveguide/slotline double-balanced mixer, in: *IEEE 1989 MTT-S International Microwave Symposium Digest*, 13–15 June 1989, Long Beach, CA, vol. 3, 1989, pp. 967–968.
67. S.-A. Kravchuk and Y.-I. Yakimenko, Hybrid integrated circuits of balance diode mixers for the UHF band, *Radioelectronics Commun. Syst.*, vol. 32, no. 10, 1989, pp. 81–83.
68. M. Riaziat, E. Par, G. Zdasiuk, S. Voly, and M. Glenn, Monolithic millimeter wave CPW circuits, in: *IEEE 1989 MTT-S International Microwave Symposium Digest*, 13–15 June 1989, Long Beach, CA, vol. 2, 1989, pp. 525–528.

69. H.-Y. Lee and T. Itoh, GaAs traveling-wave optical modulator using a modulated coplanar strip electrode with periodic cross-tie overlay, *Int. J. Infrared Millimeter Waves*, vol. 10, no. 3, 1989, pp. 321–335.
70. T. Hirota and H. Ogawa, Uniplanar monolithic frequency doublers, *IEEE Trans. Microwave Theory Tech.*, vol. 37, no. 8, August 1989, pp. 1249–1254.
71. S. S. Bharj, S. P. Tan, and B. Thomson, A 2–18 GHz 180 degree phase splitter network, in: *IEEE MTT-S International Microwave Symposium Digest*, 1989, pp. 959–962.
72. F. Ishitsuka, M. Hosoya, H. Tomimuro, N. Takachio, and K. Onodera, 0.5–23 GHz MMIC amplifier modules, in: *12th Annual GaAs IC Symposium. Technical Digest*, 7–10 Oct. 1990, New Orleans, LA, pp. 109–112.
73. R. Majidi-Ahy, C. Nishimoto, M. Riaziat, Y.-C. Pao, S.-L. Weng, S. Silverman, M. Glenn, G. Zdasiuk, S. Voly, and Z. Tan, 100 GHz high-gain InP MMIC cascode amplifier, in: *12th Annual GaAs IC Symposium Technical Digest*, 7–10 Oct. 1990, New Orleans, LA, pp. 173–176.
74. R. Majidi-Ahy, M. Riaziat, C. Nishimoto, M. Glenn, S. Silverman, S. Weng, and Y.-C. Pao, 5–100 GHz InP CPW MMIC 7-section distributed amplifier, in: *IEEE 1990 Microwave and Millimeter-Wave Monolithic Circuits Symposium Digest of Papers*, 7–8 May 1990, Dallas, TX, 1990, pp. 31–34.
75. C. Y. Chang and T. Itoh, Microwave active filters based on coupled negative resistance method, *IEEE Trans. Microwave Theory Tech.*, MTT-38, Dec. 1990, pp. 1879–1884.
76. R. Majidi-Ahy, M. Riaziat, C. Nishimoto, M. Glenn, S. Silverman, S. Weng, and Y.-C. Pao, 94 GHz InP MMIC five-section distributed amplifier, *Electronics Lett.*, vol. 26, no. 2, 1990, pp. 91–92.
77. W. Bischof, W. Ehrlinger, M. Berroth, and W. Reinert, Modelling and realization of a monolithic 27 GHz HEMT amplifier in coplanar waveguide technology, in: *Proceedings 20th European Microwave Conference*, Budapest, H, September 10–13, 1990, vol. 2, 1990, pp. 943–948.
78. H.-O. Vickes, Comments on "A novel MMIC active filter with lumped and transversal elements," *IEEE Trans. Microwave Theory Tech.*, MTT-38, Oct. 1990, pp. 1530–1532.
79. J. Browne, Coplanar circuits arm limiting amp with 100-dB gain, *Microwaves & RF*, vol. 29, no. 4, 1990, pp. 213–214, 216, 218, 220.
80. A.-R. Kerr, S. Pan, S. Whiteley, M. Radparvar, and S. Faris, A fully integrated SIS mixer for 75–110 GHz, in: *1990 IEEE MTT-S Internat. Microwave Symposium Digest*, 8–10 May 1990, Dallas, TX, vol. 2, 1990, pp. 851–854.
81. S. E. Sussman-Fort, A realization of a GaAs FET microwave active filter, *IEEE Trans. Microwave Theory Tech.*, MTT-38, Oct. 1990, pp. 1524–1526.
82. T.-H. Chen, T.-N. Ton, G.-S. Dow, K. Nakano, L.-C-T. Liu, and J. Berenz, A Q-band monolithic balanced diode mixer using AlGaAs/GaAs HEMT and CPW hybrid, in: *1990 IEEE MTT-S International Microwave Symposium Digest*, 8–10 May 1990, Dallas, TX, vol. 2, 1990, pp. 895–898.
83. T.-H. Chen, T.-N. Ton, G.-S. Dow, K. Nakano, L.-C.-T. Liu, and J. Berenz, A Q-band monolithic balanced resistive HEMT mixer using CPW/slot-line balun, in: 12th Annual GaAs IC Symposium. Technical Digest 1990, 7–10 Oct. 1990, New Orleans, LA, pp. 177–180.

84. A.-R. Kerr and S.-K. Pan, Some recent developments in the design of SIS mixers, *Int. J. Infrared and Millimeter Waves*, vol. 11, no. 10, 1990, pp. 1169–1187.

85. R.-A. Marsland, C.-J. Madden, Der-Weide-D-W. Van, M.-S. Shakouri, and D.-M. Bloom, Monolithic integrated circuits for mm-wave instrumentation, in: *12th Annual GaAs IC Symposium Technical Digest 1990*, 7–10 Oct. 1990, New Orleans LA, pp. 19–22.

86. C.-H. Bulmer, W.-K. Burns, and C.-W. Pickett, Linear, 1*2 directional coupler modulator with 10 GHz bandwidth, in: *Conference Digest. LEOS Summer Topical on Broadband Analog Optoelectronics—Devices and Systems*, 23–25 July 1990, Monterey, CA, pp. 31–32.

87. M.-A.-R. Violas and D.-J. Heatley, Microstrip and coplanar design considerations for microwave bandwidth PIN-HEMT optical receivers, in: *IEE Colloquium on "Microwave Optoelectronics,"* 26 Oct. 1990, London, vol. 139, 1990, pp. 8/1–8/8.

88. T. Tokumitsu, S. Hara, T. Takenaka, and M. Aikawa, Divider and combiner line-unified FET's as basic circuit function modules—part I, *IEEE Trans. Microwave Theory Tech.*, vol. 38, no. 9, Sept. 1990, 1210–1217.

89. T. Tokumitsu, S. Hara, T. Takenaka, and M. Aikawa, Divider and combiner line-unified FET's as basic circuit function modules—part II, *IEEE Trans. Microwave Theory Tech.*, vol. 38, no. 9, Sept. 1990, 1218–1226.

90. R. Majidi-Ahy et al., 5–100 GHz InP coplanar waveguide MMIC distributed amplifier, *IEEE Trans. Microwave Theory and Tech.*, vol. 38, no. 12, 1990, pp. 1986–1993.

91. T. Hiraoka, T. Tokumitso, and M. Akaike, A miniaturized broad-band MMIC frequency doubler, *IEEE Trans. Microwave Theory Tech.*, vol. 38, no. 12, December 1990, pp. 1932–1937.

92. T. Tsukii and M. J. Schindler, 2–20 GHz MMIC frequency doubler, in: *The 3rd Asia-Pacific Microwave Conference Proceedings*, Tokyo, 1990, pp. 857–860.

93. A. A. Sweet, *MIC & MMIC Amplifier and Oscillator Circuit Design*, Boston: Artech House, 1990.

94. B. Roth and A. Beyer, An advanced CAD-procedure for the design and simulation of autonomous systems applicated to an integrated microwave oscillator, in: *Proceedings, 20th European Microwave Conference*, Budapest, vol. 1, 1990, pp. 45–56.

95. Y. Deville, High-frequency and high-*Q* tunable active filters, *IEEE Trans. Circuits Systems*, vol. 36, no. 8, Aug. 1989, pp. 1128–1133.

96. Y. Deville, A high-selectivity continuous-time GaAs balanced filter, *IEEE J. Solid-State Circuits*, vol. 25, no. 3, June 1990, pp. 889–892.

97. S. E. Sussmann-Fort, Inductance simulation for microwave active filter design, in: *IEEE MTT-S Internat. Microwave Symposium Digest*, Boston, June 1991, Workshop D, UHF and Active Filter Technology.

98. R. Majidi-Ahy, C. Nishimoto, M. Riaziat, M. Glenn, S. Silverman, S.-L. Weng, Y.-C. Pao, G. Zdasiuk, S. Voly, and Z. Tan, 100-GHz high-gain InP MMIC cascode amplifier, *IEEE J. Solid-State Circuits*, vol. 26, no. 10, 1991, pp. 1370–1378.

99. J. Eisenberg, J. Panelli, and W. Ou, A new planar double-double balanced MMIC mixer structure, in: *1991 IEEE MTT-S International Microwave Symposium Digest*, 10–14 June 1991, Boston, vol. 1, 1991, pp. 81–84.

100. Y. Yamamoto and T. Itoh, Voltage control of planar active band-pass filters, *IEEE MTT-S International Microwave Symposium Digest*, Boston, MA, Workshop D, UHF and Active Filter Technology, June 1991.

101. Y. Yamamoto, J. Lin, and T. Itoh, Laser tuning of a planar active band-pass filter using MESFETs, in: *Proceedings, 21th European Microwave Conference*, Stuttgart, Sept. 1991, pp. 1317–1322.

102. J. Eisenberg, I. Panelli, and W. Ou, A new planar double-double balanced MMIC mixer structure, in: *IEEE 1991 Microwave and Millimeter-Wave Monolithic Circuits Symposium Digest of Papers*, 10–11 June 1991, Boston, pp. 69–72.

103. T.-H. Chen, T.-N. Ton, G.-S. Dow, K. Nakano, L.-C.-T. Liu, and J. Berenz, A *Q*-band monolithic balance resistive HEMT mixer using CPW/slotline balun, *IEEE J. Solid-State Circuits*, vol. 26, no. 10, 1991, pp. 1389–1394.

104. M. J. Schindler, MMIC active filters using lumped and transversal elements, in: *1991 IEEE MTT-S International Microwave Symposium Digest*, Boston, MA, Workshop D, UHF and Active Filter Technology, June 1991.

105. T.-H. Chen, K.-W. Chang, H. Wang, G.-S. Dow, L.-C.-T. Liu, S.-B.-T. Bui, and T.-S. Lin, Broadband monolithic passive baluns and monolithic double-balanced mixer, in: *1991 IEEE MTT-S International Microwave Symposium Digest*, 10–14 June 1991, Boston, MA, vol. 2, 1991, pp. 861–864.

106. A. Nesic, Integrated uniplanar antenna mixer and local oscillator, *Electronics Lett.*, vol. 27, no. 1, 1991, pp. 56–58.

107. A. Bangert and M. Ludwig, A direct optical injection locked 8 GHz MMIC oscillator, in: *1991 IEEE MTT-S International Microwave Symposium Digest*, 10–14 June 1991, Boston, vol. 2, 1991, pp. 499–502.

108. P. Ho et al., Extremely high gain 0.15 μm gate-length InAlAs/InGaAs/InP HEMTs, *Electronics Lett.*, vol. 27, no. 4, 1991, pp. 325–327.

109. J.-A. Navarro, Y.-H. Shu, and K. Chang, A novel varactor tunable coplanar waveguide-slotline Gunn VCO, in: 1991 *IEEE MTT-S International Microwave Symposium Digest*, 10–14 June 1991, Boston, MA, vol. 3, 1991, pp. 1187–1190.

110. S. Koßlowski, R. Bertenburg, N.-H.-L. Koster, I. Wolff, F.-J. Tegude, P. Narozny, J. Wenger, and H. Dämbkes, Advances in CPW-design applied to monolithic integrated Ka-band MESFET and HEMT-amplifiers on GaAs, in: *14th Annual IEEE GaAs IC Symposium, Technical Digest*, 1992, Miami Beach, FL, Oct. 4–7, 1992, pp. 119–122.

111. M. Schlechtweg, W. Reinert, P.-J. Tasker, R. Bosch, J. Braunstein, A. Hulsmann, and K. Kohler, Design and characterization of high performance 60 GHz pseudomorphic MODFET LNAs in CPW-technology based on accurate S-parameter and noise models, *IEEE Trans. Microwave Theory Tech.*, vol. 40, no. 12, 1992, pp. 2445–2451.

112. T. Takenaka and H. Ogawa, Miniaturized MMIC mixers; image rejection and balanced mixers using multilayer microstrip lines and line-unified HEMT modules, *IEICE Trans. Electronics*, Japan, vol. E75-C, no. 6, 1992, pp. 689–697.

113. R. Klieber, R. Ramisch, A.-A. Valenzuela, R. Weigel, and P. Russer, A coplanar transmission line high-T (ind c), superconductive oscillator at 6.5 GHz on a single substrate, *IEEE Microwave and Guided Wave Lett.*, vol. 2, no. 1, 1992, pp. 22–24.

114. P.-A. Stimson, R.-J. Dengler, P.-H. Siegel, and H.-G. LeDuc, A planar quasi-optical SIS receiver for array applications, in: *Proceedings of the 3rd International Symposium on Space Terahertz Technology*, Ann Arbor, MI, March 24–26, 1992, pp. 235–242.

115. J. Dickmann et al., Influence of surface layers on the RF-performance of AlInAs/InGaAs HFETs, *IEEE Microwave and Guided Wave Lett.*, vol. 2, no. 12, 1992, pp. 472–474.

116. T. Suzaki, M. Soda, T. Morikawa, H. Tezuka, C. Ogawa, S. Fujita, H. Takemura, and T. Tashiro, Si bipolar chip set for 10-Gb/s optical receiver," *IEEE J. Solid-State Circuits*, vol. 27, no. 12, 1992, pp. 1781–1786.

117. J. Browne, Broadband amp drops through noise floor, *Microwaves & RF*, vol. 31, no. 2, Feb. 1992, pp. 141–144.

118. M. Abdo Tuko and I. Wolff, Novel 36 GHz GaAs frequency doublers using (M)MIC coplanar technology, in: *IEEE MTT-S International Microwave Symp. Digest*, 1992, pp. 1167–1170.

119. H. Wang et al., A D-band monolithic low noise amplifier, in: *Proceedings of the 14th GaAs IC Symposium*, Oct. 1992, pp. 23–26.

120. S. Kosslowski, et al., Advances in CPW-Design Applied to Monolithic Integrated Ka-band Amplifiers on GaAs, in: *Proceedings of the 14th. GaAs IC Symposium*, Oct. 1992, pp. 119–122.

121. H. Wang et al., A D-band monolithic low noise amplifier, *IEEE Trans. Microwave Theory and Tech.*, vol. 40, 1992, pp. 417–428.

122. M. Abdo Tuko and I. Wolff, Ka-band FET frequency doublers in CPW-techniques, in: *IEEE MTT/AP-German Chapter Workshop on Integrated Nonlinear Microwave Circuits Digest*, 1992, University of Duisburg, pp. 201–215.

123. M. R. Moazzam and A. H. Aghvami, Design concepts of a novel microwave active filter, *Int. J. Microwave Millimeter-Wave Computer-Aided Eng.*, vol. 2, no. 1, Jan. 1992, pp. 28–33.

124. R. M. Bertenburg, *Technologie koplanarer monolithisch integrierter Mikrowellenschaltungen auf Galliumarsenid unter Verwendung von Submikron-Feldeffekttransistoren*, Doctoral Thesis, Duisburg University, Duisburg, Germany, 1992.

125. M. Abdo Tuko and I. Wolff, Ka-band FET frequency doublers in CPW-techniques, in: *IEEE MTT/AP-German Chapter Workshop on Integrated Nonlinear Microwave Circuits Digest*, 1992, University of Duisburg, pp. 201–215.

126. D.-J. Newson, P. Birdsall, P.-H. Loosemore, and I.-D. Henning, 39 GHz monolithic coplanar waveguide amplifier using doped channel HFET technology on InP, *Electronics Lett.*, vol. 29, no. 13, 1993, pp. 1177–1178.

127. K. Minot, B. Nelson, and W. Jones, A low noise phase linear distributed coplanar waveguide amplifier, *IEEE Trans. Microwave Theory and Tech.*, vol. 41, no. 9, 1993, pp. 1650–1653.

128. M. Schlechtweg, P.-J. Tasker, W. Reinert, J. Braunstein, W. Haydl, A. Hülsmann, and K. Köhler, High gain 70–80 GHz MMIC amplifiers in coplanar waveguide technology, *Electronics Lett.*, vol. 29, no. 12, 1993, pp. 1119–1120.

129. R. Heilig, D. Hollmann, and G. Baumann, A monolithic 1–55 GHz HEMT distributed amplifier in coplanar waveguide technology, in: *Proceedings, 23rd European Microwave Conference*, Madrid, E, Sept. 6–9, 1993, vol. 23, 1993, pp. 364–367.

130. M. Schlechtweg, P.-J. Tasker, W. Reinert, J. Braunstein, W.-H. Haydl, and Lohrmann, High gain CPW MMIC amplifiers and waveguide modules for V-band applications, in: *Proceedings, 23rd European Microwave Conference*, Madrid, E, Sept. 6–9, 1993, vol. 23, 1993, pp. 144–146.

131. J. Braunstein, M. Schlechtweg, P.-J. Tasker, W. Reinert, A. Hülsmann, and K. Köhler, High performance narrow and wide bandwidth amplifiers in CPW-technology up to W-band, in: *IEEE Gallium Arsenide Integrated Circuit Symposium Digest., Tech. Digest* 1993, San Jose, CA, Oct 10–13, 1993, vol. 15, 1993, pp. 277–280.

132. J. Dickmann, S. Koßlowski, B.-E. Maile, H. Haspeklo, S. Heuthe, A. Geyer, and M. Riepe, High gain 28 GHz coplanar waveguide monolithic amplifier on InP substrate, *Electronics Lett.*, vol. 29, no. 5, 1993, pp. 493–495.

133. S. Koßlowski, R. Bertenburg, N. H. L. Koster, I. Wolff, F.-J. Tegude, P. Narozny, J. Wenger, and H. Dämbkes, Advances in CPW-design applied to monolithic integrated Ka-band MESFET and HEMT-amplifiers on GaAs, in: *IEEE GaAs Integrated Circuits Symposium Digest*, Sept. 1992, pp. 119–122.

134. S. Kosslowski, J. Dickmann, P. Narozny, A. Geyer, B.-E. Maile, A. Schurr, H. Haspeklo, C. Wölk, H. Dämbkes, and I. Wolff, InP based high gain Ka-band three stage coplanar waveguide monolithic amplifier, in: *Proceedings, 23rd European Microwave Conf.*, Madrid, E, Sept. 6–9, 1993, vol. 23, 1993, pp. 147–150.

135. M. Abdo Tuko, M. Naghed, and I. Wolff, Novel 18/36 GHz (M)MIC GaAs FET frequency doublers in CPW-techniques under the consideration of coplanar discontinuity effects, *IEEE Trans. Microwave Theory Tech.*, vol. 41, no. 8, August 1993, pp. 1307–1315.

136. D. Leistner, Low noise amplifier at L- and Ku-band for space applications in coplanar technology, in: *Proceedings, 23rd European Microwave Conference*, Madrid, E, Sept. 6–9, 1993, vol. 23, 1993, pp. 823–827.

137. G. David, S. Redlich, W. Mertin, R.-M. Bertenburg, S. Koßlowski, F.-J. Tegude, E. Kubalek, and D. Jäger, Two-dimensional direct electro-optic field mapping in a monolithic integrated GaAs amplifier, in: *Proceedings, 23rd European Microwave Conference*, Madrid, E, Sept. 6–9, 1993, vol. 23, 1993, pp. 497–499.

138. J. Braunstein, P.-J. Tasker, A. Hülsmann, M. Schlechtweg, W. Reinert, and K. Köhler, Very broad band TWAs to 80 GHz on GaAs substrate, in: *Proceedings 23rd European Microwave Conference*, Madrid, E, Sept. 6–9, 1993, vol. 23, 1993, pp. 372–373.

139. S.-A. Maas and K.-W. Chang, A broadband planar, doubly balanced monolithic Ka-band diode mixer, *IEEE Trans. Microwave Theory Tech.*, vol. 41, no. 12, 1993, pp. 2330–2335.

140. W. Gross, Balanced resistive MESFET mixer in uniplanar technology, in: *Proceedings, 23rd European Microwave Conference*, Madrid, E, Sept. 6–9, 1993, vol. 23, 1993, pp. 405–406.

141. T. Becks, B. Hopf, D. Köther, R. Kulke, P. Pogatzki, M. Rittweger, and T. Sporkmann, Activities on coplanar waveguide structures and circuits at the IMST, in: *Analysis, Design and Applications of Coplanar Waveguides*, London, GB, Oct. 19, 1993, IEE Colloquium, vol. 186, 1993, pp. 5/1–5/10.

142. M. Schlechtweg, P.-J. Tasker, and W. Haydl, Design and characterisation of MMICs in coplanar waveguide technology up to 80 GHz, in: *Analysis, Design and Appli-*

cations of Coplanar Waveguides, London, GB, Oct. 19, 1993, IEE Colloquium, vol. 186, 1993, pp. 9/1–9/7.

143. V. Radisic, V. Jevremovic, and Z.-B. Popovic, CPW oscillator configuration for an electro-optic modulator, *IEEE Trans. Microwave Theory Tech.*, vol. 41, no. 9, 1993, pp. 1645–1647.
144. M.-J. Vaughan and R.-C. Compton, Resonant-tee CPW oscillator and the application of the design to a monolithic array of MESFETs, *Electronics Lett.*, vol. 29, no. 16, 1993, pp. 1477–1479.
145. C.-C. Ling and G.-M. Rebeiz. A 94-GHz planar monopulse receiver, *IEEE Microwave and Guided Wave Lett.*, vol. 3, no. 10, 1993, pp. 358–360.
146. G. Gauthier, T.-P. Budka, Ahmad.-W-Y. Ali, D.-F. Filipovic, and G.-M. Rebeiz, A low-noise 86–90 GHz uniplanar Schottky-receiver, in: *1993 IEEE MTT-S International Microwave Symposium Digest*, Atlanta, GA, June 14–18, 1993, vol. 1, pp. 325–328.
147. E. Bayar and J.-C-R. Simpson, S-band high isolation switch for space and satellite applications, in: *Recent Advantages in Microwave Sub-Systems for Space and Satellite Applications*, London, GB, 18 March 1993, IEE Colloquium, vol. 62, 1993, pp. 3.1–3.4.
148. I. Wolff, Design rules and realization of coplanar circuits for communication applications, *Proceedings of 23rd European Microwave Conference*, Sept. 1993, pp. 36–41.
149. M. Abdo Tuko, M. Naghed, and I. Wolff, Novel 18/36 GHz (M)MIC GaAs FET frequency doublers in CPW-techniques under the consideration of coplanar discontinuity effects, *IEEE Trans. Microwave Theory Tech.*, vol. 41, no. 8, Aug. 1993, pp. 1307–1315.
150. B. Roth, *Methoden zur Simulation, Realisierung und Messung integrierter Mikrowellenoszillatoren*, Doctoral Thesis, Duisburg University, Duisburg, Germany, 1993.
151. M.-H. Murgulescu, A. Boulouard, M. LeRouzic, P. Legaud, E. Penard, and I. Zaquine, 38 GHz, coplanar waveguide GaAs MMIC amplifier, *Electronics Lett.*, vol. 30, no. 21, 1994, pp. 1768–1770.
152. M. Berroth, Analog, digital and mixed-signal integrated circuits for high speed applications, in: *Gallium Arsenide and Related Compounds 1993, Proceedings of the 12th International Symposium*, Freiburg, D, Aug. 29–Sept. 2, 1993, Institute of Physics Conference Series, vol. 136, 1994, pp. 1–8.
153. M. Abdo Tuko, *Theoretical and experimental investigation of the design and realization of microwave frequency multipliers in coplanar (M)MIC techniques*, Doctoral Thesis, Duisburg University, Duisburg, Germany, 1994.
154. P. Pogatzki, D. Köther, R. Kulke, B. Hopf, T. Sporkmann, and I. Wolff, Coplanar hybrids based on an enhanced inductor model for mixer applications up to mm-wave frequencies, in: *Proceedings, 24th European Microwave Conference*, Cannes, F, Sept. 1994, pp. 258–262.
155. A. Hülsmann, W. Bronner, P. Hofmann, K. Köhler, B. Raynor, J. Schneider, J. Braunstein, et al., Fabrication of high speed MMICs and digital ICs using T-Gate technology on pseudomorphic-HEMT structures, in: *Gallium Arsenide and Related Compounds 1993, Proceedings of the 12th International Symposium*, Freiburg, D, Aug. 29–Sept. 2, 1993, Institute of Physics Conference Series, vol. 136, 1994, pp. 467–472.

156. D.-W. Tu, W.-P. Berk, S.-E. Brown, N.-E. Byer, S.-W. Duncan, A. Eskandarian, E. Fischer, D.-M. Gill, et al., High gain monolithic *p*-HEMT W-band four-stage low noise amplifiers, in: MMWMC, in: *1994 IEEE Microwave and Millimeter-Wave Monolithic Circuits Symposium Digest of Papers*, San Diego, CA, May 23–25, 1994, pp. 29–32.

157. D.-W. Tu, S.-W. Duncan, A. Eskandarian, B. Golja, B.-C. Kane, S.-P. Svensson, S. Weinreb, and N.-E. Byer, High gain monolithic W-band low noise amplifiers based on pseudomorphic high electron mobility transistors, *IEEE Trans. Microwave Theory Tech.*, vol. 42, no. 12, part 2, 1994, pp. 2590–2597.

158. B. Hopf, I. Wolff, and M. Guglielmi, Coplanar MMIC active bandpass filters using negative resistance circuits, *IEEE Trans. Microwave Theory Tech.*, vol. 42, Dec. 1994, pp. 2598–2602.

159. S. E. Sussman-Fort, An NIC-based negative resistance circuit for microwave active filters, *Int. J. Microwave Millimeter-Wave Computer-Aided Eng.*, vol. 4, no. 2, April 1994, pp. 130–139.

160. R. Heilig, D. Hollmann, and G. Baumann, A monolithic 2–52 GHz HEMT matrix distributed amplifier in coplanar waveguide technology, in: *MMWMC, 1994 IEEE Microwave and Millimeter-Wave Monolithic Circuits Symposium Digest of Papers*, San Diego, CA, May 23–25, 1994, pp. 191–194.

161. R. Lang, Modelling and characterisation of a millimeterwave monolithic amplifier, in: *MMIC's (Monolithic Microwave Integrated Circuits)*, London, GB, Nov. 15, 1994, IEE Colloquium, vol. 216, pp. 12.1–12.5.

162. D. Köther, B. Hopf, T. Sporkmann, and I. Wolff, Active CPW MMIC circulator for the 40 GHz band, *Proceedings, 24th European Microwave Conference*, Cannes, 5–8 Sept. 1994, vol. 1, pp. 542–547.

163. H.-S. Tsai, M.-J-W. Rodwell, and R.-A. York, Planar amplifier array with improved bandwidth using folded-slots, *IEEE Microwave and Guided Wave Lett.*, vol. 4, no. 4, 1994, pp. 112–114.

164. Y. Kalayci, R. Tempel, B. Hopf, J.-J. Borkes, R. Gründler, and I. Wolff, Miniaturising of K-band coplanar MMIC-amplifiers by using lumped elements, in: *Proceedings, 24th European Microwave Conference*, Cannes, Sept. 5–8, 1994, vol. 24, no. 1, pp. 343–348.

165. Y. Kalayci, R. Tempel, W. Lütke, M. Akpinar, and I. Wolff, A miniaturized Ka-band MMIC high gain medium power amplifier in coplanar line technique by using lumped elements, in: *1995 IEEE MTT-S International Microwave Symposium Digest*, Orlando, FL, May 1995, pp. 1623–1626.

166. D. Köther, B. Hopf, T. Sporkmann, and I. Wolff, MMIC Wilkinson couplers for frequencies up to 110 GHz, in: *1995 IEEE MTT-S International Microwave Symposium Digest*, Orlando, FL, May 1995, pp. 663–666.

167. M. Boheim, L.-P. Schmidt, J. Ritter, V. Brankovic, R. Beyer, F. Arndt, U. Klein, K. Künzi, et al., A new 140 GHz planar diode finline mixer for radiometer applications, in: *Proceedings, 24th European Microwave Conference*, Cannes, Sept. 5–8, 1994, vol. 24, no. 1, pp. 664–669.

168. A. Bangert, M. Schlechtweg, M. Lang, W. H. Haydl, W. Bronner, T. Fink, K. Köhler, and B. Raynor, Monolithic integrated 75 GHz oscillator with high output power using a pseudomorphic HFET, in: *IEEE Microwave and Millimeter-Wave Monolithic Circuits Symposium Digest*, San Diego, CA, 1994, pp. 135–138.

169. K. Noguchi, H. Miyazawa, and O. Mitomi, 75-GHz Ti:LiNbO3 optical modulator, in: *OFC 94, Conference on Optical Fiber Communication*, San Jose, CA, Feb 20–25, 1994, Technical Digest Series, Conference Edition, vol. 94, no. 4, 1994, pp. 76–77.

170. B. Roth, D. Köther, M. Coady, and T. Sporkmann, A combined on-wafer measurement stand for linear and nonlinear microwave measurements, *Proceedings, 24th European Microwave Conference*, Cannes, F, Sept. 1994, pp. 962–967.

171. V.-M. Hietala, S.-H. Kravitz, M.-G. Armendariz, G.-A. Vawter, R.-F. Carson, and R.-E. Leibenguth, High-performance GaAs/AlGaAs optical phase modulators for microwave/photonic integrated circuits, in: *Optoelectronic Signal Processing for Phased-Array Antennas IV*, Los Angeles, CA, Jan. 26–27, 1994, Proceedings, SPIE, vol. 2155, 1994, pp. 29–36.

172. W. Ehrlinger, W. Bischof, K. Haug, A. Huelsmann, and M. Schlechtweg, Wideband monolithic modulator in GaAs/AlGaAs-technology, in: *Proceedings, 24th European Microwave Conference*, Cannes, Sept. 5–8, 1994, vol. 24, no. 1, pp. 560–566.

173. W. Lütke, M. Akpinar, and R. Tempel, and I. Wolff, Nonlinear modeling for MMIC design, in: *8th GaAs Simulation Workshop*, Duisburg, Germany, Oct. 1994.

174. G. Baumann, D. Hollmann, and R. Heilig, A 29 GHz DRO in coplanar waveguide configuration with an AlGaAs HEMT, in: *INMMC 94, Integrated Nonlinear Microwave and Millimeter-Wave Circuits Digest*, 3rd International Workshop, Duisburg, Germany, Oct. 5–7, 1994, pp. 237–242.

175. S. Lucyszyn, I.-D. Robertson, and H. Aghvami, 24 GHz serrodyne frequency translator using a 360 deg of angle analog CPW MMIC phase shifter, *IEEE Microwave and Guided Wave Lett.*, vol. 4, no. 3, 1994, pp. 71–73.

176. R. Tempel and B. Roth, Standardized concepts for microwave oscillators and their realization by new MMICs in coplanar line technique, in: *International Conference on Millimeter and Submillimeter Waves and Applications*, San Diego, CA, Jan. 10–14, 1994.

177. S. Lucyszyn, I.-D. Robertson, and A.-H. Aghvami, Monolithic analogue phase shifter and cascode FET amplifier using uniplanar techniques, in: *Proceedings 24th European Microwave Conference*, Cannes, F, Sept. 5–8, 1994, vol. 24, no. 1, pp. 554–559.

178. C.-C. Ling and G.-M. Rebeiz, A 94 GHz planar monopulse tracking receiver, *IEEE Trans. Microwave Theory Tech.*, vol. 42, no. 10, 1994, pp. 1863–1871.

179. B. Hopf, D. Köther, T. Sporkmann, and I. Wolff, Foundry cell oriented design of active coplanar power dividers and combiners for MMIC applications, in: *Proceedings 24th European Microwave Conference*, Cannes (France), 5–8 Sept. 1994, vol. 2, pp. 1489–1494.

180. B. Hopf, S. Koßlowski, D. Köther, T. Sporkmann, and I. Wolff, Active CPW circulators for the 40 GHz band, *Proceedings, 24th European Microwave Conf.*, Cannes, F, Sept. 5–8, vol. 24, no. 1, 1994, pp. 254–257.

181. B. Hopf, I. Wolff, and M. Guglielmi, Coplanar MMIC active bandpass filters using negative resistance circuits, in: *IEEE 1994 Microwave and Millimeter-Wave Monolithic Circuits Symposium*, San Diego, CA, May 1994, pp. 229–231.

182. B. Hopf, I. Wolff, and M. Guglielmi, Coplanar MMIC active bandpass filters using negative resistance circuits, *IEEE Trans. Microwave Theory Tech.*, vol. 42, Dec. 1994, pp. 2598–2602.

183. B. Hopf and I. Wolff, Narrow band MMIC active filter using negative resistance circuits in coplanar line technique, *Proceedings, 25th European Microwave Conference*, Bologna, Italy, Sept. 1995, pp. 1110–1112.

184. R. Tempel and B. Roth, Standardized concepts for microwave oscillators and their realization by new MMICs in coplanar line technique, in: *International Conference on Millimeter and Submillimeter Waves and Applications*, San Diego, CA, Jan. 10–14, 1994.

185. P. Nevermann, *Modellbildung und Simulation der temperaturabhängigen Rauscheigenschaften von GaAs-MESFET*, Doctoral Thesis, Duisburg University, Duisburg, Germany, 1994.

186. Y. Kalayci, R. Tempel, and I. Wolff, 30 GHz coplanar MMIC high-gain amplifier for medium power applications, in: *MIOP 95, Mikrowellen und Optronik, 8. Kongreßmesse für Hochfrequenztechnik Digest*, Sindelfingen, D, 30 May–1 June, 1995, pp. 298–301.

187. M. Schefer, B.-U. Klepser, H.-P. Meier, W. Patrick, U. Lott, and W. Bächtold, Comparison of coplanar matching networks for V-band amplifiers, *Electronics Lett.*, vol. 31, no. 16, 1995, pp. 1351–1353.

188. Y. Baeyens, D. Schreurs, B. Nauwelaers, M. van. Hove, W. de Raedt, and M. Rossum, Ultra-broadband distributed variable gain amplifiers using lattice matched and pseudomorphic InP cascode HEMTs, in: *MIOP 95, Mikrowellen und Optronik, 8. Kongre ßmesse für Hochfrequenztechnik Digest*, Sindelfingen, D, 30 May–1 June 1995, pp. 345–349.

189. K.-M. Strohm, J.-F. Luy, F. Schäffler, H. Jorke, H. Kibbel, C. Rheinfelder, R. Doerner, J. Gerdes, F.-J. Schmückle, and W. Heinrich, Coplanar Ka-band SiGe-MMIC amplifier, *Electronics Lett.*, vol. 31, no. 16, 1995, pp. 1353–1354.

190. M. Schlechtweg, W.-H. Haydl, J. Braunstein, P.-J. Tasker, A. Bangert, and W. Reinert, 110 GHz amplifiers based on compact coplanar W-band receiver technology, in: *1995 IEEE GaAs IC Symposium, 17th Annual IEEE Gallium Arsenide Integrated Circuit Symposium, Technical Digest*, San Diego, CA, Oct. 29–Nov. 1, 1995, pp. 214–217.

191. Y.-I. Ryu, K.-W. Kobayashi, and A.-K. Oki, A monolithic broadband doubly balanced EHF HBT star mixer with novel microstrip baluns, in: *1995 IEEE MTT-S International Microwave Symposium Digest*, Orlando, FL, May 16–20, 1995, vol. 1, pp. 119–122.

192. A. Nesic, S. Jovanovic, and I. Radnovic, Integrated uniplanar oscillator-transmitter and mixer with active antennas, in: *Proceedings, 25th EUMC 95, European Microwave Conference*, Bologna, I, Sept. 4–7, 1995, vol. 1, pp. 324–328.

193. W. Wang, D. Chen, H.-R. Fetterman, Y. Shi, W.-H. Steier, and L.-R. Dalton, 60 GHz polymer electro-optic phase modulators, in: *IEEE/LEOS 1995 Digest of the LEOS Summer Topical Meetings, RF Optoelectronics*, Keystone, CO, Aug. 7–11, 1995, pp. 4–6.

194. M. Yu, A. Gopinath, Seventeen-GHz directional coupler optical modulator, *IEEE Trans. Microwave Theory Tech.*, vol. 43, no. 9, part II, 1995, pp. 2198–2202.

195. D. Jäger, R. Kremer, and A. Stöhr, High-speed travelling-wave photodetectors, modulators and switches, in: *IEEE/LEOS 1995 Digest of the LEOS Summer Topical Meetings, RF Optoelectronics*, Keystone, CO, Aug. 7–11, 1995, pp. 57–58.

196. K.-W. Kobayashi, L.-T. Tran, A.-K. Oki, T. Block, and D.-C. Streit, A coplanar waveguide InAlAs/InGaAs HBT monolithic Ku-band VCO, *IEEE Microwave and Guided Wave Lett.*, vol. 5, no. 9, 1995, pp. 311–312.

197. J. Lin, Y.-K. Chen, D.-A. Humphrey, R.-A. Hamm, R.-J. Malik, A. Tate, R.-F. Kopf, and R.-W. Ryan, Ka-band monolithic InGaAs/InP HBT VCO's in CPW structure, *IEEE Microwave and Guided Wave Lett.*, vol. 5, no. 11, 1995, pp. 379–381.

198. U. Güttich, Microstrip DROs: MESFET, PMHEMT, and HBT compared as active oscillator devices, in: *MIOP 95, Mikrowellen und Optronik, 8. Kongreßmesse für Hochfrequenztechnik Digest*, Sindelfingen, D, 30 May–1 June 1995, pp. 341–344.

199. M.-J. Vaughan and R.-C. Compton, 28 GHz oscillator for endfire quasioptical power combining arrays, *Electronics Lett.*, vol. 31, no. 17, 1995, pp. 1453–1455.

200. T. Felgentreff, P. Selmeier, and G. Olbrich, Rauschverhalten eines 15-GHz-AlGaAs/GaAs-HEMT-Oszillators, in: *MIOP 95, Mikrowellen und Optronik, 8. Kongreßmesse für Hochfrequenztechnik Digest*, Sindelfingen, D, 30 May–1 June 1995, pp. 306–312.

201. S. Mollenkopf, L.-P.-B. Katehi, and G.-M. Rebeiz, A low-cost 20–22 GHz MIC active receiver/radiometer, *IEEE Trans. Microwave Theory Tech.*, vol. 43, no. 4, Part II, 1995, pp. 989–993.

202. S.-M. Mollenkopf, S. Raman, L.-P.-B. Katehi, and G.-M. Rebeiz, Low-cost receivers for millimeter-wave imaging array applications, in: *Smart Structures and Materials 1995, Smart Electronics*, San Diego, CA, March 2–3, 1995, Proceedings, SPIE, vol. 2448, pp. 175–180.

203. G.-P. Gauthier, Ahmad.-W.-Y. Ali, T.-P. Budka, D.-F. Filippovic, and G.-M. Rebeiz, A uniplanar 90-GHz Schottky-diode millimeter-wave receiver, *IEEE Trans. Microwave Theory Tech.*, vol. 43, no. 7, part II, 1995, pp. 1669–1672.

204. K.-W. Kobayashi, Tran Liem, A.-K. Oki, and D.-C. Streit, A 50 MHz–30 GHz broadband co-planar waveguide SPDT PIN diode switch with 45-dB isolation, *IEEE Microwave and Guided Wave Lett.*, vol. 5, no. 2, 1995, pp. 56–58.

205. S.-E. Saddow and C.-H. Lee, Optical control of microwave-integrated circuits using high-speed GaAs and Si photoconductive switches, *IEEE Trans. Microwave Theory Tech.*, vol. 43, no. 9, part II, 1995, pp. 2414–2420.

206. S. Chouteau, J. Boussey, B. Cabon, and A. Iliadis, Optoelectronic microswitch on SOI based structure, in: *Proceedings, 1995 IEEE International SOI Conference*, Tucson, AZ, Oct. 3–5, 1995, pp. 40–41.

207. R. Tempel, Transistor switches in coplanar line technique suited for MMIC applications, in: *MIOP 95, Mikrowellen und Optronik, 8. Kongreßmesse für Hochfrequenztechnik Digest*, Sindelfingen, D, 30 May–1 June, 1995, pp. 403–407.

208. D. Köther, B. Hopf, T. Sporkmann, I. Wolff, and S. Koßlowski, New types of MMIC circulators, in: *1995 IEEE MTT-S International Microwave Symposium Digest*, Orlando, FL, 16–20 May 1995, vol. 2, pp. 877–880.

209. W. H. Haydl, A. Tessmann, K. Zuefle, H. Massler, T. Krems, L. Verweyen, and J. Schneider, Models of coplanar lines and elements over the frequency range 0–120 GHz, in: *European Microwave Conference Digest*, Prague, 1996, pp. 996–1000.

210. T. Krems, W. Haydl, H. Massler, and J. Rüdiger, Millimeter-wave performance of chip interconnection using wire bonding and flip chip, in: *IEEEMTT-S International Microwave Symposium Digest*, San Francisco, 1996, pp. 247–250.

211. K. Maruhashi, M. Madihian, L. Desclos, K. Ouda, and M. Kuzuhara, A K-band monolithic oscillator integrated with a buffer amplifier using a device-circuit interaction design concept, *IEEE Trans. Microwave Theory Tech.*, vol. 44, 1996, no. 8, pp. 1424–1428.

212. R. Tempel, W. Lütke, J. Herrmann, and I. Wolff, MMICs for sensor applications, in: *MMWMC-Symposium Digest*, June 1996, pp. 209–212.

213. K.-W. Kobayashi, L.-T. Tran, J.-C. Cowles, T.-R. Block, A.-K. Oki, and D.-C. Streit, Extending the bandwidth performance of heterojunction bipolar transistor-based distributed amplifiers, *IEEE Trans. Microwave Theory Tech.*, vol. 44, no. 5, 1996, pp. 739–748.

214. M. Nakamura, Y. Imai, S. Yamahata, and Y. Umeda, Over-30-GHz limiting amplifier IC's with small phase deviation for optical communication systems, *IEEE J. Solid-State Circuits*, vol. 31, no. 8, 1996, pp. 1091–1099.

215. K.-W. Kobayashi, L.-T. Tran, M.-D. Lammert, A.-K. Oki, and D.-C. Streit, Transimpedance bandwidth performance of an HBT loss-compensated coplanar waveguide distributed amplifier, *Electronics Lett.*, vol. 32, no. 24, 1996, pp. 2287–2288.

216. W. Ehrlinger, U. Banhahrdt, W. Bischof, T. Diercks, M. Glauner, K. Haug, R. Keller, R. Lohrmann, and M. Ottka, 60 GHz Bausteine für MMIC-Modulatoren in koplanarer Wellenleitertechnik, in: *Report: Forsch.-Ber. Bundesministerium für Bildung, Wissenschaft, Forsch. und Technol.*, Germany, 1996, Report-Nr. 01BM104/4, pp. 1–19.

217. M. Schefer, U. Lott, H. Benedickter, B.-U. Klepser, W. Patrick, and W. Bächtold, Monolithic coplanar, varactor tunable V-band HEMT oscillator with injection locking capability, *Electronics Lett.*, vol. 32, no. 20, 1996, pp. 1899–1900.

218. M. Schlechtweg, W. H. Haydl, A. Bangert, J. Braunstein, P. J. Tasker, L. Verweyen, H. Massler, W. Bronner, A. Hülsmann, and K. Koehler, Coplanar millimeter-wave ICs for W-band applications using 0.15 μm pseudomorphic MODFETs, in: *IEEE J. Solid-State Circuits*, vol. 31, no. 10, 1996, pp. 1426–1434.

219. M. Berg, T. Hackbarth, B. E. Maile, J. Dickmann, D. Köther, and B. Hopf, Active circulator MMIC in CPW technique using quarter micron gate-length InAlAs/InGaAs/InP HFETs, in: *Proc. 8th Internat. Conference on Indium Phosphide and Related Materials*, Schwäbisch Gmünd, Germany, April 1996.

220. C. Rheinfelder, K. Strohm, F. Beißwanger, J. Gerdes, F.-J. Schmückle, J.-F. Luy, and W. Heinrich, 26 GHz coplanar SiGe MMICs, in: *1996 IEEE MTT-S International Microwave Symposium Digest*, vol. 1, San Francisco, June 17–21, 1996, pp. 273–276.

221. J.-E. Müller, T. Grave, H.-J. Siweris, M. Kärner, A. Schäfer, H. Tischer, H. Riechert, L. Schleicher, L. Verweyen, A. Bangert, W. Kellner, and T. Meier, A GaAs HEMT MMIC chip set for automotive radar systems fabricated by optical stepper lithography, in: *IEEE J. Solid-State Circuits*, vol. 32, no. 9, 1997, pp. 1342–1349.

222. S. Raman, F. Rucky, and G.-M. Rebeiz, A high-performance W-band uniplanar subharmonic mixer, *IEEE Trans. Microwave Theory Tech.*, vol. 45, no. 6, 1997, pp. 955–962.

223. Y.-C. Chen and G.-M. Rebeiz, Design of Lange-couplers and single-sideband mixers using micromachining techniques, *IEEE Trans. Microwave Theory Tech.*, vol. 45, no. 2, 1997, pp. 291–294.

224. I. Toyoda, K. Nishikawa, T. Tokumitsu, K. Kamogawa, C. Yamaguchi, M. Hirano, and M. Aikawa, Three-dimensional masterslice MMIC on Si substrate, *IEEE Trans. Microwave Theory Tech.*, vol. 45, no. 12, pt. 2, 1997, pp. 2524–2530.

225. U. Banhardt, W. Bischof, W. Ehrlinger, K. Haug, R. Lohrmann, R. Keller, M. Schlechtweg, F. Arndt, and J. Ritter, 60 GHz-Bausteine für MMIC-Modulatoren in koplanarer Wellenleitertechnik, in: *MIOP 97, Mikrowellen und Optronik, 9. Kongreßmesse für Hochfrequenztech., Funkkommunikation und elektromagnetische Verträglichkeit Digest*, Sindelfingen, D, 22–24 April 1997, pp. 492–497.

226. M. Schefer, *Integrated coplanar millimeter wave circuits using InP HEMTs*, Doctoral Thesis, Germany, 1997, pp. 1–109.

227. W.-H. Kwok, Y.-W. Bo, and K.-S. Chiang, Design and analysis of coplanar waveguide optical modulator using finite element method, *Int. J. Infrared Millimeter Waves*, vol. 18, no. 4, 1997, pp. 875–887.

228. P. Cameron, W. Pan, C. Hanz, R. Nicklaus, A flip-chip high efficiency X-band HPA, in: 1997 *IEEE MTT-S Int. Microwave Symposium Digest*, June 1997, pp. 889–892.

229. Z.-G. Wang, M. Berroth, A. Thiede, M. Motzer-Rieger, T. Jakobus, A. Hülsmann, K. Köhler, and B. Raynor, 40 GHz monolithically-integrated fully-balanced VCO using 0.3 micron HEMTs, *Electronics Lett.*, vol. 33, no. 5, 1997, pp. 422–423.

230. L. Verweyen, A. Bangert, H. Massler, T. Fink, M. Neumann, R. Osorio, T. Krems, T. Jakobus, W. Haydl, and M. Schlechtweg, Compact integrated coplanar T/R-modules for automotive applications, in: *IEEE MTT-S International Microwave Symposium Digest*, Denver, 1997, pp. 243–246.

231. S. Maas, Broadband planar monolithic balanced mixers and frequency multipliers, in: *INMMC 98, 5th International Workshop on Integrated Nonlinear Microwave and Millimeterwave Circuits*, Duisburg, Germany, Oct. 1–2, 1998, pp. 129–137.

232. T. Takeuchi, T. Nakata, K. Fukuchi, K. Makita, and K. Taguchi, A high-efficiency waveguide photodiode for 40-Gb/s optical receivers, in: *Recent Progress in Optoelectronics and Communications*, 12–16 July 1998, Chiba, Japan, *IEICE Trans. Electronics*, vol. E82-C, no. 8, 1999, pp. 1502–1508.

233. E. Alekseev, D. Pavlidis, V. Ziegler, M. Berg, and J. Dickmann, 77 GHz high-isolation coplanar transmit-receive switch using InGaAs/InP PIN diodes, in: *20th IEEE Gallium Arsenide Integrated Circuit Symposium Technical Digest*, 1998, 1–4 Nov. 1998, Atlanta, 1998, pp. 177–180.

234. J. Schönauer, Hochfrequenz-Relais in Micro Strip oder Coplanar Wave Guide-Konstruktion, *Elektron. Ind.*, vol. 29, no. 11, 1998, pp. 44, 46, 48–50.

235. T. Krems, A. Tessmann, W. H. Haydl, C. Schmelz, and P. Heide, Avoiding cross talk feedback effects in packaging coplanar millimetre-wave circuits, *IEEE Trans. Microwave Theory Tech.*, vol. 46, no. 1, 1998, pp. 1–4.

236. V. Hoel, S. Boret, B. Grimbert, G. Aperce, S. Bollaert, H. Happy, X. Wallart, and A. Cappy, 94-GHz low noise amplifier on InP in coplanar technology. in: *Proceedings, 1999 European Gallium Arsenide and Related III-V Compounds, Applications Symposium*, Munich, D, 4–5 Oct. 1999, pp. 257–262.

237. A. Tessmann, W.-H. Haydl, M. Neumann, S. Kudszus, and A. Hülsmann, A coplanar W-band power amplifier MMIC using dual-gate HEMTs, in: *MIOP 99, 29th European Microwave Conference, Conference Proceedings*, vol. 1, Munich, D, Oct. 5–7, 1999, pp. 246–249.

238. S. Kudszus, W.-H. Haydl, M. Neumann, and A. Hülsmann, W-Band HEMT-oscillator with stabilization by phase controlled subharmonic injection locking, in: *MIOP 99, 29th European Microwave Conf., Conf. Proc.*, vol. 1, Munich, D, Oct. 5–7, 1999, pp. 198–201.

239. M.-G. Keller, A.-P. Freundorfer, and Y.-M.-M. Antar, A single-chip coplanar 0.8-μm GaAs MESFET K/Ka-band DRO, *IEEE Microwave and Guided Wave Lett.*, vol. 9, no. 12, 1999, pp. 526–528.

240. A.-S. Nagra, Xu. Jian, E. Erker, and R.-A. York, Monolithic GaAs phase shifter circuit with low insertion loss and continuous 0–360 degrees phase shift at 20 GHz, *IEEE Microwave and Guided Wave Lett.*, vol. 9, no. 1, 1999, pp. 31–33.

241. F. Steinhagen, H. Massler, W.-H. Haydl, A. Hülsmann, and K. Köhler, Coplanar W-band SPDT and SPTT resonated PIN diode switches, in: *MIOP 99, 29th European Microwave Conference, Conference Proceedings*, vol. 2, Munich, Oct. 5–7, 1999, pp. 53–56.

242. K.-M. Strohm, C.-N. Rheinfelder, A. Schurr, and J.-F. Luy, SIMMWIC capacitive RF switches, in: *MIOP 99, 29th European Microwave Conference, Conference Proceedings*, vol. 2, Munich, Oct. 5–7, 1999, pp. 411–414.

243. A. Werthof, T. Grave, and W. Kellner, 90 GHz amplifier fabricated by a low cost PHEMT technology, in: *GAAS 99 Conf. Proc., 1999 European Gallium Arsenide and Related III–V Compounds, Applications Symp.*, Munich, 4–5 Oct. 1999, pp. 263–266.

244. M. Stotz, S.-D. Fritze, H. Downar, and J. Wenger, Thermally controlled coplanar microwave switches, in: *MIOP 99, 29th European Microwave Conference, Conference Proceedings*, vol. 2, Munich, Oct. 5–7, 1999, pp. 415–418.

245. A. Bessemoulin, W. Marsetz, Y. Baeyens, R. Osorio, H. Massler, A. Hülsmann, and M. Schlechtweg, Design of coplanar power amplifiers for mm-wave system applications including thermal aspects, *CD-ROM: Europ. Microwave Week 2000, 30th European Microwave Conference, GAAS 2000, European Conference on Wireless Technol. 2000, Conference Proceedings*, Paris, Oct. 2–6, 2000, \GAAS_6_2.pdf, pp. 1–4.

246. R. Follmann, J. Borkes, P. Waldow, and I. Wolff, Extraction and modeling method for FET devices, *IEEE Microwave Mag.*, vol. 1, no. 3, 2000, pp. 49–55.

247. R. Follmann, J. Berben, D. Köther, and I. Wolff, A universal method for calculating and extracting the LF and RF noise behaviour of nonlinear devices, in: *22nd IEEE GaAs IC Symposium Digest*, Seattle, vol. 1, 2000, pp. 47–50.

248. S. Kudszus, M. Neumann, T. Berceli, and W.-H. Haydl, Fully integrated 94-GHz subharmonic injection-locked PLL circuit, *IEEE Microwave and Guided Wave Letters*, vol. 10, no. 2, 2000, pp. 70–72.

249. E. Hirt, M. Scheffler, G. Tröster, W. Wendel, and R. Epple, A 9:4 satellite receiver switch in MCM-C/D technology, in: *Microtec 2000, Applications–Trends–Visions, VDE World Microtechnology Congress Proceedings*, vol. 2, Hannover, Sept. 25–27, 2000, pp. 137–141.

250. M. Ulm, T. Walter, J. Schier, F.-R. Mueller, and E. Kasper, Microelectromechanical capacitive RF switches on high resistivity silicon substrates, in: *Microtec 2000, Applications–Trends–Visions, VDE World Microtechnology Congress, Proceedings*, vol. 2, Hannover, Sept. 25–27, 2000, pp. 93–96.

251. C. Yongming and L.-P.-B. Katehi, Wide band series switch fabricated using metal as sacrificial layer, *CD-ROM: Europ. Microwave Week 2000, 30th European Microwave Conference, GAAS 2000, European Conference on Wireless Technology 2000, Conference Proceedings*, Paris, Oct 2–6, 2000, \Eumc\EuMC_2_4.pdf, pp. 1–3.

252. Leng-Tan Guan and G.-M. Rebeiz, Microwave absorptive MEMS switches, *CD-ROM: European Microwave Week 2000, 30th European Microwave Conference, GAAS 2000, European Conference on Wireless Technology 2000, Conference Proceedings*, Paris, F, Oct. 2–6, 2000, \Gaas\EuMC_2_2.pdf, pp. 1–4.

253. B. Hopf, *Aktive monolithisch integrierte Mikrowellen-Kopplungsbandfilter in Koplanarleitungstechnik*, Doctoral Thesis, Duisburg University, Duisburg, Germany, 2000.

254. A. Fujihara, H. Miyamoto, K. Yamanoguchi, E. Mizuki, and N. Samoto, 60 GHz coplanar low-noise amplifier fabricated using AlAs/InAs superlattice-inserted InP-based heterojunctions FETs, *NEC Res. Dev.*, vol. 42, no. 3, 2001, pp. 304–308.

255. K. Uchida, H. Matsuura, T. Yakihara, S. Kobayashi, S. Oka, T. Fujita, and A. Miura, A series of InGaP/InGaAs HBT oscillators up to D-band, *IEEE Trans. Microwave Theory Tech.*, vol. 49, no. 5, 2001, pp. 858–865.

256. R. Mehran, M.-M. Bokatius, D. Lumbeck, R. Reber, and H.-P. Feldle, A highly integrated X-band MMIC SPDT switch in CPW technique, in: *MIOP, the German Wireless Week, Mikrowellen und Optronik/Microwaves and Optronics, 11th Conference and Exhibition on Microwaves, Radio Communication and Electromagnetic Compatibility, Conference Proceedings*, Stuttgart, 8–10 May 2001, pp. 222–225.

257. J.-Y. Chen, L.-S. Huang, C.-H. Chu, I.-Y. Li, Y.-H. Kuo, and P. Chang, A new assembly-transferred microwave switch with suspended CMOS-compatible coplanar waveguides, in: *Micro-Electro-Mechanical Systems (MEMS), the 2001 ASME International Mechanical Engineering Congress and Exposition*, Nov. 11–16 2001, New York, MEMS, vol. 3, 2001, pp. 65–69.

258. C. Delong, H. Shawn, D. Pavlidis, P. Chin, and T. Block, A Ka-band monolithic low phase noise coplanar waveguide oscillator using InAlAs/InGaAs HBT, *Solid State Electronics*, vol. 46, no. 2, 2002, pp. 249–253.

259. R. Follmann, G. Langgartner, J. Borkes, I. Wolff, and H.-P. Feldle, Influence of backside metallization on a coplanar X-band LNA, in: *Proceedings European Microwave Conference*, Milan, 2002.

260. Daimler Benz: GaAs Gallium Arsenide Foundry Service Design Rules.

261. Ferdinand-Braun-Institut für Höchstfrequenz-technik, Berlin.

262. EEsof Inc.: Program Touchstone/Libra, Version 3.1 and Manual.

263. HP-EEsof Inc., Libra 4.0, Series IV.

264. Agilent. ADS™.

INDEX

Accuracy, finite difference method, 49
Accuracy, modeling, 312
Active band pass filter, coplanar, first order, 460
Active band pass filter, coplanar, second order, 460
Active filter, tunable, coplanar, 463
Active filters, coplanar, 457
Active inductor, coplanar, 458
Active power divider, coplanar, 444
Active power combiner, coplanar, 444
Additional charge, discontinuities, 158
Admittance per unit length matrix, coupled coplanar waveguides, 92
Air bridge, coplanar, model, C_AIR, 359
Air bridges, 183
Amplifier, coplanar, high gain, medium power, 488
Amplifier, low noise, influence of backside metallization, 481
Amplifiers, coplanar, 471
—lumped elements, 477
—waveguide design, 471
Attenuation coefficient, coplanar waveguide, 18, 66
—dependence on width of ground plane, 22
Attenuation coefficient, coplanar-like waveguides, 82
Attenuation coefficient, microstrip line, 66
Attenuation coefficient, surface wave, 33

Backside metallization, coplanar waveguide, 108
influence on low noise amplifier, 481
Balanced frequency doubler, coplanar, 495
Band reject filter, coplanar, 394
Band-pass filter, active, coplanar, first order, 460
—active, coplanar, second order, 460
—lumped element, coplanar, 388
Bend, coplanar transmission line, model, C_BEND, 361
Bend, coplanar waveguide, 192

Coplanar Microwave Integrated Circuits, by Ingo Wolff.

Bridge data definitions, coplanar, C_AIRTYP, 331
Broadside-coupled coplanar spiral transformer, 301
Broadside-coupled coplanar waveguides, 109
Buttress, coplanar-like waveguides, 87

C_AIR, coplanar air bridge, model, 359
C_AIRTYP, bridge data definitions, coplanar, 331
C_BEND, bend, coplanar transmission line, model, 361
C_CAPLIN, capacitor, MIM, to ground, coplanar, model, 349
C_CROSS, coplanar waveguides crossing, model, 364
C_GAP, coplanar waveguide, gap, model, 354
C_GRID, grid data definitions, coplanar, 333
C_IDC, capacitor, interdigital, coplanar, model, 367
C_LAYER, layer data definitions, 338
C_LIN, coplanar transmission line, 344
C_LINTYP, linetype data definition, coplanar, 327
C_METVIA, intermetal via, coplanar, model, 346
C_MIM, capacitor, MIM, coplanar, model, 372
C_NLTYP, coupled lines data definitions, coplanar, 329
C_OPEN, coplanar open ended transmission line, model, 351
C_PORT, waveguide RF-port, coplanar, model, 343
C_PROCES, process, used for fabrication, 335
C_RIND, coplanar rectangular inductor, model, 368
C_SHORT, coplanar short circuited transmission line, model, 353
C_STEP, step, coplanar transmission line, model, 355
C_SUB, substrate definition, coplanar, 325
C_TAPER, coplanar waveguide, taper, model, 357
C_TECH, technological data definitions, 336
C_TEE, coplanar waveguide T-junction, model, 363
C_TFG, transmission line, resistively loaded, coplanar, model, 347
C_TFR, resistor, thin film, coplanar, model, 370
Cache management, 321
Capacitance matrix, coupled coplanar waveguides, 90
Capacitance per unit length matrix, coupled coplanar waveguides, 92
Capacitive feedback, common-source, 457
Capacitor, interdigital, coplanar, 250
—model, C_IDC, 367
Capacitor, MIM, coplanar, model, C_MIM, 372
—to ground, coplanar, model, C_CAPLIN, 349
Characteristic impedance, conventional coplanar waveguide, 106
—coplanar waveguide, 19, 64
—coplanar waveguide, influence of losses, 30
—dependence on ground plane width, 20
—open coplanar waveguide, 30
—supported coplanar waveguide, 98
Characteristic parameters, coplanar waveguides, 63
Charge per unit line length, 60
Charge, discontinuities, 158
Chebychev band-pass filter, coplanar, 389
Circuit design program, 309
Circulator, coplanar, electronic, 491
Closed formula static analysis, coplanar waveguides, 95
Common-gate inductive feedback, 457
Common-source capacitive feedback, 457
Components, coplanar, models, 339
Conductor backed coplanar waveguide, 4, 77
Conductor losses, coplanar waveguide, 18
Conformal mapping technique, 96

Convergence, modeling, 312
Coplan for ADS, 315
Coplanar active filter, tunable, 463
Coplanar active inductor, 458
Coplanar active power combiner, 444
Coplanar active power divider, 444
Coplanar air bridge, model, C_AIR, 359
Coplanar band reject filter, 394
Coplanar bridge data definitions, C_AIRTYP, 331
Coplanar components, models, 339
Coplanar coupled lines data definitions, C_NLTYP, 329
Coplanar couplers, 377
Coplanar data items, 322
Coplanar discontinuities, models, 340, 341
Coplanar edge coupled line structures, 404
Coplanar electronic circulator, 491
Coplanar element library, 309
Coplanar end coupled structures, 409
Coplanar filters, 377
Coplanar frequency doubler, 495
—balanced, 495
—realization, 497
—wideband, 497
Coplanar grid data definition, C_GRID, 333
Coplanar hybrid band pass filter, 390
Coplanar inter metal via, model, C_METVIA, 346
Coplanar interdigital capacitor, 250
—model, C_IDC, 367
—electric field, 253
—equivalent circuit, 252
—millimeter wave frequencies, 269
—Q-factor, 262
—scattering parameters, 266
—technological realization, 251
—to ground, 271
Coplanar line type data definitions, C_LINTYP, 327
Coplanar lumped element amplifiers, 477
Coplanar lumped element band-pass filter, 388
Coplanar lumped element filters, 377
Coplanar lumped elements, 249
—models, 342
Coplanar meander line, 197
Coplanar microwave active filters, 457
Coplanar microwave amplifiers, 471
—coplanar waveguide design, 471
—lumped element design, 477
Coplanar microwave integrated circuits, 439
Coplanar microwave oscillator, 508
Coplanar millimeter wave filter, 398
Coplanar MIM capacitor, 272
—model, C_MIM, 372
—technological realization, 273
—to ground, 276
—to ground, model, C_CAPLIN, 349
Coplanar open end, 162
—model, C_OPEN, 351
Coplanar phase shifting filter, 386
Coplanar resistively loaded transmission line, model, C_TFG, 347
Coplanar RF switches, design and realization, 453
Coplanar short circuited end, 167
—model, C_SHORT, 353
Coplanar spiral inductor, 276
—as filter, 377
—electric field, 281
—equivalent circuit, 279
—millimeter wave frequencies, 290
—model, C_RIND, 368
—open ended, 380
—Q-factor, 286
—short circuited, 380
—surface current density, 283
—technological realization, 278
—coupled, 291
Coplanar spiral transformer, 295
—broadside coupled, 301
—electric field, 298
—scattering parameters, 301
—technological realization, 300
Coplanar strips, 3
Coplanar substrate data definition C_SUB, 325
Coplanar switch circuits, 446
Coplanar switches, 444
Coplanar thin film resistor, 303
—model, C_TFR, 370
Coplanar times five frequency multiplier, 504

Coplanar T-junction, 202
—as mode converter, 225
—electric field, 209
—model, C_TEE, 363
—mode transfer, 222
—odd mode excitation, 221
—surface current distribution, 205
Coplanar transistors, 444
Coplanar voltage controlled oscillator, 511, 514
Coplanar waveguide bend, 192
Coplanar waveguide couplers, 420
Coplanar waveguide crossing, 234
Coplanar waveguide crossing, model, C_CROSS, 364
Coplanar waveguide filters, 392
Coplanar waveguide gap, 169
Coplanar waveguide gap, electric field, 171
Coplanar waveguide mode, 4, 12
Coplanar waveguide step, 175
Coplanar waveguide taper, 182
Coplanar waveguide T-junction, 202
Coplanar waveguide T-junction, model, C_TEE, 363
Coplanar waveguide T-junction, surface current distribution, 205
Coplanar waveguide, 4
—attenuation coefficient, 18, 66
—attenuation coefficient, width of ground-planes, 22
—backside metallization, 108
—characteristic impedance, 19
—characteristic impedance, 64
—characteristic impedance, dependence on ground plane width, 20
—characteristic impedance, influence of losses, 30
—characteristic parameters, 63
—closed formula static analysis, 95
—conductor backed, 4
—conductor backed, 77
—conductor losses, 18
—covered, 96
—field distribution, odd mode, 13
—discontinuities, 145, 154
—dispersion, 15
—effective dielectric constant, 14
—effective dielectric constant, 65
—effective dielectric constant, 97
—effective dielectric constant, width of ground planes, 21
—even mode, 4, 13
—field distribution, even mode, 14
—field distribution, odd mode, 15
—filling factor, 100
—gap, model, C_GAP, 354
—generalized, 95
—influence metallization thickness, 72, 107
—influence of shielding, 75, 106
—influence of the ground strip width, 74
—losses, 18
—measured field distribution, 23
—odd mode, 4, 13
—open, 26, 96
—open, characteristic impedance, 30
—open, electric field, 26
—open, magnetic field, 26
—phase velocity, 14
—power concentration ratio, 17
—quasistatic analysis technique, 46, 55
—resistance per unit line length, 61
—single center strip, finite ground plane, 12
—single center strip, infinite ground plane, 26
—slotline mode, 4, 13
—special forms, 76
—supported, 4
—supported, 96
—supported, characteristic impedance, 98
—surface wave, 26, 31
—taper, model, C_TAPER, 357
—transmission properties, 11
Coplanar waveguides, asymmetric, supported, broadside coupled, 115
—broadside-coupled, 109
—broadside-coupled, even/odd mode, 112
—coupled, 34, 89
—coupling, 405
—orthogonal, coupling, 411
Coplanar Wilkinson coupler, 426

Coplanar, high gain, medium power amplifier, 488
Coplanar-like waveguides, 78
—attenuation coefficient, 82
—buttress, 87
—current density distribution, 80
—dispersion effective dielectric constant, 83
—effective dielectric constant, 82
—electric and magnetic field, 80
Coplanar transmission line model, C_LIN, 344
Coupled coplanar line couplers, 414
Coupled coplanar line filters, 414
Coupled coplanar waveguides, 34, 89
—admittance per unit length matrix, 92
—capacitance matrix, 90
—capacitance per unit length matrix, 92
—coupling coefficient, 41
—current matrix, 38
—current transformation matrix, 94
—impedance matrix, 39
—impedance per unit length matrix, 92
—inductance per unit length matrix, 92
—isolation, 41
—mode currents, 39
—mode voltages, 39
—power matrix, 37
—reflection coefficient, 41
—resistance per unit length matrix, 92
—scattering matrix, 36
—transmission line voltages, 39
—voltage matrix, 39
—voltage transformation matrix, 94
Coupled lines data definitions, coplanar, C_NLTYP, 329
Coupled microstrip lines, 3, 44
Coupled microstrip lines, coupling coefficient, 44
—isolation, 44
—transmission coefficient, 44
Coupled spiral inductors, coplanar, 291
Coupler, coupled coplanar line, 414
Coupler, Wilkinson, coplanar, 426
—conventional, 427
—coplanar waveguide technique, 431
—coplanar, with lumped elements, 429
—with discrete elements, 427
Couplers, coplanar waveguide, 377, 420
Coupling coefficient, coupled coplanar waveguides, 41
—coupled microstrip lines, 44
Coupling to orthogonal coplanar waveguide, 411
Coupling, coplanar waveguides, 234, 405
Crossing, coplanar waveguides, model, C_CROSS, 364
Current matrix, coupled coplanar waveguides, 38
Current transformation matrix, coupled coplanar waveguides, 94

Data definitions, layers, C_LAYER, 338
Data definitions, technological, C_TECH, 336
Data items, coplanar, 322
Device scaling, transistors, 450
Dielectric losses, coplanar waveguide, 19
Dirichlet wall, finite difference method, 49
Discontinuities, models, 155
—additional charge, 158
—electric field strength, 147
—inductance, 161
—magnetic field strength, 150
—magnetic flux, 160
—model parameter extraction, 157
—surface current density, 152
—coplanar waveguides, 145, 154
—coplanar, models, 340, 341
Discretization, finite difference method, 49
Dispersion, coplanar waveguide, 15
—effective dielectric constant, coplanar waveguide, 58
—effective dielectric constant, coplanar-like waveguides, 83
effective dielectric constant, microstrip line, 58
—surface wave, 32

Edge coupled line structures, coplanar, 404
Effective dielectric constant, coplanar waveguide, 14, 65, 97
—coplanar waveguide, width of ground planes, 21
—coplanar-like waveguides, 82
—dispersion, coplanar waveguide, 58
—dispersion, microstrip line, 58
—dispersion, coplanar-like waveguides, 83
—surface wave, 32
Efficiency, finite difference method, 49
Electric field strength, 50
—discontinuities, 147
Electric field, coplanar spiral inductor, 281
—coplanar spiral transformer, 298
—coplanar T-junction, 209
—coplanar waveguide gap, 171
—coplanar-like waveguides, 80
—even mode, open coplanar waveguide, 26
—interdigital capacitor, 253
—open end, 163
End coupled coplanar structures, 409
Even mode, coplanar waveguide, 4, 13

Feedback, capacitive, common-source, 457
Feedback, inductive, common-gate, 457
Field distribution, coplanar waveguide, even mode, 13
—coplanar waveguide, odd mode, 13
—coplanar waveguide, parasitic even mode, 13
—measured, coplanar waveguide, 23
Filling factor, coplanar waveguides, 100
Filter, active, tunable, coplanar, 463
Filter, band-pass, first order, active, coplanar, 460
—hybrid, coplanar, 390
—lumped element, coplanar, 388
—second order, active, coplanar, 460
—coplanar, Chebychef, 389
Filter, coplanar, 377
—coplanar waveguide, 392
—coplanar, band reject, 394
—coupled coplanar line, 414
—interdigital, coplanar, 414
—millimeter-wave, coplanar, 398
—phase shifting, coplanar, 386
—active, coplanar, 457
Finite difference method, 46
—accuracy, 49
—Dirichlet wall, 49
—discretization, 48
—efficiency, 49
—mesh size, 49
—Neumann wall, 49
—shielding, 49
—three-dimensional, 146
Frequency doubler, coplanar, 495
—balanced, 495
—realization, 497
—wideband, 497
Frequency multiplier, coplanar, times five, 504

Gap, coplanar waveguide, 169
—model, C_GAP, 354
Gauss-Seidel iteration technique, 52, 148
Generalized coplanar waveguides, 95
Grid data definitions, coplanar, C_GRID, 333
Ground strip width, influence on coplanar waveguides, 74

Hybrid mode, 56

Impedance matrix, coupled coplanar waveguides, 39
Impedance per unit length matrix, coupled coplanar waveguides, 92
Inductance per unit line length matrix, coupled coplanar waveguides, 90
Inductance, discontinuities, 161
Inductive feedback, common-gate, 457
Inductor, active, coplanar, 458
—rectangular, coplanar, model, C_RIND, 368
Interdigital capacitor, coplanar, 250
—model, C_IDC, 367
—electric field, 253
—equivalent circuit, 252
—Q-factor, 262
—coplanar, scattering parameters, 266

—technological realization, 251
—to ground, 271
Interdigital coplanar filter, 414
Intermetal via, coplanar, model, C_METVIA, 346
Isolation, coupled coplanar waveguides, 41
—coupled microstrip lines, 44

Laplace equation, planar and coplanar line structures, 48
—three-dimensional, 148
Layer data definitions, C_LAYER, 338
Layout, coplanar circuits, 321
Library elements, 319
Linteype data definition, coplanar, C_LINTYP, 327
Losses, coplanar waveguide, 18
Low noise amplifier, influence of backside metallization, 481
Lumped element band-pass filter, coplanar, 388
Lumped element filters, coplanar, 377
Lumped element modeling, 250
Lumped elements, coplanar, 249
Lumped elements, coplanar, models, 342

Magnetic field strength, discontinuities, 150
Magnetic field, coplanar-like waveguides, 80
Magnetic field, open coplanar waveguide, 26
Magnetic flux, discontinuities, 160
Meander line, coplanar, 197
Mesh size, finite difference method, 49
Metallization thickness, coplanar waveguide, 72, 107
Microstrip line, 2
—attenuation coefficient, 66
—suspended, 2
Microstrip lines, coupled, 3
Microwave integrated circuits, coplanar, 439
Millimeter wave filter, coplanar, 398
MIM capacitor, coplanar, model, C_MIM, 372
MIM-capacitor to ground, coplanar, model, C_CAPLIN, 349

Mode converter, coplanar T-junction, 225
Mode currents, coupled coplanar waveguides, 39
Mode transfer, coplanar T-junction, 222
Mode voltages, coupled coplanar waveguides, 39
Model parameter extraction, discontinuities, 157
Modeling, convergence, accuracy, 312
Models of discontinuities, 155
Models, coplanar components, 339

Negative conductance circuit, 509, 511
Negative resistance, 457
Neumann wall, finite difference method, 49

Odd mode excitation, coplanar T-junction, 221
Odd mode, coplanar waveguide, 4, 13
Open end, coplanar, 162
—electric field, 163
Open ended transmission line, coplanar, model, C_OPEN, 351
Oscillator, coplanar, 508
—voltage controlled, 514
Over-relaxation technique, 148

Phase shifting filter, coplanar, 386
Phase velocity, coplanar waveguide, 14
Power combiner, active, coplanar, 444
Power concentration ratio, coplanar waveguide, 17
Power divider, active, coplanar, 444
Power matrix, coupled coplanar waveguides, 37
Process, used for fabrication, C_PROCES, 335

Q-factor, coplanar interdigital capacitor, 262
—coplanar spiral inductor, 286
Quasistatic analysis technique, coplanar waveguides, 55

Rectangular Inductor, coplanar, model, C_RIND, 368
Reflection coefficient, coupled coplanar waveguides, 41

Resistance per unit length matrix, coupled coplanar waveguides, 92
—coplanar waveguide, 61
Resistance, negative, 457
Resistively loaded transmission line, coplanar, model, C_TFG, 347
Resistor, thin film, coplanar, 303
—coplanar, model, C_TFR, 370
Resonant circuit, coplanar, parallel, fixed frequency, 510
—parallel, tunable, 510
—series, fixed frequency, 512
—series, tunable frequency
RF switches, coplanar, design and realization, 453
RF-port, coplanar, model, C_PORT, 343

Scattering matrix, coupled coplanar waveguides, 36
Scattering parameters, coplanar interdigital capacitor, 266
Shielding, coplanar waveguide, 106
—effect on modeling, 440
—finite difference method, 49
—influence on coplanar waveguides, 75
Short circuited end, coplanar waveguide, 167
Skin effect, 29
Slotline mode, 4, 13
Spiral inductor, coplanar, 276
—as filter, 377
—electric field, 281
—equivalent circuit, 279
—coplanar, millimeter-wave frequencies, 290
—open ended, 380
—Q-factor, 286
—short circuited, 380
—surface current density, 283
—technological realization, 278
Spiral inductors, coplanar, coupled, 291
Spiral transformer, coplanar, 295
—broadside coupled, 301
—scattering parameters, 301
—technological realization, 300
Step, coplanar waveguide, 175
—model, C_STEP, 355
Strip line, 2
Substrate definition, coplanar, C_SUB, 325
Successive over-relaxation method, 54
Surface current distribution, coplanar T-junction, 205
Surface current density, 61
—coplanar spiral inductor, 283
—discontinuities, 152
Surface wave, attenuation coefficient, 33
Surface wave, coplanar waveguide, 26, 31
—dispersion, 32
—effective dielectric constant, 32
Suspended microstrip line, 2
Switches, coplanar, 444
—RF, coplanar, design and realization, 453

Taper, coplanar waveguide, 182
—model, C_TAPER, 357
Technological data definitions, C_TECH, 336
Thin film resistor, coplanar, 303
—model, C_TFR, 370
T-junction, coplanar waveguide, 202
—model, C_TEE, 363
Transistors, coplanar, 444
—device scaling, 450
Transmission coefficient, coupled microstrip lines, 44
Transmission line , coplanar, model C_LIN, 344
—open ended, model, C_OPEN, 351
—resistively loaded, model, C_TFG, 347
—short circuited, model, C_SHORT, 353
—step, model, C_STEP, 355
Transmission line voltages, coupled coplanar waveguides, 39

Voltage matrix, coupled coplanar waveguides, 39
Voltage transformation matrix, coupled coplanar waveguides, 94

Waveguide properties, coplanar, microstrip, 441

Waveguide RF-port, coplanar, model, C_PORT, 343
Wilkinson coupler, conventional, 427
—with discrete elements, 427
Wilkinson coupler, coplanar, 426
—waveguide technique, 431
—with lumped elements, 429

Printed and bound by CPI Group (UK) Ltd, Croydon, CR0 4YY
16/09/2026
14963414-0004